The 9th International Conference on Time Series and Forecasting-Volume I

The 9th International Conference on Time Series and Forecasting-Volume I

Editors

Ignacio Rojas
Hector Pomares
Luis Javier Herrera
Fernando Rojas
Olga Valenzuela

MDPI

Basel • Beijing • Wuhan • Barcelona • Belgrade • Novi Sad • Cluj • Manchester

Editors
Ignacio Rojas
University of Granada
Granada, Spain

Hector Pomares
University of Granada
Granada, Spain

Luis Javier Herrera
University of Granada
Granada, Spain

Fernando Rojas
University of Granada
Granada, Spain

Olga Valenzuela
University of Granada
Granada, Spain

Editorial Office
MDPI
St. Alban-Anlage 66
4052 Basel, Switzerland

This is a reprint of articles from the Proceedings published online in the open access journal *Engineering Proceedings* (ISSN 2673-4591) (available at: https://www.mdpi.com/2673-4591/39/1).

For citation purposes, cite each article independently as indicated on the article page online and as indicated below:

Lastname, A.A.; Lastname, B.B. Article Title. *Journal Name* **Year**, *Volume Number*, Page Range.

Volume I
ISBN 978-3-0365-9728-7 (Hbk)
ISBN 978-3-0365-9729-4 (PDF)
doi.org/10.3390/books978-3-0365-9729-4

Set
ISBN 978-3-0365-9726-3 (Hbk)
ISBN 978-3-0365-9727-0 (PDF)

Contents

engineering proceedings

MDPI

Editorial

New Developments in Time Series and Forecasting, ITISE-2023 †

Olga Valenzuela [1], Fernando Rojas [2], Luis Javier Herrera [2], Hector Pomares [2] and Ignacio Rojas [2,*]

[1] Department of Applied Mathematics, University of Granada, 18071 Granada, Spain; olgavc@ugr.es
[2] Department of Computer Engineering, Automation and Robotics, CITIC-UGR, University of Granada, 18071 Granada, Spain; frojas@ugr.es (F.R.); jherrera@ugr.es (L.J.H.); hector@ugr.es (H.P.)
* Correspondence: irojas@ugr.es
† All papers are presented at the 9th International Conference on Time Series and Forecasting, Gran Canaria, Spain, 12–14 July 2023.

1. Introduction

The ITISE 2023 (9th International Conference on Time Series and Forecasting) sought to provide a discussion forum for scientists, engineers, educators and students about the latest ideas and realizations in the foundations, theory, models and applications for interdisciplinary and multidisciplinary research encompassing the disciplines of computer science, mathematics, statistics, forecaster, econometric, etc., in the field of time series analysis and forecasting.

ITISE 2023 solicited high-quality original research papers (including significant work-in-progress) on any aspect of time series analysis and forecasting in order to motivate the generation and use of knowledge and new computational techniques and methods on forecasting in a wide range of fields.

As in previous editions, ITISE 2023 was held in Gran Canaria (Spain), with the dates being 12–14 July 2023.

2. Main Topics of ITISE

As is well known, ITISE aims to provide a friendly discussion forum for scientists, engineers, educators and students to discuss the latest ideas and achievements in the fundamentals, theory, models and applications in the field of time series analysis and forecasting. More specifically, the main topics of ITISE are:

1. Time series analysis and forecasting
    - Nonparametric and functional methods;
    - Vector processes;
    - Probabilistic approaches to modeling macroeconomic uncertainties;
    - Uncertainties in forecasting processes;
    - Nonstationarity;
    - Forecasting with many models. Model integration;
    - Forecasting theory and adjustment;
    - Ensemble forecasting;
    - Forecasting performance evaluation;
    - Interval forecasting;
    - Data preprocessing methods: data decomposition, seasonal adjustment, singular;
    - Spectrum analysis, detrending methods, etc.
2. Econometrics and forecasting
    - Econometric models;
    - Economic and econometric forecasting;
    - Real macroeconomic monitoring and forecasting;

Citation: Valenzuela, O.; Rojas, F.; Herrera, L.J.; Pomares, H.; Rojas, I. New Developments in Time Series and Forecasting, ITISE-2023. *Eng. Proc.* **2023**, *39*, 101. https://doi.org/10.3390/engproc2023039101

Published: 19 September 2023

- Advanced econometric methods.

3. Advanced methods and on-line learning in time series
    - Adaptivity for stochastic models;
    - On-line machine learning for forecasting;
    - Aggregation of predictors;
    - Hierarchical forecasting;
    - Forecasting with computational intelligence;
    - Time series analysis with computational intelligence;
    - Integration of system dynamics and forecasting models.
4. High-dimension and complex/big data
    - Local vs. global forecasts;
    - Dimension reduction techniques;
    - Multiscaling;
    - Forecasting complex/big data.
5. Forecasting in real problems
    - Health forecasting;
    - Atmospheric science forecasting;
    - Telecommunication forecasting;
    - Hydrological forecasting;
    - Traffic forecasting;
    - Tourism forecasting;
    - Marketing forecasting;
    - Modelling and forecasting in power markets;
    - Energy forecasting;
    - Climate forecasting;
    - Financial forecasting and risk analysis;
    - Forecasting electricity load and prices;
    - Forecasting and planning systems.

3. Special Session in ITISE-2023

During ITISE-2023, several Special Sessions were carried out. Special Sessions are a very useful tool in order to complement the regular program with new and emerging topics of particular interest for the participating community. Special Sessions that emphasized multidisciplinary and transversal aspects, as well as cutting-edge topics, were especially encouraged and welcomed, and in this edition of ITISE-2023, the following were received:

- SS1. Advances in time series analysis and forecasts in Engineering Sciences.

It is well-known that time series analysis is time-consuming when large data sets are used, and soft computing methods are recommended for obtaining a balance between the models' accuracy and speed of solving the problem at hand. Therefore, this Special Session aimed to present the advances in the fields of time series modeling and forecasting of large series issued from measurements and experiments in different engineering fields. Submissions were expected to reflect theoretical methods and experimental works in statistical analysis and applications to modeling such time series.

Suggested topics of this Special Session include but are not limited to:

- Parametrical versus non-parametric approaches for data series modeling in engineering sciences;
- Critical evaluation and comparisons of alternative approaches for experimental time series modeling;
- New techniques for spatial data analysis;
- New software for data analysis—development and applications for solving engineering problems;
- Soft computing and fuzzy techniques for engineering time series modeling;

- Environmental time series modeling (precipitation, temperature, pollution).

The participants will be invited to submit their extended articles to the following journals:

- Advances in Water, Air and Soil Pollution Monitoring, Modeling and Restoration
- Hydrology (MDPI)—indexed within Scopus, ESCI (Web of Science)—tracked for Impact Factor—CiteScore 3.6—https://www.mdpi.com/journal/hydrology (12 July 2023)

Organizers: Prof. Dr. Hab. Alina Bărbulescu, Transilvania University of Brașov, Romania.

- SS2. Advanced econometric methods for Economic analysis and Finance

This Special Session focused on the application of advanced econometric methods to economic analysis and finance. Three authors presented in this session. Their presentations included advanced methods for macroeconomic data decomposition, a microeconometric assessment of tightening monetary policy impact on firm debt overhang, and a financial econometric analysis of the relation between the energy sector and the financial markets.

Organizer: Prof. Paolo Canofari, an Associate Professor in Economic Policy at the Università Politecnica delle Marche (UNIVPM) in Italy.

- SS3. Cryptocurrency time series modelling and forecasting

The cryptocurrency market, or rather more generally the cryptoasset market, is a field that is rapidly and steadily growing. Similarly, the literature regarding this topic has also enormously grown over the last few years. What distinguishes this market from the traditional financial market is that it operates 24/7 and every day, contrarily to the majority of traditional financial markets that operate within particular hours and only on business days. Moreover, what we already know about cryptoasset market is that it is characterized by wide heterogeneity, in terms of different aspects—both from the perspective of the supply and the demand. Most of the studies and most of the available methodologies for the analysis of the behavior of cryptoasset prices are focused on the data of daily frequency. Therefore, the literature using daily cryptoasset data is already quite exhausted. On the other hand, there are not many studies that make use of high-frequency data. It is important to distinguish the aggregated high-frequency data (e.g., the data that contain weighted-average of prices/volumes from multiple major cryptocurrency exchanges, available for instance on coinpaprika.com up to 5 min frequency) and the tick-to-tick on-exchange data that are also openly available. Therefore, such availability of high-frequency tick-to-tick data provides a great opportunity to not only develop the cryptoasset research field but also the field of high-frequency financial data in general, which has constantly growing been over the last decade or two.

Therefore, the aim of this session was to discuss the studies that focus on the usage of high-frequency cryptoasset data, since this field requires more attention in the literature because of the large gaps up to this point.

Organizer: Prof. Damian Zięba, University of Warsaw, Faculty of Economic Sciences, Department of Quantitative Finance, Warsaw, Poland.

- SS4. Artificial Intelligence and Sustainability

I encourage you to contribute a research or comprehensive review article for consideration for publication in *Sustainability*, an international Open Access journal which provides an advanced forum for research findings in areas related to sustainability and sustainable development. *Sustainability* publishes original research articles, review articles and communications. I am confident you will find that the journal contributes to enhancing understanding of sustainability and fostering initiatives and applications of sustainability-based measures and activities.

Organizers: Prof. Ebrahim Ghaderpour, Assistant Professor in Department of Earth Sciences at Sapienza University of Rome

4. Plenary Talk in ITISE-2023

In this edition of ITISE, we are honored to have hosted the following invited speakers:

1. Prof. Eamonn Keogh, Distinguished Professor, Department of Computer Science and Engineering University of California Riverside. Title of the presentation: Irrational Exuberance: Has Deep Learning Contributed Anything to Time Series problems?
2. Prof. Martin Wagner, Professor of Economics at the University of Klagenfurt. Chief Economic Advisor at the Bank of Slovenia and Fellow of the Macroeconomics and Economic Policy group at the Institute for Advanced Studies, Vienna. Title of the presentation: Sources and Channels of Nonlinearities and Instabilities of the Phillips Curve: Results for the Euro Area and Its Member States
3. Prof. Daniel Peña Sanchez De Rivera, Professor at Universidad Carlos III de Madrid. Department of Statistics. Madrid (Spain). Title of the presentation: Finding the Number of Clusters in Time Series

These plenary lectures strengthened the aim of this conference for the diffusion and discussion of high-quality research from some of the most recognized scientists in their fields.

5. Peer-Review Statement and MDPI *Engineering Proceedings*

As in previous editions, for ITISE 2023, a selection of papers were published in *Engineering Proceedings* (ISSN2673-4591). The first volume of *Engineering Proceedings* with the contributions of the congress was published for ITISE 2021 [1], as well as for ITISE 2022 [2].

In submitting conference proceedings to *Engineering Proceedings*, the volume editors of the proceedings certify to the publisher that all papers published in this volume have been subjected to peer review performed by the volume editors. Reviews are conducted by expert referees adhering to the professional and scientific standards expected of a proceedings journal. The type of peer review was single-blind, and the conference submission management system was Easychair (319 contributions were submitted), presenting in these proceedings a selection of such contributions which, on average, have been reviewed by at least two expert reviewers.

Funding: This research received no external funding.

Conflicts of Interest: The editors declare no conflict of interest.

References

1. Rojas, I.; Rojas, F.; Herrera, L.; Pomares, H. *The 7th International Conference on Time Series and Forecasting*; MDPI Proceedings; MDPI: Basel, Switzerlands, 2022; ISBN 978-3-0365-1732-2; ISSN 2504-3900.
2. Rojas, I.; Pomares, H.; Valenzuela, O.; Rojas, F.; Herrera, L.J. *The 8th International Conference on Time Series and Forecasting*; Engineering Proceedings; MDPI: Basel, Switzerlands, 2022; ISBN 978-3-0365-5452-5; ISSN 2673-4591.

engineering proceedings

MDPI

Proceeding Paper

Inventory Improvement in Tyre Retail through Demand Forecasting †

Diana Neves [1], Magda Monteiro [1,2,*] and Maria José Felício [1,3]

1 ESTGA-Águeda School of Technology and Management, University of Aveiro, 3750-127 Águeda, Portugal; dcn@ua.pt (D.N.); mjfelicio@ua.pt (M.J.F.)
2 CIDMA–Center for Research and Development in Mathematics and Applications, University of Aveiro, 3810-193 Aveiro, Portugal
3 GOVCOPP-Research Unit on Governance, Competitiveness and Public Policies, University of Aveiro, 3810-193 Aveiro, Portugal
* Correspondence: msvm@ua.pt

† Presented at the 9th International Conference on Time Series and Forecasting, Gran Canaria, Spain, 12–14 July 2023.

Abstract: The aim of this study is to develop the inventory planning system of a Portuguese tyre retailer based on forecasting sales models. Using sales history up to 2020, tyres were grouped into three levels of sales aggregation and different quantitative forecasting models were applied. The comparison of these models resorted to various evaluation measures to choose the most suitable one for each group. The study shows that for items with sales grouped monthly and for items with sales grouped by semester, Holt's method had a better performance on determining sales forecasts, while for tyres with sales grouped quarterly, it was Croston's method that stood out. The inventory policy outlined for each group of items reflects the results of the forecasted demand, and the review period depends on the sales group under analysis. In agreement with previous studies, the usefulness of statistical methods is corroborated. Additionally, the advantage of combining the said methods proved helpful, particularly as a starting point for tyre retail inventory planning.

Keywords: iventory; inventory planning; forecasting models; sales forecasting; intermittent demand

Citation: Neves, D.; Monteiro, M.; Felício, M.J. Inventory Improvement in Tyre Retail through Demand Forecasting. *Eng. Proc.* **2023**, *39*, 1. https://doi.org/10.3390/engproc2023039001

Academic Editors: Ignacio Rojas, Hector Pomares, Luis Javier Herrera, Fernando Rojas and Olga Valenzuela

Published: 25 June 2023

1. Introduction

A retailer's inventory is extremely critical. Low liquidity caused by high inventory backlog or poor customer experiences ensuing from a shortage of inventory arises from decisions taken during inventory management [1]. Hence, it is necessary to improve the levels of inventory held, which can be achieved by forecasting the sales of the items sold or needed to provide the service.

The European Union represents 20% of the global market of light and commercial tyres [2]. However, this market is highly sensitive to external factors: namely, the advancement of the automobile industry or improvements in the economy and road transportation.

Mahama-Musah et al. [3] observed that the independent aftermarket (private workshops) is most popular for purchasing tyres and that the internet is widely consulted to search for locations to replace tyres and information about brands, prices, and timings. Concerning the choice of tyre brand, price and quality are the most common, with this choice also being highly influenced by the mechanic's opinion.

In Portugal, tyre retailing is quite fragmented and occurs mainly in small businesses, and, consequently, there is a high level of competition. Thus, this case study aimed to develop sales forecast models for tyres sold by a Portuguese retailer so as to improve the planning system and the inventory policy in place while also maintaining a compromise between the reduction of inventory related costs and the satisfaction of consumers.

To date, inventory planning in this retailer only resorts to human judgment without assistance from any forecasting models. Therefore, the aim was to develop simple models that will allow a user to forecast the sales of each tyre size so as as to upgrade the company's tyre inventory levels and lower the associated costs.

The study focused on applying several quantitative forecasting methods to each tyre size. Afterwards a comparative analysis was made to select the most adequate size for each one. The novelty of the study relies on the field of application, bringing the use of simple and commonly used quantitative forecasting methods, which allows the improvement of the current tyre inventory levels without much effort due to their easy implementation either in the usual demand context or in a context of intermittent demand.

This paper is organized as follows. The next section is devoted to a literature review followed by the presentation of the case study. Afterwards, the practical work carried out is described, namely, the selection of the data used, the methods followed, the results obtained, and the main conclusions drawn. Finally, the general conclusions are shown.

2. Literature Review

Inventory management is responsible for ensuring that the right amount of each item is always in stock and in a cost-effective way. To do so, it resorts to inventory control, which supports operational decisions on when and how much to replenish for each of the stock keeping units, as well as the parts and materials used to produce them. The operating costs involved also include inventory, holding, ordering, and stockout costs [4].

The time to place an order can follow one of three approaches [5,6]:

- Carry out a periodic review (R) and place orders of variable size at regular time intervals, bringing the inventory to a certain level (S) ((R,S) policy).
- Conduct an ongoing review and place a fixed quantity (S) order as soon as inventory levels fall below the defined threshold (s) ((s,S) policy).
- Link supply to demand by ordering sufficient stock to meet expected demand in a specific time period ((R,s,S) policy).

Once the frequency and the size of the orders are decided, aspects such as the average inventory level, safety inventory, and level of customer service are automatically defined.

The methods developed for making predictions can be divided into two broad groups: qualitative (by judgment) and quantitative.

The most common judgment forecasting methods include the manager's opinion, the panel of executive opinion, the sales force opinion, the market survey to consumers, the historical analogy, and the Delphi method [5,7].

These methods are valuable when there is little or no historical data or when changes in the market turn existing data unsuitable for forecasting purposes. Strategies for their improvement include combining with other judgmental or quantitative estimates [8].

In retail sales forecast, statistical methods are the most utilized—namely, simple moving averages, exponential smoothing, autoregressive integrated moving averages, and regression, applying the Box and Jenkins approach type [9–11].

The advantages of exponential smoothing methods are their simplicity, low cost, and easy implementation. Fildes et al. [12] concluded that the performance of these methods depends on how the smoothing parameters are estimated and on how they are initialized, so information should be taken from the time series itself.

Moreover, there are several situations where items in inventory are infrequently requested and show great variability in demand values, which results in sporadic demand with a high risk of obsolescence. Simple exponential smoothing (SES) has proven to be a robust forecasting method and is probably the most used among statistical approaches to forecast intermittent demand [11,13]. However, Croston [14] observed that SES obtained negatively biased forecasts immediately before demand occurred and positively afterwards, which resulted in excessive inventory levels, and thus created a method where the forecast results from the ratio between the smoothed demand and the smoothed time between demands, using it for both SES with the same smoothing parameter [13,15,16].

Syntetos and Boylan [17] revised Croston's method, having developed a modified version (Syntetos–Boylan approximation, SBA) that theoretically eliminates the positive bias of the forecast.

Sbrana [18] has also suggested an intermittent model that considers that a time series switches between the state of a local level plus a constant and zero, reflecting the intermittency of demand. By doing this, it derives prediction intervals surpassing Croston's theory and its lack of an underlying stochastic model.

Nikolopoulos et al. [19] applied the aggregate-disaggregate intermittent demand approach, based on the theory that forecasts with higher levels of aggregation are more accurate and less variable. If after aggregation there is no demand equal to zero, any forecasting method can be used and the estimates can be disaggregated for a detailed analysis.

Artificial Neural Networks (ANN) are also involved in forecasting intermittent demand as they can model time series without assuming function models *a priori* [20,21].

The forecasting method to be used depends on several factors, from the forecast time window to the demand behaviour or its causes. Therefore, it is highly unlikely that a single model remains the best fit over time. Thus, estimates resulting from different methods may provide useful information, so combining them can be advantageous, as proven by Petropoulos et al. [22] and Hibon and Evgeniou [23].

It is also standard to assess the performance of forecasting methods by measuring their errors. These measures are useful to determine the model that best fits a time series and help in choosing and optimizing parameters, such as smoothing constants. However, choosing one measure over others may lead to completely different conclusions, and may even lead to disregarding one model that might be perfectly suitable for forecasting [24].

Syntetos and Boylan [13] argue that the presence of zeros needs to be considered in intermittent demand. So, studying the effect of the forecasts on inventory control parameters is more adequate, specifically on the resulting inventory and service levels [25,26].

Wallström and Segerstedt [27] compared several forecasting methods to show that a single error measure is not representative. Among others, they determined the number of stockouts and introduced the number of periods in inventory (PI) that considers the total number of periods in which the forecasted units remain in stock. Beyond the error, this also evaluates the time it takes to correct it.

As stated by several authors [28–30], the better accuracy of one forecasting model over others does not translate into better efficiency in inventory control, as what is crucial is how to use the forecast to achieve the targeted level of consumer service or to minimize the cost. A holistic understanding of the specific (and joint) nature of the inventory forecasting problem is required as it is furthered [4].

Regarding the forecast of tyre sales, the existing literature is very limited as performing these has been difficult, to some extent due to the impact that human psychology has on the decision of purchasing tyres [31]. The models used to this purpose are univariate, and rely only on past sales and estimates by experts [32].

More recent studies on forecasting retail sales have included macroeconomic factors, ANN, data mining models, hybrid models, or even extreme learning machine. Some papers have also combined expert judgement and statistical forecasts [9,32].

3. Data and Methods

This section describes the process of choosing the tyre sizes that will be targeted for inventory improvement by forecasting their sales volume as well as the methods used.

3.1. Selection of Items for Analysis

The company started selling tyres in May 2011, and by October 2020 had sold a total of 225 tyre sizes, including tyres for light, commercial, 4×4, and heavy vehicles. Currently there are only 181 of these references in inventory. The sizes corresponding to single orders from specific customers were also excluded.

For each of the 181 tyre sizes, monthly sales were aggregated in a total of 114 months. There were 99 references with highly intermittent sales (less than 20 non-zero observations over 114 months) and a very low impact on profit, reasons that led to their exclusion from the analysis. For each of the remaining 82 references, the percentage of months with non-zero observations was determined.

Items with less than 30% of non-zero observations were aggregated in semesters (28 references), items ranging from 30% to 50% non-zero observations were trimestral aggregated (23 references), and items with more than 50% of the months with non-zero observations were considered in this time unit (31 references).

The advantage of data aggregation is that it broadens the spectrum of forecasting methods that can be employed, as advocated by Nikolopoulos et al. [19].

3.2. Methods

To predict sales, exponential smoothing methods, such as SES and Holt's method, were tested, and so were Croston's method and SBA. Multiple linear regression (MLR) and generalized linear models (GLM) were also examined.

A training set was used that included the data, since sales first occurred in February 2020, and so different sales conditions were showcased. The test set comprised data from March 2020 until February 2021 with the aim of including all 12 months.

3.2.1. Smoothing Methods

The prediction formula for SES is given by:

$$P_{t+1} = \alpha Y_t + (1-\alpha)P_t, \quad (1)$$

with Y_t representing the sales of the item at period t, P_{t+1} the forecast obtained at the next period and α the smoothing parameter, ranging from 0 to 1. To initialize the method, the first prediction was considered equal to the first observation [33].

Regarding the smoothing parameter, according to Hyndman and Athanasopoulos [33], it is more accurate to estimate it from the observed data, so the parameter was obtained by minimizing the root mean squared error (RMSE) using Microsoft's Excel Solver.

The application of Holt's method requires the following equations [34]:

$$\hat{a}_t = \alpha Y_t + (1-\alpha)\left(\hat{a}_{t-1} + \hat{b}_{t-1}\right), \quad (2)$$

$$\hat{b}_t = \beta(\hat{a}_t + \hat{a}_{t-1}) + (1-\beta)\hat{b}_{t-1}, \quad (3)$$

$$P_{t+h} = \hat{a}_t + \hat{b}_t \times h, \quad (4)$$

where the first and second equations define the level and slope at each time point, respectively. The smoothing parameters, α and β, range from 0 to 1. The third equation corresponds to the sales forecast h-step-ahead.

This method also needs to be initialized, both in level and slope, so the average method was used, considering the recommendation of Fildes et al. [12]. For the tyre sizes where monthly forecasts were made, the first semester was used to this purpose. The first quarter was used for quarterly forecasts and two years for half-yearly forecasts.

The smoothing parameters were obtained through the minimization of the RMSE.

In Croston's method, the forecast is made using the following equations [14]:

$$P_{t+1} = \frac{a_t}{T_t}, \quad (5)$$

$$\begin{cases} a_{t+1} = \alpha Y_t + (1-\alpha)a_t, & Y_t > 0 \\ a_{t+1} = a_t, & Y_t = 0 \end{cases}{}' \quad (6)$$

$$T_{t+1} = \alpha q_t + (1-\alpha)T_t, \quad (7)$$

with P_t being the forecast of the demand at time t, a_t the correspondent level, T_t the time between the occurrence of two demands, and q_t the number of successive periods since the last demand occurrence. Thus, T_t is only updated when demand occurs.

Regarding the initialization of the method, a_1 corresponds to the first demand occurrence Y_1 and T_1 is equal to $Y_1 + 1$. The parameter α is the smoothing parameter, ranging from 0 to 1, and was obtained by minimizing the RMSE.

The SBA correction, which multiplies the forecasts resulting from the previous method (P_{t+1}) by $1 - \frac{\alpha}{2}$, was also tested.

3.2.2. Multiple Linear Regression and Generalized Linear Models

Having detected an apparent increase in tyre purchases in the months before and when rainfall typically occurs (September, October, November, and December) and in the months before summer travels (June, July, and August), simple linear regression (SLR) models were employed with time as the independent variable (t):

$$Y_t = \beta_0 + \beta_1 t + \varepsilon_t, \tag{8}$$

and MLR models, which also considered dummy variables:

$$Y_t = \beta_0 + \beta_1 t + \beta_2 c_t + \beta_3 v_t + \varepsilon_t, \tag{9}$$

that considered purchases in the rainy months (c_t, where 1 identifies September/October, October/November, or November/December, and 0 the remaining months) and others that account for purchases before summer trips (v_t, where 1 identifies June/July or July/August and 0 the remaining months), following Waters [5].

The parameter estimates of the models that minimize the MSE were obtained via SPSS software, version 27. The addition of the dummy variables to the SLR model increases the coefficient of determination, R^2, but the significant dummy variables are different between tyre sizes.

The use of these models in forecasting tyre sales should only be undertaken if the assumptions associated with them are met: namely, the normal distribution, homoscedasticity, and independence of the errors. In cases where these were compromised, GLM with Poisson and negative binomial distributions were considered due to the discrete nature of tyre demand. For the above distributions, the link function used is the logarithm [35]:

$$f(\mu) = log(\mu). \tag{10}$$

For a set of covariates, the tyre sales are conditionally independent with one of the above distributions whose mean, μ, relates with these covariates through the formula:

$$log(\mu) = \beta_0 + \beta_1 x_1 + \beta_2 x_2 + \beta_3 x_3. \tag{11}$$

Note that these models have a set of premises that need to be validated. The response variable must be non-negative, its observations independent, and the log mean a linear function of the covariates. In Poisson's distribution, equidispersion must also be validated.

3.2.3. Assessment Measures

To compare the accuracy and efficiency of the tested forecasting models, several measures were determined in accordance with Wallström and Segerstedt's [27] conclusion. The first was the RMSE, which measures the deviation of the estimates from the real value, advocated as appropriate by Bretschneider [36].

The average final inventory level, the percentage of shortages (which occurs whenever the final inventory of the period is equal to or lower than zero,) and PI were also calculated. The formulas applied were:

$$\text{RMSE} = \sqrt{\frac{1}{n}\sum_{t=1}^{n}(Y_t - \hat{Y}_t)^2}; \tag{12}$$

$$\text{Average Final Inventory: } \overline{FI} = \tfrac{1}{n}\textstyle\sum_{t=1}^{n} FI_t \times \mathbf{1}_{FI_t \geq 0}, \quad \text{where } FI_t = FI_{t-1} + \hat{Y}_t - Y_t, \text{ and } FI_1 = \hat{Y}_1 - Y_1; \tag{13}$$

$$\text{Shortage Percentage: } SP = \frac{1}{n}\sum_{t=1}^{n} \mathbf{1}_{FI_t \leq 0} \times 100\%; \tag{14}$$

$$\text{Average Period in Inventory: } \overline{PI} = \tfrac{1}{n}\textstyle\sum_{t=1}^{n} PI_t, \quad \text{where } PI_t = PI_{t-1} + \textstyle\sum_{i=1}^{t}(\hat{Y}_i - Y_i), \text{ and } PI_1 = \hat{Y}_1 - Y_1. \tag{15}$$

Finally, the estimates of the two forecasting methods with lower RMSE were combined through simple average according to Makridakis and Winkler [37], and following Aiolfi and Timmermann [38], a weighted combination of the same two methods was also performed for each tyre size according to the following formula:

$$Weighted\ Prediction_t = \frac{\text{RMSE}_2}{\text{RMSE}_1 + \text{RMSE}_2} Prediction_{1t} + \frac{RMSE_1}{\text{RMSE}_1 + \text{RMSE}_2} Prediction_{2t} \tag{16}$$

4. Results

This section presents the results obtained, divided into the levels of the aggregation of the tyre. First, the values determined for a specific tyre size are revealed, and then a global analysis of the remaining sizes is carried out.

4.1. Sales Grouped by Month

As for tyres with sales grouped monthly, the number of observations stands between 93 and 106, while the percentage of non-zero demand varies from 57% to 100%.

4.1.1. The Case of the 185/55R15 Tyre Size

The 185/55R15 tyre size has a total of 391 units sold, showing sales in 83% of the months. Table 1 shows the results obtained with each of the methods tested and in relation to each evaluation measure. The values of the combination of the two methods with the lowest RMSE value are also presented.

Table 1. Results of the evaluation measures for five forecasting methods for the 185/55R15 tyre size.

Measures	SES	Holt	GLM	Croston	SBA	Combined Forecast	
						Simple Average	Weighted Average
RMSE	3.039	3.170	**2.978**	3.252	3.156	3.004	3.004
$\overline{FI}$	16.248	**10.406**	19.535	34.347	12.822	16.248	16.248
SP	7.767%	17.476%	**6731%**	14.563%	15.534	7.767%	7.767%
$\overline{PI}$	**187**	−1127	−232	−708	−821	−22	−4

The method with the lowest RMSE value is the Poisson regression model, followed by SES and SBA. In this case, the estimated GLM model (ANOVA p-value < 5%) is:

$$log(\mu_t) = 1.266 + 0.298c_t, \tag{17}$$

with the independent variable months of October/November (c_t) being statistically significant, considering a 5% significance level. Time (t) and months of July/August (v_t) are not statistically significant.

The premises associated with the Poisson model were assessed and validated.

Figures 1 and 2 allow the visualization of the adjusted *p*-value of the determined models to the real observations, reflecting some differences, which consequently translate into different values in the evaluation measures, as shown in Table 1.

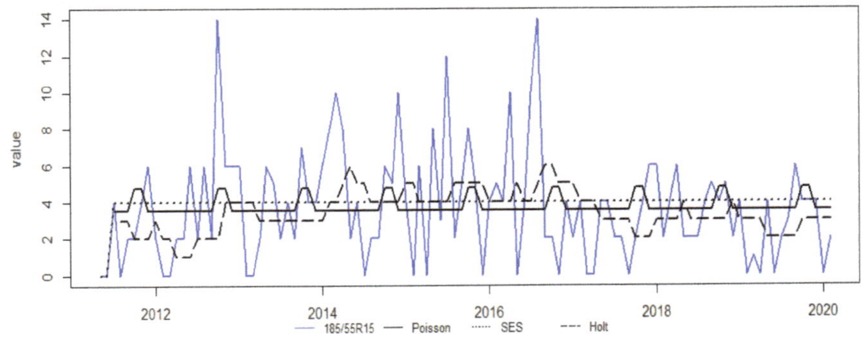

Figure 1. Poisson, SES and Holt's models for the 185/55R15 tyre size.

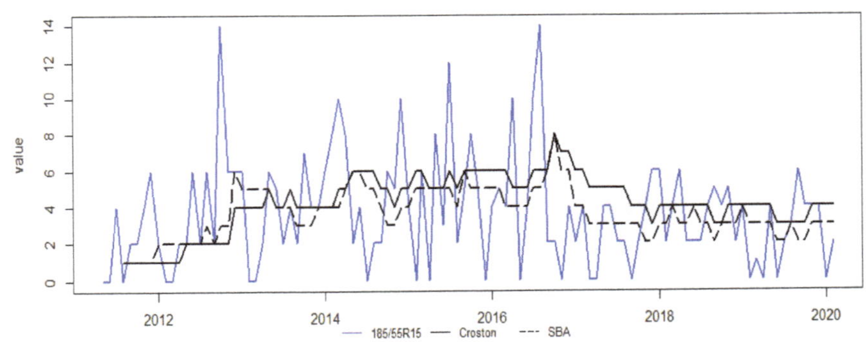

Figure 2. Intermittent models for the 185/55R15 tyre size.

Regarding the remaining assessment measures, the lowest $\overline{FI}$ value is obtained with Holt's method, but this method shows the highest SP, as seen in Table 1. Poisson's model, on the other hand, has the lowest SP, and SES shows the lowest $\overline{PI}$ value (not considering the weighted forecast).

Thus, based on Table 1, the model with the appropriated adjustment for the 185/55R15 tyre size results from the weighted combination of two methods with the lowest RMSE value, SES and GLM, as suggested by Petropoulos et al. [22]. In fact, the weighted model presents the second smallest value of RMSE and SP, the third smallest $\overline{FI}$ and the lowest $\overline{PI}$ value, being considered suitable to forecast.

Through the analysis of Table 2, one can observe that Holt's method and SES are the methods with the lowest value of RMSE when making estimates. Furthermore, the table shows that the optimal period for reviewing the data is monthly, which is expected since this item's sales are high and show variability, requiring constant revision.

Given the above, the most suitable inventory policy for the 185/55R15 tyre size is the (R,s,S) policy, where the review period (R) must be monthly and the quantity to order (S) will aim to satisfy the expected demand, which is around 3 units (average S), as shown in Table 3, considering the final inventory of the previous month and the safety inventory (s). The latter must be established in two units in order to satisfy unforeseen demand because of the variability observed in sales, ensuring that it is possible to satisfy one customer's needs (in general, two tyres are always supplied at a time).

Table 2. RMSE values from March 2020 to February 2021, for the 185/55R15 tyre size, without reviewing the previously obtained model and updating it every 1, 2, 4, and 6 months with real data.

		SES	Holt	GLM	Croston	SBA
DETERMINED MODEL		3.039	3.170	2.978	3.252	3.156
REVISION	No review	3.109	2.692	3.132	3.004	2.842
	Monthly	3.109	**2.475**		3.141	3.140
	2 months	3.109	2.795		3.187	3.226
	4 months	3.109	2.934		3.169	3.161
	6 months	3.109	2.737		2.955	2.883

Table 3. Determination of the monthly quantity to order (S) for the 185/55R15 tyre size.

	Real Sales	Forecast		
		Holt	SES	Weighted
MAR/20	0	2	4	3
APR/20	4	2	4	3
MAY/20	2	2	4	3
JUN/20	0	2	4	3
JUL/20	2	2	4	3
AUG/20	4	2	4	3
SEP/20	10	2	4	3
OCT/20	2	3	4	4
NOV/20	0	3	4	4
DEC/20	0	3	4	3
JAN/21	2	2	4	3
FEB/21	4	2	4	3

4.1.2. Global Analysis of Remaining Tyre Sizes

The results of the forecasting methods allow us to conclude that SES is the method with the lowest RMSE in 48% of tyre sizes, followed by Holt's method in 26% of the sizes, as shown in Table 4.

Table 4. Summary of the evaluation measures applied to each forecasting method, considering tyre sizes with sales grouped monthly.

	SES	Holt	MLR/GLM	Croston	SBA
LOWEST RMSE	**15 (48%)**	8 (26%)	2 (6%)	2 (6%)	4 (13%)
2ND LOWEST RMSE	8 (26%)	**9 (29%)**	0 (0%)	**9 (29%)**	5 (16%)
SMALLEST $\overline{FI}$	6 (19%)	7 (23%)	0 (0%)	6 (19%)	**12 (39%)**
LARGEST $\overline{FI}$	7 (23%)	**13 (42%)**	0 (0%)	11 (35%)	0 (0%)
LOWER *SP*	**17 (55%)**	11 (35%)	1 (3%)	1 (3%)	1 (3%)
HIGHER *SP*	2 (6%)	4 (13%)	0 (0%)	11 (35%)	**14 (45%)**
LOWEST $\overline{PI}$ (ABSOLUTE VALUE)	7 (23%)	**16 (52%)**	2 (6%)	2 (6%)	4 (13%)
$\overline{PI} < 0$	14 (45%)	17 (55%)	12 (39%)	**30 (97%)**	**30 (97%)**

The small presence of MLR and GLM in this general analysis results from the fact that once the validity of the assumptions of the different models and the statistical significance of the independent variables was analysed, it was determined that in only two tyre sizes, one with normal and other with Poisson distribution, can these models be considered valid, and so they were abandoned for the remaining sizes.

Croston's method and SBA only performed better in six sizes, which show demand in less than 70% of the months, and in the remaining four sizes with the same percentage of demand, SES and Holt's method show lower RMSE.

According to Table 4, concerning the average inventory, SBA and Holt's method present the lowest $\overline{FI}$ in 39% and 23% of the sizes, respectively. However, Holt's and

Croston's methods show, in 42% and 35% of the analysed sizes, the largest $\overline{FI}$. On the other hand, the highest SP is verified in Croston's method and SBA (35% and 45%, respectively), with SES being the method with the most intermediate values of $\overline{FI}$ and SP.

Observing the $\overline{PI}$ values, in absolute value, the smallest numbers are found with Holt's method in 52% of the sizes. The greatest underestimation of inventory occurs with Croston's method and SBA in 97% of the sizes, with SES and Holt's method showing a similar percentage of underestimation and overestimation of demand.

When the validated models with the two smallest RMSE are grouped, the combination of SES and Holt's method is the winner in 45% of the sizes, followed by Croston's method and SBA in 16% of the sizes, according to Table 5.

Table 5. Crossover of the models with the lowest RMSE for tyre sizes with sales grouped monthly.

	Holt	GLM	MLR	Croston	SBA
SES	**14 (45%)**	1 (3%)	1 (3%)	3 (10%)	4 (13%)
HOLT		0 (0%)	0 (0%)	3 (10%)	0 (0%)
CROSTON		0 (0%)	0 (0%)		5 (16%)

After obtaining the forecasts by weighting the two methods with the lowest RMSE, one observes that the resulting model presents the lowest RMSE, average inventory, SP, and $\overline{PI}$ values in 97%, 39%, 65%, and 100% of the sizes, respectively.

Once the forecasts from March 2020 to February 2021 were determined, the accuracy of the forecasting methods regarding the RMSE (and shown in Table 6) was studied.

Table 6. Evaluation of the RMSE value obtained for each model in different review periods, considering tyre sizes with sales grouped monthly.

Forecast	SES	Holt	MLR/GLM	Croston	SBA
SAME AS MARCH 2020	10 (32%)	**13 (42%)**	0 (0%)	3 (10%)	5 (16%)
MONTHLY REVIEW	6 (19%)	**22 (71%)**	1 (3%)	0 (0%)	2 (6%)
2-MONTH REVIEW	8 (26%)	**20 (65%)**	0 (0%)	0 (0%)	3 (10%)
4-MONTH REVIEW	13 (42%)	**14 (45%)**	0 (0%)	1 (3%)	3 (10%)
6-MONTH REVIEW	10 (32%)	**16 (52%)**	0 (0%)	3 (10%)	2 (6%)

Note that if the forecast calculated for March 2020 does not change during the following year, the method with the smallest error becomes Holt's method, with SES being the second-best option. If the forecasts are revised monthly, Holt's method remains the one with the lowest RMSE, as when the revision is carried out every 2, 4, or 6 months. The second method with the lowest RMSE in all revision periods remains SES.

As for the review period, the lowest RMSE value is obtained when the data is reviewed monthly (77% of the sizes), followed by the constant forecast equal to March 2020 or revision every 6 months (10% of the sizes) and the 4-month review (3% of the sizes).

Thus, for tyre sizes with sales grouped by month, one concludes that the smallest error is achieved by combining the weighted forecasts resulting from SES and Holt's method, with monthly data revision.

Table 7 presents the parameters of the (R,2,S) inventory policy specific to three tyre sizes. It is possible to verify the variability of the sales units foreseen for each month and between sizes.

Table 7. (R,2,S) Inventory policy parameters for 175/65R14, 195/65R15 and 385/65R22.5 tyre sizes.

R–Monthly		175/65R14	195/65R15	385/65R22.5
AVERAGE S		13	17	15
Weighted Forecast	20 March	14	17	12
	20 April	13	16	12
	20 May	13	16	13
	20 June	13	16	13
	20 July	13	17	14
	20 August	14	17	14
	20 September	14	18	14
	20 October	13	18	15
	20 November	13	19	16
	20 December	13	17	16
	21 January	14	17	18
	21 February	13	17	18

4.2. Sales Grouped by Quarter

Considering the tyre sizes with sales grouped quarterly, the number of observations ranges between 31 and 35, while the percentage of demand different from zero fluctuates from 54% to 89%.

A similar analysis was carried out for the monthly cases, and the results of the forecasting methods show that SES is the method with the lowest RMSE in 30% of the sizes, followed by Croston's method in 26% of the sizes.

The MLR and GLM models, except for one tyre size, were abandoned since the premises of the models were not met.

Regarding the $\overline{FI}$, SBA presents the lowest inventory level in 43% of the sizes, contrary to SES and Croston's method, with 48% and 35%, respectively, of the largest inventory. The highest SP happens with Croston's method and SBA (in 39% of the sizes).

In absolute value, the lowest $\overline{PI}$ numbers are obtained with Holt's method. The largest underestimation of inventory occurs with Croston's method and SBA.

When combining the validated models with the two smallest RMSE, grouping SES and Holt's method show the best values in 43% of the sizes, followed by Croston's method and SBA in 39% of the sizes.

After obtaining the forecasts and weighing the two methods with the lowest RMSE, the resulting model shows the lowest RMSE, $\overline{FI}$, SP, and $\overline{PI}$ values in most tyre sizes.

Once the forecasts from the second quarter of 2020 to the first quarter of 2021 were determined, one verifies that when keeping the forecast for the second quarter of 2020 constant during the analysis period, the method with the lowest MSE continues to be SES, followed by Croston's method. If the forecasts are revised quarterly, SES remains the method with the minimum error and the second place is divided by Croston's method and SBA. However, when the review is performed every two quarters, in 30% of the sizes, SES obtains the lowest RMSE, as does SBA.

As for the review period, the lowest RMSE value is obtained when there is no revision, followed by the review every two quarters and, finally, the quarterly review.

Therefore, the weighted combination of SES and Croston's method appears to be the most adequate for data grouped quarterly, keeping the forecast constant for one year.

4.3. Sales Grouped by Semester

Concerning sales grouped every six months, the number of observations ranges between 11 and 19, while the percentage of non-zero demand varies from 57% to 100%.

As in the previous cases, a thorough analysis of the forecasting methods under study was undertaken. After applying the forecasting methods, Holt's method reveals the lowest RMSE in 46% of the tyre sizes, followed by SES in 29% of the sizes.

Regarding the $\overline{FI}$, SBA presents the lowest inventory level in 71% of the sizes, contrary to SES and Holt's method, with 39% and 46%, of the largest inventory. However, the highest SP is also verified with SBA.

Concerning the $\overline{PI}$, in absolute value, the lowest numbers are obtained with Holt's method, followed by SES and MLR/GLM. The greatest underestimation of inventory occurs with Croston's method and SES.

When combining the forecasting models with the two smallest RMSE, the combination of SES and Holt's method is preferred in 46% of sizes, followed by SES and Croston's method in 14% of sizes.

Having determined the weighted forecasts with the two methods with the lowest RMSE, the resulting model achieves the lowest RMSE, $\overline{FI}$, SP, and $\overline{PI}$ values in the majority of the sizes.

After obtaining the forecasts from the second half of 2020 to the first half of 2021, one notices that when keeping the forecast for the second semester of 2020 constant during the year, the method with the smallest error becomes SES, followed by Holt's. If the forecasts are revised every six months, SES remains the method with the lowest error, with Holt's method following.

As for the review period, the lowest RMSE value is obtained when no review is performed, compared to the biannual review (in 54% and 46% of sizes, respectively). Note that there are only two forecasts under analysis.

In summary, concerning the tyre sizes with sales grouped by semester, the smallest RMSE is achieved by combining the weighted forecasts resulting from SES and Holt's method. Keeping the forecast constant for one year or revising it every semester presents a similar error value.

5. Discussion and Conclusions

The importance of having a clearly defined inventory policy in a company is revealed by the present work. In fact, determining when to place orders and what is the optimal quantity to order for each item not only improves the use of the retailer's financial resources, but also guarantees consumer satisfaction, as it is possible to answer their needs in a faster and more diversified way.

However, it is not always clear which of the sales forecasting method is the most suitable—the one that allows to reduce the uncertainty of demand—and it may even be difficult to find one that can be adjusted.

After testing five forecasting models with 82 tyre sizes, some differences arose according to the level of sales aggregation. Specifically, for tyre sizes with sales grouped monthly, the method considered the most suitable when determining the model was SES, which was surpassed by Holt's method after determining the forecasts, which agrees with some authors about being unrealistic that a single model is predominant over time.

Likewise, it appears that while the second method with the lowest RMSE in the models determined for tyre sizes with sales grouped quarterly was Holt's method, once the forecasts were calculated, it was surpassed by Croston's method.

It is also determined that the revision periods are different and that the quantities to order are quite diverse between sizes and, in multiple situations, between months/quarters/semesters for the same size.

For the most sold tyre sizes (those grouped by month), the review period must be monthly. For sizes gathered by quarter, the review period should be annual. As for the tyres grouped by semester, it seems to be more prudent to review sales every semester.

Regarding the quantity to order, and since it was possible to adjust demand forecasting models to the different tyre sizes, supply must be related to demand—that is, one ought to order a sufficient quantity in order to meet the expected demand during the revision period and ensure that there is always a safety inventory to satisfy unexpected demand (at least, a pair of tyres of each size sold frequently).

This is defined as an (R,s,S) inventory policy, considered adequate for items with faster output, but it is also slower, and the model to be used results from the weighted combination of the forecasts of the two methods that presented the lowest RMSE for each level of data aggregation.

In summary, for data grouped monthly and semesterly, the forecasts obtained with SES and Holt's method should be combined, while for data grouped quarterly, the estimates resulting from SES and Croston's method ought to be weighted.

SES seems to be suitable to forecast items with intermittent demand. However, it is worth mentioning the importance of analysing the models at each review period and of adjusting accordingly to new information that is acquired. The need for evaluating the error measures is also evident in order to detect biases as soon as possible.

As for future work, it can be useful to carry out a survey to assess the consumer's purchasing behaviour and decision-making process regarding tyres and to collect complementary information, whether concerning economic aspects or the proximity of competitors, so as to add practical information and make the forecasting models more accurate.

Furthermore, the application of other forecasting methods, such as neural networks and bagged forecasts, should be evaluated, analysing the possible benefits of improved inventory system versus increased complexity in forecasting methods given the retail sector we are working in.

Consequently, the practical implications of this study ought to be analysed after the implementation of the proposed inventory plan in the company's inventory levels and associated costs.

Author Contributions: Conceptualization, all authors; methodology, software, visualization and writing—original draft preparation, D.N. and M.M.; validation, formal analysis and data curation, D.N.; supervision and writing—review and editing, M.M. and M.J.F. All authors have read and agreed to the published version of the manuscript.

Funding: This research was funded by The Center for Research and Development in Mathematics and Applications (CIDMA) through the Portuguese Foundation for Science and Technology (FC-Fundação para a Ciência e a Tecnologia), grant number UIDB/04106/2020.

Institutional Review Board Statement: Not applicable.

Informed Consent Statement: Not applicable.

Data Availability Statement: The data presented in this study are available on request from the corresponding author.

Conflicts of Interest: The authors declare no conflict of interest.

References

1. Tian, X.; Wang, H.; Erjiang, E. Forecasting intermittent demand for inventory management by retailers: A new approach. *J. Retail. Consum. Serv.* **2021**, *62*, 102662. [CrossRef]
2. European Tyre and Rubber Manufacturers' Association. *The ETRMA Statistics Report*; ETRMA: Brussels, Belgium, 2019.
3. Mahama-Musah, F.; Vanhaverbeke, L.; Gillet, A. The impact of personal, market- and product-relevant factors on patronage behaviour in the automobile tyre replacement market. *J. Retail. Consum. Serv.* **2020**, *57*, 102206. [CrossRef]
4. Goltsos, T.E.; Syntetos, A.A.; Glock, C.H.; Ioannou, G. Inventory–forecasting: Mind the gap. *Eur. J. Oper. Res.* **2022**, *299*, 397–419. [CrossRef]
5. Waters, D. *Inventory Control and Management*; Wiley: Hoboken, NJ, USA, 2003.
6. Babiloni, E.; Cardós, M.; Albarracín, J.M.; Palmer, M. Demand Categorisation, Forecasting, and Inventory Control for Intermittent Demand Items. *South Afr. J. Ind. Eng.* **2012**, *21*, 101–113. [CrossRef]
7. Hillier, F.; Lieberman, G. *Introduction to Operations Research*; McGraw-Hill: New York, NY, USA, 2001.
8. Blattberg, R.C.; Hoch, S.J. Database models and managerial intuition: 50% model + 50% manager. *Perspect. Promot. Database Mark. Collect. Work. Robert C Bl.* **2010**, *36*, 228–240. [CrossRef]
9. Beheshti-Kashi, S.; Karimi, H.R.; Thoben, K.-D.; Lütjen, M.; Teucke, M. A survey on retail sales forecasting and prediction in fashion markets. *Syst. Sci. Control Eng.* **2015**, *3*, 154–161. [CrossRef]
10. Ma, S.; Fildes, R. Retail sales forecasting with meta-learning. *Eur. J. Oper. Res.* **2021**, *288*, 111–128. [CrossRef]

11. Willemain, T.R.; Smart, C.N.; Schwarz, H.F. A new approach to forecasting intermittent demand for service parts inventories. *Int. J. Forecast.* **2004**, *20*, 375–387. [CrossRef]
12. Fildes, R.; Hibon, M.; Makridakis, S.; Meade, N. Generalising about univariate forecasting methods: Further empirical evidence. *Int. J. Forecast.* **1998**, *14*, 339–358. [CrossRef]
13. Syntetos, A.A.; Boylan, J.E. The accuracy of intermittent demand estimates. *Int. J. Forecast.* **2005**, *21*, 303–314. [CrossRef]
14. Croston, J.D. Forecasting and Stock Control for Intermittent Demands. *Oper. Res. Q. (1970–1977)* **1972**, *23*, 289. [CrossRef]
15. Gardner, E.S., Jr. Exponential smoothing: The state of the art—Part II. *Int. J. Forecast.* **2006**, *22*, 637–666. [CrossRef]
16. Kourentzes, N. On intermittent demand model optimisation and selection. *Int. J. Prod. Econ.* **2014**, *156*, 180–190. [CrossRef]
17. Syntetos, A.; Boylan, J.E. On the bias of intermittent demand estimates. *Int. J. Prod. Econ.* **2001**, *71*, 457–466. [CrossRef]
18. Sbrana, G. Modelling intermittent time series and forecasting COVID-19 spread in the USA. *J. Oper. Res. Soc.* **2022**, *74*, 465–475. [CrossRef]
19. Nikolopoulos, K.; Syntetos, A.A.; Boylan, J.E.; Petropoulos, F.; Assimakopoulos, V. An aggregate–disaggregate intermittent demand approach (ADIDA) to forecasting: An empirical proposition and analysis. *J. Oper. Res. Soc.* **2011**, *62*, 544–554. [CrossRef]
20. Gutierrez, R.S.; Solis, A.O.; Mukhopadhyay, S. Lumpy demand forecasting using neural networks. *Int. J. Prod. Econ.* **2008**, *111*, 409–420. [CrossRef]
21. Mitrea, C.A.; Lee, C.K.M.; Wu, Z. A comparison between neural networks and traditional forecasting methods: A case study. *Int. J. Eng. Bus. Manag.* **2009**, *1*, 19–24. [CrossRef]
22. Petropoulos, F.; Makridakis, S.; Assimakopoulos, V.; Nikolopoulos, K. 'Horses for Courses' in demand forecasting. *Eur. J. Oper. Res.* **2014**, *237*, 152–163. [CrossRef]
23. Hibon, M.; Evgeniou, T. To combine or not to combine: Selecting among forecasts and their combinations. *Int. J. Forecast.* **2005**, *21*, 15–24. [CrossRef]
24. Koutsandreas, D.; Spiliotis, E.; Petropoulos, F.; Assimakopoulos, V. On the selection of forecasting accuracy measures. *J. Oper. Res. Soc.* **2022**, *73*, 937–954. [CrossRef]
25. Strijbosch, L.W.; Syntetos, A.A.; Boylan, J.E.; Janssen, E. On the interaction between forecasting and stock control: The case of non-stationary demand. *Int. J. Prod. Econ.* **2011**, *133*, 470–480. [CrossRef]
26. Teunter, R.H.; Duncan, L. Forecasting intermittent demand: A comparative study. *J. Oper. Res. Soc.* **2009**, *60*, 321–329. [CrossRef]
27. Wallström, P.; Segerstedt, A. Evaluation of forecasting error measurements and techniques for intermittent demand. *Int. J. Prod. Econ.* **2010**, *128*, 625–636. [CrossRef]
28. Eaves, A.H.C.; Kingsman, B.G. Forecasting for the ordering and stock-holding of spare parts. *J. Oper. Res. Soc.* **2004**, *55*, 431–437. [CrossRef]
29. Syntetos, A.A.; Boylan, J.E. On the stock control performance of intermittent demand estimators. *Int. J. Prod. Econ.* **2006**, *103*, 36–47. [CrossRef]
30. Syntetos, A.A.; Boylan, J.E. Demand forecasting adjustments for service-level achievement. *IMA J. Manag. Math.* **2008**, *19*, 175–192. [CrossRef]
31. Natarajan, K.; Soundararajan, K.; Jayakrishnan, J. Buyer Behaviour Patterns and Satisfaction Trends of Commercial Vehicles Owners in Replacement of Tyres. *Annamalai Int. J. Bus. Stud. Res.* **2013**, *5*, 79–86.
32. Sagaert, Y.R.; Aghezzaf, E.-H.; Kourentzes, N.; Desmet, B. Temporal big data for tactical sales forecasting in the tire industry. *INFORMS J. Appl. Anal.* **2018**, *48*, 121–129. [CrossRef]
33. Hyndman, R.; Athanasopoulos, G. *Forecasting: Principles and Practice*; Springer: Berlin/Heidelberg, Germany, 2018.
34. Holt, C.C. Author's retrospective on 'Forecasting seasonals and trends by exponentially weighted moving averages'. *Int. J. Forecast.* **2004**, *20*, 11–13. [CrossRef]
35. Turkman, M.A.A.; Silva, G.L. *Modelos Lineares Generalizados: Da Teoria à Prática*; Edições SPE: Lisboa, Portugal, 2000.
36. Bretschneider, S. Estimating forecast variance with exponential smoothing Some new results. *Int. J. Forecast.* **1986**, *2*, 349–355. [CrossRef]
37. Makridakis, S.; Winkler, R.L. Averages of Forecasts: Some Empirical Results. *Manag. Sci.* **1983**, *29*, 987–996. [CrossRef]
38. Aiolfi, M.; Timmermann, A. Persistence in forecasting performance and conditional combination strategies. *J. Econ.* **2006**, *135*, 31–53. [CrossRef]

engineering proceedings

Proceeding Paper

Medium-Term Horizon Time Photovoltaic Power Generation Prediction for an Island Zone †

Harry Ramenah [1,*], Camel Tanougast [1], Nidhal Rezg [2] and Abdel Khoodaruth [3]

1 Laboratoire de Conception, Optimisation et Modélisation des Systèmes (LCOMS), University of Lorraine, 57070 Metz, France; camel.tanougast@univ-lorraine.fr
2 Laboratoire de Génie Informatique, de Production et de Maintenance, University of Lorraine, 57000 Metz, France; nidhal.rezg@univ-lorraine.fr
3 Department of Mechanical and Production Engineering, University of Mauritius, Réduit 80837, Mauritius; a.khoodaruth@uom.ac.mu
* Correspondence: harry.ramenah@univ-lorraine.fr
† Presented at the 9th International Conference on Time Series and Forecasting, Gran Canaria, Spain, 12–14 July 2023.

Abstract: This article presents a Johansen test assessing the predetermined long-term relationships of datasets from a photovoltaic (PV) plant to predict the power output of an island zone. The goal was to use Johansen's model to predict the PV power generation in the island of Mauritius. In this article, time series using an on-site measurement dataset have been used to design an original prediction model, the Johansen model for PV power output. This model is trained to predict random monthly, weekly, and daily PV power outputs in different seasons and years. The experimental results demonstrate that the Johansen model is a powerful medium-term predicting tool.

Keywords: predicting; cointegration; Johansen

Citation: Ramenah, H.; Tanougast, C.; Rezg, N.; Khoodaruth, A. Medium-Term Horizon Time Photovoltaic Power Generation Prediction for an Island Zone. *Eng. Proc.* **2023**, *39*, 2. https://doi.org/10.3390/engproc2023039002

Academic Editors: Ignacio Rojas, Hector Pomares, Luis Javier Herrera, Fernando Rojas and Olga Valenzuela

Published: 25 June 2023

1. Introduction

Islands close to the Indian Ocean require imports of primary energy resources such as natural gas, coal, and fossil energy to meet everyday requirements. Therefore, in response to the severe climate change and global energy crisis, the government of Mauritius has outlined its intention to oversee and promote solar photovoltaic (PV) technologies through the local electricity public utility company. However, the accurate prediction and integration of PV energy must be preceded by better planning and distribution strategies to ensure the stability of electrical grid energy.

Systems transform solar energy into electric power. Accurate photovoltaic power generation is linked to the accurate forecasting of solar irradiance. Thus, most research studies are based on solar irradiance forecasting, and several forecasting methods have been cited in the literature [1–4], ranging from physical [5,6] to machine learning methods [7,8] via classical statistical techniques. The latter have been classified as linear stationary models, which is mainly where Auto-Regressive Moving Average (ARMA) models and Vector Auto Regressive (VAR) are concerned, or linear non-stationary models, e.g., Auto-Regressive Integrated Moving Average (ARIMA) [9] techniques, which are stochastic process combining Auto-Regressive (AR) and Moving Average (MA) components. The downside of these statistical models is that forecast accuracy decreases with increasing independent variables such as for the PV power output, which is a function that depends on various weather parameters. Moreover, the forecast of PV power generation has also been classified [10–12] in time horizon depending on the needs of the PV production and electrical transport management. The proposed time horizon is defined as follows:

- Very short-term forecast horizon: a few seconds to one hour; used for electricity dispatch in real-time and energy smoothing.

- Short-term forecast: for one or several hours up to the day ahead; guarantees system commitment and scheduling.
- Medium-term forecast: multiple days to months ahead; ensures power system planning.
- Long-term forecast: months to one or several years; finds and assesses potentially resourceful sites.

For this research paper, we used a robust statistical cointegration approach, the Johansen vector error correction model (VECM), to determine long-term equilibrium relationships. It is important to note that the Johansen approach for cointegration has been a popular tool in applied economics but it has never been applied to PV systems for forecasting. A unique feature of the resulting model is that it can be applied for the medium-term horizon time PV power output prediction in an island zone. The measured dataset is compared to the model one for different seasonal horizon times. The obtained results prove the model's efficacy and show promise for application in further studies aiming to investigate short-term horizon times.

The rest of this paper is organized as follows: Section 2 contains a short historical overview of the experimental conditions and methods used in our study. Section 3 presents a brief introduction of the Johansen test. Section 4 presents the obtained results. Finally, Section 5 provides our conclusions and presents ideas for future studies.

2. Research History

2.1. European Zone

Firstly, we studied a linear relation analysis of time series data collected from a PV system from the GREEN lab in Metz, France [13]. The PV system is a grid-connected system equipped with solar irradiance and temperature sensors. Therefore, the power output is the dependent variable, while the solar irradiance and PV module temperature are the explanatory variables. We performed time series, correlograms at level, Augmented Dickey–Fuller (ADF) [14,15] stationarity tests, and residuals analysis such as the Goldfeld–Quandt (GQ) and Durbin–Watson tests. When outliers were suspected, we applied the Engle–Granger (EG) method [16,17] to determine the most appropriate model. The determined power output EG cointegration relationship was then applied to other experimental years for the Green lab in the European zone and compared to the measured power output data.

2.2. Island Zone

The EG model was applied to a PV system in the Reunion Island in the Indian Ocean, and results were not very conclusive. Other weather parameters (such as wind, humidity, dust) must be considered in such zones as they have an impact on the PV power output. Spurious regression appeared. Consequently, we investigated whether the Johansen VECM was a model that could be fitted to the zone, as more than two explanatory variables should be considered. Indeed, the Johansen VECM cointegration test can be considered as a multivariate generalization of the ADF test, enabling the estimation of all cointegrating vectors when more than two variables are considered. The PV system in Reunion island is equipped with solar irradiance, module temperature, and wind and humidity sensors, and data is recorded at a sample time of 15 min. The Johansen VECM cointegration test has already been applied to Reunion Island, as reported elsewhere [18]. Johansen cointegration Trace and Eigen value tests, as well as lag criteria had been carried out to determine the number of cointegration equations with the corresponding error correction coefficients. The Wald test was applied to exclude short-term relationships, and white noise tests, such as the Lagrange multiplier and Jarque–Bera test, were applied to ensure normal residual distribution. The final cointegration relationship was then validated to determine long-term horizon times [18].

The obtained resulting Johansen cointegration relationship was applied to a PV plant in the island of Mauritius in the southwest Indian Ocean close to Reunion island. These islands have similar weather conditions, and the year is divided in two main seasons: summer and winter. The summer and the intermediate seasons run from November to

April, while the winter and the intermediate seasons run from May to October. In this paper, the used dataset for the island of Mauritius and the Johansen model spans over a one-year period (in this case, from July 2019 to July 2020 and from October 2021 to December 2022). The PV technology was silicon polycrystalline with a rated power of 320 Wp (north-facing orientation), and the PV total power was 20 kWp. The recorded data was filtered so that data between 4:30 P.M. and 7:30 A.M. for the following day and missing data were removed.

3. Johansen Procedure

If time series are nonstationary, it is not possible to use ordinary least square (OLS) to estimate their long-run linear relationships because this would lead to spurious regression. The technique of cointegration was introduced according to which models contained non-stationary stochastic variables and can be constructed in a way that ensures that the results are statistically meaningful. Therefore, the cointegration test and a vector error correction model were established to distinguish between short-term and long-term equilibrium. The Johansen procedure involves cointegrated variables that are directly constructed based on maximum likelihood estimation instead of depending on OLS estimation. The Johansen procedure [19,20] is simply a multivariate generalization of the ADF test and proposes two different likelihood ratio tests: (a) Trace test, (b) maximum Eigen values test. The Trace test is the null hypothesis tests of r cointegrating vectors against the alternative hypothesis of n cointegrating vectors, whereas the maximum eigen-value test is the null hypothesis tests of r cointegrating vectors against the alternative hypothesis of (r + 1) cointegrating vectors. A more complete description of the Johansen procedure is given in [18]. Therefore, the procedure required when conduction a Johansen test can be summarized as follows:

- Perform series stationarity tests to determine the existence of cointegration relationships.
- If the previous step has been followed correctly, then the series are of the same order of integration, and cointegration is likely; therefore, the VECM model can be estimated. The lag length should be determined using Akaike and Schwarz's criteria.
- The Johansen Trace and Eigen value tests should then be performed to determine the number of cointegration relationships.
- Then, identify the long-term relationships or cointegration relationships between variables.
- Subsequently, via the maximum likelihood method, the VECM model can be estimated, and validation tests can be performed for residuals and white noise.

4. Results

4.1. Long-Horizon Term: Yearly

The methodology described in the previous section was applied to develop a long-term prediction model for a grid-connected PV plant in Reunion island. Hence, the results reported in this section are based on the Johansen model that was initially trained on a dataset from 2013. The dataset consisted of a 15 by 15 min series of simultaneous solar irradiance, wind speed, humidity, and PV system parameters (mainly power output and module temperature) covering a whole year. The model had been trained [18] and compared to a dataset from 2014 for validation regarding yearly horizon time.

The considered model was also applied to the island of Mauritius, where the experimental data employed in this current analysis were obtained from a PV plant located on the rooftop of the University of Mauritius. The experimental conditions were listed in Section 2.2. Figure 1 illustrates a scatter plot of the measured power output, determined via the Johansen model, on a yearly basis from July 2019 to July 2020 for each solar day hour between 09:00 A.M. to 04:00 P.M. The very high linear regression value (96%), reveals that the on-site measurement of PV power generation and the Johansen model are fundamental in accurately predicting the PV power output at different horizon times. This is elaborated in the following sections. Moreover, the model can be applied on a yearly basis for long-term horizon predicting, meaning that it can be used for unit commitment, load balancing and scheduling, and for planning infrastructure.

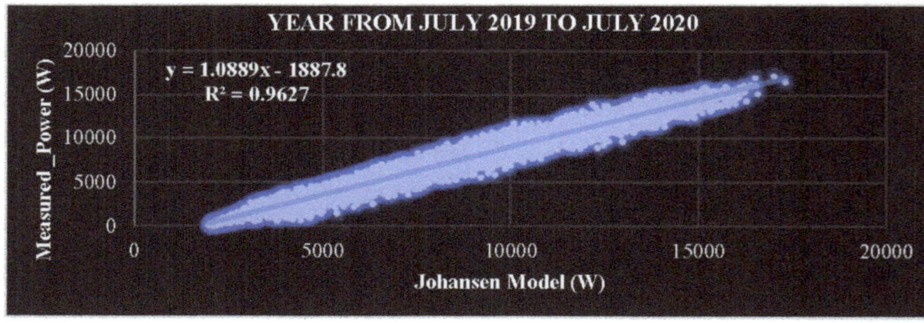

Figure 1. Validation of Johansen model for one dataset between July 2019 to July 2020 in the island of Mauritius.

4.2. Long-Horizon Term: Monthly

In the literature, periods of a month to a year are considered as long horizon term. These predicted horizons are suitable for long-term power generation and feed into power grids while playing a role in determining seasonal trends. Therefore, taking into consideration the weather conditions described in Section 2.2, we compared the on-site power output measurements to the Johansen model for different months in different seasons. We have chosen one particular month for each summer and winter season and two others for the intermediate season for the years 2019 and 2020. In Figure 2, the left-hand side plots represent the measured power output and the corresponding Johansen model for the month of January 2020 in the summer season, July 2019 for the winter season, and the months of April 2020 and October 2019 for the intermediate season. The high R^2 values of the linear regressions and the fact that most of these values are greater than 92% mean that there is good agreement between the measured power and the Johansen model. The right-hand side of Figure 2 are scatter plots representing the predicting power output. Indeed, we are comparing the results for the same period on the left-hand side but for years N + 1 or N + 2, as if we were predicting the same corresponding months.

The R^2 values are greater than 90%, indicating that the Johansen model is positively and strongly correlated with on-site measurements. For the island of Mauritius and in agreement with the local energy supplier, this monthly long horizon term can be essential for maintaining the solar schedule.

4.3. Medium-Horizon Term: Weekly

Although the prior results seem adequate, weather fluctuations spanning one or few days may reduce predicting accuracy. Therefore, we decided to extend the study to shorter periods, that is, weekly medium-horizon term. For this study, we selected random weeks for each season and the intermediate seasons. The results are given in Figure 3. The left-hand side are the scattered plots of random weeks for the weeks in 2020 or 2019. For the summer season, the random week selected was 19 to 25 January 2020. For the intermediate seasons, the selected weeks were 12 to 18 April 2020 and 10 to 17 September 2019. Finally, for the winter season, the selected week was 10 to 17 August for 2019. The vertical axis is the on-site measurements, and the horizontal axis is the Johansen model. The R^2 values of the linear regressions are greater than 95%, showing that the model is well adapted for predictions. Therefore, we applied the model to the same periods but for the year of 2022, as if we were led to make predictions for these similar weeks. The scattered plots on the right-hand side of Figure 3 show the predicting results. The vertical axis is the Johansen model, and the horizontal axis is the measured power output. Figure 3 shows that the R^2

values are more than 90%, again indicating a good correlation and proving suitable for this horizon term.

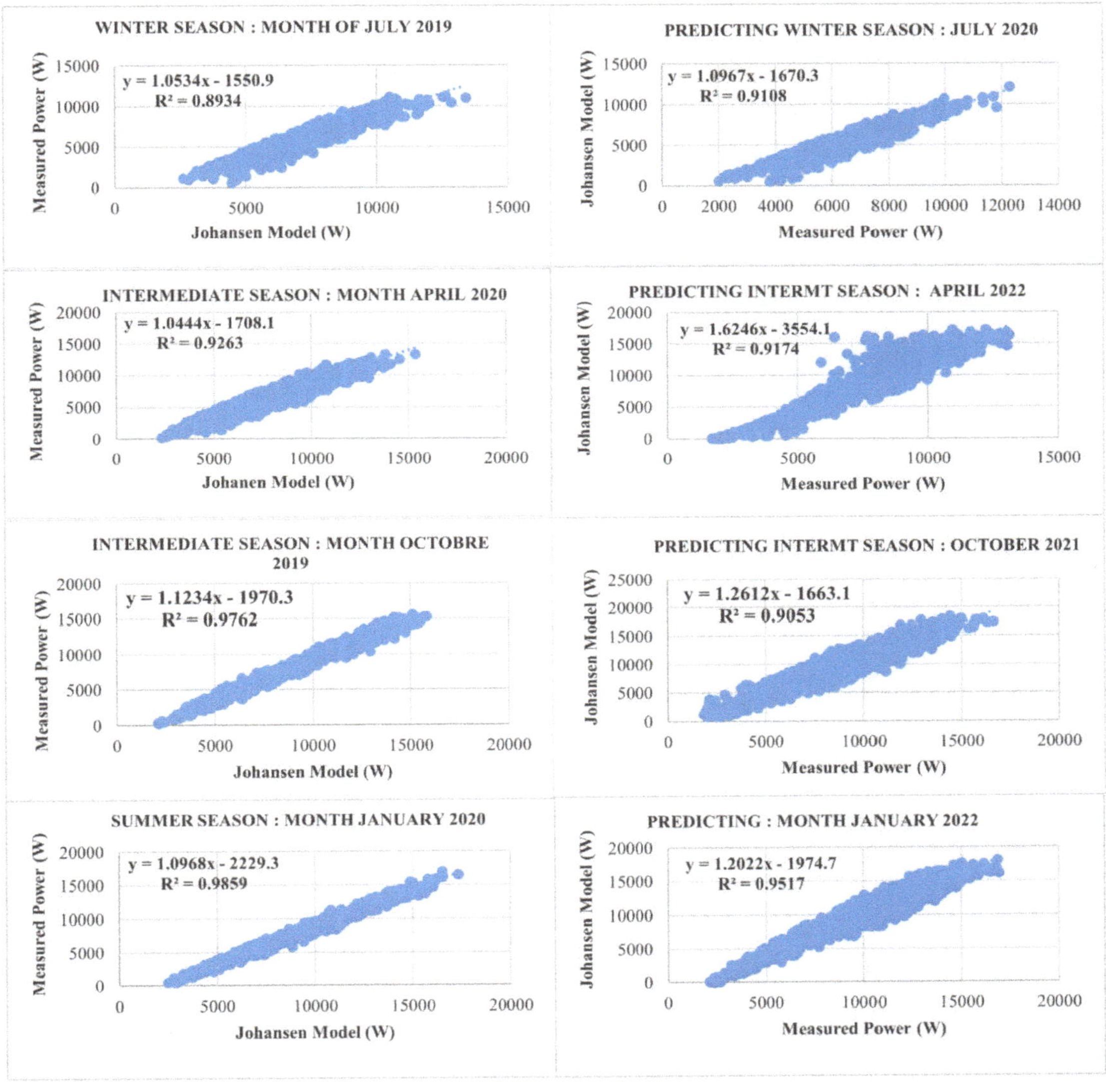

Figure 2. On the left: testing the Johansen model for random months in 2019/2020. On the right: predicting for the same random months in 2021/2022.

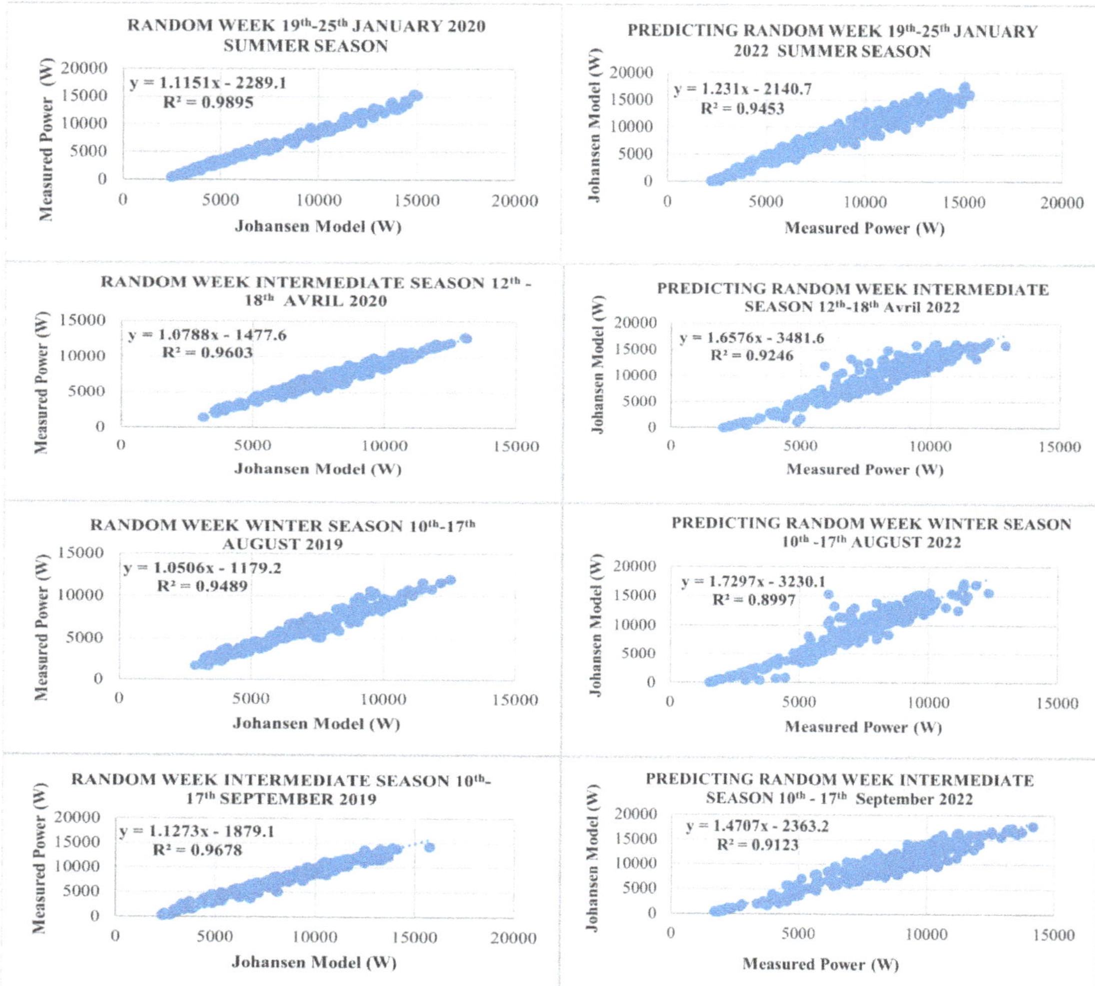

Figure 3. Left side: testing Johansen model random weeks for year 2019/2020. Right side: predicting same random weeks for 2022.

4.4. Medium-Horizon Term: Daily

This sub-section is concerned with a daily medium-horizon term, although in the literature, the daily period is sometimes considered as short-horizon term. We performed this last experiment by using a dataset consisting of daily 15 by 15 min data from on-site measurements between 08:00 A.M. to 04:00 P.M. as indicated in Figure 4.

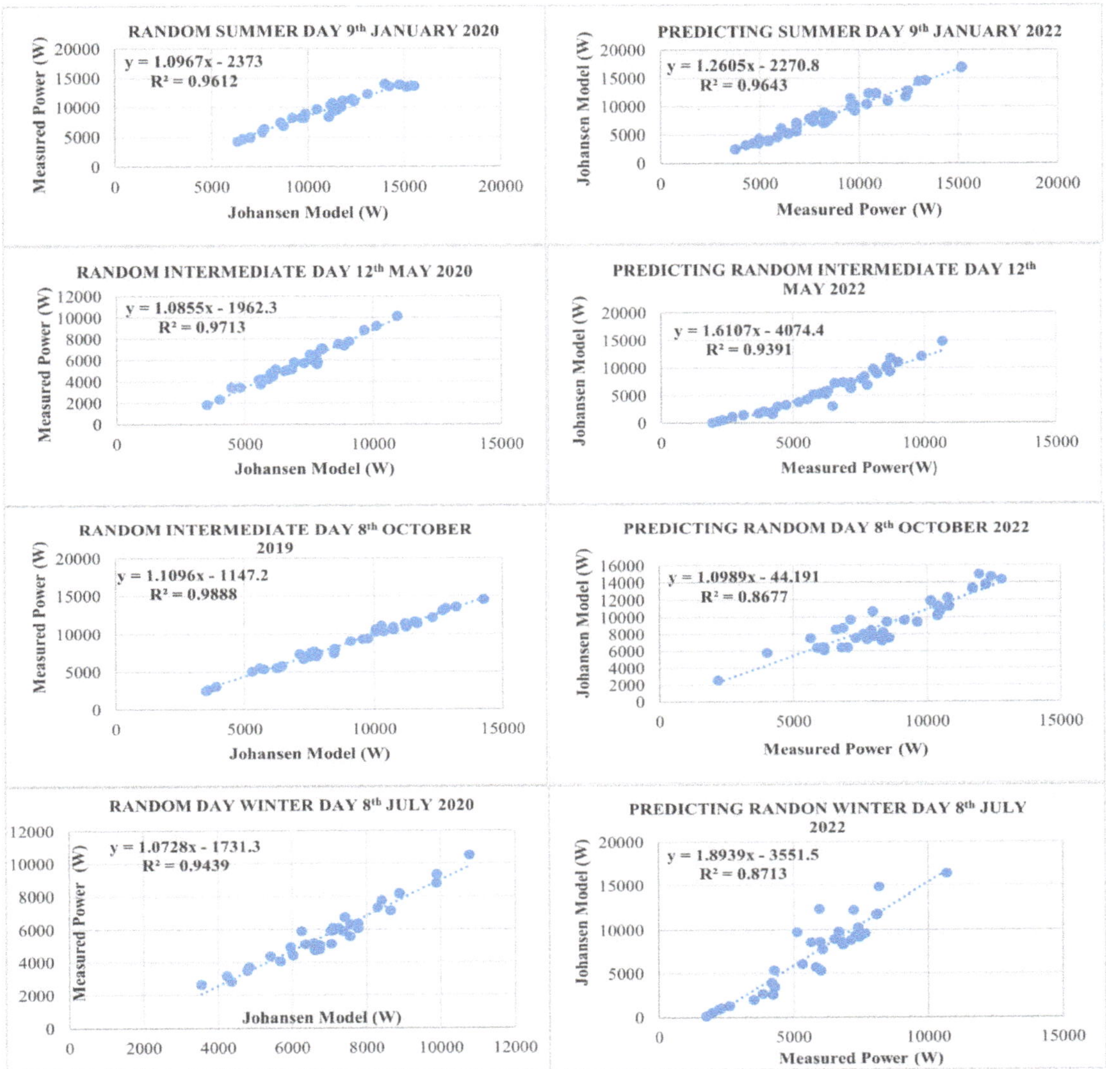

Figure 4. Left side: testing Johansen model for random days in 2019/2020. Right side: predicting same random days for 2022.

5. Conclusions

Predicting PV information is crucial for fulfilling energy efficiency requirements, electricity grid stability, and solar energy management. We focused on the use of a robust statistical technique in the form of the Johansen VECM to predict PV generation in the island zone of Mauritius. We proved the efficiency of the Johansen model for PV power generation by comparing our experimental results to on-site measurements. Results are not only in agreement with the aforementioned measurements but are also compliant with respect to the time horizon conditions of grid stability and where PV energy management as imposed by the local government. Indeed, we showed that the Johansen technique is a powerful tool for monthly and daily PV medium horizon time prediction. Further tests are in progress for the forecasting of very short-term horizon times, i.e., hourly to every 15 min. However, this method still needs to be improved by considering the climate vagaries. Therefore, our future work will focus on incorporating artificial intelligence to optimize the model.

Author Contributions: Conceptualization, H.R.; methodology, H.R.; validation, H.R. and C.T.; formal analysis, N.R.; investigation, H.R. and A.K.; resources, H.R. and A.K.; data curation, H.R. and A.K.; writing—original draft preparation, H.R.; writing—review and editing, C.T. and N.R.; visualization, A.K.; supervision, A.K.; project administration, H.R. and A.K; funding acquisition, A.K. All authors have read and agreed to the published version of the manuscript.

Funding: The authors thank the University of Mauritius for supporting this research.

Institutional Review Board Statement: Not applicable.

Informed Consent Statement: Not applicable.

Data Availability Statement: The data for this this paper can be found at the University of Mauritius.

Acknowledgments: This work was supported by the University of Mauritius.

Conflicts of Interest: Abdel Khoodaruth is the project manager of the following project at the University of Mauritius: Development of a framework for predicting photovoltaic plant power output considering temperature effects under real conditions to enhance grid stability in Mauritius.

References

1. Qing, X.; Niu, Y. Hourly day-ahead solar irradiance prediction using weather forecasts by LSTM. *Energy* **2018**, *148*, 461–468. [CrossRef]
2. Cao, J.C.; Cao, S.H. Study of forecasting solar irradiance using neural networks with preprocessing sample data by wavelet analysis. *Energy* **2006**, *31*, 3435–3445. [CrossRef]
3. Sharma, V.; Yang, S.; Walsh, W.; Reindl, T. Short term solar irradiance forecasting using a mixed wavelet neural network. *Renew. Energy* **2016**, *90*, 481–492. [CrossRef]
4. Lima, F.; Martins, J.; Periera, R.; Lorenz, B.; Heinemann, D. Forecast for surface solar irradiance at the Brazilian Northeastern region using NWP model and Artificial Neural Networks. *Renew. Energy* **2016**, *87*, 807–818. [CrossRef]
5. Lonij, V.P.A.; Brooks, A.E.; Cronin, A.D.; Leuthol, M.; Koch, K. Intra-hour forecasts of solar power production using measurements from a network of irrandiance sensors. *Sol. Energy* **2013**, *97*, 58–66. [CrossRef]
6. Pedro, H.T.C.; Coimbra, C.F.M. Assessement of forecasting techniques for solar power production with no exogenous inputs. *Sol. Energy* **2012**, *86*, 2017–2028. [CrossRef]
7. Voyant, C.; Notton, G.; Kalogirou, S.; Nivet, M.-L.; Paoli, C.; Motte, F.; Fouilloy, A. Machine learning methods for solar radiation forecasting: A review. *Renew. Energy* **2017**, *105*, 569–582. [CrossRef]
8. Sharma, N.; Sharma, P.; Irwin, D.; Shenoy, P. Predicting solar generation from weather forecasts using machine learning. In Proceedings of the IEEE International Conference on Smart Grid Communications, Brussels, Belgium, 17–20 October 2011; pp. 528–533.
9. Colak, I.; Yesilbudak, M.; Genc, N.; Bayindir, R. Multi-period prediction of solar radiation using ARMA and ARIMA models. In Proceedings of the IEEE 14th International Conference on Machine Learning and Applications, Miami, FL, USA, 9–11 December 2015; pp. 1045–1049.
10. Boroojeni, K.G.; Amini, M.H.; Bahrami, S.; Iyengar, S.S.; Sarwat, A.I.; Karabasoglu, O. A novel multi-time scale modeling for electric power demand forecasting: From short-term to medium-term horizon. *Electr. Power Syst. Res.* **2017**, *142*, 58–73. [CrossRef]
11. Lundquist, K.; Chow, F.; Lundquist, J. An Immersed Boundary Method for the Weather Research and Forecasting Model. *Am. Meteorol. J.* **2010**, *138*, 796–817. [CrossRef]
12. Chaturvedi, D.K. Solar Power Forecasting: A Review. *Int. J. Comput. Appl.* **2016**, *145*, 28–50.
13. Ramenah, H.; Casin, P.; Ba, M.; Benne, M.; Tanougast, C. Accurate determination of parameters relationship for photovoltaic power output by Augmented Dickey Fuller test and Engle & Granger method. *AIMS Energy* **2018**, *6*, 19–48.
14. Dickey, D.; Fuller, W. Distribution of the estimates for the autoregressive time series with a unit root. *J. Am. Stat. Assoc.* **1979**, *74*, 427–431.
15. Gujarati, D.N. *Basic of Econometrics*, 4th ed.; McGraw-Hill Econometrics: New York, NY, USA, 2004.
16. Engle, R.F.; Granger, C.W.J. Co-integration and error correction: Representation, estimation and testing. *Econometrica* **1987**, *55*, 251–276. [CrossRef]
17. Ender, W. *Applied Economic Time Series*; Wiley Series in Probabilty and Statistics; Wiley: New York, NY, USA, 1995.
18. Fanchette, Y.; Ramenah, H.; Tanougast, C.; Benne, M. Applying Johansen VECM cointegration approach to propose a forecast model of photovoltaic power output plant in Reunion Island. *AIMS Energy* **2020**, *8*, 179–213. [CrossRef]

19. Johansen, S. Statistical Analysis of Cointegration Vectors. *J. Econ. Dyn. Control.* **1987**, *12*, 231–254. [CrossRef]
20. Johansen, S.; Juselius, K. Maximum likelihood estimation and inferences on cointegration with applications to the demand for money. *Oxf. Bull. Econ. Stat.* **1990**, *52*, 169–210. [CrossRef]

engineering proceedings

MDPI

Proceeding Paper

Forecasting the Case Number of Infectious Diseases Using Type-2 Fuzzy Logic for a Diphtheria Case Study †

Wiwik Anggraeni [1,*], Maria Firdausiah [1] and Muhammad Ilham Perdana [2]

[1] Department of Information Systems, Institut Teknologi Sepuluh Nopember, Surabaya 60111, Indonesia; mariafrdsh@gmail.com
[2] Department of Informatics, Universitas Muhammadiyah Malang, Malang 65145, Indonesia; ilhamperdana@umm.ac.id
* Correspondence: wiwik@is.its.ac.id

† Presented at the 9th International Conference on Time Series and Forecasting, Gran Canaria, Spain, 12–14 July 2023.

Abstract: Diphtheria is an infectious disease with a high mortality rate. In Indonesia, the number of diphtheria cases has remained relatively high in recent years, so efforts to prevent and control diphtheria are needed. Forecasting of the number of diphtheria cases was carried out in this study by applying a type-2 fuzzy logic systems method. Forecasting in this study was carried out by involving the variables of the number of diphtheria sufferers, the percentage of immunization coverage comprising four immunization types, and population density. Regions are grouped into three clusters based on the number of cases that have occurred. Each cluster is taken and sampled in the form of one region to acquire a robust model for other regions. The forecasting results for the next 24 periods show that the performance of the type-2 fuzzy logic systems method is quite good, with accuracy values in the Malang area showing an MSE of 8.785 and an SMAPE of 54.91%. In the Surabaya area, the forecasting accuracy results have an MSE value of 14.940 and an SMAPE of 35.51%. In the Sumenep area, the forecasting accuracy results show an MSE value of 2.188 and an SMAPE of 67.63%. The results of the forecasting of the number of cases can be used as a guide in planning and making decisions regarding the prevention and management of diphtheria.

Keywords: forecasting; diphtheria; type-2 fuzzy logic; infectious disease

Citation: Anggraeni, W.; Firdausiah, M.; Perdana, M.I. Forecasting the Case Number of Infectious Diseases Using Type-2 Fuzzy Logic for a Diphtheria Case Study. *Eng. Proc.* **2023**, *39*, 3. https://doi.org/10.3390/engproc2023039003

Academic Editors: Ignacio Rojas, Hector Pomares, Luis Javier Herrera, Fernando Rojas and Olga Valenzuela

Published: 25 June 2023

1. Introduction

Diphtheria is a disease caused by the bacterium *Corynebacterium diptheriae* [1]. This disease is classified as a contagious disease and can cause death in sufferers. Diphtheria can be transmitted directly through physical contact with sufferers or through a patient's aerosol fluids [2]. The disease primarily affects the nose, throat, and airways, resulting in difficulty breathing, fever, and the formation of a thick coating in the throat [1,2]. In addition, diphtheria is a type of communicable disease that requires surveillance activities for prevention [3] and control as soon as possible; it is necessary to study how vaccination affects diphtheria [4].

In recent years, Asia has seen outbreaks and an increase in the prevalence of diphtheria [5]. According to the WHO, Indonesia is in the top ten countries with the most diphtheria cases. In terms of case numbers, Indonesia is in third position, after India and Nepal [6]. In Indonesia, the number of diphtheria cases is high, sometimes leading to outbreaks. The province with the largest number of sufferers is East Java, where there is a fairly high mortality rate [7].

The high number of diphtheria cases in Indonesia, especially East Java, requires efforts to prevent and control diphtheria to reduce the number of diphtheria cases. In order to carry out good planning in efforts to prevent and control diphtheria, forecasting of the number of cases is carried out. The results of this forecasting can later be used as the basis for decision-making related to efforts to prevent and control diphtheria. So far, the prevention efforts of the Health Service have come in the form of immunizations to minimize the number of occurrences of diphtheria. However, increases in numbers of cases are still common.

So far, there are still very few studies that have predicted the number of diphtheria cases, especially those involving influential variables. Research related to diphtheria is more often focused on analyzing the impact of vaccines on public health [4,8]. Analyses are usually distinguished based on various characteristics. Our past research has tried to predict the number of diphtheria cases [9], but our best model only involved one variable, namely the number of cases in the previous period. In fact, research developments state that the risk of diphtheria can be influenced by various other factors [10].

For this reason, this study will provide forecasts of the number of diphtheria cases in various regions with different characteristics involving various influential factors, such as population density and the coverage of various types of vaccination treatments. The approach used is a fuzzy type-2 approach. This method is considered very effective in dealing with uncertainties such as linguistic uncertainty [11]. The fuzzy type-2 approach also has the ability to model problems with more complex situations [11,12]. Type-2 fuzzy systems can help to reduce the difficulties faced in modeling a system based on rules, and they make it possible to tune and increase our understanding of rule-based systems [12]. The proposed method is expected to improve system performance [13].

2. Related Works

At this time, many studies related to diphtheria have been carried out, but only a few are related to case number forecasting. So far, research related to diphtheria has focused more on analyzing the impact of diphtheria vaccine administration. Related to research on the factors involved, previous studies have stated that the risk factors that influence the occurrence of diphtheria cases are demographics [10,14], the administration of vaccines [1,14,15], and familial wealth [14]. For this reason, this research will forecast the number of diphtheria cases by involving the variables of various vaccines that have been obtained, the number of cases in the previous period, and population density. There is still very little research that predicts the number of diphtheria cases involving population density variables and the number of cases in the previous period. So far, research related to diphtheria has focused more on analyzing the impact of vaccines on public health [4,8], distinguished by age group [5,15,16], population [14], demographics [10], area characteristics [17], changes in social behavior [18], and the size of the given country's income [19].

Regarding the method used herein to predict the number of diphtheria cases, thus far, it has rarely been used specifically to research numbers of diphtheria cases; very few alternative methods have been proposed. Past research has proposed using the radial basis function network method to predict the number of diphtheria cases, but the best model is said to involve only the number of cases in the previous period [9]. This is somewhat different from the recent findings previously mentioned. For this reason, in this study, we will forecast the number of cases using a time series approach. The method used in forecasting is the type-2 fuzzy logic systems method. This method is also considered excellent in dealing with complex situations [11,12], and has been widely used for forecasting in various fields. In previous research, the type-2 fuzzy model has been compared to the artificial neural network model and the type-1 fuzzy logic systems model in forecasting coal production capacity; the type-2 fuzzy logic systems model was considered better in terms of stability and consistency [11]. Other studies have shown a low number of errors in prediction when using a type-2 fuzzy method. Prediction of clinical data using type-1 fuzzy and type-2 fuzzy models was carried out in [13]. The study [13] stated that forecasting results produced using the type-2 fuzzy model were superior to those of the type-1 fuzzy model. Currently, there are many applications of the fuzzy type-2 model, including decision-making [11], pattern recognition, classification, and control [12]. However, to the best of the author's knowledge, until now, there has still only been occasional use of the fuzzy type-2 model for forecasting time series data, particularly cases related to the spread of disease. With that in mind, this study uses a type-2 fuzzy model to forecast numbers of diphtheria cases.

3. Methodology

3.1. Data

The data to be used in this study include the number of diphtheria sufferers, population density, and immunization coverage, including the diphtheria-1 immunization (DPT-1), diphtheria-2 immunization (DPT-2), diphtheria-3 immunization (DPT-3), and diphtheria-4 immunization (DPT-4). The data periods used were monthly, from 2013 to 2018. Data were obtained from the East Java Provincial Health Office and the East Java Central Bureau of Statistics. The data obtained are data from all cities/districts in the province of East Java. District/City data in the following provinces are grouped based on the number of cases. Group 1 consists of five cities/districts, while Group 2 and Group 3 each have seventeen cities/districts. The descriptive statistics data used are shown in Table 1. The data in Table 1 indicate that the data are not normally distributed. This can be seen from the skewness value, which is different from zero. The range of data and the standard deviation are also very large.

Table 1. Descriptive statistics data.

	Minimum Statistic	Maximum Statistic	Mean Statistic	Std. Deviation Statistic	Skewness Statistic
Case_Number	0	17	2.09	2.722	2.304
Population_ Density	529	8232	3127.5	3.543.534	0.711
DPT_1	3.13	17.1	84.352	113.736	2.433
DPT_2	3.23	16.29	83.390	110.404	1.626
DPT_3	3.51	16.35	83.053	116.969	1.931
DPT_4	0	127.02	52.462	1.184.238	7.710

3.2. Methodology

The experimental stages used in this study are shown in Figure 1.

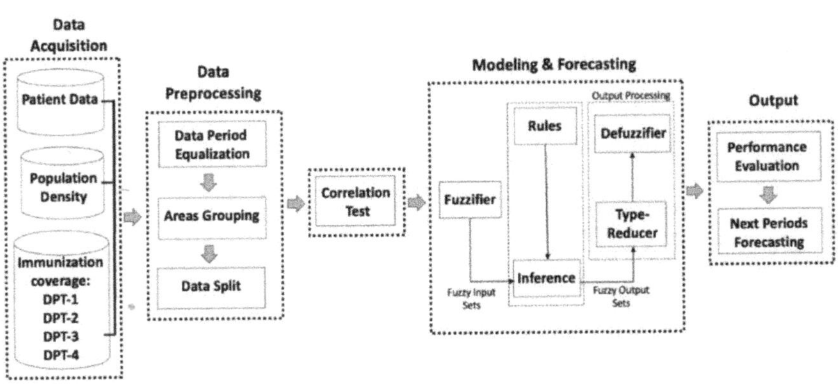

Figure 1. Proposed methodology.

3.2.1. Data Preprocessing

The acquired data still need to be processed, so that they become structured data. Data that are not in the monthly period format will be made so. Regency/city data are grouped based on data on the highest number of sufferers that they have. In addition, the data will also be divided into training data and testing data, with a ratio of 75:25 [9].

3.2.2. Correlation Test

The correlation test is a statistical method used to determine the relationship between two or more variables [9]. The variables analyzed are the independent variables and the dependent variables. In this study, the correlation test was used to see the effect of seven input variables in the tth period on the output variable, namely the number of diphtheria sufferers in the $(t + 1)$th period.

3.2.3. Modeling and Forecasting

Modeling is carried out using type-2 fuzzy logic systems (T2FLS). The characteristics of the type-2 fuzzy model lie in the membership function [12]. In a type-2 set, the degree of membership for each element is a fuzzy type-1 set in [0, 1]. Type-2 fuzzy logic has two membership degrees: primary and secondary membership [11]. In a type-2 fuzzy interval, there are limits on the membership functions, namely the upper membership function (UMF) and lower membership function (LMF). The membership function used is of the Gaussian type. Using the Gaussian primary membership function, the antecedent and the number of rules in the membership function are expressed in Equation (1); the upper membership function is defined in Equation (2). The lower membership function is defined in Equation (3). Furthermore, to generate fuzzy rules from input–output pairs, the lookup table scheme is used.

$$\mu_k^l(x) = \exp\left[-\frac{1}{2}\left(\frac{x_k - m_k^l}{\sigma_k^l}\right)^2\right] \sigma_k^l \in \left[\sigma_{k1}^l, \sigma_{k2}^l\right] \quad (1)$$

$$\mu_k^l(x_k) = N\left(m_k^l,\ \sigma_{k2}^l;\ x_k\right) \quad (2)$$

$$\mu_k^l(x_k) = N\left(m_k^l, \sigma_{k1}^l;\ x_k\right) \quad (3)$$

The input and output variables used in modeling are shown in Table 2. Experiments were carried out in various scenarios, which were combinations of input variables. The scenarios in this study are shown in Table 3. There are 12 subsequent scenarios denoted by scenarios A3, A5, A7, B3, and so on, where 3, 5, and 7 show the many linguistic categories of each group of variables. The distribution of training and testing data used is 75% and 25%, respectively. Furthermore, an example of the linguistic category 5 membership function is shown in Table 3.

Table 2. Input and output variables.

Input Variable	Output Variable
Number of Diphtheria Sufferers Period t	Number of Diphtheria Sufferers Period $t + 1$
Population Density Period t	
Diphtheria-1 Immunization (DPT-1) Coverage Period t	
Diphtheria-2 Immunization (DPT-2) Coverage Period t	
Diphtheria-3 Immunization (DPT-3) Coverage Period t	

Next, twelve scenarios will each be applied to the three selected areas, which are the areas with the highest diphtheria case numbers. These areas are Surabaya City, Malang Regency, and Sumenep Regency. The selection was based on areas in which the number of cases had dominated in the previous year.

Table 3. Input variables combination.

Combination	MF	Input Variable
A	3	Number of Diphtheria Sufferers Period t
	5	
	7	
B	3	Number of Diphtheria Sufferers Period t Population Density Period t
	5	
	7	
C	3	Number of Diphtheria Sufferers Period t Diphtheria-1 Immunization (DPT-1) Coverage Period t Diphtheria-2 Immunization (DPT-2) Coverage Period t Diphtheria-3 Immunization (DPT-3) Coverage Period t Diphtheria-4 Immunization (DPT-4) Coverage Period t
	5	
	7	
D	3	Number of Diphtheria Sufferers Period t Population Density Period t Diphtheria-1 Immunization (DPT-1) Coverage Period t Diphtheria-2 Immunization (DPT-2) Coverage Period t Diphtheria-3 Immunization (DPT-3) Coverage Period t Diphtheria-4 Immunization (DPT-4) Coverage Period t
	5	
	7	

3.2.4. Model Performance Calculations

The model's performance is found using symmetric mean absolute percentage error (SMAPE) and mean square error (MSE). The SMAPE is depicted in Equation (4) [9], while the standard MSE is obtained by using Equation (5), where n shows the number of periods, X_i is the actual value in period ith, and F_i is the predicted value in ith period.

$$\text{SMAPE} = \frac{100\%}{n} \sum_{i=1}^{n} \frac{|X_i - F_i|}{|X_i| + |F_i|} \tag{4}$$

$$\text{MSE} = \frac{1}{n} \sum_{i=1}^{n} [X_i - F_i]^2 \tag{5}$$

3.3. *Model's Robustness Test*

The model robustness test is used to see how well the model performs in each group when it is used to forecast data in other regions.

3.4. *Case Number Forecasting in the Next Several Periods*

The model that has proven to be robust is then used to forecast the number of cases in the coming period. The Health Service has stated that the forecast that needs to be made concerns the next 24 months.

4. Results and Discussion

Modeling in each sample area is carried out by involving many variables that are considered to have an effect on the number of cases of this disease. Tables 4–6 show the results of the correlation test between the variables involved for Surabaya, Malang, and Sumenep, respectively.

Table 4. Correlation test results for independent variables in Surabaya.

Variable	P_{t+1}	P_t	DPT-1	DPT-2	DPT-3	DPT-4	K
P_t	0.485	1.000					
DPT-1	0.091	0.052	1.000				
DPT-2	0.052	−0.003	0.930	1.000			
DPT-3	0.042	−0.033	0.894	0.944	1.000		
DPT-4	−0.033	−0.080	0.139	0.167	0.150	1.000	
K	−0.111	−0.165	0.088	0.177	0.148	0.146	1.000

Table 5. Correlation test results for independent variables in Malang.

Variable	P_{t+1}	P_t	DPT-1	DPT-2	DPT-3	DPT-4	K
P_t	0.110	1.000					
DPT-1	−0.086	−0.114	1.000				
DPT-2	−0.035	−0.063	0.959	1.000			
DPT-3	−0.023	−0.085	0.947	0.962	1.000		
DPT-4	−0.083	−0.140	−0.026	0.028	−0.005	1.000	
K	0.959	0.026	−0.200	−0.190	−0.196	0.675	1.000

Table 6. Correlation test results for independent variables in Sumenep.

Variable	P_{t+1}	P_t	DPT-1	DPT-2	DPT-3	DPT-4	K
P_t	0.168	1.000					
DPT-1	−0.039	0.062	1.000				
DPT-2	0.049	0.101	0.877	1.000			
DPT-3	0.072	0.112	0.889	0.908	1.000		
DPT-4	−0.002	−0.028	−0.125	−0.129	−0.141	1.000	
K	0.117	0.066	−0.328	−0.363	−0.371	0.120	1.000

Tables 4–6 shows that the value of the correlation coefficient in the city of Surabaya ranges from −0.033 to 0.485. The variables that have a negative correlation are the DPT-4 immunization coverage variable and the population density variable, which means that when the value of these variables is greater, the value of the number of cases in the next period will decrease, while the remaining variables have a positive correlation, which means that changes in the values of these two variables are directly proportional to changes in the value of case numbers in period of t. In Surabaya city, it was found that the most influential independent variable was the number of sufferers in the t period, with a correlation coefficient value of 0.485, which is included in the sufficient criteria [9]. The other independent variables in the Surabaya city data have a very weak correlation with the dependent variable. From the analysis of the correlation results in Malang and Sumenep, it emerges that the variable that has the highest correlation value is also the number of sufferers in the period t.

The results of determining the range values for each variable, taken from the lowest and highest values of the training data for each variable, can be seen in Table 7. Meanwhile, the parameters of each model in each region are listed in Table 8. An example of the formed rule fragments ("L" is for "Low Number", "MN" is "Medium Number", "MA" is "Many", "LD" is "Low Density", "FD" is "Fair Density", "HD" is "High Density", "U" is "Uneven", "FE" is "Fairly Even", and "E" is "Even") is shown in Table 9.

Table 7. Variable ranges.

Variable	Malang		Surabaya		Sumenep	
	Upper Limit	Lower Limit	Upper Limit	Lower Limit	Upper Limit	Lower Limit
Number of Diphtheria Sufferers Period *t*	0	6	0	16	0	4
Population Density Period *t*	706	728	8008	8183	528	540
DPT-1 Immunization Coverage Period *t*	6.92	17.1	6.48	14.44	6.7	11.07
DPT-2 Immunization Coverage Period *t*	6.97	16.29	6.21	12.17	6.43	11.04
DPT-3 Immunization Coverage Period *t*	7.16	16.35	6.17	13.35	6.89	11.75
DPT-4 Immunization Coverage Period *t*	0	10.54	0	100	0	127.02
Number of Diphtheria Sufferers Period *t* + 1	0	6	0	16	0	4

Table 8. Parameter model C.3 in Sumenep.

Variables	MF Label	Standard Deviation: Lower	Standard Deviation: Upper	Average
Sufferers	Low Number	0.637	0.8493	0
	Medium Number			2
	Many			4
DPT-1 Immunization Coverage	Uneven	0.6959	0.9279	6.7
	Fairly Even			8.885
	Even			11.07
DPT-2 Immunization Coverage	Uneven	0.7341	0.9788	6.43
	Fairly Even			8.73
	Even			11.04
DPT-3 Immunization Coverage	Uneven	0.7739	1.032	6.89
	Fairly Even			9.32
	Even			11.75
DPT-4 Immunization Coverage	Uneven	20.23	26.97	0
	Fairly Even			63.51
	Even			127.02

The results of the model's performance in the studies of twelve scenarios in the districts of Malang, Surabaya and Sumenep are shown sequentially in Tables 10–12. In Malang, the model that has the lowest SMAPE value is the model of scenario C.7. Specifically, C.7 is the scenario in which all independent variables and seven membership functions are used.

Table 9. Model results in Malang, using scenario D.3.

P_t	K	DPT-1	DPT-2	DPT-3	DPT-4	P_{t+1}
L	LD	U	U	U	U	
L	LD	E	E	E	U	B
L	FD	FE	U	FE	U	LD
L	HD	U	U	U	FE	LD
MN	FD	U	U	U	FE	LD
MN	FD	U	U	U	FE	B
MN	HD	U	U	U	FE	MN
B	LD	U	U	U	U	MN
B	FD	U	U	U	U	LD

Table 10. Model accuracy in Malang.

MF	Model	Train		Test	
		MSE	SMAPE	MSE	SMAPE
3	A.3	2.503	61.9%	8.785	54.91%
	B.3	2.643	57.8%	11.674	99.83%
	C.3	3.077	63.2%	8.580	67.75%
	D.3	3.418	49.8%	7.300	54.37%
5	A.5	2.475	46.4%	6.368	51.56%
	B.5	1.645	45.7%	8.129	63.92%
	C.5	1.854	43.1%	6.327	49.75%
	D.5	1.863	44.2%	5.837	46.78%
7	A.7	2.248	62.1%	9.785	67.70%
	B.7	1.411	41.1%	7.579	55.02%
	C.7	2.286	44.4%	6.269	45.83%
	D.7	3.681	51.92%	5.971	49.52%

Table 11. Model accuracy in Surabaya.

MF	Model	Train		Test	
		MSE	SMAPE	MSE	SMAPE
3	A.3	7.705	40.20%	14.940	35.51%
	B.3	8.967	64.31%	45.637	99.98%
	C.3	6.414	37.28%	20.017	34.48%
	D.3	6.309	36.01%	21.039	35.57%
5	A.5	6.633	40.43%	20.037	34.80%
	B.5	5.211	32.95%	29.928	52.35%
	C.5	9.133	38.85%	13.657	31.02%
	D.5	5.967	34.93%	27.693	36.71%
7	A.7	6.296	35.98%	20.401	39.17%
	B.7	3.780	30.38%	-	-
	C.7	7.072	36.39%	18.256	34.40%
	D.7	4.997	33.43%	25.150	35.67%

Table 12. Model accuracy in Sumenep.

MF	Model	Train		Test	
		MSE	SMAPE	MSE	SMAPE
3	A.3	1.285	76.9%	2.188	67.63%
	B.3	0.838	75.4%	2.126	70.87%
	C.3	1.130	85.6%	1.980	68.86%
	D.3	1.264	84.0%	2.130	78.09%
5	A.5	0.802	86.7%	2.739	84.61%
	B.5	0.618	85.7%	2.971	91.27%
	C.5	0.586	75.8%	1.985	76.35%
	D.5	0.572	77.0%	1.891	71.02%
7	A.7	1.310	90.6%	2.527	84.88%
	B.7	0.809	85.5%	3.261	77.18%
	C.7	1.078	79.0%	2.545	70.86%
	D.7	1.079	90.3%	2.954	75.07%

In the Surabaya city model, the model that has the best accuracy/smallest SMAPE is the model with scenario C.5, in which the number of sufferers and the coverage of DPT immunization are the variables used. The best performance in the Surabaya model is 31.02%. In scenario B.7 of the Surabaya city model, the testing results cannot be obtained. This is because in scenario B.7, the variables used are the number of sufferers and population density, where the population density variable shows an upward trend. In scenario B.7, the standard deviation value used in the model is smaller than in B.5 and B.3, so the model cannot reach variable values that are far from those of the predetermined range.

In the model for the Sumenep area, the model with the best performance is the model with scenario A.3. In Table 12, it can be seen that model A.3 has the lowest SMAPE value, which is equal to 67.63%.

The SMAPE value calculated for each model can be used to determine the best model. However, it transpires that the model that has the lowest SMAPE score in each city has a different number of membership functions. In the Malang regency model, the model with the lowest SMAPE score is that which has a scenario with seven membership functions. In the Surabaya city model, the model with the lowest SMAPE score is the model with five membership functions. Finally, in the Sumenep model, the model with the lowest SMAPE score is the model that has three membership functions. If the models with low SMAPE scores are used, they may lead to differences in linguistic categories. Therefore, the selection of the best model to be used in the next process is carried out to equalize the number of membership functions.

Looking at the graphs of the actual data forecasting results, for the city of Surabaya and the Malang regency, the forecasting chart that follows the actual data pattern is the model with a total of three membership functions, while for Sumenep, the model with a good data pattern is the model with a total of five membership functions. So, the model chosen for forecasting is the model with a membership function of three, which only uses the variable of the number of cases.

Next, to find model with the best robustness for each group, the models will be tested on data from other cities/regencies in the same group. The model robustness test was carried out in other regions. Comparisons of the actual data with the results of the forecast by the robustness model in the three regions are shown in Figures 2–4.

Figure 2 shows the results of testing the model on data from other cities/districts. The trial results of the Surabaya city model using Blitar data graphs show forecast results that follow a pattern. However, the graph seems to shift. Within the Blitar data, the trend of increasing in the mid-period is not captured in the forecast results. Both of these trends occur because of the combination of basic rules used in the model.

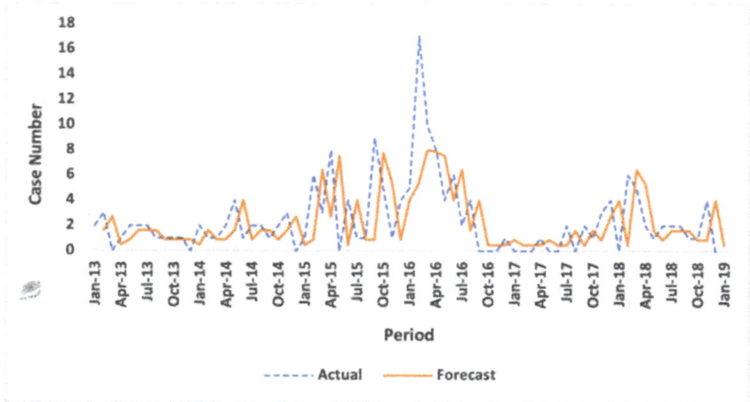

Figure 2. Results of the Surabaya model trial using Blitar data.

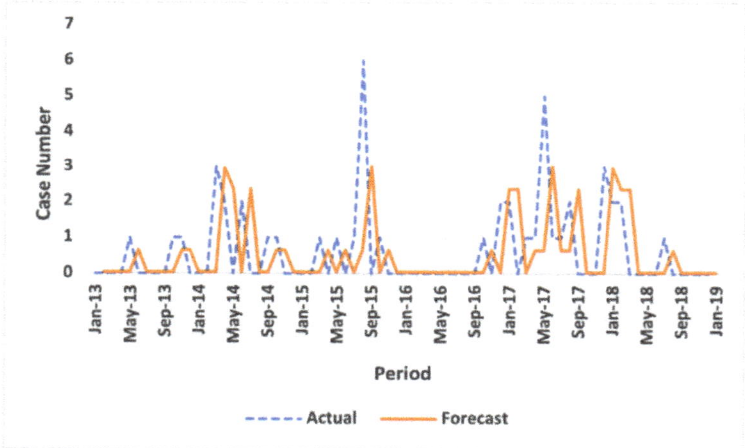

Figure 3. Results of the Malang model trial using Batu data.

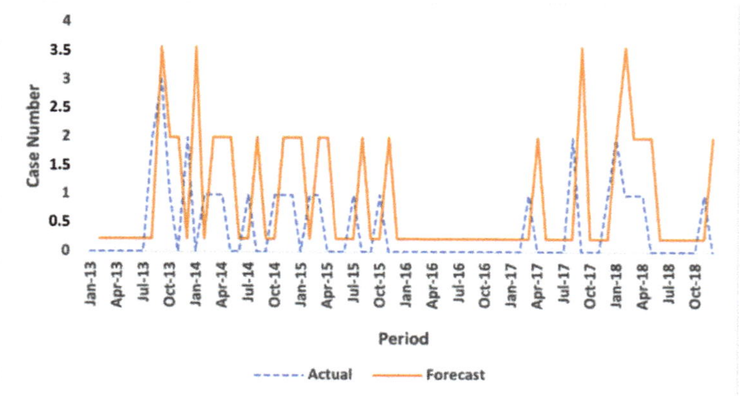

Figure 4. Results of the Sumenep model trial using Magetan data.

The results of the trials for the Malang regency and the Sumenep models in Figures 3 and 4 show the same result: the graphs follow a pattern, but there is a 'delay' in the pattern. The reason for this is the same, namely the basic rules used. From the trials conducted, it can be concluded that the rules cannot be used optimally, because the data do not have a strong linear correlation; therefore, they are less able to capture patterns.

After we know how the model performs against other data, we may forecast the case numbers for the next several periods. The forecasting results for the next several periods using model A.3 in the Malang region are shown in Figure 5. The future forecasting results for the Malang regency show a straight graph with the same value, without any up or down pattern. This is due to the rules used in the model.

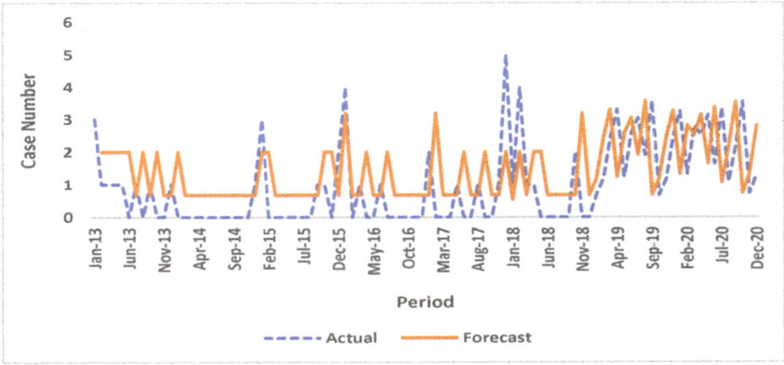

Figure 5. The next period forecasts results in Sumenep.

The actual value used in the January 2019 period is included in the first category, namely "Low Number", and according to the rules used, if there is a value that is in the "Low Number" category, the result is "Low Number". This is why the value of the forecasting results does not increase or decrease; the value remains in the "Low Number" category.

5. Conclusions

Accurate forecasting of the number of diphtheria cases is very important, because forecasting numbers are needed as a basis for making decisions regarding preventive measures. The type-2 fuzzy approach used herein produces different performances when it involves different independent variables and membership functions. The results of the correlation test show that not all the independent variables involved have a significant effect on the forecasting results. The type-2 fuzzy method is more suitable for application to data that show a strong relationship between variables. Forecasting involving the number of sufferers in the previous period produces the best forecasting. The relationship between these variables has an impact on the basic rules generated in the fuzzy model.

Experiments and robustness model tests show that the Malang group model in several regions produces forecasting results with patterns similar to those of actual data; however, it experienced time delays. However, specifically in the Batu region, the forecasting results were less able to follow the actual data pattern, because there are actual data whose values are outside the range of variables used in the training model. The variable number of patients in the training model reached a maximum of six, but other regional data have a value of more than six. This condition means the rules used on the model are unable to capture patterns. The same is true of the results of the Surabaya group model. The forecasting results in the Blitar region showed that the increasing trend in the middle of the period was not captured in the forecasting results. This result was also influenced by the rules used in the model. The increment value is very high, and lies outside the variable range; thus, the basic rules do not capture this value. This was a similar result to that captured using the models of the Sumenep model trials in other regions. The forecasting results in several test areas, including Magetan, follow the actual data pattern, but with delays. The model can effectively capture the value of the increase; this is because the data value in this district is still within the pool of variable values used in the training model.

The basic rules used greatly affect the forecasting results, as do the values of the upper limit and lower limit. Determining the value of the upper limit and lower limit for each range of variables using the min–max method on training data transpired to be less than optimal. With min–max, the model cannot capture values that are far from the range of values that existed before. Thus, this method is not suitable for application to data that shows a trend. In future research, this type-2 fuzzy model will be developed in terms of the basic rules used. In addition, it is necessary to develop this type-2 fuzzy method so that the range of variables may be dynamic, and may capture all existing data patterns.

Author Contributions: Conceptualization, W.A. and M.I.P.; methodology, W.A.; validation, W.A., M.F. and M.I.P.; formal Analysis, W.A., M.F. and M.I.P.; data curation, M.F.; experiment, M.F.; writing—original draft preparation, W.A.; editing, W.A., M.F. and M.I.P. All authors have read and agreed to the published version of the manuscript.

Funding: This research was funded by Institut Teknologi Sepuluh Nopember through the Scientific Research scheme with funding number 1748/PKS/ITS/2023.

Institutional Review Board Statement: Not applicable.

Informed Consent Statement: Informed consent was obtained from all subjects involved in the study.

Data Availability Statement: Data available on request due to restrictions e.g., privacy or ethical.

Acknowledgments: This research was supported by the Public Health Office of East Java Province, Indonesia.

Conflicts of Interest: The authors declare no conflict of interest.

References

1. Badell, E.; Alharazi, A.; Criscuolo, A.; Almoayed, K.A.A.; Lefrancq, N.; Bouchez, V.; Guglielmini, J.; Hennart, M.; Carmi-Leroy, A.; Zidane, N.; et al. Ongoing diphtheria outbreak in Yemen: A cross-sectional and genomic epidemiology study. *Lancet Microbe* **2021**, *2*, e386–e396. [CrossRef]
2. Diphtheria | CDC. 9 September 2022. Available online: https://www.cdc.gov/diphtheria/index.html (accessed on 23 February 2023).
3. van Seventer, J.M.; Hochberg, N.S. Principles of Infectious Diseases: Transmission, Diagnosis, Prevention, and Control. In *International Encyclopedia of Public Health*, 2nd ed.; Quah, S.R., Ed.; Academic Press: Cambridge, MA, USA, 2017; pp. 22–39. [CrossRef]
4. Reid, M.C.; Peebles, K.; Stansfield, S.E.; Goodreau, S.M.; Abernethy, N.; Gottlieb, G.S.; Mittler, J.E.; Herbeck, J.T. Models to predict the public health impact of vaccine resistance: A systematic review. *Vaccine* **2019**, *37*, 4886–4895. [CrossRef]
5. Nicholson, L.; Adkins, E.; Karyanti, M.R.; Ong-Lim, A.; Shenoy, B.; Huoi, C.; Vargas-Zambrano, J.C. What is the true burden of diphtheria, tetanus, pertussis and poliovirus in children aged 3–18 years in Asia? A systematic literature review. *Int. J. Infect. Dis.* **2022**, *117*, 116–129. [CrossRef]
6. Clarke, K. Review of the Epidemiology of Diphtheria 2000–2016. Available online: https://cdn.who.int/media/docs/default-source/immunization/sage/2017/sage-meeting-of-april-2017/background-docs/session-diphtheria/1.-review-of-the-epidemiology-of-diphtheria{-}{-}-2000-2016-pdf-829kb.pdf?sfvrsn=9ba4f061_3 (accessed on 1 June 2023).
7. Sujarwo, E. 460 Kasus dan 16 Orang Meninggal, KLB Difteri di Jatim Belum Dicabut [460 Cases and 16 People Died, Diphtheria Outbreak in East Java Has Not Been Re-voked]. Detiknews. Available online: https://news.detik.com/berita-jawa-timur/d-3855026/460-kasus-dan-16-orang-meninggal-klb-difteri-di-jatim-belum-dicabut (accessed on 29 April 2023).
8. Bellavite, P. Factors that influenced the historical trends of tetanus and diphtheria. *Vaccine* **2018**, *36*, 5506. [CrossRef]
9. Anggraeni, W.; Nandika, D.; Mahananto, F.; Sudiarti, Y.; Fadhilla, C.A. Diphtheria Case Number Forecasting using Radial Basis Function Neural Network. In Proceedings of the 2019 3rd International Conference on Informatics and Computational Sciences (ICICoS), Semarang, Indonesia, 29–30 October 2019; pp. 1–6. [CrossRef]
10. Leidere-Reine, A.; Kolesova, O.; Kolesovs, A.; Viksna, L. Seroprevalence of diphtheria and measles antibodies and their association with demographics, self-reported immunity, and immunogenetic factors in healthcare workers in Latvia. *Vaccine X* **2022**, *10*, 100149. [CrossRef]
11. Runkler, T.; Coupland, S.; John, R. Interval type-2 fuzzy decision making. *Int. J. Approx. Reason.* **2017**, *80*, 217–224. [CrossRef]
12. Mittal, K.; Jain, A.; Vaisla, K.S.; Castillo, O.; Kacprzyk, J. A comprehensive review on type 2 fuzzy logic applications: Past, present and future. *Eng. Appl. Artif. Intell.* **2020**, *95*, 103916. [CrossRef]

13. Najafi, A.; Amirkhani, A.; Mohammadi, K.; Naimi, A. A novel soft computing method based on interval type-2 fuzzy logic for classification of celiac disease. In Proceedings of the 2016 23rd Iranian Conference on Biomedical Engineering and 2016 1st International Iranian Conference on Biomedical Engineering (ICBME), Tehran, Iran, 24–25 November 2016; pp. 257–262. [CrossRef]
14. Gonzales, A.; Choque, D.; Marcos-Carbajal, P.; Salvatierra, G. Factors associated with diphtheria vaccination completion among children under five years old in Peru 2010–2019: A cross-sectional population-based study. *Heliyon* **2022**, *8*, e11370. [CrossRef]
15. Khetsuriani, N.; Zaika, O.; Slobodianyk, L.; Scobie, H.M.; Cooley, G.; Dimitrova, S.D.; Stewart, B.; Geleishvili, M.; Allahverdiyeva, V.; O'Connor, P.; et al. Diphtheria and tetanus seroepidemiology among children in Ukraine, 2017. *Vaccine* **2022**, *40*, 1810–1820. [CrossRef]
16. Verdier, R.; Marchal, C.; Belhassen, M.; Pannerer, M.L.; Guiso, N.; Cohen, R. Coverage rates for diphtheria, tetanus, poliomyelitis, and pertussis age-specific booster recommendations in France: 2018 update of the real-world cohort analysis. *Infect. Med.* **2022**, *2*, 51–56. [CrossRef]
17. Pruitt, S.L.; Tiro, J.A.; Kepka, D.; Henry, K. Missed Vaccination Opportunities Among U.S. Adolescents by Area Characteristics. *Am. J. Prev. Med.* **2022**, *62*, 538–547. [CrossRef]
18. Dadari, I.; Ssenyonjo, J.; Anga, J. Effective vaccine management through social behavior change communication: Exploring solutions using a participatory action research approach in the Solomon Islands. *Vaccine* **2020**, *38*, 6941–6953. [CrossRef]
19. Li, X.; Mukandavire, C.; Cucunuba, Z.M.; Londono, S.E.; Abbas, K.; Clapham, H.E.; Jit, M.; Johnson, H.L.; Papadopoulos, T.; Vynnycky, E.; et al. Estimating the health impact of vaccination against ten pathogens in 98 low-income and middle-income countries from 2000 to 2030: A modelling study. *Lancet* **2021**, *397*, 398–408. [CrossRef]

engineering proceedings

MDPI

Proceeding Paper

Yearly Residential Electricity Forecasting Model Based on Fuzzy Regression Time Series in Indonesia †

Riswan Efendi [1,*], Noor Wahida Md Yunus [1], Sri Rahayu Widyawati [2], Rika Susanti [2], Erol Egrioglu [3], Muhammad Syahri [4], Emansa Hasri Putra [5] and Amir Hamzah [6]

1 Mathematics Department, Faculty of Sciences and Mathematics, Universiti Pendidikan Sultan Idris, Tanjung Malim 35900, Malaysia; wahida@fsmt.upsi.edu.my
2 Faculty of Science and Technology, UIN Sultan Syarif Kasim, Pekanbaru 28293, Indonesia; sri.rahayu.widyawati08@gmail.com (S.R.W.); rika.susanti@uin-suska.ac.id (R.S.)
3 Statistics Department, Giresun University, Giresun 28200, Turkey; erol.egrioglu@giresun.edu.tr
4 Power Distribution Unit Pekanbaru, State Electricity Company Indonesia, P. Sekaki, Pekanbaru 28291, Indonesia; muhammad.syahri@pln.co.id
5 Electrical Engineering Department, Politeknik Caltex Riau, Rumbai 28265, Indonesia; emansa@pcr.ac.id
6 Electrical Department, Universitas Riau, Pekanbaru 28291, Indonesia; amir.hamzah@eng.unri.ac.id
* Correspondence: riswanefendi@fsmt.upsi.edu.my
† Presented at the 9th International Conference on Time Series and Forecasting, Gran Canaria, Spain, 12–14 July 2023.

Abstract: Triangular fuzzy numbers (TFNs) are used to express the weights of criteria and alternatives to account for the ambiguity and uncertainty inherent to subjective evaluations. However, the proposed method can easily be extended to other fuzzy settings depending on the uncertainty facing managers and decision-makers. Triangular fuzzy number (TFN) is a critical component in building fuzzy models such as fuzzy regression and fuzzy autoregressive. Many symmetrical triangular fuzzy numbers have been proposed to improve the scale's linguistic accuracy. Additionally, Sturges' rule is a well-known approach to determining criteria or intervals of grouped data. However, some existing TFN methods are challenging despite being considered in building fuzzy regression models. The increase in electricity distribution is caused by the number of customers and the amount of installed capacity factors in Indonesia. The identified factors are uncertainty, inexactness, and random nature. This paper investigates the residential electricity distribution model using fuzzy regression time series. In the beginning step, the integration between conventional TFN and Sturges' rule was proposed to determine the criteria or scale of linguistic terms. The secondary data was collected from BPS Indonesia from 2000 to 2021. The dependent variable was denoted as electric power distribution (Y_{RT}). On the other hand, the number of customers and the amount of installed capacity were grouped as independent variables (X_{PL} and X_{KT}). The results showed that the best forecasting model is an FLR right upper limit without constant. This proposed model also has higher MAPE accuracy at 1.44% compared to classical models. Additionally, the proposed triangular fuzzy number could improve the accuracy of the proposed model significantly. Interestingly, both dependent and independent factors were initially forecasted using a basic time series model, namely exponential smoothing.

Keywords: fuzzy linear regression; exponential smoothing; triangular fuzzy number; Sturges rule; electricity power distribution

Citation: Efendi, R.; Yunus, N.W.M.; Widyawati, S.R.; Susanti, R.; Egrioglu, E.; Syahri, M.; Putra, E.H.; Hamzah, A. Yearly Residential Electricity Forecasting Model Based on Fuzzy Regression Time Series in Indonesia. *Eng. Proc.* **2023**, *39*, 4. https://doi.org/10.3390/engproc2023039004

Academic Editors: Ignacio Rojas, Hector Pomares, Luis Javier Herrera, Fernando Rojas and Olga Valenzuela

Published: 26 June 2023

1. Introduction

The conventional ordinary regression method requires very strict statistical assumptions such as linearity of variables, no multicollinearity among independent variables, homoskedasticity, reliability of measurement, error should be normally distributed and independently [1]. All assumptions above should be provided completely to attain the best regression model.

Additionally, the input information related to data quality is a highly indispensable component that should be considered for this method. However, this regression method will not be effective and is not recommended for limited data size and linguistic variables. Based on a systematic review paper, multiple linear regression, general linear regression, polynomial regression, exponential regression, and multivariate adaptive regression spline are frequently implemented for electricity load consumption forecasting models [2].

In previous studies, some non-statistical methods, such as fuzzy regression, fuzzy autoregressive, general regression neural network, kernel regression with *k*-nearest neighbors, and fuzzy time series, have been integrated with ordinary regression to handle the previously-mentioned limitations [3–6]. Its applications are commonly employed for electricity forecasting [2]. For example, one of them is the integration between fuzzy and regression methods in handling some issues like linguistic data, small-size data, and normality data. Fuzzy regression estimates parameters using the fuzzy optimization approach more effectively than the ordinary least square [7,8]. Some fuzzy regression methods consider the triangular fuzzy number (TFN) for data pre-processing [9].

In each country, electricity forecasting and its models are the main components to be managed and projected by state and private companies for efficient operations of power distribution systems in supporting daily life activities [10,11]. The conventional models have been discussed and implemented by previous researchers to investigate electricity power distribution and its factors using conventional regression or time series. However, the highest forecasting accuracy is an arduous task since various unpredictable factors may influence electricity power distributions.

Hybrid models have been introduced to improve elements, such as forecasting accuracy and data size. Fuzzy regression is one of the hybrid model types in electricity forecasting [12–14]. This model deals with the triangular fuzzy number (TFN) of fuzzy form data and is not strictly vital in terms of statistical assumptions [15–17]. In this paper, time series analysis is proposed to support fuzzy regression in predicting the value of each variable (dependent and independent) by following a series of times (yearly data). Because the fuzzy regression model is suitable for estimating the significant relationship between dependent and independent variables using fuzzy parameters, it is not a recommended model to forecast future values of variables, especially time series data. Thus, an exponential smoothing model is more practical for such forecasting purposes. Essentially, there are two forecasting phases in this paper.

2. Fundamental Concept

2.1. Triangular Fuzzy Number (TFN)

In 1965, the concepts of fuzzy set and membership function were first proposed by Zadeh [18–20]. Some basic notions on fuzzy sets and numbers are included below:

Definition 1. *Fuzzy sets*

A fuzzy set A of a universal set X is defined as follows:

$$A = \{(x, \mu_A(x)) \,\|\, x \in X\}$$

where $\mu_A(x) : X \to [0,1]$ *is the membership function of the set A. The membership value* $\mu_A(x)$ *indicates the degree of membership of* $x \in X$ *to the set A.*

Definition 2. *Triangular Fuzzy Number (TFN)*

Let a, b and c be real numbers with, a < b < c. Then, the triangular fuzzy number (TFN), A = (a, b, c) is the fuzzy number (FN) with a membership function [18].

$$y = m(x) = \begin{cases} \frac{x-a}{b-a}, \; x \in [a,b] \\ \frac{c-x}{c-b}, \; x \in [b,c] \\ 0, \; x < a \, and \, x > c \end{cases} \tag{1}$$

Thus, Equation (1) is interpreted as membership functions as shown in Figure 1.

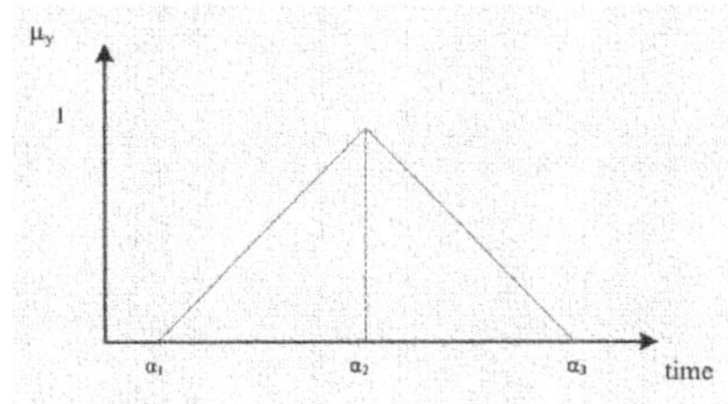

Figure 1. Triangular fuzzy number $A = (\alpha_1, \alpha_2, \alpha_3)$.

Based on Equation (1), a TFN can be defined as:

$$\text{TFN} = y = [\alpha_l, c, \alpha_r] \tag{2}$$

Based on Equation (2), if the TFN is symmetrical, $\alpha_2 - \alpha_1 = \alpha_3 - \alpha_2$, then y is denoted as

$$y = [c, \alpha] \tag{3}$$

where a is a spread of TFN and y is a non-fuzzy number if $a = 0$.

2.2. Fuzzy Regression Model (FRM)

Fuzzy least square and fuzzy linear regression models have been introduced by Tanaka in 1982. Both models used fuzzy forms in terms of input, process, and output, respectively. Mathematically, FRM with and without intercept is written as [21]:

$$\widetilde{Y} = \widetilde{A}_0 + \widetilde{A}_1 X_{1i} + \cdots + \widetilde{A}_N X_{iN} = \widetilde{A}X \tag{4}$$

and

$$\widetilde{Y} = \widetilde{A}_1 X_{1i} + \cdots + \widetilde{A}_N X_{iN} = \widetilde{A}X \tag{5}$$

From Equations (4) and (5), $\widetilde{A} = (p_j; c_j), j = 1, 2, \cdots, m$, while p_j is a mid-value of j and c_j reveals a spread value of j. Both equations are detailed in Table 1.

Table 1. General FRM based on intercept and bound functions.

Bound Function	FRM with Intercept	FRM without Intercept
Lower bound function	$Y_i^l = \sum_{j=0}^{m} (p_j - c_j) X_{ij}$	$Y_i^l = \sum_{j=1}^{m} (p_j - c_j) X_{ij}$
Mid-value function	$Y_i^h = \sum_{j=0}^{m} p_j X_{ij}$	$Y_i^h = \sum_{j=1}^{m} p_j X_{ij}$
Upper bound function	$Y_i^u = \sum_{j=0}^{m} (p_j + c_j) X_{ij}$	$Y_i^u = \sum_{j=1}^{m} (p_j + c_j) X_{ij}$

Based on Table 1, some extension models have been proposed by previous researchers [21–23] to handle the limitation and minimize the spread of the triangular fuzzy number (TFN) from FRM, as presented in Table 2.

Table 2. Extended FRM based on [11–13].

FRM with Intercept [21]	FRM without Intercept [21]
$Z = min\sum_{j=0}^{m} c_j$ $\sum_{j=0}^{m} p_j X_{ij} + (1-h)\sum_{j=0}^{m} c_j \lvert X_{ij} \rvert \geq Y_i + (1-h)e_i$ $\sum_{j=0}^{m} p_j X_{ij} + (1-h)\sum_{j=0}^{m} c_j \lvert X_{ij} \rvert \leq Y_i + (1-h)e_i$ which $c_j \geq 0,\ p \in R; 0 \leq h \leq 1$	$Z = min\sum_{j=1}^{m} c_j$ $\sum_{j=1}^{m} p_j X_{ij} + (1-h)\sum_{j=1}^{m} c_j \lvert X_{ij} \rvert \geq Y_i + (1-h)e_i$ $\sum_{j=1}^{m} p_j X_{ij} + (1-h)\sum_{j=1}^{m} c_j \lvert X_{ij} \rvert \leq Y_i + (1-h)e_i$ which $c_j \geq 0,\ p \in R; 0 \leq h \leq 1$
FRM with intercept [22]	FRM without intercept [22]
$Z = min\sum_{j=0}^{m} c_j \sum_{i=0}^{n} \lvert X_{ij} \rvert$ $\sum_{j=0}^{m} p_j X_{ij} + (1-h)\sum_{j=0}^{m} c_j X_{ij} \geq Y_i + (1-h)e_i$ $\sum_{j=0}^{m} p_j X_{ij} + (1-h)\sum_{j=0}^{m} c_j X_{ij} \leq Y_i + (1-h)e_i$	$Z = min\sum_{j=1}^{m} c_j \sum_{i=1}^{n} \lvert X_{ij} \rvert$ $\sum_{j=1}^{m} p_j X_{ij} + (1-h)\sum_{j=1}^{m} c_j X_{ij} \geq Y_i + (1-h)e_i$ $\sum_{j=1}^{m} p_j X_{ij} + (1-h)\sum_{j=1}^{m} c_j X_{ij} \leq Y_i + (1-h)e_i$
FRM with intercept [23]	FRM without intercept [23]
$Z = \underset{c_j}{min}(\sum_{i=0}^{n} \sum_{j=0}^{m} c_j X_{ij})$ $Y_i \leq \sum_{j=0}^{m} p_j X_{ij} + (1-h)\sum_{j=0}^{m} c_j X_{ij}.$ $Y_i \geq \sum_{j=0}^{m} p_j X_{ij} - (1-h)\sum_{j=0}^{m} c_j X_{ij}.$	$Z = \underset{c_j}{min}(\sum_{i=1}^{n} \sum_{j=1}^{m} c_j X_{ij})$ $Y_i \leq \sum_{j=1}^{m} p_j X_{ij} + (1-h)\sum_{j=1}^{m} c_j X_{ij}$ $Y_i \geq \sum_{j=1}^{m} p_j X_{ij} - (1-h)\sum_{j=1}^{m} c_j X_{ij}$

From Table 2, the general FRM has been extended in terms of objective and constraint functions, respectively. All extended models will be used to estimate the significant factors that contribute to the electricity power distribution for residential sectors in Indonesia from 2000–2016.

2.3. Exponential Smoothing Model (ESM)

In time series data analysis, ESM is widely used for estimating in the light of more recent in an exponentially decreasing manner. The most recent observation receives the most weight, α (where $0 < \alpha < 1$); the second most recent observation receives less weight, $\alpha(1-\alpha)$; the observation of two time periods in the past receives even less weight, $\alpha(1-\alpha)^2$; and so forth. Formally, ESM is written mathematically as below [24,25]:

$$\hat{Y}_{t+1} = \alpha Y_t + (1-\alpha)\hat{Y}_t \quad (6)$$

From Equation (6), $\hat{Y}_{t+1}$ is a new smoothed value or the forecast value for the next period, α is the smoothing constant, Y_t is a new observation or the actual value of the series in period t, and $\hat{Y}_t$ is the old-smoothed value or the forecast for period t. This model is also frequently applied to the forecast of electricity load demand data.

3. Forecasting Model for Electricity Power Distribution

In this paper, we considered three phases on electricity power distribution and its factors were proposed based on three different forecasting models as illustrated in Figure 2.

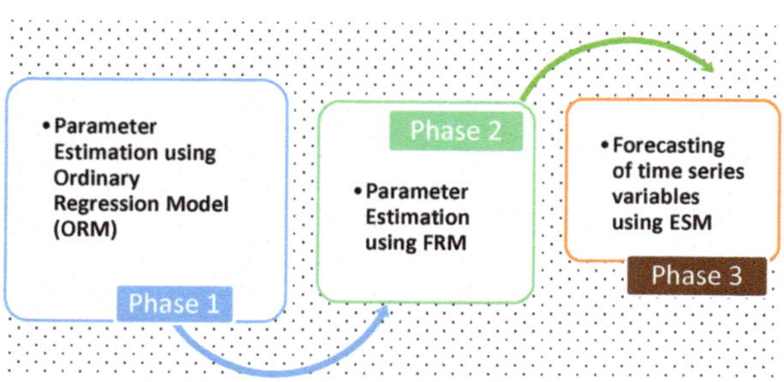

Figure 2. Phase on building an electricity forecasting model.

Based on Figure 2, FRM can be applied if the ORM has been established in the first phase. All slopes or parameters obtained from the ORM will be used to formulate a fuzzy linear programming model. From the FRM, some possibilistic models will be obtained, as described in Table 2, respectively. Measurement error using mean absolute percentage error (MAPE) through data training-testing will be considered to choose the best FR. At the end of the process, ESM will be implemented to forecast the electricity power distribution and its factors for the residential sector by following a series of times (yearly data).

4. Empirical Study

In this section, the implementation of the suggested phases is attempted in two case studies as follows:

Case study A: Electricity power distribution

Step 1: Build ORM for electricity power distribution using secondary data [26] as presented in Table 3.

Table 3. ORM for electricity power distribution data.

Assumptions	Statistical Test	Output
Normality	p-value = 0.200	p-value > α, variables are normally distributed.
Linearity	Scatterplot	Linear trends between dependent and independent factors occurred.
Autocorrelation	p-value = 1.00	p-value > α, no autocorrelation in variable.
Heteroskedastic	p-value x_{PL} = 0.00 p-value x_{KT} = 0.35	p-value > α, there is no heteroscedastic occurred among variables.
Multicollinearity	VIF = 22.302	VIF > 10

Step 2: Transform single data into symmetrical TFN forms for electricity power distribution and its factors using Sturges rule as follows:

- Determine range (R) data for each dependent and independent variable.
- Determine $k = 1 + 3.322\log(n)$.
- Determine $I = \frac{R}{k}$.
- Determine lower and upper limits of intervals.
- Provide a distribution table.

For example, the transformation value of customer numbers is illustrated in Figure 3.

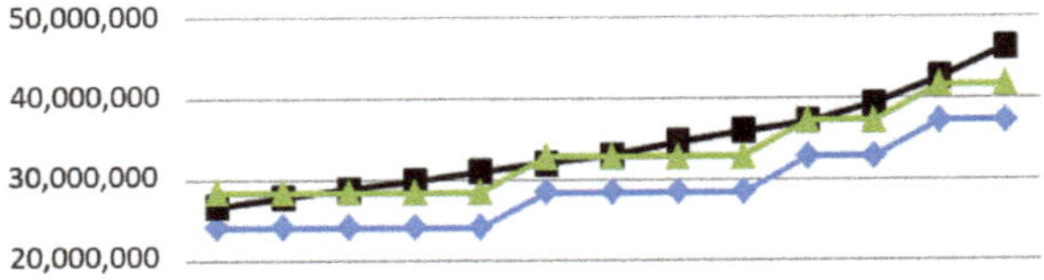

Figure 3. Number of customers in TFN form using Sturges.

Step 3: The estimates of fuzzy parameters presented in Table 4 illustrate the building of fuzzy optimization.

Table 4 shows the minimization of the spread function (c_j) from the mid value (p_j) using fuzzy intervals to left-right constraints.

Step 4: Based on parameters obtained in Step 3, build FRMs as presented in Table 5.

Table 5 shows that the left and right sides have three different FRMs, respectively. Furthermore, these models will be used for forecasting purposes using training and testing data in Step 5.

Step 5: Forecast electricity power distribution using all possible FRM as expressed in Table 6, respectively.

Step 6: Evaluate and validate all possible FRMs using MAPE of training and testing data, respectively, as presented in Table 7.

Table 4. Fuzzy parameters estimation.

Side	Model	h	$(p_0; c_0)$	$(p_1; c_1)$	$(p_2; c_2)$
Left	With intercept	0.1	$(0; 6254.65)$	$(0.0009; 0)$	$(0.4744; 0)$
	Without intercept	0.1	-	$(0.0011; 0)$	$(0.1689; 0.2478)$
Right	With intercept	0.1	$(0; 5566.48)$	$(0.0014; 0)$	$(0; 0)$
	Without intercept	0.1	-	$(0.0014; 0)$	$(0; 0.1761)$

Table 5. FRM for left and right sides.

Side	Bound	FRM with Intercept	FRM without Intercept
Left	Lower	$\tilde{y}_{RT} = (-6254.653) + (0.0009)x_{PL} + (0.4744)x_{KT}$	$\tilde{y}_{RT} = (0.0011)x_{PL} + (-0.0789)x_{KT}$
	Middle	$\tilde{y}_{RT} = (0.0009)x_{PL} + (0.4744)x_{KT}$	$\tilde{y}_{RT} = (0.0011)x_{PL} + (0.1689)x_{KT}$
	Upper	$\tilde{y}_{RT} = (6254.653) + (0.0009)x_{PL} + (0.4744)x_{KT}$	$\tilde{y}_{RT} = (0.0011)x + (0.4167)x_{KT}$
Right	Lower	$\tilde{y}_{RT} = (-5566.489) + (0.00145)x_{PL}$	$\tilde{y}_{RT} = (0.00142)x_{PL} + (-0.1761)x_{KT}$
	Middle	$\tilde{y}_{RT} = (0.00145)x_{PL}$	$\tilde{y}_{RT} = (0.00142)x_{PL}$
	Upper	$\tilde{y}_{RT} = (5566.489) + (0.0014)x_{PL}$	$\tilde{y}_{RT} = (0.00142)x_{PL} + (0.1761)x_{KT}$

Table 6. Forecast values using left-right sides FRM.

Side		Year	FRM with Intercept		
			Lower Forecast (GWh)	Middle Forecast (GWh)	Upper Forecast (GWh)
Left	Training	2000	27,753.59	34,008.25	40,262.90
		2001	28,847.35	35,102.00	41,356.66
		...	...	...	...
		2012	56,522.18	62,776.84	69,031.49
	Testing	2013	60,169.24	66,423.89	72,678.54
		2014	66,913.75	73,168.41	79,423.06
		2015	71,008.86	77,263.51	83,518.17
Right	Training	2000	33,288.69	38,855.18	44,421.67
		2001	34,896.46	40,462.95	46,029.44
		...	...	...	...
		2012	61,502.51	67,069.00	72,635.49
	Testing	2013	67,144.44	72,710.93	78,277.41
		2014	71,795.22	77,361.71	82,928.20
		2015	76,574.60	82,141.09	87,707.58

Table 7. MAPE training-testing of FRM.

		MAPE Training	
Side	Bound	FRM with Intercept	FRM without Intercept
Left	Lower	25.21%	87.87%
	Mid	10.22%	11.45%
	Upper	10.27%	8.49%
Right	Lower	8.17%	10.15%
	Mid	10.74%	10.16%
	Upper	16.99%	13.73%

Table 7. *Cont.*

		MAPE Training	
Side	**Bound**	**FRM with Intercept**	**FRM without Intercept**
		MAPE Testing	
Left	Lower	26.71%	85.80%
	Mid	15.69%	24.11%
	Upper	6.44%	4.50%
Right	Lower	16.35%	25.07%
	Mid	7.96%	10.24%
	Upper	1.05%	1.44%

Based on Table 6, scatter plots between actual and forecast values are illustrated in Figure 4.

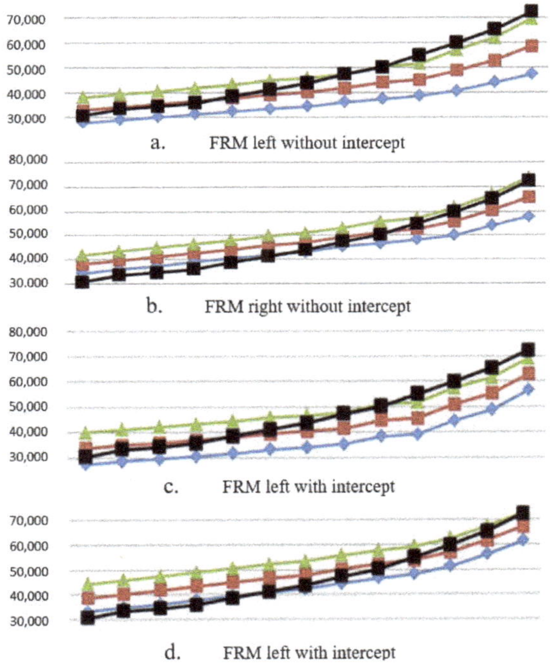

Figure 4. Actual and forecasted values using various FRM.

Step 7: Forecast electricity power distribution for 2016–2021 using the best FRM model (smallest MAPE) without intercept and exponential smoothing (ES) as presented in Tables 8 and 9, respectively.

Table 8. Testing data for 2016.

α	x_{PL}	x_{KT}	y_{RT} (FLR)	y_{RT} (ES)
0.1	39,323,118	33,265	[49,980.83; 61,696.83]	56,462.07
0.5	53,328,382	51,360	[66,681.89; 84,770.72]	83,418.89
0.9	56,283,050	55,075	[70,223.31; 89,620.56]	88,212.52

Table 9. Testing data for 2017–2021.

Year	Actual Load (GWh)	FLR	ES
2017	94,457.38	93,788.67	92,600.35
2018	97,832.28	96,988.25	95,566.78
2019	103,733.43	103,300.98	102,342.23
2020	112,155.85	110,256.76	109,432.34
2021	115,370.04	114,399.12	112,578.35

Based on Table 8, electricity power distribution (y_{RT}) is predicted using FRM right without intercept, as expressed in Table 7. On a regular basis, a regression model is not directly practical for forecasting purposes. In this case, each variable was gathered and measured by considering time intervals (yearly time series data). Thus, they should be predicted separately using time series models such as exponential smoothing (ES). Additionally, each forecasted value was obtained from the ES model, respectively.

In the final stage, the prediction of power distribution can be substituted into FRM right model as written in Equation (7):

$$\widetilde{y}_{RT} = (0.00142)x_{PL} + (0.1761)x_{KT} \tag{7}$$

From this table, the predicted $\widetilde{y}_{RT}$ values were obtained using Equation (7) and ES model. Actual y_{RT} power distribution was 93,634.63 GWh in 2016. On a note, the predicted and actual values revealed immense differences because the State Electricity Company of Indonesia offered a power subsidy for the residential sector for that year. Additionally, the national championship sports of Indonesia were also conducted in 2016. Therefore, the electricity distribution exceeded the actual amount. In this case, two forecasting parts, namely parameter estimation using fuzzy regression and future amount estimation, were already taken into account using the exponential smoothing technique. Unlike some previous studies [11–14], the researchers were only concerned with the fuzzy regression part.

The State Electricity Company of Indonesia offers subsidies for their customers every year. Thus, the proposed model lacks the ability to capture the actual amount. Occasionally, the difference is also significant between forecasted and actual amounts.

Case study B: Palm oil production

By following the same steps given in Case study A, the comparison between actual and forecast values can be shown for palm oil data from January–December 2012 in Tables 10–12, respectively.

Table 10. Forecast values of monthly palm oil production using left-right sides FRM.

Side		Month	FRM without Intercept		
			Lower Forecast (Ton)	Middle Forecast (Ton)	Upper Forecast (Ton)
Left	Training	January	94,798.74	142,018.68	189,209.00
		February	103,945.71	151,423.05	198,870.76
		...	...	...	...
		July	107,295.64	160,669.99	214,014.74
	Testing	October	91,942.59	141,056.79	190,141.37
		November	123,030.85	180,891.52	238,722.57
		December	89,362.53	145,768.25	202,144.36
Right	Training	January	110,503.70	142,588.11	188,983.20
		February	119,041.69	151,316.91	197,902.81
		...	...	...	...
		July	122,625.62	159,480.26	210,645.57
	Testing	October	107,851.67	141,462.56	189,384.12
		November	137,206.26	177,686.27	232,476.96
		December	105,871.75	145,206.90	198,852.74

Table 11. MAPE training-testing data with and without intercept using FRM.

		Left			Average Left	Right			Average Right
		Bound Lower	Bound Middle	Bound Upper		Bound Lower	Bound Middle	Bound Upper	
Training	With intercept	32.45%	10.28%	25.95%	22.90%	21.68%	11.22%	36.73%	23.21%
	Without intercept	29.21%	8.86%	37.91%	25.33%	18.86%	8.84%	36.70%	21.46%
Testing	With intercept	23.32%	22.31%	53.57%	33.07%	17.42%	32.06%	67.20%	38.89%
	Without intercept	21.51%	24.77%	61.64%	35.97%	16.17%	25.36%	61.20%	34.24%

Table 12. Testing data for January 2013.

α	X_1	X_2	X_3	Y (FLR)	Y (ES)
0.1	24,281.6	28,352.1	217.4	[134,795.7; 176,277.5]	153,420.7
0.5	24,220.2	28,398.1	266.0	[135,332.0; 186,087.2]	159,424.9
0.9	26,089.9	28,362.0	245.9	[11,321,307.1; 11,368,226.9]	154,326.2

5. Conclusions

In this paper, the parameters (intercept and slopes) of ordinary regression in building fuzzy linear regression were implemented. Both parameters were employed for fuzzy optimization purposes, namely objective function and left-right constraints. Furthermore, the Sturges rule was used to determine the symmetrical TFN and the number of fuzzy intervals when the total number of observations was specified.

In application, FRM without intercept was considered to capture the actual electricity data precisely. Each variable from FRM was predicted using a basic time series technique known as exponential smoothing. Therefore, two types of forecasting strategies have been employed to estimate yearly electricity power distribution in Indonesia from 2000 to 2021 and palm oil production. In this paper, we also considered the effectiveness between with and without intercepts in the forecasting models.

Author Contributions: Conceptualization, R.E., S.R.W. and E.E.; methodology, N.W.M.Y.; software, E.H.P. and R.S.; data analysis, R.E., A.H. and M.S.; data interpretation, A.H. and M.S.; writing—preparation, R.E. and S.R.W.; writing—editing, S.R.W. and R.S.; visualization, E.H.P. All authors have read and agreed to the published version of the manuscript.

Funding: This research received no external funding.

Institutional Review Board Statement: Not applicable.

Informed Consent Statement: Not applicable.

Data Availability Statement: No new data are created and analyzed in this paper. Data sharing is not allowed.

Acknowledgments: The authors acknowledge all possible administrative and financial support provided by our institute, i.e., "Academic Affair and International Relation, Universiti Pendidikan Sultan Idris, Malaysia" to conduct and publish this research work.

Conflicts of Interest: The authors declare no conflict of interest.

References

1. Franzco, R.J.C.D.P.; Farmer, L.D.M. Review understanding and checking the assumptions of linear regression: A primer for medical researchers. *Clin. Exp. Ophthalmol.* **2014**, *42*, 590–596.
2. Zhang, L.; Wen, J.; Li, Y.; Chen, J.; Ye, Y.; Fu, Y.; Livingood, W. A Review of Machine Learning in Building Load Prediction. *Appl. Energy* **2021**, *285*, 116452. [CrossRef]

3. Ismail, Z.; Efendi, R.; Deris, M.M. Interquartile Range Approach to Length-Interval Adjustment of Enrolment Data in Fuzzy Time Series Forecasting. *Int. J. Comp. Intell. Appl.* **2012**, *12*, 1350016. [CrossRef]
4. Efendi, R.; Arbaiy, N.; Deris, M.M. A New Procedure in Stock Market Forecasting Based on Fuzzy Random Auto-Regression Time Series Model. *Inf. Sci.* **2018**, *441*, 113–132. [CrossRef]
5. Efendi, R.; Deris, M.M. Non-probabilistic inverse fuzzy model in time series forecasting. *Int. J. Uncertain. Fuzziness Knowl.-Based Syst.* **2018**, *26*, 855–873. [CrossRef]
6. Sturges, H. The Choice of A Class-Interval. *J. Am. Stat. Assoc.* **1926**, *21*, 65–66. [CrossRef]
7. Tanaka, H. Fuzzy Data Analysis by Possibilistic Linear Models. *Fuzzy Sets Syst.* **1987**, *24*, 363–375. [CrossRef]
8. Tanaka, H.; Watada, J. Possibilistic linear systems and their application to the linear regression model. *Fuzzy Sets Syst.* **1988**, *27*, 275–289. [CrossRef]
9. Wang, H.M.; Lee, M.J. Fuzzy regression model with fuzzy input and output data. *Inf. Sci.* **2007**, *177*, 2049–2067.
10. Rosadi, M.; Syamsul, A.B. Faktor-Faktor yang Mempengaruhi Konsumsi Listrik di Indonesia. *J. Kaji. Ekon. Dan Pembang.* **2019**, *1*, 273–286. [CrossRef]
11. Nazarko, J.; Zalewski, W. The fuzzy regression approach to peak load estimation in power distribution systems. *IEEE Trans. Power Syst.* **1999**, *14*, 809–814. [CrossRef]
12. Purwareta, H.P.; Gusti, N.I.; Nuri, W. Model Peramalan Pasokan Energi Primer dengan Pendekatan Metode Fuzzy Linear Regression (FLR). *J. Sains Dan Seni ITS* **2012**, *1*, A34–A39.
13. Khairudin, M.; Nursusanto, U.; Ismara, K.I.; Arifin, F.; Fahrurrozi, D.B.; Yahya, A.; Prabuwono, A.S.; Mohamed, Z. Estimated Use of Electrical Load Using Regression Analysis and Adaptive Neoro Fuzzy Inference System. *J. Eng. Sci. Technol.* **2021**, *16*, 4452–4467.
14. Lee, G.-C. Regression-Based Methods for Daily Peak Load Forecasting in South Korea. *Sustainability* **2022**, *14*, 3984. [CrossRef]
15. Lah, M.S.C.; Arbaiy, N.; Efendi, R. Stock Market Forecasting Model Based on AR(1) with Adjusted Triangular Fuzzy Number Using Standard Deviation Approach for ASEAN Countries. *Intell. Interact. Comput.* **2019**, *67*, 103–114.
16. Efendi, R.; Imandari, A.N.; Rahmadhani, Y.; Suhartono; Samsudin, N.A.; Arbai, N.; Deris, M.M. Fuzzy autoregressive time series model based on symmetry triangular fuzzy numbers. *New Math. Nat. Comput.* **2021**, *17*, 387–401. [CrossRef]
17. Azadeh, A.; Khakestani, M.; Saberi, M.A. Flexible Fuzzy Regression Algorithm for Forecasting Oil Consumption Estimation. *J. Energy Policy* **2009**, *37*, 5567–5579. [CrossRef]
18. Mir, A.A.; Alghassab, M.; Ullah, K.; Khan, Z.A.; Lu, Y.; Imran, M. A Review of Electricity Demand Forecasting in Low and Middle Income Countries: The Demand Determinants and Horizons. *Sustainability* **2020**, *12*, 5931. [CrossRef]
19. Zadeh, L.A. Fuzzy sets. *Inf. Control* **1965**, *8*, 338–353. [CrossRef]
20. Afrasiabi, A.; Tavana, M.; Caprio, D.D. An Extended Hybrid Fuzzy Multi-Criteria Decision Model for Sustainable and Resilient Supplier Selection. *Environ. Sci. Pollut. Res.* **2022**, *29*, 37291–37314. [CrossRef]
21. Tanaka, H.; Uejima, S.; Asia, K. Linear Regression Analysis with Fuzzy Model. *IEEE Trans. Syst. Man Cybern.* **1982**, *12*, 903–907.
22. Tanaka, H.; Hayashi, I.; Watada, J. Possibilistic Linear Regression Analysis For Fuzzy Data. *Eur. J. Oper. Res.* **1989**, *40*, 389–396. [CrossRef]
23. Chang, Y.H.O.; Ayyub, B.M. Fuzzy Regression Methods-A Comparative Assessment. *Fuzzy Sets Syst.* **2001**, *199*, 187–203. [CrossRef]
24. Wooldridge, M. *Introductory Econometrics A Modern Approach*, 3rd ed.; South-Western: Mason, OH, USA, 2006.
25. Hanke, J.E.; Wichern, D.W. *Business Forecasting*, 9th ed.; Pearson/Prentice Hall: Hoboken, NJ, USA, 2009.
26. BPS. Available online: https://www.bps.go.id/subject/7/energi.html#subjekViewTab3 (accessed on 20 June 2023).

engineering proceedings

MDPI

Proceeding Paper

Long Lead ENSO Forecast Using an Adaptive Graph Convolutional Recurrent Neural Network †

Jahnavi Jonnalagadda * and Mahdi Hashemi

Department of Information Sciences and Technology, George Mason University, Fairfax, VA 22030, USA; mhashem2@gmu.edu

* Correspondence: jjonnala@gmu.edu

† Presented at the 9th International Conference on Time Series and Forecasting, Gran Canaria, Spain, 12–14 July 2023.

Abstract: El Niño-Southern Oscillation (ENSO), a natural phenomenon in the Pacific Ocean, is caused by cyclic changes in sea-surface temperature (SST) and the overlying atmosphere in the tropical Pacific. The impact of ENSO varies, ranging from slightly warmer or colder temperatures to extreme weather events such as flash floods, droughts, and hurricanes, affecting various regions around the globe. Therefore, ENSO forecasting has paramount importance in the atmospheric and oceanic sciences. The Oceanic Niño Index (ONI), a three-month running mean of SST anomalies over the east–central equatorial Pacific region, is the commonly used metric for measuring ENSO events. However, the literature shows that the forecasting accuracy of ONI for lead times exceeding one year is low. This study aims to improve the forecast accuracy of ONI for up to 18 months lead time by applying an Adaptive Graph Convolutional Recurrent Neural Network (AGCRNN). The graph-learning module adaptively learns the spatial structure of features during training, while the graph convolution in hidden layers of the recurrent neural network captures the temporal relationships of features with ONI. Experiments conducted on simulation and reanalysis datasets demonstrate that AGCRNN outperforms state-of-art statistical and eight dynamical models for forecasting ONI with up to 18 months' lead time.

Keywords: graph networks; deep learning; climate anomalies; spatial-temporal data mining

Citation: Jonnalagadda, J.; Hashemi, M. Long Lead ENSO Forecast Using an Adaptive Graph Convolutional Recurrent Neural Network. *Eng. Proc.* **2023**, *39*, 5. https://doi.org/10.3390/engproc2023039005

Academic Editors: Ignacio Rojas, Hector Pomares, Luis Javier Herrera, Fernando Rojas and Olga Valenzuela

Published: 26 June 2023

1. Introduction

El Niño-Southern Oscillation (ENSO) represents the climate variability in the tropical Pacific Ocean caused by coupled ocean–atmosphere interactions and is associated with severe rainfalls, floods, and droughts affecting both tropics and subtropics [1,2]. El Niño in ENSO refers to the periodic warming of sea-surface temperature (SST) across the central and east–central Equatorial Pacific Ocean. Southern Oscillation in ENSO refers to the atmospheric component coupled with sea-surface temperature changes. The warm phase of ENSO is known as El Niño, and the cold phase of ENSO is known as La Niña.

The commonly used metric for measuring ENSO events is the Oceanic Niño Index (ONI). This is a three-month running mean of SST anomalies over the east–central equatorial Pacific region (aka Niño 3.4 region) spanning between 5° N–5° S and 120° W–170° W. An El Niño event is observed in the Niño 3.4 region if the ONI is +0.5 °C or higher, which means that surface waters in the east–central equatorial Pacific are warmer than average by +0.5 °C or higher for five consecutive months. Conversely, a La Niña event is observed in Nino 3.4 region if the ONI is −0.5 °C or lower, which means that surface waters in the east–central equatorial Pacific are colder than average by −0.5 °C or lower for five consecutive months.

ENSO has a profound impact on global climate, with different impacts observed for the warm (El Niño) and the cold phases (La Niña) of ENSO and the region being considered [3,4]. For instance, El Niño is typically linked with warm and dry conditions in the southeastern areas of Australia, Indonesia, the Philippines, and the central Pacific Islands. Furthermore, prolonged El Niño episodes have caused droughts in India, Indonesia and Australia, and flash floods in the southern United States [5]. In contrast, La Niña is often associated with wetter conditions in eastern Australia and severe rainfall in Indonesia the Philippines, and Thailand. Coastal Ecuador and northwestern Peru experience drier than average conditions during La Niña. The prolonged La Niña episodes have caused severe rainfall in India, Indonesia, and Australia and droughts in the southern United States [6]. The impacts differ for the two opposite phases of ENSO and their evolutionary patterns. For example, the transition from El Niño to La Niña or La Niña to El Niño caused flash floods in the northeast regions of Asia. Overall, understanding the impact of ENSO events is essential to predicting weather patterns and mitigating the risk of natural disasters

The conventional approach to machine learning (ML) involves using the input features of each observation to predict the output, which, in this case, determines ONI. However, these features alone are insufficient for spatial–temporal (ST) prediction problems since time and location are critical factors. In addition, forecasting ONI in a given period requires multiple data points across the tropical Pacific, not just a single data point. Therefore, making predictions across the tropical Pacific rather than relying on a single data point is essential. In this study, we structured input features, such as SST and upper oceanic heat content at various points across the Pacific region, in a proper spatial–temporal format over the past three months to forecast ONI for an up to 18-month lead time.

ENSO events occur due to interactions between SST and the overlying atmosphere. SST is a measurement of the ocean's surface temperature and is defined only for the ocean, so SST values are unavailable for land. However, deep learning models such as convolutional neural networks (CNNs) require an image or grid-like input. To maintain this grid-like input, missing SST values for land points are estimated using interpolation methods. Unfortunately, this interpolation approach may result in incorrect predictions due to errors being introduced in the training data. To address this challenge, we apply an adaptive graph convolutional recurrent neural network (AGCRNN), a graph-based approach to handling non-grid input: SST and upper oceanic heat content for the past three months to forecast ONI with up to 18 months' lead time.

In a graph network, a node represents a data point, and an edge represents the spatial connectivity between a pair of data points. An adjacency matrix represents the spatial distance or proximity between any two nodes, which, in this study, is learned adaptively from the data during training. Experiments on simulation and reanalysis datasets indicate that our proposed approach is superior to previous methods in terms of the correlation coefficient (CC) and coefficient of determination (R^2) for all lead times up to 18 months. The rest of the paper is organized as follows. Section 2 reviews the selected literature on ENSO forecasting, followed by a description of our dataset, given in Section 3. Section 4 elaborates on the proposed methodology, and Section 5 discusses the results of the experiments. Finally, Section 6 concludes the paper with a summary of the findings and directions for future research.

2. Literature Review

Although not specifically used for forecasting ENSO events, the significance of location and time in predicting and forecasting geographical phenomena has been underscored in the literature, both theoretically [7,8] and experimentally [9,10]. The literature on ENSO forecasting can be broadly classified into dynamical and statistical methods. The dynamical methods represent the physical processes of ENSO forecasting, such as coupled ocean–atmosphere interactions using complex numerical equations. These methods require large-scale computational resources and can take several hours to generate predictions. In contrast, statistical methods such as ours can extract meaningful patterns from historical data and require fewer computational resources than dynamical models.

2.1. Dynamical Methods

Zhang et al. [11] examined the changes in forecasting skills caused by changes in SST anomalies in the Pacific after the year 2000. The ensemble of SST forecasts from five dynamical models is evaluated. Another dynamical model, Climate Forecast System version 2 (CFSv2) [12], forecasts SST at a 9-month lead time by considering the initial conditions at the 0th, 6th, 12th, and 18th hour of every 5th day from 1982 to 2010. Therefore, an ensemble of 24 (6 days $\times$ 4 timesteps) forecasts for each month is referred to as the forecast for that month. The CC for a 9-month lead time is 0.34.

2.2. Statistical Methods

Linear models, such as least regression [13] and support vector machine [14,15], and non-linear models, such as decision trees [16] and random forest [17], are used for forecasting ENSO events. For example, in [13], the upper oceanic heat content and the thermocline depth at 20 °C isotherm are the variables used for forecasting. Ensemble learning is applied in both dynamical methods and statistical methods [17]. The three models used in the ensemble classifier are the artificial neural network, random forest, and nearest neighbor. The ensemble classifier uses the voting scheme to generate the final prediction. It correctly predicted 12 out of 13 central Pacific El Nino, 20 out of 20 eastern Pacific El Nino, 19 out of 26 La Nina, and 63 out of 64 neutral events. In a recent review [18], the application of machine learning algorithms and their role in improving the prediction skill of ENSO are discussed.

Recent studies applied artificial neural networks [19–21] convolutional neural networks (CNNs) [22–24], and recurrent neural networks [25–27] to ENSO forecasting. For example, in [22], CNN outperformed state-of-the-art dynamical models and achieved a CC of 0.5 for forecasting ONI for lead times of up to 17 months. Furthermore, the enhanced version of CNN, called all season-CNN (A-CNN), proposed in [23] improves the CC from 0.3 to 0.4 for a lead time of 23 months. A variant of CNN, such as the dense convolutional long short-term memory (DC-LSTM), is used in [28] for forecasting ENSO events. A dense convolutional layer and a transposed convolutional layer are used to extract the spatial features from the input, and multi-layer casual L-shaped LSTM is used to capture the temporal dynamics of SST anomalies. In addition to SST, T300 (vertically averaged oceanic temperature above 300 m), zonal wind, and meridional wind are used as predictors. Their results conclude that additional predictors showed no correlation with ENSO events. Although the performance of DC-LSTM is superior to that of CNN [22], the number of trainable parameters is relatively higher. Other variants, such as deformable CNN [29] and residual CNN [30], are also applied to forecast ENSO events.

Graph Neural Networks (GNNs) are the generalized version of CNNs that can handle non-Euclidean/non-uniform data. GNNs work well for spatial data as they can model the relationships between variables as graph edges. They are widely used in applications such as intelligent transportation, earthquake prediction, recommendation systems, and social media data-mining. However, it is surprising that GNNs are rarely applied in climate science and weather forecasting. Therefore, this study applies a graph-based approach, mainly AGCRNN, to forecast ONI for up to 18 months' lead time. The main difference between our work and the other graph-based ENSO prediction [27] is that we learn the graph structure from the dataset during training, rather than learning a predefined graph structure that can be incomplete and inaccurate. Then, we modify feed-forward connections in the gated recurrent unit (GRU) with graph convolutions to capture the features' spatial structure and temporal relationships with ONI.

3. Data Description

This study uses historical simulation and reanalysis datasets to forecast ONI, which is a three-month average of SST anomalies over the east–central equatorial Pacific region ranging from 5° N–5° S to 120° W–170° W. The historical simulation data were collected from 21 selected Coupled Model Intercomparison Project 5 (CMIP5) models. A single ensemble member from each CMIP5 model was chosen for 1860–2001. The reanalysis datasets were collected from Simple Ocean Data Assimilation (SODA) and Global Ocean Data Assimilation System (GODAS) for 1871–1973 and 1984–2017, respectively.

The study region contained both ocean and land data points (1728) collected across 0–360° E and 55° S–60° N with a 5° × 5° resolution, as shown in Figure 1. However, since SST and upper oceanic heat content are defined only for the ocean, we excluded land data points from our analysis, which resulted in 1345 data points, as shown in Figure 2. The features used in the study were three months' SST and upper oceanic heat content anomalies measured at each point, and output was the ONI for a lead time of up to 18 months.

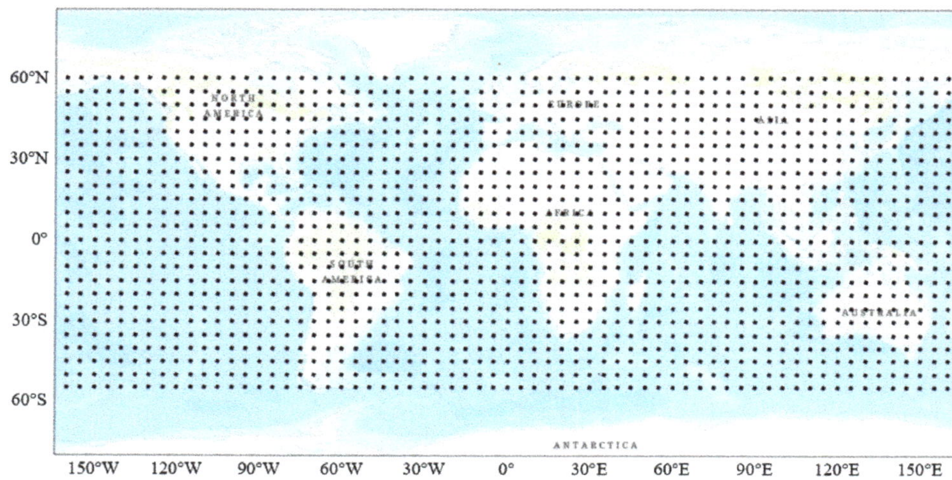

Figure 1. Data points are separated by 5° across latitude and longitude.

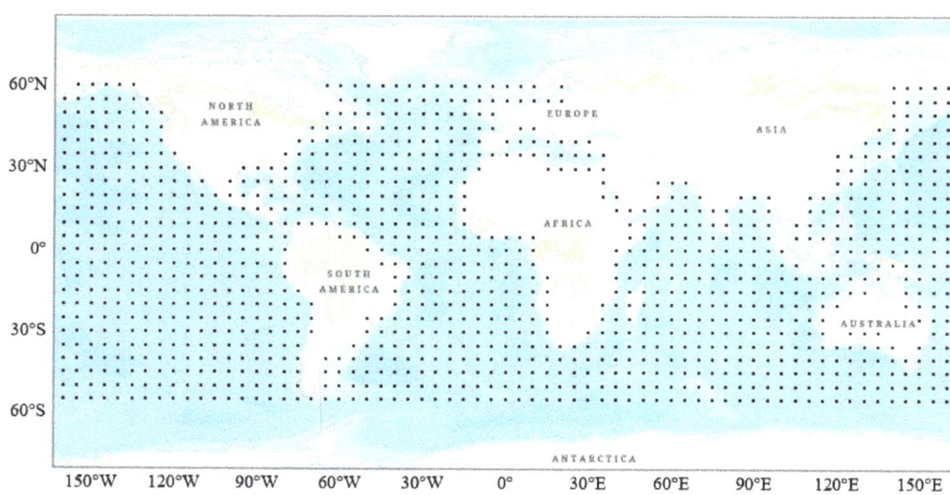

Figure 2. Data points are separated by 5° across latitude and longitude after excluding points on land.

4. Methodology

Data standardization is a common process to standardize features in ML. Since both features (SST and upper oceanic heat content) have different ranges, we standardize them using the below equation, where x denotes a feature and x_i denotes the value of that feature at a single data point.

$$\hat{x}_i = \frac{x_i - mean(x)}{standard\ deviation(x)} \tag{1}$$

4.1. Problem Statement

Consider a time series $X = \{X_{:,1}, \ldots, X_{:,t}, \ldots\}$, where $X_{:,t} = \{x_{1,t}, x_{2,t}, \ldots x_{N,t}\}^T \in R^{N\times 1}$ represents the observations recorded at N data points for time step t. Our goal is to predict the future values of the time series based on historical values. We formulated a spatial–temporal prediction problem to find a function F to forecast the next τ steps based on previous T historical values:

$$\{X_{:,t+1}, X_{:,t+2}, \ldots, X_{:,t+\tau}\} = F_\theta(X_{;,t}, X_{:,t-1}, \ldots, X_{t-T+1}) \tag{2}$$

where θ denotes model parameters. The spatial correlations between different time series were formulated as a graph network. In an undirected graph network $G(V, E, A)$, $V_1, V_2, \ldots V_N$ are the nodes of the graph, and E is the set of edges connecting the pairs of nodes. A node represents a data point based on latitude and longitude, and an edge represents the spatial connectivity between pair of data points. The adjacency matrix $A \in [0, 1]$ representing the proximity between nodes is an $N \times N$ matrix where each entry A_{ij} is the weighted representation of spatial similarity between nodes V_i and V_j. If an edge exists between nodes V_i and V_j, then $0 < A_{ij} < 1$; otherwise, $A_{ij} = 0$. Therefore, the spatial–temporal prediction problem is modified by adding graph network G as follows:

$$\{X_{:,t+1}, X_{:,t+2}, \ldots, X_{:,t+\tau}\} = F_\theta(X_{;,t}, X_{:,t-1}, \ldots, X_{t-\tau+1}; G) \tag{3}$$

4.2. Graph Learning Module (GLM)

Graph convolutional networks (GCNs) are applied in many domains, such as traffic prediction, urban anomaly prediction, recommendation systems, molecular biology, and social network analysis. A multi-layer GCN with layer-wise propagation is approximated using the first-order Chebyshev polynomial given by

$$H^{l+1} = \sigma(\tilde{D}^{-\frac{1}{2}}\tilde{A}\tilde{D}^{-\frac{1}{2}}H^l W^l) \tag{4}$$

Here, $\tilde{A} = A + I_N$ is the adjacency matrix of the graph G with self-connections, I_N is the identity matrix, D is the degree matrix and $\tilde{D}_{ii} = \sum_j \tilde{A}_{ij}$. W^l is a layer-specific trainable weight matrix. $\sigma(\cdot)$ denotes an activation function, such as the $ReLU(\cdot) = max(0, \cdot)$. $H^l \in R^{N\times D}$is the output of the lth layer and $H(0) = X$. More details of the first-order approximation are given in [31]. Most GCNs for prediction tasks rely on a predefined graph structure/adjacency matrix (A) computed using node distances or similarities to perform graph convolutions. Nevertheless, this predefined structure might not encompass all spatial dependencies, which can introduce bias into the model.

In this study, we used a graph learning module (GLM) to automatically infer the hidden dependencies from the data. First, the AGLM randomly initializes a learnable node-embedding matrix ($E_A \in R^{N\times d_e}$) for all nodes, where each row of E_A represents the embedding of a node and d_e represents the dimension of the node-embedding. Then, we can infer the spatial dependencies between pairs of nodes by multiplying E_A and E_A^T. To reduce the computational cost, we directly generated the Laplacian Matrix $\tilde{D}^{-\frac{1}{2}}\tilde{A}\tilde{D}^{-\frac{1}{2}}$ using the following equation, where the SoftMax function normalizes the adaptive matrix.

$$\tilde{D}^{-\frac{1}{2}}\tilde{A}\tilde{D}^{-\frac{1}{2}} = SoftMax(ReLU(E_A.E_A^T)) \tag{5}$$

During training, E_A was updated automatically to learn the hidden dependencies between data points and generate the adaptive matrix for graph convolutions. Finally, by incorporating GLM into GCN, we obtained an adaptive graph convolution network (AGCN), given by the below equation:

$$H^{l+1} = SoftMax(ReLU(E_A.E_A^T))H^l W^l \tag{6}$$

4.3. AGCRNN

The local and global spatial patterns are captured hierarchically using AGCN layers However, to capture the temporal relationships, we proposed the use of AGCRNN by replacing fully connected Gated Recurrent Unit (GRU) layers with AGCN. The equations for AGCRNN are given by

$$\tilde{A} = SoftMax(ReLU(EE^T)) \tag{7}$$

$$z_t = \sigma(\tilde{A}[X_{:,t}, h_{t-1}]W_z + b_z) \tag{8}$$

$$r_t = \sigma(\tilde{A}[X_{:,t}, h_{t-1}]W_r + b_r) \tag{9}$$

$$\hat{h}_t = tanh(\tilde{A}[X_{:,t}, r \odot h_{t-1}]W_{\hat{h}} + b_{\hat{h}}) \tag{10}$$

$$h_t = z \odot h_{t-1} + (1 - z) \odot \hat{h}_t \tag{11}$$

where $[\cdot]$ represents the concatenate operation, $X_{:,t}$ h_t are input and output at time step t, $\odot$ represents element wise multiplication, and z and r represent the reset and update gate, respectively. The trainable parameters of AGCRNN are embedding matrix E, weights $W_Z, W_r, W_{\hat{h}}$, and bias $b_Z, b_r, b_{\hat{h}}$. All parameters are trained using the backpropagation algorithm.

4.4. Implementation Details

The AGCRNN architecture comprises an embedding layer, two AGCRNN layers, and a convolutional layer. The node-embedding layer takes the randomly initialized graph as input and assigns each node a low-dimensional embedding vector. Additionally, the node embeddings capture information about each node's features and relationships with other nodes in the graph. The AGCRNN layers perform graph convolution on the node embeddings and dynamically adjust the graph structure during the convolution operation, enabling the network to learn the spatial structure of SST and upper-ocean heat content and their temporal relationships with ONI. Finally, a single convolutional filter with a kernel size of (nodes × hidden dimension of the previous layer) maps the output of the last AGCRNN layer to generate output. The AGCRNN's network parameters are an embedding vector of size 10 and hidden layer of size 64. We optimized the AGCRNN using the Adam optimizer with a learning rate of 0.001 for a maximum of 100 epochs. Additionally, we implemented an early-stopping algorithm that halts training if the validation loss does not improve in the last 15 epochs. All parameters were selected through hyperparameter-tuning on the validation set.

For CMIP5 and SODA datasets, 60% of the data are used for training, 20% for validation, and 20% for testing. For the GODAS dataset, due to the relatively low number of samples (which might lead to overfitting), we followed the same procedure as detailed in [23]. Therefore, we used CMIP5, SODA, and GODAS for training, validation, and testing.

4.5. Model Evaluation

The performance of AGCRNN was compared with CNN [22] and eight dynamical models. Unfortunately, the results for the dynamical models are only available from 1984 to 2017. Therefore, for CMIP5 and SODA datasets, we compared the performance of AGCRNN with CNN alone, and for the GODAS dataset, we compared AGCRNN with CNN and dynamical models. Notably, only AGCRNN and CNN models were implemented in this study, while the results of the dynamical models are borrowed from [22].

We deployed two widely used regression metrics—correlation coefficient (CC) and coefficient of determination (R^2)—to compare the performance of the predictive models. CC measures the strength of the relationship between actual and predicted values, while R^2 provides the variation in the actual values explained by the predictions. $CC^2 = R^2$ only if the prediction model is a linear regression and the mean of the predicted values equals the mean of the actual values. The mathematical formulation of these metrics is given by the following equations, where y_i and p_i denote the values of observed and predicted output for data point i and $\bar{y}$ and $\bar{p}$ indicate the mean of the output variable y and predicted variable p, respectively.

$$R^2 = 1 - \frac{\sum_i (p_i - \bar{y})^2}{\sum_i (y_i - \bar{y})^2} \tag{12}$$

$$CC = \frac{\sum_i (p_i - \bar{p})(y_i - \bar{y})}{\sum_i (p_i - \bar{p})^2 \sum_i (y_i - \bar{y})^2} \tag{13}$$

5. Results and Discussion

5.1. CMIP5 Dataset

In Figure 3, we compare the performance of CNN and AGCRNN for forecasting ONI using the CMIP5 dataset. Our proposed model outperforms CNN for all lead times, with a CC of greater than 0.9 for up to five months compared to CNN's performance of up to only two months. The rapid decrease in CNN's performance after five months is due to its poor handling of irregular data, such as the missing land points in our dataset, which CNN fills with an average value. This approach can lead to a significant loss of information and bias in the model. In contrast, our proposed model is specifically designed to handle irregular data and uses graph-based operations to process the data effectively. Overall, these results highlight the advantages of our proposed AGCRNN model over CNN for forecasting ONI in the CMIP5 dataset.

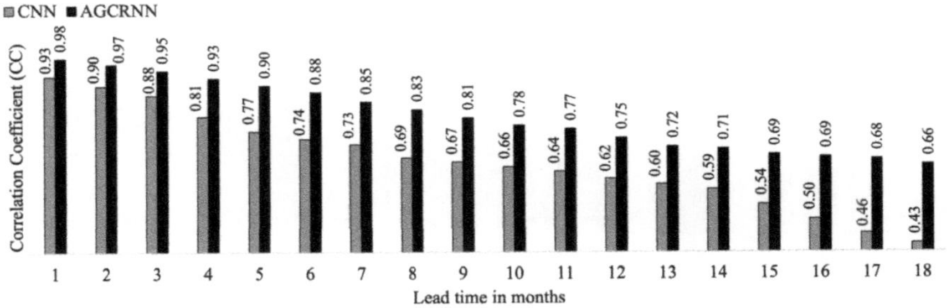

Figure 3. Correlation Coefficient (CC) of CNN and AGCRNN in forecasting ONI for CMIP5 dataset.

In Table 1, we also compare the R^2 values of our model with CNN. Once again, our AGCRNN model outperforms CNN for all lead times, maintaining an R^2 value of 0.5 for up to 13 months' lead time.

Table 1. Coefficient of determination (R^2) of CNN and AGCRNN in forecasting ONI for CMIP5 dataset.

	Lead Time in Months								
Model	**1**	**2**	**3**	**4**	**5**	**6**	**7**	**8**	**9**
CNN	0.86	0.81	0.79	0.70	0.65	0.61	0.55	0.53	0.51
AGCRNN	0.97	0.93	0.90	0.85	0.81	0.78	0.73	0.69	0.65
	10	11	12	13	14	15	16	17	18
CNN	0.46	0.42	0.41	0.37	0.33	0.32	0.30	0.25	0.21
AGCRNN	0.61	0.58	0.55	0.50	0.49	0.48	0.47	0.45	0.42

5.2. SODA Dataset

Figure 4 compares the performance of AGCRNN and CNN for forecasting ONI using the SODA dataset. Our proposed model outperforms CNN for all lead times. However, the *CC* of both models on the SODA dataset is poorer than that of the CMIP5 dataset. This can be attributed to the smaller sample size of the SODA dataset, making models more susceptible to overfitting.

To further evaluate both models, we compare their R^2 values in Table 2. Both models perform similarly in terms of R^2, indicating that a complex network structure does not improve the model performance for smaller datasets. Based on the results from both datasets, we conclude that the performance of AGCRNN improves for larger datasets. In other words, the ACGRNN's performance is proportional to the sample size, as is the case for most deep-learning models.

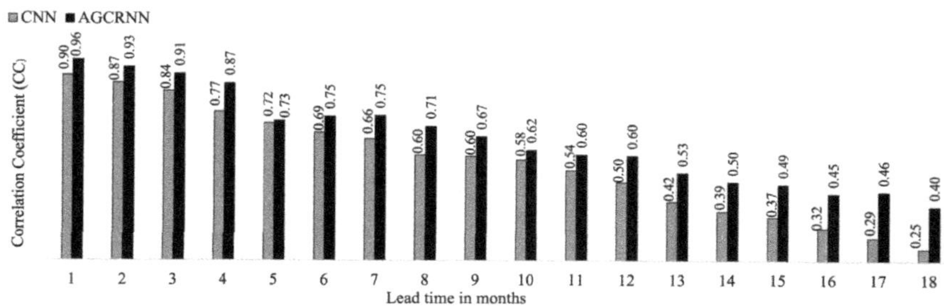

Figure 4. Correlation Coefficient (*CC*) of CNN and AGCRNN in forecasting ONI for SODA dataset.

Table 2. Coefficient of determination (R^2) of CNN and AGCRNN in forecasting ONI for SODA dataset.

	Lead Time in Months								
Model	**1**	**2**	**3**	**4**	**5**	**6**	**7**	**8**	**9**
CNN	0.72	0.64	0.60	0.55	0.47	0.42	0.41	0.45	0.42
AGCRNN	0.81	0.67	0.65	0.67	0.51	0.55	0.48	0.46	0.43
	10	11	12	13	14	15	16	17	18
CNN	0.36	0.34	0.28	0.18	0.18	0.17	0.15	0.16	0.09
AGCRNN	0.38	0.36	0.32	0.19	0.20	0.19	0.18	0.19	0.12

5.3. GODAS Dataset

Figure 5 shows the *CC* of statistical models (AGCRNN and CCN) and dynamical models (all other models) when forecasting ONI for various lead times using the GODAS dataset from 1984 to 2017. Among all models, AGCRNN exhibits the highest forecast skill for all lead times, surpassing both CNN and all other state-of-the-art dynamical models. CNN is the second-best performer, followed by SINTEX-F. AGCRNN achieves a *CC* of 0.68 for an 18-month lead time, while CNN and SINTEX-F, the leading statistical and dynamical models, achieve 0.48 and 0.345, respectively. Consequently, we can conclude that AGCRNN can provide accurate ONI forecasts for up to 18 months of lead time. This superior performance of AGCRNN can be attributed to its proper representation of input features and adaptive graph structure, which capture the spatial structure of features and their temporal relationships with ONI.

Table 3 compares the R^2 values of statistical models since only *CC* values are available for dynamical models. The R^2 values of AGCRNN are close to those of CNN for lead times up to 11 months. However, a substantial difference in R^2 values for higher lead times indicates that AGCRNN outperforms CNN in ONI forecasting.

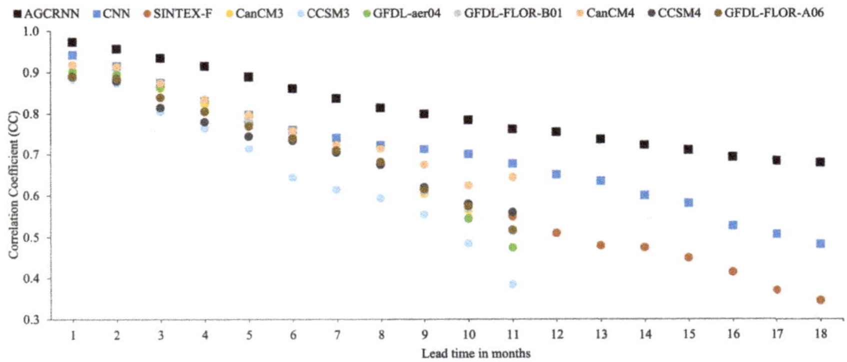

Figure 5. Correlation Coefficient (*CC*) of statistical models (CNN and AGCRNN) versus dynamical models for GODAS dataset.

Table 3. Coefficient of determination (R^2) of CNN and AGCRNN in forecasting ONI for GODAS dataset.

	Lead Time in Months								
Model	**1**	**2**	**3**	**4**	**5**	**6**	**7**	**8**	**9**
CNN	0.92	0.89	0.83	0.76	0.71	0.67	0.63	0.59	0.55
AGCRNN	0.94	0.90	0.86	0.82	0.78	0.73	0.69	0.66	0.63
	10	11	12	13	14	15	16	17	18
CNN	0.52	0.48	0.46	0.43	0.38	0.36	0.35	0.31	0.28
AGCRNN	0.61	0.58	0.57	0.54	0.52	0.50	0.47	0.46	0.45

6. Conclusions and Future Directions

This study uses an adaptive graph convolutional recurrent neural network (AGCRNN), a graph-based approach for forecasting ONI for up to 18 months' lead time. Experiments on reanalysis and simulation datasets demonstrate that AGCRNN outperforms CNN [18] and eight dynamical models for all lead times. Specifically, for the 1984–2017 evaluation period, AGCRNN achieves a *CC* of 0.68 for an 18-month lead time, while CNN and SINTEX-F, the leading statistical and dynamical models, achieve 0.48 and 0.345, respectively.

Our future work involves including other variables, such as southern oscillation index, warm water volume, and thermocline depth, as features, along with SST and upper ocean heat content. We also plan to explore better GCN architectures to understand the spatial structure of these features and their temporal relationships with ENSO.

Author Contributions: Conceptualization, J.J. and M.H.; methodology, J.J.; formal analysis, J.J. and M.H.; writing—original draft preparation, J.J.; writing—review and editing, J.J. and M.H.; visualization, J.J. and M.H.; supervision, M.H.; project administration, J.J. All authors have read and agreed to the published version of the manuscript.

Funding: This research received no external funding.

Institutional Review Board Statement: Not applicable.

Informed Consent Statement: Not applicable.

Data Availability Statement: The data presented in this study are openly available in Zenodo at https://doi.org/10.5281/zenodo.8067393.

Conflicts of Interest: The authors declare no conflict of interest.

References

1. Hardiman, S.C.; Dunstone, N.J.; Scaife, A.A.; Smith, D.M.; Ineson, S.; Lim, J.; Fereday, D. The impact of strong El Niño and La Niña events on the North Atlantic. *Geophys. Res. Lett.* **2019**, *46*, 2874–2883. [CrossRef]
2. Bjerknes, J. Atmospheric teleconnections from the equatorial Pacific. *Mon. Weather. Rev.* **1969**, *97*, 163–172. [CrossRef]
3. Cai, W.; Van Rensch, P.; Cowan, T.; Sullivan, A. Asymmetry in ENSO teleconnection with regional rainfall, its multidecadal variability, and impact. *J. Clim.* **2010**, *23*, 4944–4955. [CrossRef]
4. Yu, J.Y.; Fang, S.W. The distinct contributions of the seasonal footprinting and charged-discharged mechanisms to ENSO complexity. *Geophys. Res. Lett.* **2018**, *45*, 6611–6618. [CrossRef]
5. McPhaden, M.J.; Zhang, X. Asymmetry in zonal phase propagation of ENSO sea surface temperature anomalies. *Geophys. Res Lett.* **2009**, *36*. [CrossRef]
6. Cole, J.E.; Overpeck, J.T.; Cook, E.R. Multiyear La Niña events and persistent drought in the contiguous United States. *Geophys Res. Lett.* **2003**, *29*, 25-1–25-4. [CrossRef]
7. Hashemi, M.; Karimi, H. Weighted machine learning. *Stat. Optim. Inf. Comput.* **2018**, *6*, 497–525. [CrossRef]
8. Hashemi, M.; Karimi, H.A. Weighted machine learning for spatial-temporal data. *IEEE J. Sel. Top. Appl. Earth Obs. Remote. Sens.* **2020**, *13*, 3066–3082. [CrossRef]
9. Hashemi, M.; Alesheikh, A.A.; Zolfaghari, M.R. A spatio-temporal model for probabilistic seismic hazard zonation of Tehran *Comput. Geosci.* **2013**, *58*, 8–18. [CrossRef]
10. Hashemi, M.; Alesheikh, A.A.; Zolfaghari, M.R. A GIS-based time-dependent seismic source modeling of Northern Iran. *Earthq. Eng. Eng. Vib.* **2017**, *16*, 33–45. [CrossRef]
11. Zhang, S.; Wang, H.; Jiang, H.; Ma, W. Evaluation of ENSO prediction skill changes since 2000 based on multimodel hindcasts *Atmosphere* **2000**, *12*, 365. [CrossRef]
12. Saha, S.; Moorthi, S.; Wu, X.; Wang, J.; Nadiga, S.; Tripp, P.; Behringer, D.; Hou, Y.T.; Chuang, H.Y.; Iredell, M.; et al. The NCEP climate forecast system version 2. *J. Clim.* **2014**, *27*, 2185–2208. [CrossRef]
13. Lima, C.H.; Lall, U.; Jebara, T.; Barnston, A.G. Machine learning methods for ENSO analysis and prediction. In *Machine Learning and Data Mining Approaches to Climate Science: Proceedings of the 4th International Workshop on Climate Informatics, Boulder, CO, USA, 25–26 September 2014*; Springer: Cham, Switzerland, 2015; pp. 13–21.
14. Pal, M.; Maity, R.; Ratnam, J.V.; Nonaka, M.; Behera, S.K. Long-lead prediction of ENSO modoki index using machine learning algorithms. *Sci. Rep.* **2020**, *10*, 365. [CrossRef]
15. Aguilar-Martinez, S.; Hsieh, W.W. Forecasts of tropical Pacific sea surface temperatures by neural networks and support vector regression. *Int. J. Oceanogr.* **2009**, *2009*, 167239. [CrossRef]
16. Silva, K.A.; de Souza, Rolim, G.; de Oliveira Aparecido, L.E. Forecasting El Niño and La Niña events using decision tree classifier. *Theor. Appl. Climatol.* **2022**, *148*, 1279–1288. [CrossRef]
17. Maher, N.; Tabarin, T.P.; Milinski, S. Combining machine learning and SMILEs to classify, better understand, and project changes in ENSO events. *Earth Syst. Dyn.* **2022**, *13*, 1289–1304. [CrossRef]
18. Dijkstra, H.A.; Petersik, P.; Hernández-García, E.; López, C. The application of machine learning techniques to improve El Niño prediction skill. *Front. Phys.* **2019**, *7*, 153. [CrossRef]
19. Baawain, M.S.; Nour, M.H.; El-Din, A.G.; El-Din, M.G. El Niño southern-oscillation prediction using southern oscillation index and Niño3 as onset indicators: Application of artificial neural networks. *J. Environ. Eng. Sci.* **2005**, *4*, 113–121. [CrossRef]
20. Tangang, F.T.; Hsieh, W.W.; Tang, B. Forecasting regional sea surface temperatures in the tropical Pacific by neural network models, with wind stress and sea level pressure as predictors. *J. Geophys. Res. Ocean.* **1998**, *103*, 7511–7522. [CrossRef]
21. Nooteboom, P.D.; Feng, Q.Y.; López, C.; Hernández-García, E.; Dijkstra, H.A. Using network theory and machine learning to predict El Niño. *Earth Syst. Dyn.* **2018**, *9*, 969–983. [CrossRef]
22. Ham, Y.G.; Kim, J.H.; Luo, J.J. Deep learning for multi-year ENSO forecasts. *Nature* **2019**, *573*, 568–572. [CrossRef] [PubMed]
23. Ham, Y.G.; Kim, J.H.; Kim, E.S.; On, K.W. Unified deep learning model for El Niño/Southern Oscillation forecasts by incorporating seasonality in climate data. *Sci. Bull.* **2021**, *66*, 1358–1366. [CrossRef] [PubMed]
24. Hashemi, M. Forecasting El Nino and La Nina Using Spatially and Temporally Structured Predictors and A Convolutional Neural Network. *IEEE J. Sel. Top. Appl. Earth Obs. Remote. Sens.* **2021**, *14*, 3438–3446. [CrossRef]
25. Broni-Bedaiko, C.; Katsriku, F.A.; Unemi, T.; Atsumi, M.; Abdulai, J.D.; Shinomiya, N.; Owusu, E. El Niño-Southern Oscillation forecasting using complex networks analysis of LSTM neural networks. *Artif. Life Robot.* **2019**, *24*, 445–451. [CrossRef]
26. Jonnalagadda, J.; Hashemi, M. Spatial-Temporal Forecast of the probability distribution of Oceanic Niño Index for various lead times. In Proceedings of the 33rd International Conference on Software Engineering and Knowledge Engineering, Virtual, 1–10 July 2021; p. 309.
27. Jonnalagadda, J.; Hashemi, M. Feature Selection and Spatial-Temporal Forecast of Oceanic Nino Index Using Deep Learning. *Int. J. Softw. Eng. Knowl. Eng.* **2022**, *32*, 91–107. [CrossRef]
28. Geng, H.; Wang, T. Spatiotemporal model based on deep learning for ENSO forecasts. *Atmosphere* **2021**, *12*, 810. [CrossRef]
29. Nielsen, A.H.; Iosifidis, A.; Karstoft, H. Forecasting large-scale circulation regimes using deformable convolutional neural networks and global spatiotemporal climate data. *Sci. Rep.* **2022**, *12*, 8395. [CrossRef]

30. Hu, J.; Weng, B.; Huang, T.; Gao, J.; Ye, F.; You, L. Deep residual convolutional neural network combining dropout and transfer learning for ENSO forecasting. *Geophys. Res. Lett.* **2021**, *48*, e2021GL093531. [CrossRef]
31. Defferrard, M.; Bresson, X.; Vandergheynst, P. Convolutional Neural Networks on Graphs with Fast Localized Spectral Filtering. *Adv. Neural Inf. Process. Syst.* **2016**, *29*, 3844–3852.

engineering proceedings

MDPI

Proceeding Paper

An Application of Ensemble Spatiotemporal Data Mining Techniques for Rainfall Forecasting †

Shanthi Saubhagya [1,*], Chandima Tilakaratne [1], Musa Mammadov [2] and Pemantha Lakraj [1]

1 Department of Statistics, University of Colombo, Colombo P.O. Box 1490, Sri Lanka; cdt@stat.cmb.ac.lk (C.T.); pemantha@stat.cmb.ac.lk (P.L.)
2 School of Info Technology, Faculty of Science, Engineering and Built Environment, Geelong Waurn Ponds Campus, Deakin University, Geelong P.O. Box 423, Australia; musa.mammadov@deakin.edu.au
* Correspondence: shanthi@stu.cmb.ac.lk

† Presented at the 9th International Conference on Time Series and Forecasting, Gran Canaria, Spain, 12–14 July 2023.

Abstract: The study proposes an ensemble spatiotemporal methodology for short-term rainfall forecasting using several data mining techniques. Initially, Spatial Kriging and CNN methods were employed to generate two spatial predictor variables. The three days prior values of these two predictors and of other selected weather-related variables were fed into six cost-sensitive classification models, SVM, Naïve Bayes, MLP, LSTM, Logistic Regression, and Random Forest, to forecast rainfall occurrence. The outperformed models, SVM, Logistic Regression, Random Forest, and LSTM, were extracted to apply Synthetic Minority Oversampling Technique to further address the class imbalance problem. The Random Forest method showed the highest test accuracy of 0.87 and the highest precision, recall and an F1-score of 0.88.

Keywords: deep learning; spatial kriging; ensemble; cost-sensitive; data mining; imbalance learning

Citation: Saubhagya, S.; Tilakaratne, C.; Mammadov, M.; Lakraj, P. An Application of Ensemble Spatiotemporal Data Mining Techniques for Rainfall Forecasting. *Eng. Proc.* **2023**, *39*, 6. https://doi.org/10.3390/engproc2023039006

Academic Editors: Ignacio Rojas, Hector Pomares, Luis Javier Herrera, Fernando Rojas and Olga Valenzuela

Published: 27 June 2023

1. Introduction

Rainfall is identified as one of the most chaotic and dynamic phenomena that varies spatiotemporally [1]. Heavy and extreme rainfall creates a serious threat to human lives and properties through severe flooding. Therefore, an accurate rainfall nowcasting is of great significance in preventing devastating consequences.

The data mining techniques optimally capture the hidden spatiotemporal patterns among largely available weather-related data [2]. Many researchers achieved high prediction accuracy in rainfall classification using techniques such as Random Forest, Artificial Neural Network (ANN), K—Nearest Neighbour (KNN) and Support Vector Machine (SVM). As a recent trend, deep learning methods such as Convolutional Neural Network (CNN) and Long-Short-Term-Memory (LSTM) are employed to explore the meteorological big data due to its promising technical advantages and performance [3]. The study performed in [4] used K-means clustering to predict rainfall states. The identified clusters were used as predictands for training the Classification and Regression Tree (CART) model with five climate input variables and obtained a satisfactory value of goodness-of-fit. Another study performed CART and C4.5 models with thirteen input variables to predict the chance of rain and gained average accuracies of 99.2% (CART) and 99.3% (C4.5) [5]. Moreover, [6] modeled weekly rainfall with weather variables using ANN and produced higher prediction accuracy than multiple linear regression model. The summer precipitation patterns over eastern China were modeled using multinomial logistic regression (MLR) by [7] and gained a prediction accuracy range of (60–70%). Authors of [8] compared several machine learning models in classifying month of a year as dry or wet. The rainfall classification carried out by [9] concluded that Decision Trees and Random Forests could perform well even with a low proportion of training data. A similar study conducted by [10] extracted the Adaboost algorithm, which produced F1-score of 0.9726. Authors of [11] carried out rainfall classification addressing the class imbalance through over and under-sampling techniques, and the results indicated varying performance with different inputs generated by resampling techniques.

Our study employed several cost-sensitive machine learning models, including Penalized SVM and Complement Naïve Bayes followed by a resampling technique to address the natural rarity of extreme rainfall events. Prior to that two spatial input variables were generated by modeling satellite data with deep learning method (CNN) and rainfall at nearby rain gauging stations with Spatial Kriging. Therefore, the study focused on three solutions proposed by the literature [12] for imbalance learning. Moreover, we have performed a comparative study between machine learning and deep learning models suggested by many researchers [11,12].

The next section describes the materials and methods used during the study while Sections 3 and 4 line up the results and discussion, and conclusions, respectively.

2. Materials and Methods

2.1. Description of Data

The weather data was obtained from the Meteorology Department of Sri Lanka based on the Kalu River basin over the period from 2015 to 2019. It includes daily data on 28 variables including rainfall at target rain gauging station (Rathnapura), rainfall of six nearby gauging stations, relative humidity, mean sea level pressure, wind speed, temperature, sunshine hours, evaporation, and Southern Oscillation Index. Additionally, the study collected daily satellite images (with a size of 500 × 512 pixels) covering the river basin from China Meteorological Administration National Satellite Meteorological Center for the same time period.

2.2. Methods

The main objective of the study is to forecast rainfall occurrences from highly imbalanced spatiotemporal time series data by using machine learning and deep learning methods. Initially, the rainfall classes needed to be identified. Therefore, with the cutoff levels established by Meteorology Department of Sri Lanka and a comparison of flood occurrence with respect to rainfall at Rathnapura, the rainfall values were categorized into three classes, 'No rainfall', 'Normal rainfall', and 'Extreme rainfall'. Following the norm of the Department of Meteorology, Sri Lanka, the cutoff level for extreme events was set at 110 mm of rainfall. Then, the methodology illustrated in Figure 1 was carried out.

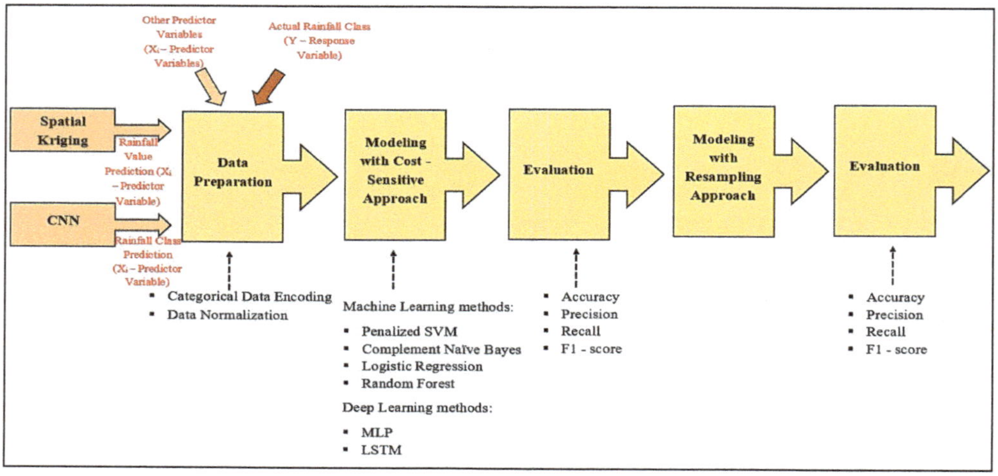

Figure 1. Ensemble Spatiotemporal Data Mining Approach.

The Spatial Kriging was applied to predict current day (t) rainfall at Rathnapura using current day rainfall values of six nearby stations to incorporate spatial correlation between nearby stations and target station to final model. The previous day (t − 1) satellite image was modeled to predict the current day (t) rainfall class through CNN model. Then, the predicted rainfall class of the day t with the predicted rainfall value of the same day were brought as predictor variables to the final dataset along with other selected variables. The values of these predictor variables of last three consecutive days (t, t − 1, t − 2) were fed into six classifiers from different model families (Linear Classifier, Ensemble, and deep and sequential learning) for forecasting next day (t + 1) actual rainfall class of target variable or the response variable.

2.2.1. Spatial Kriging

Let the rainfall value in Rathnapura on a particular day j be y_j. In Spatial Kriging estimates of y_j, $\hat{y}_j$ is modeled through the rainfall values of m neighboring sample locations X_i, i.e., $Z(X_i)$. It gives an optimal linear combination of $Z(X_i)$ with weights w_i, which are taken according to covariance values [13].

$$\hat{y}_j = k + \sum_{i=1}^{n} w_i Z(X_i) = k + w^T \mathbf{Z} \tag{1}$$

2.2.2. Multi-Layer Perceptron (MLP)

MLP is a feed forward neural network that consists of three types of layers, the input layer, hidden layer (s), and output layer [14]. Let us consider a MLP model (see Figure 2) with one hidden layer.

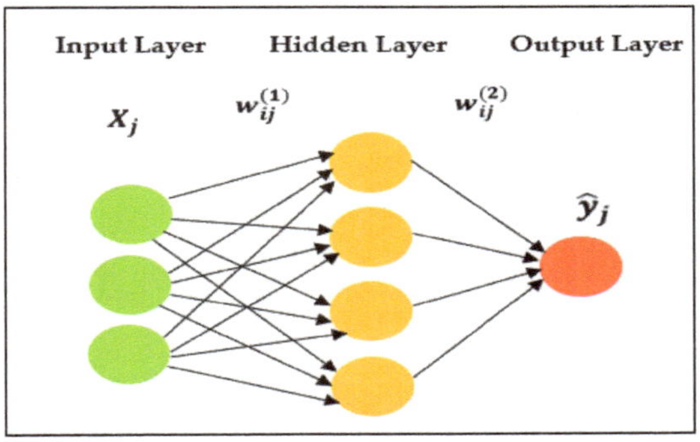

Figure 2. A multilayer perceptron model with one hidden layer.

Here, the X_j is the input to the jth neuron of the input layer, the $w_{ij}{}^{(1)}$ is the weight of the link connecting j^{th} neuron of the input layer to i^{th} neuron of the hidden layer, b_i is the bias associated with the ith neuron of the hidden layer, and $\varnothing^{(1)}$ is the activation function associated with the hidden layer. Then, the net output from i^{th} neuron of the hidden layer is given by h_i (see Equation (2)). $w_{ij}{}^{(2)}$ is the weight of the link connecting i^{th} neuron of the hidden layer to j^{th} neuron of the output layer, b_j is the bias associated with the jth neuron of the output layer, and $\varnothing^{(2)}$ is the activation function associated with the output layer. In this case, $j = 1$. The output of that j^{th} neuron of the output layer (or, in our case, the final rainfall prediction at Rathnapura by MLP) will be $\hat{y}_j$ [15]. Since the network is fully connected, each unit has its own bias, and there is a weight for every pair of units in two consecutive layers. Then, the MLP network computations can be written as:

$$h_i = \varnothing^{(1)}\left(\sum_i w_{ij}^{(1)} X_j + b_i\right) \text{ Here, } i = 1, 2, 3, 4 \tag{2}$$

$$\hat{y}_j = \varnothing^{(2)}\left(\sum_i w_{ij}^{(2)} h_i + b_j\right) \text{ Here, } i = 1,\ 2,\ 3,\ 4 \tag{3}$$

2.2.3. Long Short-Term Memory (LSTM)

LSTM has four neural network layers interacting in a very special way. Its memory cell consists of a forget gate, input gate, and output gate [16].

As shown in Figure 3, the output of the last moment and current input value are fed into the forget gate to obtain the following output at the forget gate.

$$f_t = \sigma\left(w_f \,.[h_{t-1},\ x_t] + b_f\right) \tag{4}$$

where $f_t \in (0,\ 1)$, w_f—weight at forget gate, b_f—bias at forget gate, x_t- current input value, h_{t-1}—output at previous moment. Then, the same previous output and current input value are inputted to the input gate, and the output value and candidate cell state at the input gate are calculated as below.

$$i_t = \sigma(w_i \,.[h_{t-1},\ x_t] + b_i) \tag{5}$$

$$\widetilde{C}_t = tanh(w_c \,.[h_{t-1},\ x_t] + b_c) \tag{6}$$

where $i_t \in (0,\ 1)$, w_i—weight at input gate, b_i—bias at input gate, w_c—weight at candidate input gate, b_c—bias at candidate input gate. Update the current cell state using the following formulae.

$$C_t = f_t * C_{t-1} + i_t * \widetilde{C}_t \tag{7}$$

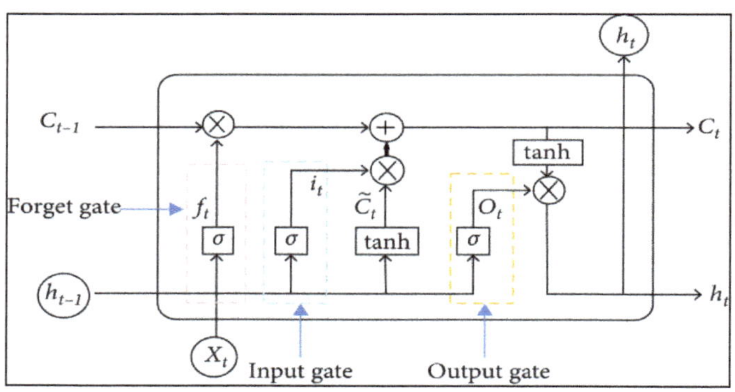

Figure 3. LSTM structure diagram (Adapted with permission from Ref. [17]. 2020, Wenjie Lu et al.).

The h_{t-1} and x_t are then fed into output gate at time t and obtain output O_t at output gate as follows. Here, w_o—weight at output gate and b_o—bias at input gate.

$$O_t = \sigma(w_o \,.[h_{t-1},\ x_t] + b_o) \tag{8}$$

Finally, the output of the LSTM was obtained using the current cell state and output at the output gate using the following formulae.

$$h_t = O_t * tanh(C_t) \tag{9}$$

2.2.4. Convolutional Neural Network (CNN)

CNN is very popular for image processing and computer vision. It consists of three layers as seen in Figure 4. Convolution layer performs linear convolution operation, and the features of the data are extracted. Since the feature dimensions are very high, a pooling layer is added after the convolution layer. To make a final forecast, a fully connected (FC) layer (or dense layer) is added, and inputs to this layer are the flattened features resulted from convolutional and pooling layers [18,19].

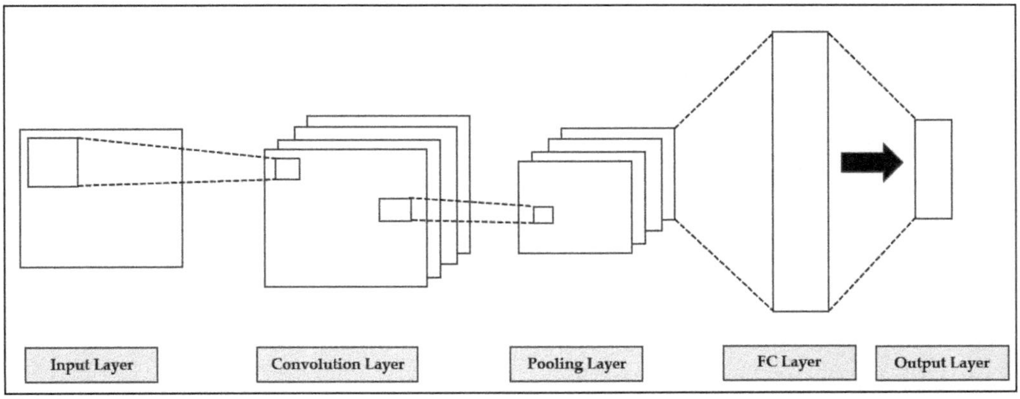

Figure 4. The architecture of CNN.

2.2.5. Random Forest (RF)

RF is known as a supervised ensemble learning method. The method constructs a multitude of decision trees with controlled variation at the training phase. Then, using bagging, each tree in the ensemble is constructed (with sample with replacement) from training data. In the classification problem, each tree in ensemble is a base classifier to identify the class of the unlabeled observation. Through voting of each classifier for their predicted classes, the final class is obtained computing the majority votes [20,21].

2.2.6. Support Vector Machine (SVM)

SVM classifier finds a hyperplane to segregate the nodes for classification. The following optimization problem is solved when deriving the optimal hyperplane which separates two classes.

$$\underset{w,\ b,\varepsilon_j}{\text{Minimize}}\frac{1}{2}w^T w + C\sum_j \varepsilon_j \tag{10}$$

$$\text{Subject to } y_j\left(w^T\Phi(X_j) + b\right) \geq 1 - \varepsilon_j \tag{11}$$

$$\varepsilon_j \geq 0 \tag{12}$$

where y_j is the class label, w is the weights vector, X_j is the input feature vector, Φ is the transformation function, ε_j is the degree of misclassification corresponding to X_j, C is the regularization parameter and b is the bias.

The optimal hyperplane (a maximum marginal hyperplane (MMH)) is learnt by training the samples using several kernel functions such as linear, radial basis function (rbf) and polynomial (poly) [22]. The Penalized SVM (PSVM or Cost-sensitive SVM) is a modification of SVM that weighs the margin proportional to the class importance which can be applied to an imbalanced dataset [23,24].

2.2.7. Naïve Bayes (NB)

The Naïve Bayes Algorithm is based on the Bayes Theorem of probability.

$$Pr(C|X) = \frac{Pr(X|C)Pr(C)}{Pr(X)} \tag{13}$$

where, $Pr(C|X)$: The posterior probability of the class of interest given predictor, $Pr(C)$: The prior probability of the class of interest, $Pr(X|C)$: The probability of predictor given class C, and $Pr(X)$: The prior probability of predictor.

NB method calculates the probability of an observation belonging to a certain class. The Complement Naïve Bayes (CNB) method computes the probability of the observation belonging to all the classes. Thus, CNB is more suitable in dealing with imbalanced datasets [25].

2.2.8. Multinomial Logistic Regression (MLR)

Multinomial Logistic Regression is the generalization of Logistic regression which allows more than two categories of output variable. It also obtains maximum likelihood estimation to evaluate the probability of categorical membership [26]. The formula of MLR is as follows:

$$\log\left[\frac{Pr_i(Y)}{Pr_j(Y)}\right] = \alpha_i + \beta_i^T X \tag{14}$$

where $Pr_i(Y)$ is the probability of Y in i^{th} class, $Pr_j(Y)$ is the probability of Y in j^{th} class, α_i is the intercept, β_i^T is the vector of covariates for the ith class of the output variable Y, and X is the input feature vector [27].

2.3. *Imbalanced Learning*

The data mining techniques will produce biased classification if the dataset is imbalanced [11]. Moreover, it can lead to a problem in ignoring the minority class entirely in the case where the predictions on the minority class are most important. This is a major issue found in rainfall forecasting. Certain methodologies can deal with the class imbalance of the data.

2.3.1. Cost—Sensitive Learning

This tactic uses penalized learning algorithms which give higher misclassification costs (or weights) for instances of the minority class and lower misclassification costs for the majority class [12,24].

2.3.2. Resampling Techniques

The resampling techniques are applied to obtain more balanced datasets. In this study, Synthetic Minority Oversampling Technique (SMOTE) which synthesizes new samples from the minority class was applied.

2.4. *Model Evaluation*

The metrics, Accuracy, Precision, Recall and F1-Score were used to evaluate the classification models, and the Spatial Kriging model results were evaluated using Mean Absolute Error (MAE), Root Mean Squared Error (RMSE), and R^2 value.

Firstly, the cost-sensitive approach was followed in rainfall class prediction. Through the model evaluation results, the best set of models were chosen to apply the resampling technique. The final evaluation based on resampling was taken into consideration in selecting the best model for rainfall classification. Before applying the classification models, all the input variables were normalized. The machine learning and deep learning algorithms were run in Python.

3. Results and Discussion

As mentioned previously, initially the Spatial Kriging method was applied to find the daily rainfall prediction at Rathnapura gauging station.

The results shown in Table 1 indicate that the Spatial Kriging model fitted using the rainfall values of nearby stations cannot be solely used to explain the variation of the rainfall at target station, yet, they have an influence on the target station's rainfall.

Table 1. Performance of Spatial Kriging method.

Metric	RMSE	MAE	R^2
Error	16.2	6.64	30.3%

Then, the previous day's (t − 1) satellite image was modeled with the current day's (t) rainfall occurrence in Rathnapura using the CNN model. The best model parameters obtained via 60 trials of training the models CNN, MLP and LSTM are presented in Table 2.

Table 2. Parameter specification of CNN, MLP and LSTM models.

Parameters	CNN	MLP	LSTM
Activation function	Softmax	Softmax	Softmax
Epochs	50	50	50
Number of hidden layers	7	2	1
Number of neurons in final dense function	3	3	3
Batch size	72	72	72
Kernel size	(3,3)	-	-
Learning rate	0.01	0.01	0.01
Optimizer	Adam	Adam	Adam
Loss Function	Categorical Cross Entropy	Categorical Cross Entropy	Categorical Cross Entropy

The CNN model showed 64.9% of Accuracy and Recall with 59.7% of Precision and 52.8% of F1-Score. The results also suggest the same conclusion produced by the Spatial Kriging method.

However, applying Spatial Kriging reduced the dimensions (number of input variables) of the final model. This method along with satellite images analysis are set to incorporate the spatial variation of the rainfall data to the final model.

The predictions obtained from the above two spatial models were incorporated as new predictor variables to the final dataset. Then, there were 23 predictor variables. The values of the past three days (t, t − 1, t − 2 on (2)) spatial correlation between nearby stations and target station to final model. t − 2) of each predictor variables were modeled with the next day (t + 1) actual rainfall class since through a preliminary data analysis we could identify that the past three days rainfall values have much impact on the next day (t + 1) rainfall occurrence.

To address the class imbalance, cost-sensitive models were applied. The entire data set was split as 80% for training and 20% for testing. For the training set, Repeated Stratified 5—Fold Cross Validation (which repeats the cross-validation procedure multiple times) was applied to further address the class imbalance problem. Tables 3 and 4 show model performance.

Table 3. The performance of cost-sensitive models in training sets.

Metric	Cost-Sensitive Model						
	PSVM (Kernel = rbf)	PSVM (Kernel = Poly)	CNB	MLR	RF	MLP	LSTM
Accuracy	0.75	0.73	0.71	0.74	0.73	0.73	0.73
Recall	0.75	0.74	0.72	0.74	0.74	0.74	0.74
Precision	0.75	0.73	0.71	0.74	0.73	0.73	0.73
F1-Score	0.75	0.73	0.71	0.73	0.73	0.73	0.73

Table 4. The performance of cost-sensitive models in test set.

Metric	Cost-Sensitive Model						
	PSVM (Kernel = rbf)	PSVM (Kernel = poly)	CNB	MLR	RF	MLP	LSTM
Accuracy	0.71	0.77	0.68	0.81	0.76	0.70	0.71
Precision	0.72	0.75	0.75	0.79	0.75	0.72	0.75
Recall	0.71	0.77	0.68	0.81	0.76	0.70	0.71
F1-Score	0.72	0.76	0.71	0.78	0.75	0.71	0.73

The training and testing performances indicate that cost-sensitive SVM, Random Forest, Multinomial Logistic Regression, and LSTM models have performed better in terms of metrics, especially precision, recall, and F1-score, which are more suitable in evaluating class-imbalanced problems [28]. The selected models depict more than 70% Accuracy, Precision, Recall, and F1-Score.

Then, for the selected models, the SMOTE resampling technique was applied. The performance was evaluated after refitting the balanced dataset using the selected best set of models. The following Tables 5 and 6 illustrate the final performance results.

Table 5. The performance of cost-sensitive resampled models in training set.

Metric	Cost-Sensitive Resampled Model				
	PSVM (Kernel = rbf)	PSVM (Kernel = Poly)	MLR	RF	LSTM
Accuracy	0.82	0.83	0.80	0.87	0.77
Precision	0.82	0.82	0.79	0.87	0.78
Recall	0.82	0.83	0.80	0.87	0.79
F1-Score	0.82	0.83	0.80	0.87	0.77

Table 6. The performance of cost-sensitive resampled models in test set.

Metric	Cost-Sensitive Resampled Model				
	PSVM (Kernel = rbf)	PSVM (Kernel = Poly)	MLR	RF	LSTM
Accuracy	0.82	0.82	0.79	0.87	0.78
Precision	0.82	0.83	0.79	0.88	0.79
Recall	0.82	0.83	0.80	0.88	0.79
F1-Score	0.82	0.83	0.79	0.88	0.78

It can be observed that after two operations, the performance of all selected models has improved. Out of them, the Random Forest method gives the best and consistent performance in both the training set and in the final evaluation (in test set) of the selected models (nearly 88% of Accuracy, Precision, Recall, and F1-Score).

Overall, the study results indicate the importance of incorporating spatial variation of the rainfall data in predicting future events and highlight the effectiveness of step-wise imbalance learning to obtain consistent and more accurate predictions which could not be attained in some previous studies.

4. Conclusions

In this study, we proposed a novel ensemble spatiotemporal data mining approach to forecast rainfall occurrence at the Rathnapura gauging station. Spatial Kriging and a Deep Learning model (CNN) were employed to capture the spatial variation over the selected grid. The temporal variation of the rainfall data was brought to the model by modeling with the three past consecutive days' values of the variables. Five cost-sensitive models were further improved to address the imbalanced problem found in rainfall classes through a resampling technique. The final performance summary emphasizes the outperformance of the cost-sensitive resampled Random Forest method (nearly 88% of accuracy, precision, recall, and F1-score) in forecasting future rainfall occurrences.

During the study, we found the complexity of working with high number of predictor variables. Therefore, our future studies are expected to enhance further by focusing on feature selection and application of dimension reduction prior to the model application. Collecting data for an extended period (e.g., 30 years) and selecting novel approaches will also be taken into consideration when dealing with highly imbalanced rainfall data.

Author Contributions: Conceptualization, C.T. and S.S.; methodology, S.S. and C.T.; software, S.S.; validation, S.S.; formal analysis, S.S.; investigation, C.T., M.M. and P.L.; resources, S.S.; data curation, S.S.; writing—original draft preparation, S.S.; writing—review and editing, C.T. and M.M.; visualization, S.S.; supervision, C.T., M.M. and P.L.; project administration, C.T.; funding acquisition, C.T. All authors have read and agreed to the published version of the manuscript.

Funding: This research was funded by the Research Grant of the University of Colombo, Sri Lanka, grant number AP/3/2019/CG/30.

Institutional Review Board Statement: Not applicable.

Informed Consent Statement: Not applicable.

Data Availability Statement: The data that support the findings of this study are available from the authors upon reasonable request.

Acknowledgments: The authors wish to acknowledge Irrigation Department of Sri Lanka for providing flood related data and the China Meteorological Administration National Satellite Meteorological Center (NSMC) for providing the satellite data for the study.

Conflicts of Interest: The authors declare no conflict of interest.

References

1. Hussein, E.A.; Ghaziasgar, M.; Thron, C.; Vaccari, M.; Jafta, Y. Rainfall Prediction Using Machine Learning Models: Literature Survey. In *Artificial Intelligence for Data Science in Theory and Practice*; Studies in Computational, Intelligence; Alloghani, M., Thron, C., Subair, S., Eds.; Springer: Cham, Switzerland, 2022; Volume 1006. [CrossRef]
2. Parmar, A.; Mistree, K.; Sompura, M. Machine Learning Techniques For Rainfall Prediction: A Review. In Proceedings of the 2017 International Conference on Innovations in information Embedded and Communication Systems (ICIIECS), Coimbatore, India, 17–18 March 2017.
3. Sun, D.; Wu, J.; Huang, H.; Wang, R.; Liang, F.; Xinhua, H. Prediction of Short-Time Rainfall Based on Deep Learning. *Math. Probl. Eng.* **2021**, *2021*, 6664413. [CrossRef]
4. Kannan, S.; Ghosh, S. Prediction of daily rainfall state in a river basin using statistical downscaling from GCM output. *Stoch. Environ. Res. Risk Assess.* **2010**, *25*, 457–474. [CrossRef]
5. Ji, S.-Y.; Sharma, S.; Yu, B.; Jeong, D.H. Designing a rule-based hourly rainfall pre-diction model. In Proceedings of the 2012 IEEE 13th International Conference on Information Reuse & Integration (IRI), Las Vegas, NV, USA, 8–10 August 2012; Volume I. [CrossRef]
6. Sharma, M. Comparative Study of rainfall forecasting models MA Sharma. *JB Singh N. Y. Sci. J.* **2011**, *4*, 115–120.
7. Gao, L.; Wei, F.; Yan, Z.; Ma, J.; Xia, J. A Study of Objective Prediction for Summer Precipitation Patterns Over Eastern China Based on a Multinomial Logistic Regression Model. *Atmosphere* **2019**, *10*, 213. [CrossRef]
8. Aguasca-Colomo, R.; Castellanos-Nieves, D.; Méndez, M. Comparative Analysis of Rainfall Prediction Models Using Machine Learning in Islands with Complex Orography: Tenerife Island. *Appl. Sci.* **2019**, *9*, 4931. [CrossRef]
9. Zainudin, S.; Jasim, D.; Abu Bakar, A. Comparative Analysis of Data Mining Techniques for Malaysian Rainfall Prediction. *Int. J. Adv. Sci. Eng. Inf. Technol.* **2016**, *6*, 1148. [CrossRef]
10. Singh, G.; Kumar, D. Hybrid Prediction Models for Rainfall Forecasting. In Proceedings of the 2019 9th Inter-national Conference on Cloud Computing, Data Science & Engineering (Confluence), Noida, India, 10–11 January 2019; pp. 392–396. [CrossRef]
11. Oswal, N. Predicting rainfall using machine learning techniques. *arXiv* **2019**, arXiv:1910.13827.
12. Katrakazas, C.; Antoniou, C.; Yannis., G. Time Series Classification Using Imbalanced Learning for Real-Time Safety Assessment. In Proceedings of the Transportation Research Board 98th Annual Meeting, Washington, DC, USA, 13–17 January 2019.
13. Cuenca, J.; Correa-Flórez, C.; Patino, D.; Vuelvas, J. Spatio-Temporal Kriging Based Economic Dispatch Problem Including Wind Uncertainty. *Energies* **2020**, *13*, 6419. [CrossRef]

14. Abirami, S.P.; Chitra, P. Chapter Fourteen—Energy-efficient edge based real-time healthcare support system. *Adv. Comput.* **2020**, *117*, 339–368.
15. Grosse, R. Lecture 5: Multilayer Perceptrons. 2018. Available online: https://www.cs.toronto.edu/~rgrosse/courses/csc321_2018/readings/L05%20Multilayer%20Perceptrons.pdf (accessed on 1 February 2023).
16. Understanding LSTM Networks. Available online: https://colah.github.io/posts/2015-08-Understanding-LSTMs/ (accessed on 1 February 2023).
17. Lu, W.; Li, J.; Li, Y.; Sun, A.; Wang, J. A CNN-LSTM-Based Model to Forecast Stock Prices. *Complexity* **2020**, *2020*, 6622927. [CrossRef]
18. How Do Convolutional Layers Work in Deep Learning Neural Networks? Available online: https://machinelearningmastery.com/convolutional-layers-for-deep-learning-neural-networks/ (accessed on 1 February 2023).
19. Torres, J.F.; Hadjout, D.; Sebaa, A.; Martínez-Álvarez, F.; Troncoso, A. Deep Learning for Time Series Forecasting: A Survey. *Big Data. Feb.* **2018**, *2021*, 3–21. [CrossRef]
20. Goehry, B.; Yan, H.; Goude, Y.; Massart, P.; Poggi, J.M. Random Forests for Time Series. *REVSTAT Stat. J.* **2021**. *accepted*. Available online: https://revstat.ine.pt/index.php/REVSTAT/article/view/400 (accessed on 1 February 2023).
21. Fawagreh, K.; Gaber, M.; Elyan, E. Random Forests: From Early Developments to Re-cent Advancements. *Syst. Sci. Control. Eng* **2014**, *2*, 602–609. [CrossRef]
22. Yu, N.; Haskins, T. KNN, An Underestimated Model for Regional Rainfall Forecasting. *arXiv* **2021**, arXiv:2103.15235.
23. Cost-Sensitive SVM for Imbalanced Classification. Available online: https://machinelearningmastery.com/cost-sensitive-svm-for-imbalanced-classification/ (accessed on 2 February 2023).
24. How to Handle Imbalanced Classes in Machine Learning. Available online: https://elitedatascience.com/imbalanced-classes (accessed on 2 February 2023).
25. Complement Naive Bayes (CNB) Algorithm. Available online: https://www.geeksforgeeks.org/complement-naive-bayes-cnb-algorithm/ (accessed on 2 February 2023).
26. Kwak, C.; Clayton-Matthews, A. Multinomial logistic regression. *Nurs. Res.* **2002**, *51*, 404–410. [CrossRef] [PubMed]
27. Hashimoto, E.M.; Ortega, E.M.M.; Cordeiro, G.M.; Suzuki, A.K.; Kattan, M.W. The multinomial logistic regression model for predicting the discharge status after liver transplantation: Estimation and diagnostics analysis. *J. Appl. Stat.* **2019**, *47*, 2159–2177. [CrossRef] [PubMed]
28. Andersson, M. Multi-Class Imbalanced Learning for Time Series Problem: An Industrial Case Study. Master's Dissertation, Uppsala University, Uppsala, Sweden, 2020. Available online: http://urn.kb.se/resolve?urn=urn:nbn:se:uu:diva-412799 (accessed on 2 February 2023).

engineering proceedings

MDPI

Proceeding Paper

Musical Aptitude Screening: A Brazilian Experience under Construction †

Fabiana Oliveira Koga [1,*], Rosemeire de Araújo Rangni [1] and Rafael Pereira [2]

1 Department of Psychology, Federal University of São Carlos, São Carlos 13565-905, Brazil; rose.rangni@ufscar.br

2 Department of Technology, Federal Institute of Bahia, Vitória da Conquista 45078-900, Brazil; contatorafaeldutra@outlook.com

* Correspondence: fabianapsicopedagogiamusical@gmail.com

† Presented at the 9th International Conference on Time Series and Forecasting, Gran Canaria, Spain, 12–14 July 2023.

Abstract: In Brazil, Law n. 9394/96 ensures rights such as the identification of talented students in order to offer specialized educational attention; in this sense, the Protocol for Screening of Musical Abilities was elaborated with 54 items and its complementary instruments (scales and questionnaires) in order to collaborate with the survey of students with indicators of musical talent. This work, therefore, aims to present the instruments and the evidence of effectiveness and usability found in a preliminarily manner. It is an investigation in progress and incorporates experimental psychometric (elaboration of scales) and psycho-physical (peer comparison method) methods. In total, 800 individuals, including children from six to eleven years old, their guardians (family members) and teachers, took part in the research. The results have indicated that participants with higher scores remain with the same indices in the later stage of evaluation; however, only from the statistical tests intended for validation, standardization and reliability, as well as exploratory factor analysis will it be possible to attest the validity, standardization of scores and prepare the final version for wide use of the instruments.

Keywords: music aptitude; measurement; screening

Citation: Koga, F.O.; de Araújo Rangni, R.; Pereira, R. Musical Aptitude Screening: A Brazilian Experience under Construction. *Eng. Proc.* **2023**, *39*, 7. https://doi.org/10.3390/engproc2023039007

Academic Editors: Ignacio Rojas, Hector Pomares, Luis Javier Herrera, Fernando Rojas and Olga Valenzuela

Published: 28 June 2023

1. Introduction

The Brazilian legislation Law n. 9394/96, states that talented students are part of the Special Education audience and that rights and resources are ensured for them, such as: identification, specialized educational assistance, permission to leave school, supplementation and curriculum planning, as well as acceleration. These legal rights extend from kindergarten to Higher Education, and they are considered in all areas of human, academic and creative knowledge, including for Arts such as Music [1]. After all, talent can manifest itself in different individuals and in any period of life, regardless of socioeconomic and cultural level [2].

In the last Brazilian school census, in 2022, released by the National Institute of Educational Studies and Research Anísio Teixeira (INEP), the total enrolled students identified as talented were 26,589 (regular classes) and 226 (special classes), and in the State of São Paulo, the locus of this research, the numbers were 2918 (regular classes) and 9 (special classes). From these indices, it is possible to conclude the small number of students that are identified and referred to educational care in a country such as Brazil, a country with a continental dimension and a significant number of inhabitants. The studies by Rangni, Rossi and Koga [3] and, mainly, those by Koga and Rangni [4], showed the neglection of talented Brazilian students, especially in the area of Music, with evidence also of a lack of instruments available for selection on the poll.

Some initiatives are widely known, including Carl Seashore's aptitude and musical talent test, and Edwin E. Gordon's tracking instruments, among others, as summarized by Gagné and McPherson [5]. Recently, the battery of scales and questionnaires by Haroutounian [6]; the Spanish scale for probing aptitude and musical talent by Hernández and Pérez [7] and the scale by Mullensiefen, Gingras, Musil and Stewart [8] have been recognized. Although these instrumental survey possibilities exist, they have not been validated for a Brazilian cultural diversity. It should be noted that this cultural structure relies on the miscegenation of many peoples, especially making it difficult to culturally validate instruments [9]. It is in this context that Koga [9] elaborated the Musical Abilities Screening Protocol (PSHM) and complementary instruments in order to track possible cases of musical talent in Brazilian schools and in other realities.

It is inferred that musical talent is a complex and abstract phenomenon with multiple variables, such as aptitude (eminent musical potential or intelligence) [6], motivation, rhythmic and aesthetic sense, imagination and creativity (giftedness) [10], audiation and musical memory [11], precocious musicality [12] and personality and over-excitability [13]. Moreover, for Gagné and McPherson [5] aptitude is synonymous with the concept of endowment, and for Haroutounian [6], Kirnarskaya [10] and Gordon [11] it is an important basis or variable, but not the only one, used to infer the presence of musical talent in a preliminary way. This route was chosen because it unites the cognitive system with the affective and creative one through the internal, symbolic or abstract representations of Music, which generate emotional meaning in the search for meaning when composing or performing a piece [14], with the dialectical and mediated interactions of these variables making a composer or interpreter unique or original [14].

Under these aspects, the following question arises: What evidence about the effectiveness of the PSHM and complementary instruments is already observed in relation to the survey of the presence of musical talent, based on the musical aptitude indices? What are the steps required to be completed for validity, standardization and normalization?

The objective outlined was to present the PSHM and its complementary instruments and the preliminary evidence of effectiveness.

1.1. Method

This is ongoing research which is based on the psychometric model "Item Response Theory" (validity of tests) and psychophysics (peer comparison method). Aptitude is considered a measurable construct through performance of the individual in certain tasks (items); thus, performance is treated as an effect and the construct (latent trait) as the cause in which individuals with more aptitude are more likely to present high performance in a set of items. It is worth mentioning that only the PSHM is based on psychophysics, as it deals with pairs of sound items [15,16].

A sample of approximately 800 individuals (students from six to eleven years old), teachers/adults and legal guardians/adults) from public, private, indigenous schools and institutions dedicated to serving talented students identified in Brazil and Mexico participated in the research.

The PSHM and other instruments were applied in two scenarios, the first with all students (PSHM and complementary instruments), and the second with those who had higher, medium and lower levels of musical aptitude, via random sampling; the students were to be evaluated musically and confirmation of musicality indices given (consecrated validated instruments and observations in musical activities). The research has ethical approval CAEE: 5 2224021.0.0000.5504.

Figure 1 outlines the research design, data collection and analysis procedure.

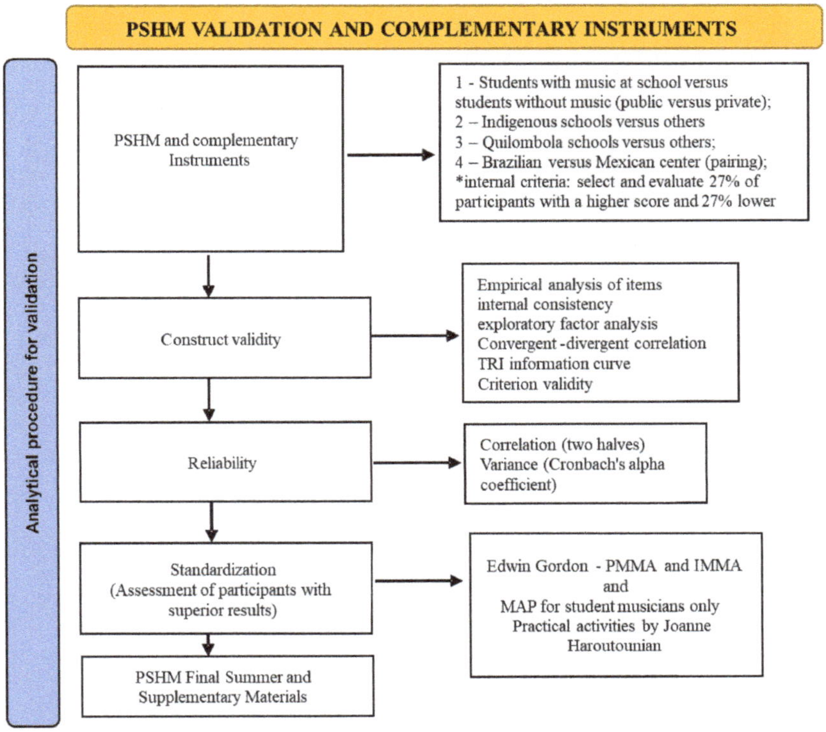

Figure 1. Research design and analytical procedures for validation.

Data collection was carried out during the year 2022, post-pandemic. In the first stage, pairs of sounds were presented to the participants, and they had five seconds to respond. At the end, they filled out a questionnaire with questions to mark, two of which were dissertations. Teachers and family members also completed the musical evaluation, with the teachers completing it in the classroom and family members taking it home and returning the results the next day. The interviews with teachers and students were audio-recorded. Due to the existence of bilingual schools, immigration, indigenous peoples and participation of Mexican students, the musical evaluation has versions in English, Spanish and Portuguese as well as guidance for translation (carried out at the time of the survey by indigenous teachers) in Guarany and Terena indigenous languages.

Each school received a researcher for an average of two to six days to carry out the musical aptitude survey. Completing the musical evaluation took an average of 50 min with students and 20 min with teachers. Figures 2 and 3 exemplify data collection in educational institutions. Figure 2 illustrates students responding in person and, in Figure 3, three participants responding virtually (completed directly on the PSHM platform after sending the access link).

Data analysis and validation of the PSHM toolkit are currently underway. The next steps include item analysis, internal consistency analysis (correlation), exploratory factor analysis, establishing cutoff scores in the psychometric test through distribution curve analysis, reliability assessment, and other procedures as guided by Pasquali [15] and Vieira [16].

Figure 2. Data collection in a public school (Brazil).

Figure 3. Data collection in a private school (Brazil).

1.2. Results and Discussion

The PSHM and complementary instruments could be given to groups and individuals, as was the case for headphones or amplified boxes for sound reproduction. There was no need for soundproofing; however, if it would have helped with the quality of the results, it was available to give. Tabulation was standard and by age group. There is the extraction of raw data and its conversion into scores. The complete material of the PSHM was developed to carry out an initial identification or screening, that is, there was a need for evaluation procedures as a way of confirming the individual's condition. In addition, it was essential to plan the intervention or musical enrichment later [9]. The structure of the PSHM and complementary materials followed the same method as the musical evaluations of Haroutounian [6], Hernández and Pérez [7], Kirnarskaya [10] and Gordon [11]. Figure 4 contemplates the description of the structure of the instruments; for more details, see Koga [9].

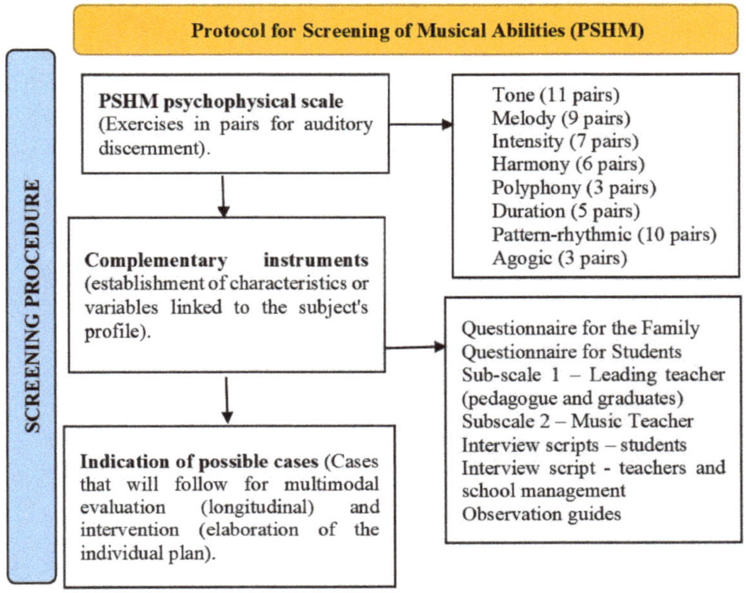

Figure 4. Complete PSHM structure and complementary instruments.

The PSHM and complementary instruments were restricted to school spaces. In the case of the online version, it could be carried out in computer labs or classrooms, and navigation in the system was intuitive and simple (self-application). Additionally, they had resources adapted for people with disabilities. The application worked on electronic devices—computers, tablets and cell phones—and responses were marked individually. Figure 5 illustrates the system design.

Figure 5. Design of the Professora Fabi system (teacher Fabi), which hosts the PSHM and its complementary instruments and was created by Rafael Pereira with the contribution of a Theme Selection design.

The printed version follows the same standards as the online version, with changes in manual filling out. Figure 6 exposes the printed format.

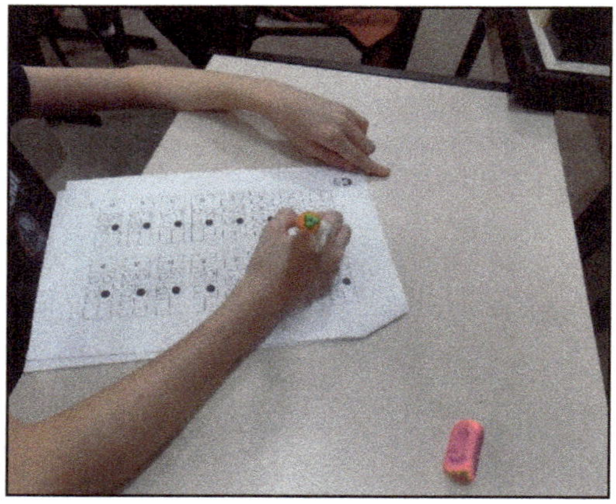

Figure 6. Printed design.

Since the construction reported in Koga [9] and Koga and Rangni [3], the PSHM and complementary instruments have shown satisfactory indices ($p \leq 0.05$ in the chi-square test) in their measurements. The authors point out that there was a significant correlation between the items and dimensions ($p \leq 0.05$ in Kendell's W). Differences were observed between schools which had Music as a subject and those without. To exemplify: in the Mann–Whitney statistical test, $p = 0.0479$ was obtained, considered statistically significant for the 1st years; $p = 0.0031$, very significant for the 2nd years; $p = 0.0001$, considered extremely significant for the 3rd years; $p = 0.0487$, significant for the 4th year; and $p = 0.0040$, very significant for the 5th year. Finally, the exploratory factor analysis among public schools (without Music as a subject and private schools with the subject) showed different factor distributions. Private school students generalize more about musical elements, which is why they presented only one factor (analytical ear), and public school students presented two factors (expressive ear and rhythmic sense), indicating a more specific perception of musical elements. In this sense, scores are assigned differently, as guided by Pasquali [15]. Based on the studies by Kirnarskaya [9] and Gordon [10], students considered superior in the PSHM and complementary instruments are those who present, in the private network, an architectural ear (score equal to or greater than 80) and, in the private network public, an analytical ear (score equal to or greater than 70).

Although these results were promising, as pointed out by Koga and Rangni [3], the binomial study (items less than and equal to $p \leq 0.05$) carried out by the authors showed a discrepancy among the items (some very easy and others difficult) and the amount of items, making it necessary to expand the study. Pasquali [15] and authors of international scales [6,7,11] demonstrate the importance of validation and the search for reliability as a way of adjusting the instrument to calibrate its measurement sensitivity. Kirnarskaya [10] emphasizes the importance of psychometric instruments in the musical evaluation process and how accuracy benefits the correct indication of possible cases of talent. Haroutounian [6] and Gordon [11] point out that these are processes that decide the type of educational attention, which is the culmination and reason why survey procedures and evaluations are carried out.

Even though the tabulation of data is in progress, an improvement in the usability and completion of the instruments has preliminarily been observed based on the sample of the first and second stages, it was necessary to determine whether the participants experienced fatigue while performing the musical tracking task. Pasquali [15] recommends that scales should have a maximum of 30 items. Regarding usability with children and young people, there are instruments such as those developed by Haroutounian [6], Hernández and Pérez [7], Kirnarskaya [10], and Gordon [11], which demonstrate practicality for both administrators and participants. These instruments are quick to apply. However, the limitation is that they are designed specifically for music specialists, thus their use is restricted. In contrast, PSHM can be used by multiple educators. Additionally, after comparing the results, it will be possible to improve the cut-off scores of the items, including the score that will be assigned to indigenous schools [15]. Abramo and Natalie-Abramo [13] alert us to the specificities in the musicality of individuals, as well as the possibilities of manifestation of musical talent, which is why the cut-off notes or scores need to pass through the appropriate statistical tests, such as, for example, using linear transformations and even score matching [15]. Perception itself is also a challenge; after all, what are at stake are the elaboration and presentation of the stimulus and the individual's response, which passes through the sieve of understanding and sound representation in the brain [16].

Furthermore, with the tabulation of raw data completed, it will be possible to analyze whether the distribution will be normal or not, thus allowing us to collaborate in decision making regarding the use of parametric or non-parametric statistical tests. Based on studies by Gordon [11] and observing the scale by Hernández and Pérez [7], both paths allow for validation.

When analyzing the students who participated in the first and second stages, the following observations can be made: those who obtained results above 70 (indicative of average/superior auditory aptitude) were considered superior; those who scored between 31 and 69 were considered to have average musical aptitude; participants who scored below 30 were considered to have below-average scores; individuals who performed similarly in both stages were considered to have more reliable results [15].

With the data analysis completed and adjustments made, the instruments will be made available for use in Brazil and abroad.

1.3. Final Considerations

Limitations have occurred when carrying out the construction of the PSHM; however, for Brazilian cultural–musical diversity, it has been shown to be effective in a way that it will make it possible to know the aptitude and signs of musical talent for several cultures. The PSHM and complementary instruments, until the moment of data analysis, show evidence of contribution to the identification of children with musical aptitude and who may be referred to specialized Music Education. Collaborative work between teachers and specialist musicians is suggested. It is expected, through future research, to expand the sample of participants to contemplate other cultural realities and age groups, as well as to refine the capacity of these instruments in carrying out the musical aptitude survey, in addition to reaching reliable estimation indices, inferring the reliability and allowing score standards.

Author Contributions: Conceptualization, F.O.K. and R.d.A.R.; methodology, F.O.K. and R.d.A.R.; software, R.P.; validation, F.O.K., R.d.A.R. and R.P.; formal analysis, F.O.K.; investigation, F.O.K.; resources, F.O.K.; data curation, F.O.K.; writing—original draft preparation, F.O.K.; writing—review and editing, F.O.K. and R.d.A.R.; visualization, F.O.K. and R.d.A.R.; supervision, F.O.K. and R.d.A.R.; project administration, F.O.K. and R.d.A.R.; funding acquisition, F.O.K. All authors have read and agreed to the published version of the manuscript.

Funding: This research was funded by the Research Support Foundation of the State of São Paulo, São Paulo, Brazil grant number [Process: 2019/14466-8].

Institutional Review Board Statement: The study was conducted in accordance with the Ethics and Research Committee of the Federal University of São Carlos with protocol code 52224021.0.0000.5504, 29 August 2022.

Informed Consent Statement: Informed consent was obtained from all subjects involved in the study.

Data Availability Statement: This research is in progress and the set of PSHM instruments is in the process of patenting. Therefore, at the moment it will not be possible to share the data from the survey. Soon they will be made available on the page of the Research Group for the Development of Human Potential (GRUPOH, UFSCAR) with access at https://altashabilidadesgrupoh.com.br/ (accessed on 6 June 2023).

Conflicts of Interest: The authors declare no conflict of interest.

References

1. Koga, F.; Rangni, R.A.; Pereira, J.D.S. Talento Musical: Identificação, avaliação e enriquecimento. In *VIII Congresso Nacional de Educação*; Castro, P.A., Lima, E.W., Eds.; Realize editora: Campina Grande, Brazil, 2022; pp. 624–643.
2. Renzulli, J.S. A concepção de superdotação no modelo dos três anéis: Um modelo de desenvolvimento para a promoção da produtividade criativa. In *Altas Habilidades/Superdotacao, Inteligência e Criatividade: Uma Visao Multidisciplinar*; Virgolim, A.M.R., Konkiewitz, E.C., Eds.; Papirus: Campinas, Brazil, 2014.
3. Rangni, R.A.; Rossi, C.S.; Koga, F.O. Estudantes com altas habilidades ou superdotação: Desdobramentos dos índices da sinopse estatística e dos microdados da região sudeste do Brasil. *Res. Soc. Dev.* **2021**, *10*, 1–15. [CrossRef]
4. Koga, F.O.; Rangni, R.A. Preliminary evidences of musical talent in basic education children. *Música Hodie* **2020**, *20*, 1–22.
5. Gagné, F.; Mcpherson, G. Analyzing musical prodigiusness using Gagné's integrative model of talent development. In *Musical Prodigies: Interpretations from Psychology, Education, Musicology and Ethnomusicology*; Mcpherson, G., Ed.; Reino Unido, Oxford University Press: Oxford, UK, 2016; pp. 3–114.
6. Haroutounian, J. *Artes Talent ID: A Framework for the Identification of Talented Students in the Arts*; Royal Fireworks Press: New York, NY, USA, 2014.
7. Hernández, C.A.; Pérez, L.G. Normativización de una escala para la detección temprana de talentos musicales (TAMU). *Sobredotação* **2021**, *17*, 193–220.
8. Müllensiefen, D.; Gingras, B.; Musil, J.; Stewart, L. The Musicality of Non-Musicians: An Index for Assessing Musical Sophistication in the General Population. *PLoS ONE* **2014**, *9*, e89642. [CrossRef] [PubMed]
9. Koga, F. *Protocol for Screening Music Abilities*; Cultura Acadêmica: São Paulo, Brazil, 2021.
10. Kirnarskaya, D. Why workaholics are not Mozarts? Musical abilities in post-cognitive era. *J. Mod. Foreign Psychol. Moscou* **2021**, *10*, 64–72. [CrossRef]
11. Gordon, E.E. *Dissonance in Space Audiation*; GIA Publication: Chicago, IL, USA, 2015.
12. Hallam, S. Musicality. In *The Child as Musician: A Handbook of Musical Development*; Mcpherson, G.E., Ed.; Reino Unido, Oxford Press: Oxford, UK, 2016; pp. 67–80.
13. Abramo, J.M.; Natale-Abramo, M. Reexamining "Gifted and Talented" in Music Education. *Music. Educ. J.* **2020**, *106*, 38–46. [CrossRef]
14. Sloboda, J.A. *A Mente Musical: Psicologia Cognitiva da Música*; EDUEL: Londrina, Brazil, 2008.
15. Pasquali, L. *Psicometria: Teoria dos Testes na Psicologia e na Educação*; Vozes: Petrópolis, Brazil, 2013.
16. Vieira, S. *Estatística Básica*; Cengage: São Paulo, Brazil, 2018.

engineering proceedings

MDPI

Proceeding Paper

Methods and Scenario Analysis into Regional Area Participatory Planning of Sustainable Development: The "Roses Valley" in Southern Morocco, A Case Study †

Antonio Bertini, Immacolata Caruso and Tiziana Vitolo *

Institute of Mediterranean Studies (ISMed), National Research Council (CNR), 80134 Naples, Italy; bertini@ismed.cnr.it (A.B.); caruso@ismed.cnr.it (I.C.)

* Correspondence: vitolo@ismed.cnr.it

† Presented at the 9th International Conference on Time Series and Forecasting, Gran Canaria, Spain, 12–14 July 2023.

Abstract: As the global environmental crisis grows in scale and complexity, land protection experts and policy makers are increasingly called upon to make decisions, despite high levels of uncertainty, limited resources and insufficient or, conversely, available but unintegrated data. Efforts to protect biodiversity at the national and, especially, the local level, which aim to achieve sustainable development in territories and local communities, require the incorporation of social, economic and political considerations to ensure that participatory planning of strategies is adopted and undertaken. With this issue in mind, the geographical focus chosen for this contribution is the territory of the Valley of Roses located in the southern area of Morocco. From a methodological point of view, this paper will address the state of the existing literature on sustainable development and the good practices implemented in studied territories. The final objective, which is related to the application and resolution of real problems, concerns, on the one hand, the possibility of valorizing the material and immaterial cultural heritage of the area and, on the other hand, identifying the steps to be taken as part of a long-term vision aimed at identifying concrete actions for the valorization and development of the area.

Keywords: sustainable development; scenario analysis; southern Morocco; backcasting real problems; territorial planning

Citation: Bertini, A.; Caruso, I.; Vitolo, T. Methods and Scenario Analysis into Regional Area Participatory Planning of Sustainable Development: The "Roses Valley" in Southern Morocco, A Case Study. *Eng. Proc.* **2023**, *39*, 8. https://doi.org/10.3390/engproc2023039008

Academic Editors: Ignacio Rojas, Hector Pomares, Luis Javier Herrera, Fernando Rojas and Olga Valenzuela

Published: 28 June 2023

1. Introduction

The data concerning the southern area of Morocco are alarming [1]. The difficult conditions that the populations living on the border endure, which are caused by the inexorably advancing desert [2,3], are endangering the permanence of local people. These communities have lived for hundreds of years in areas that are particularly disadvantaged in terms of water, climate and material conditions, with these areas being dense with ancient and largely preserved cultures. In addition to the well-known and adverse climatic conditions, which put a strain on residents' lives, there are the precarious economic conditions that constitute the most pressing threat to individuals who are tied to those lands, as well as to the ways of life of peoples who want to live a dignified existence in their places of origin. The analyzed areas, which are still being investigated through interdisciplinary research projects organized by the National Research Council [4–6], are located inland, being defined as peripheral [7] from the country's most important centers of activity and economy; being a local resident is characterized by belonging to a settlement system that exists along three valleys.

The local people, who settled there over time and who have lived there permanently for over a century, can now also count, from a hydrological point of view, on the presence of three '*ouadi*', which create a complex of oases of extreme charm and low environmental impact. In this context, particular attention was dedicated to the area defined and recognised as the 'Valley of Roses', where there are rose cultivations, particularly of the Damask variety, whose potential is yet to be fully exploited. Moreover, the fact that the Valley of Roses belongs to the Unesco Mab Biosphere Reserve called 'Oasis du Sud Marocain'[1] is not insignificant. The techniques used and the products obtained, i.e., the methods used to build and maintain the oases;, as well as the procedures used to create houses, i.e., the kasbah, the ksour, the agadir, the ighrem and the links, be they material and immaterial, all have a very low environmental impact, thus constituting a model of good practice for the entire world to follow to create a truly 'sustainable' human habitat.

By themselves, these characteristics, in our opinion, are sufficient to make the 'oasis–city system' of southern Morocco worth promoting, studying, disseminating and enhancing. Many attempts were and are being made by the Moroccan Government to revive the fortunes of these lands, though, so far, the results have been less than encouraging. With this paper, in which the results of studies and exchanges of ideas, research and opinions, many of which are now decades old,[2] are illustrated in synthesis, the aim is to contribute to defining, as best as possible, the reality about which we are talking by highlighting its great potential and the problems of the system, as well as analysing, valuing and fielding existing and future opportunities. A further objective is to contribute to the composition of one or more scenarios aimed at supporting decision makers and managers in the planning of cities and territories.

2. Method and Scenario Analysis into Regional Area: The Territorial Approach

This study was based on a territorial approach [8], which we believed could provide a "global" vision for the area in question, offering, at the same time, a basis upon which to define a local action plan that is representative of the area's characteristics. In particular, the methodology used envisaged that, once the "case study territory" was selected, its profile would be defined by assessing its "territorial capital" [9], thus producing a sort of "diagnosis" capable of offering indications for a suitable local development strategy. The choice of the territorial approach is linked to the importance of local resources used to achieve sustainable development, albeit without neglecting the contribution of ancient know-how rooted in the territory that, once reintroduced and professionalized, could create new activities and added value. These endogenous resources could be physical, environmental, cultural, human, economic and financial, as well as institutional and administrative, in nature. In other words, territorial capital corresponds to the elements that constitute the territory's wealth and can be summarized as follows:

- physical resources, in particular natural resources, facilities and infrastructure, and historical, architectural, urban and landscape heritage, as well as and their management;
- human resources (residents in the area, such as people who move there and people who leave it, i.e., demographic characteristics and the social structure of the population);
- activities (enterprises and the related industry sector, their weight within the sector, size, geographical concentration, etc.) and employment;
- know-how and skills in the area, culture and identity (the values generally shared by the players in the territory, their interests, attitudes, forms of recognition, customs, etc.);
- the level of "governance" (local institutions and administrations, rules, collective operators, relationships between these stakeholders, the degree of autonomy in managing development, including financial resources and forms of consultation and participation, etc.);
- the area's image and perception (both among the inhabitants themselves and externally), communication within the area and relationships with the outside world (in particular, the area's positioning in the various markets, contacts with other areas, exchange networks, etc.).

These elements could constitute strengths or limitations depending on the aspects considered and represent a complex whole that is part of a wider spatio-temporal logic. The concept of 'territorial capital' was used in a dynamic sense because the territory itself represents an entity with multiple facets, which evolves and is the result of a link between past, present and future. Constantly evolving, it is enriched and defined ever-more precisely thanks to elements drawn from the past (history), the future (the development project), the territory's internal components (interactions between the various players, institutions, local networks, etc.) and its relations with the outside world (exchanges with markets, institutions and external networks). Consequently, the 'territorial' approach enabled local players to define a development policy based on the realities, advantages (strengths), limitations (weaknesses), needs and opportunities of a given area, as identified through the analyses conducted.

Among the methodological approaches used to analyze territorial capital and, thus, proceed to the elaboration of a territory's development project, the initial diagnosis, in this case study, was an essential step required to ensure the success of the development approach, especially in the case of rural territories in inland and fragile areas, such as those analysed in Morocco. Moreover, by examining the existing links between sectors of activity, operators and areas, the initial diagnosis could lead local stakeholders to discover various, often unexpected, ways to revitalize their territory. At the same time, the analysis of alternative scenarios provided indications of long-term risks and opportunities, highlighting various possible paths. The participation of representatives of numerous local interests remained essential at subsequent stages. The use of participatory methods (animation, information, training, "sweeping" research into potential projects, etc.), even as soon as the initial diagnosis was drawn up, could foster a feeling of ownership of the approach to the development process and, subsequently, the achievement of a consensus regarding the measures to be implemented. The territorial approach, moreover, proposed a new concept regarding the population's relationship with the development project by encouraging local involvement, consultation and participation.

3. The Case Study: "The Roses Valley"

There is no lack of existing literature that consider the analysed area from landscape and social–economic perspectives [10]. The area is much studied [11] and was the subject, along with the entire province, of plans and programmes issued by the Moroccan Government [12] that attempted to revive the fortunes of a particularly peripheral region with little economic activity [13].

3.1. The Ribbon Oases

The study area is part of one of the largest and most important UNESCO Man and Biosphere (MaB) reserves in Africa—the 'Oasis du Sud Marocain'[3]—which extends over a sizeable area of approximately 7 million hectares (Table 1), where there is a wide range of bioclimatic zones, such as the Mediterranean wetlands and the hyper-arid Sahara.[4]

Table 1. Zoning and dimensions of UNESCO Mab site 'Oasis de Sud Marocain', designated in 2000. Source: MaB UNESCO 'Oasis de Sud Marocain'.

Map Surface ha	Core Area(s) ha	Buffer Zone(s) ha	Transition Zone(s) ha
7,185,371	908,581	4,619,230	1,657,560

The Reserve is the main place of endemism in North Africa, as well as hosting the majority of Moroccan species, such as 63% of birds, 60% of reptiles and 66% of mammals present in the country. Its highly effective oasis system protects the land from desertification and nurtures a very rich biodiversity. A skillful combination of crop-based production systems and sheep and goat breeding ensures high prolificacy, even in difficult environments. The cultivation systems practised are generally organised in three layers (date palm, fruit trees and crops below) or, in some cases, only in two layers, without the date palm. Ecologically, the oases are threatened by salinity and siltation. These forms of degradation are exacerbated by the harshness of the climate, water scarcity, a lack of control over irrigation, the grazing of spontaneous vegetation and the excessive removal of woody vegetation in pastoral areas.

Over the years, the oases were subjected to increasingly severe droughts and affected by bayoud disease,[5] as well as strong migration flows. The combined action of all of these factors gradually transformed a natural area into a heavily anthropised zone. Date palms also form the basic agronomic structure of Moroccan oases, in particular by creating a microclimate that is indispensable for the good development of crops growing at the below level. All of these crops condition the economy of the oases and the stability of the populations living there. Date palm cultivation, which constitutes one of the main agricultural economies in the area, occupied a large area for a long time, though the area occupied by palm groves has shrunk considerably since the beginning of the century. A Ministry of Agriculture study [14] details the regional distribution of the national phoenicultural heritage, highlighting its concentration in three main areas: Ouarzazate (41%), Tafilalet (28%) and Skoura (20%). The problem of siltation has become a threat and, since the 1970s, has also progressively endangered homes, farmland, irrigation canals and road infrastructure in palm groves. These worrying events, combined with the poor economic diversification of the oases, pose a serious threat to these areas, a threat that risks generating long-term exclusion problems in these areas of strong identity-based and geostrategic importance for Morocco.

In this context, the "Roses Valley" is part of a millenary historical system that is a fragment of the caravan routes that both crossed the area studied longitudinally north–south and transversely west–east. The north–south axis connected the Mediterranean with the city of Timbuktu [15] in Mali,[6] while the west–east transverse axis connected the Atlantic and all of the coastal centres with the Arabian Peninsula, which is the site of Mecca and the main centre of pilgrimage for Muslims. Over the centuries, along these caravan routes, many Berber tribes, through skillful use of the scarce water resources, along the Ouedas created a continuous set of oases that follow one another, forming a 'ribbon' to make it possible, in a suitable habitat for permanent settlement. This succession of oases forms a narrow band of varying size, never exceeding 10 kilometres, but winding for hundreds of miles, along which the Berber tribes settled. In part of this ribbon system, two valleys along the Dadès and M'Gouna oases show the conspicuous presence of rows of Damascene roses and a considerable number of kasbah, ksour and granaries that constitute a material heritage of considerable historical, architectural, landscape, urban and environmental interest.

3.2. *The Kelaat M'Gouna Urban Area*

The nearest population centres to the area are Agadir, which is an important port on the southern coast of Morocco, and Marrakesh, which is a famous southern Moroccan city with a history as a capital. The connection between Marrakesh and Kelaat M'Gouna is via national road No. 10, which crosses the "Tizi 'n Tichka pass" (at an altitude of 2260 m). The M'Gouna Valley, known as the 'Valley of Roses', is about 30 kilometres long and stretches at the foot of the Ighil M'Goun massif in the middle of the central High Atlas mountain range, which reaches altitudes of over four thousand metres (4071 m). The main centre of the area is Kelaat M'Gouna, which rises at the confluence of the Dadès and the ouadi Asif M'Goun. It is only a few kilometres, about 20, away from the other centre of interest for the roses: Boumalne Dadès. Kelaat M'Gouna is a town in the province of Tinghir Drâa–Tafilalet [16,17]. Like the major cities of southern Morocco, it is made up of several neighbourhoods, forming a polynuclear settlement that administratively gravitates around a more densely urbanised central area. In particular, the central part of the city is formed of the following districts: Ait Aissi, R'kon, Elkelaa, Zawiyt nAguerd, Ait Baamran, Hay Annahda, Ait Boubker, Mirna and Taltnamart. Moreover, not physically distant from the central area are many douars (rural villages) that surround the town but are not administratively part of the municipality of Kelaat M'Gouna (Figure 1).

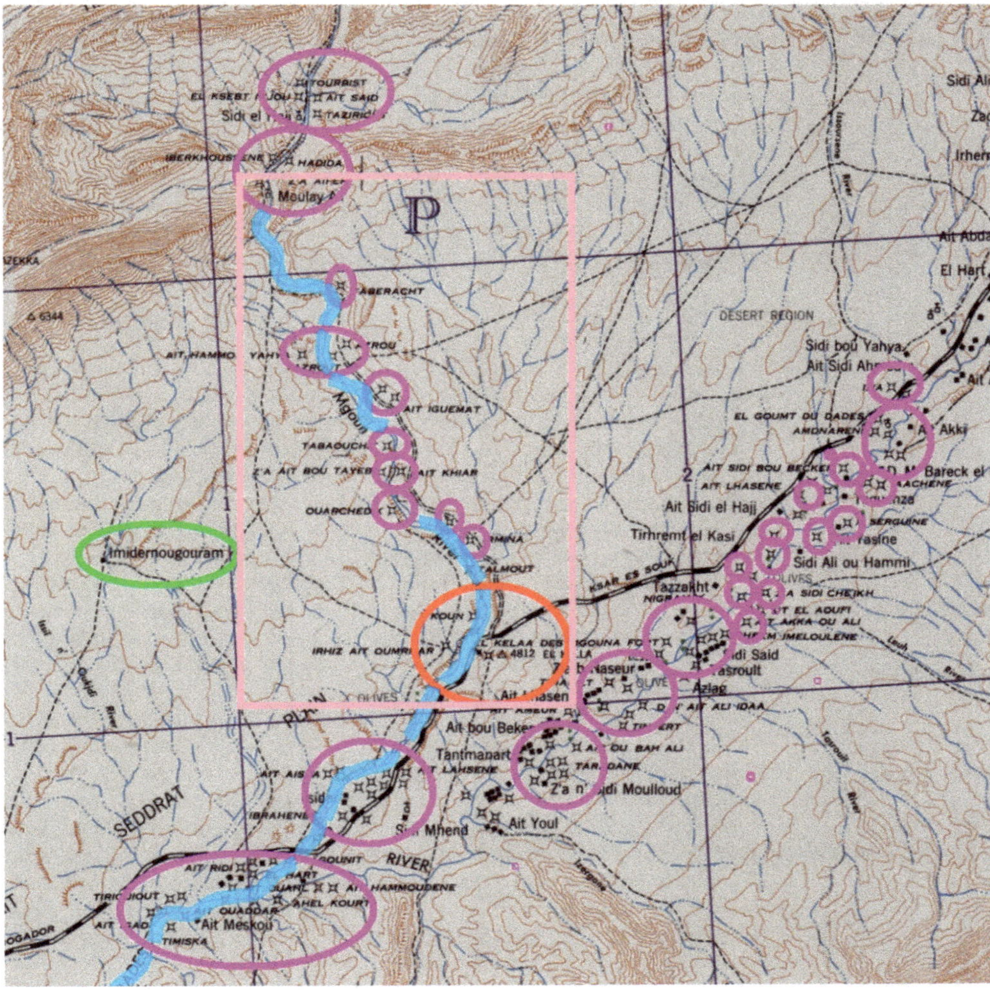

Figure 1. Map of Valley of Roses. Legend: pink square locates the Valley of Roses; the red circle highlights centre of Kelaat M'Gouna, purple circles highlight the kasbah; green oval highlights granary; light blue line locates course of Dades river (Cartographic base: Army Map Service U.S. Army, Washington D.C., 1943, Southern Morocco 1: 125,000 Aine Toumert sheet; cartographic elaboration performed by Antonio Bertini).

These villages include Aït Sidi Boubker, Ifri, Zawiyt Elbir, Amdnagh, Sarghin, Timskelt, Ait Boukidour, Tazzakht, Tawrirt and Tasswit. School facilities are also located in the central part of the city. In addition to primary schools, there are three high schools: Al Woroud (Roses), which is named after the roses of the Dades Valley, M'gGun and an Moulay Baamrane. There are two main souk (market) days in this city: Tuesdays involve the livestock trade, and Wednesdays involve the sale of food and other goods. Regardig secondary activities, there is a factory for the distillation of roses and the production of rose water (eau de rose), essential oils and cosmetic products. The route through the valley is dotted with traditional adobe kasbah that, once renovated, are available as small hotels along the road between Kelaat M'Gouna and Bou Thrarar. Due to its particularly lively nature, Kelaat M'Gouna is a very important economic and social centre for the entire region. According to the most recent 'Recensement Général de la Population et de l'Habitat (RGPH)' study, which dates back to 2014, the urban area has a population of 16,956. The old part of the city is also characterised by typical constructions in pisé, i.e., the traditional technique of southern Morocco, with kasbah being of particular architectural interest as they constitute one of the most important elements on which to focus this programme, as well as to plan and manage a sustainable development of the territory in a touristic key. In this direction, there are a number of good practices already implemented in the Dades and Draa valleys, where a considerable number of "land architectures" of historical and environmental interest are concentrated [18–21], from which we can draw ideas and methods that were previously successfully tested.

3.3. *The Roses Garden of Kelaat M'Gouna*

The cultivation of roses is spread over two areas: the first areas begins about 10 km from the village of Kelaat M'Gouna and extends for 10 km, ending near Boumalne Dades, while the second cultivation area begins in Kelaat M'Gouna and continues along a provincial road that climbs northwards, ultimately reaching the village of Bou Tharar, representing about 30 uninterrupted kilometres of rose gardens along the edges of the ouadi. The inhabitants practice subsistence farming in these locations, using small plots of land, that is based mainly on the cultivation of wheat, alfalfa for livestock and fruit trees. Birch trees are also planted, which are used as construction timber. These plots of land are watered by many irrigation canals. All strips of land are bordered by hedges of wild Damask roses, specifically of the botanical species 'Rosa damascena', which provide excellent protection from ruminants; thus, without the rose hedges, ruminants could ruin the cultivated fields. One of the first functions given to rose bushes by the Berbers was creating a separation barrier more than two metres high (a height sufficient to prevent goats from eating crops) between plots of land.

It was in the 1930s that the French realised the commercial value of these fragrant flowers and specifically started to cultivate them. To obtain the best olfactory yield, the rose petals are harvested in the absence of sunlight, i.e. at sunrise or sunset. After sorting, the rose buds are dried for two days on the earthen roofs of the kasbah. The very dry climate, which is characteristic of these altitudes, allows for excellent drying of the flowers. The cultivation of roses, together with other activities in the area, is of great importance to the local populations, since even if it is seasonal and not adequately paid, it constitutes one of their main sources of income. Cultivation takes place around the city of M'Gouna, while 30% of the processing of roses into finished and derived products takes place in Kelaat M'Gouna, Marrakech, Casablanca and Fez.

4. Results and Discussion

The territorial approach pursued in the study of the Valley of Roses highlighted the issues related to the development model pursued so far in this area. The latter model is essentially based, on one hand, on the desire to relaunch and promote 'inland tourism' that is linked to the richness of the landscape and cultural heritage and, on the other hand, on the production and processing of roses. These types of tourism, being valorised as local identity products (and as such labelled PDO-Protected Designation of Origin in 2014), are considered capable of acting as a flywheel in the complex articulation of the relationship between the dynamics of the production process and the relaunch of rural tourism in the area. In this regard, however, the SWOT analysis (Table 2) conducted on the potentialities and limitations detected in the area allowed a better definition of these scenarios; these definitions are useful for the elaboration not only of the initial diagnosis, but also, in particular, of sustainable development projects that are more in line with the needs of the territory and the communities living there.

Table 2. Swot Analysis of Valley of Roses.

Strengths	**Weaknesses**
Biosphere Reserve Raw earth constructions, i.e., kasbah, igherem and entire neighbourhoods (ksars) Local handicrafts Tourist equipment Rural tourism Rose picking Rose processing Rose processing techniques Identity and typical landscape in the area Climatic conditions suitable for diversifying tourism Distance from the polluted urban environment Healthy environment Well-maintained oases Date production Wrought iron craftsmanship Identity-based production—mantice Identity-based production—rural ceramics Identity-based production—Weaving Identity-based production—shoemaking Identity-based production—daggers (typical of and exclusive to Kelaat M'Gouna)	Kettara in desiccation Peripherals Reduced agricultural production and date production Few economic activities Low employment rates in general, and high youth employment in particular Low attractiveness for residents Low attractiveness for tourists and travellers Most economic activities and, to a large extent, those related to the tourism industry are in the hands of external companies, which have little involvement in local communities Income from tourism activities contributes little to improving the living standards of local communities A lack of ecological education that can make people understand that the old way of life is better than the contemporary one Innovative and non-traditional techniques are problematic because they do not use natural products that stem from the production of rose by-products The world's largest solar power plant located just 10 km from Ouarzazate, inaugurated in 2016, does not power the Draa Tafilalet region
Opportunities	**Threats**
World Heritage Earthen Architecture Programme (WHEAP) [22] Federation Interprofessionelle Marocainede la Rose a Parfume New Development Model (NMD) New Action Plan for the tourism sector for the period 2023–2026 with an allocation of 6.1 billion Moroccan dinars (the tourism sector has a new action plan, with a new strategic roadmap for the period 2023–2026. More than DH 6.1 billion will be allocated to the implementation of this roadmap. For the implementation of this roadmap is scheduled to take 4 years. The plan aims to attract 17.5 million tourists in 2026, after reaching 11 million last year. The plan has multipe specific aims: (1) to reach USD 120 billion in foreign exchange earnings; and (2) to create 200,000 new direct and indirect jobs by 2026. To achieve these objectives, the sector will have to achieve the following aims: (a) creating a new diversified structure of the tourism offer; (b) strengthening air connections; (c) supporting and promoting e-marketing; (d) diversifying the cultural and free-time offers; (e) redeveloping the existing hotel heritage and creating new hotel facilities; (f) strengthening human capital via an attractive training framework (Joint Note No. 42-INAC-HCP—April 2023).) Développement Rural et des Eaux et Forêts à la zone d'action de ORMVA Tafilalet, 2020	Architectural and natural heritage under threat Desertification Drying out of khettara irrigation systems Palm disease (BAYARD) Migration processes of local populations Falsification in the production of rose derivatives Lack of brand protection for the processing and production chains of rose derivatives

Source: own elaboration.

More specifically, reflecting on the link between the development of rural tourism and the valorization of rose products, one element that emerged from the survey is the lack of co-ordination between the stakeholders involved at various levels (local, regional, national) in the two key sectors of the local economy (tourism and roses). Tourism remains a priority, as does the need for better economic structuring of the rose chain, in particular at the level of the marketing derived products, such as essential rose oil that is destined for the foreign market and, therefore, has a strong added value. In this regard, the New Model of Development (NMD),[7] which was recently promoted by the government, suggests strengthening integration and territorial dynamics using the financial, human and natural resources available in the area. In particular, it aims to develop new ways of governing the territory in favor of State–Region complementarity, the development of integrated economic ecosystems, the valorization of living space and the conservation of natural resources [23].

5. Conclusions

The scenario identified in order to compose a framework of possible interventions in the area was intended to balance the potential and needs for valorizations in both the 'rose route' and 'kasbah' route tourist basins and create a model of sustainable balance for areas with particularly harsh climates for human beings [24]. It is a model that was proposed precisely to create a contrast with those more recurrent and in vogue in recent decades in the Arab world, particularly in the Arabian Peninsula (Dubai, Kuwait, etc.), where the relationship between man and nature is being irreparably devastated.

The Valley of Roses' natural and cultural capital offers strong potential for the national and international tourism sector, particularly cultural tourism or wellness tourism, alongside niches in sports tourism and eco-tourism. Its diversified natural capital, its historical and cultural heritage, the hospitality of its population, its culinary expertise and its safe and secure environment are all assets that make it a popular tourist destination. Proximity, particularly to a European market that will be increasingly sensitive to the environmental impacts of and low-carbon footprint travel, further strengthens its potential. However, in line with the recommendations of the "New Development Model", to fully harness this potential, it is necessary to develop a diversified tourism offer that highlights the country's cultural and heritage assets and promotes them fully in all territories, as well as enhancing tourist entertainment. To achieve these objectives, close co-ordination among all sectors involved is necessary, in particular agriculture (for eco-tourism, gites, natural parks), culture (for the enhancement of tangible and intangible heritage), health (for wellness tourism) and sport (for activities with a high potential for exposure, such as surfing, mountaineering, etc.). This viewpoint calls for a systemic and partner-based approach, which embraces the growing complexity and interdependence of issues and decompartmentalizes the barriers to allow all potential to be unleashed through co-construction. This systemic approach necessarily entails the acceptance of a new mode of governance, setting up the conditions for cross-functionality and co-ordination (information sharing, multi-stakeholder implementation bodies, etc.). In this regard, one of the methodological trajectories to be pursued next is backcasting (especially participatory backcasting, with broad stakeholder involvement), which is an innovative planning methodology that gained ground in recent years, particularly in sustainability studies, but whose use as a strategic tool to arrive at a shared local development project is yet to be sufficiently tested. In contrast to forecasting scenarios, the main characteristic of backcasting scenarios is that they do not deal with forecasts of how the future might evolve, but focus on the definition of a desirable future and the development of strategies required to achieve it. This methodology requires backcasting from the end point to the current situation, making it possible to determine the feasibility of that future and the definition of what is needed to achieve the set goals. In particular, as stated by Robinson [25], backcasting aims to suggest the implications of different future situations, which are not chosen based on their probability, but based on other criteria, such as social or environmental desirability criteria, and emphasises the importance of a clear relationship between goal setting and planning. Ultimately, backcasting studies concern social interaction processes involving various stakeholders; thus, the pluralistic character of a society cannot be overlooked. The results of such studies also highlight the advantage of fostering a social learning process, precisely because of the interaction between the different subjects involved, who may change their views in this process of exchange with others to develop a shared vision and actions. Participation also makes it possible to give a voice to all citizens and make them responsible for the projects that affect them, thus fostering their success. In the light of these reflections, therefore, the continuation of this line of research envisages the use of participatory backcasting to support the elaboration, by local actors, of a systemic and shared sustainable development vision.

Author Contributions: Conceptualization, A.B., I.C. and T.V.; methodology, A.B., I.C. and T.V.; formal analysis, A.B., I.C. and T.V.; writing original draft preparation, A.B., I.C. and T.V.; writing review and editing, A.B., I.C. and T.V. All authors have read and agreed to the published version of the manuscript.

Funding: This research received no external funding.

Informed Consent Statement: Not applicable.

Data Availability Statement: Some or all data, models or code that support the findings of this study are available from the corresponding author upon reasonable request.

Acknowledgments: Paolo Pironti (CNR-ISMed).

Conflicts of Interest: The authors declare no conflict of interest.

Notes

1. The Biosphere Reserve 'Oasis du Sud Marocain was designated in 2000. It has rich agronomic diversity, including vast areas with date palms, acacia forests and other high-altitude forests covered with Juniperus thurifera, J. phoenicea and Quercus rotundiforia, among other species. In particular, the area is home to dorcas gazelles, gazelles de cuvier (Gazella cuvieri), barbarian sheep (Ammotragus lervia), ubara otards (Chlamydotis undelata) and cobras (Naja haje). The population of the Oasis du Sud Marocain has lived in this area for thousands of years. Most of their economic activities come from agriculture, particularly the production of cereals, potatoes, olives and, above all, dates, which are the backbone of the local economy. Recently, the cultivation and processing of new products, such as apples, roses and henna, significantly increased the farmers' incomes.
2. Project "Moroccan migration to Italy in time of global economic crisis" in partnership with Moulay Ismail University, Department of Sociology, Faculty of Arts and Humanities, Meknes, Morocco on the basis of the Bilateral Agreement for Scientific and Technological Cooperation CNR-Italy/CNRST-Morocco 2012–2013.
3. https://en.unesco.org/biosphere/arab-states/oasis-sud-marocain (accessed on 16 March 2023).
4. The region is classified by UNESCO as being part of the Biosphere Reserve of the Southern Moroccan Oases ReBOSuM. The ecological, environmental and socio-economic importance of the region faces several challenges, as anthropogenic impacts on the oasis ecosystem are changing the trajectory of ecosystem services.
5. Bayoud disease is an epiphytic fungal disease of date palms. The disease was first reported in Morocco in 1870. The term 'bayoud' is derived from the Arabic abiadh ('white') and refers to the whitish colouring of diseased fronds.
6. Timbuctu (Tumbuktu), originating as a cultural centre, became a trading centre in the 15th century. Caravan routes connected it to Sudan, Egypt, Tunis and Morocco. It cultivated trade relations with Italy and, in particular, with Florence.
7. https://www.hcp.ma/Developpement-regional_r614.html (accessed on 28 March 2023).

References

1. Royaume du Maroc, Haut-Commissariat au Plan, Direction Régionale de Draa Tafilalet. Monographie Régionale 2021, La région Draa Tafilalet. Available online: https://www.hcp.ma/draa-tafilalet/docs/monographies/Monographie%20Regionale%20de%201%27%27annee%202021.pdf (accessed on 28 April 2023).
2. Salem, A.B. *Changements Climatiques Dans les Oasis de Tafilalet Sud Est du Maroc*; Editions Universitaires Européennes: Sarrebruck, Germany, 2016.
3. Kuisma, J.; Haanpera, H. Water quality pressures and water management. In *Encounters across the Atlas: Fieldtrip in Morocco 2011*; Minoia, P., Kaakinen, I., Eds.; Unigrafia: Helsinki, Finland, 2012; pp. 58–71. Available online: https://helda.helsinki.fi/handle/10138/37939 (accessed on 28 April 2023).
4. Bertini, A.; Caruso, I.; Cuturi, C.; Vitolo, T. Preserving and valorising the settlement system of South Morocco. In Proceedings of the 6th International Conference on Heritage and Sustainable Development, Granada, Spain, 12–15 June 2018; pp. 1895–1906.
5. Bertini, A. Sviluppo sostenibile nel Sud del Marocco. In *Percorsi Migranti*; Bruno, G.C., Caruso, I., Sanna, M., Vellecco, I., Eds.; McGraw-Hill: Milano, Italy, 2011; pp. 143–160. ISBN 978-88-386-7296-5.
6. Caruso, I.; Vitolo, T.; Bertini, A. Cultural heritage and territorial development: A comparative analysis between Italy and Morocco. In Proceedings of the 6th International Conference on Heritage and Sustainable Development, Granada, Spain, 12–15 June 2018; pp. 719–730.
7. Troin, J.F.; Berriane, M.; Guitouni, A.; Laouina, A.; Kaioua, A. *Maroc: Régions, Pays, Territoires*; Maisonneuve et Larose: Paris, France, 2002.

8. Camagni, R. Territorial capital and regional development. In *Handbook of Regional Growth and Development Theories*; Capello, R., Nijkamp, P., Eds.; Edward Elgar: Cheltenham, UK, 2009; pp. 118–132.
9. Lacquement, G.; Chevalier, P. Capital territorial et développement des territoires locaux, enjeux théoriques et méthodologiques de la transposition d'un concept de l'économie territoriale à l'analyse géographique. *Ann. De Géographie* **2016**, *5*, 490–518. [CrossRef]
10. Berriane, M. *Dynamiques Territoriales et Politiques Publiques: Territoires Fonctionnels et Territoires Officiels In Le Maroc au Présent: D'une Époque à L'Autre, une Société en Mutation*; Dupret, B., Rhani, Z., Boutaleb, A., Ferrié, J.-N., Eds.; Centre Jacques-Berque: Casablanca, Morocco, 2015. Available online: http://books.openedition.org/cjb/998 (accessed on 26 April 2023).
11. Michon, G.; Berriane, M.; Aderghal, M.; Landel, P.-A.; Medina, L.; Stephane, G. Construction d'une destination touristique d'arrière-pays: La « Vallée des roses » (Maroc). *GÉOgraphie Et DÉVeloppement Au Maroc.* **2018**, *5*, 1–13.
12. Sadki, A. Environnement, paysage et projet de territoire. Vers une approche territoriale pour la sauvegarde et la mise en valeur de la réserve de biosphère des oasis du sud marocain. Master's Thesis, Université Senghor, Alexandrie, Egypt, 2007. [CrossRef]
13. Boulanouar, B.; Kradi, C. *Actes du Symposium International Sur: Le developpement Durable des Systemes Oasiens*; Institut National de la Recherche Agronomiqu: Erfoud, Morocco, 2005. Available online: https://www.inra.org.ma/sites/default/files/publications/ouvrages/oasis.pdf (accessed on 3 May 2023).
14. Office Régionaux de la Mise en Valeur dé l'Agricol, OMRVA 2019. Available online: https://www.ormvatafilalet.ma/ (accessed on 2 May 2023).
15. Hammer, J. La biblioteca segreta di Timbuctù. In *La Vera Storia Degli Uomini Che Salvarono Trecentomila Libri Dalla Furia Della Jihad*; Rizzoli: Milan, Italy, 2017.
16. Sadki, A. Urbanisme et Dégradation de L'Habitat Traditionnel des Oasis du Sud-Est Marocain: L'Exemple des Ksour du Tafilalet (Province d'Errachidia). Available online: http://www.archi-mag.com/essai_18.php (accessed on 5 May 2023).
17. Sadiki, M.; Essami, A.; Bouaouinate, A. Les ksour du Tafilalet: Un patrimoine culturel et touristique en déclin. *J. Political. Orbits* **2019**, *3*, 76–94.
18. Boussalh, M.; Cancino, C.; Marcus, B.; Wong, L. The development of a conservation and rehabilitation plan (CRP) for the earthen Kasbah of Taourirt in southern Morocco. *J. Mater. Environ. Sci.* **2016**, *7*, 3579–3583.
19. Ginex, G. Un sistema interattivo per la conoscenza e la gestione del patrimonio culturale in area mediterranea. La Kasbah di Ait Ben Haddou, un progetto di rappresentazione, (Cultural heritage Documentation. The Kasbah of Ait Ben Haddou in Morocco). *Disegnare Con* **2010**, 64–76. [CrossRef]
20. Kagermeier, A.; Amzil, L.; Elfasskoui, B. Touristification of the Moroccan Oasis Landscape: New Dimensions, New Approaches, New Stakeholders and New Consumer Formulas. Available online: https://www.academia.edu/37245318/Touristification_of_the_Moroccan_oasis_landscape_new_dimensions_new_approaches_new_stakeholders_and_new_consumer_formulas (accessed on 4 May 2023).
21. Cancino, C.; Marcus, B.; Boussalh, M. *Conservation and Rehabititation Plan for Tighermt (Kasbah) Taourirt, Southern Morocco*; The getty Conservation Institues, Los Angees, Centre de Conservation et de rehabilitation du Patrimoine Architectural atlasique et subatlasique (CERKAS): Ouarzazate, Morocco, 2016; pp. 1–400.
22. World Heritage Earthen Architecture Programme (WHEAP). Available online: https://whc.unesco.org/en/earthen-architecture (accessed on 1 May 2023).
23. Bertini, A.; Caruso, I.; Vitolo, T. Inland Areas, Protected Natural Areas and Sustainable Development. *Eng. Proc.* **2022**, *18*, 20. [CrossRef]
24. Dale, R. Lightfoot, Moroccan Khettara: Traditional Irrigation and Progressive Desiccation. *Geoforum* **1996**, *27*, 261–273.
25. Robinson, J.B. Futures under glass: A recipe for people who hate to predict. *Futures* **1990**, *22*, 820–842. [CrossRef]

engineering proceedings

MDPI

Proceeding Paper

Time-Frequency Varying Estimation of Okun's Law in the European Union: A Wavelet-Based Approach †

Roman Mestre

Montpellier Recherche en Économie (MRE), University of Montpellier, 34960 Montpellier, France; roman.mestre@umontpellier.fr

† Presented at the 9th International Conference on Time Series and Forecasting, Gran Canaria, Spain, 12–14 July 2023.

Abstract: In this paper, we use the time-frequency wavelet estimators to analyze the robustness of Okun's Law in the European Union across time and within various economic cycles. We extend the Okun's Law literature as we focus on Europe, directly estimating the time-frequency varying Okun's coefficient. We observe that Okun's coefficient in Europe is unstable at short run (infra and annual cycles). The strength of Okun's Law is time dependent at short run as linkages between growth and unemployment are stronger only during crisis times. Such instability is explained as unemployment predominates growth, leading to a positive coefficient and weaker strength. However, as the frequencies increase, the coefficient is more stable over time and the strength is higher and homogenous over time.

Keywords: Okun's Law; time-frequency estimator; wavelets transform; unemployment; growth

Citation: Mestre, R. Time-Frequency Varying Estimation of Okun's Law in the European Union: A Wavelet-Based Approach. *Eng. Proc.* **2023**, *39*, 9. https://doi.org/10.3390/engproc2023039009

Academic Editors: Ignacio Rojas, Hector Pomares, Luis Javier Herrera, Fernando Rojas and Olga Valenzuela

Published: 25 June 2023

1. Introduction

Okun's Law [1] established a negative relationship between unemployment rate variations and the output gap. This relationship estimates that 1% variation in the real output from the potential leads to a 0.33% change in the unemployment rate from the natural rate. For the US, Okun's Law studies estimated that each 1% deviation in real output leads to an unemployment reduction of 0.4% or a heuristic value of 0.5%. Okun's Law is a pilar of macroeconomics policy tools due to its ability to easily examine unemployment dynamics according to basic economics conditions (crisis time or expansion period).

The stability of the Okun parameter across time is discussed in the literature focusing on macroeconomic aspects explaining, or not, the effect of structural breaks on the Okun's Law results. Okun's coefficient can be considered as stable over time for various countries while different values are noted across countries [2,3]. The stability hypothesis of the Okun parameter is also supported [4,5] as no significant differences were found between parameters estimated over different periods (before and after 1984—Great Moderation). However, some studies highlight an unstable Okun parameter value across time [6–11]. The unemployment rate seems react more to real output changes during recessions than in expansion periods or subject to structural breaks [10,12–14]. Changes in the Okun coefficients have been observed during the Great Moderation [15,16] and in the 2001 recession [17]. The question of the Okun coefficient stability is still debated today; for example, some authors continue to estimate a static parameter to study the relationships [18] or simply use the intuitive value of 0.5 [19].

Recent studies highlight additional weaknesses of Okun's Law. For instance, the estimation is not conditional on the frequency, here associated to the period of the business cycles [20,21]. Previous literature indicates that the Okun's Law debate on stability is related to the different kinds of cycles considered in the studies. Then, Okun's Law can be stable for some cycles (frequencies) while being unstable for others depending on the speed of recovery and the strength of a shock/recession. The output growth volatility transfer from short run cycles (1–4 years) to long run cycles (8–16 years) during the Great Moderation period has been observed [22,23]. These results can explain some variability in Okun's coefficients across time while supporting the frequency hypothesis. Relevant results were found using time-frequency analysis with a continuous wavelet coherence phase [20]. The Okun coefficient is time-frequency varying but not asymmetric as the value is not dependent on crisis and expansion periods. At the same time, the heuristic value of 0.5% is observed at specific frequencies. Using phase analysis, it has been found that output (and growth) is predominantly variable of the relation. The time-frequency variation of Okun's coefficient and the output gaps leads on unemployment are also supported [21]. However few papers have analyzed Okun's Law in Europe, assuming that results observed for the US are applicable to European situations. But the European business cycles and labor market are different, potentially leading to different results for Europe. Since the 2008–2009 crisis unemployment has been a persistent and structural issue in many European countries (especially in Mediterranean countries such as France, Spain, and Italy).

Time-frequency analysis is then suitable to study the relationship between variables, particularly the coherence and phase. Multiple studies focus on the application of wavelets in finance to highlight linkages between commodities, stock indices, and financial assets [24–28] to outline risk transmission mechanisms or contagion effects between major stock indices. The notion of wavelet gain is developed [20,29] in order to assess the absolute value of a parameter's regression model (applied for the Okun framework, among others) and analyze the phase to assess the sign and the lead–lag relationship. In parallel, the concept of wavelet gain is extended by time-frequency estimators [30,31] to directly estimate the beta at each time and frequency with its sign (positive or negative).

In this paper, we use the time-frequency wavelet estimators to analyze Okun's Law in the European Union. We extend the Okun's Law literature as we focus on Europe and estimate the time-frequency varying Okun's coefficient. We first present the technical improvement of wavelet time-frequency estimators, and we then discuss and analyze the results.

2. Materials and Methods

Wavelets are an extension of the spectral analysis of time series, allowing the decomposition of a chronic in the time-frequency space [32–37]. We distinguish two forms of wavelet decomposition processes: the discrete process, called maximal overlap discrete wavelets transform (MODWT); and the continuous process, called continuous wavelet transform (CWT). The main difference between MODWT and CWT is the type of wavelets used, the accuracy in the frequency scale, and the computational time. MODWT is based on a dyadic scale decomposition in which frequency components are ranked into wavelet frequency bands of a length equal to a multiple of 2 only. The CWT is finer and can more accurately distinguish between frequencies. In addition, CWT provides cross-transformation of two variables useful to analysis interaction and between them across time and frequencies.

The CWT is based on a wavelet mother $\psi(t)$ serving as a filter to extract information from a time series $x(t)$ of length N. This function will be shifted by a time parameter and dilated by s, a frequency scale parameter, generating the wavelet family $\psi_{\tau,s}(t)$ composed by "wavelet daughters" representing each version of $\psi(t)$ according to τ and s.

$$\psi_{\tau,s}(t) = \frac{1}{\sqrt{s}}\psi\left(\frac{t-\tau}{s}\right) \tag{1}$$

The initial chronic $x(t)$ will be projected in the $\psi_{\tau,s}(t)$ generating the wavelet coefficients, $W(s,\tau)$. These coefficients represent the variations in $x(t)$ around an area around $t \mp \tau$ with a frequential length s.

$$W(s,\tau) = \int_{-\infty}^{+\infty} x(t)\frac{1}{\sqrt{s}}\psi^*\left(\frac{t-\tau}{s}\right)dt \tag{2}$$

where $\psi^*\left(\frac{t-\tau}{s}\right)$ is the complex conjugate of $\psi_{\tau,s}(t)$.

Then, the CWT of $x(t)$ generates multiple sub-chronics of the same length, N, at each frequency scale. The inverse process, called inverse continuous wavelet transform (ICWT), allows the reconstruction of the initial $x(t)$ from all of the wavelet coefficients.

$$x(t) = \frac{1}{C_\psi}\int_{-\infty}^{+\infty}\int_{-\infty}^{+\infty}\psi_{\tau,s}(\mathrm{t})W(s,\tau)\frac{\mathrm{d}\tau\mathrm{ds}}{\mathrm{s}^2} \tag{3}$$

The previous equation of ICWT assumes that the expression C_ψ is non-null and less than $+\infty$. This condition ensures the condition of the existence or admissibility of the wavelet mother [37]:

$$C_\psi = \int_0^{+\infty}\frac{\left|\widehat{\Psi}(f)\right|^2}{f}df < +\infty \tag{4}$$

where f is the frequency and $\widehat{\Psi}(f)$ is the Fourier transform of the wavelet mother.

This condition indicates and is respected when the wavelet mother is a zero-mean and square norm function, then the variance of $x(t)$ is preserved during the process of decomposition and reconstruction. In this paper, we use the complex Morlet wavelet $\psi_M(t)$ as it provides a good balance between time and frequency representations.

$$\psi_M(t) = \pi^{-1/4}e^{if_0t}e^{\left(-\frac{t^2}{2}\right)} \tag{5}$$

where $i^2 = -1$ and f_0, the non-dimensional frequency, equals 6 to satisfy the condition on C_ψ.

From a CWT based on the Morlet wavelets, we define the notion of wavelet coherence, which is similar to the squared correlation and useful to visualize co-movement between two chronics, $x(t)$ and $y(t)$ [38]. The two time series are decomposed by CWT providing their respective wavelet coefficients $W_x(s,\tau)$; $W_y(s,\tau)$. So, we can compute time-frequency covariance called cross-wavelet decomposition:

$$SW_{xy}(s,\tau)\text{: } SW_{xy}(s,\tau) = W_x(s,\tau)W_y^*(s,\tau) \tag{6}$$

$W_y^*(s,\tau)$ is the complex-conjugate of $W_y(s,\tau)$.

The formula of the wavelet coherence, $WQ(s,\tau)$, between $x(t)$ and $y(t)$ is like the R^2 one in terms of the ratio of covariance and variance products.

$$WQ(s,\tau) = \frac{\left|G\left(s^{-1}.SW_{xy}(s,\tau)\right)\right|^2}{G\left(s^{-1}.|W_x(s,\tau)|^2\right).G\left(s^{-1}.|W_y(s,\tau)|^2\right)} \tag{7}$$

G is a smoothing time-frequency operator.

Time-frequency smoothing is required because the coherence coefficients are complex [39]. The coherence coefficient $WQ(s,\tau)$ is between 0 and 1 at each time t and frequency scale s.

The wavelet phase difference (or simply phase) is a complementary notion to the coherence as it contains information on the leadership and sign of the relationship between $x(t)$ and $y(t)$. The phase difference function, $\theta_{x,y}(s,\tau)$, is the arctangent of the ratio of the imaginary part and real part of the cross-transform $SW_{xy}(s,\tau)$:

$$\theta_{x,y}(s,\tau) = \arctan\left(\frac{\Im\left(SW_{xy}(s,\tau)\right)}{\Re\left(SW_{xy}(s,\tau)\right)}\right) \tag{8}$$

According to the value of the phase difference between $-\pi$ and π, we can study the sign and the leadership of the relationship between $x(t)$ and $y(t)$ as follows:

- $\theta_{x,y}(s,\tau) \in [0, \frac{\pi}{2}]$: $x(t)$ and $y(t)$ move together in phase so they are positively correlated. In this case, $x(t)$ leads $y(t)$.
- $\theta_{x,y}(s,\tau) \in [\frac{\pi}{2}, \pi]$: $x(t)$ and $y(t)$ move together in anti-phase so they are negatively correlated. In this case, $y(t)$ leads $x(t)$.
- $\theta_{x,y}(s,\tau) \in [-\frac{\pi}{2}, 0]$: $x(t)$ and $y(t)$ move together in phase so they are positively correlated. In this case, $y(t)$ leads $x(t)$.
- $\theta_{x,y}(s,\tau) \in [-\pi, -\frac{\pi}{2}]$: $x(t)$ and $y(t)$ move together in anti-phase so they are negatively correlated. In this case, $x(t)$ leads $y(t)$.

We define the time-frequency estimator as follows:

$$\beta_{s,\tau} = \vartheta_{s,\tau} * WQ(s,\tau)^{\frac{1}{2}} * \frac{G\left(s^{-1}.|W_y(s,\tau)|^2\right)^{\frac{1}{2}}}{G\left(s^{-1}.|W_x(s,\tau)|^2\right)^{\frac{1}{2}}} \quad (9)$$

where $\vartheta_{s,\tau}$ is a phase parameter providing the sign of the correlation ([30,31] $\vartheta = 1$ when chronics are in phase $|\theta_{x,y}(s,\tau)| \in [0, \frac{\pi}{2}]$ and $\vartheta = -1$ when chronics are in anti-phase $|\theta_{x,y}(s,\tau)| \in [\frac{\pi}{2}, \pi]$..

We use official data for the European Union quarterly real GDP (Y) and unemployment rate (U) for 2001—Q1 to 2019—Q4. Previously, we computed the GDP growth (ΔY_t) and unemployment rate variations ΔU_t to estimate the first difference Okun's Law equation:

$$\Delta U_t = \alpha + \beta \Delta Y_t + \varepsilon_t \quad (10)$$

The wavelet coherence indicates the strength (R^2) of Okun's Law while the phase shows which variable leads the relationships. Equation (10) in the time-frequency space is then:

$$\Delta U_{t,s} = \alpha + \beta_{s,\tau}\, \Delta Y_{t,s} + \varepsilon_{t,s} \quad (11)$$

Finally, we apply the time-frequency estimators' formula (Equation (9)), and we study the time and frequency dynamic of the Okun coefficient.

3. Results and Discussion

Figure 1 shows the coherence, phase difference, and time frequency varying Okun coefficient results. Figure 1a is a 2D presentation of the time-frequency coherence (R^2 coefficients) associated with the phase difference represented by arrows. The color code shows the value of the coherence, with red for high coherence and blue for low coherence between the two variables. The direction of the arrows indicates both the sign of the relationship—negative by arrows pointing left and positive by arrows pointing right—and the variable leader. Figure 1b shows a 3D representation of the coherence between the two variables. The analysis of the leader variable is more easily performed from a representation of the phase for a given frequency, which is performed hereafter. The frequency unit is the period related to the cycle. Then, period 2 represents a cycle of 6 months (two quarters), period 4 represents a cycle of 1 year (four quarters), etc.

The coherence is strong at high-frequencies between periods 2 and 4 (here, the short-run horizon) only during 2007–2010 (Global Financial Crisis and 2009 Recession). In the long term, for cycles of more than three years, the coherence is very high across time and seems independent of crisis and expansion periods. We can thus conclude that the intensity of the Okun relationship is dependent of the European economic condition (recession expansion) for short cycles (less than 1 year) but remains homogeneous in the longer term. These results are consistent with current literature mainly focused on the US economy as we confirm the frequency hypothesis of Okun's Law. However, we note that the strength of Okun's Law is more dependent on short run economics situations, especially crisis times, in Europe than in the US. Then, we notify a frequency asymmetric relation. Okun's Law is consistent and relevant in the long run but sensitive to economic conditions in the short term.

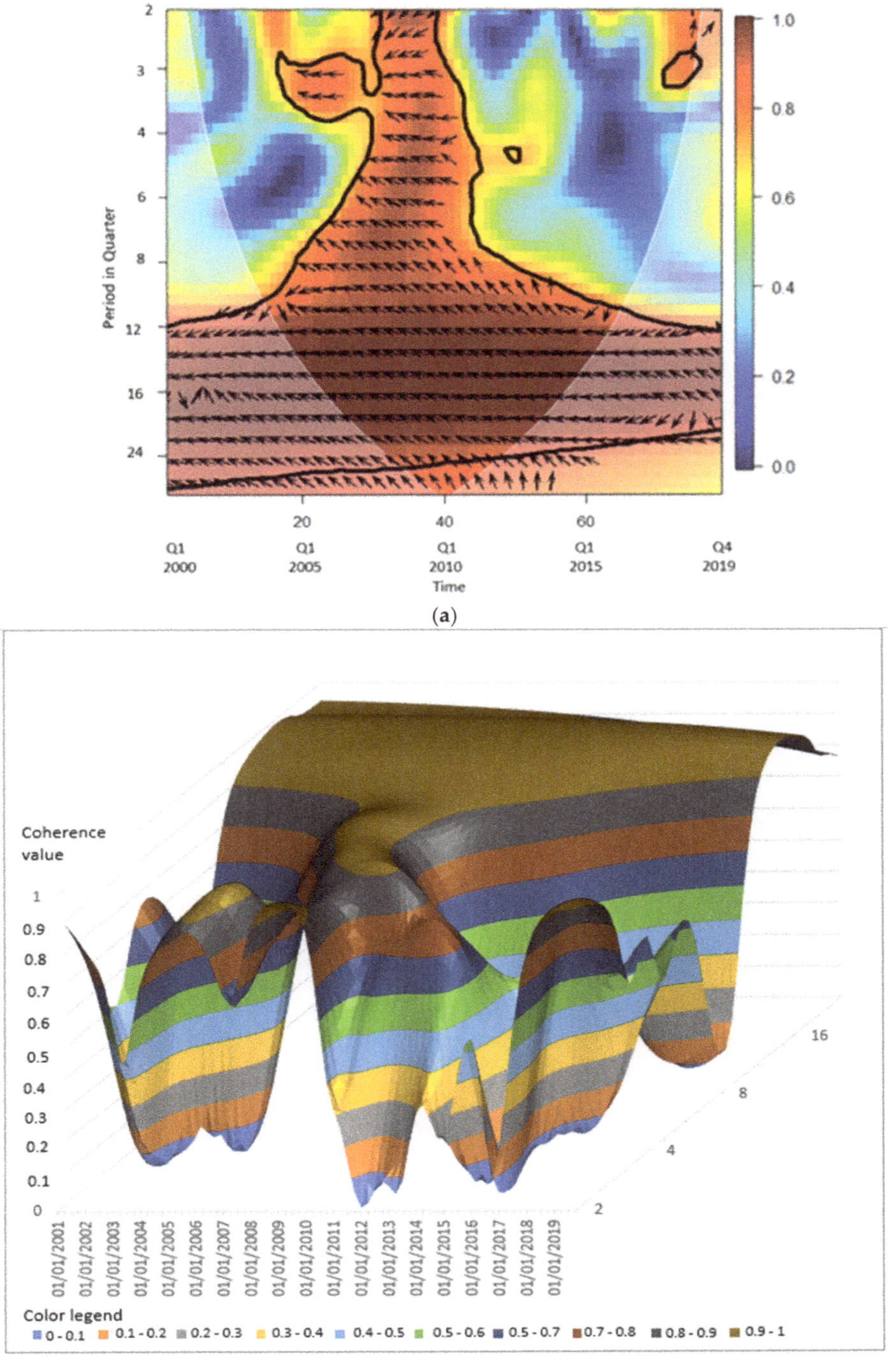

(a)

(b)

Figure 1. *Cont.*

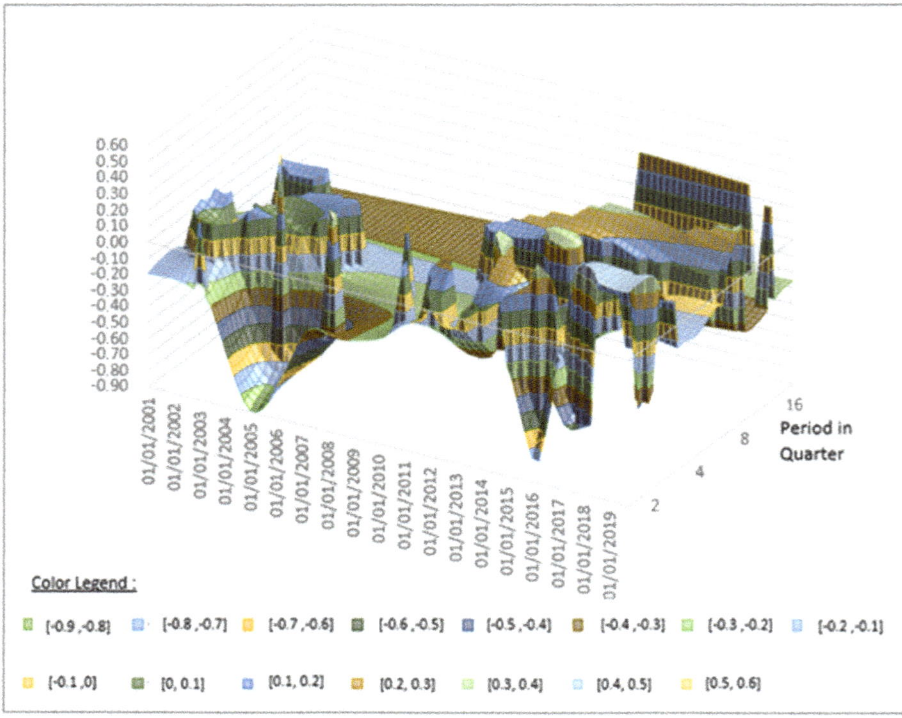

(c)

Figure 1. (**a**) Two-dimensional representation of wavelet coherence and phase between growth and unemployment rate variations; (**b**) three-dimensional representation of wavelet coherence; (**c**) three-dimensional representation of time-frequency varying the Okun coefficient. Corresponding color legends describing coefficents values are put on each figure.

In addition, the estimation of the Okun coefficient in the time-frequency space extends our interpretation. Figure 1c shows that the Okun coefficient is not stable around the commonly acceptable value of −0.5. Then, the coefficient is more erratic and volatile, especially in the very short term. However, in the long run, the coefficient seems to be more stable as the strength of the relation increases. We also note an unusual result with positive Okun coefficients while a negative one is expected.

In the short term for cycles between 6 months and less than 1 year in length, the coherence is relatively high and the Okun coefficients are decreased and negative from 2001 to Q1 of 2005, reaching its lowest value at −0.9. Growth is the predominant variable, but after 2005 the coefficient is increasing to stabilize around −0.3 until 2010. Here; we note that unemployment leads growth. During the European debt crisis (2011–2012), the Okun coefficient increased to 0 and the coherence sharply decreased to 0. From 2014 to Q1 of 2016, the coherence slowly increases while being non-significant and the Okun coefficient returns to the negative zone at −0.8. Note that unemployment still predominates growth. However, from Q1 of 2016 to mid-2018, the coherence sharply increases and growth leads unemployment, as expected by Okun's Law. Afterward, we note a positive Okun coefficient with unemployment as the leading variable. For cycles between 1 and 3 years in length, the results show that Okun's Law is relevant only during crisis times from 2007 to 2012 with growth as the predominant variable. The Okun coefficient is stable at around −0.27. As previously observed, the coherence is non-significant after the debt crisis and sharply decreases as unemployment dominates growth. It seems that when unemployment predominates growth, coherence tends to decrease sharply and the Okun coefficient (usually negative) increases, frequently exceeding zero. For long run cycles with a period above 3 years, we note a stable and negative Okun coefficient around [−0.37, −0.3] with growth as the predominant variable. However, since mid-2015, there has been a slow decline in coherence, a slow increase in the Okun coefficient, and a more frequent predominance of unemployment over growth.

The leadership of unemployment indicates standard unilateral causal representation from growth to unemployment of Okun's Law is not fully observed, supposing a feedback relationship. This unusual aspect is observed over all business cycles, regardless of the frequencies, and coincides with a decline in the strength of Okun's Law and its coefficient (turning from negative to positive). The literature mainly focusing on the US economy indicates that this fact is only concentrated on specific frequencies.

We find that, for Europe, the unemployment changes play a greater role especially after the 2008 crash and debt crisis (2011–2012). Consequently, unemployment changes lead over growth, reducing the strength of Okun's Law, and then tends to affect the coefficient, equating to or exceeding zero more frequently than in the USA. Such results can be explained by the structural unemployment rate being more important in Europe than in the US, for which the labor market is more liquid. Many European countries (for example, France, Spain, and Italy) have suffered high unemployment rate since the 2008–2009 crisis.

4. Conclusions

In this article, we discuss the strength of Okun's Law for Europe as well as the stability of its coefficient across time and frequencies. We use a wavelet approach distinguishing Okun's relationship with various economic cycles of different amplitudes. The literature recommends the estimation of the gain in wavelets (the absolute value of the parameter), but our results propose the direct estimation of the time-frequency varying Okun's coefficient.

We find that the value of Okun's coefficient and the coherence are then unstable in the short run as unemployment variations more frequently lead growth for infra-annual or annual cycles. However, we assume that the short-run coefficient is relatively stable and negative across time and frequencies in crisis times only (for example, the 2008 crash and debt crisis). In this context, growth is predominant over unemployment, satisfying the Okun's Law specification. However, when unemployment leads the relation, the coefficient tends to be equal to zero or positive while the coherence declines. This fact is observed for relatively quiet periods or post-crisis periods. We note that the coefficient gradually stabilizes in value as the frequencies increase (i.e., for long-run cycles). For cycles of more than 3 years, the coefficient is stable and insensitive to economic conditions. Okun's Law is then more relevant for long-run cycle than for short-run cycles in Europe.

Compared with the US, the European Okun's Law is frequency asymmetric as the coefficient is dependent on crises in the short term but not in the long run. This result is consistent with a previous one indicating that the short-run Okun's Law is less robust across time than the long-run one. In addition, the standard value of −0.5 is discussed for Europe as the long-run coefficient is close to −0.3, the initial value provided by Okun. This result suggests that the structural unemployment in Europe is stronger and then less affected by economic growth than in the US. This conclusion is also consistent with phase analysis showing a more frequent leadership of unemployment in Europe than commonly observed in the literature for the US.

Funding: This research received no external funding.

Institutional Review Board Statement: Not applicable.

Informed Consent Statement: Not applicable.

Data Availability Statement: The data presented in this study are openly available in the Eurostat database or FRED data website.

Conflicts of Interest: The authors declare no conflict of interest.

References

1. Okun, A.M. Potential GNP: Its Measurement and Significance. In *Proceedings of the Business and Economic Statistics Section of the American Statistical Association*; American Statistical Association: Alexandria, VA, USA, 1962; Volume 7, pp. 98–104.
2. Galí, J.; Smets, F.; Wouters, R. Slow recoveries: A structural interpretation. *J. Money Credit Bank.* **2012**, *44*, 9–30. [CrossRef]
3. Ball, L.; Leigh, D.; Loungani, P. Okun's Law: Fit at 50? *J. Money Credit Bank.* **2017**, *49*, 1413–1441. [CrossRef]
4. Fernald, J.G.; Hall, R.E.; Stock, J.H.; Watson, M.W. *The Disappointing Recovery of Output after 2009*; Working Paper No. w23543; National Bureau of Economic Research: Cambridge, MA, USA, 2017.
5. Daly, M.C.; Fernald, J.G.; Jordà, O.; Nechio, F. *Shocks and Adjustments*; Working Paper 2013-32; Federal Reserve Bank of San Francisco: San Francisco, CA, USA, 2017.
6. Lee, J. The robustness of Okun's law: Evidence from OECD countries. *J. Macroecon.* **2000**, *22*, 331–356. [CrossRef]
7. Cuaresma, J.C. Okun's law revisited. *Oxf. Bull. Econ. Stat.* **2003**, *65*, 439–451. [CrossRef]
8. Silvapulle, P.; Moosa, I.A.; Silvapulle, M.J. Asymmetry in Okun's law. *Can. J. Econ.* **2004**, *37*, 353–374. [CrossRef]
9. Holmes, M.J.; Silverstone, B. Okun's law, asymmetries and jobless recoveries in the United States: A Markov-switching approach. *Econ. Lett.* **2006**, *92*, 293–299. [CrossRef]
10. Gordon, R.J. Okun's law and productivity innovations. *Am. Econ. Rev.* **2010**, *100*, 11–15. [CrossRef]
11. Lucchetta, M.; Paradiso, A. Sluggish US employment recovery after the Great Recession: Cyclical or structural factors? *Econ. Lett.* **2014**, *123*, 109–112. [CrossRef]
12. Knotek, E.S., II. How useful is Okun's law? *Econ. Rev.-Fed. Reserve Bank Kans. City* **2007**, *92*, 73–103.
13. Meyer, B.; Tasci, M. An unstable Okun's Law, not the best rule of thumb. *Fed. Reserve Bank Clevel. Econ. Comment.* **2012**, *8*. [CrossRef]
14. Grant, A.L. The Great Recession and Okun's law. *Econ. Model.* **2018**, *69*, 291–300. [CrossRef]
15. Owyang, M.T.; Sekhposyan, T. Okun's law over the business cycle: Was the great recession all that different? *Fed. Reserve Bank St. Louis Rev.* **2012**, *94*, 399–418. [CrossRef]
16. Valadkhani, A.; Smyth, R. Switching and asymmetric behaviour of the Okun coefficient in the US: Evidence for the 1948–2015 period. *Econ. Model.* **2015**, *50*, 281–290. [CrossRef]
17. Berger, T.; Everaert, G.; Vierke, H. Testing for time variation in an unobserved components model for the US economy. *J. Econ. Dyn. Control* **2016**, *69*, 179–208. [CrossRef]
18. Kamber, G.; Morley, J.; Wong, B. Intuitive and reliable estimates of the output gap from a Beveridge Nelson filter. *Rev. Econ. Stat.* **2018**, *100*, 550–566. [CrossRef]
19. Arai, N. Evaluating the Efficiency of the FOMC's New Economic Projections. *J. Money Credit Bank.* **2016**, *48*, 1019–1049. [CrossRef]
20. Aguiar-Conraria, L.; Martins, M.; Soares, M.J. Okun's Law across time and frequencies. *J. Econ. Dyn. Control* **2020**, *116*, 103897. [CrossRef]
21. Krüger, J.; Neugart, M. Dissecting Okun's law beyond time and frequency. *Appl. Econ. Lett.* **2021**, *28*, 1744–1749. [CrossRef]
22. Crowley, P.M.; Hallett, A.H. Great moderation or "Will o' the Wisp"? A time-Frequency decomposition of GDP for the US and UK. *J. Macroecon.* **2015**, *44*, 82–97. [CrossRef]
23. Crowley, P.M.; Hallett, A.H. What causes business cycles to elongate, or recessions to intensify? *J. Macroecon.* **2018**, *57*, 338–349. [CrossRef]
24. Aguiar-Conraria, L.; Soares, M.J. The Continuous Wavelet Transform: Moving beyond Uni-and Bivariate Analysis. *J. Econ. Surv.* **2014**, *28*, 344–375. [CrossRef]
25. Bekiros, S.; Nguyen, K.D.; Salah Uddin, G.; Sjo, B. On the time scale behavior of equity-commodity links: Implications for portfolio management. *J. Int. Financ. Mark. Inst. Money* **2016**, *41*, 30–46. [CrossRef]
26. Kahraman, E.; Unal, G. Multiple wavelet coherency analysis and forecasting of metal prices. *arXiv* **2019**, arXiv:1602.01960.
27. Mestre, R. Stock profiling using time-frequency-varying systematic risk measure. *Financ. Innov.* **2023**, *9*, 52. [CrossRef]
28. Mestre, R. A wavelet approach of investing behaviors and their effects on risk exposures. *Financ. Innov.* **2021**, *7*, 24. [CrossRef]
29. Aguiar-Conraria, L.; Martins, M.; Soares, M.J. Estimating the Taylor Rule in the time-frequency domain. *J. Macroecon.* **2018**, *57*, 122–137. [CrossRef]
30. Mestre, R.; Terraza, M. Time-Frequency varying beta estimation-a continuous wavelets approach. *Econ. Bull.* **2018**, *38*, 1796–1810.
31. Mestre, R.; Terraza, M. Time-frequency varying estimations: Comparison of discrete and continuous wavelets in the market line framework. *J. Bank. Financ. Technol.* **2019**, *3*, 97–111. [CrossRef]
32. Grossmann, A.; Morlet, J. Decomposition of Hardy functions into square integrable wavelets of constant shape. *SIAM J. Math. Anal.* **1984**, *15*, 723–736. [CrossRef]
33. Meyer, Y. Ondelettes et algorithmes concurrents. *Actual. Mathématiques Sci. Arts* **1990**, *7*, 217–381.
34. Mallat, S. A theory for multiresolution signal decomposition: The wavelet representation. *IEEE Trans. Pattern Anal. Mach. Intell.* **1989**, *11*, 674–693. [CrossRef]
35. Mallat, S. *A Wavelet Tour of Signal Processing: The Sparce Way*, 3rd ed.; AP Professional: London, UK, 2009.
36. Mallat, S. *Une Exploration des Signaux en Ondelettes*; Ecole Polytechnique: Paris, France, 2000.
37. Daubechies, I. *Ten Lectures on Wavelets*; Conference Series of Applied Mathematics; Society for Industrial and Applied Mathematics: Philadelphia, PA, USA, 1992.

38. Grinsted, A.; Moore, J.C.; Jevrejeva, S. Application of the cross wavelet transform and wavelet coherence to geophysical time series. *Nonlinear Process. Geophys.* **2004**, *11*, 561–566. [CrossRef]
39. Torrence, C.; Compo, G. A practical guide to wavelet analysis. *Bull. Am. Meteorol. Soc.* **1998**, *79*, 61–78. [CrossRef]

engineering proceedings

MDPI

Proceeding Paper

Optimizing the Spatial-Temporal Extent of Environmental Factors in Forecasting El Niño and La Niña Using Recurrent Neural Network †

Jahnavi Jonnalagadda * and Mahdi Hashemi

Department of Information Sciences and Technology, George Mason University, Fairfax, VA 22030, USA; mhashem2@gmu.edu

* Correspondence: jjonnala@gmu.edu

† Presented at the 9th International Conference on Time Series and Forecasting, Gran Canaria, Spain, 12–14 July 2023.

Abstract: El Niño-Southern Oscillation (ENSO) is caused by periodic fluctuations in sea surface temperature and overlying air pressure across the Equatorial Pacific region. ENSO has a global impact on weather patterns and can cause severe weather events, such as heat waves, floods, and droughts, affecting regions far beyond the tropics. Therefore, forecasting ENSO with longer lead times is of great importance. This study utilizes Long Short-Term Memory (LSTM) network to predict ENSO events in the coming year based on environmental variables from previous years, including sea-surface temperature, sea level pressure, zonal wind, meridional wind, and zonal wind flux. These environmental variables are collected only inside certain spatial and temporal windows and used to forecast ENSO events. Furthermore, this study investigates how the size of these spatial and temporal windows influences the generalization accuracy of forecasting ENSO events. The size of spatial and temporal windows is optimized based on the generalization accuracy of the LSTM network in forecasting ENSO events. Our results indicated that the accuracy of the ENSO forecast is significantly sensitive to the extent of spatial and temporal windows. Specifically, increasing the temporal window size from one to nine years and the spatial window from 0 to 17.7 geographical degrees resulted in generalization accuracies, ranging from 40.1% to 83% in forecasting Central Pacific ENSO and 39.2% to 65% in forecasting Eastern Pacific ENSO.

Keywords: climate anomalies; machine learning; spatial–temporal data mining; deep learning; ENSO events

Citation: Jonnalagadda, J.; Hashemi, M.; Optimizing the Spatial-Temporal Extent of Environmental Factors in Forecasting El Niño and La Niña Using Recurrent Neural Network. *Eng. Proc.* **2023**, *39*, 10. https://doi.org/10.3390/engproc2023039010

Academic Editors: Ignacio Rojas, Hector Pomares, Luis Javier Herrera, Fernando Rojas and Olga Valenzuela

Published: 28 June 2023

1. Introduction

El Niño-Southern Oscillation (ENSO) refers to the cyclic variations in sea-surface temperature (SST) and air pressure in the overlying atmosphere in the Equatorial Pacific Ocean. It has two phases—El Niño and La Niña—characterized by warm and cold conditions. El Niño events bring above-average SST in the central and east-central Equatorial Pacific. These conditions typically result in warmer temperatures over the western and northern United States and wetter than average conditions over the US Gulf Coast and Florida during winter. Conversely, La Niña events cause below-average SST in the east-central Equatorial Pacific. During La Niña, winter temperatures are warmer in the southeast and colder than average in the northwestern regions of the United States. Approximately every four years, the SST of the tropical Pacific becomes exceedingly warm, leading to abnormal global climate patterns [1].

The National Oceanic and Atmospheric Administration (NOAA) uses the Oceanic Niño Index (ONI) to identify ENSO events. ONI determines ENSO events by averaging the SST anomalies for three consecutive months in the Niño 3.4 region (5° S–5° N, 170° W–120° W). An ONI greater than 0.5 °C for five consecutive months indicates El Niño, while an ONI less than −0.5 °C for five consecutive months indicates La Niña. Studies have shown that ENSO events concentrated in the Central Pacific (CP) and Eastern Pacific (EP) regions differ, not only in the SST anomaly patterns but also in the oceanic surface currents and their impact on the global climate [2,3].

EP El Niño has a larger SST anomaly centered at 120° W than CP El Niño (also known as El Niño Modoki). In contrast, CP El Niño has a weaker SST anomaly and tends to shift westwards (150° W) during its mature phase [4]. EP La Niña and CP La Niña show differences in the physical mechanism of SST anomaly development [5,6], tropical climate responses, and global environmental impact [2]. Forecasting the type of ENSO events could help prepare for their global repercussions.

State-of-the-art research has applied both dynamical and statistical models to forecast ENSO events. Dynamical models utilize physical laws to conserve the ocean, land, and atmosphere and their interactions to predict ENSO events, while statistical models rely on statistical learning patterns from historical data [7]. While dynamical models require scientists to develop specific mathematical and physical laws to model the relationship between different parameters and ENSO events, statistical models learn those relationships automatically from historical data. The downside is that statistical models, especially deep learning models, do not provide a clear and easy-to-understand mathematical representation of the patterns they have learned. A Long Short-Term Memory (LSTM) network is a recurrent neural network (RNN) tailored for sequential prediction problems with the help of memory gates. This study applies the LSTM network for forecasting ENSO events, given its proven capability in effectively capturing the non-linear complexities of multi-variate time series [8–10].

The first step of this study is to optimize the size of spatial and temporal windows, where environmental features related to forecasting ENSO events are extracted. The environmental features include SST, sea level pressure (SLP), zonal wind (ZW), meridional wind (MW), and zonal wind flux (WF). Next, the environmental features, extracted from varying spatial and temporal window sizes, are fed to an LSTM network to forecast ENSO events in the next year to determine which window size results in the highest generalization accuracy. It was shown that the generalization accuracy significantly depends on the spatial and temporal window sizes.

The remainder of this paper is organized as follows. Section 2 reviews select literature in ENSO forecasting. Section 3 describes the data collection and preprocessing. Section 4 outlines the methodology to optimize the size of spatial and temporal windows, and Section 5 discusses the experimental results. Section 6 concludes the paper with a summary of findings and future directions.

2. Literature Review

While ENSO events have been forecasted using statistical [11–13] and dynamical models [14,15] in the literature, this section focuses on reviewing statistical methods since our proposed model is also statistical.

Early studies utilized Artificial Neural Networks (ANNs) with the leading empirical orthogonal functions of wind stress or SLP as input [16]. These studies yielded a correlation coefficient (CC) of 0.6 for the prediction period of 1980–1990 and 0.1 for 1950–1970. Baawain et al. [17] used a multi-layer perceptron to predict ENSO events with a one-year lead time. Ham et al. [18] employed transfer learning to train a Convolutional Neural Network (CNN) on both Coupled Model Intercomparison Project Phase 5 (CMIP5) data and reanalysis data for the training period (1871–1973). Their results showed that the correlation coefficient (CC) of the observed and forecasted Niño 3.4 index is above 0.5 for a lead time of 17 months. Martínez-Alvarado et al. [19] used Support Vector Regressor (SVR) and Bayesian neural networks to forecast the tropical SST anomalies for a lead time of up to 15 months. The results demonstrated that the Bayesian neural network model outperformed the SVR. Noteboom et al. [20] proposed a hybrid model using autoregressive integrated moving average and ANN to predict ENSO events with a one-year lead time.

Guo et al. [21] proposed a hybrid neural network model that combines ensemble empirical mode decomposition with a CNN and LSTM network. Finally, Peter et al. [22] attempted to forecast ENSO events using Gaussian density neural networks and quantile regression neural network ensembles. These models can assess the predictive uncertainty of the forecast by predicting a Gaussian distribution and the quantiles of the forecasts, respectively.

Several studies have explored the effectiveness of combining statistical and dynamical models for forecasting ENSO events. Hong et al. [23] developed a dynamical-statistical forecast model for predicting SST anomalies, which achieved a CC of 0.8 for a lead time of 12 months. Meanwhile, Zhang et al. [24] used Bayesian model averaging (BMA) to combine the results of three statistical and one dynamical model to forecast ONI. They applied an expectation-maximization algorithm to derive the maximum likelihood estimation for model parameters. Their combined statistical–dynamical model outperformed either statistical or dynamical models when used alone. Finally, Ha et al. [25] proposed an encoder–decoder structure for predicting river flow using ENSO. They achieved an R^2 of 0.8 with the CLSTM encoder–decoder and an LSTM network in forecasting Yangtze River flow.

Although not specifically in forecasting ENSO events, the significance of location and time in predicting and forecasting geographical phenomena has been underscored in the literature both theoretically [26,27] and experimentally [28,29]. Here we aim to explore how the extent of spatial and temporal windows, where the environmental features for forecasting ENSO events are extracted, influences the generalization accuracy of an LSTM network.

3. Data Description

Our data consists of seven environmental variables: SST, SLP, ZW, MW, WF, CP's ONI, and EP's ONI, recorded at 4697 geographical locations (Figure 1) between 1949 and 2014. These variables were obtained from the following sources:

- SST data in °C was retrieved from NOAA extended reconstructed SST version 3b. The spatial resolution is 1° × 1°. Since this resolution differs from other variables, we used geographical interpolation to convert it to a 2.5° × 2.5° resolution, consistent with the rest of the variables. Additionally, 25% of SST values are missing in this dataset. Therefore, we replaced the missing values of SST with the mean value across the entire region, which is zero.
- SLP in Hecto Pascals was obtained from National Centre for Environmental Climate Prediction (NCEP) reanalysis version 1 [30]. The spatial resolution is 2.5° × 2.5°.
- The horizontal (ZW) and vertical wind (MW) components at 10m depth were obtained from NCEP reanalysis version 1. The spatial resolution is 2.5° × 2.5°.
- The horizontal wind flux (WF) was obtained from NCEP reanalysis version 1. The spatial resolution is 2.5° × 2.5°.
- Details on calculating CP and EP indices from Niño indices are available in [31].

The environmental variables in the dataset represent the three-month averages over December, January, and February. For instance, the SLP data referring to the time 1949-01-01 in the dataset is the average of SLP over December 1948, January 1949, and February 1949.

The continuous values of CP's ONI and EP's ONI are transformed into three categories, El Niño where ONI is greater than +0.5, La Niña where ONI is less than −0.5, and Neutral where ONI is between −0.5 and +0.5. There are 24 neutral, 22 El Niño, and 20 La Niña events that occurred in CP, and 34 neutral, 18 La Niña, and 14 El Niño events that occurred in EP between 1949 and 2014. We aim to forecast the ONI class in the next year based on SST, SLP, ZW, MW, and WF from previous years, making this a three-way classification problem.

Feature standardization is a prerequisite for machine learning and deep learning models. Therefore, all variables are standardized to have a zero mean and unit variance.

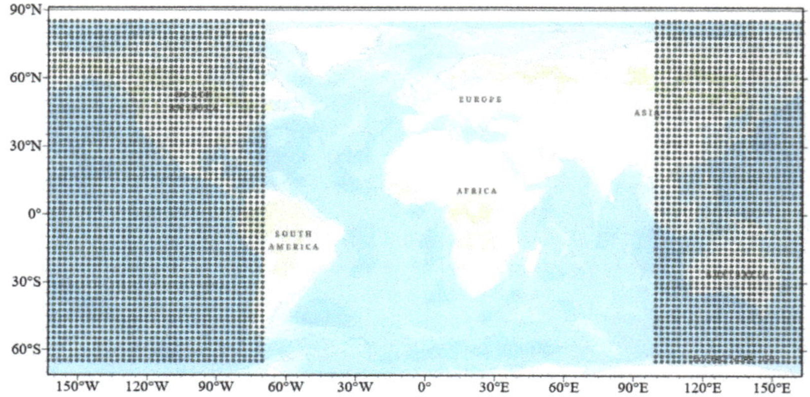

Figure 1. Data points are separated by 2.5° across latitude and longitude.

4. Methodology

An important property of spatial–temporal phenomena is that samples are not independent but rather spatially and temporally auto-correlated. Spatial (temporal) autocorrelation refers to the phenomenon where samples recorded at nearby locations (time) display similar patterns to those recorded further apart. As a result, careful consideration should be given to the spatial–temporal autocorrelation among observations when designing a prediction model.

Every spatial–temporal observation (st_i) contains the values of a set of variables at a specific location (s) and time (t). Therefore, a spatial–temporal dataset contains recorded values of features at different locations and times. We define spatial–temporal forecast here as predicting the value of a variable at time t and location s based on the variable's value and other contributing variables at different locations and previous time stamps. The question we are posing is at what locations (how far from the point where the forecast is being made) and at how many previous timestamps the values of those variables would positively contribute to forecasting the target variable. We refer to the size of the geographical neighborhood as the spatial window size and the number of time stamps going back as the temporal window size. We will measure the forecast generalization accuracy at varying spatial and temporal window sizes. The optimal window sizes are those resulting in the highest generalization accuracy.

4.1. Temporal Window

A temporal window is the number of previous timesteps the environment variable's values would contribute to forecasting the target variable. For example, a temporal window of one year means environmental variables, SST, SLP, ZW, MW, and WF, recorded from one year ago are used to forecast the occurrence of El Niño, La Niña, or a neutral event in the current year. Similarly, a temporal window of two years means environmental variables recorded from the past two years are used to forecast the occurrence of El Niño, La Niña, or a neutral event in the current year. Figure 2 illustrates the use of a two-year temporal window for forecasting ENSO events. The temporal window size is optimized based on the generalization accuracy of the LSTM network in forecasting ENSO events.

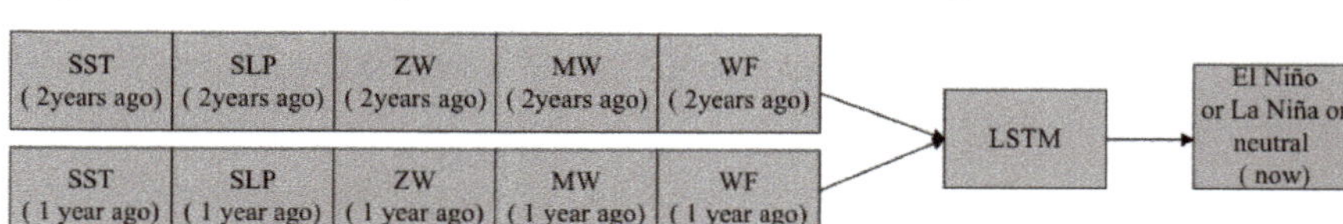

Figure 2. A temporal window with a size of two years.

4.2. Spatial Window

A spatial window is a distance in geographical degrees from the point where the forecast is being made. The values of the environmental variables that fall within the spatial window would contribute to forecasting the target variable. The spatial window size varies from one point to a circle of radius nr, where $r = \sqrt{2.5^2 + 2.5^2} = 3.54$ geographical degrees. The coefficient n is treated as a hyperparameter and optimized during training. For example, when the spatial window size is zero, the values of environmental variables SST, SLP, ZW, MW, and WF, recorded only at the forecast point (Figure 3a), are used for forecasting ENSO events in the following year. On the other hand, if the spatial window size is r, the values of those variables recorded at the forecast point and within a distance of 3.54 geographical degrees (Figure 3b) are employed for forecasting ENSO events in the following year. Similarly, a spatial window of size $2r$ encompasses the values of the environmental variables recorded at the forecasting point and within the distance of 7.08 geographical degrees (Figure 3c) for forecasting ENSO events in the following year.

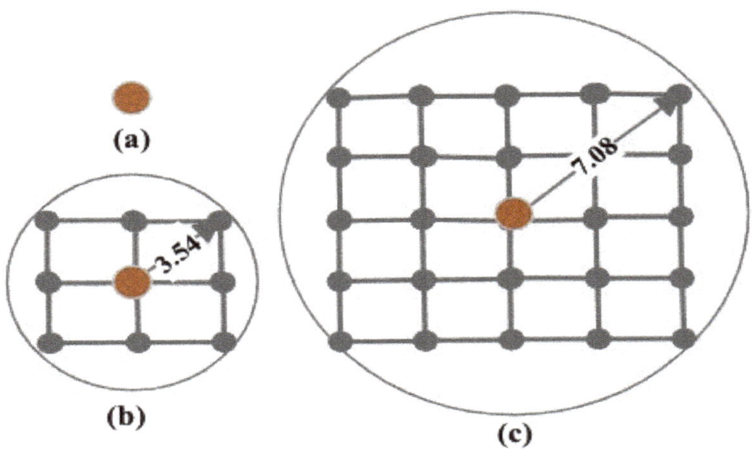

Figure 3. Varying spatial window sizes: (**a**) spatial window is one point, (**b**) spatial window is a circle with a radius of $3.54 = \sqrt{2.5^2 + 2.5^2}$ geographical degrees, and (**c**) spatial window is a circle with a radius of 7.08 geographical degrees.

4.3. LSTM

LSTM overcomes the limitations of traditional RNNs in capturing long-term temporal dependencies in sequential data. Unlike standard RNN, which suffers from the vanishing gradient problem, LSTM uses a gating mechanism to regulate the flow of information through the network. The main component of an LSTM unit is the memory cell, which serves as a storage unit to retain information for longer durations. The memory cell has three gates, input, forget, and output. These gates regulate the flow of information into and outside of the cell, allowing the LSTM to retain or discard information. The input gate decides how much information should be added to the memory cell, while the forget gate decides what information to discard from the memory cell. The output gate regulates the amount of information to be outputted from the memory cell to the next time step or the final prediction. This gating mechanism allows LSTM to capture long-term temporal dependencies from the sequential data. The mathematical equations for information flow in LSTM are given in [32].

The LSTM network consists of an input layer, an LSTM layer, and an output layer. The LSTM layer consists of 32 memory units, ReLU activation, and a dropout of 0.2. Dropout [33] is used as regularization to drop a fraction of randomly selected memory units in the LSTM layer. This means that the information passing through these units is not considered in the forward pass and is not updated during backpropagation, allowing the network to learn robust and generalized data representations. The output layer is a fully connected dense layer with three units, where each unit represents each class. SoftMax is the activation function used in the output layer to generate the probability of each class, and the class with the highest probability is chosen as the target class. The loss function employed in the LSTM network is categorical cross-entropy, and the network is optimized using Adam with a learning rate of 0.001. The model is trained for 100 epochs, and an early stopping algorithm [34] is implemented if the validation loss does not improve for 10 epochs. The network parameters of LSTM are determined by hyperparameter tuning. Notably, adding more LSTM layers to the network led to overfitting.

4.4. Benchmark Models and Evaluation Metrics

To evaluate the performance of the LSTM, we employ Multi-Layer Perceptron (MLP), Random Forest (RF), and CLSTM encoder–LSTM decoder [25] as baselines. The MLP consists of an input layer, two fully connected hidden layers, and an output layer. Two hidden layers comprise 100 units each and ReLU activation. The output layer has three units representing three classes. SoftMax is the activation used in the output layer to generate a probability for each class, and the class with the highest probability is selected as the target class. The loss function employed in the MLP is categorical cross-entropy, and the network is optimized using Adam with a learning rate of 0.001. The model is trained for 100 epochs, and early stopping is implemented if the validation loss does not improve for 10 epochs. The maximum number of features required for splitting in the RF varies with the spatial window size. The number of trees in the RF is 50. We followed the same architecture shown in [25] for the CLSTM encoder–LSTM decoder.

We split the data into 80% for training and 20% for testing. Five-fold cross-validation is implemented on the training dataset to determine the optimal spatial and temporal window sizes and for hyperparameter tuning.

All experiments are evaluated using the $F1$ score, which measures the model's accuracy on a dataset. $F1$ score is the harmonic mean of precision and recall and is calculated as follows. $F1$ score reaches its best value at one and worst value at zero. The best value for the $F1$ score is achieved for perfect precision (100%) and recall (100%).

$$F1 = \frac{\sum_{i=1}^{M} F1(w_i)}{M} \tag{1}$$

$$F1(w_i) = \frac{2 \times precision(w_i) \times recall(w_i)}{precision(w_i) + recall(w_i)} \tag{2}$$

$$Precision(w_i) = \frac{TP(w_i)}{TP(w_i) + FP(w_i)} \tag{3}$$

$$Recall(w_i) = \frac{TP(w_i)}{TP(w_i) + FN(w_i)} \tag{4}$$

Here, $TP(w_i)$, $TN(w_i)$, $FP(w_i)$, and $FN(w_i)$ denotes true positives, true negatives, false positives, and false negatives for class w_i and M denotes the number of classes. True positives and true negatives are the test samples that are correctly classified, and false positives and false negatives are the test samples that LSTM wrongly classifies.

5. Results and Discussions

Figures 4 and 5 present the results of five-fold cross-validation accuracies, expressed as percentages of $F1$ scores, for predicting ENSO events in CP and EP regions, respectively. Figure 4 indicates that increasing the temporal window from 1 year to 9 years results in an improvement in accuracy of approximately 50%, with the highest accuracy observed at 7 years. Similarly, when varying the spatial window from 0 to $5r$, the accuracy increases until $2r$ and decreases from $2r$ to $5r$. The combination of spatial window size at $2r$ (7.08 geographical degrees) and temporal window size of 7 years yields the highest cross-validation accuracy and, therefore, is optimal.

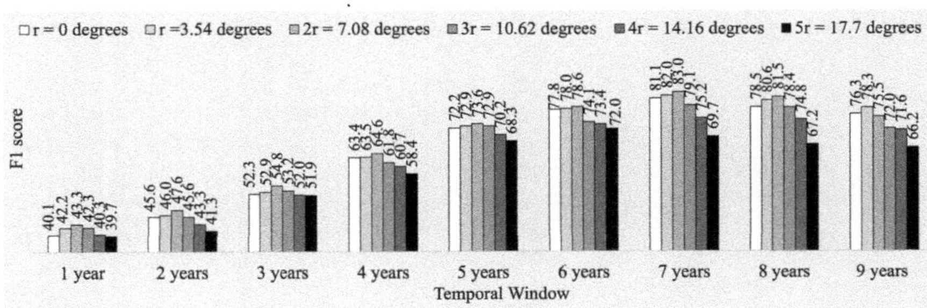

Figure 4. Five-fold cross validation accuracy ($F1$ score) in forecasting El Niño and La Niña in CP using LSTM for various spatial and temporal window sizes (values shown in percentage).

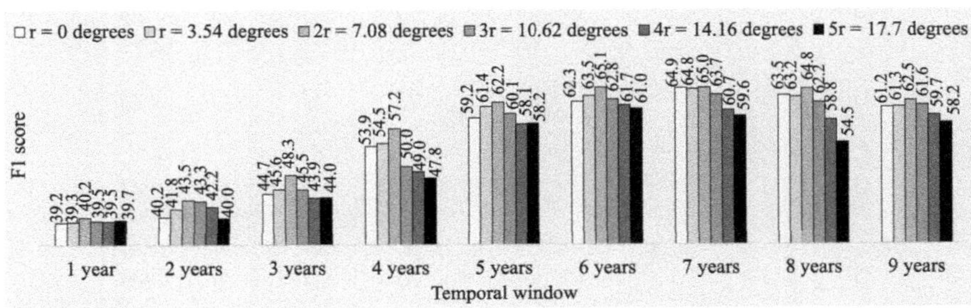

Figure 5. Five-fold cross validation accuracy ($F1$ score) in forecasting El Niño and La Niña in EP using LSTM for various spatial and temporal window sizes (values shown in percentage).

For the EP region, the five-fold cross-validation accuracy improves as the temporal window increases from 1 year to 7 years, as depicted in Figure 5. However, when the temporal window is increased from 7 years to 9 years, the accuracy declines. Similarly, regardless of the temporal window size, an increase in the spatial window size from 0 to $5r$ leads to a rise in accuracy until $2r$ and a drop in accuracy from $2r$ to $5r$. The optimal spatial and temporal window sizes for the EP region are the same as those for the CP region, a spatial window size of $2r$ (7.08 geographical degrees) and a temporal window size of 7 years, respectively.

The performance of different models in forecasting ENSO events for a one-year lead time is compared in Table 1. The results indicate that LSTM performs better than the other three baseline models, achieving $F1$ scores of 0.83 and 0.65 for the CP and EP regions, respectively. The second best-performing models are the CLSTM encoder–LSTM decoder and RF, which exhibit similar accuracies for both regions. The inferior performance of the CLSTM encoder–decoder LSTM model, when compared to LSTM, can be attributed to the flattening of the encoder output, resulting in the loss of spatial structure in environmental variables and their temporal relationships with ENSO events. On the other hand, the limitations of RF lie in its inability to capture the temporal dependencies between environmental variables and ENSO. The MLP performs poorly compared to other models, as it cannot handle the spatial and temporal relationships between environmental variables and ENSO. In summary, LSTM, with optimized spatial and temporal windows, demonstrates its superiority over other baseline models. Specifically, LSTM outperforms RF and CLSTM encoder–decoder LSTM by 1.3 and 1.08 times, respectively, in forecasting ENSO events in the CP and EP regions.

Table 1. Comparison of different models in forecasting ENSO events (measured in terms of *F*1 score).

	MLP	RF	CLSTM Encoder-Decoder LSTM	LSTM (Ours)
CP	0.60	0.62	0.63	0.83
EP	0.56	0.58	0.60	0.65

Table 2 presents the precision, recall, and *F*1 scores for each type of ENSO event, providing insight into the performance of LSTM. Higher precision indicates that the classifier misclassified fewer negative samples as positive samples, and a higher recall indicates that the classifier misclassified fewer positive samples as negative samples. Notably, the recall values for ENSO events in both CP and EP exceed the precision values, indicating that LSTM is suitable for ENSO event forecasting. Further investigation reveals that most misclassified ENSO events have ONI values close to ±0.5.

Table 2. Highest accuracies in forecasting El Niño and La Niña in CP and EP using LSTM with spatial window of 7.08 geographical degrees and temporal window of 7 years.

	CP			EP		
	Precision	*Recall*	*F*1	*Precision*	*Recall*	*F*1
El Niño	1.00	1.00	1.00	0.5	1.00	0.66
La Niña	0.71	0.83	0.77	0.60	0.60	0.60
Neutral	0.80	0.66	0.72	0.78	0.64	0.70

Additionally, the proposed model successfully identifies one of the strongest El Niño events from 1997 to 1998. It is worth mentioning that the *F*1 score of LSTM for EP is slightly reduced compared to CP due to the high-class imbalance ratio, which was not addressed in this study. The experimental results highlight that varying spatial and temporal window sizes, ranging from 0 to 17.7 degrees and 1 year to 9 years, yield forecasting accuracy ranging from 40.1% to 83% for CP and from 39.2% to 65% for EP regions.

6. Conclusions and Future Directions

Our findings demonstrated that the accuracy of forecasting CP (or EP) ENSO events in the coming year could be improved by up to 40% (or 25%) by optimizing the spatial and temporal extent of the environmental variables (i.e., SST, SLP, ZW, MW, and WF) which are used as features in deep learning. Furthermore, it was shown that these optimal spatial and temporal window sizes are 7.08 geographical degrees and 7 years. In our future work, we explore the impact of including additional climate features, such as warm water volume and upper ocean heat content, on the ENSO forecast accuracy of an LSTM. We will also focus on integrating dynamic models into deep learning and studying their impact on forecast accuracies.

Author Contributions: Conceptualization, J.J. and M.H.; methodology, M.H.; formal analysis, J.J. and M.H.; writing—original draft preparation, J.J.; writing—review and editing, J.J. and M.H.; visualization, J.J. and M.H.; supervision, M.H.; project administration, J.J. All authors have read and agreed to the published version of the manuscript.

Funding: This research received no external funding.

Institutional Review Board Statement: Not applicable.

Informed Consent Statement: Not applicable

Data Availability Statement: The data presented in this study are openly available in GitHub at https://github.com/jahnavijo/ENSO-Dataset.git.

Conflicts of Interest: The authors declare no conflict of interest.

References

1. Philander, S. El Niño Southern Oscillation phenomena. *Nature* **1983**, *302*, 295–301. [CrossRef]
2. Hsun-Ying, K.; Jin-Yi, Y. Contrasting Eastern-Pacific and Central-Pacific Types of ENSO. *J. Clim.* **2009**, *22*, 615–632.
3. Jin-Yi, Y.; Tae, K.S. Identifying the types of major El Niño events since 1870. *Int. J. Climatol.* **2013**, *33*, 2105–2112.
4. Mingcheng, C.; Tim, L. ENSO evolution asymmetry: EP versus CP El Niño. *Clim. Dyn.* **2021**, *56*, 3669–3579.
5. Ashok, K.; Behera, S.K.; Rao, S.A.; Weng, H.; Yamagata, T. El Niño Modoki and its possible teleconnection. *J. Geo Phys. Lett.* **2007**, *112*. [CrossRef]
6. Yuan, Y.; Song, Y. Impacts of Different Types of El Niño on the East Asian Climate: Focus on ENSO Cycles. *J. Clim.* **2012**, *25*, 7702–7722. [CrossRef]
7. Barnston Anthony, G.; Tippett Michael, K.; L'Heureux Michelle, L.; Li, S.; DeWitt David, G. Skill of Real-Time Seasonal ENSO Model Predictions during 2002–11: Is Our Capability Increasing? *Bull. Am. Meteorol. Soc.* **2012**, *93*, 631–651. [CrossRef]
8. Qianlong, W.; Yifan, G.; Lixing, Y.; Pan, L. Earthquake Prediction Based on Spatio-Temporal Data Mining: An LSTM Network Approach. *IEEE Trans. Emerg. Top. Comput.* **2020**, *8*, 148–158.
9. Jahnavi, J.; Hashemi, M. Forecasting Atmospheric Visibility Using Auto Regressive Recurrent Neural Network. In Proceedings of the 2020 IEEE 21st International Conference on Information Reuse and Integration for Data Science (IRI), Las Vegas, NV, USA, 11–13 August 2020.
10. Jahnavi, J.; Hashemi, M. Feature Selection and Spatial-Temporal Forecast of Oceanic Nino Index Using Deep Learning. *Int. J. Softw. Eng. Knowl. Eng.* **2022**, *32*, 91–107.
11. David, C.; Cane Mark, A.; Naomi, H.; Eun, L.D.; Chen, C. A Vector Auto regressive ENSO Prediction Model. *J. Clim.* **2015**, *28*, 8511–8520.
12. Ren, H.L.; Zuo, J.; Deng, Y. Statistical predictability of Niño indices for two types of ENSO. *Clim. Dyn.* **2018**, *52*, 5361–5382. [CrossRef]
13. Hashemi, M. Forecasting El Niño and La Niña Using Spatially and Temporally Structured Predictors and A Convolutional Neural Network. *IEEE J. Sel. Top. Appl. Earth Obs. Remote Sens.* **2021**, *14*, 3438–3446. [CrossRef]
14. Ren, H.L.; Scaife, A.A.; Dunstone, N.; Tian, B.; Liu, Y.; Ineson, S.; Lee, J.Y.;, Smith, D.; Liu, C.; Thompson, V.; et al. Seasonal predictability of winter ENSO types in operational dynamical model predictions. *Clim. Dyn.* **2018**, *52*, 3869–3890. [CrossRef]
15. Larson, S.M.; Kirtman, B.P. Drivers of coupled model ENSO error dynamics and the spring predictability barrier. *Clim. Dyn.* **2017**, *48*, 3631–3644. [CrossRef]
16. Tangang, F.T.; Hsieh, W.W.; Benyang, T. Forecasting regional sea surface temperatures in the tropical Pacific by neural network models, with wind stress and sea level pressure as predictors. *J. Geophys. Res. Ocean.* **1998**, *103*, 7511–7522. [CrossRef]
17. Baawain, M.S.; Nour, M.H.; El-Din, A.G.; El-Din, M.G. El Niño southern-oscillation prediction using southern oscillation index and Niño3 as onset indicators: Application of artificial neural networks. *J. Environ. Eng. Sci.* **2005**, *4*, 113–121. [CrossRef]
18. Ham, Y.-G.; Kim, J.-H.; Luo J.-J. Deep learning for multi-year ENSO forecasts. *Nature* **2019**, *573*, 568–572. [CrossRef]
19. Aguilar-Martinez, S.; Hsieh, W.W. Forecasts of Tropical Pacific Sea Surface Temperatures by Neural Networks and Support Vector Regression. *Int. J. Oceanogr.* **2009**, *2009*, 167239. [CrossRef]
20. Nooteboom, P.D.; Feng, Q.Y.; López Cristóbal, H.-G.E.; Dijkstra Henk, A. Using Network Theory and Machine Learning to predict El Nino. *Phys.-Atmos. Ocean. Phys.* **2018**, *9*, 969–983. [CrossRef]
21. Guo, Y.; Cao, X.; Liu, B.; Peng, K. El Niño Index Prediction Using Deep Learning with Ensemble Empirical Mode Decomposition. *Symmetry* **2020**, *12*, 893. [CrossRef]
22. Petersik, P.J.; Dijkstra, H.A. Probabilistic Forecasting of El Niño Using Neural Network Models. *Geophys. Res. Lett.* **2020**, *47*, e2019GL086423. [CrossRef]
23. Hong, M.; Chen, X.; Zhang, R.; Wang, D.; Shen, S.; Singh, V.P. Forecasting experiments of a dynamical–statistical model of the sea surface temperature anomaly field based on the improved self-memorization principle. *Ocean. Sci.* **2018**, *14*, 301–320. [CrossRef]
24. Zhang, H.; Chu, P.S.; He, L.; Unger, D. Improving the CPC's ENSO Forecasts using Bayesian model averaging. *Clim. Dyn.* **2019**, *53*, 3373–3385. [CrossRef]
25. Ha, S.; Liu, D.; Mu, L. Prediction of Yangtze River stream flow based on deep learning neural network with El Niño-Southern Oscillation. *Sci. Rep.* **2021**, *11*, 11738. [CrossRef]
26. Hashemi, M.; Karimi, H. Weighted machine learning. *Stat. Optim. Inf. Comput.* **2018**, *6*, 497–525. [CrossRef]
27. Hashemi, M.; Karimi, H.A. Weighted machine learning for spatial-temporal data. *IEEE J. Sel. Top. Appl. Earth Obs. Remote Sens.* **2020**, *13*, 3066–3082. [CrossRef]
28. Hashemi, M.; Alesheikh, A.A.; Zolfaghari, M.R. A spatio-temporal model for probabilistic seismic hazard zonation of Tehran. *Comput. Geosci.* **2013**, *58*, 8–18. [CrossRef]
29. Hashemi, M.; Alesheikh, A.A.; Zolfaghari, M.R. A GIS-based time-dependent seismic source modeling of Northern Iran. *Earthq. Eng. Eng. Vib.* **2017**, *16*, 33–45. [CrossRef]
30. Saha, S.; Moorthi, S.; Wu, X.; Wang, J.; Nadiga, S.; Tripp, P.; Behringer, D.; Hou, Y.T.; Chuang, H.Y.; Iredell, M.; et al. The NCEP Climate Forecast System version 2. *J. Clim.* **2014**, *27*, 2185–2208. [CrossRef]
31. Vimont, D.J.; Alexander, M.A.; Newman, M. Optimal growth of central and east Pacific ENSO events. *Geophys. Res. Lett.* **2014**, *41*, 4027–4034. [CrossRef]
32. Hochreiter Sepp and Schmidhuber Jürgen. Long short-term memory. *Neural Comput.* **1997**, *9*, 1735–1780. [CrossRef] [PubMed]

33. Srivastava, N.,; Hinton, G.; Krizhevsky, A.; Sutskever, I.; Salakhutdinov, R. Dropout: A simple way to prevent neural networks from overfitting. *J. Mach. Learn. Res.* **2014**, *15*, 1929–1958.
34. Lutz, P. Early Stopping—However, When? In *Neural Networks: Tricks of the Trade*, 2nd ed.; Springer: Berlin/Heidelberg, Germany, 2012; pp. 53–67.

engineering proceedings

Proceeding Paper

Quality-Aware Conditional Generative Adversarial Networks for Precipitation Nowcasting †

Jahnavi Jonnalagadda * and Mahdi Hashemi

Department of Information Sciences and Technology, George Mason University, Fairfax, VA 22030, USA; mhashem2@gmu.edu
* Correspondence: jjonnala@gmu.edu

† Presented at the 9th International Conference on Time Series and Forecasting, Gran Canaria, Spain, 12–14 July 2023.

Abstract: Accurate precipitation forecasting is essential for emergency management, aviation, and marine agencies to prepare for potential weather impacts. However, traditional radar echo extrapolation has limitations in capturing sudden weather changes caused by convective systems. Deep learning models, an alternative to radar echo extrapolation, have shown promise in precipitation nowcasting. However, the quality of the forecasted radar images deteriorates as the forecast lead time increases due to mean absolute error (MAE, a.k.a L1) or mean squared error (MSE, a.k.a L2), which do not consider the perceptual quality of the image, such as the sharpness of the edges, texture, and contrast. To improve the quality of the forecasted radar images, we propose using the Structural Similarity (SSIM) metric as a regularization term for the Conditional Generative Adversarial Network (CGAN) objective function. Our experiments on satellite images over the region 83° W–76.5° W and 33° S–40° S in 2020 show that the CGAN model trained with both L1 and SSIM regularization outperforms CGAN models trained with only L1, L2, or SSIM regularizations alone. Moreover, the forecast accuracy of CGAN is compared with other state-of-the-art models, such as U-Net and Persistence. Persistence assumes that rainfall remains constant for the next few hours, resulting in higher forecast accuracies for shorter lead times (i.e., <2 h) measured by the critical success index (CSI), probability of detection (POD), and Heidtke skill score (HSS). In contrast, CGAN trained with L1 and SSIM regularization achieves higher CSI, POD, and HSS for lead times greater than 2 h and higher SSIM for all lead times.

Keywords: weather prediction; precipitation prediction; deep learning; rainfall forecasting

Citation: Jonnalagadda, J.; Hashemi, M. Quality-Aware Conditional Generative Adversarial Networks for Precipitation Nowcasting. *Eng. Proc.* **2023**, *39*, 11. https://doi.org/10.3390/engproc2023039011

Academic Editors: Ignacio Rojas, Hector Pomares, Luis Javier Herrera, Fernando Rojas and Olga Valenzuela

Published: 28 June 2023

1. Introduction

In recent years, rapid climate change and global warming have led to catastrophic weather events, such as heavy rainfall and flash floods in various parts of the world [1]. Heavy rains, followed by strong winds, can cause significant economic and social losses. "Precipitation nowcasting" refers to forecasting rainfall intensity in a region over a relatively short period (between 0 and 4 h). Precipitation nowcasting provides multiple sectors, including emergency management services, flood warning systems, and flight and marine operations, with warnings before heavy rainfalls. Additionally, short-term weather forecasts are critical for outdoor activities, such as construction, roadworks, sports, and community gatherings.

Deep Learning (DL) has revolutionized the field of computer vision due to its ability to capture semantic information from images and videos [2]. Despite the promising performance of current state-of-the-art models in predicting future radar echoes, the quality of the predictions generated by these models is poor, often referred to as blurry predictions in the literature [3,4]. For instance, using Mean Squared Error (MSE, a.k.a L2) or Mean Absolute Error (MAE, a.k.a L1) losses in image-to-image prediction problems produces low-quality images due to uniform distribution assumption, which is not valid for multi-modal data distribution. In addition, both L1 and L2 losses force the model to generate a mean prediction of all possible outcomes, thereby losing sharp details such as edges and color transitions between pixels.

The main contribution of this study is to improve the quality of forecasted images by introducing an image quality metric, Structural Similarity (SSIM), as a regularization term for the objective function of the Conditional Generative Adversarial Network (CGAN). The SSIM is a visual image quality metric that measures structural changes such as local structure, luminosity, and contrast between the ground truth and generated images. This study conducts several experiments to demonstrate the superiority of the SSIM regularization over other pixel-wise regularization approaches, such as L1 and L2, measured in terms of the critical success index (CSI), false alarm ratio (FAR), Heidtke Skill Score (HSS), probability of detection (POD), and SSIM.

The rest of the paper is organized as follows. Section 2 reviews the relevant literature on precipitation nowcasting. Section 3 describes the data and preprocessing, followed by the proposed methodology in Section 4. Finally, Section 5 discusses the experimental results, and Section 6 concludes the paper with a summary of findings and future directions.

2. Literature Review

Both theoretical [5,6] and practical [7–11] application of ML in forecasting and predicting spatial–temporal phenomena has been underscored in the literature. Random forest and decision tree performed better in rainfall forecasting for shorter lead times [12,13]. Prudden et al. [14] comprehensively reviewed existing methods in radar-based nowcasting. This study sheds light on the limitations of optical flow-based and persistence-based methods and the scope of ML methods in precipitation nowcasting.

DL models are widely used in precipitation nowcasting. U-Net is a convolutional neural network (CNN) commonly used for precipitation nowcasting [15]. A Small Attention U-Net (SmaAt-UNet) was developed by adding an attention module and depth-wise convolutions to the original U-Net architecture [16]. Although SmaAt-UNet performs the same as U-Net, the number of trainable parameters is reduced to one-quarter. RainNet, another member of the U-Net family, is inspired by SegNet and used in radar-based nowcasting [17]. The attention mechanism in deep learning has shown promising results in computer vision. Yan et al. [18] proposed multi-head attention in a dual-channel network to highlight the critical areas of precipitation. Results indicate that the addition of multi-head attention and residual connections to the CNN can precisely extract the local and global spatial features of the radar image.

Some studies have used ConvLSTM for precipitation nowcasting [19–21]. Shi et al. proposed two DL models, Trajectory Gated Recurrent Unit (TrajGRU) [22] and ConvLSTM [23], for precipitation nowcasting. TrajGRU overcomes the location invariance problem in ConvLSTM using fewer training parameters. Most recent studies have applied generative adversarial networks (GANs) [24] for weather forecasting [25]. For instance, Choi and Kim [26] trained Radar CGAN on radar images with a spatial resolution of 128 × 128 km and a temporal resolution of 10 min. They showed that Rad-CGAN outperforms U-Net and ConvLSTM in terms of CSI. Few studies use satellite images for precipitation nowcasting. Hayatbini et al. [27] used satellite images and applied CGANs to generate forecasts over the contiguous United States for up to four hours of lead times. Their model outperformed Precipitation Estimation from Remotely Sensed Information using Artificial Neural Networks for Cloud Classification System (PERSIAN-CSS) in terms of CSI, POD, FAR, MSE, bias, and correlation coefficient. Despite the use of advanced DL models, such as U-Net, and GANs in precipitation nowcasting, the predicted images often turn out to be blurry due to the L1 or L2 losses [3,4].

3. Data Description

The dataset used in this study was collected from the National Aeronautics and Space Administration's (NASA's) Integrated Multi-Satellite Retrievals for Global Precipitation Measurement (IMERG) algorithm [28], which provides precipitation estimates with a spatial resolution of 0.1° × 0.1° (10 km × 10 km) all over the globe. We used half-hourly final IMERG data from 1 January 2020–31 December 2020, positioned over the east coast of the United States (−83° W–76.5° W, 33° S–40° S) with a spatial resolution of 0.1° × 0.1° (10 km × 10 km) and precipitationCal as the variable of interest.

Our dataset is multidimensional, with latitude, longitude, and timestamp as dimensions. Our program reads through the dataset for each timestamp and constructs a two-dimensional matrix (also called a rain map) based on latitude and longitude. A rain map represents the snapshot of the precipitation collected over the study region. The rain maps have a dimension of 64 × 64, with each pixel representing average rainfall in the last 30 min over 10 km × 10 km. Most rain maps contain pixels with low or no rain, so we created two balanced datasets, Dataset-A and Dataset-B, by ensuring that each rain map has a minimum number of rainy pixels, following a similar approach to that shown in [24]. Dataset-A contains at least 50% of pixels with rain, while Dataset-B contains at least 20%. Both datasets have gaps in the timestamps because the rain maps that do not qualify for the threshold were dismissed. We handle the gaps in both datasets by creating an empty set and adding samples to it sequentially until the first gap is identified. We follow the same process through the end of the dataset, generating several small subsets. Finally, we treat each subset as a separate dataset and create input and output sequences. The input to the model is a sequence of four rain maps (past 2 h) stacked along the channel dimension, and the output from the model is a single rain map at a lead time of 30 min, 1 h, 2 h, or 4 h. Therefore, the input and output dimensions are (64 × 64 × 4) and (64 × 64 × 1), respectively.

Dataset-A and Dataset-B are balanced datasets containing 944 and 3736 samples, respectively. Although they have fewer samples than the original dataset (17,548), they are suitable for this study to predict rainfall. Therefore, we evaluate our proposed model on these two datasets. In the results section of this paper, we provide a detailed comparison of CGAN's performance on both datasets.

4. Methodology

4.1. Problem Statement

The radar echo extrapolation task can be viewed as a sequence-to-sequence prediction problem as it considers the historical radar echo maps to forecast future radar echo maps. The radar echo map/rain map at any time t is a tensor χ of shape $A \times B$, where A and B represent a rain map's height (rows) and width (columns). In this study, $A = B = 64$. The sequence of tensors from $t = 1$ to T is obtained by collecting rain maps at fixed time intervals over T. Therefore, the spatial–temporal prediction problem can be obtained by finding a function F to forecast a rain map at l time steps ahead given the sequence of j rain maps, as shown below:

$$\chi_{t+1} = F_\theta(\chi_{t-j+1} \cdots \chi_t) \tag{1}$$

where θ denotes model parameters. In this study, the temporal interval between rain maps is 30 min. The size of the rain map is 64 × 64, the value of j is empirically chosen to be 4, and l is varied up to 8. For example, when $j = 4$ and $l = 8$, the past two hours of rain maps are used to forecast a rain map after four hours.

4.2. Conditional Generative Adversarial Network (CGAN)

This study applies CGANs for precipitation nowcasting. GANs are generative models that learn a mapping from a random noise vector z to output image y, G: $z \rightarrow y$ [29]. In contrast, CGANs learn a conditional generative model, i.e., instead, CGANs learn a mapping from both an auxiliary information x and noise random vector z to output image y, G: $\{x, z\} \rightarrow y$ [30]. The following equation gives the objective function of CGAN:

$$G^* = arg \min_G \max_D L_{CGAN} \tag{2}$$

where the generator (G) tries to minimize the objective against its opponent discriminator (D), whereas D tries to maximizes the objective against its adversary G, and the loss function of CGAN denoted as L_{CGAN} is given by

$$L_{CGAN} = \mathbb{E}_y[\log D(y)] + \mathbb{E}_{x,z}[\log(1 - D(x, G(x,z))]. \tag{3}$$

The use of pixel-wise loss functions such as L1 or L2 in the objective function of CGAN is common in precipitation nowcasting literature [25–27]. This allows the generator to produce images with pixel values close to the ground truth. However, a common limitation of L1 or L2 loss functions is blurriness in the forecasts. This is because L1 or L2 loss functions assume that noise is independent of the local image characteristics, whereas the sensitivity of the Human Visual System (HVS) depends on the structural changes such as local contrast, luminance, and structure. Other studies [4] support this argument, suggesting that L1 or L2 loss functions assume global similarity but do not capture local structures or intricate characteristics of the HVS, such as edges and color transitions between pixels.

4.3. CGAN with SSIM Regularization

To improve the quality of forecasted images, this study introduces the SSIM loss as a regularization to the objective function of CGAN, which is given by the below equation.

$$G^* = arg \min_G \max_D E_y[\log D(y)] + E_{x,z}[\log(1 - D(x, G(x,z)))] + \lambda[1 - \frac{1}{M}\sum_{k=1}^{M} SSIM(y_k, y_k')]. \tag{4}$$

where $SSIM(y_k, y_k')$ is the SSIM of the kth kernel window of the output image (y) and generated image (y'), respectively, and M is the number of kernels. The penalty factor λ is set empirically. When λ is set to zero, CGAN is trained using only the adversarial loss. If λ is set to one, both the adversarial loss and SSIM loss are used in equal proportions during training. However, if λ is set to a much larger value, the training of CGAN is primarily optimized to minimize errors caused by the loss of structural information.

The Structural Similarity (SSIM) [31] is a perceptual metric that measures the quality of an image. Unlike other metrics, SSIM requires both a reference image and a processed image to calculate the similarity between the two. We use the following formula to compute the SSIM value for two equal sized windows i and j of reference and processed image. For simplicity, we use the notations i and j instead of y_k and y_k'.

$$SSIM(i,j) = \frac{(2\mu_i\mu_j + C_1)(2\sigma_{ij} + C_2)}{(\mu_i^2 + \mu_j^2 + C_1)(\sigma_i^2 + \sigma_j^2 + C_2)} \tag{5}$$

where the symbols μ_i, μ_j, σ_i, and σ_j represent the mean and standard deviation of pixel values within windows i and j, respectively. The symbol σ_{ij} represents the covariance between the pixel values of windows i and j. The constants C_1 and C_2 are given by $(k_1L)^2$ and $(k_2L)^2$, respectively, where k_1 and k_2 are default values of 0.01 and 0.03, and L is the dynamic range of pixels, which is $2^{Numberofbits} - 1$.

Studies show that training deep learning models with a combined SSIM loss function and pixel-wise losses such as L1 or L2 loss can improve the quality of forecasts for image-to-image translation problems [32,33]. This is because L1 or L2 loss functions preserve colors and luminance (local structure), whereas the SSIM loss function preserves contrast. Therefore, we analyze the performance of CGAN by training the generator using different combinations of SSIM and L1 or L2 loss functions. Excluding random noise (z) from Equation (4), the generator can still learn a mapping from x to y, but it is highly sensitive to randomness because slight changes in x can significantly vary y. Our experimental results found that the noise vector z has no impact on generator training because the same results were produced with and without the noise vector. Therefore, we ignore the noise vector z from the generator and instead use dropout as noise in the layers of the generator during training and testing [30]. In pix2pix CGAN, U-Net is used for the generator and Patch-GAN is used for the discriminator. It is worth noting that we utilized the pix2pix version of CGAN [30], but with a slightly modified U-Net generator designed to fit the size of our input image (64 × 64). Adam is the optimizer used with a learning rate of 0.0002, and momentum parameters are 0.5 and 0.99. We determined the optimal hyperparameter values for epochs, batch size, and penalty factor (λ) through hyperparameter tuning on the validation set. The resulting values are as follows: 100 epochs, a batch size of 16, and a penalty factor of 100. The patch size of the discriminator (a.k.a discriminator receptive field) is set to 1 × 1 (PixelGAN).

4.4. Baselines (U-Net and Persistence)

To compare the performance of CGAN, we utilized two state-of-the-art baselines: U-Net and Persistence. The architecture used for U-Net is identical to that of the CGAN generator. We optimized U-Net for 100 epochs using Adam with a learning rate of 0.001 and implemented an early stopping algorithm that halts the training process if the validation loss fails to improve in the last ten epochs. We performed several experiments to train U-Net using all combinations of loss functions for a fair comparison. However, due to space constraints, we only reported the best values in Figure 1. Another commonly used baseline in nowcasting is Persistence. It assumes that rainfall after a few hours is the same as the present rainfall. Unfortunately, this baseline is challenging to surpass because it aligns with the fact that weather does not change much in the span of few hours. In other words, there is a possibility that rainfall remains the same during the observed and forecasted periods.

4.5. Evaluation Metrics

We evaluate all models in this study using several metrics, including CSI, FAR, POD, HSS, and SSIM. To accomplish this, we begin by rescaling the pixel values of both the predicted and ground truth images. Then, we convert these values to binary values using a rainfall rate of 0.5 mm/h as a threshold, as outlined in previous research [19,23]. The SSIM equation is already shown in Equation (5), while the mathematical formulas for the other metrics are given in [21].

5. Results and Discussions

Table 1 shows the test accuracy of CGAN on Dataset-A for various regularizations. The values highlighted in the bold indicate the best values for that lead time. It is noticeable that the accuracy of CGAN decreases as the lead time increases, regardless of the regularization used. However, CGAN trained using L1 and SSIM regularization outperformed other regularizations in all metrics except POD. In other words, CGAN trained with L1 and SSIM regularization has better CSI, FAR, HSS, and SSIM for all lead times than CGAN trained with other regularizations. The CGAN model trained with the L2 regularization has higher POD than models trained with other regularizations. This is unsurprising, as the L2 regularization minimizes the mean square of pixel differences between observed and forecasted images, resulting in pixel values closer to the average and fewer false negatives. However, the FAR of the L2 regularization is higher than other regularizations as it predicts the majority of non-rainy pixels as rainy pixels. Moreover, the CSI and HSS of the L2 regularization are lower than other regularizations due to more false positives. Lastly, the SSIM of the CGAN model trained with the L2 regularization is less than models trained with the SSIM regularization because L2 estimates rainfall intensity globally, while SSIM estimates it locally. The model trained with L1 and SSIM regularization has less FAR than other regularizations for all lead times. Additionally, the HSS of the CGAN model trained with L1 and SSIM regularization for a lead time of four hours is 0.25, which means the CGAN model is 25% better than the random forecast.

Table 1. Test accuracy of CGAN on Dataset-A for different regularizations (the values in the bold indicate the best values for that lead time).

Metric	Regularization	Lead Time in Hours			
		0.5	1	2	4
CSI	L1	0.481	0.574	0.546	0.442
	L2	0.562	0.575	0.547	0.481
	SSIM	0.650	0.613	0.556	0.497
	L2 + SSIM	0.677	0.624	0.559	0.495
	L1 + L2 + SSIM	0.711	0.637	0.557	0.496
	L1 + SSIM	**0.713**	**0.642**	**0.598**	**0.505**
POD	L1	0.811	0.775	0.724	0.634
	L2	**0.932**	**0.886**	**0.855**	**0.851**
	SSIM	0.917	0.880	0.839	0.837
	L2 + SSIM	0.900	0.875	0.836	0.809
	L1 + L2 + SSIM	0.822	0.785	0.738	0.652
	L1 + SSIM	0.829	0.786	0.749	0.660
FAR	L1	0.186	0.238	0.308	0.405
	L2	0.411	0.371	0.406	0.455
	SSIM	0.303	0.332	0.369	0.446
	L2 + SSIM	0.265	0.312	0.368	0.442
	L1 + L2 + SSIM	0.163	0.229	0.305	0.406
	L1 + SSIM	**0.160**	**0.221**	**0.300**	**0.404**
HSS	L1	0.455	0.578	0.443	0.239
	L2	0.397	0.438	0.337	0.186
	SSIM	0.454	0.510	0.399	0.212
	L2 + SSIM	0.452	0.537	0.402	0.219
	L1 + L2 + SSIM	0.702	0.596	0.455	0.249
	L1 + SSIM	**0.704**	**0.605**	**0.463**	**0.251**
SSIM	L1	0.848	0.831	0.808	0.775
	L2	0.835	0.810	0.780	0.748
	SSIM	**0.880**	**0.849**	0.825	0.801
	L2 + SSIM	**0.880**	0.847	**0.826**	0.803
	L1 + L2 + SSIM	**0.880**	**0.849**	**0.826**	0.804
	L1 + SSIM	**0.880**	**0.849**	**0.826**	**0.806**

Table 2 shows the test accuracy of CGAN on Dataset-B for various regularizations. The values highlighted in the bold indicate the best values for that lead time. For Dataset-B, we observe similar results as Dataset-A. That is, the forecast accuracy of CGAN decreases as lead time increases, regardless of the choice of the regularization. In other words, as the lead time increases, CSI, POD, SSIM, and HSS gradually decrease while FAR rises. Furthermore, the forecast accuracy of CGAN improves when trained with combined regularization. As a result, for all lead times, models trained with combined regularization have a lower FAR than models trained with a single regularization. The CGAN model's CSI, POD, and FAR on Dataset-B are less than those on Dataset-A for all lead times because Dataset-B has fewer rainy pixels. However, the SSIM values for Dataset-B are higher and vary slightly across lead times because 80% of pixels contain no rain. The HSS scores for both datasets are greater than 0.5 for lead times up to 2 h, suggesting that our proposed model is 50% better than the random model. Overall, the CGAN model trained with L1 and SSIM regularization performs well for four out of five metrics for both datasets. These results demonstrate that combining adversarial loss with L1 and SSIM regularization can produce quality predictions similar to previous studies [32].

Table 2. Test accuracy of CGAN on Dataset-B for different regularizations (the values in the bold indicate the best values for that lead time).

Metric	Regularization	Lead Time in Hours			
		0.5	1	2	4
CSI	L1	0.607	0.522	0.409	0.310
	L2	0.484	0.453	0.382	0.281
	SSIM	0.587	0.512	0.409	0.306
	L2 + SSIM	0.559	0.512	0.413	0.292
	L1 + L2 + SSIM	0.623	0.525	0.415	0.311
	L1 + SSIM	**0.624**	**0.564**	**0.428**	**0.335**
POD	L1	0.753	0.668	0.566	0.421
	L2	0.858	**0.852**	**0.822**	**0.708**
	SSIM	0.857	0.785	0.713	0.632
	L2 + SSIM	**0.863**	0.764	0.681	0.601
	L1 + L2 + SSIM	0.766	0.702	0.598	0.452
	L1 + SSIM	0.750	0.696	0.571	0.451
FAR	L1	0.234	0.297	0.406	0.549
	L2	0.470	0.531	0.542	0.602
	SSIM	0.347	0.422	0.504	0.592
	L2 + SSIM	0.377	0.425	0.505	0.589
	L1 + L2 + SSIM	0.235	0.305	0.422	0.565
	L1 + SSIM	**0.202**	**0.235**	**0.354**	**0.403**
HSS	L1	0.676	0.581	0.429	0.218
	L2	0.498	0.418	0.333	0.182
	SSIM	0.593	0.521	0.385	0.206
	L2 + SSIM	0.602	0.548	0.402	**0.221**
	L1 + L2 + SSIM	0.691	0.582	0.433	0.211
	L1 + SSIM	**0.696**	**0.618**	**0.442**	0.212
SSIM	L1	0.933	0.912	0.914	0.905
	L2	0.928	0.924	0.913	0.905
	SSIM	0.942	0.922	**0.918**	0.902
	L2 + SSIM	0.946	0.921	0.916	0.903
	L1 + L2 + SSIM	0.945	0.921	**0.918**	0.902
	L1 + SSIM	**0.948**	**0.926**	**0.918**	**0.906**

Figure 1 depicts the input and forecasted images generated using different regularization terms on Dataset-A, specifically for a four-hour lead time. Both L1 and L2 losses accurately identified the rain location but lost intricate details such as sharp edges, especially for pixels with low rainfall. On the other hand, SSIM loss preserved color and brightness but fell short in predicting intensity in some pixels if used alone instead of combining L1 or L2 regularization. Finally, the combination of L1 and SSIM regularization looks close to the target, whereas adding L2 to the SSIM regularization generated smooth edges.

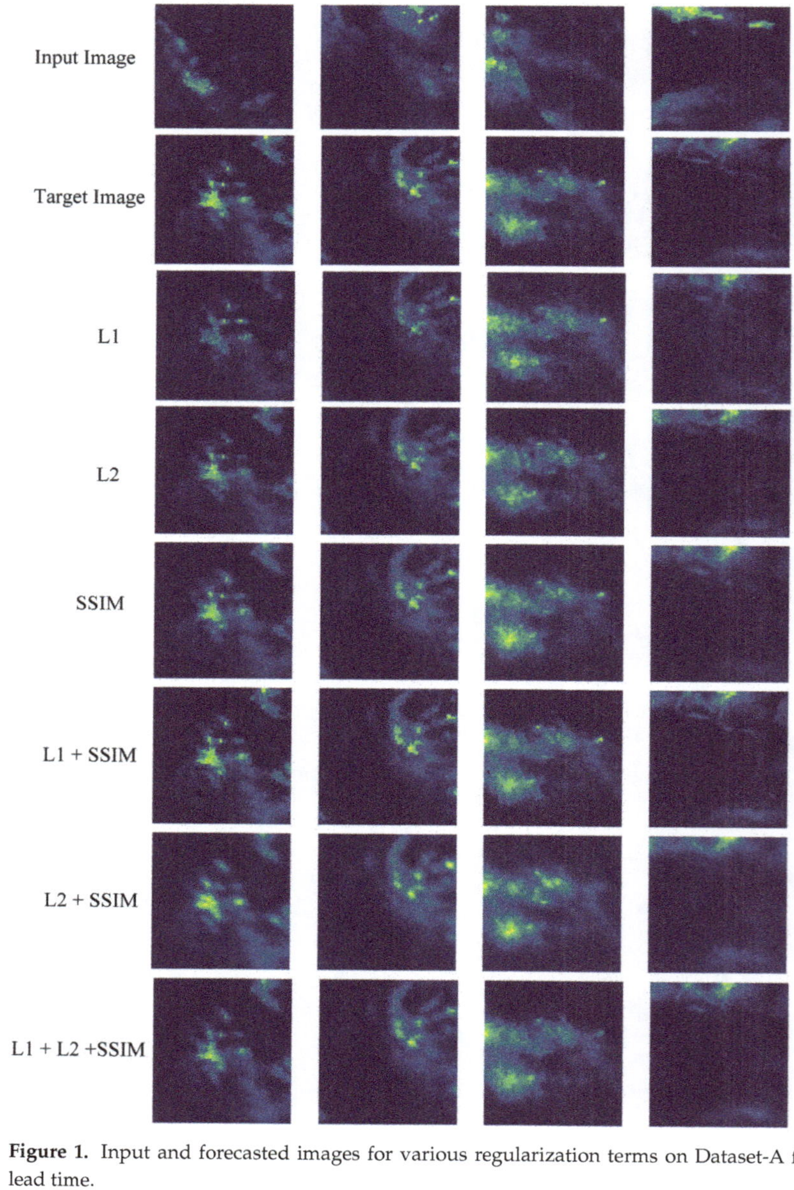

Figure 1. Input and forecasted images for various regularization terms on Dataset-A for 4 h of lead time.

We compare the forecast accuracy of CGAN with the baselines U-Net and Persistence. As mentioned earlier, U-Net is also evaluated using different regularizations, but only the best values are reported in Figure 2 for space constraints. Figure 2a provides the test accuracy of models on Dataset-A. Notably, the reported results for CGAN are based on the L1 + SSIM regularization. Persistence shows a high POD, U-Net has a low FAR, and CGAN has a high SSIM on average. Persistence performs slightly better than CGAN in CSI, POD, and HSS for shorter lead times (up to one hour) because precipitation does not vary significantly in one hour, making it challenging to outperform Persistence. However, our CGAN model outperforms U-Net and Persistence in three out of five metrics for lead times greater than one hour. Additionally, CGAN and Persistence beat U-Net for all lead times in all metrics except FAR. Overall, CGAN scored 8, Persistence scored 7, and U-Net scored 5 out of 20 (five metrics × four lead times), demonstrating the capability of CGAN and the SSIM regularization.

Figure 2b compares the test accuracy of models on Dataset-B. These results follow the same trend as Figure 2a. Overall, Persistence has a high POD, and CGAN has a low FAR and high SSIM. Furthermore, CGAN and Persistence outperform U-Net for almost all lead times and across all metrics. CGAN scored 12, and Persistence scored eight out of 20 (five metrics × four lead times), highlighting the capability of CGAN and SSIM regularization for precipitation nowcasting. An important observation from Figure 2a,b is that the POD (recall) for Dataset-B is lower than for Dataset-A at all lead times because of the fewer rainy pixels in Dataset-B than Dataset-A. This biases the CGAN model to forecast the majority of the pixels as no rain, indicating that CGAN can achieve higher accuracy in forecasting precipitation for longer lead times if more samples from the positive class (rainy pixels) are available.

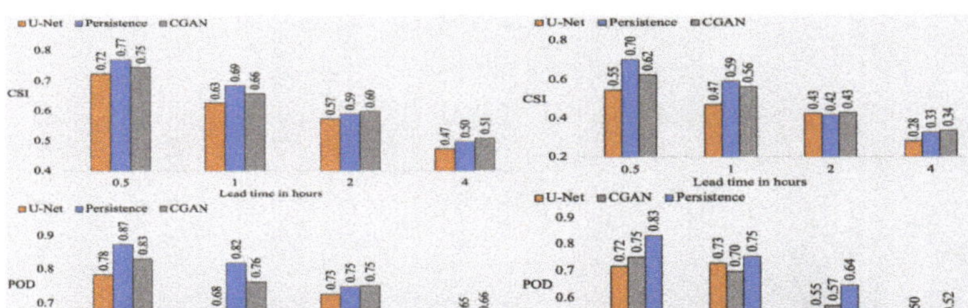

(a) (b)

Figure 2. Test accuracy of models trained with L1 and SSIM regularization for (**a**) Dataset-A, (**b**) Dataset-B measured using CSI, POD, FAR, HSS, and SSIM.

6. Conclusions and Future Directions

This study demonstrates that using SSIM and mean absolute error (MAE, also known as L1) regularization in CGAN's objective function can result in higher-quality predicted images, as measured by CSI, POD, FAR, HSS, and SSIM. Although the CSI, POD, and HSS for Dataset-B are lower than those for Dataset-A due to the smaller number of rainy pixels in Dataset-B, the proposed model can achieve higher forecast accuracy for larger datasets with more positive samples.

To evaluate the effectiveness of CGAN, we compared it with other baselines such as U-Net and Persistence. Persistence assumes that rainfall remains constant after a few hours making it a difficult baseline to outperform in weather prediction, particularly for shorter lead times. Therefore, for up to one hour lead times, the CSI, POD, and HSS of Persistence are greater than those of CGAN and U-Net. However, for two hours and longer lead times, CGAN outperformed Persistence in terms of CSI, POD, and SSIM, demonstrating the superiority of CGAN and SSIM regularization in rainfall forecasting for longer lead times

Author Contributions: Conceptualization, J.J. and M.H.; methodology, J.J.; formal analysis, J.J and M.H.; writing—original draft preparation, J.J.; writing—review and editing, J.J. and M.H.; visualization, J.J. and M.H.; supervision, M.H.; project administration, J.J. All authors have read and agreed to the published version of the manuscript.

Funding: This research received no external funding.

Institutional Review Board Statement: Not applicable.

Informed Consent Statement: Not applicable.

Data Availability Statement: The data presented in this study are openly available in Github at https://github.com/jahnavijo/Data-for-Precipitation-Nowcasting.git.

Conflicts of Interest: The authors declare no conflict of interest.

References

1. Guhatakurta, P.; Sreejith, O.P.; Menon, P.A. Impact of climate change on extreme rainfall events and flood risk in India. *J. Earth Syst. Sci.* **2011**, *120*, 359. [CrossRef]
2. Srivastava, N.; Mansimov, E.; Salakhutdinowv, R. Unsupervised Learning of Video Representations using LSTMs. In Proceedings of the 32nd International Conference on Machine Learning, Lille, France, 6–11 July 2015; Volume 37, pp. 843–852.
3. Tran, Q.-K.; Song, S.-K. Computer Vision in Precipitation Nowcasting: Applying Image Quality Assessment Metrics for Training Deep Neural Networks. *Atmosphere* **2019**, *10*, 244. [CrossRef]
4. Klein, B.; Wolf, L.; Afek, Y. A Dynamic Convolutional Layer for Short Range Weather Prediction. In Proceedings of the IEEE Conference on Computer Vision and Pattern Recognition (CVPR), Boston, MA, USA, 7–12 June 2015.
5. Mahdi, H.; Hassan, K. Weighted machine learning. *Stat. Optim. Inf. Comput.* **2018**, *6*, 497–525.
6. Mahdi, H.; Hassan, K. Weighted machine learning for spatial-temporal data. *IEEE J. Sel. Top. Appl. Earth Obs. Remote Sens.* **2020**, *13*, 3066–3082.
7. Jonnalagadda, J.; Hashemi, M. Feature Selection and Spatial-Temporal Forecast of Oceanic Nino Index Using Deep Learning. *Int. J. Softw. Eng. Knowl. Eng.* **2022**, *32*, 91–107. [CrossRef]
8. Jonnalagadda, J.; Hashemi, M. Forecasting Atmospheric Visibility Using Auto Regressive Recurrent Neural Network. In Proceedings of the 2020 IEEE 21st International Conference on Information Reuse and Integration for Data Science (IRI), Las Vegas, NV, USA, 11–13 August 2020.
9. Jonnalagadda, J.; Jahnavi; Hashemi, M. Spatial-Temporal Forecast of the probability distribution of Oceanic Niño Index for various lead times. In Proceedings of the 33rd International Conference on Software Engineering and Knowledge Engineering, Online, 1–10 July 2021; p. 309.
10. Hashemi, M.; Alesheikh, A.A.; Zolfaghari, M.R. A GIS based time-dependent seismic source modeling of Northern Iran. *Earthq. Eng. Eng. Vib.* **2017**, *16*, 33–45. [CrossRef]
11. Hashemi, M.; Alesheikh, A.A.; Zolfaghari, M.R. A spatio-temporal model for probabilistic seismic hazard zonation of Tehran. *Comput. Geosci.* **2013**, *58*, 8–18. [CrossRef]
12. Balamurugan, M.S.; ManojKumar, R. Study of short term rain forecasting using machine learning based approach. *Wirel. Netw.* **2021**, *27*, 5429–5434. [CrossRef]
13. Sangiorio, M.; Barindelli, S.; Guglieri, V.; Biondi, R.; Solazzo, E.; Realini, E.; Venuti, G.; Guariso, G. A Comparative Study on Machine Learning Techniques for Intense Convective Rainfall Events Forecasting. In *Theory and Applications of Time Series Analysis: Selected Contributions from ITISE 2019*; Springer: Cham, Switzerland, 2020; Volume 6, pp. 305–317.

14. Pruden, R.; Adams, S.; Kangin, D.; Robinson, N.; Ravuri, S.; Mohammed, S.; Alberto, A. A review of radar-based nowcasting of precipitation and applicable machine learning techniques. *arXiv* **2020**, arXiv:2005.04988.
15. Agarwal, S.; Barrington, L.; Bromberg, C.; Burge, J.; Gazen, C.; Hickey, J. Machine Learning for Precipitation Nowcasting from Radar Images. In Proceedings of the 33rd Conference on Neural Information Processing Systems (NeurIPS 2019), Vancouver, BC, Canada, 8–14 December 2019.
16. Trebing, K.; Tomasz, S.; Siamak, M. SmaAt-UNet: Precipitation nowcasting using a small attention-UNet architecture. *Pattern Recognit. Lett.* **2021**, *145*, 178–186. [CrossRef]
17. Ayzel, G.; Scheffer, T.; Heistermann, M. RainNet v1.0: A con- volutional neural network for radar-based precipitation nowcasting. *Geosci. Dev.* **2020**, *13*, 2631–2644. [CrossRef]
18. Yan, Q.; Ji, F.; Miao, K.; Wu, Q.; Xia, Y.; Teng, L. Convolutional Residual-Attention: A Deep Learning Approach for Precipitation Nowcasting. *Adv. Meteorol.* **2020**, *2020*, 6484812. [CrossRef]
19. Kumar, A.; Islam, T.; Sekimoto, Y.; Mattmann, C. Convcast: An embedded convolutional LSTM based architecture for precipitation nowcasting using satellite data. *PLoS ONE* **2020**, *15*, e0230114. [CrossRef]
20. Lei, C.; Yuan, C.; Ma, L.; Junping, Z. A Deep Learning-Based Methodology for Precipitation Nowcasting With Radar. *Earth Space Sci.* **2020**, *7*, e2019EA000812.
21. Jeong, C.; Kim, W.; Joo, W.; Jang, D.; Yi, M. Enhancing the Encoding-Forecasting Model for Precipitation Nowcasting by Putting High Emphasis on the Latest Data of the Time Step. *Atmosphere* **2021**, *12*, 261. [CrossRef]
22. Shi, X.; Chen, Z.; Wang, H.; Yeung, D.; Wong, W.; Woo, W. Convolutional LSTM Network: A Machine Learning Approach for Precipitation Nowcasting. *Adv. Neural Inf. Process. Syst.* **2015**, *1*, 802–810.
23. Shi, X.; Gao, Z.; Lausen, L.; Wang, H.; Yeung, D.; Wong, W.; Woo, W. Deep Learning for Precipitation Nowcasting: A Benchmark and A New Model. In Proceedings of the 31st Conference on Neural Information Processing Systems, LongBeach, CA, USA, 4–9 December 2017.
24. Zhang, C.; Yang, X.; Tang, Y.; Zhang, W. Learning to Generate Radar Image Sequences Using Two-Stage Generative Adversarial Networks. *IEEE Geosci. Remote Sens. Lett.* **2020**, *17*, 401–405. [CrossRef]
25. Bihlo, A. A generative adversarial network approach to (ensemble) weather prediction. *Neural Netw.* **2021**, *139*, 1–16. [CrossRef]
26. Choi, S.; Kim, Y. Rad-cGAN v1.0: Radar-based precipitation nowcasting model with conditional Generative Adversarial Networks for multiple domains. *Geosci. Model Dev.* **2020**, *15*, 5967–5985. [CrossRef]
27. Hayatbini, N.; Kong, B.; Hsu, K.; Nguyen, P.; Sorooshian, S.; Stephens, G.; Fowlkes, C.; Nemani, R. Conditional generative adversarial networks (cGANs) for near real-time precipitation estimation from multispectral GOES-16 satellite imageries-PERSIANN-cGAN. *Remote Sens.* **2019**, *11*, 2193. [CrossRef]
28. Huffman, G.J.; Stocker, E.F.; Bolvin, D.T.; Nelkin, E.J.; Tan, J. GPM IMERG Final Precipitation L3 Half Hourly 0.1 degree × 0.1 degree V06. Goddard Earth Sciences Data and Information Services Center (GES DISC): Greenbelt, MD, USA. Available online: https://disc.gsfc.nasa.gov/datasets/GPM_3IMERGHH_06/summary (accessed on 10 January 2022).
29. Goodfellow, I.; Pouget-Abadie, J.; Mirza, M.; Xu, B.; Warde-Farley, D.; Ozair, S.; Courville, A.; Bengio, Y. Generative Adversarial Nets. *Commun. ACM* **2014**, *63*, 139–144. [CrossRef]
30. Isola, P.; Zhu, J.; Zhou, T.; Efros, A.A. Image-to-Image Translation with Conditional Adversarial Networks. In Proceedings of the 2017 IEEE Conference on Computer Vision and Pattern Recognition (CVPR), Honolulu, HI, USA, 21–26 July 2017; pp. 5967–5976. [CrossRef]
31. Wang, Z.; Bovik, A.C.; Sheikh, H.R.; Simoncelli, E.P. Image quality assessment: From error visibility to structural similarity. *IEEE Trans. Image Process.* **2004**, *13*, 600–612. [CrossRef] [PubMed]
32. Zhao, H.; Gallo, O.; Frosio, I.; Kautz, J. Loss functions for image restoration with neural networks. *IEEE Trans. Comput.* **2016**, *3*, 47–57. [CrossRef]
33. Kancharla, P.; Channappayya, S.S. Quality Aware Generative Adversarial Networks. In Proceedings of the 33rd Conference on Neural Information Processing Systems, Vancouver, BC, Canada, 8–14 December 2019.

engineering proceedings

MDPI

Proceeding Paper

Forecasting of Signals by Forecasting Linear Recurrence Relations †

Nina Golyandina * and Egor Shapoval

Faculty of Mathematics and Mechanics, St. Petersburg State University, 199034 St. Petersburg, Russia; st063753@student.spbu.ru
* Correspondence: n.golyandina@spbu.ru

† Presented at the 9th International Conference on Time Series and Forecasting, Gran Canaria, Spain, 12–14 July 2023.

Abstract: The forecasting of a signal that locally satisfies linear recurrence relations (LRRs) with slowly changing coefficients is considered. A method that estimates the local LRRs using the subspace-based method, predicts their coefficients and constructs a forecast using the LRR with the predicted coefficients is proposed. This method is implemented for time series that have the form of a noisy sum of sine waves with modulated frequencies. Linear and sinusoidal frequency modulations are considered. The application of the algorithm is demonstrated with numerical examples.

Keywords: time series; signal; forecasting; singular spectrum analysis; linear recurrence relation

1. Introduction

Let us consider the problem of forecasting time series using singular spectrum analysis (SSA) [1–6]. The theory of SSA is quite well-developed; there are many papers with applications of SSA to real-life time series (see, for example, [6] and [7] (Section 1.7) with short reviews). SSA does not require a time-series model to construct the decomposition into interpretable components such as the trend, periodicities, and noise. However, for prediction in SSA, it is assumed that the signal (the deterministic component of the time series) $S_N = (s_1, \ldots, s_N)$ satisfies some model, in particular, a linear recurrence relation (LRR) with constant coefficients (maybe, approximately):

$$s_{i+d} = \sum_{k=1}^{d} a_k s_{i+d-k} \quad \text{for } i = 1, \ldots, N-d. \tag{1}$$

This assumption is valid for signals in the form of a sum of products of polynomial, exponential, and sinusoidal time series, in particular, a sum of exponentially modulated periodic components. SSA also works for the case of trend extraction and the general case of amplitude-modulated harmonics, where the model is satisfied approximately. However, SSA is not applicable if the signal locally satisfies a changing LRR. An example of such a signal is sinusoidal frequency-modulated time series. This paper aims to construct a method for the prediction of time series locally governed by changing LRRs, staying within the framework of SSA.

Let us consider the model of time series in the form of a noisy signal, where the signal is locally governed by LRRs with slowly time-varying coefficients. A local version of SSA has already been considered earlier for signal estimation [8]. However, it results in different approximations of the segments of the time series and the prediction can be performed based on the last segment only. In this paper, a local modification of the recurrent SSA prediction, based on the construction of a prediction of the coefficients of the local LRRs, is proposed. This modification was applied to time series in which the signal is a sum of sinusoids with time-varying frequencies having non-intersecting frequency ranges, where the instantaneous frequency of each summand is slowly varying.

Citation: Golyandina, N.; Shapoval, E. Forecasting of Signals by Forecasting Linear Recurrence Relations. *Eng. Proc.* **2023**, *39*, 12. https://doi.org/10.3390/engproc2023039012

Academic Editors: Ignacio Rojas, Hector Pomares, Luis Javier Herrera, Fernando Rojas and Olga Valenzuela

Published: 28 June 2023

2. Basic Notions

2.1. Linear Recurrence Relations

Let us introduce several definitions. By time series of length N we mean a sequence of real numbers $\mathsf{X}_N = (x_1, \dots, x_N) \in \mathbb{R}^N$. Consider a time series in the form of the sum of a signal and noise $\mathsf{X}_N = \mathsf{S}_N + \mathsf{R}_N$ and state the problem of signal forecasting.

A time series S_N satisfies a linear recurrence relation (LRR) of order d if there exists a sequence $\{a_i\}_{i=1}^{d}$ such that $a_d \neq 0$ and (1) takes place.

A time series governed by an LRR satisfies a set of LRRs, one of which has minimal order; we will call it minimal LRR. Among the set of the governing LRRs, there is the so-called min-norm LRR with the minimum norm of coefficients, which suppresses noise in the best way [5] and is used for the recurrent SSA forecasting.

The characteristic polynomial of LRR (1) is defined as

$$P_d(\mu) = \mu^d - \sum_{k=1}^{d} a_k \mu^{d-k}.$$

The roots of the characteristic polynomial of the minimal LRR are called signal roots. The characteristic polynomial corresponding to an LRR governing the signal includes the signal roots, among others.

Remark 1. *Since* $P_m(\mu) = \prod_{k=1}^{m} (\mu - \mu_k) = \mu^m - \sum_{k=1}^{m} a_k \mu^{m-k}$*, the roots of the characteristic polynomial provides the LRR coefficients, and vice versa.*

The following result [9] together with Remark 1 shows how to find the roots of the governing minimal LRR $s_{i+r} = \sum_{k=1}^{r} a_k s_{i+r-k}$ using the common term of the time series S_N. Let $\mu_1, \dots, \mu_p$ be the roots of the LRR characteristic polynomial with multiplicities $k_1, \dots, k_p$. The time series S_N satisfies the LRR if and only if $s_n = \sum_{m=1}^{p} \left(\sum_{j=0}^{k_m - 1} c_{m\,j} n^j \right) \mu_m^n$, where $c_{m\,j} \in \mathbb{C}$ depends on $s_1, \dots, s_r$.

Example 1. *Consider the time series* $\mathsf{S}_N = (s_1, \dots, s_N)$ *with* $s_n = A\cos(2\pi\omega n + \phi)$, $\omega \in \left(0, \frac{1}{2}\right)$*. Since* $A\cos(2\pi\omega n + \phi) = Ae^{-\mathrm{i}\phi} e^{-\mathrm{i}\cdot 2\pi\omega n}/2 + Ae^{\mathrm{i}\phi} e^{\mathrm{i}\cdot 2\pi\omega n}/2$*, we have* $\mu_1 = e^{-\mathrm{i}\cdot 2\pi\omega}$, $\mu_2 = e^{\mathrm{i}\cdot 2\pi\omega}$*. Therefore, the characteristic polynomial is* $P_r(\mu) = (\mu - \mu_1)(\mu - \mu_2) = (\mu - e^{-\mathrm{i}\cdot 2\pi\omega n})(\mu - e^{\mathrm{i}\cdot 2\pi\omega n}) = \mu^2 - 2\mu\cos 2\pi\omega + 1$*. Thus,* $a_1 = 2\cos 2\pi\omega$, $a_2 = -1$ *and* $s_{i+2} = 2s_{i+1}\cos 2\pi\omega - s_i$.

Remark 2. *The method for constructing the min-norm LRR of a given order with the given signal roots is described in [10]. This method will be used in the algorithm proposed in Section 3.2.*

2.2. Harmonic Signal with Time-Varying Frequency: Instantaneous Frequency

In this paper, the basic form of signals will be the discrete-time version of

$$s(t) = \sum_{i=1}^{p} \cos(2\pi\omega_i(t) + \phi_i), \tag{2}$$

where $\omega_i(t)$, $i = 1, \dots, p$, are slowly changing functions. Note that if $\omega_i(t)$ are linear functions, the signal satisfies an LRR; see [4] (Section 2.2) and Remark 1.

Let $s(t) = \cos(2\pi\psi(t) + \phi)$. The instantaneous frequency of the signal $s(t)$ is defined as $\omega_{\mathrm{ins}}(t) = \psi'(t)$. The instantaneous period is the function $T_{\mathrm{ins}}(t) = \frac{1}{\omega_{\mathrm{ins}}(t)}$. If $\omega_{\mathrm{ins}}(a) = 0$ for some a, we put $T_{\mathrm{ins}}(a) = +\infty$. The frequency range of the signal S_N is the range $\omega_{\mathrm{ins}}([1, N])$, that is, the image of $[1, N]$.

Example 2. *For the signal* $s(t) = \cos\left(2\pi\left(\frac{t}{100}\right)^2\right)$, *the instantaneous frequency* $\omega_{\text{ins}}(t) = \frac{t}{5000}$ *is a linear function; the frequency range equals* $\omega_{\text{ins}}([1, N]) = \left[\frac{1}{5000}, \frac{N}{5000}\right]$. *For the signal* $s(t) = \cos\left(\frac{2\pi}{20}\left(t + 5\sin\frac{2\pi t}{100}\right)\right)$, *the instantaneous frequency equals* $\omega_{\text{ins}}(t) = \frac{1}{20}\left(1 + \frac{5\cdot 2\pi}{100}\cos\frac{2\pi t}{100}\right)$ *and is a periodic function with period equal to* $T = 100$; *the frequency range is*

$$\omega_{\text{ins}}([1, N]) \subset \left[\frac{1}{20}\left(1 - \frac{\pi}{10}\right), \frac{1}{20}\left(1 + \frac{\pi}{10}\right)\right].$$

We assume that on short segments of time, the signals considered in Example 2 are well-enough approximated by a sinusoid with a frequency equal to the instantaneous frequency in the middle of the segment. This assumption will be used for the construction of the signal-forecasting algorithm.

2.3. SSA, Signal Subspace and Recurrent SSA Forecasting

In the version for signal extraction, SSA has two parameters, the window length L, $1 < L < N$, where N is the time-series length, and the number of elementary components r. At the first step of SSA, the trajectory matrix of size $L \times K$, where $K = N - L + 1$, is constructed and then decomposed into elementary components using the SVD. The leading r SVD components are used for the estimation of the signal and the signal subspace basis, which is used for constructing the forecasting LRR. Here is the scheme of the method given in [6]:

$$\mathsf{X} \xrightarrow[\boxed{L}]{\mathcal{T}} \mathbf{X} = \begin{pmatrix} x_1 & x_2 & \cdots & x_K \\ x_2 & x_3 & \cdots & x_{K+1} \\ \vdots & \vdots & \ddots & \vdots \\ x_L & x_{L+1} & \cdots & x_N \end{pmatrix} \xrightarrow[\boxed{r}]{\text{SVD:}(\sqrt{\lambda_m}, U_m, V_m),\ \Pi_r}$$

$$\left\{\begin{array}{l} \mathcal{L}_r = \text{span}(U_1, \ldots, U_r) \\ \text{is the signal space;} \\ \Pi_r \text{ is the projector on } \mathcal{L}_r; \\ \widehat{\mathbf{S}} = \sum_{m=1}^r U_m (\mathbf{X}^{\mathrm{T}} U_m)^{\mathrm{T}} = \Pi_r \mathbf{X}. \end{array}\right. \quad \xrightarrow{\Pi_{\mathcal{H}}} \widetilde{\mathbf{S}} = \begin{pmatrix} \widetilde{s}_1 & \widetilde{s}_2 & \cdots & \widetilde{s}_K \\ \widetilde{s}_2 & \widetilde{s}_3 & \cdots & \widetilde{s}_{K+1} \\ \vdots & \vdots & \ddots & \vdots \\ \widetilde{s}_L & \widetilde{s}_{L+1} & \cdots & \widetilde{s}_N \end{pmatrix} \xrightarrow{\mathcal{T}^{-1}} \widetilde{\mathsf{S}}.$$

Thus, a concise form of the SSA algorithm for signal extraction is

$$\widetilde{\mathsf{S}} = \mathcal{T}^{-1} \circ \Pi_{\mathcal{H}} \circ \Pi_r \circ \mathcal{T}(\mathsf{X}),$$

where $\Pi_{\mathcal{H}}$ is the projector to the set of Hankel matrices, that is, the set of trajectory matrices.

The L-rank of a time series is the rank of its trajectory matrix $\mathcal{T}(\mathsf{X})$, or, equivalently, the dimension of the column space of $\mathcal{T}(\mathsf{X})$. For infinite time series and $L > r$, the rank is equal to the order of the minimal LRR governing the time series. For example, the rank of the signal $\mathsf{S}_\infty = (s_1, s_2, \ldots)$ with the common term $s_n = A\cos(2\pi\omega n + \phi)$ and $\omega \in \left(0, \frac{1}{2}\right)$ equals two.

The construction of the forecasting min-norm LRR of order $L - 1$ based on the SSA decomposition follows the formula for the LRR coefficients $\mathcal{R}$ [4] (Equation (2.1)):

$$\mathcal{R} = \frac{1}{1 - \nu^2} \sum_{i=1}^{r} \pi_i \underline{U_i} =: (a_{L-1}, \ldots, a_1), \tag{3}$$

where $\underline{U_i} \in \mathbb{R}^{L-1}$ is the vector U_i without the last coordinate, which is denoted by π_i. The forecast is constructed as $\widetilde{s}_{N+1} = \sum_{i=1}^{L-1} a_i \widetilde{s}_{N+1-i}$.

In addition to forecasting, the SSA decomposition allows one to estimate the signal roots. Let us describe the ESPRIT (see [7] (Algorithm 3.3) and [11]) for signal-root estimation. We define $\mathbf{U} := [U_1 : \ldots : U_r]$, $\overline{\mathbf{U}}$ the matrix $\mathbf{U}$ without the first row and $\underline{\mathbf{U}}$ the matrix $\mathbf{U}$ without the last row. The ESPRIT estimates of the signal roots $\{\mu_k\}_{k=1}^r$ are the eigenvalues of a matrix that is an approximate solution of the equation $\underline{\mathbf{U}}\mathbf{D} = \overline{\mathbf{U}}$; e.g., $\mathbf{D} = (\underline{\mathbf{U}}^{\mathrm{T}}\underline{\mathbf{U}})^{-1}\underline{\mathbf{U}}^{\mathrm{T}}\overline{\mathbf{U}}$ for the LS-ESPRIT version.

3. Signal Forecasting by Forecasting of Local LRRs

3.1. General Model of Signals

Let us describe a general model of time series, for which the developed approach to forecasting will be applied. Consider the signal S_N and take some natural Z, $1 < Z < N$. For a time series Y, we denote $\mathsf{Y}_{A,B}$ the series $(y_A, \ldots, y_B)$.

The signal model is

$$s_n = \sum_{j=1}^{p} \rho_j^n(n) \cos(2\pi\psi_j(n) + \varphi_j), \tag{4}$$

where we assume that

1. For the time series S, on its sequential segments $\mathsf{S}_{i,i+Z-1}$ of length Z, $i = 1, \ldots, N-Z+1$, every summand in (4) is well-approximated by a series in the form $s_n^{\text{approx}} = \rho_j^n(n_0)\cos(2\pi\omega_{j,0} \cdot n + \phi)$, where $\omega_{j,0} = \psi_j'(n_0)$ and $n_0 = n_0(i)$ is the middle point of the segment.
2. The series $\rho_j(n)$ and $\psi_j'(n)$ behave regularly in n, $0 < \psi_j'(n) < 0.5$ and there exist methods that can forecast such kinds of series.

To construct a forecast s_{N+1} one needs to find the instantaneous frequencies at the point $N+1$.

3.2. Algorithm LocLRR SSA Forecast

Hereinafter, we will consider the model $\mathsf{X}_N = \mathsf{S}_N + \mathsf{R}_N$, where $\mathsf{S}_N = (s_1, \ldots, s_N)$ satisfies the conditions described above and R_N is white Gaussian noise with zero mean and standard deviation σ. Let $\mathsf{Z}_{(i)} = \mathsf{X}_{i,i+Z-1}$, $i = 1, \ldots, W$, where $W = N - Z + 1$. The estimates of instantaneous roots, frequencies, moduli and LRR coefficients will be enumerated according to the middle of the local segment. In particular, denote $\{a_k^{(i)}\}_{k=1}^r$ the coefficients of the minimal LRR that approximates the local segment of the series $\mathsf{S}_{i-[Z/2],i+[Z/2]-1}$ with the center in s_i.

The local segment of length Z has the structure depicted in Figure 1, where the middle of $\mathsf{Z}_{(i_0)}$ is $N+1$.

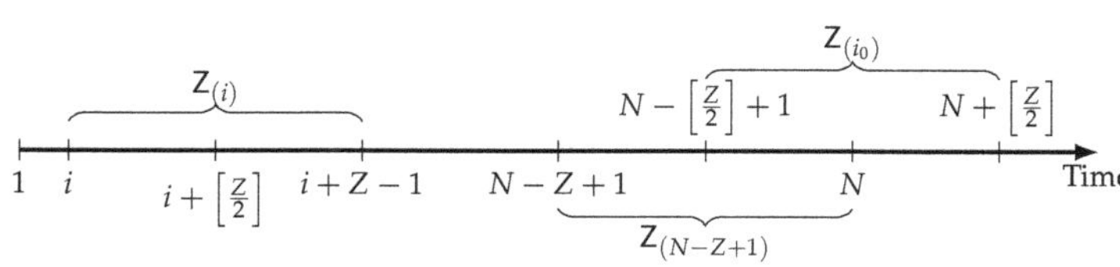

Figure 1. Scheme of moving segments.

3.2.1. Scheme

In the scheme below, we consider $j = 1, \ldots, p'$, where p' is the number of signal roots with nonnegative imaginary parts (note that signal roots with negative imaginary parts are conjugate to signal roots with positive imaginary parts):

$$\begin{array}{ccccc}
\mathsf{X}_N & & \{\rho_j^{(i+[Z/2])}\}_{i=1}^W & \xrightarrow{\text{FOR MODs}} & \{\widetilde{\rho}_j^{(N+l)}\}_{l=1}^M \\
\downarrow \mathsf{Z} & & \uparrow & & \downarrow \\
\{\mathsf{Z}_{(i)}\}_{i=1}^W & \xrightarrow[L,r]{ESPRIT} & \{\mu_j^{(i+[Z/2])}\}_{i=1}^W & & \{\widetilde{\mu}_j^{(N+l)}\}_{l=1}^M \rightarrow \{\widetilde{\mathcal{R}}_{N+l}\}_{l=1}^M. \\
 & & \downarrow & & \uparrow \\
 & & \{\omega_j^{(i+[Z/2])}\}_{i=1}^W & \xrightarrow{\text{FOR ARGs}} & \{\widetilde{\omega}_j^{(N+l)}\}_{l=1}^M
\end{array}$$

Here $\{\widetilde{\mathcal{R}}_{N+l}\}_{l=1}^M$, where $\widetilde{\mathcal{R}}_k = (\tilde{a}_r, \ldots, \tilde{a}_1)$ is the sequence of coefficients of the forecasting minimal LRRs. If the required length m of the forecasting LRRs is larger than r, then we lengthen each $\widetilde{\mathcal{R}}_{N+l}$, $l = 1, \ldots, M$, to $\widetilde{\mathcal{R}}_{N+l}^{(m)}$ of the min-norm LRR of order m; see Remark 2.

The result is a sequence of coefficients of the forecasting min-norm LRRs $\{\widetilde{\mathcal{R}}^{(m)}_{N+l}\}_{l=1}^{M}$ of order m. To obtain the results, the series of moduli and frequencies should be forecasted using some algorithms FOR MODs and FOR ARGs for each j.

After the sequence of coefficients is constructed, the forecasting values $\widetilde{\mathsf{S}}_{N+1,N+M} = (y_{N+1}, \ldots, y_{N+M})$ are taken from the time series

$$y_n = \begin{cases} \tilde{s}_n, & n = 1, \ldots, N, \\ \sum\limits_{k=1}^{m} \widetilde{a}_k^{(n)} y_{n-k}, & n = N+1, \ldots, N+M, \end{cases}$$

where $\tilde{s}_n$, $n = 1, \ldots, N$, is the signal estimate obtained using SSA.

3.2.2. Algorithm in Detail

Let us formally describe the algorithm for forecasting the minimal LRRs (Algorithm 1).

Algorithm 1: LocLRR SSA Forecast

Input:
- time series $\mathsf{X}_N = (x_1, \ldots, x_N)$,
- forecast length M,
- local segment length Z,
- window length L,
- number r of leading eigentriples of the trajectory matrices of the local segments that are used to find the estimates of the signal roots,
- length m of the forecasting LRRs,
- algorithm for forecasting the root moduli FOR MODs,
- algorithm for forecasting the instantaneous frequencies FOR ARGs.

Steps:

1. For each segment $\mathsf{Z}_{(i)}$, $i = 1, \ldots, W$, where $W = N - Z + 1$, estimate the signal roots $\{\mu_j^{(i+[Z/2])}\}_{j=1}^{r}$ of the approximating series using ESPRIT with window length L and signal rank r. Suppose that the estimates of the roots form complex-conjugate pairs. Choose the roots with positive argument $\{\mu_j^{(i+[Z/2])}\}_{j=1}^{p}$, $i = 1, \ldots, W$, $p = \frac{r}{2}$.
2. Arrange the first set of roots $\{\mu_j^{(1+[Z/2])}\}_{j=1}^{p}$ in descending argument order.
3. Order the sets of roots $\{\mu_j^{(i+[Z/2])}\}_{j=1}^{p}$, $i = 2, \ldots, W$, so that the sum
$$\sum_{j=1}^{p} \left| \mu_j^{(i+[Z/2])} - \mu_j^{(i+[Z/2]-1)} \right|$$
is minimal among all possible permutations of $\{\mu_j^{(i+[Z/2])}\}_{j=1}^{p}$.
4. For each $i = 1, \ldots, W$ and $j = 1, \ldots, p$, calculate the moduli $\rho_j^{(i+[Z/2])} = |\mu_j^{(i+[Z/2])}|$ and the frequencies $\omega_j^{(i+[Z/2])} = \frac{\operatorname{Arg}\mu_j^{(i+[Z/2])}}{2\pi}$. For $j = 1, \ldots, p$, set the series $\mathsf{P}_{(j)} = \left(\rho_j^{(1+[Z/2])}, \ldots, \rho_j^{(N-[Z/2]+1)}\right)$ and $\Omega_{(j)} = \left(\omega_j^{(1+[Z/2])}, \ldots, \omega_j^{(N-[Z/2]+1)}\right)$.
5. Using the algorithms FOR MODs and FOR ARGs, for each $j = 1, \ldots, p$, construct the forecast $\left(\widetilde{\rho}_j^{(N+1)}, \ldots, \widetilde{\rho}_j^{(N+M)}\right)$ of the series $\mathsf{P}_{(j)}$ and the forecast $\left(\widetilde{\omega}_j^{(N+1)}, \ldots, \widetilde{\omega}_j^{(N+M)}\right)$ of $\Omega_{(j)}$, respectively.
6. Using the obtained forecasts of frequencies and moduli, calculate the roots $\widetilde{\mu}_j^{(n)} = \widetilde{\rho}_j^{(n)} \exp(\mathrm{i} 2\pi \widetilde{\omega}_j^{(n)})$, $n = N+1, \ldots, N+M$, $j = 1, \ldots, p$, supplement them by their complex conjugates and then, using the relation between characteristic polynomials and LRRs (see Remark 1), find the sequence $\{\widetilde{\mathcal{R}}_{N+j}\}_{j=1}^{M}$ of the LRR coefficients, $\widetilde{\mathcal{R}}_{N+j} = \left(\widetilde{a}_r^{(N+j)}, \ldots, \widetilde{a}_1^{(N+j)}\right)$.

Output: The sequence $\{\widetilde{\mathcal{R}}_{N+l}\}_{l=1}^{M}$ of coefficients of minimal LRRs of order r approximately governing the future signal segments $\mathsf{S}_{l+N-[Z/2]-1, l+N+[Z/2]}$.

Remark 3. *If real-valued roots are obtained on some segments of* $\mathsf{Z}_{(i)}$, $i = 1, \ldots, W$, *we replace the values of the roots with missing values. That is, for* $i = 1, \ldots, W$ *and* $j = 1, \ldots, r$ *such that* $\mu_j^{(i)} \in \mathbb{R}$ *we put* $\mu_j^{(i)} := \mathrm{NA}$ *(not available). Possible gaps in the series of frequency estimates can be filled in; e.g., one can fill them with the iterative gap-filling method [12]; see also the description of the igapfill algorithm in [7] (Algorithm 3.7).*

An appropriate choice of the algorithms FOR MODs and FOR ARGs depends on the form of the frequency modulation.

Finally, each LRR of $\{\widetilde{\mathcal{R}}_{N+j}\}_{j=1}^{M}$ is enlarged to a min-norm LRR of length m with coefficients $\{\widetilde{\mathcal{R}}_{N+l}^{(m)}\}_{l=1}^{M}$ (see Remark 2) and this enlarged LRR is used for the prediction of s_{N+l}.

4. Examples

4.1. Description

In this section, we demonstrate the forecasting using the LocLRR SSA Forecast algorithm. The following types of time series were considered.

4.1.1. Sinusoid with Linearly Modulated Frequency

The signal has the form

$$s(t) = \cos(2\pi(at)^2), \quad a \neq 0.$$

The instantaneous frequency is $\omega_{\mathrm{ins}}(t) = 2a^2t$; the frequency range is $\omega_{\mathrm{ins}}([1,N]) = [2a^2, 2a^2N]$. We will consider such values of the parameter a and series length N at which the frequency range $\omega_{\mathrm{ins}}([1,N]) \subset (0,\ 0.5)$. Since the instantaneous frequency is a linear function, we will take the linear-regression-prediction algorithm as FOR ARGs.

4.1.2. Sinusoid with Sinusoidal Frequency

The signal is

$$s(t) = \cos(2\pi\omega_{\mathrm{ext}}(t + b\sin 2\pi\omega_{\mathrm{int}}t)), \quad b > 0,$$

$\omega_{\mathrm{ext}}, \omega_{\mathrm{int}} \in (0, 0.5)$, where ω_{int} is much smaller than ω_{ext}. The instantaneous frequency equals $\omega_{\mathrm{ins}}(t) = \omega_{\mathrm{ext}}(1 + 2\pi\omega_{\mathrm{int}}b\cos 2\pi\omega_{\mathrm{int}}t)$; the frequency range is $\omega_{\mathrm{ins}}([1,N]) = [\omega_{\mathrm{ext}}(1 - 2\pi\omega_{\mathrm{int}}b), \omega_{\mathrm{ext}}(1 + 2\pi\omega_{\mathrm{int}}b)]$. We will consider values of signal parameters and series lengths such that $\omega_{\mathrm{ins}}([1,N]) \subset (0,\ 0.5)$. The rank of the time series of the series Ω_N of instantaneous frequencies with terms $\omega_n = \omega_{\mathrm{ins}}(n)$, $n = 1, \ldots, N$, is equal to 3. Therefore, we take the recurrent SSA forecasting algorithm with $r = 3$ as FOR ARGs; the window length L is chosen to be half of the length of the frequency-estimation series following the general recommendations.

Since we do not consider time-varying amplitude modulation in the examples, we take the forecast of moduli series using the average value over local intervals as algorithm FOR MODs.

4.1.3. Sum of Sinusoids

The signal is

$$s(t) = \sum_{j=1}^{p_1} \cos(2\pi(a_j t)^2) + \sum_{k=1}^{p_2} \cos\left(2\pi\omega_k^{\mathrm{ext}}\left(t + b_k \sin 2\pi\omega_k^{\mathrm{int}}t\right)\right), \tag{5}$$

$a_j \neq 0$, $j = 1, \ldots, p_1$, $\omega_k^{\mathrm{ext}}, \omega_k^{\mathrm{int}} \in (0, 0.5)$, $b_k > 0$, $k = 1, \ldots, p_2$. We will consider the signal parameters and the time-series length N such that the frequency ranges of the summands are not mutually intersected and the frequency range of each summand belongs to the interval $(0,\ 0.5)$.

The frequencies corresponding to the signal summands of $s(t)$ will be forecasted either with linear regression or with SSA, based on the type of obtained estimates of the instantaneous frequencies on the local segments. The moduli will be predicted using the average of the estimates over the local segments.

In numerical examples, we will consider the signal parameters such that the instantaneous frequencies are slow-varying functions. Since the examples under consideration satisfy the general time-series model (Section 3.1), the LocLRR SSA-forecast algorithm can be used for forecasting.

In real-life problems, the value of the optimal LRR order m can be chosen based on the training data. In the model examples, the value of the optimal LRR order m will be chosen by trying all possible values of m in the range from r to Z and comparing the mean squared errors (MSE) of the predictions.

The following approach was used for choosing the length Z of local segments. We take such a value of the parameter Z that most of the local segments $Z_{(i)}, i = 1, \dots, W$, contain at least 2–3 instantaneous periods of each summand of the signal satisfying model (5). For small Z, we obtain estimates of the instantaneous frequency with large variability, whereas for large Z, we have a considerable bias. The necessary condition for appropriate values of Z is that there are no (or a few) segments providing real-valued roots.

4.2. Numerical Experiments

Consider the following numerical examples:

- Sinusoid with linearly modulated frequency,

$$s_n = \cos\left(2\pi\left(\frac{n}{100}\right)^2\right)$$

  (denoted by $\cos(\mathsf{n}^2)$);
- Sinusoid with sinusoidal frequency modulation,

$$s_n = \cos\left(\frac{2\pi}{20}\left(n + 5\sin\frac{2\pi n}{100}\right)\right)$$

  (denoted by $\cos\sin(\mathsf{n})$);
- Sum of sinusoids with linear and sinusoidal frequency modulations,

$$s_n = \cos\left(2\pi\left(\frac{n}{100}\right)^2\right) + \cos\left(\frac{2\pi}{10}\left(n + \sin\frac{2\pi n}{100}\right)\right) \tag{6}$$

  (denoted by $\cos(\mathsf{n}^2) + \cos\sin(\mathsf{n})$);
- Sum of two sinusoids with sinusoidal frequency modulation,

$$s_n = \cos\left(\frac{2\pi}{20}\left(n + \sin\frac{2\pi n}{100}\right)\right) + \cos\left(\frac{2\pi}{10}\left(n + 2\sin\frac{2\pi n}{140}\right)\right)$$

  (denoted by $\cos\sin(\mathsf{n}) + \cos\sin(\mathsf{n})$).

We compare the proposed LocLRR SSA-forecast algorithm (denoted by "alg") with two simple methods:

- Forecasting by constant, which forecasts by zero, since we consider time series with zero average (denoted by 'by 0').
- Forecasting using the last local segment, which is performed with the min-norm LRR computed using the roots of the last local segment $Z_{(N-Z+1)} = (x_{N-Z+1}, \dots, x_N)$ (denoted by 'last').

Let us consider accuracies of $M = 30$ step ahead forecasts for time series of length $N = 300$. For each example, the prediction is performed for the pure signal and the noisy signal, where the noise is white Gaussian with standard deviation $\sigma = 0.25$. In the noisy case, a sample of size $P = 100$ is used for the estimation of accuracy.

Let $\{\mathsf{X}_N^{(i)}\}_{i=1}^P$ be the time-series sample. The prediction error is estimated as RMSE $= \sqrt{\frac{1}{P}\sum_{i=1}^{P} \mathrm{MSE}(\widetilde{\mathsf{S}}^{(i)}_{N+1,N+M}, \mathsf{S}_{N+1,N+M})}$.

The results are shown in Table 1, where the best results are highlighted in bold. They confirm the advantage of the proposed method over the simple methods under consideration.

Table 1. RMSE of forecasts; 'alg' is the proposed algorithm; m is the optimal length of the forecasting LRRs.

Signal S_N	σ	by 0	Last		Alg	
			RMSE	m	RMSE	m
$\cos(n^2)$	0	0.689	0.717	2	**0.014**	3
	0.25	0.733	0.754	5	**0.135**	11
$\cos\sin(n)$	0	0.698	0.309	2	**0.097**	2
	0.25	0.741	0.438	5	**0.232**	6
$\cos(n^2) + \cos\sin(n)$	0	1.060	0.880	6	**0.184**	4
	0.25	1.089	0.958	10	**0.295**	12
$\cos\sin(n) + \cos\sin(n)$	0	0.873	0.587	4	**0.191**	5
	0.25	0.908	0.656	28	**0.291**	15

4.3. Detailed Example

To demonstrate the approach more clearly, let us consider the example (6) without noise; see Figure 2. Take $Z = 61$, $L = 30$, $r = 4$.

The results of forecasting the signal root corresponding to the first summand are shown in Figure 3, and the results of forecasting the signal root corresponding to the second summand can be seen in Figure 4. The frequency ranges of modulations in the summands are approximately $[0, 0.06]$ and $[0.094, 0.106]$, respectively. The forecasts are depicted in Figure 5 (forecasting with the last segment) and Figure 6 (forecasting with the proposed algorithm). Since there is no noise, the optimal LRR length m is small; here, $m = 4$.

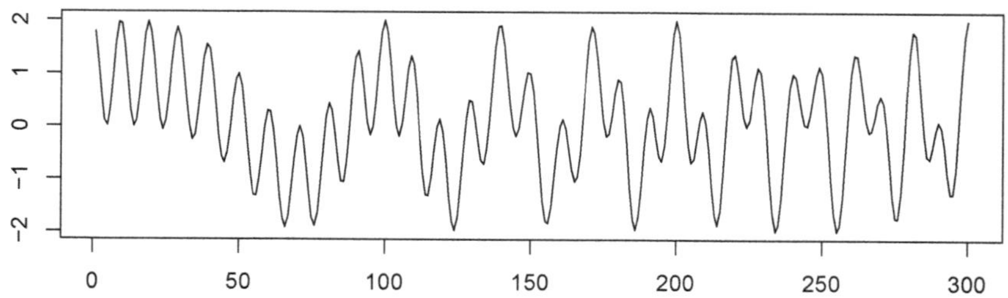

Figure 2. Initial signal $s_n = \cos\left(2\pi\left(\frac{n}{100}\right)^2\right) + \cos\left(\frac{2\pi}{10}\left(n + \sin\frac{2\pi n}{100}\right)\right)$.

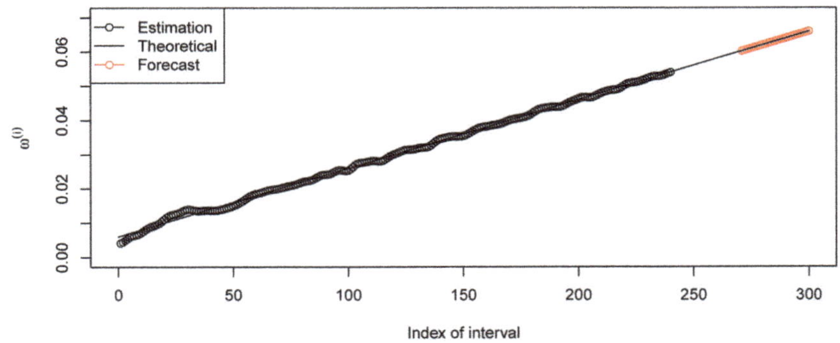

Figure 3. Forecasting the series of linear instantaneous frequencies for the summand $\cos\left(2\pi\left(\frac{n}{100}\right)^2\right)$.

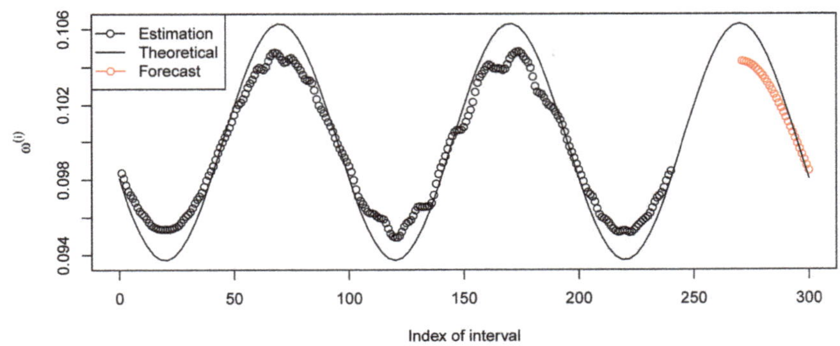

Figure 4. Forecasting the series of sinusoidal instantaneous frequencies for the summand $\cos\left(\frac{2\pi}{10}\left(n + \sin\frac{2\pi n}{100}\right)\right)$.

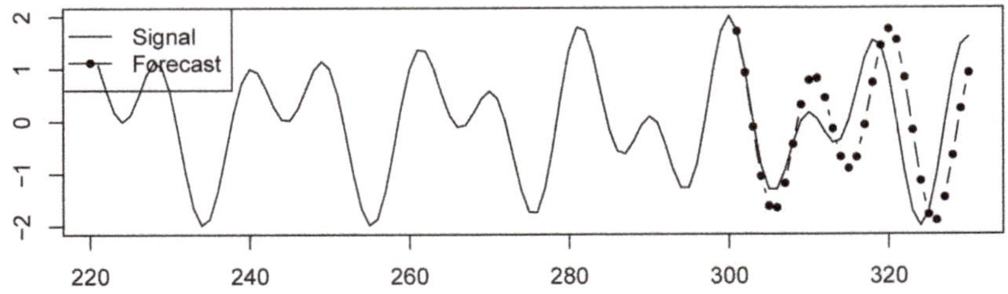

Figure 5. Forecasting using the last local segment, RMSE = 0.88. A shift is clearly seen.

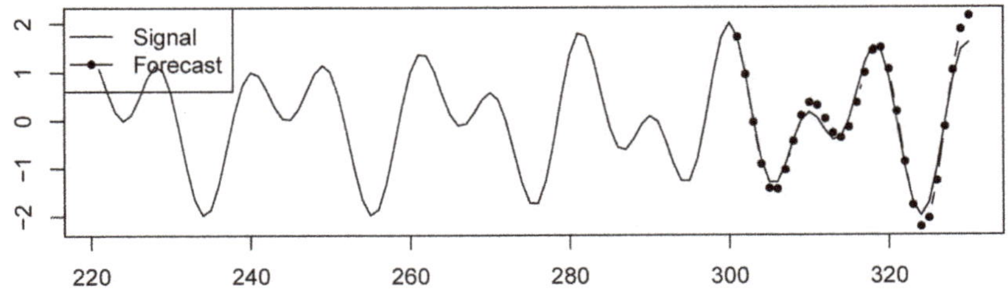

Figure 6. Forecasting using LocLRR SSA forecast, RMSE = 0.2. There is no shift.

5. Conclusions

In this paper, we proposed a method for forecasting time series that extends the capabilities of SSA and allows one to predict time series in which the signal just locally satisfies LRRs. A regular behaviour of the coefficients of the governed LRRs was assumed. In [6] (page 9), it was stated for the considered type of time series that "[t]he problem is how to forecast the extracted signal, since its local estimates may have different structures on different time intervals. Indeed, by using local versions of SSA, we do not obtain a common nonlinear model but instead we have a set of local linear models". In this paper, we proposed an answer to the problem of prediction of local structures of time series for some class of signals.

We constructed an algorithm for predicting local structures of time series that are the sum of frequency-modulated sinusoids and showed that the proposed forecasting method gives reasonable results for the cases of linear and sinusoidal frequency modulations.

Certainly, the considered comparison with a couple of simple methods is not enough; a more extensive comparison should be performed in the future. However, the results of this work show that the proposed approach based on the prediction of the coefficients of LRRs is promising.

Author Contributions: Conceptualization, N.G.; methodology, N.G. and E.S.; software, E.S.; validation, N.G. and E.S.; formal analysis, N.G. and E.S.; investigation, N.G. and E.S.; resources, N.G.; data curation, N.G. and E.S.; writing—original draft preparation, N.G. and E.S.; writing—review and editing, N.G. and E.S.; visualization, N.G. and E.S.; supervision, N.G.; project administration, N.G.; funding acquisition, N.G. All authors have read and agreed to the published version of the manuscript.

Funding: The work is supported by the Russian Science Foundation (project No. 23-21-00222).

Institutional Review Board Statement: Not applicable.

Informed Consent Statement: Not applicable.

Data Availability Statement: The data presented in this study are openly available. The R-code for data generation and result replication can be found at https://zenodo.org/record/8087608, doi 10.5281/zenodo.8087608.

Conflicts of Interest: The authors declare no conflict of interest.

References

1. Broomhead, D.; King, G. Extracting qualitative dynamics from experimental data. *Physica D* **1986**, *20*, 217–236. [CrossRef]
2. Vautard, R.; Ghil, M. Singular spectrum analysis in nonlinear dynamics, with applications to paleoclimatic time series. *Physica D* **1989**, *35*, 395–424. [CrossRef]
3. Elsner, J.B.; Tsonis, A.A. *Singular Spectrum Analysis: A New Tool in Time Series Analysis*; Plenum Press: New York, NY, USA, 1996.
4. Golyandina, N.; Nekrutkin, V.; Zhigljavsky, A. *Analysis of Time Series Structure: SSA and Related Techniques*; Chapman & Hall/CRC: Boca Raton, FL, USA, 2001.
5. Golyandina, N.; Zhigljavsky, A. *Singular Spectrum Analysis for Time Series*, 2nd ed.; Springer Briefs in Statistics; Springer: Berlin/Heidelberg, Germany, 2020.
6. Golyandina, N. Particularities and commonalities of singular spectrum analysis as a method of time series analysis and signal processing. *Wiley Interdiscip. Rev. Comput. Stat.* **2020**, *12*, e1487. [CrossRef]
7. Golyandina, N.; Korobeynikov, A.; Zhigljavsky, A. *Singular Spectrum Analysis with R*; Springer: Berlin/Heidelberg, Germany, 2018.
8. Leles, M.; Sansão, J.; Mozelli, L.; Guimarães, H. Improving reconstruction of time-series based in Singular Spectrum Analysis: A segmentation approach. *Digit. Signal Process.* **2018**, *77*, 63–76. [CrossRef]
9. Hall, M.J. *Combinatorial Theory*; Wiley: New York, NY, USA, 1998.
10. Usevich, K. On signal and extraneous roots in Singular Spectrum Analysis. *Stat. Interface* **2010**, *3*, 281–295. [CrossRef]
11. Roy, R.; Kailath, T. ESPRIT-estimation of signal parameters via rotational invariance techniques. *IEEE Trans. Acoust. Speech Signal Process.* **1989**, *37*, 984–995. [CrossRef]
12. Kondrashov, D.; Ghil, M. Spatio-temporal filling of missing points in geophysical data sets. *Nonlinear Process. Geophys.* **2006**, *13*, 151–159. [CrossRef]

engineering proceedings

MDPI

Proceeding Paper

Automated Approach for Generating and Evaluating Traffic Incident Response Plans †

Adel Almohammad * and Panagiotis Georgakis

School of Architecture and the Built Environment, University of Wolverhampton, Wolverhampton WV10 0JP, UK; p.georgakis@wlv.ac.uk

* Correspondence: a.almohammad@wlv.ac.uk

† Presented at the 9th International Conference on Time Series and Forecasting, Gran Canaria, Spain, 12–14 July 2023.

Abstract: Traffic incidents usually have negative effects on transportation systems such as delays and traffic jams. Therefore, a traffic incident response plan can guide management actors and operators to take action effectively and timely after traffic incidents. In this paper, an approach has been proposed to generate and evaluate traffic incident response plans automatically when a traffic incident is detected. In this approach, a library of response action templates has been constructed beforehand to be used in the real-time generation process of a response plan template. According to the type and severity of the detected and confirmed traffic incident, a combination of relevant response action templates will provide a set of response plans. In addition, we have developed a simulation model for the study area by using Aimsun Next software (version 20.0.3), developed by Aimsun, to evaluate the performance of the generated response plans. Therefore, the simulation outcomes determine the rank of the generated response plans including the optimal response plans. The proposed approach considers the characteristics of input traffic incidents and transport road networks to generate response plans. Furthermore, the choice of the optimal response plan considers the characteristics of the input traffic incident. The implementation results show that the generated response plans can enhance and improve the overall network performance and conditions efficiently. In addition, the response plan ranking is considered to be a supportive tool in the network operators' decision-making process in terms of the optimal response plan to be implemented or propagated.

Keywords: road traffic incident; traffic management system; response plan; response action; road traffic simulation; simulation network statistics

Citation: Almohammad, A.; Georgakis, P. Automated Approach for Generating and Evaluating Traffic Incident Response Plans. *Eng. Proc.* **2023**, *39*, 13. https://doi.org/10.3390/engproc2023039013

Academic Editors: Ignacio Rojas, Hector Pomares, Luis Javier Herrera, Fernando Rojas and Olga Valenzuela

Published: 28 June 2023

1. Introduction

Traffic incidents may cause property damage, injuries, and fatalities. Furthermore, they can quickly lead to congestion and associated travel delay, increased pollutant emissions, and wasted fuel. An incident represents any unpredictable occurrence that disrupts traffic flow and reduces roadway capacity such as a broken-down vehicle, accidents and collisions, fires, or hazardous material spills. Traffic management represents a crucial tool in minimizing the impact of incidents and the negative consequences on network safety and efficiency [1]. Moreover, traffic incident management represents an important component of intelligent transportation systems for reducing the durations and impacts of traffic incidents through systematic, coordinated, and pre-planned use of human resources and equipment [2]. Therefore, once an incident is detected and confirmed, the incident response management system must operate in order to maintain a reasonable level of safety for all road users, preserve and protect human life, and minimise travel delay [3].

Generally, a traffic incident response plan can be defined as a preprepared artificial document containing some guidance and instructions for responders and traffic network actors in order to make a decision rapidly, accurately, and orderly during incident response [4].

Therefore, road users must be supplied with timely, useful, and accurate information to minimise the traffic impact of an incident (e.g., advise them to avoid the problem area by using variable message signs—VMS). In addition, the traffic management operator must review and approve the response plan before sending the messages (e.g., encouraging diversion) out to the signs [3]. An emergency response plan may include the following elements: traffic diversion, traffic flow regulation, and the dynamic dissemination of information (e.g., traffic conditions, traffic speeds, routing, and changes in roadway geometry) to road users [1].

Emerging situations on the different transportation services (e.g., traffic incident) can be defined and will delineate scenarios by using various predictive analysis and anomaly detection algorithms. For each scenario, different response plans can be generated in a dynamic manner. Response plans may contain multiple device level mitigation measures ranging from a "light" response to a "severe" response, and it is expected that the "intensity" of the response plans depends on or is proportional to the severity of the traffic incident. The mitigation measures within each response can be different, and the response variables can be configured according to the severity of the traffic incident. Each response plan can be evaluated with the use of simulations, or data-driven analytics, and the results of the evaluations can be presented to the network's operators for further arbitration and decision-making.

In this paper, we describe an automated approach for generating and evaluating traffic incident response plans that is being developed in the context of FRONTIER, an EU funded research project. The proposed approach involves two main components: the Response Plans Generation Module (RPG) and the Response Plans Scoring Module (RPS). The first component will generate a list of response plans based on the scenarios that may be enacted on the network. Essentially, each response plan represents a configuration of a pre-defined template that includes a sequence of network management actions such as the opening and closing of road lanes, variable speed limiting, and traffic redirection. However, the second component, RPS, measures and assesses the impact of each response plan on the network performance (e.g., traffic flow, speed, density, and travel time). This response plan evaluation may support the decision-making process that needs to be done at some point by traffic management and network's operators. We have developed and calibrated a simulation model for the study area using Aimsun Next software to be used for simulating and assessing the impact of different response actions of each response plan generated.

The remainder of this paper is organised as follows. In Section 2, we discuss related work, and in Section 3, we present the proposed approach and describe the details of its elements. Then, in Section 4, we discuss the implementation of the proposed approach, describe some example experiments, and demonstrate the results of these experiments. Finally, in Section 5, we conclude with a summary of the current status and future work regarding the proposed methodology.

2. Related Work

Traffic incident response plans guide traffic management actors and operators to take actions effectively and timely after traffic incidents. Considering and using such response plans will alleviate traffic congestion, save many losses, and increase safety.

An innovative real-time incident management platform has been introduced in [5] to detect and then classify incidents into two types, recurrent and non-recurrent, based on their frequency and characteristics. When an accident is detected and confirmed by the system, the platform triggers either a data-driven machine learning module (if the incident is recurrent) or traffic simulation modules (if the incident is non-recurrent). Therefore, for the non-recurrent incidents, the process of choosing the most appropriate response plan will mainly depend on the simulation output.

In another study, a real-time management system for traffic incident response plans has been developed in [4] in order to automatically and timely generate and manage traffic incident response plans. Three indicators were used to measure the system performance:

precision, *P*; recall, *R*; and indicator, *F*. The method contains four procedures: case representation, case retrieval, case revision, and case learning. Therefore, when a traffic incident is verified, the response plans database will be triggered by the emergency management centre to generate response plans.

Additionally, a traffic management system has been proposed in [6] at a traffic control centre to support traffic management authorities in the traffic flow management task in Beijing. The system consists of a traffic modelling system, a traffic plan builder, and a traffic plan selector. It generates optimum traffic response plans based on traffic incidents, and for a given traffic incident, the traffic operations are analysed and simulated to provide the quantitative evaluation results for these alternatives. A traffic plan builder provides several features to help the operator to select the needed traffic measures (traffic diversion, entry gating, of promoting traffic flows of diversion routes) in response to a specific incident and automatically suggests the corresponding traffic control aspects.

To minimise an incidents impact (i.e., congestion, queues, and travel delays) on traffic, [1] designed a system to support traffic control operators when they select traffic incident response measures. The system uses a set of parameters such as demand, incident severity, and duration to select the most suitable traffic response plan. It predicts the duration of the input incident and estimates the impact area extent and the travel delays that will be caused by the incident in order to select a response plan. Each response plan includes a number of strategies to manage traffic flow such as diversion traffic volumes, diversion points, termination points for diversion, diversion routes, VMSs state, and new timing for traffic signals.

In order to achieve a state-of-the-art traffic management software system, ref. [7] implemented traffic incident response strategies using a real-time knowledge-based expert system. The real-time incident response plan generation subsystem provides response plans that are customised for the characteristics of the given incident. Response plans can be generated from incident data and can support various actions such as messaging signs and sending emergency vehicles to the location of the incident. Generating device states to support a response plan represents a key function of this subsystem. Six types of devices were involved: lane use signals (LUS), variable message signs (VMS), variable speed limit signs (VSLS), traffic signals (TS), blankout signs (BOS), and highway advisory radio (HAR).

In addition to this, some of the earlier studies that address traffic incidents, response plans and strategies, incident impacts, and evaluation of response plans have been reviewed and are described in Table 1.

Table 1. Examples of some studies related to traffic incidents and response plans.

Reference	Description
[8]	Designed a simulation and evaluation framework to model an emergency response system for highway traffic safety and security or for minimising the average response times associated with different accidents on the road.
[9]	Assessed and enhanced traffic incident management techniques by using a scenario-based assessment methodology involving macroscopic traffic simulations, traffic performance calculations, and cost–benefit analysis.
[10]	Evaluated the impacts of incident management strategies such as ramp metering, VMS, and VSLS using a large-scale microsimulation model.
[11]	Analysed the impact of incidents in an urban traffic system by using flow data collected from loop detectors.
[12]	Analysed a wide range of traffic incidents and responses for the set of critical locations on a test network by using VISSIM microsimulation.
[13]	Introduced a method for deriving an optimal dispatching strategy for incident response by integrating GIS, traffic simulation, and optimisation systems.
[14]	Introduced a discrete time train and passenger simulation engine for urban railway networks and a mixed integer programming formulation for the problem of finding an optimal action plan as a response to an incident.
[15]	Examined and improved indicators for better incident prioritisation and development of rapid incident response plans.

The existing research efforts regarding real-time generation of traffic incident response plans as part of traffic incident management systems are somehow limited. In addition, most of these efforts address the problem without considering the effects of some traffic incident characteristics, such as incident location, start time, duration, type, and severity, on the generated response plans. Therefore, the performance and effectiveness of the existing real-time methods are uncertain. Therefore, our suggested approach for generating and evaluating traffic incident response plans represents a contribution to address the limitations of current practices by considering the following:

- The library of response action templates used to generate the appropriate response plans has been principally constructed by considering the characteristics of the road network (e.g., motorway or urban area) of the study area.
- The characteristics of the input traffic incident (i.e., incident type and severity) have been used to determine which response actions will be included in the response plan template.
- The impact of input traffic incident characteristics (i.e., incident location, time, and duration) has been considered in the response plan application and evaluation process. Therefore, ranking and suggesting optimal response plans will be largely affected by the traffic incident characteristics.

3. Approach

The proposed approach uses the anomalies and critical situations detected in the traffic network as input to generate many response plans aiming to optimise the transport network. Each response plan contains a set of actions that can adjust supply or/and demand features of a transport network in an optimal way. In addition, this approach uses Aimsun Next simulation framework to assess the response plans generated. Therefore, this approach aims to be enacted after an incident has been identified in the network to provide capabilities for the generation of appropriate response plans along with their scores. Each response plan includes a number of measures that aim to alleviate the network-wide congestion and restore the network conditions to normal. The overview diagram of the response plans generation and evaluation approach is depicted in Figure 1. It contains two main components: Response Plans Generation Module (RPG) and Response Plans Scoring Module (RPS).

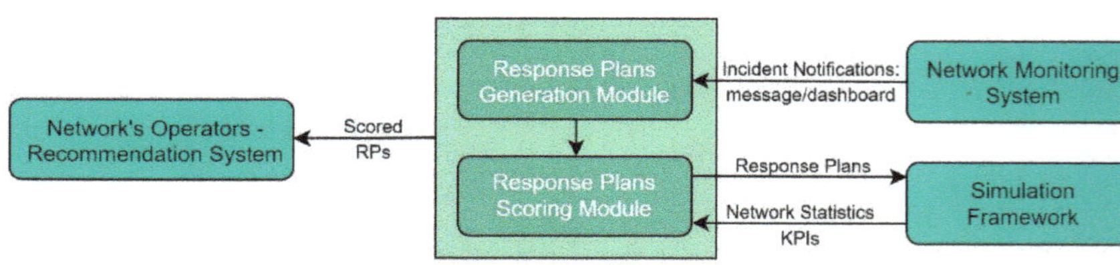

Figure 1. Overview of the proposed approach.

3.1. *Response Plans Generation Module—RPG*

This component is responsible for building and constructing the traffic incident response plans when a traffic incident is detected and verified (Figure 1). It has been assumed that the input traffic incident can be identified through automatic (e.g., via a dedicated incident detection system) or manual means (e.g., the network operators insert a traffic incident in the corresponding traffic management system). Figure 1 illustrates that the Network Monitoring System—NMS—(for example) is responsible for detecting critical network anomalies that require specific remedial actions and response plans. In addition, NMS is responsible for making the detected incidents (i.e., timestamp, duration, location, type, and severity) available to the RPG (e.g., via a message or through a dedicated dashboard). Therefore, the RPG will be triggered by traffic incidents detected and verified by the NMS. Figure 2 shows an example for the main characteristics of a verified traffic incident. For each detected and verified traffic incident, the RPG will generate a collection of response plans that will assist in the mitigation of the incident's impact on the network capacity or demand conditions. It has also been considered in the proposed approach that the types of input traffic incidents can be categorised into five key categories (Table 2), and the severities of the input traffic incidents can be categorised into three key categories (Table 3).


```
{
  "incident_id" : 222223,
  "incident_location" : [37.956261, 23.886498],
  "incident_start" : "02/02/2023 8:40:00",
  "incident_end" : "02/02/2023 18:40:00",
  "incident_duration" : 36000,
  "incident_severity" : "severe",
  "incident_type" : "road accident",
  "incident_status" : "resolved"
}
```


Figure 2. An example of verified traffic incidents.

Table 2. Types of input traffic incidents.

ID	Description	Examples
U1	Unplanned: road-based incident	Overturning of a truck, road accident, accident, traffic congestion, broken down vehicle, closed road, obstacle on the roadway, impassable water, pedestrians crossing the road, live animal on the roadway.
U21	Unplanned: public transport-based incident—road	Disrupted bus service, diverted bus service.
U22	Unplanned: public transport-based incident—rail	Metro station closure, broken down train, voltage drop in railway network, obstacle on the railway.
U3	Unplanned: other incidents	severe weather conditions, flood, fire, strong wind, tornado.
P	Planned incident	Sport event, recurring congestion, concert, exhibition, demonstrations, protests, road work, forecasted extreme weather.

Table 3. Severities of input traffic incidents.

ID	Description	Examples
M	Minor incident	Minor, low impact, slight, small, insignificant, trivial, negligible.
J	Major incident	Major, medium, substantial, great, considerable, significant.
C	Critical incident	Critical, high, severe, extreme, acute, crucial, serious, strong.

Figure 3 shows the main subcomponents of Response Plans Generation Module (RPG), including the Response Plans Template Manager (RPTM) and the Response Plans Building Module (RPBM). Effectively, the library of response action templates (RPTM) must be constructed in advance and before it can be utilised by the RPBM. Therefore, the RPBM will use this built library of response action templates every time an input traffic incident triggers the RPG.

In the proposed approach, a library of pre-defined templates of response actions is required to facilitate the operation of the RPG and the real-time generation of appropriate response plans. Additionally, these pre-defined templates need to be categorised according to the characteristics of potential input traffic incidents to support the process of selecting the suitable response actions. Each template will specify what required or optional actions are applicable for each incident and some default values for the initialisation of the response plan. For example, a template that can be used as a response plan for a bomb threat at a metro station will have a required action to 'suspend a specific metro service (or all services)'. The default initialisation value would be to suspend all services that pass through the station for which the thread has been made.

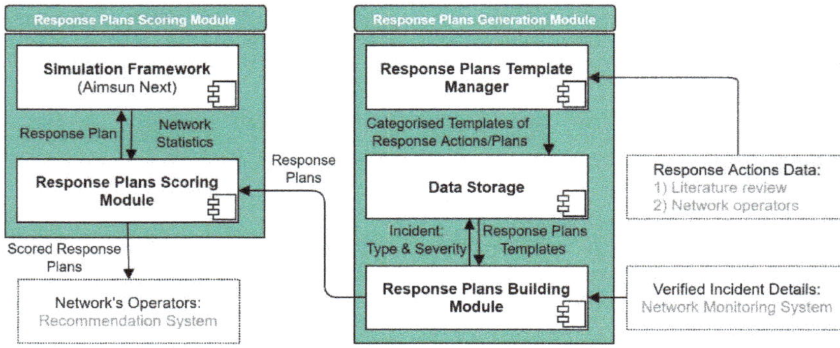

Figure 3. Operational workflow of the response plans generation and evaluation approach.

3.1.1. Response Plans Template Manager—RPTM

This subcomponent is responsible for creating a library of categorised templates of response actions that mainly construct the templates of response plans as a basis for generating traffic incident response plans. In the proposed approach, constructing this library considered the literature review of the related studies and the response logic (i.e., traffic management systems) of network operators in the study area, which is the Athens Airport—Metamorfosis corridor. The response logic algorithms and heuristics express the qualitative knowledge for the handling of traffic incidents by each actor. Therefore, the traffic management procedures of network operators (Attiki Odos, Attiko Metro, and OASA) of the Athens corridor have been considered and utilised for developing and constructing the mentioned library of categorised response action templates. As a result of extensive consideration of the related literature and the concept of operation for the different network operators, twenty-nine abstract response action templates have been identified. These templates include informative, executable, and simulatable response actions. Despite the templates having a unified format, the values of their variables may differ from one response action template to another.

Each response action template is labelled by both of the related incident types (i.e., U1, U21, U22, U3, and/or P in Table 2) and the related incident severities (i.e., M, J, and/or C in Table 3). These categorised templates of response actions are stored in a dedicated data storage (or a file system) to be accessed and used by the Response Plans Building Module—RPBM. Table 4 represents some examples of these identified templates of response actions. As mentioned above, constructing the library of response action templates needs to be accomplished just once for a given study area before the approach can deal with any input traffic incident and generate the appropriate response plans (RPBM in Figure 3).

Table 4. Examples of identified response action templates.

ID	Response Action Name	Type					Severity		
		U1	U21	U22	U3	P	M	J	C
2	Inform and advise the road users via a specific VMS (i.e., to redirect or change mode).	✓	✓	✓	✓	✓	✓	✓	✓
8	Close an upstream corridor/motorway entrance.	✓							✓
10	Close some or all lanes of a section/road.	✓			✓	✓	✓	✓	✓
15	Send more staff to a metro station for assistance.			✓				✓	✓
23	Temporary alteration and redirection of a bus service route.	✓			✓	✓			✓

3.1.2. Response Plans Building Module—RPBM

This subcomponent is responsible for generating a set of response plans for each detected and verified traffic incident. According to the type and severity of the input incident, this subcomponent picks up the relevant templates of response actions (that were constructed earlier and stored in the data storage) to build a response plan template. Therefore, the generated response plan template represents a combination of all relevant response action templates, and each response plan represents a configuration of this created response plan template. Response Plans Template Manager—RPTM—identified all possible configurations for each template of response actions and possible values of its variables according to the incident severity. Therefore, the number of possible configurations for a response plan template determines the number of response plans that can be generated for a given traffic incident. Having the input traffic incident shown in Figure 2, the RPBM generates the response plan template demonstrated in Table 5. This template includes 21 templates of response actions that can provide 128 response plans (combinations of all possible values of variables).

Table 5. The generated response plan template for the traffic incident in Figure 2.

ID	Response Action	Values	Simulation
1	Providing information to the road users via media and call centre.		
2	Inform and advise the road users via a specific VMS (i.e., to redirect or change mode).		
3	Inform and advise metro/train passengers via a specific VMS at metro stations.		
4	Inform and advise passengers via app & DMS at bus stops and in-vehicle screens.		
5	Contact the traffic police and exchange information.		
6	Activate a VSLS in a specific road/location.		
7	Divert/redirect traffic from a specific corridor/road to other major arterials.	75, 90	✓
8	Close an upstream corridor/motorway entrance.		✓
9	Temporary application of variable speed limiting at specific roads/sections.	75, 90	✓
10	Close some or all lanes of a section/road.		✓
11	Open the toll in a specific interchange.		
12	Apply dynamic toll pricing.	60, 80	
13	Divert/redirect some road users to a P&R station/facility.	5	✓
14	Send patrols, ambulance, and traffic police for assistance.		
17	Increase the frequency of a metro service.	50, 100	
18	Increase the frequency of a bus service.	50, 100	
19	Deployment of an on-demand bus service.	5	
21	Divert/redirect m% of traffic/users to bus services.	3, 4	✓
22	Divert/redirect n% of traffic/users to metro/train services.	3, 4	✓
23	Temporary alternation and redirection of a bus service route.		
29	Control of a specific signalised intersection.		

3.2. *Response Plans Scoring Module—RPS*

This component is responsible for assessing and evaluating the impact of the response plans generated by the RPG component. It uses the simulation framework (i.e., Aimsun Next) to calculate Key Performance Indicators (KPIs) for each response plan. For a given response plan, the RPS calculates and compares two scenarios: the baseline scenario, where only the actions representing the traffic incident (e.g., closing two lanes) are considered and simulated, and the response scenario, where all actions representing both the traffic incident and response plan are considered and simulated. Therefore, Aimsun Next framework simulates these scenarios for each response plan and produces relevant KPIs. The latter are used to assess the traffic conditions both at baseline conditions, as well as assess generated response plans. A list with all possible KPIs is specified within the online Aimsun Next manual [16]. Therefore, amongst the types of output that the Aimsun Next can provide, statistical data is of special relevance and, more specifically, a selection of relevant KPIs will be consumed by the RPS. Such KPIs may, but are not limited to, include delay time, density, flow, speed, mean queue, and travel time.

It has been assumed that some of the response actions that construct the generated response plan can be simulated (tic sign in Table 5) while other actions are either non-simulatable (e.g., contact the traffic police and exchange information) or have a trivial effect on the simulation outcomes (e.g., deployment of an on-demand bus service). Considering the response plan template in Table 5, sixteen response plans can be simulated and evaluated (a combination of actions 7, 9, 21, and 22). Here, the baseline scenario represents the simulation of action 10 while the response scenario represents the simulation of actions 7, 8, 9, 10, 13, 21, and 22.

As a result, these calculated KPIs may be used to rank the response plans generated for a given traffic incident. This ranking of response plans can support the stakeholders or network operators to select and follow the optimal response plan. Alternatively, these KPIs can be used by another component or service (recommendation system in Figures 1 and 3) that can further analyse and process these KPIs along with other aspects of the response plans (e.g., safety, environmental impact, and passengers' satisfaction) to more precisely suggest and recommend the optimal response plan.

4. Results and Discussion

The proposed approach is being implemented using a micro-services architecture in order to ensure that it can be integrated with other potential components and secure the interoperability of the suite of integrated services. Various traffic conditions, strategies, and mechanisms have been used with Aimsun Next software to simulate different traffic response actions (defined as simulatable actions) of response plans. At this stage, this mainly includes lane closure, speed reduction, forced turn, and demand reduction. Therefore, the simulatable response actions of a given response plan have been represented within the simulation network and model using the python scripting approach that is offered and supported by Aimsun Next software.

We have tested the proposed approach through a number of experiments in order to investigate its efficiency and performance. Various input traffic incidents (with different types and severities) have been applied in these experiments to demonstrate the details of the generated response plans. Furthermore, the impact of the generated response plans on the network performance has been assessed by using Aimsun Next software and a network simulation model for the Athens Attiki Odos corridor, which we have developed and calibrated.

After constructing the library of response action templates, the response plan template (e.g., Table 5) and all generated response plans represent the main outcome of the proposed approach. However, the results of simulating these generated response plans represent their evaluation (or rank), which needs to be considered in the process of selecting the optimal response plan. Figure 4 shows the results of simulating the generated response plans from Table 5. Each variable value (e.g., the most top left circle on the blue polyline—speed) for each response plan (e.g., speed for response plan 1) represents the percentage by which this variable value has been changed from the baseline scenario (do-nothing) to the response scenario (response plan application). For example, the most top left circle indicates that the application of the first response plan (RP1) will increase the traffic speed by 38.30% compared to the "do-nothing" or baseline scenario. Figure 4 shows that almost all of the variables seem to be rather invariant across the different response plans. This is because the differences among the generated response plans are trivial and each response plan generated is similar to another one after slightly tweaking the value of one variable related to one of this response plan's actions. For example, the difference between RP1 and RP2 is that the variable value of response action 21 is 3 in RP1 while it is 4 in RP2. Therefore, these minor differences among response plans show just a slight change in the simulation results.

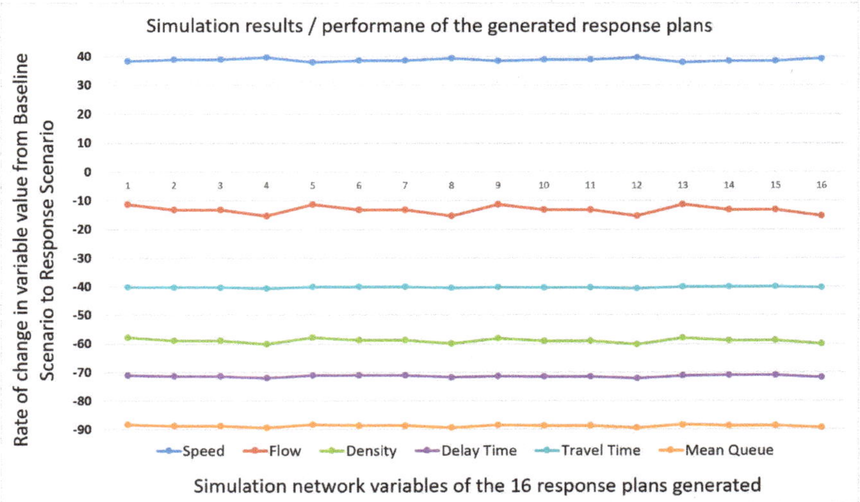

Figure 4. The results of simulating the generated response plans (Section 3.2).

Table 6 represents the aggregated results for all response plans generated and simulated when five different traffic incidents are used as input to the proposed approach. For instance, the incident 3: U1-C in this table represents the aggregated results shown in Figure 4, as the "Speed" value represents the average value of the 16 values depicted on the top blue polyline in Figure 4.

Table 6. The results of the proposed approach using five input traffic incidents.

Incident	Speed	Flow	Density	Delay Time	Travel Time	Mean Queue
1: U22-C	−5.67	5.45	−12.80	−39.04	−7.03	−42.69
2: P-C	−10.18	1.66	−20.64	−17.52	−10.58	−25.12
3: U1-C	38.82	−13.27	−58.90	−71.44	−40.20	−88.75
4: U3-C	11.19	3.47	−29.02	−27.90	−11.80	−57.73
5: P-J	−4.00	4.33	−16.34	−18.16	−10.00	−24.75

To find the optimal response plans out of the different response plans generated for a particular traffic incident (e.g., 16 response plans in Figure 4), a particular ranking approach can be used. In addition, such ranking methods may consider various characteristics or attributes of the generated response plans (e.g., network performance, environmental aspects, easiness to apply, and user satisfaction). For instance, if only the simulation results and network statistics (i.e., Speed, Flow, Density, Delay Time, Travel Time, and Mean Queue) have been considered and have the same weight, the response plans in Figure 4 can be ranked from the optimal one as follows: 12, 16, 4, 8, 10, 11, 2, 3, 14, 15, 6, 7, 9, 13, 1, 5. However, alterative ranking methods can be investigated and implemented by selecting and considering some of these network variables along with other characteristics of response plans.

The proposed approach generated and evaluated different response plans for each input traffic incident. Moreover, the assessment process of the generated response plans revealed that application of these generated response plans improves the network performance and conditions. Table 6 shows that the delay time and travel time have decreased significantly for all tested incidents as a result of using the proposed approach for generating the related response plans.

5. Conclusions

In this paper, we presented an automated approach for generating and evaluating traffic incident response plans. Firstly, a library of response action templates had been constructed, taking into account the characteristics of the road network of study area that was the Athens Attiki Odos corridor. For a given traffic incident, the approach used the constructed library of response actions to build a response plan template and, therefore, generate a set of response plans according to the incident characteristics. Aimsun Next software and a developed simulation model for the Athens corridor had been used to evaluate the generated response plans. Effectively, the outcomes of simulating and evaluating response plans reflected the different characteristics of the input traffic incident. Thus, the implementation of the generated response plans demonstrated how the network performance and conditions had been generally improved when compared to the baseline (do-nothing) scenario. In a later stage, the evaluated response plans had been ranked to define the optimal response plans to be considered and propagated by the network traffic management operators. However, future research directions can involve:

- The integration of the proposed solution as a part of traffic management system, acquiring some feedback, and improving the performance of the approach.
- The development of a general approach for generating and evaluating traffic incident response plans that considers all types of road networks through extending the library of response actions and utilising additional traffic incident characteristics to define the response plan template.
- The improvements of modelling and the representation of different response actions in the simulation framework to correctly measure the impact of real response plans on the network performance.
- The investigation of ranking methods for the generated response plans by considering the outcomes of implementing different response plans and their potential impact on users' satisfaction, network performance, and environment.

Author Contributions: Conceptualisation and methodology, A.A. and P.G.; software, A.A and P.G.; validation, A.A. and P.G.; writing—original draft preparation, A.A. and P.G.; writing—review and editing, A.A. and P.G.; visualisation, A.A. All authors have read and agreed to the published version of the manuscript.

Funding: This research is part of the Project "FRONTIER" (next generation traffic management for empowering CAVs integration, cross-stakeholders collaboration and proactive multi-modal network optimization). This project has received funding from the European Union's Horizon 2020 research and innovation programme under grant agreement No. 955317. This publication only reflects the authors' view, and the European Union is not liable for any use that may be made of the information contained therein.

Institutional Review Board Statement: Not applicable.

Informed Consent Statement: Not applicable.

Data Availability Statement: The data are available on request from the corresponding author.

Conflicts of Interest: The authors have no conflict of interest to declare that are relevant to the content of this paper.

References

1. Mitrovich, S.; Valenti, G.; Mancini, M. A decision support system (DSS) for traffic incident management in roadway tunnel infrastructure. In Proceedings of the 34th European Transport Conference, Strasbourg, France, 18–20 September 2006; p. 240.
2. Ma, Y.; Zhang, W.; Lu, J.; Yuan, L. Automated generation of traffic incident response plan based on case-based reasoning and Bayesian theory. *Discret. Dyn. Nat. Soc.* **2014**, *2014*, 920301. [CrossRef]
3. Korpal, P.R. Incident management: The key to successful traffic management in Toronto. *ITE J.* **1992**, *62*, 58–61.
4. Ma, Y.; Zhang, W.; Xie, J.; Lu, J. Application of real-time automated traffic incident response plan management system: A web structure for the regional highway net-work in China. *Adv. Mech. Eng.* **2014**, *6*, 489623. [CrossRef]

5. Wen, T.; Mihăiţă, A.; Nguyen, H.; Cai, C.; Chen, F. Integrated incident decision-support using traffic simulation and data-driven models. *Transp. Res. Rec.* **2018**, *2672*, 247–256. [CrossRef]
6. Vermeulen, M.J. A traffic management system for Beijing. In Proceedings of the 28th Annual Southern African Transport Conference 2009, Pretoria, South Africa, 6–9 July 2009.
7. Stack, R. Boston central artery/tunnel traffic management using an expert system. In Proceedings of the Conference on Intelligent Transportation Systems, Boston, MA, USA, 12 November 1997; pp. 76–81.
8. ur Rehman, M.A.; Yan, H.Y. Modelling and Simulation on Traffic Emergency Response Based on Unity3D. *Int. J. Future Comput. Commun.* **2019**, *8*, 68–71. [CrossRef]
9. Nitsche, P.; Olstam, J.; Taylor, N.; Reinthaler, M.; Ponweiser, W.; Bernhardsson, V.; Mocanu, I.; Uittenbogaard, J.; van Dam, E. Pro-active management of traffic incidents using novel technologies. *Transp. Res. Procedia* **2016**, *14*, 3360–3369. [CrossRef]
10. Dia, H.; Gondwe, W.; Panwai, S. Traffic impact assessment of incident management strategies. In Proceedings of the 2008 11th International IEEE Conference on Intelligent Transportation Systems, Beijing, China, 12–15 October 2008; IEEE: Piscataway, NJ, USA; pp. 441–446.
11. Huang, B.; Zhao, T.; Zhang, Y. Traffic incident impact analysis with random matrix theory and cluster analysis. In Proceedings of the 2010 IEEE International Conference on Emergency Management and Management Sciences, Beijing, China, 8–10 August 2010; IEEE: Piscataway, NJ, USA; pp. 198–201.
12. Tasic, I. Using microsimulation to evaluate traffic incident responses for Traffic Operations Center decision making. In *Masters Abstracts International*; The University of Utah: Salt Lake City, UT, USA, 2012; Volume 51.
13. Huang, B.; Pan, X. GIS coupled with traffic simulation and optimization for incident response. *Comput. Environ. Urban Syst.* **2007**, *31*, 116–132. [CrossRef]
14. Szabó, J.; Blandin, S.; Brett, C. Data-driven simulation and optimization for incident response in urban railway networks. In Proceedings of the 16th Conference on Autonomous Agents and MultiAgent Systems, São Paulo, Brazil, 8–12 May 2017; pp. 819–827.
15. bin Kabit, M.R.; Charles, P.; Ferreira, L.; Hojati, A.T. *Examination of Traffic Incident Records and Development of a Rapid Incident Response Plan*; Universiti Malaysia Sarawak: Kota Samarahan, Malaysia, 2011.
16. Aimsun Next Manual. Available online: https://docs.aimsun.com/next/22.0.1/UsersManual/DataOutputs.html (accessed on 20 February 2023).

engineering proceedings

MDPI

Proceeding Paper

Econometric Modeling of the Impact of the COVID-19 Pandemic on the Volatility of the Financial Markets †

Abdessamad Ouchen

National School of Business and Management (ENCG) Fez, Sidi Mohamed Ben Abdellah University, Fez 30000, Morocco; abdessamad.ouchen@usmba.ac.ma; Tel.: +212-667085262

† Presented at the 9th International Conference on Time Series and Forecasting, Gran Canaria, Spain, 12–14 July 2023.

Abstract: The purpose of this paper is to identify econometric models likely to highlight the impact of the COVID-19 pandemic on the financial markets. The Markov-switching "GARCH and EGARCH" models are suitable for analyzing and forecasting the series of daily returns of the major global stock indices (i.e., SSE, S&P500, FTSE100, DAX, CAC40, and NIKKEI225) during the pre-COVID-19 period, from 1 June to 30 November 2019, and the post-COVID-19 period, from 31 December 2019, to 1 June 2020. The Markov-switching "GARCH and EGARCH" models allow good modeling of the conditional variance. The estimated conditional variance values by these models highlight the increase in volatility for the stock markets in our sample, during the post-COVID-19 period compared to that pre-COVID-19, with a peak in volatility in "early January 2020" for the Chinese stock market and in "March 2020" for the other five stock markets (i.e., New York, Paris, Frankfurt, London, and Tokyo). The stock exchange of Frankfurt has shown great resilience compared to other international stock exchanges (i.e., the stock exchanges in Paris, London, and New York). The modeling of the impact of the COVID-19 pandemic on the financial markets by the Markov-switching "GARCH and EGARCH" models makes it possible to simultaneously take into consideration the nonlinearity at the level of the mean and the variance, and to obtain the results of the transition probabilities, the unconditional probabilities and the conditional anticipated durations during the pre-COVID-19 period and the post-COVID-19 period.

Keywords: financial markets; Markov-switching GARCH models; COVID-19 pandemic; volatility

Citation: Ouchen, A. Econometric Modeling of the Impact of the COVID-19 Pandemic on the Volatility of the Financial Markets. *Eng. Proc.* **2023**, *39*, 14. https://doi.org/10.3390/engproc2023039014

Academic Editors: Ignacio Rojas, Hector Pomares, Luis Javier Herrera, Fernando Rojas and Olga Valenzuela

Published: 28 June 2023

1. Introduction

The prices of the main international financial market portfolios experienced a plunge in March 2020 due to the COVID-19 pandemic. Pandemics can also have a substantial impact on financial systems due to their enormous economic costs [1]. It is true that the previous literature remains limited as to how pandemics affect financial markets. However, some research has advanced the impact of the COVID-19 pandemic on financial volatility [2–4]. It should be noted that other forms of natural disasters, such as earthquakes and volcanoes; air disasters; as well as acts of terrorism, have a negative impact on financial markets [1,5–9]. Since the appearance of the first case of COVID-19 in Wuhan in December 2019, the virus has quickly spread to all corners of the world. On 11 March 2020, when it has already affected more than 100,000 people and killed thousands of people in over 100 countries, the World Health Organization (WHO) declared the coronavirus epidemic (COVID-19) as a global pandemic. The global spread of COVID-19, which has saturated healthcare systems, has forced societies and economies to shut down, causing social and economic disruption. The negative repercussions of the COVID-19 pandemic on foreign trade, tourism, transport, and industry were evident [10]. Its economic consequences are likely to exceed those of the global financial crisis of 2007–2009. In fact, in its April 2020 report, the International Monetary Fund forecasts a global growth rate of −3% in 2020, which is lower than the lowest rate of −1.7% recorded in 2009 during the global financial crisis of 2007–2009 [11]. The negative, substantial, and sudden impact of the COVID-19 pandemic on major stock markets was evident in March 2020 [1,10,12,13]. Indeed, the price of the S&P 500 stock index, which reached a high value of 3386.15 points on 19 February 2020, experienced, after almost a month, a decline of around 34% and recorded a low value of 2237.40 points on 23 March 2020. The price of FTSE 100 also registered a decline of about 30% for a single month and fell from 7403.9 points on 16 February 2020 to 5190.8 points on 15 March 2020. Likewise, the price of CAC 40 fell from 6111.24 points on 19 February 2020 to 3754.84 points on 18 March 2020, a drop of around 39% for a single month. The literature explains financial volatility through factors related to economic conditions, institutional problems, market uncertainty, good or bad announcements, and economic policy uncertainty [10,14–23]. There is a fair amount of research that has focused on estimating and forecasting the economic and financial costs of pandemics [1]. For example, the study of the economic costs of the HIV pandemic [24], the impact of the HIV pandemic on development [25], the costs of growing global obesity and diabetes [26], the work on forecasting the economic costs for possible future pandemics which has highlighted the importance of good health system management to address the people affected and tackling outbreaks as well as the negative impact of social distancing on economic activity [27], with the study indicating the need to prepare for pandemics and estimating the value of the annual losses due to a possible pandemic at around 500 billion US dollars, or 0.6% of global income [28]-currently considered to be underestimated [1]-the work highlighting the need for economic risk management versus potential probability future pandemics [29,30]. The objective of this paper is, therefore, to identify econometric models likely to model the processes of the series of daily returns of the main world stock market indices: SSE, S&P500, FTSE 100, DAX, CAC40, and NIKKEI 225, during the pre-COVID-19 period, from 1 June to 30 November 2019, and the post-COVID-19 period, from 31 December 2019 to 1 June 2020, in order to highlight the substantial impact of the COVID-19 pandemic on the financial markets. Since these series experience phases of calm or low volatility and phases of crisis or high volatility, the Markov-switching "GARCH and EGARCH" models constitute the econometric methods adequate to model their volatility during the period pre-COVID-19 and the post-COVID-19 one [31–36].

2. Results and Concluding Remarks

Both the graphical examination of our variables of interest and the unit root tests, i.e., the increased Dickey–Fuller, Phillips–Perron, and KPSS (Kwiatkovski, Phillips, Schmidt and Shin), show that the daily prices of the main world stock indices: SSE, S&P500, FTSE 100, DAX, CAC40, NIKKEI 225 are not stationary, while the series of daily returns of the same indices: RSSE, RS&P500, RFTSE 100, RDAX, RCAC40 and RNIKKEI 225, are stationary, during the pre-COVID-19 period and the post-COVID-19 one.

Tables 1–4 below present the results of the estimation of the Markov-switching "GARCH and EGARCH" models during the pre-COVID-19 and post-COVID-19 periods.

Given the results of the individual significance test of the coefficients, the information criteria (i.e., Bayesian Information Criteria (BIC)), and the Log-likelihood (Log(L)), the models suitable for modeling conditional volatility are, on the one hand, the Markov-switching EGARCH model, with normal distribution, for the RS&P500, RSSE, RDAX, RFTSE, and RNIKKEI series, the Markov-switching EGARCH model, with a generalized distribution of errors, for the RCAC series, during the pre-COVID-19 period; and, on the other hand, the Markov-switching EGARCH models, with normal distribution, for the RS&P500, RCAC, RFTSE, and RNIKKEI series, the Markov-switching EGARCH model, with a Student distribution, for the RDAX series, and the Markov-switching GARCH model, with normal distribution, for the RSSE series, during the post-COVID-19 period. Figure 1 below illustrates the graphical representations of the conditional volatilities estimated by the models indicated above, during the pre-COVID-19 period and the post-COVID-19 one, for the six series of our interest: RS&P500, RSSE, RDAX, RCAC, RFTSE, and RNIKKEI.

Table 1. Estimation of Markov-switching GARCH models during the pre-COVID-19 period from 1 June to 30 November 2019. $h_t^{(i)} = \alpha_0^{(i)} + \alpha_1^{(i)}\varepsilon_{t-1}^2 + \beta^{(i)}h_{t-1}$. Where h_{t-1} is the independent state of the past conditional variance; $\alpha_0^{(i)} > 0, \alpha_1^{(i)} \geq 0 \wedge \beta^{(i)} \geq 0$, with $i \in \{1;2\}$.

	RS&P500			RSSE			RDAX			RCAC			RFTSE			RNIKKEI 225		
	MS-nGARCH (Normal Distribution)	MS-sGARCH (Student Distribution)	MS-gedGARCH (GED Distribution)	MS-nGARCH	MS-sGARCH	MS-gedGARCH	MS-nGARCH	MS-sGARCH	MS-gedGARCH	MS-nGARCH	MS-sGARCH	MS-gedGARCH	MS-nGARCH	MS-sGARCH	MS-gedGARCH	MS-nGARCH	MS-sGARCH	MS-gedGARCH
$\alpha_0^{(1)}$	0.0397	0.0432	0.0413	0.0000	0.0142	0.0194 *	0.0594 *	0.0559	0.0583	0.0399	0.0300	0.0340	0.0219	0.1495	0.0196	0.2008 **	0.2442	0.2685
$\alpha_0^{(2)}$	0.2507 **	0.2244	0.2421 *	0.0000	0.0003	0.0001	0.1674 **	0.1194	0.1380	0.2722 *	0.1586	0.2093	0.0266	0.0001	0.0198	0.0000 ***	0.0001	0.0001
$\alpha_1^{(1)}$	0.0397	0.0432	0.0413	0.0004	0.0142 *	0.0194	0.0594	0.0559	0.0583	0.0399 *	0.0300	0.0340	2.3633	1.1209	3.4992	0.2008 ***	0.2442	0.2686
$\alpha_1^{(2)}$	0.2507 *	0.2244 *	0.2421	0.0367	0.0003	0.0001	0.1674 *	0.11943	0.1380	0.2722 **	0.1586	0.2093	0.0009	0.0032	0.0013	0.0000 ***	0.0001	0.0001
$\beta_1^{(1)}$	0.6836 ***	0.6959 ***	0.6860 ***	0.9999 ***	0.9800 ***	0.9724 ***	0.7545 ***	0.8025 ***	0.7770 ***	0.7090 ***	0.8119 ***	0.7623 ***	0.8962 ***	0.4959	0.9139 ***	0.6985 ***	0.6582 ***	0.5945 **
$\beta_1^{(2)}$	0.6837 ***	0.8787 ***	0.6860 ***	0.9632 ***	0.9800 ***	0.9724 ***	0.7545 ***	0.8025 ***	0.7770 ***	0.7090 ***	0.8119 ***	0.7624 ***	0.0229	0.1810	0.0920	0.6985 ***	0.6582 ***	0.5945 *
$v^{(1)}$		8.4613 *	1.5233 *		6.7499 **	1.4285 ***		4.8566	1.3430 **		3.9446 **	1.1658		99.7120 ***	3.7656 ***		4.6154 ***	1.3526 ***
$v^{(2)}$		8.4615 *	1.5233 *		6.7499 ***	1.4285 ***		4.8564 *	1.3430 ***		3.9447	1.1659 *		6.9040	19.9959 ***		4.6154 **	1.3525 ***
P_{11}	0.9708 ***	0.9204 ***	0.9708 ***	0.4511 ***	0.0000	0.5745 ***	0.9635 ***	0.9135 ***	0.9135 ***	0.9174 ***	0.9481 ***	0.9174 ***	0.9675 ***	0.9687 ***	0.9623 ***	0.5000 ***	0.5936 ***	0.5000 **
P_{21}	0.0796 ***	0.0292	0.0796 **	1.0000	0.4255 ***	1.0000 ***	0.0865 ***	0.0365	0.0365	0.0519 **	0.0826	0.0519	0.3088 **	0.1067 ***	0.6949 ***	0.4064 ***	0.5000 ***	0.4064 *
Log(L)	−133.3526	−131.4676	−131.982	−145.6943	−149.1706	−149.2006	−153.8733	−149.0492	−150.3224	−151.7521	−142.9516	−144.7371	−131.3101	−132.3429	−128.8778	−147.0999	−144.4282	−144.4264
BIC	305.3954	311.2981	312.3268	329.8209	346.3814	346.4414	346.5001	346.5403	349.0866	342.3827	334.5014	338.0724	301.4365	313.2061	306.276	332.632	336.8965	336.893

*: indicates the significance of the coefficient at the statistical threshold of 10%; **: indicates the significance of the coefficient at the statistical threshold of 5%; and ***: indicates the significance of the coefficient at the statistical threshold of 1%.

Table 2. Estimation of Markov-switching GARCH models during the post-COVID-19 period from 31 December 2019 to 1 June 2020. $h_t^{(i)} = \alpha_0^{(i)} + \alpha_1^{(i)}\varepsilon_{t-1}^2 + \beta^{(i)}h_{t-1}$. Where h_{t-1} is the independent state of the past conditional variance; $\alpha_0^{(i)} > 0, \alpha_1^{(i)} \geq 0 \wedge \beta^{(i)} \geq 0$, with i $\in\{1;2\}$.

	RS&P500			RSSE			RDAX			RCAC			RFTSE			RNIKKEI 225		
	MS-nGARCH	MS-sGARCH	MS-gedGARCH	MS-nGARCH	MS-sGARCH	MS-gedGARCH	MS-nGARCH	MS-sGARCH	MS-gedGARCH	MS-nGARCH	MS-sGARCH	MS-gedGARCH	MS-nGARCH	MS-sGARCH	MS-gedGARCH	MS-nGARCH	MS-sGARCH	MS-gedGARCH
$\alpha_0^{(1)}$	0.1409	0.0878	0.1153	1.0390 **	0.0252	1.6457	0.2328 **	0.0282	0.1110	0.1191 *	0.0000	0.3499	0.1872	0.0000	0.5019	0.2429	0.2229	4.2051
$\alpha_0^{(2)}$	0.2992 *	0.2813	0.2932	0.9999 ***	0.0708	0.9999 ***	0.1970 **	0.1047	0.1470	0.5735 *	0.1619	0.2245	0.2845	0.1756	0.2298	0.1816 **	0.1743	0.9998 ***
$\alpha_1^{(1)}$	0.1409	0.0878	0.1153	1.0390 *	0.0252 **	1.6460 **	0.2329	0.0282	0.1110	1.1660 *	0.0000	0.3499	0.1874 *	0.0000	0.5019	0.2429	0.2229	4.2051
$\alpha_1^{(2)}$	0.2992 **	0.2813 *	0.2932	0.9999 ***	0.0708	0.9999 ***	0.1970	0.1048	0.1470	0.1796	0.1619	0.2245	0.2852 **	0.1757	0.2298 *	0.1816	0.1743	0.9998 ***
$\beta_1^{(1)}$	0.6498 ***	0.6764 ***	0.6601 ***	0.0000 ***	0.9143 ***	0.0000	0.7566 ***	0.8787 ***	0.8172 ***	0.3134 ***	0.8380 ***	0.7214 ***	0.6637 ***	0.8242 ***	0.7074 ***	0.7526 ***	0.7641 ***	0.0001
$\beta_1^{(2)}$	0.6498 ***	0.6764 ***	0.6601 ***	0.0000 ***	0.9143 ***	0.0000	0.7566 ***	0.8787 ***	0.8172 ***	0.7291 ***	0.8380 ***	0.7215 ***	0.6631 ***	0.8242 ***	0.7074 ***	0.7526 ***	0.7641 ***	0.0001
$v^{(1)}$		11.1271 **	1.6108 ***		3.2933 **	0.7000 ***		4.5435 ***	1.0829		4.8549 ***	0.7000 ***		6.6031 ***	0.7000 ***		11.3866	0.7000 ***

Table 2. *Cont.*

	RS&P500			RSSE			RDAX			RCAC			RFTSE			RNIKKEI 225		
	MS-nGARCH	MS-sGARCH	MS-gedGARCH	MS-nGARCH	MS-sGARCH	MS-gedGARCH	MS-nGARCH	MS-sGARCH	MS-gedGARCH	MS-nGARCH	MS-sGARCH	MS-gedGARCH	MS-nGARCH	MS-sGARCH	MS-gedGARCH	MS-nGARCH	MS-sGARCH	MS-gedGARCH
$v^{(2)}$		11.1273 **	1.6108 ***		3.2936	0.7000 ***		4.5434	1.0829 ***		4.8545	0.7000 ***		6.5993	0.7000 ***		11.3867	0.7000 ***
P_{11}	0.9747 ***	0.9747 ***	0.9820 ***	0.9624 ***	0.9624 ***	0.9013	0.9992 ***	0.9992 ***	0.9715 ***	0.9834 ***	0.9996	0.9996 ***	0.9724 ***	0.9996	0.9724 ***	0.9813 ***	0.9596 ***	0.9596
P_{21}	0.0180 **	0.0180	0.0253 ***	0.0987 ***	0.0987 ***	0.0376	0.0285 ***	0.0285 ***	0.0008	0.0120	0.0276	0.0276 ***	0.0004	0.0276	0.0004	0.0404 ***	0.0187	0.0187
Log(L)	−216.1008	−214.935	−215.2672	−164.8534	−156.4963	−161.0797	−233.2177	−224.3603	−225.0192	−218.9026	−218.1442	−227.0016	−215.6416	−210.3256	−220.6903	−199.1081	−198.6219	−206.969
BIC	469.2794	476.2174	476.8816	366.3045	358.7397	367.9064	503.5132	495.068	496.3858	474.9604	482.7323	500.4472	468.3611	466.9985	487.7279	434.8139	442.991	459.6852

*: indicates the significance of the coefficient at the statistical threshold of 10%; **: indicates the significance of the coefficient at the statistical threshold of 5%; and ***: indicates the significance of the coefficient at the statistical threshold of 1%.

Table 3. Estimation of Markov-switching EGARCH models during the pre-COVID-19 period from 1 June to 30 November 2019. $log\left(h_t^{(i)}\right) = \alpha_0^{(i)} + \alpha_1^{(i)}\left|\frac{\varepsilon_{t-1}}{h_{t-1}}\right| + \alpha_2^{(i)}\frac{\varepsilon_{t-1}}{h_{t-1}} + \beta^{(i)}log(h_{t-1})$. Where $i \in \{1,2\}$ and h_{t-1} is the independent state of the past conditional variance.

	RS&P500			RSSE			RDAX			RCAC			RFTSE			RNIKKEI 225		
	MS-nEGARCH	MS-sEGARCH	MS-gedEGARCH	MS-nEGARCH	MS-sEGARCH	MS-ged-EGARCH	MS-nEGARCH	MS-sEGARCH	MS-ged-EGARCH	MS-nEGARCH	MS-sEGARCH	MS-ged-EGARCH	MS-nEGARCH	MS-sEGARCH	MS-ged-EGARCH	MS-nEGARCH	MS-sEGARCH	MS-ged-EGARCH
$\alpha_0^{(1)}$	−0.0311	−0.0334	−0.0326	−0.1773	−0.1448	−0.1549	−0.0116 ***	−0.0023 ***	−0.0093	−0.0153 ***	−0.0022	−0.0114 ***	−1.1020 ***	−0.0620 ***	−0.0654 ***	−0.1106 ***	−0.0637 ***	−0.0813 ***
$\alpha_0^{(2)}$	−0.0311	−0.0334	−0.0325	−0.1773	−0.1448	−0.1549 ***	−0.0115 ***	−0.0023 ***	−0.0085	−0.0153 ***	−0.0022	−0.0114 ***	−0.2666 **	−0.0606 ***	−0.0737 ***	−0.1101 ***	−0.0637 ***	−0.0813 ***
$\alpha_1^{(1)}$	0.2485 ***	0.2491 ***	0.2502 ***	−0.3709 ***	−0.3586 **	−0.3712 *	−0.2009 ***	−0.2288 ***	−0.1436 ***	−0.2602 ***	−0.2275 ***	−0.2284 ***	0.4673 ***	−0.3216 ***	−0.3363 ***	−0.5506 ***	−0.5615 ***	−0.4395 ***
$\alpha_1^{(2)}$	0.2485 ***	0.2491 ***	0.2502 ***	−0.3709 ***	−0.3586 *	−0.3712 **	−0.2009 ***	−0.2288 ***	−0.1438 ***	−0.2602 ***	−0.2275 ***	−0.2284 ***	−1.4492 ***	−0.2780 ***	−0.3501 ***	−0.5509 ***	−0.5615 ***	−0.4395 ***
$\alpha_2^{(1)}$	−0.3162 ***	−0.3259 ***	−0.3204 ***	−0.1616**	−0.1374	−0.1467	−0.2628 ***	−0.2543 ***	−0.2025 ***	−0.3680 ***	−0.4195 ***	−0.3489 ***	0.3644 ***	−0.2453 ***	−0.2596 ***	−0.1819 ***	−0.2344 ***	−0.1823 ***
$\alpha_2^{(2)}$	−0.3162 ***	−0.3259 ***	−0.3204 ***	−0.1616**	−0.1374	−0.1467	−0.2628 ***	−0.2543 ***	−0.2026 ***	−0.3680 ***	−0.4195 ***	−0.3489 ***	−1.2816 ***	−0.2606 ***	−0.3138 ***	−0.1820 ***	−0.2344 ***	−0.1823 ***
$\beta_1^{(1)}$	0.9138 ***	0.9103 ***	0.9123 ***	0.5603 ***	0.6049 ***	0.5893 ***	0.9427 ***	0.9491 ***	0.9504 ***	0.9345 ***	0.9268 ***	0.9336 ***	0.0895 *	0.9088 ***	0.9071 ***	0.7769 ***	0.8058 ***	0.8172 ***
$\beta_1^{(2)}$	0.9138 ***	0.9103 ***	0.9123 ***	0.5603 ***	0.6049 ***	0.5893 ***	0.9427 ***	0.9491 ***	0.9504 ***	0.9345 ***	0.9268 ***	0.9336 ***	0.3108 ***	0.9079 ***	0.8950 ***	0.7769 ***	0.8058 ***	0.8172 ***
$v^{(1)}$		24.3579 ***	1.8826 ***		7.4575 ***	1.4773 ***		4.8542 ***	0.9504 ***		4.3069 ***	1.5106 ***		6.3084 ***	1.4212 ***		5.6409 ***	1.7891 ***
$v^{(2)}$		24.3579 ***	1.8826 ***		7.4575 ***	1.4773 ***		4.8542 ***	0.9504 ***		4.3069 ***	1.5106 ***		6.3110 ***	1.3899 ***		5.6409 ***	1.7891 ***
P_{11}	0.9708 ***	0.9708 ***	0.9708 ***	0.5745 ***	0.5745 ***	0.0000	0.9635 ***	0.9635 ***	0.9135 ***	0.9481 ***	0.9174 ***	0.9481 ***	0.9148 ***	0.5002 ***	0.5014 ***	0.5000 ***	0.5936 ***	0.5000 ***
P_{21}	0.0796 ***	0.0796 ***	0.0796 ***	1.0000 ***	1.0000 ***	0.4255 ***	0.0865 ***	0.0865 ***	0.0365 ***	0.0826 ***	0.0519 ***	0.0826 ***	0.3550 ***	0.5002 ***	0.5009 ***	0.4064 ***	0.5000 ***	0.4064 ***
Log(L)	−123.5262	−123.4272	−123.476	−148.5441	−146.9717	−147.1036	−140.0383	−139.6108	−140.959	−138.0418	−133.0186	−132.0914	−126.9224	−128.758	−128.2535	−131.8982	−132.0324	−132.3902
BIC	295.4153	304.8897	304.9874	345.1284	351.5917	351.8555	328.5185	337.3519	340.0481	324.6817	324.3549	322.5006	302.365	315.7403	314.7314	311.8367	321.7131	322.4287

*: indicates the significance of the coefficient at the statistical threshold of 10%; **: indicates the significance of the coefficient at the statistical threshold of 5%; and ***: indicates the significance of the coefficient at the statistical threshold of 1%.

Table 4. Estimate of EGARCH regime change models during the post-COVID-19 period from 31 December 2019 to 1 June 2020. $log\left(h_t^{(i)}\right) = \alpha_0^{(i)} + \alpha_1^{(i)}\left|\frac{\varepsilon_{t-1}}{h_{t-1}}\right| + \alpha_2^{(i)}\frac{\varepsilon_{t-1}}{h_{t-1}} + \beta^{(i)}log(h_{t-1})$. Where $i \in \{1,2\}$ and h_{t-1} is the independent state of the past conditional variance.

	RS&P500			RSSE			RDAX			RCAC			RFTSE			RNIKKEI 225		
	MS-nEGARCH	MS-sEGARCH	MS-gedEGARCH	MS-nEGARCH	MS-sEGARCH	MS-ged-EGARCH	MS-nEGARCH	MS-sEGARCH	MS-ged-EGARCH	MS-nEGARCH	MS-sEGARCH	MS-ged-EGARCH	MS-nEGARCH	MS-sEGARCH	MS-ged-EGARCH	MS-nEGARCH	MS-sEGARCH	MS-ged-EGARCH
$\alpha_0^{(1)}$	−0.0158 ***	−0.0118 *	−0.0151 ***	0.0257	0.0385	0.0261	−0.0006	−0.0008	−0.0001	−0.0167 ***	−0.0139	−0.0101 ***	−0.0134 ***	−0.0117 ***	−0.0099 ***	0.0149 ***	0.0082 ***	0.0141 ***
$\alpha_0^{(2)}$	−0.0157 ***	−0.0116 *	−0.0151 ***	0.0258	0.0385	0.0261	−0.0006	−0.0008	−0.0001	−0.0167 ***	−0.0139	−0.0101 ***	−0.0134 ***	−0.0117 ***	−0.0099 ***	0.0151 ***	0.0084 ***	0.0142 ***
$\alpha_1^{(1)}$	−0.4035 ***	−0.3505 *	−0.3677 ***	0.0542	0.0405	0.0464	−0.1609 ***	−0.1560 ***	−0.1461 ***	−0.2236 ***	−0.2022 ***	−0.2031 ***	−0.1686 ***	−0.1764 ***	−0.2013 ***	−0.1915 ***	−0.1706 ***	−0.2003 ***
$\alpha_1^{(2)}$	−0.4035 ***	−0.3505 **	−0.3677 ***	0.0542	0.0405	0.0464	−0.1609 ***	−0.1560 ***	−0.1461 ***	−0.2236 ***	−0.2022**	−0.2031 ***	−0.1686 ***	−0.1764 ***	−0.2013 ***	−0.1915 ***	−0.1707 ***	−0.2003 ***
$\alpha_2^{(1)}$	−0.7525 ***	−0.6938 ***	−0.7034 ***	−0.7902 ***	−0.6965 **	−0.7399 **	−0.2850 ***	−0.2914 ***	−0.2875 ***	−0.3796 ***	−0.3974 ***	−0.3351 ***	−0.316 ***	−0.3271 ***	−0.3036 ***	−0.2588 ***	−0.2667 ***	−0.2965 ***
$\alpha_2^{(2)}$	−0.7521 ***	−0.6932 ***	−0.7034 ***	−0.7902 ***	−0.6965	−0.7399	−0.2850 ***	−0.2914	−0.2875 ***	−0.3796 ***	−0.3974 ***	−0.3351 ***	−0.316 ***	−0.3271 ***	−0.3036 ***	−0.2587 ***	−0.2665 ***	−0.2965 ***
$\beta_1^{(1)}$	0.9561 ***	0.9609 ***	0.9597 ***	0.7145 ***	0.7429 ***	0.7251 ***	0.9801 ***	0.9858 ***	0.9839 ***	0.9761 ***	0.9778 ***	0.9810 ***	0.9825 ***	0.9841 ***	0.9868 ***	0.9699 ***	0.9778 ***	0.9679 ***
$\beta_1^{(2)}$	0.9561 ***	0.9609 ***	0.9597 ***	0.7145 ***	0.7429 ***	0.7251 ***	0.9801 ***	0.9858 ***	0.9839 ***	0.9761 ***	0.9778 ***	0.9810 ***	0.9825 ***	0.9841 ***	0.9868 ***	0.9699 ***	0.9778 ***	0.9679 ***
$v^{(1)}$		55.0694 ***	2.3187 ***		5.3726 ***	1.3823 ***		5.1847 ***	1.3514 ***		6.2915 ***	1.5078 ***		7.0457 ***	1.5115 ***		29.1137 ***	1.9323 ***
$v^{(2)}$		55.0694 ***	2.3187 ***		5.3726 ***	1.3823 ***		5.1847 ***	1.3514 ***		6.2915 ***	1.5078 ***		7.0457 ***	1.5115 ***		29.1137 ***	1.9323 ***
P_{11}	0.9747 ***	0.9747 ***	0.9820 ***	0.9624 ***	0.9624 ***	0.9624 ***	0.9715 ***	0.9992 ***	0.9715 ***	0.9724 ***	0.9724 ***	0.9724 ***	0.9724 ***	00.9724 ***	0.9724 ***	0.9813 ***	0.9596 ***	0.9596 ***
P_{21}	0.0180 ***	0.0180 ***	0.0253 ***	0.0987 ***	0.0987 ***	0.0987 ***	0.0008 ***	0.0285 ***	0.0008 ***	0.0004 ***	0.0004	0.0004 ***	0.0004 ***	0.0004 ***	0.0004 ***	0.0404 ***	0.0187 ***	0.0187 ***
Log(L)	−201.2783	−201.4655	−201.0586	−151.7309	−150.2307	−150.0951	−216.7804	−214.6281	−215.3432	−209.9184	−209.109	−209.8444	−202.4837	−201.7522	−201.9135	−188.9213	−188.9126	−188.6964
BIC	448.9038	458.5478	457.734	349.2089	355.3579	355.0866	479.9082	484.8729	486.3032	466.2807	473.9507	475.4216	451.3147	459.1211	459.4438	423.5896	432.7216	432.2894

*: indicates the significance of the coefficient at the statistical threshold of 10%; **: indicates the significance of the coefficient at the statistical threshold of 5%; and ***: indicates the significance of the coefficient at the statistical threshold of 1%.

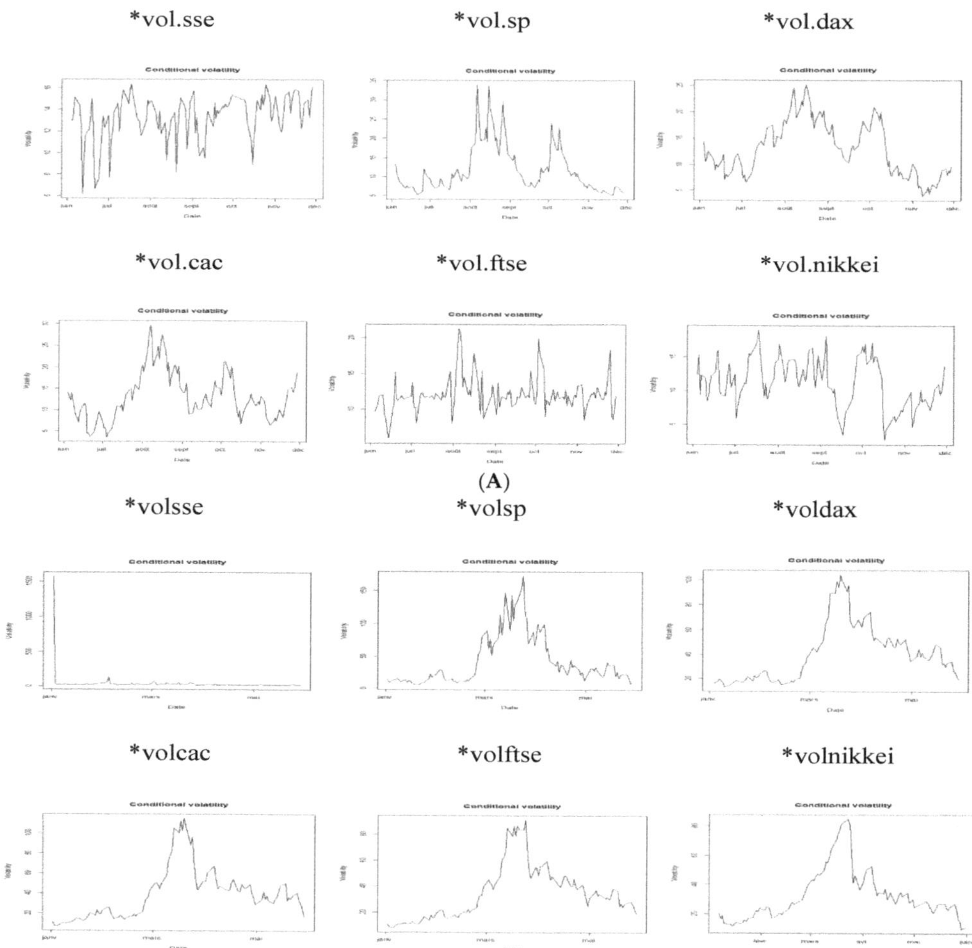

(B)

Figure 1. Conditional volatility during the pre-COVID-19 and post-COVID-19 period. (**A**) Pre-COVID-19 period from 1 June to 30 November 2019. (**B**) Post-COVID-19 period from 31 December 2019 to 1 June 2020.

From this figure, we see that the extent (maximum-minimum) is very large during the post-COVID-19 period compared to the pre-COVID-19 period. Apart from the stock market of the epicenter country of the COVID-19 pandemic (the Chinese stock market), for which conditional volatility (denoted "volsse") reached its peak in "late December-early January" during the post-COVID-19 period, volatilities conditions of the other five stock markets (the New York, Paris, Frankfurt, London, and Tokyo stock exchanges) reached their climax in March 2020. This finding clearly illustrates the increase in volatility in the main stock markets due to the pandemic of COVID-19.

The indicators of central tendency and position (the first quartile, the median, the third quartile, the arithmetic mean), the minimum value, the maximum value, and the dispersion indicators (the standard deviation, the coefficient of variation, and the interquartile range) of conditional volatility (Tables 5 and 6) increased during the post-COVID-19 period compared to the pre-COVID-19 period. This increase clearly illustrates the increase in volatility in major stock markets due to the COVID-19 pandemic.

Table 5. Statistical indicators of conditional volatility during the pre-COVID-19 period.

	Min.	1st Qu.	Median	Mean	3rd Qu.	Max.	Std. Dev.	Interquartile Coefficient
S&P	5.134	7.262	9.663	11.77	14.275	33.858	6.263305	0.725758046
SSE	6.156	12.171	13.616	13.218	14.766	16.375	2.089393	0.190584606
DAX	3.962	8.628	12.042	12.961	17.328	25.193	5.18288	0.72247135
CAC	3.682	9.25	12.389	13.273	16.346	29.703	5.59991	0.572766164
FTSE	5.849	10.977	11.667	11.963	12.457	21.238	2.40386	0.126853518
NIKKEI	2.672	8.664	11.171	11.195	14.086	18.974	3.578993	0.485363889

Table 6. Statistical indicators of conditional volatility during the post-COVID-19 period.

	Min.	1st Qu.	Median	Mean	3rd Qu.	Max.	Std. Dev.	Interquartile Coefficient
S&P	4.89529	13.23112	27.1176	42.77574	62.66145	172.00951	39.30484	1.822813597
SSE	16.12	16.9	19.17	40.42	26.43	1569	157.4888	0.497130934
DAX	13.31	19.29	37.62	40.35	52.3	103.76	23.27155	0.877458799
CAC	7.773	17.917	34.91	38.874	49.572	114.521	25.16885	0.906760241
FTSE	8.316	17.32	31.563	34.389	45.535	90.947	20.76365	0.893926433
NIKKEI	10.06	21.05	27.23	32.47	39.47	85	17.08098	0.676459787

According to the values of the transition probability P_{21}, we see that the chances of passing from the state of crisis at t-1 to the state of stability at t have greatly decreased, during the post-COVID-19 period compared to that pre-COVID-19, for all stock markets in our sample (Tables 7 and 8). For example, this probability goes, between the pre-COVID-19 and post-COVID-19 period, from 0.355 to 0.0004 for the stock exchange of London, from 0.4064 to 0.0404 for the stock exchange of Tokyo, and from 0.0826 to 0.0004 for the stock exchange of Paris. It should be noted that the stock exchange of Frankfurt recorded the smallest decrease in this probability between the pre-COVID-19 and the post-COVID-19 period, a drop from 0.0865 to 0.0285. Based on the values of the unconditional probability in the stable state π_1 and the unconditional probability in the crisis state π_2, we see that the COVID-19 pandemic has had a negative impact on the stock exchanges of New York, Paris, and London. In other words, the proportion of observations that should be in a state of crisis (π_2) increased significantly, during the post-COVID-19 period compared to the pre-COVID-19 period, for these three stock exchanges. Between the pre-COVID-19 and post-COVID-19 period, this proportion rose from 19.35% to 98.57% for the stock exchange of London, from 38.59% to 98.57% for the stock exchange of Paris, and from 26.84% to 58.43% for the stock exchange of New York. In return, the stock exchange of Frankfurt showed great resilience, compared to other international stock exchanges, with the recording of a very low value of the unconditional probability of the state of crisis π_2 and a very high value of the unconditional probability of the state of stability π_1, during the post-COVID-19 period compared to the pre-COVID-19 period. In fact, the proportion of observations that should be in the state of crisis increased from 38.59%, during the pre-COVID-19 period, to 2.73%, during the post-COVID-19 period, for this stock exchange. It should also be noted that, to a lesser extent, the stock exchange of Tokyo has also shown a certain resilience, with a decrease in the proportion of observations that should be in the state of crisis during the post-COVID-19 period compared to the pre-COVID-19, or 31.64% against 55.16%. These latest results are supported by the variations in the values of the expected duration conditional on the state of crisis and the expected duration conditional on the state of stability, namely: the significant increase in the expected period of high volatility ($1/1 - P_{22}$) during the post-COVID-19 period compared to that pre-COVID-19 for the stock exchanges of Paris, London, and New York, as well as the significant increase in the expected period of low volatility ($1/1 - P_{11}$) during the post-COVID-19 period compared to that pre-COVID-19 for the stock exchange of Frankfurt. On the one hand, we can expect a period of high volatility equal to 2500 days, or 1.6 years, during the post-COVID-19 period against a period of high volatility equal, respectively, to almost 12 days and 3 days during the pre-COVID-19 period for stock exchanges in Paris and London. On the other hand, we can expect a period of high volatility equal to only 35 days during the post-COVID-19 period against a period of high volatility equal to almost 11 days during the pre-COVID-19 period for the stock exchange of Frankfurt. The resilience of this stock exchange in relation to the COVID-19 pandemic is illustrated by the expected period of low volatility equal to 1250 days, or 0.8 years, during the post-COVID-19 period, against an expected period of low volatility equal to 27 days, during the pre-COVID-19 period, for the German stock market.

Table 7. Transition probabilities, unconditional probabilities, and conditional anticipated duration (the pre-COVID-19 period).

Transition Probabilities, Unconditional Probabilities and Conditional Anticipated Duration	RS&P500	RSSE	RDAX	RCAC	RFTSE	RNIKKEI225
P_{11}	0.9708	0.5745	0.9635	0.9481	0.9148	0.5
$P_{22} = 1 - P_{21}$	0.9204	0	0.9135	0.9174	0.645	0.5936
$P_{12} = 1 - P_{11}$	0.0292	0.4255	0.0365	0.0519	0.0852	0.5
P_{21}	0.0796	1	0.0865	0.0826	0.355	0.4064
$\pi_1 = \frac{1-P_{22}}{2-P_{22}-P_{11}}$	0.7316	0.7015	0.7033	0.6141	0,8065	0.4484
$\pi_2 = \frac{1-P_{11}}{2-P_{22}-P_{11}}$	0.2684	0.2985	0.2967	0.3859	0.1935	0.5516
Conditional anticipated duration on the state of crisis = $1/1 - P_{22}$	12.5628	1.0000	11.5607	12.1065	2.8169	2.4606
Conditional anticipated duration on the state of stability = $1/1 - P_{11}$	34.2466	2.3502	27.3973	19.2678	11.7371	2.0000

Table 8. Transition probabilities, unconditional probabilities, and conditional anticipated duration (the post-COVID-19 period).

Transition probabilities, Unconditional Probabilities, and Conditional Anticipated Duration	RS&P500	RSSE	RDAX	RCAC	RFTSE	RNIKKEI225
P_{11}	0.9747	0.9624	0.9992	0.9724	0.9724	0.9813
$P_{22} = 1 - P_{21}$	0.982	0.9013	0.9715	0.9996	0.9996	0.9596
$P_{12} = 1 - P_{11}$	0.0253	0.0376	0.0008	0.0276	0.0276	0.0187
P_{21}	0.018	0.0987	0.0285	0.0004	0.0004	0.0404
$\pi_1 = \frac{1-P_{22}}{2-P_{22}-P_{11}}$	0.4157	0.7241	0.9727	0.0143	0.0143	0.6836
$\pi_2 = \frac{1-P_{11}}{2-P_{22}-P_{11}}$	0.5843	0.2759	0.0273	0.9857	0.9857	0.3164
Conditional anticipated duration on the state of crisis = $1/1 - P_{22}$	55.5556	10.1317	35.0877	2500.0000	2500.0000	24.7525
Conditional anticipated duration on the state of stability = $1/1 - P_{11}$	39.5257	26.5957	1250.0000	36.2319	36.2319	53.4759

Funding: This research received no external funding.

Institutional Review Board Statement: Not applicable.

Informed Consent Statement: Not applicable.

Data Availability Statement: Data available on request from the author.

Conflicts of Interest: The author declares no conflict of interest.

References

1. Goodell, J.W. COVID-19 and finance: Agendas for future research. *Financ. Res. Lett.* **2020**, *35*, 101512. [CrossRef] [PubMed]
2. Albulescu, C.T. Coronavirus and financial volatility: 40 days of fasting and fear. *Financ. Res. Lett.* **2020**, *11*, 454–462. [CrossRef]
3. Bakas, D.; Triantafyllou, A. Commodity price volatility and the economic uncertainty of pandemics. *Econ. Lett.* **2020**, *193*, 109283. [CrossRef]
4. Zaremba, A.; Kizys, R.; Aharon, D.Y.; Demir, E. Infected markets: Novel coronavirus, government interventions, and stock return volatility around the globe. *Financ. Res. Lett.* **2020**, *35*, 101597. [CrossRef]
5. Chesney, M.; Reshetar, G.; Karaman, M. The impact of terrorism on financial markets: An empirical study. *J. Bank. Financ.* **2011**, *35*, 253–267. [CrossRef]
6. Choudhry, T. September 11 and time-varying beta of United States companies. *Appl. Financ. Econ.* **2005**, *15*, 1227–1242. [CrossRef]
7. Corbet, S.; Gurdgiev, C.; Meegan, A. Long-term stock market volatility and the influence of terrorist attacks in Europe. *Q. Rev. Econ. Financ.* **2018**, *68*, 118–131. [CrossRef]
8. Hon, M.T.; Strauss, J.; Yong, S.-K. Contagion in financial markets after September 11: Myth or reality? *J. Financ. Res.* **2004**, *27*, 95–114. [CrossRef]
9. Nikkinen, J.; Vähämaa, S. Terrorism and stock market sentiment. *Financ. Rev.* **2010**, *45*, 263–275. [CrossRef]
10. Albulescu, C.T. COVID-19 and the United States financial markets' volatility. *Financ. Res. Lett.* **2021**, *38*, 101699. [CrossRef]
11. IMF. World Economic Outlook, April 2020: The Great Lockdown. 2020. Available online: https://www.imf.org/en/Publications/WEO/Issues/2020/04/14/weo-april-2020 (accessed on 1 June 2020).
12. Ashraf, B.N. Stock markets' reaction to COVID-19: Cases or fatalities? *Res. Int. Bus. Financ.* **2020**, *54*, 101249. [CrossRef] [PubMed]
13. Zhang, D.; Hu, M.; Ji, Q. Financial markets under the global pandemic of COVID-19. *Financ. Res. Lett.* **2020**, *36*, 101528. [CrossRef] [PubMed]
14. Antonakakis, N.; Chatziantoniou, I.; Filis, G. Dynamic co-movements of stock market returns, implied volatility and policy uncertainty. *Econ. Lett.* **2013**, *120*, 87–92. [CrossRef]
15. Chen, X.; Chiang, T.C. Empirical investigation of changes in policy uncertainty on stock returns—Evidence from China's market. *Res. Int. Bus. Financ.* **2020**, *53*, 101183. [CrossRef]
16. Hartwell, C.A. The impact of institutional volatility on financial volatility in transition economies. *J. Comp. Econ.* **2018**, *46*, 598–615. [CrossRef]
17. Kalyvas, A.; Papakyriakou, P.; Sakkas, A.; Urquhart, A. What drives Bitcoin's price crash risk? *Econ. Lett.* **2019**, *191*, 108777. [CrossRef]
18. Li, T.; Ma, F.; Zhang, X.; Zhang, Y. Economic policy uncertainty and the Chinese stock market volatility: Novel evidence. *Econ. Model.* **2020**, *87*, 24–33. [CrossRef]
19. Mei, D.; Zeng, Q.; Zhang, Y.; Hou, W. Does US Economic Policy Uncertainty matter for European stock markets volatility? *Phys. A Stat. Mech. Its Appl.* **2018**, *512*, 215–221. [CrossRef]
20. Onan, M.; Salih, A.; Yasar, B. Impact of macroeconomic announcements on implied volatility slope of SPX options and VIX. *Financ. Res. Lett.* **2014**, *11*, 454–462. [CrossRef]
21. Su, Z.; Fang, T.; Yin, L. Understanding stock market volatility: What is the role of U.S. uncertainty? *N. Am. J. Econ. Financ.* **2019**, *48*, 582–590. [CrossRef]
22. Tiwari, A.K.; Jana, R.; Roubaud, D. The policy uncertainty and market volatility puzzle: Evidence from wavelet analysis. *Financ. Res. Lett.* **2019**, *31*. [CrossRef]
23. Zhu, S.; Liu, Q.; Wang, Y.; Wei, Y.; Wei, G. Which fear index matters for predicting US stock market volatilities: Text-counts or option based measurement? *Phys. A Stat. Mech. Its Appl.* **2019**, *536*, 122567. [CrossRef]
24. Haacker, M. *The Impact of HIV/AIDS on Government Finance and Public Services*; IMF: Washington, DC, USA, 2004.
25. Santaeulalia-Llopis, R. Aggregate Effects of AIDS on Development. Washington University in St. Louis Working Paper. 2008. Available online: http://www.eco.uc3m.es/temp/agenda/Santaeulalia_LlopisRaul_jmp1.pdf (accessed on 15 May 2020).
26. Yach, D.; Stuckler, D.; Brownell, K.D. Epidemiologic and economic consequences of the global epidemics of obesity and diabetes. *Nat. Med.* **2006**, *12*, 62–66. [CrossRef] [PubMed]
27. Bloom, D.E.; Cadarette, D.; Sevilla, J.P. Epidemics and economics: New and resurgent infectious diseases can have far-reaching economic repercussions. *Financ. Dev.* **2018**, *55*, 46–49.
28. Fan, V.Y.; Jamison, D.T.; Summers, L.H. Pandemic risk: How large are the expected losses? *Bull. World Health Organ.* **2018**, *96*, 129–134. [CrossRef] [PubMed]
29. Lewis, M. The Economics of Epidemics. *Georget. J. Int. Aff.* **2001**, *2*, 25–31.

30. Tam, C.C.; Khan, M.S.; Legido-Quigley, H. Where economics and epidemics collide: Migrant workers and emerging infections. *Lancet* **2016**, *388*, 1374–1376. [CrossRef] [PubMed]
31. Cai, J. A Markov Model of Switching-Regime ARCH. *J. Bus. Econ. Stat.* **1994**, *12*, 309. [CrossRef]
32. Hamilton, J.D.; Susmel, R. Autoregressive conditional heteroskedasticity and changes in regime. *J. Econ.* **1994**, *64*, 307–333. [CrossRef]
33. Ardia, D. *Financial Risk Management with Bayesian Estimation of GARCH Models: Theory and Applications*; Springer: Berlin/Heidelberg, Germany, 2008.
34. Marcucci, J. Forecasting stock market volatility with regime-switching GARCH models. *Stud. Nonlinear Dyn. Econ.* **2005**, *9*, 1558–3708. [CrossRef]
35. Hansen, P.R.; Lunde, A. A forecast comparison of volatility models: Does anything beat a GARCH(1,1)? *J. Appl. Econ.* **2005**, *20*, 873–889. [CrossRef]
36. Ouchen, A. Is the ESG portfolio less turbulent than a market benchmark portfolio? *Risk Manag.* **2022**, *24*, 1–33. [CrossRef]

engineering proceedings

MDPI

Proceeding Paper

Modeling Energy Transition in US Commercial Real Estate: A Diffusion Comparison with the Industrial Sector †

Andrea Savio

Steers Center for Global Real Estate, McDonough School of Business, Georgetown University, Washington, DC 20057, USA; andrea.savio@georgetown.edu

† Presented at the 9th International Conference on Time Series and Forecasting, Gran Canaria, Spain, 12–14 July 2023.

Abstract: This paper proposes a refinement of the multivariate diffusion UCTT model to explore the energy transition in the US commercial sector. The model analyzes the electricity market's interdependencies between coal, gas, and biomass, allowing a deeper understanding of the system. In addition, the comparison with the industrial sector electric system provides a valuable indication of how the US commercial sector has solid grounds for a more ready and suitable environment to accelerate the energy transition.

Keywords: real estate; energy transition; multivariate diffusion model; renewable energy; commercial sector

Citation: Savio, A. Modeling Energy Transition in US Commercial Real Estate: A Diffusion Comparison with the Industrial Sector . *Eng. Proc.* **2023**, *39*, 15. https://doi.org/10.3390/engproc2023039015

Academic Editor: Ignacio Rojas, Hector Pomares, Luis Javier Herrera, Fernando Rojas and Olga Valenzuela

Published: 29 June 2023

1. Introduction

An extensive structural transformation in energy systems is denoted by the term energy transition.This transition is often referred to as the *decarbonization* of the energy sector and aims to shift the system to renewable energy technologies (RETs), implying a change from centralized to decentralized energy production [1]. According to the International Renewable Energy Agency [2], the use of appropriate technology and regulations in all sectors, including real estate, may potentially reduce carbon emissions from the energy sector by 90%. In recent years, many studies have been conducted on the causal relationships between green energy consumption and economic growth in the US [3–6], emphasizing how institutional and political policies have impacted on the diffusion of US renewable energy and the decrease in fossil fuels [7]. In this context, the US commercial real estate industry has made significant strides in energy efficiency [8] and sustainability [9] using green energy sources. Many policies and initiatives have been put in place to encourage and facilitate the adoption of more environmentally friendly practices in this sector [10–12]. In this context, the Energy Performance of Buildings Directive requires all new real estate construction beginning after 2021 to adhere to the "virtually zero-energy buildings" standard in order to combat the property industry's GHG emissions' slow decline [13]. Incentives, technological advancements, and cost reductions all contribute to reducing barriers that hinder renewable energy development [14], paving the way to continue expanding the use of green energy in the future [15]. Energy efficiency and sustainability are becoming increasingly valued in the commercial sector not only for the environmental benefit aspects but also for investing in energy-efficient technologies, and infrastructure is becoming a sliding door for constructors and property managers to increase the market value of their properties [16].

Based on these premises, this project aims to analyze in depth how the energy transition path is developing in the commercial real estate sector. This study compares this green energy diffusion scenario with the one of another exemplary and significant US sector, the industrial one, focusing on examining the relationships between renewable and carbon fossil energy diffusion. Energy policies for the commercial and industrial sectors in the US are similar in many ways, as both sectors are subject to the same national- and state-level policies regarding energy efficiency and sustainability [17]. However, the specifics of these policies and regulations may differ based on several factors, such as the scale, energy intensity, and operational differences between the two sectors [18].

In the literature, diffusion models have been extremely valuable for defining and forecasting the development of an energy source, considering it as a technology that must be accepted in a market [19]. This well-established area of study [19–23] allows for analysis of the temporal dynamics of energy sources in order to comprehend the intricate dynamics of energy systems. Understanding how products or technologies compete or collaborate is essential for describing the trend of diffusion processes. Depending on the situation, the presence of competition can act both as a barrier to the growth of the innovation under consideration and as a stimulus for its development [24]. From this perspective, this project studies and compares commercial and industrial sectors' energy transitions in order to comprehend the intricate dynamics of energy systems through a refinement of the UCTT multivariate diffusion model presented in [25]. This paper examines the temporal diffusion of coal, gas, and biomass in the two sectors to identify peculiarities, similarities, and differences that characterize each energy system since the development and diffusion among different sectors can play a role in establishing technological innovation systems [26].

The rest of this paper is organized as follows. Section 2 clarifies the methodological approach based on the UCTT model refinement and presents the data of interest. In Section 3, the model is applied to the commercial and industrial sector cases, and the main outcomes are illustrated. Section 4 is left for concluding discussion about the findings. Furthermore, Appendix A reports a detailed explanation of the refinement method applied, while Appendix B discusses similarities and differences between the old and new UCTT model versions.

2. Materials and Methods

The data on the energy transition in the US commercial and industrial sectors were retrieved from the US Energy Information Administration [27] and refer to the energy consumption (billion Btus) for electricity generation and useful thermal output from 2003 to 2021 (2003, from which biomass data are available, is considered the starting date, even if the data for the other energy sources are available from 2001). The analysis focuses on three primary energy sources: *coal* (coal, petroleum liquids, petroleum coke), *gas* (natural gas), and *biomass* (wood waste biomass, landfill gas, biogenic municipal solid waste, other waste biomass). Figure 1 illustrates the observed time series for each sector separately.

Grounded on the diffusion model literature, the analysis has been implemented through a refinement of the *synchronic* form of the ODE multivariate diffusion model called *Unbalanced Competition for Three Technologies* (UCTT) [25] (for the industrial case, the constraint assumption ($\zeta = \rho + \xi$) has been considered on the model). The model is a system of differential equations in which $z'_i(t)$, $i = 1,2,3$ reflects the instantaneous consumption of the first, second, and third technologies, respectively, and $z_i(t), i = 1,2,3$ indicates the cumulative consumption of each technology i.

$$
\begin{aligned}
z'_1(t) &= m\{[p_{1d} + (q_{1d} + \zeta)\frac{z_1(t)}{m} + q_{1d}\frac{z_2(t) + z_3(t)}{m}]\}[1 - \frac{z(t)}{m}] \\
z'_2(t) &= m\{[p_{2d} + q_{2d}\frac{z_2(t)}{m} + (q_{2d} - \rho)\frac{z_1(t) + z_3(t)}{m}]\}[1 - \frac{z(t)}{m}] \\
z'_3(t) &= m\{[p_{3d} + q_{3d}\frac{z_3(t)}{m} + (q_{3d} - \xi)\frac{z_1(t) + z_2(t)}{m}]\}[1 - \frac{z(t)}{m}] \qquad (1) \\
\\
m &= m_{1d} + m_{2d} + m_{3d} \\
z(t) &= z_1(t) + z_2(t) + z_3(t)
\end{aligned}
$$

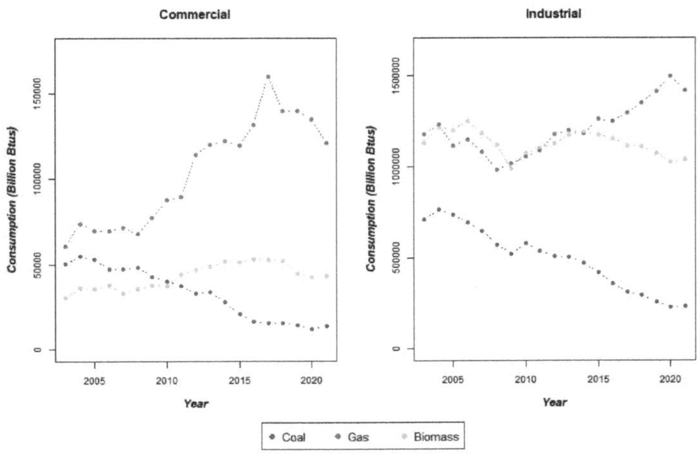

Figure 1. Observed time series of energy consumption for electricity generation and useful thermal output of coal, gas, and biomass in the US commercial and industrial sectors (2003–2021).

The description of the parameters is summarized in Table 1 (see [25] for a more extensive discussion of the model's parameters). The crucial feature of this model is its ability to capture and compare the interplay of different energy sources using the same modeling approach, testing the nature and significance of the relationship between technologies that may be in competition (when the cross-influence parameter estimate is negative) or collaboration (when this parameter estimate is positive).

Table 1. Description of the UCTT model's parameters.

$m = m_{1d} + m_{2d} + m_{3d}$	**Market potential:** max. consumption level in that market for the analyzed technologies
p_{1d}, p_{2d}, p_{3d}	**Seed coefficient of each technology:** initial dissemination of the technology
$q_{1d} + \zeta, q_{2d}, q_{3d}$	**Internal influence of each technology:** technology-specific growth after the innovation phase
$q_{1d}, q_{2d} - \rho, q_{3d} - \xi$	**Cross-influence of each technologies:** competitors' effect on the considered technology

The refinement proposed in this paper has been implemented on the $z(t)_i$ components. The previous version of the model presented in [25] considered $z(t)_i$ to be the observed cumulative series of each i series in the phase where all the three sources compete in the market, assuming an initial condition of $z(0)_i = 0$. Instead, following the lead of [28], in this work, the model implementation has been progressed by introducing an initial condition in which the first value $z(0)_i$ of the cumulative series $z(t)_i$ has to be ideally as close as possible to the cumulative historical value of the series examined at the time before the competition begins. From the standpoint of the multivariate diffusion model world, this new assumption plays a crucial role in introducing additional knowledge on the life cycle of the energy sources. This is fundamental for better estimating the parameters that reflect the scale of the process and identifying more precisely the factors that determine the trends during the competition phase. So, in this analysis, these initial conditions have been introduced to this new version of the UCTT model, considerably changing the $z(t)_i$ values. Consequently, the m_{id} values that are initialized as the Bass model market potential of each $z(t)_i$ are also different from the previous version of the model (see [29] for a detailed description of the parameter).

The data analyzed are accessible from 2001 (coal and gas) and 2003 (biomass); however, it is known that these sources were employed for generation power prior to these dates [30]. To take into account this crucial information and set up the initial condition of the model, the series of US total electricity generation (billion Btus) from 1985 to 2021 has been utilized to calculate an estimation of the initial value $z(0)_i$ for the three series analyzed (source: BP Statistical Review of World Energy [31]). Since the BP series includes all energy sources, the average proportion of electricity generated via each of the three series of interest has been calculated from 2001 (or 2003, respectively) to 2021. Based on these averages, approximations of the series from 1985 to 2001 (or 2003, respectively) have been generated. The initial condition of each source was set as the cumulative value of the new approximated series in 2002 (for coal and gas, the values of 2001 and 2002 were retrieved for the EIA's available ones to be more precise in the $z(0)_i$ definition). The specification of the method utilized to calculate the estimation of the initial value $z(0)_i$ and illustrations of the approximative historical series for both sectors are reported in Appendix A.

3. Model Application and Results

Figure 2 shows that the model can account for the energy transitions in both sectors. The models present a satisfactory result in terms of the *goodness-of-fit* and significance of parameters (see Tables A1 and A2 in Appendix B), and the relationships between the three competitors take a clear meaning. Critical parameter estimates are summarized in Table 2 and denote that coal and gas have similar parameter estimation values, whereas the different patterns for biomass emphasize a distinct evolution of the energy transition in the two sectors. The estimates for coal's internal influence are positive and relatively weak (in the commercial case, the parameter ζ is not significant, which is a reasonable outcome given its stable trend in the last years evaluated) and the cross-influence ones are negative and powerless, which is coherent with the declining direction of both series determined by the contenders' competition and coal's own weak strength on the market due to the more intense restrictions on its utilization. The gas results highlight solid internal growth and a significant competition. Competitors (especially biomass) seem to be able to wield a mighty competitive force; in the commercial sector, this is more emphasized by the highest value of cross-influence, which reflects the gas series trend decline in the last years observed. Besides this competition, the strong and positive internal coefficients show the importance of this energy source in both markets. For the green energy source, the sectors present different results. The negative internal influence estimate for the industrial one contemplates the difficulty of developing the consumption of green energy sources, which is fluctuating around an evident stable trend. The collaborative support highlighted from the cross-influence corresponds to the necessary push promoted by competitors, who, by reducing their use, force the continued pursuit of green source development. On the contrary, the commercial sector presents a strong internal influence, emphasizing the force of the incentives promoted to stimulate renewable diffusion and a negative cross-influence driven by the strong gas development within the market, which has limited biomass expansion.

Table 2. UCTT internal and cross-influence parameter estimates by sector.

	Influence	Commercial	Industrial
Coal	Internal	0.0514	0.0737
	Cross	−0.0234	−0.0291
Gas	Internal	0.7508	0.2801
	Cross	−0.5212	−0.1553
Biomass	Internal	2.0332	−0.1861
	Cross	−0.5675	0.1466

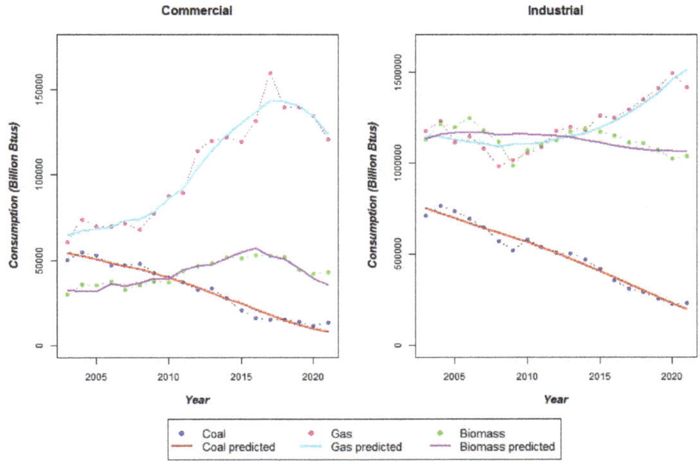

Figure 2. Observed and predicted time series of energy consumption for electricity generation and useful thermal output of coal, gas, and biomass in the US commercial and industrial sectors (2003–2021).

4. Discussion

The efficacy and distributional implications of policies and initiatives on the energy transition of the US commercial sector are still broadly open questions to date. Answering these questions with sound analyses is crucial to the future design of policies as well as the implementation of ESG regulations by private-sector operators. The analysis conducted in this paper does not consider specific policy implementation; still, it may provide a useful means to describe and explore, in a relatively simplified way, the dynamics occurring between the studied critical energy sources in the US commercial sector.

The UCTT model indicated that the grounds for the energy transition in the US commercial sector are strong: biomass is growing and sustained by a widespread internal belief towards sustainability pursuits. Still, it also suggested a significantly negative effect played by gas and coal, limiting their expansion due to their higher efficiency to date and presence in the market. On the contrary, the model identifies a negative internal component in the industrial sector that outlines biomass consumption's complex and slow diffusion. It also accounts for a positive influence of competitors that describes the need to make up for reducing fossil fuel consumption to undergo the tight CO2 emission levels and pursue environmental policies.

From this perspective, how different policies may affect the relationships among energy sources in this market remains an open question. According to [7], renewable energy policies are important tools in stimulating the deployment capacity of green energy sources for electricity, although their effectiveness varies depending on the type of policy instrument. From this standpoint, a possible future development benefit from diffusion models is to study each individual policy's impact within the relationships identified by the UCTT model; this could be a first step toward a more quantitative analysis of policy effects on the energy transition.

In conclusion, compared to the industrial sector, the US commercial sector has a more ready and suitable environment to accelerate the energy transition to renewables. In the industrial sector, progress is occurring slowly; more excellent material and energy efficiency, a more rapid uptake of renewable fuels, and a faster development and deployment of low-carbon production processes are all critical requirements. Instead, the commercial sector's trend towards adopting green energy solutions is expected to persist as policy initiatives, private sector action, and market forces continue to decrease the cost of renewable energy technologies and improve sustainability and energy efficiency in the real estate market.

Funding: This research was funded by the Steers Center for Global Real Estate, McDonough School of Business, Georgetown University, Washington, DC, USA.

Institutional Review Board Statement: Not applicable.

Informed Consent Statement: Not applicable.

Data Availability Statement: All the data used in this analysis are public. As reported in Section 2, the source is: the US Energy Information Administration [27] and refer to the energy consumption (billion Btus) for electricity generation and useful thermal output from 2003 to 2021.

Conflicts of Interest: The authors declare no conflict of interest.

Abbreviations

The following abbreviations are used in this manuscript:

Btu	British thermal unit
EIA	Energy Information Administration
GDP	Gross Domestic Product
ODEs	Ordinary Differential Equations
RET	Renewable Energy Technology
UCTT	Unbalanced Competition for Three Technologies

Appendix A

This appendix illustrates the approximate reconstruction of the coal, gas, and biomass time series in both the commercial and industrial sectors. As mentioned in Section 2, the series of US total electricity generation (TOT) [31] from 1985 to 2021 has been utilized to calculate an estimate of the initial conditions $z(0)_i$ for the three series analyzed in both sectors considered. This series aggregates all the US electricity generation, without distinctions of sectors or technology types.

For each technology series in each sector (TEC_i), the approximate series from 1985 to 2001 has been reconstructed with the basic method described below (for simplicity, the year 2001 is used to define the time division between approximate and observed series, but for biomass it is implied to be 2003 instead of 2001: see Section 2 for details).

- Calculate the average of the observed yearly proportion:
  $AV_i = mean(\frac{TEC(j)_i}{TOT(j)})$, for j = 2001, ... , 2021.
- Calculate the yearly approximate series:
  $TEC(j)_i = TOT(j) * AV_i$, for j = 1985, ... , 2001.

It is recognized that the approximate series obtained with this basic method are different from the real ones; still, since the real series are not available, this study aims to provide only an estimate of the cumulative series before 2001, and this method has been considered a sufficient starting point. Especially for the coal series, it is known from the literature that the series should be characterized by a strong decreasing trend instead of the flat ones presented in the graphs below. Still, though the scope of this work is not to replicate the real series of coal, gas, and biomass in the two sectors, this reconstruction has been used as an instrument to estimate the values to initialize the initial conditions of the model (this approximate reconstruction could be improved through a more sophisticated method, but this is not the focus of this work. Some tests were run by changing the initial conditions, and the model returned similar estimates that do not modify the interpretation of the results of this paper).

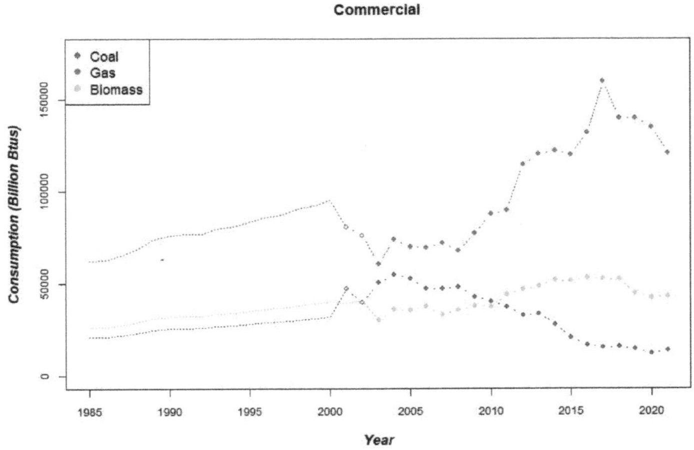

Figure A1. Approximate reconstruction through the basic method described in Appendix A of coal, gas, and biomass time series for the US commercial sector.

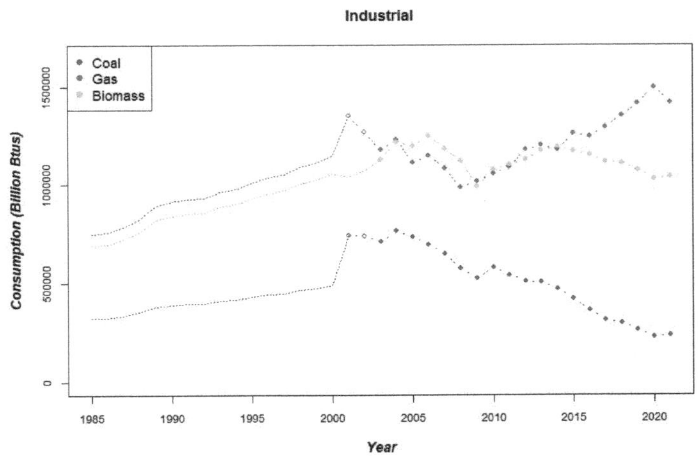

Figure A2. Approximate reconstruction through the basic method described in Appendix A of coal, gas, and biomass time series for the US industrial sector.

Appendix B

The parameter estimate results from the new versions of the UCTT model implementation on the data analyzed are reported in this Appendix.

Table A1. Parameter estimates of new version of the UCTT model for the commercial sector.

	Estimate	Std.Error	Lower	Upper	*p*-Value
m_d	6,973,179	91,091	6,794,644	7,151,714	0.0000
p_{1d}	0.0156	0.0017	0.0123	0.0190	0.0000
p_{2d}	−0.0539	0.0042	−0.0622	−0.0456	0.0000
p_{3d}	−0.0132	0.0022	−0.0175	−0.0089	0.0000
q_{1d}	−0.0234	0.0123	−0.0476	0.0008	0.0642
q_{2d}	0.7508	0.0674	0.6188	0.8828	0.0000
q_{3d}	2.0332	0.4549	1.1416	2.9247	0.0000
ζ	0.0748	0.0569	−0.0366	0.1862	0.1950
ρ	1.2720	0.1317	1.0139	1.5301	0.0000
ξ	2.6007	0.5979	1.4288	3.7726	0.0001

Table A2. Parameter estimates of the new version of the constrained UCTT model for the industrial sector.

	Estimate	Std.Error	Lower	Upper	*p*-Value
m_d	187,205,020	40,593,547	107,643,130	266,766,910	0.0000
p_{1d}	0.0075	0.0018	0.0040	0.0110	0.0001
p_{2d}	0.0012	0.0018	−0.0024	0.0047	0.5270
p_{3d}	0.0034	0.0013	0.0009	0.0060	0.0107
q_{1d}	−0.0291	0.0099	−0.0485	−0.0096	0.0051
q_{2d}	0.2801	0.0960	0.0919	0.4683	0.0054
q_{3d}	−0.1861	0.0753	−0.3336	−0.0385	0.0170
ρ	0.4354	0.1489	0.1437	0.7272	0.0052
ξ	−0.3327	0.1362	−0.5996	−0.0658	0.0183

References

1. Jäger, J.; O'Riordan, T. The history of climate change science and politics. In *Politics of Climate Change*; Routledge: London, UK, 2019; pp. 1–31.
2. Energy Transition. Available online: https://www.irena.org/energytransition (accessed on 8 February 2023).
3. Alola, A.A.; Yildirim, H. The renewable energy consumption by sectors and household income growth in the United States. *Int. J. Green Energy* **2019**, *16*, 1414–1421. [CrossRef]
4. Carmona, M.; Feria, J.; Golpe, A.A.; Iglesias, J. Energy consumption in the US reconsidered. Evidence across sources and economic sectors. *Renew. Sustain. Energy Rev.* **2017**, *77*, 1055–1068. [CrossRef]
5. Omri, A. An international literature survey on energy-economic growth nexus: Evidence from country-specific studies. *Renew. Sustain. Energy Rev.* **2014**, *38*, 951–959. [CrossRef]
6. Payne, J.E. On biomass energy consumption and real output in the US. *Energy Sources Part B Econ. Plan. Policy* **2011**, *6*, 47–52. [CrossRef]
7. Kilinc-Ata, N. The evaluation of renewable energy policies across EU countries and US states: An econometric approach. *Energy Sustain. Dev.* **2016**, *31*, 83–90. [CrossRef]
8. Ionescu, C.; Baracu, T.; Vlad, G.E.; Necula, H.; Badea, A. The historical evolution of the energy efficient buildings. *Renew. Sustain. Energy Rev.* **2015**, *49*, 243–253. [CrossRef]
9. Starr, C.W.; Saginor, J.; Worzala, E. The rise of PropTech: Emerging industrial technologies and their impact on real estate. *J. Prop. Invest. Financ.* **2021**, *39*, 157–169. [CrossRef]
10. Robinson, S.; McIntosh, M.G. A Literature Review of Environmental, Social, and Governance (ESG) in Commercial Real Estate. *J. Real Estate Lit.* **2022**, *30*, 54–67. [CrossRef]
11. Basher, S.A.; Masini, A.; Aflaki, S. Time series properties of the renewable energy diffusion process: implications for energy policy design and assessment. *Renew. Sustain. Energy Rev.* **2015**, *52*, 1680–1692. [CrossRef]
12. Popp, D.; Hascic, I.; Medhi, N. Technology and the diffusion of renewable energy. *Energy Econ.* **2011**, *33*, 648–662. [CrossRef]
13. Hirsch, J.; Spanner, M.; Bienert, S. The carbon risk real estate monitor—Developing a framework for science-based decarbonizing and reducing stranding risks within the commercial real estate sector. *J. Sustain. Real Estate* **2019**, *11*, 174–190. [CrossRef]
14. Eleftheriadis, I.M.; Anagnostopoulou, E.G. Identifying barriers in the diffusion of renewable energy sources. *Energy Policy* **2015**, *80*, 153–164. [CrossRef]
15. Norberg-Bohm, V. Creating incentives for environmentally enhancing technological change: lessons from 30 years of US energy technology policy. *Technol. Forecast. Soc. Chang.* **2000**, *65*, 125–148. [CrossRef]
16. Leskinen, N.; Vimpari, J.; Junnila, S. The impact of renewable on-site energy production on property values. *J. Eur. Real Estate Res.* **2020**, *13*, 337–356. [CrossRef]

17. Bagheri, M.; Delbari, S.H.; Pakzadmanesh, M.; Kennedy, C.A. City-integrated renewable energy design for low-carbon and climate-resilient communities. *Appl. Energy* **2019**, *239*, 1212–1225. [CrossRef]
18. Kiliccote, S.; Olsen, D.; Sohn, M.D.; Piette, M.A. Characterization of demand response in the commercial, industrial, and residential sectors in the United States. *Adv. Energy Syst. Large-Scale Renew. Energy Integr. Chall.* **2019**, 425–443. [CrossRef]
19. Bunea, A.M.; Guidolin, M.; Manfredi, P.; Della Posta, P. Diffusion of Solar PV Energy in the UK: A Comparison of Sectoral Patterns. *Forecasting* **2022**, *4*, 456–476. [CrossRef]
20. Guidolin, M.; Manfredi, P. Innovation Diffusion Processes: Concepts, Models, and Predictions. *Annu. Rev. Stat. Its Appl.* **2023**, *10*. [CrossRef]
21. Bessi, A.; Guidolin, M.; Manfredi, P. Diffusion of Renewable Energy for Electricity: An Analysis for Leading Countries. In *Theory and Applications of Time Series Analysis and Forecasting: Selected Contributions from ITISE 2021*; Springer International Publishing: Cham, Switzerland, 2022; pp. 291–305.
22. Savio, A.; Ferrari, G.; Marinello, F.; Pezzuolo, A.; Lavagnolo, M.C.; Guidolin, M. Developments in Bioelectricity and Perspectives in Italy: An Analysis of Regional Production Patterns. *Sustainability* **2022**, *14*, 15030. [CrossRef]
23. Bessi, A.; Guidolin, M.; Manfredi, P. The role of gas on future perspectives of renewable energy diffusion: Bridging technology or lock-in? *Renew. Sustain. Energy Rev.* **2021**, *152*, 111673. [CrossRef]
24. Guidolin, M.; Alpcan, T. Transition to sustainable energy generation in Australia: Interplay between coal, gas and renewables. *Renew. Energy* **2019**, *139*, 359–367. [CrossRef]
25. Savio, A.; De Giovanni, L.; Guidolin, M. Modelling energy transition in Germany: an analysis through Ordinary Differential Equations and System Dynamics. *Forecasting* **2022**, *4*, 438–455. [CrossRef]
26. Malhotra, A.; Schmidt, T.S.; Huenteler, J. The role of inter-sectoral learning in knowledge development and diffusion: Case studies on three clean energy technologies. *Technol. Forecast. Soc. Chang.* **2019**, *146*, 464–487. [CrossRef]
27. US Energy Information Admnistration: Electricity Power Annual 2021. 2013. Available online: https://www.eia.gov/electricity/annual/Last (accessed on 6 February 2023).
28. Guseo, R.; Mortarino, C. Within-brand and cross-brand word-of-mouth for sequential multi-innovation diffusions. *IMA J. Manag. Math.* **2014**, *25*, 287–311. [CrossRef]
29. Bass, F.M. A new product growth for model consumer durables. In *Management Science 15-5*; INFORMS: Catonsville, MD, USA, 1969; pp. 215–227.
30. Pérez-Lombard, L.; Ortiz, J.; Pout, C. A review on buildings energy consumption information. *Energy Build.* **2008**, *40*, 394–398. [CrossRef]
31. British Petroleum, Statistical Review of World Energy Full Report 2022. 2022. Available online: https://www.bp.com/en/global/corporate/energy-economics/statistical-review-of-world-energy/primary-energy.html (accessed on 1 April 2023).

engineering proceedings

MDPI

Proceeding Paper

A Deep Learning Model Based on Multi-Head Attention for Long-Term Forecasting of Solar Activity †

Adriana Marcucci [1,‡], Giovanna Jerse [2,*,‡], Valentina Alberti [2] and Mauro Messerotti [2]

1 Department of Physics, University of Trieste, Via A. Valerio 2, 34127 Trieste, Italy; adriana.marcucci@phd.units.it
2 Astronomical Observatory of Trieste, INAF, Via G. Tiepolo 11, 34143 Trieste, Italy; valentina.alberti@inaf.it (V.A.); mauro.messerotti@inaf.it (M.M.)
* Correspondence: giovanna.jerse@inaf.it
† Presented at the 9th International Conference on Time Series and Forecasting, Gran Canaria, Spain, 12–14 July 2023.
‡ These authors contributed equally to this work.

Abstract: The accurate long-term forecasting of solar activity is crucial in the current era of space explorations and in the study of planetary climate evolution. With timescales of about 11 years, these forecasts deal with the prediction of the very general features of a solar cycle such as its amplitude, peak time and period. Solar radio indices, continuously measured by a network of ground-based solar radio telescopes, are among the most commonly used descriptors to characterise the solar activity level. They can act as proxies for the strength of ionising radiations, such as solar ultraviolet and X-ray emissions, which directly affect the atmospheric density. In a preliminary comparative study of a selection of univariate deep-learning methods targeting medium-term forecasts of the F10.7 index, we noticed that the performance of all the considered models tends to degrade with increasing timescales and that this effect is smoother when a multi-attention module is included in the used neural network architecture. In this work, we present a multivariate approach based on the combination of fast iterative filtering (FIF) algorithm, long-short term memory (LSTM) network and multi-attention module, trained for the present solar cycle forecasting. Several solar radio flux time series, namely F3.2, F8, F10.7, F15, F30, are fed into the neural network to forecast the F10.7 index. The results are compared with the official solar cycle forecasting released by the Solar Cycle Prediction Panel representing NOAA, NASA and the International Space Environmental Services (ISES) to highlight possible discrepancies.

Keywords: solar activity forecasting; solar radio index; deep learning; multivariate prediction; time series forecasting; multi-attention; LSTM; FIF

Citation: Marcucci A.; Jerse G.; Alberti V. A Deep Learning Model Based on Multi-Head Attention for Long-Term Forecasting of Solar Activity. *Eng. Proc.* **2023**, *39*, 16. https://doi.org/10.3390/engproc2023039016

Academic Editors: Ignacio Rojas, Hector Pomares, Luis Javier Herrera, Fernando Rojas and Olga Valenzuela

Published: 29 June 2023

1. Introduction

The Sun is a moderately active G2V-type star, whose current age is 4.6 billion years. Its magnetic activity follows a quasi-periodic variation of about 11 years, after which the magnetic field completely flips. Over this time period, the number of sunspots that appear on the photosphere increases and decreases in a cyclic way, generating the so-called solar cycle, which is responsible for driving the short-to-long-term fluctuations in solar activity. Short-term solar variability includes transient events such as radiation outbursts, highly energetic particles and plasmoids, which characterise space weather (SWx) and can influence currently operating space-reliant technologies, planetary atmospheres and magnetospheres within a few days. The dynamo mechanism that works in the solar interior drives the long-term fluctuations in solar output that determine the space climate, which can impact planetary climate over decadal or longer timescales [1].

In this work, we focus on the prediction of the activity level over a period of 11 years corresponding to one solar cycle. The result can be used as input to various atmospheric models that characterise the physical state of the Earth's upper atmosphere, such as the thermosphere and ionosphere, and are used for satellite orbit determination, re-entry services, collision avoidance manoeuvres and modelling of the evolution of space debris [2]. Variations in the thermospheric density, in fact, cause variations in the atmospheric drag affecting orbiting satellites, which need to be carefully taken into account when planning spacecraft operations. Moreover, the abundance in the heliosphere of highly energetic particles, known as galactic cosmic rays, shows a periodicity that anticorrelates with solar activity. Cosmic rays constitute a hazard to space instruments and human missions. Hence, an increase in the precision at which we can predict their occurrence is critical.

Historically, long-term forecasts of solar activity have been based on the solar sunspot number (SSN), the weighted sum of the numbers of sunspots and sunspot groups on the solar disk at one time, present in records since 1600. They are good descriptors of almost all other features on the Sun, including active regions, plages, flares, prominences, and, to some extent, changes in the evolution of coronal and solar wind features [3]. Recent studies have started to consider solar radio emission at 10.7 centimetres as an alternative index to describe the solar variability. The F10.7 index is the integrated emission from the whole solar disk at the radio wavelength of 10.7 cm (2800 MHz). It results from thermal ionisation at the boundary between the photosphere and the chromosphere and from magnetic resonance above sunspots and plages [4]. In general, it is used as a proxy for the full-disk flux at the ultraviolet (UV), extreme ultraviolet (EUV), X-ray, Ca II and Mg II wavelengths, and for the total solar irradiance. Records of its values have been consistently measured by ground-based radio telescopes in Canada since 1947, on a day-by-day basis, in any weather condition. Penticton Radio Observatory in British Columbia provides a number of other radio indexes with the same cadence and duration as the F10.7, namely F3.2, F8, F15 and F30, recorded since 1957. They represent the solar flux emissions at different wavelengths measured in solar flux units, where 1 sfu $= 10^{-22}\ Wm^{-2}Hz^{-1}$. Radio waves at different frequencies are emitted at different heights in the solar chromosphere and low corona by plasma layers that have decreasing electron density with increasing altitude: short-wavelength emissions originated at lower altitudes, whereas long-wavelength ones originated at higher altitudes. Hence, we considered exploiting the "tomography" of the solar plasma layers to enhance the performances in solar activity forecasts with a multivariate approach. The majority of models adopted in the space weather framework so far to forecast the short-term trend of empirical time series such as the F10.7 index are statistical models and only a minority of the studies rely on machine learning and deep learning approaches. This number is even smaller if we consider the investigations performed to predict the solar cycle amplitude based on deep learning neural networks [5–9]. In this case, the focus is mainly on the application of classical methods such as support vector regression or single-layer feed-forward neural networks. Since the F10.7 time series is non-linear and non-stationary, we propose a viable approach to cope with its complexity, which decomposes the signal into simpler components and then predicts each of them separately. Decomposition methods, coupled with LSTM neural network, offer a significant enhancement in the field of time series prediction, allowing for a reduction in the chaotic characteristics of the original data. We use the fast iterative filtering (FIF) algorithm [10], which is a robust and stable decomposition signal technique that is suitable for analysing non-linear and non-stationary data, to identify each oscillation component, and discard the short-term ones. The LSTM network is then trained on the decomposed functions that are more correlated with the F10.7 timeseries to predict the solar cycle F10.7 values. In this work, we also adopt the attention-based architecture, which represents one of the main frontiers in deep learning and, to the best of our knowledge, has never been applied for time series forecasting in the space weather context. The attention module is an evolution of the encoder–decoder model, which has been developed to improve performance when using long input sequences.

The remainder of the paper is organized as follows. Our proposed multi-variate model for the solar cycle F10.7 prediction is illustrated in Section 2. Section 3 presents our

25th solar cycle predictions and discussion. Our results will be compared with the official forecasts given by the NOAA Space Weather Prediction Center to verify if they can be considered a good alternative to reproducing long-term variations in solar activity. Finally the conclusion of the paper is given in Section 4.

2. Deep Learning LSTM-Based Method for Long-Term F10.7 Time Series Forecasting

In this work we propose a long short-term emory (LSTM) neural network based on the Mmulti-attention architecture for multivariate time series data predictions of the 25th solar cycle. Our deep learning LSTM-based model has four processes, as shown in Figure 1.

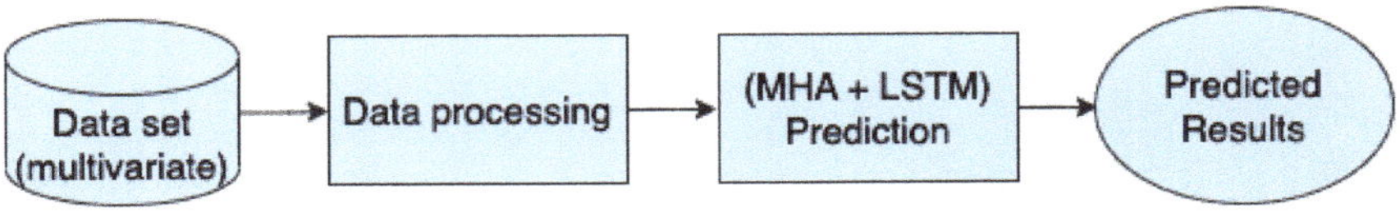

Figure 1. Model flow chart.

The data preparation and processing will be discussed in detail in Section 2.1. This part is fundamental to preparing the dataset that will be used as input to the LSTM-based neural networks. The prediction module is responsible for analyzing and predicting data based on the input data set and the attention value processed by the attention mechanism, which calculates the distribution of weights of each sequence. The output is the long-term F10.7 forecasting. The basics of the LSTM and multi-attention network that were used to develop the proposed model are revised, respectively in the Sections 2.2 and 2.3, while the model architecture will be described in Section 2.4.

2.1. Data Description and Preparation

The multivariate data set used in our paper incorporates six features: five solar radio fluxes and the International Sunspot Number from 2 November 1951 to 14 March 2023 (see Figure 2).

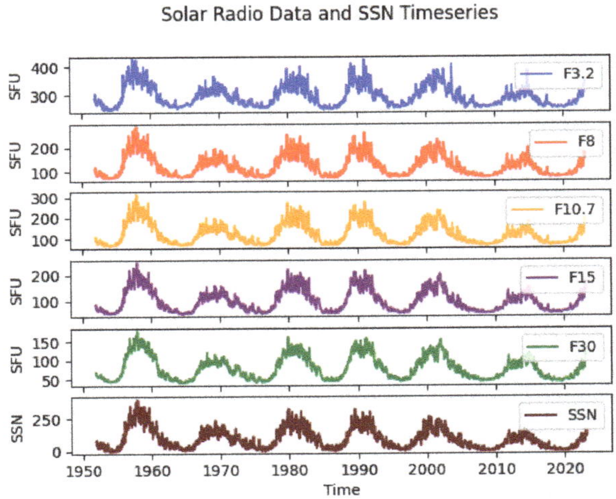

Figure 2. Solar Radio Fluxes and Sunspot Number (SSN) time series from 2 November 1951 to 14 March 2023. Data are given by LASP Interactive Solar IRradiance Datacenter (LISIRD) portal. F3.2, F8, F10.7, F15 and F30 refer, respectively, to the Solar Radio Index at 3.2, 8, 10.7, 15 and 30 cm, expressed in solar flux units (SFUs).

These data were obtained thanks to the LASP Interactive Solar IRradiance Datacenter (LISIRD) portal [11]. The service aims to facilitate solar and heliophysics studies by allowing the researcher to easily discover, visualize, and download data from a variety of space missions and ground-based facilities: the extensiveness of the provided/gathered data sets and the completeness of their description (metadata), along with the plotting capabilities and a very intuitive interface, make it a very powerful resource. The service is maintained by the Laboratory for Atmospheric and Space Physics (LASP) of the University of Colorado, Boulder [12].

Solar radio data are obtained from an external service, the Collecte Localisation Satellites (CLS), that focuses on modelling the upper atmosphere to predict low Earth satellite orbits [13]. Five different wavelengths are available and sourced by the radio telescopes situated in Toyokawa, Nobeyama, Ottawa and Penticton (see the list below for details) [13].

- F30: 30 cm radio flux from Toyokawa (historical data) and Nobeyama (recent data)
- F15: 15 cm radio flux from Toyokawa (historical data) and Nobeyama (recent data)
- F10.7: 10.7 cm radio flux from Ottawa (historical data) and Penticton (recent data)
- F8: 8 cm radio flux from Toyokawa (historical data) and Nobeyama (recent data)
- F3.2: 3.2 cm radio flux from Toyokawa (historical data) and Nobeyama (recent data)

Given the geographical position of the observatories, data were recorded at different times, namely: 03:00 UT Nobeyama and Toyokawa; 17:00 UT Ottawa (until 31 May 1991); 20:00 UT Penticton (since 1 June 1991). CLS performs the initial processing of the data to fill any potential gaps using an expectation-maximization algorithm and to replace values with the residual error above a certain threshold with one calculated via an auto-regressive model [13]. It also allows for data adjusted at 1AU to be selected, which is more convenient for the satellites' orbit prediction model. For our study, we relied on the CLs process to prepare the data set, did not attempt to resample the data to a common time grid and considered the adjusted values. The sunspot number time series is derived externally from the sunspot index and long-term solar observations (SILSO) world data Center [14] of the Royal Observatory Belgium in Brussels.

In order to improve the F10.7 long-term prediction capability of our model by reducing the complexity of the original signal, each data set was decomposed using the fast iterative filter (FIF) technique [10]. This decomposition approach decomposes the original signal into components, called intrinsic mode components (IMCs), which are basically oscillatory functions associated with intrinsic variations at various time scales and derived without leaving the time domain. The results of the FIF decomposition of F30 and SSN time-series are given as examples in Figure 3 and Figure 4, respectively.

In general lower-order IMCs show a chaotic and random signal behaviour with noise contribution while higher-order IMCs contain lower-frequency components and seem to reproduce oscillations at typical physical time scales, such as the Schwabe cycle (11 years). In order to identify the IMC that follows the Schwabe cycle trend, Pearson's-based correlation analysis was applied to each IMC of the analysed indexes and the original F10.7 signal. The results of the correlation analysis for F30 and SSN indexes are given in the correlation matrices depicted in Figure 5 (panels a and c, respectively). The IMC of any index with the highest correlation coefficient with the F10.7 time series (plotted in the panels b and d in Figure 5) was treated as a separate time series mediated over a time period of two weeks and given as an input to the multivariate LSTM network to forecast the F10.7 values. The decision to average the data over a bi-weekly interval comes from the need to balance computer performances with accurate long-term predictions.

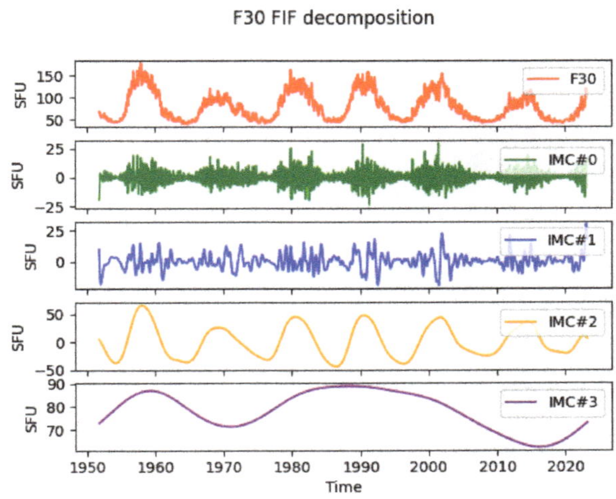

Figure 3. FIF decomposition technique results applied to the F30 radio index. A set of four IMCs, together with the F30 original signal, is plotted.

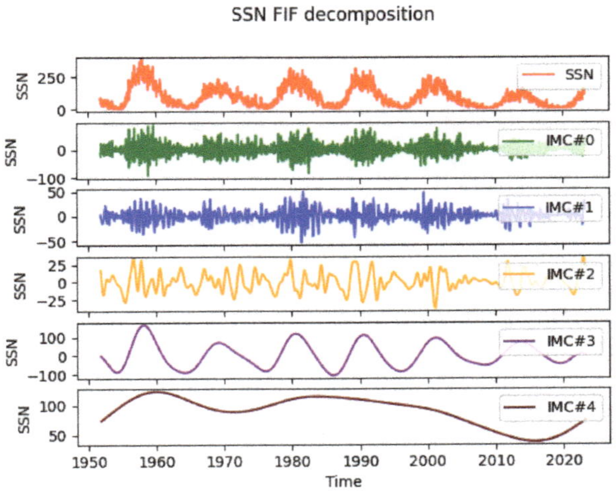

Figure 4. FIF decomposition technique results applied to the SSN index. A set of five IMCs, together with the SNN original signal, is plotted.

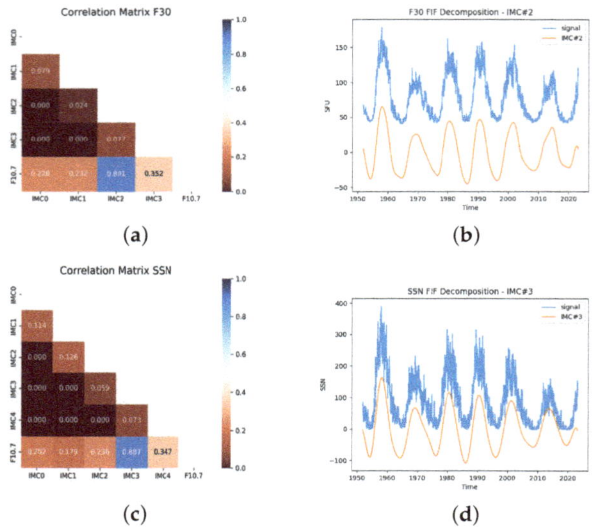

Figure 5. (**a**) Correlation coefficients between IMCs and F10.7 for F30 index. (**b**) IMC with the highest correlation coefficient with respect to F10.7 for F30 index. (**c**) Correlation coefficients between IMCs and F10.7 for SNN index. (**d**) IMC with the highest correlation coefficient with respect to F10.7 for SSN index.

2.2. LSTM Model

The LSTM networks belong to the family of *Recurrent Neural Networks* (RNNs) and they have the ability to process entire sequences of data without the memory loss that exists in classic RNN models. They are specifically designed to overcome the long-term dependency problem of the classic RNN models due to the exploding and vanishing gradient. The LSTM model is characterized by a chain-like structure (see Figure 6)—similar to the RNN architecture—of repeating memory modules, known as *LSTM cells*, with specific features to judge whether the information provided to the network is useful.

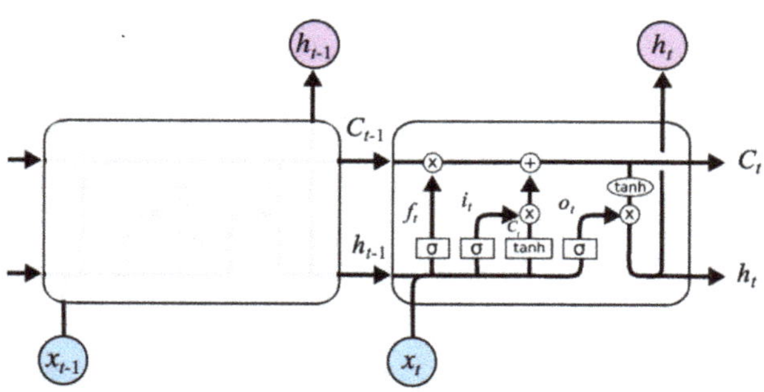

Figure 6. The repeating module in an LSTM contains four interacting layers: a sigmoid *forget gate layer* f_t to decide what information we are going to throw away from the cell state; a sigmoid layer i_t called *input gate layer* decides which values we will update; a tanh layer creates a vector of new candidate values, C_t, that could be added to the state. In the next step, we will combine these two to create an update to the state, the *output gate layer* o_t [15].

In each memory module, a cell state C_t at the time t controls the network information using its three gates: the forget, input and output gates. The input gate decides which values in input will use to modify the internal status of the cell. The output gate calculates a vector used to update the memory cell based on the other gates. The forget gate determines a vector used to choose which values of the precedent state should be maintained in the current one.

2.3. Multi-Attention Module

The attention-based architecture was first described and used in [16] and is a technique that mimics human cognitive attention. The attention mechanism is particularly useful to process multiple time series data, since it can reduce the effect of irrelevant information on the results and enhance the influence of related information by assigning different weights and improve prediction results' accuracy. In this way, it is able to keep track of long-term dependencies in data sequences and reduce the computational effort compared to recurrent or convolutional layers. It can be described as a function that maps a query vector with a set of key-value vector pairs to generate an output. The query vector represents the current state of the model, the keys are used to calculate the similarity scores with the query, and the values vector represents the different features in the input vector. In this work, we specifically used the multi-head-attention (MHA) mechanism, which permits the different heads to focus on different correlations in data and enhance different values in the sequence. Instead of using a single attention function across all the input, it runs the attention algorithm several times in parallel with different learned weights. The independent outputs of each attention module are then concatenated to obtain the final weights. This mechanism can attend to parts of the sequence differently (e.g., longer-term dependencies versus shorter-term dependencies).

2.4. Proposed Multi-Head Attention LSTM Model Architecture

The model architecture used in this work (Figure 7) consists of a three LSTM layers of 12 nodes each, an attention layer with eight heads and, finally, a dense layer, i.e., a layer that is connected deeply, in which each neuron receives input from all neurons of its previous layer. All models were implemented using Python programming language and the Tensorflow library dedicated to multiple machine learning tasks, with Keras as neural network library. In order to validate the forecasting results, we used 80% of the total values as the training set while the remaining 20% served as a validation data set. After training and testing, we used the trained LSTM model to predict F10.7 values and then compared these predicted results with the actual data. In this work, we used the root mean squared error (RMSE), namely the degree of deviation between predicted and observed values, as a performance evaluation metric, defined as follows:

$$RMSE = \sqrt{\frac{1}{N}\sum_{i=1}^{N}\left(y_i' - y_i\right)^2}, \tag{1}$$

where y_i and y_i' are the i_{th} observed and predicted values, respectively, and N is the number of data points in the testing time series. Figure 8a shows the RMSE of the model at different lead time forecasts. The model performance shows a very stable behaviour in lag size with an RMSE value (see Figure 8a) not exceeding 25%, which proves the overall model prediction effectiveness. As an example, the actual and predicted values of F10.7 index corresponding to a lead time of 80 weeks are plotted in Figure 8b. Our model tends to smooth the prediction curve due to the peculiarity of the attention module to discard less relevant features of the time series and highlight the global trend.

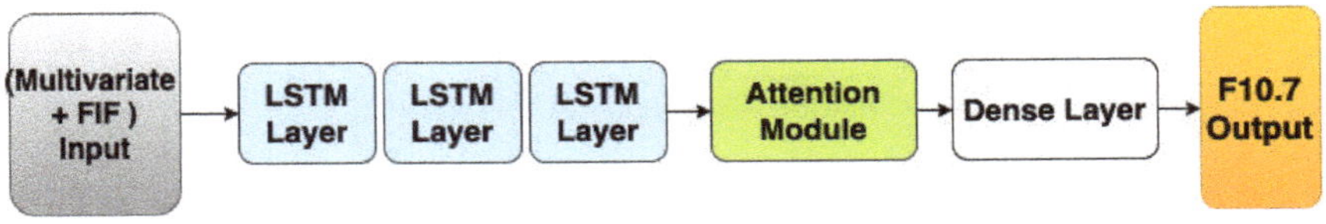

Figure 7. Multi-Head Attention LSTM Model structure diagram.

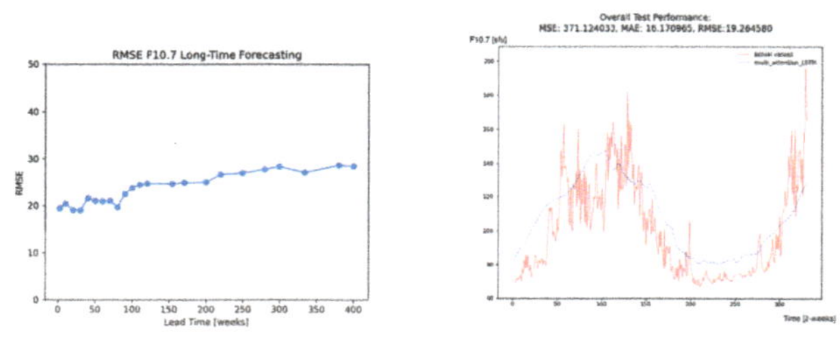

(a) (b)

Figure 8. (**a**) Performance evaluation of the model at different lead times; (**b**) Observed and predicted F10.7 values for 80 weeks ahead.

3. Experimental Results and Discussion

The forecasting result of F10.7 index for the solar cycle 25 are given in Figure 9. In order to characterize the amplitude and peak time of the predicted solar cycle, we fitted the shape of the predicted curve with a function with four free parameters proposed by [17] and [18] of the form

$$f(t) = a(t - t_0)^3 / \exp\left[(t - t_0)^2 / b^2\right] - c, \tag{2}$$

where parameter a represents the amplitude and is directly related to the rate of rise from minimum; b is related to the time in months from minimum to maximum; c gives the asymmetry of the cycle; and t_0 denotes the starting time. This well-known function reproduces both the rise and decay part of the solar cycle and has a more rapid decline after maximum. The curve fitting is given in Figure 10 with the following optimal values: $a = 124.33712279, t_0 = 16.79123391, b = 32.81850839, c = 0.6408711$. From the fitted curve, we derived that the predicted F10.7 maximum amplitude for the solar cycle 25 is about 145 ± 25, peaking in November 2023.

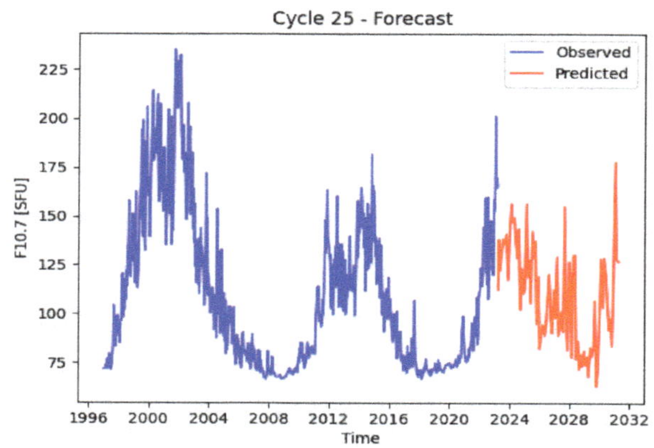

Figure 9. The forecast of F10.7 index for the solar cycle 25 (red). The Observed values for the previous solar cyles are shown in blue.

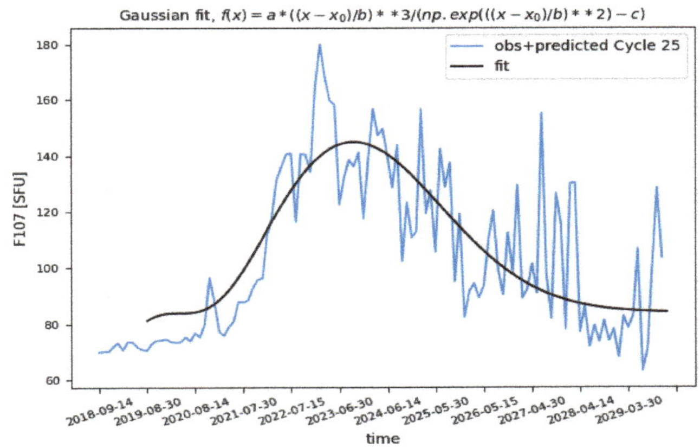

Figure 10. Fitting curve on predicted F10.7 values from 2019 and 2030.

We compared our solar cycle 25 forecasting with the official predictions released by the Solar Cycle 25 Prediction Panel and the International Space Environmental Services (ISES) in 2019 April [19]. The Solar Cycle 25 Prediction Panel is an international group of experts co-chaired by the National Aeronautics and Space Administration (NASA), National Oceanic and Atmospheric Administration (NOAA), and International Space Environmental Services (ISES), whose objective is to predict the amplitude of solar cycle 25. These predictions are a synthesis of a variety of prediction methods coming from the scientific community ranging from physical models, precursor methods to statistical inference, machine learning, and other techniques. The Prediction Panel predicted that solar cycle 25 will be similar to solar cycle 24 and will reach a maximum in July 2025 [20] with a peak F10.7 value of 135 ± 10 SFU,(see Figure 11). This prediction is in line with the current general agreement in the scientific literature, which holds that solar cycle 25 will be weaker than average, even if observed values from 2020 to 2022, the first three years of the cycle, are significantly higher than the predicted values. It is evident that our predictions are consistent with the official ones in the peak F10.7 value but our model predicts the 25th solar cycle maximum about 20 months before.

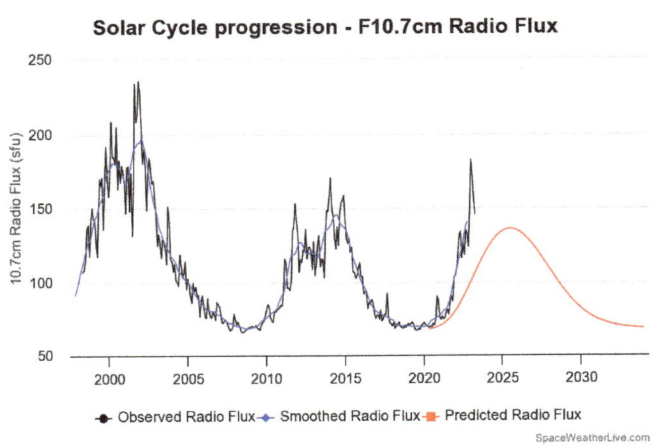

Figure 11. ISES Solar Cycle F10.7 cm Radio Flux Progression, [20].

4. Conclusions

In this paper, we present a multivariate deep-learning-based model for solar activity prediction based on a combination of decomposition algorithm, long short-term memory (LSTM) network and multi-attention module. One of the added values of this work consists of the original use of the multi-attention architecture, which has never been used before in this specific field. The study proved the prediction capability of the model using the RMSE evaluation metrics, showing a stable behaviour with increasing forecasting horizon. After the model test analysis presented in this work, we applied our prediction model to Solar cycle 25 forecasting: the peak amplitude of F10.7 is expected to be 145 SFU, while the occurrence of the solar maximum is foreseen by the end of 2023. Our results are in agreement with other studies, such as the forecasting released by the Solar Cycle 25 Prediction Panel, as regards the peak value, but its occurrence is predicted earlier than other forecasts. Future studies could include investigating the impact of other multivariate features such as solar and geomagnetic index on model performance and prediction accuracy.

Author Contributions: All authors have contributed substantially to this work. All authors have read and agreed to the published version of the manuscript.

Funding: This research received no external funding.

Institutional Review Board Statement: Not applicable.

Informed Consent Statement: Not applicable.

Data Availability Statement: The results presented in this document rely on data described in [21]. These data are available from the WDC-SILSO, Royal Observatory of Belgium, Brussels [14]. Data collected by the Solar Radio Monitoring Program with additional processing by the Space Weather Services at Collecte Localisation Satellites (CLS). These data were accessed via the LASP Interactive Solar Irradiance Datacenter (LISIRD) [11].

Acknowledgments: We would like to show our gratitude to Emanuele Papini (Astrophysical Observatory of Arcetri, INAF) for sharing his FIF algorithm in Python [22].

Conflicts of Interest: The authors declare no conflict of interest.

References

1. Messerotti, M. Space Weather and Space Climate. In *Life in the Universe. Cellular Origin and Life in Extreme Habitats and Astrobiology*; Seckbach, J., Chela-Flores, J., Owen, T., Raulin, F., Eds.; Springer: Dordrecht, The Netherlands, 2004; Volume 7. [CrossRef]
2. Petrova, E.; Podladchikova, T.; Veronig, A.M.; Lemmens, S.; Virgili, B.B.; Flohrer, T. Medium-term Predictions of F10.7 and F30 cm Solar Radio Flux with the Adaptive Kalman Filter. *Astrophys. J. Suppl. Ser.* **2021**, *254*, 9. [CrossRef]
3. Deng, L.H.; Li, B.; Zheng, Y.F.; Cheng, X.M. Relative phase analyses of 10.7cm solar radio flux with sunspot numbers. *New Astron.* **2013**, *23–24*, 1–5. [CrossRef]
4. Tobiska, W.K.; Bouwer, S.D.; Bowman, B.R. The development of new solar indices for use in thermospheric density modeling. *J. Atmos. Sol.-Terr. Phys.* **2008**, *70*, 803–819. [CrossRef]
5. Prasad, A.; Roy, S.; Sarkar, A.; Chandra Panja, S.; Narayan Patra, S. Prediction of Solar Cycle 25 using deep learning based long short-term memory forecasting technique. *Adv. Space Res.* **2022**, *69*, 798. [CrossRef]
6. Warren, H.P.; Emmert, J.T.; Crump, N.A. Linear forecasting of the F10.7 proxy for solar activity. *Space Weather.* **2017**, *15*, 1039–1051. [CrossRef]
7. Wang, Z.; Hu,Q.; Zhong, Q.; Wang, Y. Linear multistep F10.7 forecasting based on task correlation and heteroscedasticity. *Adv. Earth Space Sci.* **2018**, *5*, 863–874. [CrossRef]
8. Du, Y. Forecasting the daily 10.7 cm solar radio flux using an autoregressive model. *Sol. Phys.* **2020**, *295*, 1–23. [CrossRef]
9. Camporeale, E. The challenge of machine learning in space weather: Nowcasting and forecasting. *Space Weather.* **2019**, *17*, 1166–1207. [CrossRef]
10. Cicone, A. Iterative Filtering as a direct method for the decomposition of non-stationary signals. *arXiv* **2018**. [CrossRef]
11. Laboratory for Atmospheric and Space Physics. *LASP Interactive Solar Irradiance Datacenter*; Laboratory for Atmospheric and Space Physics: Boulder, CO, USA, 2005. [CrossRef]
12. LASP Homepage. Available online: https://lasp.colorado.edu/lisird/ (accessed on 16 May 2023).
13. CLS Homepage. Available online: https://spaceweather.cls.fr (accessed on 16 May 2023).
14. SILSO Homepage. Available online: https://www.sidc.be/silso/home (accessed on 20 May 2023).
15. LSTM Networks. Available online: http://colah.github.io/posts/2015-08-Understanding-LSTMs/ (accessed on 10 May 2023).

16. Bahdanau, D.; Cho, K.; Bengio, Y. Neural Machine Translation by Jointly Learning to Align and Translate. In Proceedings of the 3rd International Conference on Learning Representations, ICLR, San Diego, CA, USA, 7–9 May 2015. [CrossRef]
17. Hathaway, D.H.; Wilson, R.M.; Reichmann, E.J. The Shape of the Sunspot Cycle. *Sol. Phys.* **1994**, *151*, 177–190. Available online: https://ui.adsabs.harvard.edu/abs/1994SoPh..151..177H (accessed on 20 May 2023). [CrossRef]
18. Du, Z.L. The solar cycle: A modified Gaussian function for fitting the shape of the solar cycle and predicting cycle 25. *Astrophys. Space Sci.* **2022**, *367*, 20. [CrossRef]
19. Biesecker, D.A.; Upton, L. Solar Cycle 25 Consensus Prediction Update. *AGU Fall Meet. Abstr.* **2019**, *2019*, SH13B-03.
20. Solar Cycle Progression. Available online: https://www.swpc.noaa.gov/products/solar-cycle-progression (accessed on 20 May 2023).
21. Veronig, A.M.; Jain, S.; Podladchikova, T.; Pötzi, W.; Clette, F. Hemispheric sunspot numbers 1874–2020. *Astron. Astrophys.* **2021**, *652*, A56. [CrossRef]
22. Papini, E. GitHub Repository. 2022. Available online: https://github.com/EmanuelePapini/FIF (accessed on 23 March 2023).

engineering proceedings

MDPI

Proceeding Paper

Time Series Forecasting Case Study on Risk-Based Asset Integrity Management for Low-Voltage Failures of Power Distribution Systems †

A. M. Sakura R. H. Attanayake [1,2,*] and R. M. Chandima Ratnayake [2]

[1] Ceylon Electricity Board, Distribution Division 04, Dehiwala 010350, Sri Lanka
[2] Department of Mechanical and Structural Engineering and Materials Science, University of Stavanger, 4021 Stavanger, Norway; chandima.ratnayake@uis.no
* Correspondence: sakura.attanayake@ceb.lk

† Presented at the 9th International Conference on Time Series and Forecasting, Gran Canaria, Spain, 12–14 July 2023.

Abstract: The concept of risk assessment is an important tool in the asset integrity management of power distribution systems. This manuscript presents a risk-based asset integrity management (RBAIM) methodology for the optimization of power distribution assets using a time series analysis approach. This approach deals with time series forecasting on risk assessment for low-voltage-level (400/230 V) failures using the Python programming language and considering historical low-voltage (LV) fuse failure data from a case study over 44 months, starting from 2019. The proposed approach is deployed in a power distribution utility located in a densely populated area of Colombo district, Sri Lanka. The authors proposed a methodical approach for the identification of priority components for asset maintenance and repair ranking based on the risk index percentage value to enhance the predictiveness of potential defects and estimate the risk of potential failures. The results show that the proposed time series forecasting methodology for RBAIM is useful for power distribution utility asset owner organizations for continuous monitoring, the evaluation of asset conditions, and the implementation of proper maintenance and repair strategies to enable assets to perform at their optimal level. The proposed RBAIM methodology enables practicing engineers to assure the asset integrity of power distribution utilities.

Keywords: asset integrity management; low voltage failures; power distribution systems; RBAIM; risk assessment; time series forecasting

Citation: Attanayake, A.M.S.R.H.; Ratnayake, R.M.C. Time Series Forecasting Case Study on Risk-Based Asset Integrity Management for Low-Voltage Failures of Power Distribution Systems. *Eng. Proc.* **2023**, *39*, 17. https://doi.org/10.3390/engproc2023039017

Academic Editors: Ignacio Rojas, Hector Pomares, Luis Javier Herrera, Fernando Rojas and Olga Valenzuela

Published: 29 June 2023

1. Introduction

The asset integrity management (AIM) of power distribution systems is a challenging task of balancing inputs from stakeholders, such as owners, local authorities, regulatory bodies, and customers, with the decision criteria ranging from strategic to operational levels [1,2]. The necessity of effective and efficient decision-making further magnifies the complexity of AIM, depending on different planning horizons, such as the short, medium and long term [1,2]. In this study, AIM is defined as "the means of ensuring that the people, systems processes and resources which deliver the integrity, are in place, in use and fit for purpose over the whole life cycle of the asset. Whole life cycle comprises: design, construction, installation, commissioning, and operation" [3,4]. Decision-making, design, and maintenance strategies under uncertainty, whilst meeting different stakeholder requirements, are a crucial part of power distribution systems' AIM [1,2]. Asset integrity assurance (i.e., using an inspection and maintenance management process) with a limited annual budget requires effective and efficient prioritization of associated factors [5]. Hence, in order to overcome those challenges, methodical approaches for AIM are required.

Power distribution infrastructure is composed of a significant number of geographically dispersed components, including static components such as transformers, fuses, towers, poles, and power cables with rotary components such as load break switches, sectioning switches, transformer tap changers, etc. [5,6]. Hence, the main challenge faced by power distribution utilities is to maintain the integrity of the assets' complex and comprehensive infrastructure [7]. The risk of failure of those components due to degradation, external issues, manufacturing defects, etc. can reduce the integrity of the power distribution network and can also be disastrous for the power distribution utility, especially in highly populated areas [2,8,9]. Risk assessment is a vital tool in the asset and operational management of power distribution assets [1,10]. Historical failure analysis of those components is essential for power distribution system AIM to achieve high quality and reliability, while managing costs, ensuring safety, and avoiding environmental hazards [2,11]. Hence, risk-based asset integrity management (RBAIM) is required for the optimization of power distribution assets.

Time series analysis of power distribution systems' processes involves careful investigation of recorded data over time [12,13]. However, in the literature, most of the time series analysis and forecasting methods that have been discussed mainly concern electricity power distribution demand and energy forecasting [13–15]. Little research is available on time series analysis for historical failure data of power distribution system components in the literature on RBAIM [15–18]. Hence, the authors have observed that historical failure data of power distribution system components can be analyzed using time series analysis to make predictions for decision-making and scarce resource utilization. Therefore, developing a step-by-step methodology involving time series analysis and the concept of risk to facilitate the decision-making process to provide a solution to fill the gaps that exist in failure analysis is required for RBAIM.

This manuscript aims to develop a novel RBAIM approach for low-voltage (400/230 V) failures of power distribution systems, selecting failures of low-voltage (LV) fuses, applied to a regional section of a Sri Lankan power distribution utility. A Python program-based time analysis methodology has been deployed for an RBAIM approach. The proposed methodology involves a methodical approach to the identification and documentation of LV failures, the recognition of maintenance ranking based on risk indexes, resulting in the prioritization of AIM for optimum resource and spare part utilization. Hence, this paper suggests an interesting and unique time series prediction approach, providing a practical and systematic contribution to the related literature. Initially, the study's research background, the case study methodology and development are presented, considering LV failures against a quality of power supply to consumer consequence, which is represented by a reliability index called a system average interruption duration index (SAIDI) of power distribution systems. After that, the results, discussions, and conclusion from the case study are presented, including proposals for future research at the end of the manuscript.

2. Research Background

2.1. Assets, Risk, Maintenance Management, AIM, and RBAIM

The assets of electricity distribution infrastructure require inspection and maintenance to achieve goals, such as optimal allocation of yearly budget, resources, and risk-based performance assessment, which should be aligned with organizational objectives to make data-driven decisions [19,20]. The Publicly Available Specification (PAS) 55 standard, published by the British Standards Institution in 2004, necessitates the integration of organizational processes and practices to manage assets, performances, risks, and expenditures over the life cycle to achieve the organization's strategic plans in a sustainable and optimal way [4,19,21]. Using ISO 55001 Asset Management Systems, organizations can manage assets more effectively throughout the life cycle, whilst managing, improving, and implementing significant assets [22]. Furthermore, ISO 31000 provides a methodical approach to risk management through identifying risks, evaluating the probability of risk occurrence, and determining the severity of the problems caused by an event [19,23]. In addition,

ISO 9001 (2015) indicates the identification of relevant risks, controlling throughout the design with the use of a quality management system [9,24].

It is necessary to identify risks within the business process in order to implement appropriate measures to mitigate them [19,25,26]. Risk can be defined qualitatively/quantitatively as the product of the probability of an event and its consequences [19,20,26]. Risk management involves risk assessment and evaluation for decision-making [19,20]. For power distribution utilities, widely accepted business values are quality of supply (in terms of loss of electricity to consumer minutes), safety (in terms of injuries to personnel and third parties), financial impact (in terms of cost of damage to the organization), image (in terms of reputation), and compliancy (in terms of regulations and legislations) [8,10,19].

The maintenance perspective can be divided into two main categories, namely, reactive and proactive maintenance [19]. The supporting pillars for maintenance management involve process, quality, information and communication technology (ICT), as well as knowledge management [19,20]. With AIM, a set of systematic and proactive strategies can be used to enable industrial assets to operate efficiently and safely during their life cycle [3,4]. Hence, in order to achieve optimal AIM methodology, a RBAIM is required as a systematic approach to managing the physical assets of an organization, with a focus on minimizing risk and maximizing the reliability and safety of those assets.

2.2. Time Series Analysis and Forecasting

Many industries use time series forecasting for demand prediction, resource allocation, predictive maintenance, financial performance, and various other applications [27,28]. Time series forecasting is defined as a technique to predict future occurrences after analyzing past trend data [18,27,29]. In this context, models are used to fit historical data to predict future values [18,29]. Time series forecasting is a data-driven approach, which facilitates effective and efficient planning [27]. Therefore, time series forecasting can be used for an RBAIM approach to identify, assess, and manage the risks associated with physical assets.

The time series analysis approach starts with examining the historical data to check for trends, seasonal patterns, cyclical patterns, and regularity effects [27,29]. The methods used for time series analysis and forecasting include moving average (involves calculating the average of the last n observations), exponential smoothing (involves assigning exponentially decreasing weights to older observations), AutoRegressive Integrated Moving Average (ARIMA—involves modeling the dependence between observation and a number of lagged observations), Seasonal AutoRegressive Integrated Moving Average (SARIMA—involves seasonality of time series data), and machine learning methods (involve algorithms such as artificial neural networks) [18,27]. Accordingly, the proposed RBAIM methodology uses ARIMA as a time series forecasting methodology.

The steps in time analysis and forecasting can vary, depending on the method being used and the specific problem being solved [18]. A general outline of the steps for time series analysis and forecasting includes data preparation; exploratory data analysis, which involves visualizing data; summarizing statistical properties; checking for stationarity, using either an Augmented Dickey–Fuller (ADF) test, a Kwiatkowski–Phillips–Schmidt–Shin (KPSS) test or a Phillips–Perron (PP) test; appropriate preprocessing steps, such as differencing or log transformation, which should be carried out if the data set is not stationary; appropriate model selection; model fitting, evaluation, and forecasting for future values of the time series; and model refinement [18,27].

3. Case Study Methodology and Development

3.1. Use of Case Study Methodology

A case study research methodology is adopted in this research. This involves scrutinizing a particular case to obtain deeper understanding of a particular phenomenon [30]. This methodology involves steps such as defining the research question, selecting the case(s), data collection, data analysis, reporting, and interpretation [30]. The different approaches to case study methodology include the inductive approach (using data and observations to generate theories and explanations), the deductive approach (using theories and hypotheses via observation and data), and the abductive approach (using observations to come up with tentative explanations and testing, refining those via additional data collection and analysis) [30]. In this case study, an abductive approach is used, as historical time series data are used for tentative explanations and testing, with further refining of data for forecasting.

3.2. The Case Study Background

The power distribution network of the Dehiwala area of the Ceylon Electricity Board (CEB), Sri Lanka, has been selected for the application of the proposed time series forecasting case study on RBAIM for LV fuse failures of power distribution systems. The Dehiwala area has been selected as a representative area for the study, as it has a similar electricity infrastructure to other CEB areas with comparatively higher consumer density, higher consumer expectations, and a high industrial work capacity [9,10,16].

Through analysis of historical LV failure data for the overhead power distribution lines of the case study region, the general reasons for LV failures have been observed to be LV fuse failures, breakdown at the distribution transformer (DT), neutral leakage, and LV issues. The study is based on LV fuse failures, which constitute the most frequently occurring LV failure reason of all LV failures. LV fuses are used to protect the LV distribution network [31].

Although the risk of impact of LV fuse failures is comparatively low compared to medium-voltage (MV) failures (i.e., the failures at the ranges of 33/11 kV), a single LV failure causes power loss to a set of consumers in the same LV feeder, as depicted in Figure 1. Hence, an LV fuse failure results in a significant impact to a group of consumers, compared to a single consumer service connection level failure, which has an impact on only one consumer. Furthermore, LV fuse failures from 1 May 2019 to 31 December 2022 are observed and recorded for 11 kV/0.4 kV DTs of capacities 100 kVA, 160 kVA, 250 kVA, 400 kVA, 630 kVA, and 800 kVA in the selected power distribution area of the selected case study organization for the proposed RBAIM approach.

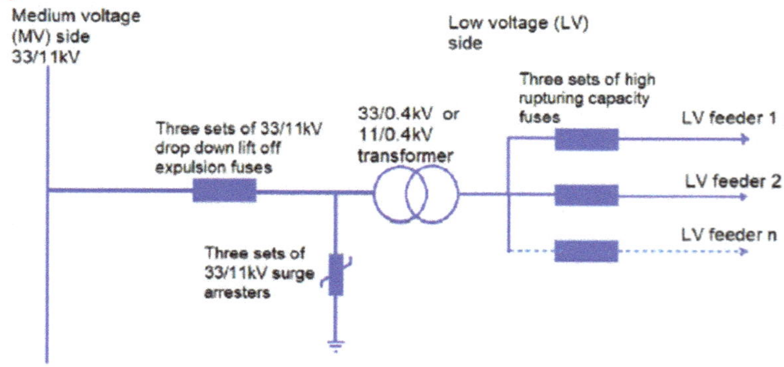

Figure 1. Typical arrangement of power distribution system in CEB [31].

3.2.1. LV Fuses

High rupturing capacity (HRC)-type current-limiting fuses of 160 A, with knife edge type as per IEC 60269, are used by CEB as LV fuses in their power distribution networks [31]. Figure 2 shows a picture of a typical HRC fuse used by CEB.

Figure 2. 160 A LV fuse used by CEB [31].

3.2.2. Probability of LV Failure

The probability of failure (PoF) of LV fuses for a particular DT capacity is obtained using Equation (1) given below:

$$\text{PoF} = \frac{\text{number of LV fuse failures in the DT category per day}}{\text{total number of DTs in the category}} \quad (1)$$

3.2.3. SAIDI and Consequence of Failure (CoF)

In order to measure the quality of the power system performance, reliability indexes are incorporated [32]. In this study, SAIDI is used as a measure of consequence for quality of power supply to consumers. For this research, SAIDI is defined as the average duration of LV fuse failure interruptions per consumer on a daily basis for different capacities of DTs. The failure rate is denoted by λi, the number of customers is given by Ni, and SAIDI is given using Equation (2), as below:

$$\text{SAIDI} = \frac{\begin{array}{c}\text{total duration of customer interruptions}\\ \text{per day due to LV fuse failures}\\ \text{per DT category}\end{array}}{\text{total number of customers served in the same DT category}} = \frac{\sum \lambda i Ni}{\sum Ni} \quad (2)$$

3.2.4. Risk Index Determination

An index of risk has been calculated for historical LV fuse failure data as PoF and SAIDI as CoF, in relation to the selected case study organization; this is given under Equation (3) below:

$$\text{Risk index} = \text{PoF} \times \text{CoF} \quad (3)$$

3.3. *Development of Time Series Analysis Forecasting Based on RBAIM Methodology*

The steps in RBAIM shall include asset identification with categorization; assessment of risk associated with each asset, based on PoF and CoF; development of a risk management plan, outlining the measures to manage or mitigate the identified risks; implementation of the risk management plan, establishing appropriate processes and procedures for AIM; monitoring and review of the risk management plan; and continuous review and improvement of the risk assessment process.

In this study, RBAIM using time series forecasting for LV fuses of power distribution systems has been investigated. Figure 3 shows the basic flow chart used for the time series forecasting for LV fuse failure data.

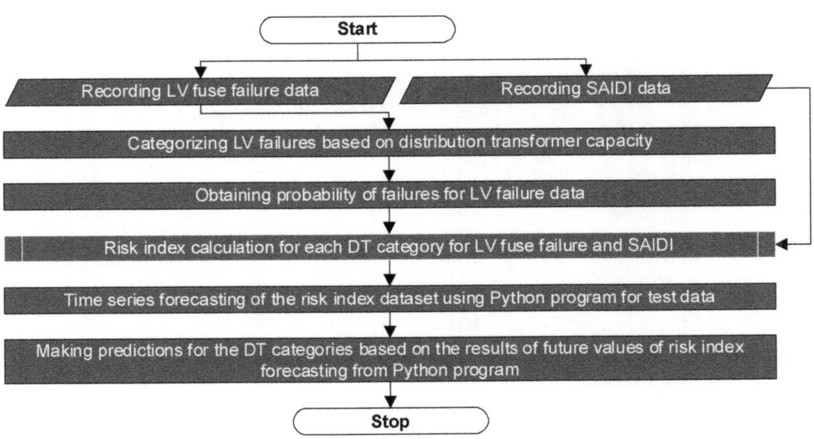

Figure 3. Basic flow chart used for the RBAIM for LV fuse failures and SAIDI.

In accordance with Figure 3, the recording of LV fuse failures was carried out to calculate the probability of LV failures. Then, daily SAIDI data for each DT category with respect to LV fuse failures were calculated. After that, daily percentage indexes of risk were calculated for each DT category, as per Equations (1)–(3) for the selected duration of 44 months. Then, time series forecasting for test data and predictions for real future based on risk index percentage values were carried out for each category of DTs.

Detailed Mathematic Modeling

Initially, historical data collection of daily LV fuse failures for 44 months were recorded for 11 kV/0.4 kV 100 kVA, 160 kVA, 250 kVA, 400 kVA, 630 kVA, and 800 kVA DT categories. Then, SAIDI data were calculated for each DT category of LV fuse failures. Risk index percentages were calculated, using PoF as LV fuse failure and CoF as SAIDI, and recorded in Comma Separated Values (CSV) files. After that, risk index percentage datasets were visualized and decomposed to check for any time series patterns, trends, and seasonality that might affect the performance of the time series model [33–35].

In the next step, stationary tests for the risk index percentage data sets were carried out, using the Augmented Dickey–Fuller (ADF) test. For ADF, the null hypothesis is non-stationarity, and the alternative hypothesis is stationarity [29,36]. The null hypothesis is not rejected if the ADF statistic is less negative than the critical value, indicating that the time series is non-stationary [29,36]. Then, the null hypothesis is rejected if the ADF statistic is more negative than the critical value, indicating that the time series is stationary [29,36].

In this study, all ADF test results have shown that the test statistic is less than the critical value for different confidence intervals. This confirms that all six categories of risk index percentage data sets are stationary. If the time series data are found to be non-stationary, it shall be transformed to stationarity using differencing, detrending, or de-seasonalizing methods.

Autoregressive Integrated Moving Average (ARIMA) with optimal set of performance parameters has been used to forecast the risk index percentages of LV fuse failures and SAIDI of all categories of DTs. ARIMA is a popular and powerful time series model which can capture the linear patterns in time series data [29,36]. The general form of ARIMA model is ARIMA(p,d,q), where p gives the order of the autoregressive component (AR), d gives the degree of differencing required to make the time series stationary (I), and q is the order of the moving average component (MA). In the AR component of ARIMA, the current value of the time series depends linearly on the past values up to lag p and is given in the following equation [29,36]:

$$y(t) = \propto + \sum_{i=1}^{p} \varphi i y(t-i) + \varepsilon t \quad (4)$$

In Equation (4), y(t) gives the current values of the time series, $\propto$ denotes a constant, φi are the autoregressive coefficients that represent the linear relationship between the current value and its past values up to lag p, and εt denotes the error term [29,36]. The MA component of ARIMA assumes that the current values of the time series depend linearly on the past errors up to lag q, and it is given in the following equation [29,36]:

$$y(t) = \mu + \sum_{i=1}^{q} \theta i \varepsilon t - i + et \quad (5)$$

In Equation (5), μ gives the mean of the time series, θi indicates the moving average coefficients that represent the linear relationship between the current values and the past errors up to lag q, εt is the current error term, and et is the white noise error term that is independent of εt [29,36]. The differencing component (I) of ARIMA indicates the time series stationarity through taking the first difference or a higher order difference, and it is given via the following equation [26,33]:

$$y(t)' = (y(t) - y(t-1)) \text{ or } y(t)'' = \left(y(t)' - y(t' - 1)\right) \quad (6)$$

In equation (6), y(t)′ and y(t)″ are the first and second differences of the time series, respectively. After determining the suitable ARIMA model for all the categories of data sets, model diagnostics for all data sets were carried out to evaluate the fit of the ARIMA model with the observed time series data, checking whether the model assumptions were met or violated [33–35]. Accordingly, residual analysis, autocorrelation, and partial autocorrelation suggested that model residuals are normally distributed for all the determined ARIMA models.

Time series forecasting was carried out for one year, from 1 January 2022 to 31 December 2022, and the forecasted values were compared with the real values to check the accuracy of the predictions. In the last step, the risk index percentages were forecast for the real future from 31 December 2022 to 100 steps ahead, and the resulting CSV files were obtained. The resulting mean squared error (MSE) and root mean squared error (RMSE) values for risk index percentages of all categories of DTs were recorded. Figure 4 indicates the risk-based time series forecasting flow chart used for the Python program for the RBAIM approach.

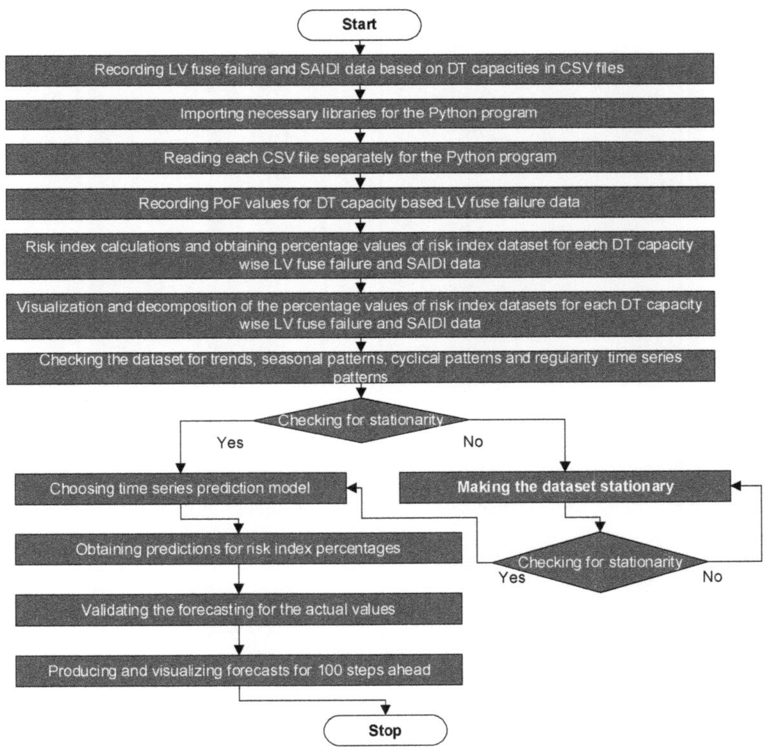

Figure 4. Flow chart for risk-based time series forecasting using Python program.

4. Data Collection and Analysis

As per Figure 4, calculated DT category-wise risk index percentages for historical LV fuse failure data and corresponding SAIDI data for 44 months were calculated, and the resulting risk index percentages values were recorded in CSV files. The resulting risk index data sets were examined to check for any time series patterns. Then, stationarity tests were carried out for all DT categories [33–35].

The resulting one-step-ahead forecasts compared with real data for one year duration and real future predictions for 100 steps ahead from 1 January 2023 are given in Figures 5–10 for each category of DT for risk index percentages calculated using LV fuse failures and corresponding SAIDI data.

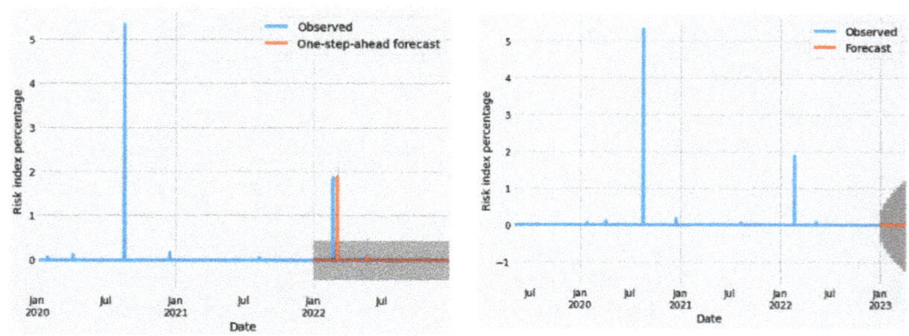

(**a**) (**b**)

Figure 5. Observed values compared to (**a**) one-step-ahead forecast and (**b**) 100-steps-ahead real future forecast of risk index percentages of LV fuse failures and SAIDI for 11 kV/0.4 kV DT rating 100 kVA.

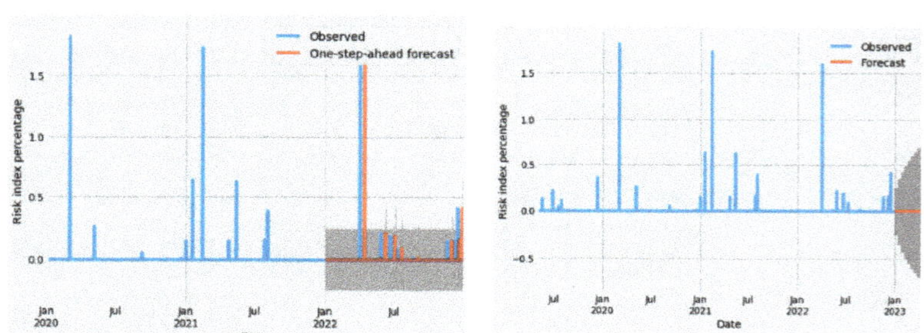

(**a**) (**b**)

Figure 6. Observed values compared to (**a**) one-step-ahead forecast and (**b**) 100-steps-ahead real future forecast of risk index percentages of LV fuse failures and SAIDI for 11 kV/0.4 kV DT rating 160 kVA.

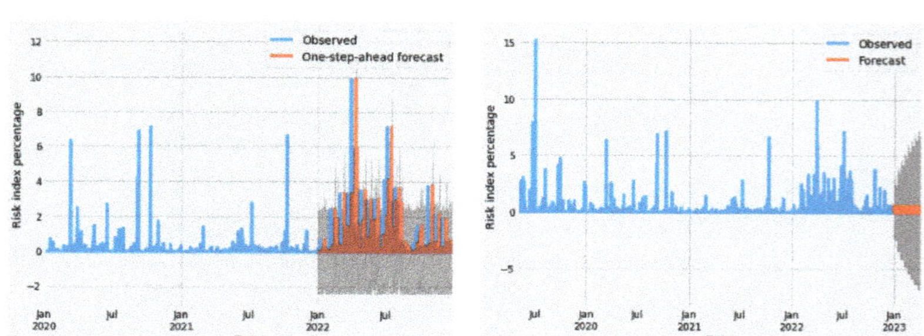

(**a**) (**b**)

Figure 7. Observed values compared to (**a**) one-step-ahead forecast and (**b**) 100-steps-ahead real future forecast of risk index percentages of LV fuses failures and SAIDI for 11 kV/0.4 kV DT rating 250 kVA.

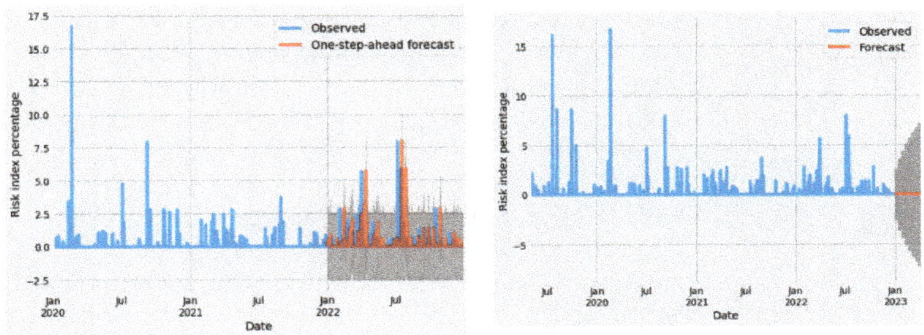

(**a**) (**b**)

Figure 8. Observed values compared to (**a**) one-step-ahead forecast and (**b**) 100-steps-ahead real future forecast of risk index percentages of LV fuse failures and SAIDI for 11 kV/0.4 kV DT rating 400 kVA.

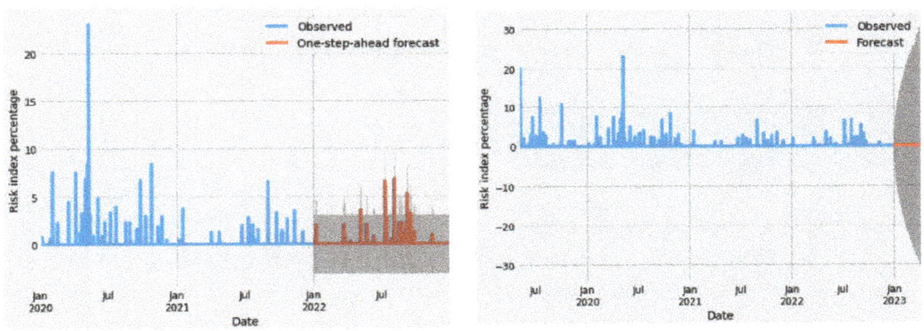

(**a**) (**b**)

Figure 9. Observed values compared to (**a**) one-step-ahead forecast and (**b**) 100-steps-ahead real future forecast of risk index percentages of LV fuse failures and SAIDI for 11 kV/0.4 kV DT rating 630 kVA.

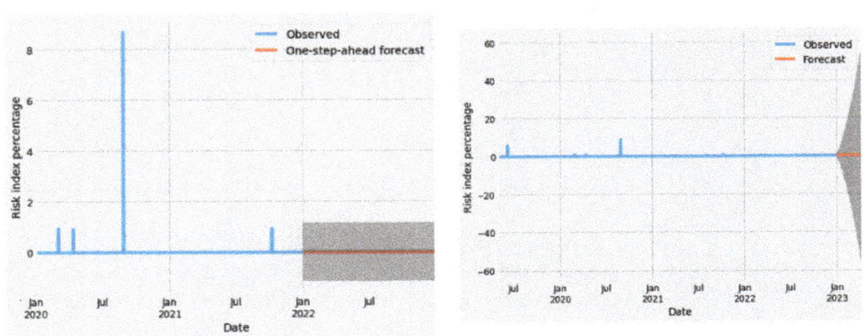

(**a**) (**b**)

Figure 10. Observed values compared to (**a**) one-step-ahead forecast and (**b**) 100-steps-ahead real future forecast of risk index percentages of LV fuse failures and SAIDI for 11 kV/0.4 kV DT rating 800 kVA.

5. Results and Discussion

Overall, Figures 5a, 6a, 7a, 8a, 9a and 10a show that the one-step-ahead forecasts approximately align with the true values for all 11 kV/0.4 kV DT 100 kVA, 160 kVA, 250 kVA, 400 kVA, 630 kVA and 800 kVA categories of LV fuse failure and SAIDI-based risk index percentage data sets, respectively. The second (b) diagrams in all Figures 5–10 show the real future predictions for 100 steps ahead for risk index percentages of all types of DT categories. The resulting average risk index percentage values are recorded in CSV files, and the data extracted from the Python program for real future forecasts are given in Table 1 for LV fuse failures and SAIDI data of all categories of DTs. Table 1 shows that 11 kV/0.4 kV 630 kVA DT has the highest risk of LV fuse failure and SAIDI consequence., Accordingly, resources and maintenance costs are to be allocated mainly for the DTs of that category. Furthermore, DTs of 11 kV/0.4 kV 250 kVA & 400 kVA have more or less resulting risk index percentage values, so resources shall be allocated similarly for those two categories. For other categories of DTs, asset managers can allocate resources according to risk index percentage predictions, as given in Table 1.

Table 1. Risk index prediction results.

11 kV/0.4 kV DT Category of LV Fuses	The Average Risk Index Prediction Values	Risk Rank
100 kVA	0.8968	3 (moderate)
160 kVA	0.5102	2 (low)
250 kVA	5.0452	4 (high)
400 kVA	5.2839	4 (high)
630 kVA	20.8804	5 (very high)
800 kVA	0.0011	1 (very low)

The resulting mean squared error (MSE) and root mean squared error (RMSE) predictions for risk index percentages of LV fuse failures of DT categories are given in Table 2. In general, lower values of MSE and RMSE are desirable for accurate predictions. Therefore, the values obtained for MSE and RMSE given in Table 2 are satisfactory for the desired predictive accuracy for time series forecasts for the RBAIM model.

Table 2. Forecasted MSE and RMSE values.

Transformer Category of LV Fuses	Mean Squared Error (MSE) of the Forecast	Root Mean Squared Error (RMSE) of the Forecast
100 kVA	0.02	0.14
160 kVA	0.02	0.13
250 kVA	1.85	1.36
400 kVA	1.06	1.03
630 kVA	0.97	0.98
800 kVA	0.0	0.0

According to the RBAIM approach, the development and implementation of the risk management plan for the identified risk index percentage predictions for each DT category and the continuous improvement of the risk assessment process shall be carried out by the organization's asset management professionals.

6. Conclusions and Recommendations

In this study, a time series forecasting approach for RBAIM has been developed, using the Python programming language, for a power distribution utility located in a densely populated area of Colombo, Sri Lanka. The established model involves historical LV fuse failure data of different categories of DTs and corresponding SAIDI values to provide risk-based predictions. Daily SAIDI values have been used as a measure of consequence for the probability of LV fuse failures. The methodology can be adopted for failures of other physical assets in a power distribution system. The ultimate goal of RBAIM is to prevent catastrophic failures that can have severe consequences, such as loss of power supply, environmental damage, danger to personnel, and financial losses. Hence, this research provides an RBAIM methodology with the support of data-driven decision-making for predictive maintenance and prioritization of scarce resources and spare parts utilization through analyzing PoF, using LV fuse failures, and CoF, using SAIDI.

Since time series analysis and forecasting can be used to make predictions about various types of data, the resulting forecasts are not always accurate, and this can be influenced by a variety of factors such as changes in the underlying data or unanticipated events.

Further research shall be carried out to investigate how to incorporate the time-dependent nature of a system's reliability for assuring asset integrity in power distribution systems, with the support of a machine learning approach. Particularly, the risk of impact on MV failures and with mean time to repair (MTTR) consequences for risk index assessment of power distribution systems for proper resource utilization and decision-making for predictive maintenance will be studied using a similar case study approach in the future.

Author Contributions: Conceptualization, A.M.S.R.H.A. and R.M.C.R.; methodology, software, validation, formal analysis, investigation, resources, data curation, A.M.S.R.H.A.; writing—original draft preparation, A.M.S.R.H.A.; writing—review and editing, visualization, supervision, R.M.C.R.; All authors have read and agreed to the published version of the manuscript.

Funding: This research received no external funding.

Institutional Review Board Statement: Not applicable.

Informed Consent Statement: Not applicable.

Data Availability Statement: Not applicable.

Conflicts of Interest: The authors declare no conflict of interest.

References

1. Nordgard, D.E.; Gjerde, K.S.O.; Catrinu, M.D.; Lassila, J.; Jartanen, J.; Bonnoit, S.; Aupied, J. A Risk based Approach to Distribution System Asset Management and A Survey of Perceived Risk Exposure Among Distribution Companies. In Proceedings of the 19th International Conference on Electricity Distribution, Vienna, Austria, 21–24 May 2007.
2. Aliyari, M.; Ayele, Y.Z. Risk analysis in low-voltage distribution systems. *Proc. Inst. Mech. Eng. Part O J. Risk Reliab.* **2018**, *233*, 118–138. [CrossRef]
3. Ratnayake, R.M.C.; Markeset, T. Asset integrity management for sustainable industrial operations: Measuring the performance. *Int. J. Sustain. Eng.* **2012**, *5*, 145–158. [CrossRef]
4. Ratnayake, R.M.C. Sustainable performance of industrial assets: The role of PAS 55-1&2 and human factors. *Int. J. Sustain. Eng.* **2013**, *6*, 198–211. [CrossRef]
5. Mirhosseini, M.; Keynia, F. Asset management and maintenance programming for power distribution systems: A review. *IET Gener. Transm. Distrib.* **2021**, *15*, 2287–2297. [CrossRef]
6. Wallnerstrom, C.J.; Bertling, L. Risk Management applied to Electrical Distribution Systems. In Proceedings of the 20th International Conference on Electricity Distribution, Prague, Czech Republic, 8–11 June 2009. [CrossRef]
7. Nordgård, D.E.; Samdal, K. Establishing risk-based maintenance strategies for electricity distribution companies. In *Reliability, Risk and Safety Three Volume Set*; CRC Press: Boca Raton, FL, USA, 2009.
8. Zhao, M.-X.; Su, J.; Liu, S.-G. Risk assessment based maintenance management for distribution network. *J. Int. Counc. Electr. Eng.* **2014**, *2*, 84–89. [CrossRef]
9. Attanayake, A.M.S.R.H.; Ratnayake, R.M.C. On the Necessity of Using Supervised Machine Learning for Risk-Based Screening of Distribution Transformers: An Industrial Case Study. In Proceedings of the IEEE International Conference on Industrial Engineering and Engineering Management (IEEM), Kuala Lumpur, Malaysia, 7–10 December 2022; pp. 0468–0473. [CrossRef]
10. Attanayake, A.M.S.R.H.; Ratnayake, R.M.C. Risk-Based Inspection and Maintenance Analysis of Distribution Transformers: Development of a Risk Matrix and Fuzzy Logic Based Analysis Approach. In Proceedings of the IEEE International Conference on Industrial Engineering and Engineering Management (IEEM), Kuala Lumpur, Malaysia, 7–10 December 2022; pp. 0457–0462. [CrossRef]
11. Gitelman, L.D.; Kozhevnikov, M.V.; Chebotareva, G.S.; Kaimanova, O.A. Asset Management of Energy Company Based on Risk-Oriented Strategy. In *Energy Production and Management in the 21st Century IV: The Quest for Sustainable Energy*; WIT Press: Southampton, UK, 2020; pp. 125–135.
12. Saha, E.; Saha, R.; Mridha, K. Short-Term Electricity Consumption Forecasting: Time-Series Approaches. In Proceedings of the 2022 10th International Conference on Reliability, Infocom Technologies and Optimization (Trends and Future Directions) (ICRITO), Noida, India, 13–14 October 2022; pp. 1–5. [CrossRef]

13. Bitos, M.; Mills, D.; Ashton, S.R. Utilising Forecasting Time Series Data and Flexibility Services to manage Distribution Networks. In Proceedings of the CIRED 2021—The 26th International Conference and Exhibition on Electricity Distribution, Online Conference, online, 20–23 September 2021; pp. 2760–2763. [CrossRef]
14. Çetinkaya, Ü.; Avcı, E.; Bayindir, R. Time Series Clustering Analysis of Energy Consumption Data. In Proceedings of the 2020 9th International Conference on Renewable Energy Research and Application (ICRERA), Glasgow, UK, 27–30 September 2020; pp. 409–413. [CrossRef]
15. Kindalkar, S.S.; Itagi, A.R.; Kappali, M.; Karajgi, S. Time Series Based Short Term Load Forecasting using Prophet for Distribution System. In Proceedings of the 2022 International Conference on Smart Generation Computing, Communication and Networking (SMART GENCON), Bangalore, India, 23–25 December 2022; pp. 1–6. [CrossRef]
16. Attanayake, A.M.S.R.H.; Ratnayake, R.M.C. Digitalization of Distribution Transformer Failure Probability Using Weibull Approach towards Digital Transformation of Power Distribution Systems. *Future Internet* **2023**, *15*, 45. [CrossRef]
17. Goswami, K.; Kandali, A.B. Electricity Demand Prediction using Data Driven Forecasting Scheme: ARIMA and SARIMA for Real-Time Load Data of Assam. In Proceedings of the 2020 International Conference on Computational Performance Evaluation (ComPE), Shillong, India, 2–4 July 2020; pp. 570–574. [CrossRef]
18. Shumway, R.H.; Stoffer, D.S. *Time Series Analysis and Its Applications with R Examples*, 3rd ed.; Springer: Berlin/Heidelberg, Germany, 2016; p. 568. [CrossRef]
19. Mehairjan, R.P.Y. *Risk-Based Maintenance for Electricity Network Organizations*; Springer International Publishing: Cham, Switzerland, 2016.
20. Gomez, J.F.; Galan, P.M.; Guillen, A.J.; Crespo, A. Risk-based criticality for network utilities asset management. *IEEE Trans. Netw. Serv. Manag.* **2019**, *16*, 755–768. [CrossRef]
21. Reyes-Picknell, J. An Introduction to PAS 55—Optimal Management of Physical Assets. Available online: https://reliabilityweb.com/articles/entry/an_introduction_to_pas_55_optimal_management_of_physical_assets (accessed on 12 February 2023).
22. ISO55001 Asset Management Systems. Available online: https://saiassurance.com.au/iso-55001/#:~:text=ISO%2055001%20Asset%20Management%20Systems%20helps%20organisations%20manage%20assets%20more,the%20value%20of%20their%20assets (accessed on 12 February 2023).
23. ISO 31000 Risk Management. Available online: https://www.iso.org/iso-31000-risk-management.html (accessed on 12 February 2023).
24. ISO 9001 and Related Standards in Quality Management. Available online: https://www.iso.org/iso-9001-quality-management.html (accessed on 12 February 2023).
25. Li, Y.; Yeddanapudi, S.; McCalley, J.; Chowdhury, A.; Jewell, W. Resource Management for Distribution System Maintenance Using Optimized Risk Reduction. In Proceedings of the 2006 International Conference on Probabilistic Methods Applied to Power Systems, Stockholm, Sweden, 11–15 June 2006. [CrossRef]
26. Khan, F.I.; Haddara, M.M. Risk-based maintenance (RBM): A quantitative approach for maintenance/inspection scheduling and planning. *J. Loss Prev. Process Ind.* **2003**, *16*, 561–573. [CrossRef]
27. Georgiou, A.D. An Introduction to Time Series Forecasting. 2021. Available online: https://www.infoworld.com/article/3622246/an-introduction-to-time-series-forecasting.html (accessed on 12 February 2023).
28. Jain, A.; Gadia, H.; Sukhdeve, T.; Sahu, S.P.; Verma, S. COVID19 Prediction using Time Series Analysis. In Proceedings of the 2021 International Conference on Artificial Intelligence and Smart Systems (ICAIS), Pichanur, India, 25–27 March 2021; pp. 1599–1606. [CrossRef]
29. Introduction to Time Series Analysis. Available online: https://www.itl.nist.gov/div898/handbook/pmc/section4/pmc4.html (accessed on 12 February 2023).
30. Yin, R.K. *Case Study Research and Applications Design and Methods*; Sage Publications, Inc.: Thousand Oaks, CA, USA, 2018.
31. Lucas, J.R.; Udayakanthi, M.V.P.G. *Identification of Causes of Distribution Transformer Failures and Introduction of Measures to Minimize Failures*; Department of Electrical Engineering, University of Moratuwa: Moratuwa, Sri Lanka, 2015.
32. *Operating Manual Area Electrical Engineer*; Ceylon Electricity Board: Colombo, Sri Lanka, 2016.
33. Jacob, S. Time Series Forecast: A Basic Introduction Using Python. 2017. Available online: https://medium.com/@stallonejacob/time-series-forecast-a-basic-introduction-using-python-414fcb963000#:~:text=Time%20series%20forecasting%20is%20basically,Handling%20Ti.me%20Series%20in%20Pandas (accessed on 2 January 2023).
34. Peixeiro, M. The Complete Guide to Time Series Analysis and Forecasting. 2019. Available online: https://towardsdatascience.com/the-complete-guide-to-time-series-analysis-and-forecasting-70d476bfe775 (accessed on 2 January 2023).
35. Li, S. An End-to-End Project on Time Series Analysis and Forecasting with Python. 2018. Available online: https://towardsdatascience.com/an-end-to-end-project-on-time-series-analysis-and-forecasting-with-python-4835e6bf050b (accessed on 2 January 2023).
36. Box, G.E.P.; Jenkins, G.M.; Reinsel, G.C.; Ljung, G.M. *Time Series Analysis: Forecasting and Control*, 5th ed.; John Wiley & Sons Incorporated: New York, NY, USA, 2015.

engineering proceedings

MDPI

Proceeding Paper

A Forecasting Model for the Prediction of System Imbalance in the Greek Power System †

Konstantinos Plakas [1,2,*], Nikos Andriopoulos [1,2], Alexios Birbas [1], Ioannis Moraitis [2] and Alex Papalexopoulos [3]

1 Electrical and Computer Engineering Department, University of Patras, 26504 Patras, Greece
2 Department of Research, Technology & Development, Independent Power Transmission Operator (IPTO) S.A, 10443 Athens, Greece
3 Ecco International Inc., San Francisco, CA 94104, USA
* Correspondence: kplakas@ece.upatras.gr
† Presented at the 9th International Conference on Time Series and Forecasting, Gran Canaria, Spain, 12–14 July 2023.

Abstract: Forecasting imbalance volumes are of great importance for the different actors in electricity markets. From a transmission system operator (TSO) perspective, balancing supply and demand in real-time is one of the main operational tasks to ensure the safe and reliable operation of the power system, while market participants also use forecasting tools to enhance their participation strategy in electricity wholesale markets. Over the last few years, the increasing integration of renewable energy sources into the power system has created additional complexity for the problem of accurately determining the imbalance volume. In the present work, a case study of the Greek balancing market is presented and analysed. Different algorithms and a set of external predictors are adopted both from the market and operational perspective and compared for two different forecasting horizons.

Keywords: system imbalance; electricity markets; time-series forecasting; machine learning; imbalance forecasting

Citation: Plakas, K.; Andriopoulos, N.; Birbas, A.; Moraitis, I.; Papalexopoulos, A. A Forecasting Model for the Prediction of System Imbalance in the Greek Power System. *Eng. Proc.* **2023**, *39*, 18. https://doi.org/10.3390/engproc2023039018

Academic Editors: Ignacio Rojas, Hector Pomares, Luis Javier Herrera, Fernando Rojas and Olga Valenzuela

Published: 29 June 2023

1. Introduction

Over the last several years, the increasing penetration of renewable generation into the energy mix has significantly changed system operations. The intermittency introduced in the system creates more frequent deviations between supply and demand, which are defined as system imbalances. In power systems, system imbalances can occur due to a number of factors, including fluctuations in demand and supply, generation outages, transmission constraints and strategic bidding. Demand fluctuations can be caused either by seasonal changes and weather conditions or by load operational constraints, while supply fluctuations are mostly due to generation constraints and commercial considerations, and, recently, due to the volatility in weather-dependent, energy-limited power sources (e.g., wind, solar, etc.). Generation outages of dispatchable power sources can also be a great source of system imbalances since they usually cover a considerable part of system demand. An indicative recent example came from France in 2022, in which the unscheduled nuclear power plant outages caused imbalances not only in the French network, but also in neighbouring countries (e.g., Belgium), due to the interconnection between the neighboring countries [1].

Finally, the transmission constraints constitute a structural—yet important—limitation of the power system. For example, if there is an over-generation situation in one region but the transmission lines are congested or at capacity, the excess energy may not be able to be transferred to another region where the demand is high. The over-generation disrupts the net position of the local area, which ideally should be kept constant. This can lead to an over-generation situation in one region and an under-generation situation in another, causing local imbalances. System operators are responsible for alleviating this problem in the spot energy markets by mainly redispatching generation to where it is needed, which is a decision that is economically not optimal and creates security challenges in some cases. Recently, measures such as dispatching economic demand response programs (managing electricity demand), and using storage plants, either as standalone systems or in combination with renewable assets (i.e., in hybrid configurations), have also been applied. In the longer timeframe, imbalances due to grid constraints can be managed by investing in transmission system upgrades (increasing the system capacity). For example, in Greece, the transmission system operator (TSO) has massively invested in transmission grid upgrades in order to increase its capacity and improve system reliability; significant RES developments are expected to be integrated into the network in forthcoming years [2].

In this paper, we propose a new methodology for predicting power system imbalances, a key factor in system reliability and energy market mechanics. The literature in the field is limited and only a few relevant publications can be found, in contrast to those addressing classical problems of load [3], RES [4] and DAM price forecast [5,6]. Nevertheless, the problem has started to attract increasing attention in the last few years, mostly due to the massive penetration of green energy technologies into the grid. Elia, the Belgian TSO, has recently published a methodology for the deterministic and probabilistic prediction of system imbalance [7]. The deterministic model is a standard linear regression with external predictors, while the probabilistic model is implemented using binomial logistic regression. A forecast is then produced and updated on a minute basis for the current and the following quarter-hour. In [8], one of the first studies in the field of system imbalance forecasting was presented, focusing on statistical techniques for the prediction process. Classical time series models (e.g., ARIMA, exponential smoothing) were employed. The presented results were of mixed promise, mainly due to the non-stationarity of the balancing data and short-term operational issues, and possibly also due to energy market conditions, such as market strategic bidding. Furthermore, [8] employed a novel data-mining technique to showcase the additional benefits when combined with classical algorithms. In [9], the most important predictors for the prediction of system imbalance in the Czech power system were presented. The predictors were divided into demand and supply variables, while the behavior of the market participants was also modelled. The results of the proposed model were compared to an ARIMA benchmark model. In [10], a random forest regressor for the prediction of the Spanish system's imbalance was presented. The analysis highlighted the benefits of an ensembling technique for system imbalance prediction, but the limited period for data analysis should also be noted. Finally, the study presented in [11] is particularly relevant as intra-hour forecasting was developed for the case of the Norwegian TSO, Statnett. The imbalance forecasting tool developed relies on quantile regression forests producing probabilistic outputs. The tool was compared to the business-as-usual approach of TSO (Planning Table) and provided significant enhancements. Apart from some theoretical analysis for the Greek case that can be found in [12,13], in which the current status and future challenges of the market are addressed, to the best of our knowledge, this work is the first that deals directly with the prediction of system imbalance in the Greek market. In [14], an approach for the probabilistic dimensioning of frequency restoration reserves was developed and compared to static dimensioning of the Greek electricity market. The main purpose of this paper is to develop a forecasting methodology for the accurate prediction of system imbalance leading to improved system reliability, decreased energy procurement costs and optimal activation of the balancing reserves.

This paper is organized as follows: In Section 2, a short description of the Greek electricity market is provided, while in Section 3, the proposed methodology is presented. In Section 4, a statistical analysis focusing on data from the Greek case study is conducted and the results obtained are given in Section 5. Finally, in Section 6, some useful conclusions are drawn, focusing on comparison of the different models.

2. The Greek Electricity Market

In Figure 1, an overview of the electricity markets in Greece is presented. Implementation of the Target Model in Greece was launched on 1 November 2020, consistent with EU policy rules (Third Energy Package, Directive 2009/72/EC [15]). The framework for the Target Model includes the forward market (FM), the day-ahead market (DAM), the intra-day market (IDM) and the balancing market (BM).

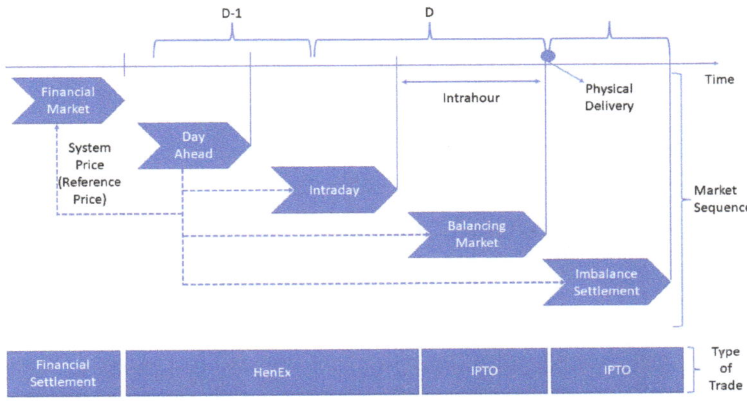

Figure 1. Overview of electricity markets.

The Hellenic Energy Exchange (HEnEx) is responsible for the operation of the FM, DAM and IDM, while the BM is operated by the Greek TSO (called the Independent Power Transmission Operator or IPTO). The forward market settlements deal with settling future contracts which determine the quantity and price of energy; FM transactions can be executed either over the counter or through the organized energy exchange and can be settled in a cash settlement or via physical delivery. The FM is an important market for hedging market risks and ensuring successful participation in sequential energy markets; market participants can mitigate their exposure to price volatility in the spot markets. In the DAM, transactions with physical delivery on a specific day D are settled. Participants (both from the supply and demand sides) submit their purchase/sale offers on the previous calendar day (D-1) for physical delivery at each market time unit. Finally, the latest market in the time sequence operated by HEnEX is the IDM where participants may change their positions, taking into account the latest market, weather and operational conditions, and submit offers to buy and/or sell on the same day D. The IDM, where energy is traded continuously, is a particularly useful market for RES participants as they can correct their DAM positions by taking into account the latest weather and market conditions to manage their deviations and risk exposure.

Since system imbalance predictions are directly connected to the BM operation, we briefly present a description of the BM. The purpose of the BM is to correct the imbalances between production and demand in real-time, maintain an uninterrupted supply of energy with predicted quality characteristics, while considering the participants' market schedule commitments in the previous markets. The BM is divided into the balancing power market, the balancing energy market and the discrepancy clearing process. In the Hellenic BM, the model of the central distribution (central dispatch) of the units is adopted by the TSO through the execution of a security-constrained unit commitment-based integrated scheduling process (ISP), which co-optimizes energy and ancillary service offers in 30 min intervals. Ancillary services are settled based on the ISP process results, while the balancing energy is settled in the subsequent real-time balancing market, which is executed every 15 min by deploying the commitment schedule of units from the ISP. BM participants are characterized as representatives of the balancing responsible parties and/or the balancing service providers. The balancing service providers, as part of their participation in the

BM, may submit the following offers: (a) upward and downward frequency conservation reserve (FCR) offers; (b) upward and downward automatic frequency restoration reserve (aFRR) offers; (c) upward and downward manual frequency restoration reserve (mFRR) offers; and (d) upward and downward balancing energy offers.

3. The Proposed Forecasting Methodology

3.1. Architecture of the Forecasting Solution

In Figure 2, the detailed architecture of the proposed solution is illustrated. For the analysis, a set of external predictors are included, which are used to predict the output of the response variable. In our study, the predictors are divided into the following groups:

- System data (system demand, RES forecast, RES actual production).
- Market data (DAM price, lagged DAM price, lagged values of system imbalance).
- Social indicators (hour index, quarter index, business day, etc.).

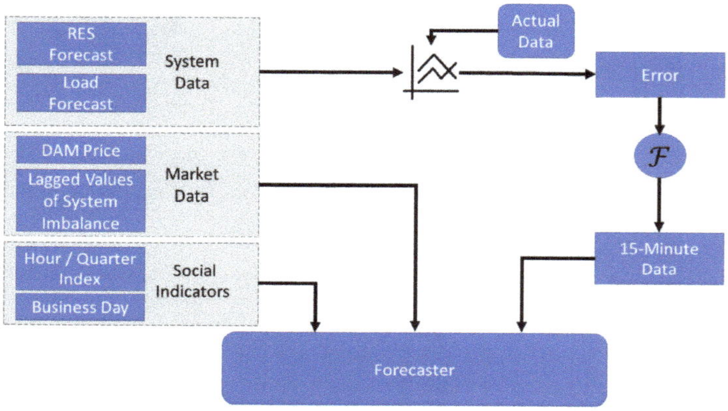

Figure 2. Architecture of the proposed forecasting solution for system imbalance prediction.

These predictors are selected due to their high predictive power in system imbalance forecasting. They not only capture the key drivers, which affect the supply and demand dynamics within the system, but also reflect information on the market side. Firstly, the module receives the forecasted system data values both for the generation and demand resources. The forecasted data is then compared with the actual values, and the respective forecast error is calculated. Then, a fast Fourier transformation (FFT) is applied to modify the sampling rate of the errors, leading to time-series data of 15 min resolution; this resolution has the desired granularity for the determination of system imbalance volumes.

The set of external predictors is not limited to the system data. Market data are also included and have been proven to be of great importance. To be more specific, the DAM price can provide useful insights since it is directly related to the market and system conditions (e.g., energy mix, RES penetration, peak load, etc.). Additionally, since there is a significant correlation of the real-time system imbalance with the past system imbalance values, the most critical system imbalance lags are also included.

Finally, in the model, some social indicator indices are also included to incorporate information related to seasonalities, which can greatly affect the accuracy of the algorithms.

3.2. Fourier Transform Resampling

As already mentioned in the methodology section, the granularity of the balancing market data differs from the granularity of the system data (system demand and RES). Therefore, a resampling technique is applied to the system data to align the granularity of both sets of data (i.e., increase the sampling rate of the system data) in order to be used in the forecasting model.

In this study, the fast Fourier transform is used for the resampling process. The three key steps of the FFT analysis which are implemented are explained below:

1. Transform the original signal to the frequency domain using the FFT:

$$X(k) = \sum_{n=0}^{N-1} x(n)e^{-j2\pi kn/N},$$

where N is the number of samples in $x(n)$.

2. Modify the frequency components to achieve the desired sampling rate. In this case, hourly frequency data are transformed to 15 min data by zero-padding the frequency components:

$$X'(k) = [X(k)\ 0\ 0\ \ldots\ 0],$$

where the number of zeros is equal to the desired increase in the number of samples.

3. Transform the modified frequency components back to the time domain using the inverse FFT:

$$x'(n) = \frac{1}{N}\sum_{k=0}^{N-1} X'(k)e^{j2\pi kn/N}.$$

In Figure 3, the FFT results for an indicative period are illustrated, in which the original 1 h resolution time series is transformed into 15 min resolution data.

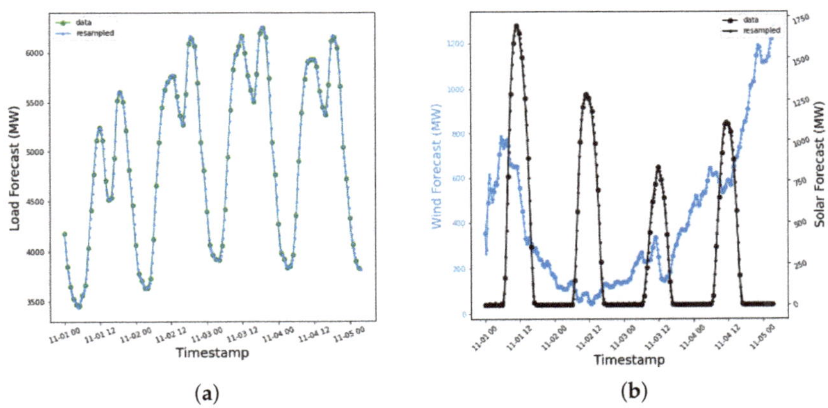

Figure 3. FFT resampling. (**a**) System load (**b**) RES (wind and solar).

3.3. Models and Metrics for System Imbalance Forecasting

In this study, three (3) different algorithms were tested for system imbalance forecasting: multivariate linear regression (LR), random forest (RF) and long short-term memory (LSTM). All the algorithms produce deterministic values of system imbalance and were tested both for 15-minute- and 1-hour-ahead forecasting horizons. The specific algorithms have been extensively studied in the literature, and were, therefore, also selected to be examined for the Greek balancing market. In all, the algorithm selection covers a wide range of predictive models, ranging from relatively simple linear models to more sophisticated and complex non-linear models. In multivariate linear regression, a linear statistical relationship between the dependent variable (system imbalance) and the set of independent variables (market data, system data, social indicators) is assumed. The other two algorithms considered in the study can both capture non-linear relationships between the predictors, but differ significantly in terms of producing the forecast. Specifically, in RF a large number of decision trees is combined to make the prediction, each trained on a random subset of the data and using a random subset of the available features. This significantly reduces the risk of overfitting, which happens quite often in ML algorithms. RF is a particularly useful prediction algorithm when dealing with datasets with complex interactions between the variables. Lastly, LSTM is a particular type of recurrent neural network (RNN), which is suited for processing sequences of data (e.g., time series). The memory cells allow an LSTM network to capture long-term dependencies in sequential data and, thus, reveal dependencies which are not captured by the standard models.

The employed metrics in this study for evaluating the performance of the forecasting algorithms were the R-squared (R^2), the root mean square error ($RMSE$), and the mean absolute error (MAE). The R^2 measures the proportion of the dependent variable's variance which is explained by the independent variables in a regression model. The $RMSE$ measures the standard deviation of the residuals (prediction errors), while the MAE measures the average of the absolute differences between the actual values and the forecasted values. The units for both the RMSE and MAE are the same as the unit of the data in the target variable, while the R^2 may not have a unit of measurement, but ranges from 0 to 1. Higher R^2 values indicate a better fit of the model to the data. The mathematical formulations of the metrics are given below:

$$R^2 = 1 - \frac{\sum_{i=1}^{n}(y_i - \hat{y}_i)^2}{\sum_{i=1}^{n}(y_i - \bar{y})^2} \quad (1)$$

$$RMSE = \sqrt{\frac{1}{n}\sum_{i=1}^{n}(y_i - \hat{y}_i)^2} \quad (2)$$

$$MAE = \frac{1}{n}\sum_{i=1}^{n}|y_i - \hat{y}_i| \quad (3)$$

where y_i denotes the true value, $\hat{y}_i$ shows the predicted value, and n indicates the number of fitted points.

4. Statistical Analysis of the Greek Case Study

The balancing market data considered in this study covered the first period of operation of the balancing market platform in Greece. The market data were downloaded from the IPTO's web portal [16], while the system data were extracted from ENTSOE's transparency platform [17]. The examined period was from 13 February 2021 until 13 November 2022. The data from November 2020 (starting date of the BM platform) until the starting date of our analysis data were disregarded since this was considered a trial period for all the participants, potentially leading to erroneous conclusions in our analysis. In Table 1, some statistical insights on the system imbalance in 2021 and 2022 are given, while in Figure 4, the respective plot for the whole period is illustrated.

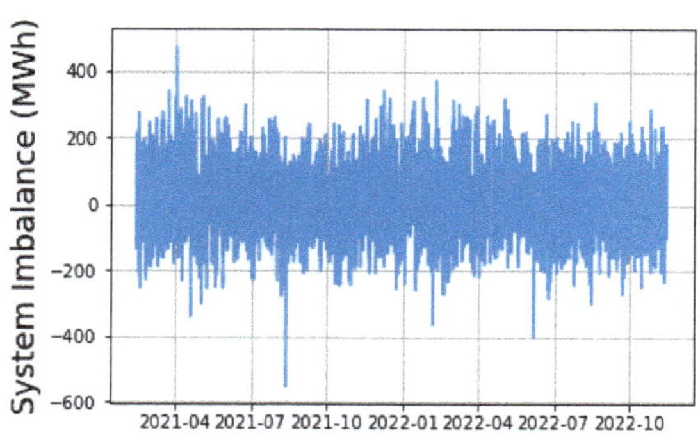

Figure 4. System imbalance in the Greek power system for the period February 2021–November 2022.

Table 1. Key statistics on the Greek system imbalance (MWh).

	2021	2022
Average	11.42	6.64
Standard Deviation	80.99	76.28
25th Quantile	−39.20	−41.31
50th Quantile	10.10	3.61
75th Quantile	59.86	51.51

In Figure 5, the autocorrelation and partial autocorrelation plot of the Greek system imbalances are presented. The red dashed line depicts the 95% confidence interval. The autocorrelation shows the correlation of the system imbalance variable at time t with the corresponding values at time t-k, while the partial autocorrelation measures the correlation between the system imbalance variable and the indicated lag without accounting for the effect of the intermediate lags. In our case, the system's imbalance volume is highly correlated with the first five lags, as illustrated by the partial autocorrelation plot. To this end, the first five lags should be added as external variables and will have a high predictive influence on the results (when applicable). Artificially generated data, which are deducted from these lags, are also suggested to be included (e.g., the average value of the first five lags). Finally, in Figure 6, the correlations of the system imbalance with the system load error, wind production error and solar production error are depicted. The error is defined as the difference between the forecasted values and the actual values, as defined in the Methodology section in Figure 2. It is clear that the correlation was the highest in the case of the wind production error ($\rho_{xy} = 0.5$), indicating that the system imbalance moves in the same direction as the error, but with lower magnitude. The respective Pearson coefficients for the load and solar error were lower, for $\rho_{xy} = -0.34$ and $\rho_{xy} = 0.21$, respectively. Finally, the 95% confidence interval for the correlation coefficients were calculated: [0.49, 0.51] for the wind production error, [−0.35, −0.33] for the load error, and [0.2, 0.22] for the solar production error [18]. The extensive size of the examined dataset, consisting of approximately 65.000 recordings, guarantees the robustness of the calculated correlation coefficients.

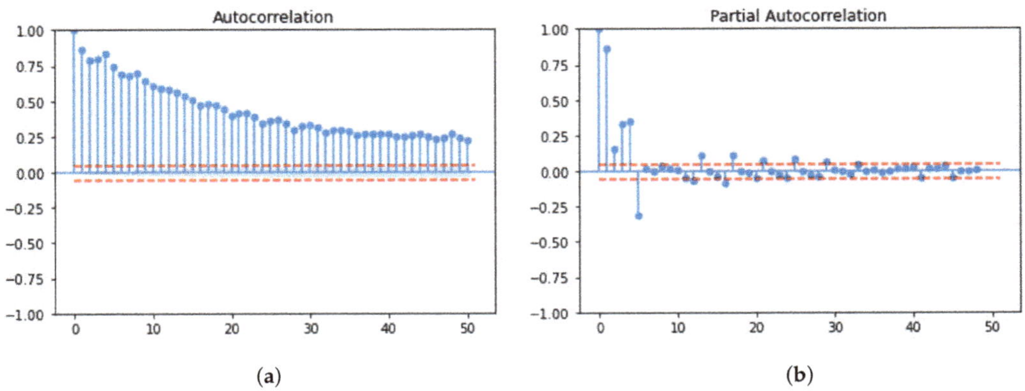

Figure 5. (**a**) Autocorrelation and (**b**) partial autocorrelation for the first 50 lags of system imbalance (the red dashed line indicates the 95% confidence interval).

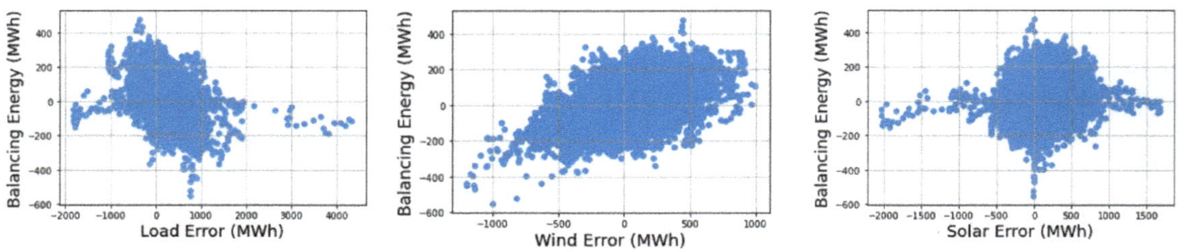

(a) (b) (c)

Figure 6. Correlation between system imbalance and (**a**) system load error, (**b**) wind production error, (**c**) solar production error.

5. Forecasting Results

For the evaluation of the different forecasting algorithms and the hyperparameter tuning of the model parameters, a time-series cross-validation methodology was employed. For cross-validation, the original dataset was split into training, validation and test sets, enabling evaluation of the model performance in a more robust way.

Two forecasting horizons were considered in the analysis: 15-minute- and 1-hour-ahead forecasts. In Table 2, the regression metrics for the three developed models are compared.

Table 2. Evaluation metrics of the forecasting algorithms.

		R^2 (-)	RMSE (MWh)	MAE (MWh)
15 min forecast	Linear Regression	0.77	33.23	24.41
	Random Forest	0.84	26.93	19.63
	LSTM	0.83	28.12	20.64
1 h forecast	Linear Regression	0.68	39.64	29.75
	Random Forest	0.72	36.53	27.01
	LSTM	0.66	40.04	30.27

The results indicate that the random forest model was the optimal selection for both the forecasting horizons. Specifically, for a 1 h forecasting horizon, random forest was more accurate by 7% compared to linear regression and by 9% compared to LSTM. In the case of the 15 min forecasting horizon, the improvement was even higher, leading to a 4.2% accuracy increase compared to LSTM and an almost 18% increase compared to linear regression.

There are several reasons for the the predictive accuracy of the random forest algorithm in system imbalance forecasting. The non-linearities which exist between the system data (demand and RES production) and the target variable can be more accurately captured in the ensembling method that the random forest methodology relies on. Additionally, there is greater resilience to outliers, which can have a large effect on the parameter estimates, and are prevalent in the first months of the training data. In terms of data requirements, the random forest methodology can perform significantly better compared to neural network approaches (e.g., LSTM), which usually require a larger amount of training data. The analysis data covers a small period of time since the balancing market in Greece is a relatively new market, thus limiting the accuracy of the LSTM methodology. Finally, the training time of the algorithms should also be accounted for as an important parameter since the application requires real-time operating actions by the TSO for the activation of balancing reserves. The training time of the random forest methodology was about 18 s, while the corresponding time for the LSTM methodology accounts was 156 s.

In Figure 7, a plot showcasing the results for the last day of the test set is presented. The specific day is presented as the volume of the system imbalance is within the indicative range for the whole period. The trend is captured by all the forecasting algorithms; the random forest methodology appears to have the smallest deviations in the extreme values, which are also reflected in the presented metric results. In any case, the results of this study can provide useful insights into the system imbalance forecasting problem and offer credence to the argument that it can be used as a complementary tool in the TSO's operational processes.

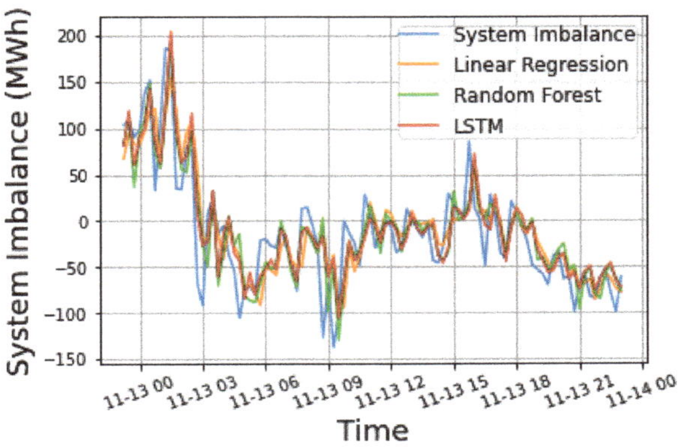

Figure 7. Forecasting results for system imbalance (13 November 2022).

6. Conclusions

In the paper, a forecasting tool for the prediction of system imbalance in the Greek power system has been presented, developed and analyzed in detail. The forecasting tool is comprised of a list of external system and market predictors. Three different algorithms have been compared for two different forecasting horizons. In both cases, the random forest algorithm proved to be the most accurate and computationally more efficient compared to the linear regression and standard neural network methodologies. Especially for the short time horizon of the study (15 min), the improvement achieved by the random forest model was considerably higher compared to the 1 h forecasting horizon.

Nevertheless, it is clear that system imbalance prediction is comprised of many different parts and prediction requires information from various sources. The error coming from the system data cannot fully explain the imbalances, and additional explanatory variables should be included to improve the accuracy of the algorithm (e.g., strategic bidding, operational, commercial and transmission constraints, outages modelling, etc.). The resulting data provide strong credence to the claim that the proposed methodology can significantly improve the system imbalance prediction process of a TSO and contribute efficiently to least-cost energy procurement and more economical dispatch of the balancing reserves.

Author Contributions: Conceptualization, K.P., N.A., A.B., I.M. and A.P.; methodology, K.P. and N.A.; formal analysis, K.P. and N.A.; writing—original draft preparation, K.P.; writing—review and editing, K.P., N.A., A.B., I.M. and A.P.; supervision, A.B., I.M. and A.P. All authors have read and agreed to the published version of the manuscript.

Funding: This work has been partially supported by the FEVER project (Horizon 2020 EU Research and Innovation Program) under Grant Agreement number 864537, and the ENERMAN project (Horizon 2020 EU research and innovation program) under Grant Agreement number 958478.

Institutional Review Board Statement: Not applicable.

Informed Consent Statement: Not applicable.

Data Availability Statement: Not applicable.

Conflicts of Interest: The authors declare no conflict of interest.

Abbreviations

The following abbreviations are used in this manuscript:

TSO	Transmission System Operator
RES	Renewable Energy Sources
DAM	Day-Ahead Market
ARIMA	Autoregressive Integrated Moving Average
IDM	Intra-Day Market
BM	Balancing Market
HEnEx	Hellenic Energy Exchange
IPTO	Independent Power Transmission Operator
ISP	Integrated Scheduling Process
FCR	Frequency Conservation Reserve
FRR	Frequency Restoration Reserve
FFT	Fast Fourier Transformation
LR	Linear Regression
RF	Random Forest
LSTM	Long Short-Term Memory
RNN	Recurrent Neural Network
R^2	R-squared
RMSE	Root Mean Squared Error
MAE	Mean Absolute Error

References

1. Zimmermann, F.; Keles, D. State or market: Investments in new nuclear power plants in France and their domestic and cross-border effects. *Energy Policy* **2023**, *173*, 113403. [CrossRef]
2. IPTO. Preliminary Draft of the Ten Year Development Plan of HETS for the Period 2024–2033. Available online: https://www.admie.gr/en/nea/anakoinosi/preliminary-draft-ten-year-development-plan-hets-period-2024-2033 (accessed on 1 February 2023).
3. Papalexopoulos, A.D.; Hesterberg, T.C. A regression-based approach to short-term system load forecasting. *IEEE Trans. Power Syst.* **1990**, *5*, 1535–1547. [CrossRef]
4. Shahid, F.; Zameer, A.; Muneeb, M. A novel genetic LSTM model for wind power forecast. *Energy* **2021**, *223*, 120069. [CrossRef]
5. Gao, F.; Guan, X.; Cao, X.R.; Papalexopoulos, A. Forecasting power market clearing price and quantity using a neural network method. In Proceedings of the Power Engineering Society Summer Meeting, Seattle, WA, USA, 16–20 July 2000; Volume 4, pp. 2183–2188.
6. Lago, J.; Marcjasz, G.; De Schutter, B.; Weron, R. Forecasting day-ahead electricity prices: A review of state-of-the-art algorithms, best practices and an open-access benchmark. *Appl. Energy* **2021**, *293*, 116983. [CrossRef]
7. Elia Group. Documentation of the Methodology Used for the System Imbalance Forecast Publications. Available online: https://www.elia.be/en/grid-data/balancing/system-imbalance-forecasts (accessed on 1 February 2023).
8. Garcia, M.P.; Kirschen, D.S. Forecasting system imbalance volumes in competitive electricity markets. *IEEE Trans. Power Syst.* **2006**, *21*, 240–248. [CrossRef]
9. Kratochvíl, Š. System Imbalance Forecast. Ph.D. Thesis, Czech Technical University, Prague, Czech Republic, 2016. Available online: https://dspace.cvut.cz/bitstream/handle/10467/67662/Disertace_Štěpan_Kratochvil_2016.pdf?sequence=1 (accessed on 1 February 2023).
10. Contreras, C. System Imbalance Forecasting and Short-Term Bidding Strategy to Minimize Imbalance Costs of Transacting in the Spanish Electricity Market. 2016. Available online: https://repositorio.comillas.edu/xmlui/handle/11531/16621 (accessed on 1 February 2023).
11. Salem, T.S.; Kathuria, K.; Ramampiaro, H.; Langseth, H. Forecasting intra-hour imbalances in electric power systems. In Proceedings of the AAAI Conference on Artificial Intelligence, Honolulu, HI, USA, 27–31 January 2019; Volume 33, pp. 9595–9600.
12. Ioannidis, F.; Kosmidou, K.; Andriosopoulos, K.; Everkiadi, A. Assessment of the Target Model Implementation in the Wholesale Electricity Market of Greece. *Energies* **2021**, *14*, 6397. [CrossRef]
13. Makrygiorgou, J.J.; Karavas C.-S.; Dikaiakos, C.; Moraitis, I.P. The electricity market in Greece: Current status, identified challenges and arranged reforms. *Sustainability* **2023**, *15*, 3767. [CrossRef]
14. Papavasiliou, A. An Overview of Probabilistic Dimensioning of Frequency Restoration Reserves with a Focus on the Greek Electricity Market. *Energies* **2021**, *14*, 5719. [CrossRef]
15. Third Energy Package. Available online: https://energy.ec.europa.eu/topics/markets-and-consumers/market-legislation/third-energy-package_en (accessed on 1 February 2023).
16. IPTO Balancing Market Data. Available online: https://www.admie.gr/en/market/market-statistics/detail-data (accessed on 1 February 2023).

17. ENTSO-E Transparency Platform. Available online: https://transparency.entsoe.eu/dashboard/show (accessed on 1 February 2023).
18. Tian, L.; Wilding, G.E. Confidence interval estimation of a common correlation coefficient. *Comput. Stat. Data Anal.* **2008**, *52*, 4872–4877. [CrossRef]

engineering proceedings

MDPI

Proceeding Paper

Sustainable Development of Renewable Energy Consumption in G7 and ASEAN-5 Countries: Panel Fixed-Effect Econometric Modelling †

Aye Aye Khin [1,*,‡], Kui Ming Tiong [1,‡], Whee Yen Wong [2,‡] and Sijess Hong [1,‡]

1 Department of Economics, Faculty of Accountancy and Management (FAM), Universiti Tunku Abdul Rahman (UTAR), Sungai Long Campus, Jalan Sungai Long, Bandar Sungai Long, Kajang 43000, Selangor, Malaysia; tiongkm@utar.edu.my (K.M.T.); sijesshong@1utar.my (S.H.)
2 Department of Internet Engineering and Computer Science, Lee Kong Chian Faculty of Engineering and Science (LKCFES), Universiti Tunku Abdul Rahman (UTAR), Sungai Long Campus, Jalan Sungai Long, Bandar Sungai Long, Kajang 43000, Selangor, Malaysia; wongwy@utar.edu.my
* Correspondence: ayekhin@utar.edu.my
† Presented at the 9th International Conference on Time Series and Forecasting, Gran Canaria, Spain, 12–14 July 2023.
‡ These authors contributed equally to this work.

Abstract: Energy is the key driver of economic growth; however, the economic leadership position of G7 countries and the rising global manufacturing hub status of the ASEAN-5 countries have yet to achieve the Sustainable Development Goals. Thus, this paper aims to examine the effects of real GDP per capita, urban population, the number of individuals using the internet, carbon dioxide emissions, total trade and net foreign direct investment (FDI) inflows on the renewable energy consumption (REC) of G7 and ASEAN-5 countries from 1990 to 2021 yearly data. Using Studenmund's and Gujarati and Porter's procedures of the panel data model, the panel fixed-effect econometric modelling held the best outcome for both G7's and ASEAN-5 countries' REC models. Based on the findings, urban population highly and positively affects REC in G7 countries. However, there is also a positive and strong relationship between net FDI inflows and REC in ASEAN-5 countries. The empirical findings prove the importance of macroeconomic, socioeconomic and environmental variables for the outcomes of REC policies across both developed and developing countries.

Keywords: sustainable development; SDG7; renewable energy consumption; G7 and ASEAN-5 countries; panel data analysis

Citation: Khin, A.A.; Tiong, K.M.; Wong, W.Y.; Hong, S. Sustainable Development of Renewable Energy Consumption in G7 and ASEAN-5 Countries: Panel Fixed-Effect Econometric Modelling. *Eng. Proc.* **2023**, *39*, 19. https://doi.org/10.3390/engproc2023039019

Academic Editors: Ignacio Rojas, Hector Pomares, Luis Javier Herrera, Fernando Rojas and Olga Valenzuela

Published: 29 June 2023

1. Introduction

Renewable energy (RE) is defined as energy derived from sources that replenish naturally but are limited in flow [1], such as geothermal, solar, hydroelectric, wind and biomass fuels (i.e., biomass waste, biofuels and wood). RE has received increasing attention globally due to its important role in reducing greenhouse gas emissions, energy imports and fossil fuel use, which are crucial to delay climate change. As the key driver of development, the sustainable use of energy is important to ensure that the investments, innovations and growth of economies allow the current and future generations to continue flourishing without affecting growth momentums in accordance with Sustainable Development Goal (SDG) 7: "Ensure Access to Affordable, Reliable, Sustainable and Modern Energy for All".

In the literature, substantial studies have been performed to identify the factors influencing the sustainable development of renewable energy consumption (REC), which can be broadly categorized into macroeconomic, socioeconomic and environmental. Since there are diverse variables in each category and different methods to estimate each variable, this produces a significant research gap, which should be addressed by narrowing the geographical scope of the research, expanding the variables used and include forecasting methods to forecast the REC rate of our target countries. This research examines the effects of real GDP per capita, urban population, the number of individual using the internet, carbon dioxide emissions, total trade and net foreign direct investment (FDI) inflows on the renewable energy consumption (REC) of G7 and ASEAN-5 countries using the panel data model.

The G7 is an informal block of the seven of the world's industrialized countries that convenes annually to discuss global issues, such as economic governance, international security and energy (Council on Foreign Relations, 2022). These countries are the United States, Canada, France, Germany, Italy, Japan and the United Kingdom, whose collective gross domestic product (GDP) contributes to about 45% of the global economy as of 2019. In contrast, ASEAN-5 is a formally instituted economic block of the top five strongest-performing ASEAN countries, which are Indonesia, Malaysia, the Philippines, Singapore and Thailand, and was established on 8 August 1967 as the Association of Southeast Asian Nations (ASEAN) [2].

In comparison with the ASEAN-5 countries, the development of RE in the G7 countries is significantly more sophisticated and began earlier, starting from hydroelectricity and biomass fuels. As technology progressed, G7's RE expanded into solar, wind and geothermal energies, although nuclear energy remained a mainstay in Japan. As for ASEAN-5, RE technology is significantly less-developed, and only Thailand, Indonesia and the Philippines can actively expand their hydro, solar, tidal and geothermal powers due to their geographical advantages [3–5]. In contrast, RE development and usage in Malaysia and Singapore is significantly lesser due to limitations in land mass and terrain construction. For example, Malaysia was heavily reliant on petroleum products until the 1980s, then natural gas usage picked up to replace petroleum products until the 2000s, which was then overtaken by coal and coke from 2000 onwards. Meanwhile, for Singapore, REC is nearly negligible—even at present, the island nation is 95% powered by natural gas despite efforts to transition to solar energy [6].

Figure 1 compares the REC of G7 and ASEAN-5 countries as a percentage share of total final energy consumption from 2016 to 2021. The REC of all G7 countries remained fairly constant in the previous 5 years due to disruptions caused by the COVID-19 virus where countries were unable to expend surplus capital to improve their technology or expand their capacity. Most notably, REC became stagnant from 2018 onwards due to COVID-19, which led to energy supply and demand disruption. Overall, among G7 countries, Canada remained consistently at the top at an average of 22.31%, while Japan remained consistently at the bottom at an average of 7.15%. In contrast, among ASEAN-5 countries, Indonesia started at 27.8348 units but was overtaken by Thailand and the Philippines in 2018 and maintained at an average of 20.86 units afterwards. During the period from 2018 to 2021, Thailand achieved an average of 23.72 units, while the Philippines had 23.22 units. Malaysia and Singapore remained at the bottom consistently at an average of 5.15 units and 0.71 units, respectively.

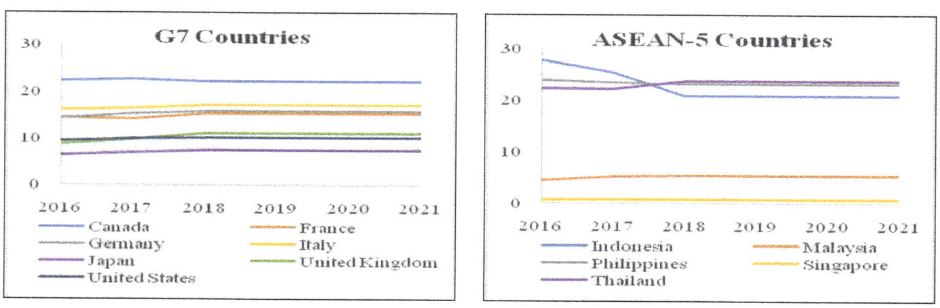

Figure 1. Renewable energy consumption of G7 and ASEAN-5 countries, 2016–2021. Source: [7].

Figure 2 compares the real GDP per capita (RGDP) (in constant USD2015 terms) of G7 and ASEAN-5 countries from 2016 to 2021. In this instance, the United States charted the highest RGDP at USD 58,811.97 in the G7 bloc on average. Meanwhile, Singapore marked an even higher RGDP across both blocs at USD 59,062.18 on average. This is due to its relatively matured economy that thrives on exports of electrical and electronics (E&E), financial services, tourism and seaport business, and its extremely small population. The rest of the G7 countries marked an average performance RGDP at an average of USD 45,641.83 by the United Kingdom, USD 43,734.42 by Canada, USD 42,209.09 by Germany, USD 37,237.84 by France, USD 35,559.48 by Japan, and USD 30,722.37 by Italy. This is followed by the rest of the ASEAN-5 economies, which marked the lowest RGDP across both blocs. Malaysia had an average RGDP of USD 10,786.49, Thailand USD 6294.97, and Indonesia USD 3695.12, while the Philippines had an average of USD 3368.50. Of all countries, the Philippines had the lowest RGDP.

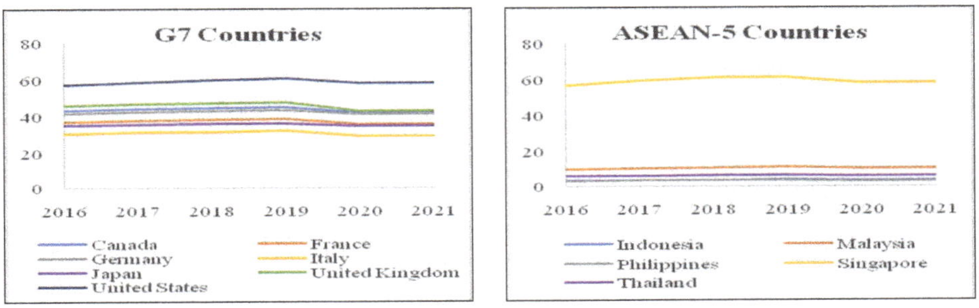

Figure 2. Real GDP per capita ('000) (constant USD 2015) of G7 and ASEAN-5 countries, 2016–2021. Source: [7].

The urban population based on the percentage of the total population of G7 and ASEAN-5 countries data were from 2016 to 2021. In this case, Singapore had the highest urbanization compared to both economic blocs at 100.00% on average. The second highest urbanization was marked by Japan at 91.65% on average, which is another high-population-density country characterized by its highly developed public transport system. Italy is the least urbanized country among G7 countries at an average of 70.54% due to its peninsula-shaped terrestrial form, which is more accommodated to fit rural towns. This caused its urbanization to fall slightly lower than Malaysia from the ASEAN-5 block, which had 76.21% on average. Meanwhile, Indonesia, Thailand, and the Philippines were the least urbanized across both economic regions at 55.54%, 50.19%, and 47.00% on average, respectively.

Figure 3 compares the individuals using the internet based on the percentage of the total population of G7 and ASEAN-5 countries from 2016 to 2021. Both economic blocks exhibited a rising pattern across the five years at different rates. In the G7 economic block, Canada achieved an average of 94.82% from 2016 to 2021, where it increased 5.81% over the years; the United Kingdom had an average of 93.01% and it increased only 0.04% across the 5 years; Japan marked an average of 91.56% but the percentage fell 2.96% within the 5 years; the United States marked an average of 88.76% with an increase of 5.36% over the 5 years; Germany marked an average of 87.23% with an increment of 5.65%; France achieved an average of 82.46% with an increment of 5.53%; and Italy marked an average of 69.31% but experienced an increase of 9.16%, the highest in the G7 economic block.

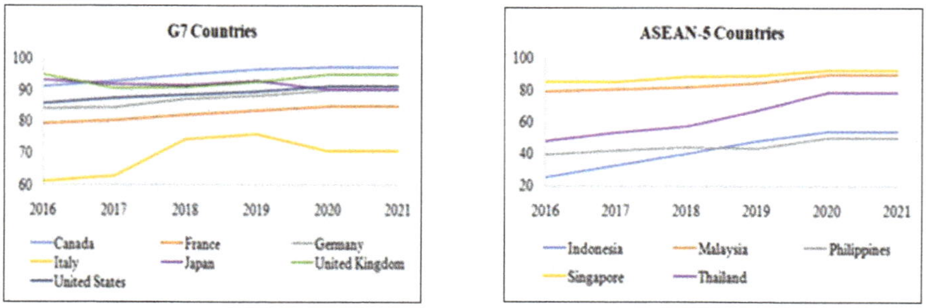

Figure 3. The individual using the internet based on % of total population of G7 and ASEAN-5 countries, 2016–2021. Source: [7].

Meanwhile, in the rest of the ASEAN-5 block, Singapore had the highest internet usage at 88.34% of the total population on average from 2016 to 2021, where the rate rose 7.55% across the five years. This is followed by Malaysia at an average of 83.90% but experienced an increase of 10.77% across the five years; it is then followed by Thailand at an average of 63.26% but increased 30.34%, which was the highest in ASEAN-5. This is then followed by the Philippines at 44.59% on average, which experienced a modest 10.60% rise across the five years; lastly, Indonesia marked 42.14% due to its extremely large population with a rise of 28.28% over the five years.

We also mentioned the carbon dioxide emissions in million tons of G7 and ASEAN-5 countries from 2016 to 2021. Although the carbon dioxide emissions of both economic blocks are much different (G7 countries produce much more carbon dioxide than ASEAN-5 countries), there is a common trend, and the United States and Indonesia are the top carbon producers of their respective blocks due to their manufacturing-intensive economic activities at an average of 4862.31 million tons and 564.82 million tons, respectively. In addition to these two countries, the rest of the countries charted fairly low carbon emissions, which are indicative of their reliance on manufacturing activities and competence in mitigating carbon emissions. In the rest of the G7 block, Japan emitted 1116.91 million tons on average; Germany 692.87 million tons; Canada 551.45 million tons; the United Kingdom 372.85 million tons; Italy 317.85 million tons; and France 289.75 million tons. Meanwhile, for the remaining ASEAN-5 economies, Malaysia emitted 255.97 million tons of carbon dioxide on average over the 5 years; followed by Thailand at 290.29 million tons; Singapore at 218.45 million tons; and lastly the Philippines at 129.28 million tons.

Figure 4 compares the total trade as a percentage of the GDP of G7 and ASEAN-5 countries from 2016 to 2021. Germany marked the highest performance in the G7 countries block at 85.04% on average but is outcompeted by Singapore at a whopping 300% and above. Meanwhile, the United States and Indonesia marked the lowest performance for the G7 and ASEAN-5 economies at 25.70% and 37.28%, respectively, meaning that their economy is very diverse, even though manufacturing is an integral component of their economic activities, such as the services and agriculture sector. For the other countries, performance averaged 63.98% for Canada, 61.36% for France, 59.48% for the United Kingdom, 57.40% for Italy, and 33.21% for Japan on average for the rest of the G7 countries; it was 124.38% for Malaysia, 111.33% for Thailand, and 64.55% for the Philippines on average for the rest of the ASEAN-5 countries.

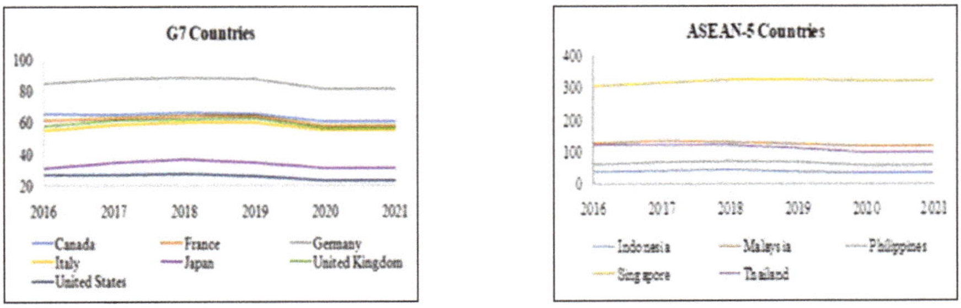

Figure 4. Total trade (% of GDP) of G7 and ASEAN-5 countries, 2016–2021. Source: [7].

Lastly, compared to the net inflow of FDI as a percentage of the GDP of G7 and ASEAN-5 countries are from 2016 to 2021. In this case, Singapore marked the highest new FDI inflows to GDP at 26.06% on average, meaning it is the most attractive foreign investment destination to its economic size. On the other hand, the FDI net inflow performance of the United Kingdom was bad and fell 10.80% across the five-year period and eventually averaged 3.01%. In general, the ASEAN-5 economies' performance was very low, similar to that of the G7 economies. For the rest of the G7 economies, Germany marked 2.74%, Canada marked 2.04%, the United States marked 1.49%, France marked 1.43%, Japan marked 0.83%, and Italy marked 0.54% on average; for the rest of the ASEAN-5 economies, Malaysia marked 2.46%, Philippines marked 2.42%, Indonesia marked 1.69%, and Thailand marked 0.70% on average. This can also be interpreted differently as ASEAN-5 economies had very regular FDI attraction performance similar to that of G7 economies, while Singapore had an excessively stellar performance in attracting FDI.

2. Methods

For the panel model analysis of renewable energy consumption (REC) in G7 and ASEAN-5 countries, this study considered macroeconomic variables, including economic growth (real GDP per capita), urbanization (urban population), trade openness (total trade), and foreign direct investment (FDI) levels (FDI, net inflows); a socioeconomic variable, which is access to the internet (individual using the internet); and an environmental variable, which is environmental degradation (carbon dioxide emissions). Annual data from 1990 to 2021 were collected from World Bank and British Petroleum (BP) Statistical Review of Energy. There are two models for analysis, one for the G7 countries while the other for ASEAN-5 countries. Figure 5 presents the proposed conceptual framework of this research.

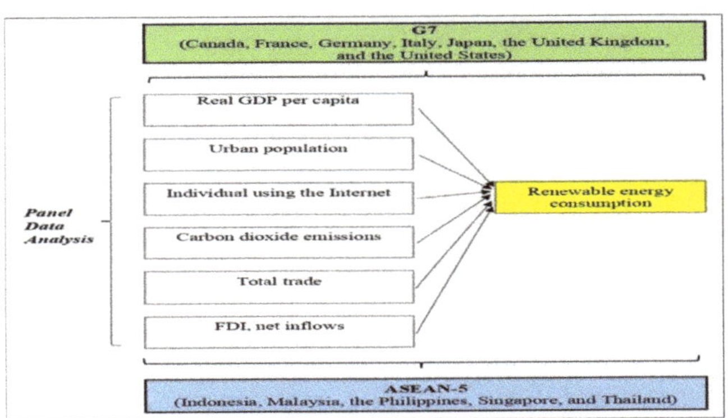

Figure 5. Proposed conceptual framework of renewable energy consumption in G7 and ASEAN-5 countries.

2.1. Research Procedure

2.1.1. Panel Data Analysis

The panel data analysis was selected for this study. It is known as a longitudinal study, i.e., a combination of cross-sectional (N) data and time-series (T) data over two or more different time periods [8,9]. In most econometric research, panel data analysis is preferable due to its ability to (1) avoid omitted variable problems, which would otherwise induce biases; (2) allow greater flexibility to study more complicated behavioral models due to its multidimensional data sample; (3) resolve heterogeneity, variability, collinearity, and degree of freedom issues; (4) and account for more dynamics of change [9].

2.1.2. Panel Unit Root Test

The unit root is acknowledged as a stochastic trend in a time series, which results in an unpredictable systematic pattern [9]. The existence of unit root in a series will cause issues in an analysis, such as spurious regressions where the R-squared value is too high even though there is no correlation between the data or erratic behavior in the model, which invalidates the ultimate results. Therefore, if the unit root exists in the series and is tested out, a series of successive differences will be made to transform the non-stationary series into a set of stationary series. There is a systematic method to depict this phenomenon. If the data are stationary at level, they are denoted as I(0); if the data are stationary at the first difference, they are denoted as I(1); and if the data are stationary at the second difference, they are denoted as I(2). The hypotheses H_{01} and H_{A1} of the unit root tests are as follows:

H_{01}: *Each time series contains a unit root (sig-p-value > α 0.05).*

H_{A1}: *Each time series is stationary (sig-p-value $\leq \alpha$ 0.05).*

The [10] test was employed for the panel model unit root tests for G7 and ASEAN-5 countries. For G7 countries, the results indicate that, at level, all variables are stationary except for REC and carbon dioxide emissions. After first differencing, all variables are stationary except for the urban population and percentage of individuals using the internet. At the second differencing, all variables are stationary except for the urbanization rate. Because the data are stationary at non-uniform difference levels, we conducted a cointegration test afterwards on the data sample. Meanwhile, for ASEAN-5 countries, a similar pattern emerged. At level, all variables are stationary except for real GDP per capita, the percentage of individuals using the internet, and total trade. After the first differencing, all variables are stationary except for the urban population. At second differencing, all variables are stationary except for urban population. Because the data are stationary at non-uniform difference levels, we also conducted a cointegration test afterwards on the data sample. The results are concluded in Table 1.

Table 1. Panel model unit root test of the renewable energy consumption of G7 and ASEAN-5 countries.

	Levin, Lin, and Chu					
		G7 Countries			ASEAN-5 Countries	
	Level	I(1)	I(2)	Level	I(1)	I(2)
REC	1.54523	−9.76818 ***	−19.1487 ***	−2.02413 **	−7.58251 ***	−3.48334 ***
RGDP	−2.50759 ***	−11.0439 ***	−14.8127 ***	0.00157	−8.86774 ***	−10.6514
URBAN	−5.24503 ***	0.98489	−2.26594 **	−12.6202 ***	2.37411	0.38139
INT	−2.70664 ***	−2.46315 ***	−12.3615 ***	4.38508	−3.22661 ***	−6.74003 ***
CO_2	3.39235	−12.3475 ***	−10.9871 ***	−1.90902 **	−5.33931 ***	−7.54595 ***
TRADE	−2.23316 ***	−11.9227 ***	−14.5881 ***	−0.31047	−6.36379 ***	−2.424899 ***
FDI	−3.83163 ***	−13.8483 ***	−14.6078 ***	−3.35995 ***	−10.5845 ***	12.3413

Note: ** and *** indicate statistical significance at α = 0.10, 0.05, and 0.01 levels, respectively.

2.1.3. Panel Cointegration Test

Non-stationarity is not necessarily a bad thing. Sometimes when two or more time series exhibit the same stochastic trend, this special case is known as cointegration [11].

Such situations can then be modelled using the vector error correction model, which can help to forecast future values. This can be achieved through the Pedroni test, Kao test, or Engle–Granger-Augmented Dickey–Fuller test. The hypotheses H_{02} and H_{A2} of the cointegration test are as follows:

H_{02}: *The time series do not have cointegration ($\alpha > 0.01$).*

H_{A2}: *The time series have cointegration ($\alpha \leq 0.01$).*

From the results in Tables 2 and 3, it is identified that all 6 cointegration tests of G7 and ASEAN-5 countries are significant at the 0.01 and 0.05 levels. Therefore, the results of the trace test and the maximum eigenvalue test indicate that 6 cointegrating equations were significant at the 0.01 and 0.05 levels, which meant that the long-term equilibrium between the variables was met.

Table 2. Johansen–Fisher panel cointegration test for the REC model of G7 countries.

Hypothesized No. of CE(s)	Fisher Stat (from Trace Test)	Prob.	Fisher Stat (from Max-Eigen Test)	Prob.
None	304.7	0.0000	183.6	0.0000
At most 1	177.9	0.0000	77.93	0.0000
At most 2	110.9	0.0000	45.97	0.0000
At most 3	73.20	0.0000	33.08	0.0028
At most 4	49.27	0.0000	30.89	0.0057
At most 5	30.95	0.0056	26.07	0.0254
At most 6	26.94	0.0196	26.94	0.0196

Table 3. Johansen–Fisher panel cointegration test for the REC model of ASEAN-5 countries.

Hypothesized No. of CE(s)	Fisher Stat (from Trace Test)	Prob.	Fisher Stat (from Max-Eigen Test)	Prob.
None	174.8	0.0000	99.44	0.0000
At most 1	104.0	0.0000	38.70	0.0000
At most 2	70.45	0.0000	21.79	0.0053
At most 3	53.56	0.0000	23.49	0.0028
At most 4	35.84	0.0000	21.67	0.0056
At most 5	22.15	0.0047	18.58	0.0173
At most 6	15.87	0.0443	15.87	0.0443

2.1.4. Panel Model Selection

There are three generic models, which are: (1) Fixed-Effects Model (FEM), (2) Random Effects Model (REM), and (3) Pooled Ordinary Least Squares Model (POLS) [8]. The REM has the advantage of avoiding bias due to omitted variables that remain constant over time and the FEM has the advantage of estimating time-invariant explanatory variables and allowing for higher degrees of freedom, whereas the POLS is the most basic OLS technique run on data that ignores all individually specific effects. To decide whether we should pursue REM, FEM or POLS, we referred to the following tests:

1. Firstly, the Redundant Fixed-Effects Test was conducted. The hypotheses H_{03} and H_{A3} of this test are as follows:

H_{03}: *POLS is preferred.*

H_{A3}: *FEM is preferred.*

If the significant p-value is lower than the α-value of 0.05, H_0 is rejected, and the fixed-effects model is preferred. If the significant p-value is higher than the α-value of 0.05, H_0 is accepted and the POLS effects model is preferred.

2. Next, the Lagrange Multiplier Test for Random Effects was conducted. The hypotheses H_{04} and H_{A4} of this test are as follows:

$\mathbf{H_{04}}$: *POLS is preferred.*

$\mathbf{H_{A4}}$: *REM is preferred.*

If the significant *p*-value is lower than the α-value of 0.05, H_0 is rejected, and the random effects model is preferred. If the significant *p*-value is higher than the α-value of 0.05, H_0 is accepted and the POLS effects model is preferred.

3. Finally, the Correlated Random Effects—Hausman Test was conducted. The hypotheses H_{05} and H_{A5} of this test are as follows:

$\mathbf{H_{05}}$: *REM is preferred.*

$\mathbf{H_{A5}}$: *FEM is preferred.*

If the significant *p*-value is lower than the α-value of 0.05, H_0 is rejected, and the fixed effects model is preferred. If the significant *p*-value is higher than the α-value of 0.05, H_0 is accepted and the random effects model is preferred.

Table 4 shows the tests for G7 panel model selection, where the fixed-effects model was identified to be the most appropriate model. Meanwhile, Table 5 shows the test for ASEAN-5 panel model selection, where the panel fixed-effects model (FEM) was also identified to be the most appropriate model.

Table 4. Panel model selection for the REC model of G7 countries.

Tests	Hypothesis	*p*-Value	Conclusion
Redundant Fixed-Effects Test	H_{03}: POLS is preferred. H_{A3}: FEM Is preferred.	$0.0000 < \alpha = 0.05$	Reject H_{03}. FEM is preferred.
Breusch–Pagan LM Test	H_{04}: POLS is preferred. H_{A4}: REM Is preferred.	$0.0000 < \alpha = 0.05$	Reject H_{04}. REM is preferred.
Hausman Test	H_{05}: REM is preferred. H_{A5}: FEM Is preferred.	$0.0000 < \alpha = 0.05$	Reject H_{05}. FEM is preferred.

Table 5. Panel model selection for the REC model of ASEAN-5 countries.

Tests	Hypothesis	*p*-Value	Conclusion
Redundant Fixed-Effects Test	H_{03}: POLS is preferred. H_{A3}: FEM Is preferred.	$0.0000 < \alpha = 0.05$	Reject H_{03}. FEM is preferred.
Breusch–Pagan LM Test	H_{04}: POLS is preferred. H_{A4}: REM Is preferred.	$0.0000 < \alpha = 0.05$	Reject H_{04}. REM is preferred.
Hausman Test	H_{05}: REM is preferred. H_{A5}: FEM Is preferred.	$0.0029 < \alpha = 0.05$	Reject H_{05}. FEM is preferred.

3. Results

3.1. *FEM of G7 Renewable Energy Consumption*

Based on Table 4, the fixed-effect panel model equation for the REC of G7 countries after obtaining the output from EViews is shown in Model (1) below:

$$\begin{aligned} \text{RECG-7}_{it} = 40.3673 + 0.1501\ \text{RGDP}_{it} + 0.3773\ \text{URBAN}_{it} + 0.07251\ \text{INT}_{it} \\ + 0.0066\ \text{CO}_{2it} + 0.1728\ \text{TRADE}_{it} + 0.2580\ \text{FDI}_{it} \end{aligned} \quad (1)$$

t = [1.8391 *] [4.4620 ***] [6.0445 ***] [7.6954 ***] [8.0767 ***] [3.3834 ***]

Note: * = at α 0.10 statistically significant level; ** = α 0.05 statistically significant level; *** = at α 0.01 statistically significant level.

Based on Model (1), the explanatory variables accounted for 92.24% of the variation in the REC of G7 countries. The results show that all the explanatory variables (namely, macroeconomic variables, economic growth (real GDP per capita), urbanization (urban population), trade openness (total trade), and foreign direct investment (FDI) levels (FDI, net inflows); the socioeconomic variable, which is the access to the internet (individuals using the internet); and the environmental variable, which is environmental degradation (carbon dioxide emissions)) are significant for the REC of G7 countries. Moreover, for the heteroskedasticity test, White test F = 1.211262, prob-value = 0.3334 and the outcome of the White test of the auxiliary regression's R2 value is 0.225228. Therefore, VIF value is 1.2907, which is less than 5. Therefore, Model (1) residuals present no multicollinearity problem. Additionally, the residuals for the normality distribution test's prob-value = 0.0741, which is more than the α 0.05 level. Thus, the residuals are also normally distributed.

A 1 unit increase in RGDP has a positive effect on the change in REC by 0.1501 units with statistical significance at the 0.10 level holding other variables constant. Next, a 1 unit increase in URBAN has a positive effect on the change in REC by 0.3773 units with statistical significance at the 0.01 level holding other variables constant. Meanwhile, a 1 unit increase in INT has a positive effect on the change in REC by 0.0725 units with statistical significance at the 0.01 level holding other variables constant. Then, a 1 unit increase in CO_2 has a positive effect on the change in REC by 0.0066 units with statistical significance at the 0.01 level holding other variables constant. Moreover, a 1 unit increase in TRADE has a positive effect on the change in REC by 0.1728 units with statistical significance at the 0.01 level holding other variables constant. Finally, a 1 unit increase in FDI has a positive effect on the change in REC by 0.2580 units with statistical significance at the 0.01 level holding other variables constant.

3.2. FEM of G7 Renewable Energy Consumption

Based on Table 5, the fixed-effects panel model equation for REC of ASEAN-5 countries after obtaining the output from EViews is shown in Model (2) below:

$$\begin{aligned} \text{RECASEAN-5}_{it} = 54.7654 + 0.1121\ \text{RGDP}_{it} + 0.2192\ \text{URBAN}_{it} + 0.2530\ \text{INT}_{it} \\ + 0.0033\ \text{CO}_{2it} + 0.1523\ \text{TRADE}_{it} + 0.6714\ \text{FDI}_{it} \end{aligned} \quad (2)$$

t = [7.9122 ***] [3.2934 ***] [8.8423 ***] [0.6126] [9.8540 ***] [2.9144 ***]

Note: *** = at α 0.01 statistically significant level.

Based on Model (2), the explanatory variables accounted for 81.17% of the variation in the REC of ASEAN-5 countries. The results show that the macroeconomic variables (economic growth (real GDP per capita), urbanization (urban population), trade openness (total trade), and foreign direct investment (FDI) levels (FDI, net inflows)) and the socioeconomic variable (which is access to the internet (individual using the Internet)) are significant factors. However, the environmental variable, which is environmental degradation (carbon dioxide emission), is not significant for the REC of ASEAN-5 countries. Furthermore, for the heteroskedasticity test, White test F = 2.150000, prob-value = 0.0828 and the outcome of the White test of the auxiliary regression's R2 value is 0.340369, Therefore, VIF value is 1.5159, which is less than 5. Therefore, Model (2) residuals present no multicollinearity problem. Additionally, the residuals for the normality distribution test's prob-value = 0.8490, which is more than α 0.05 level. Thus, the residuals are also normally distributed.

A 1 unit increase in RGDP has a positive effect on the change in renewable energy consumption (REC) by 0.1121 units with statistical significance at the 0.01 level holding other variables constant. Next, a 1 unit increase in URBAN has a positive effect on the change in REC by 0.2192 units with statistical significance at the 0.01 level holding other variables constant. Then, a 1 unit increase in INT has a positive effect on the change in REC by 0.2530 units with statistical significance at the 0.01 level holding other variables constant. Moreover, a 1 unit increase in TRADE has a negative effect on the change in REC by 0.1523 units with statistical significance at the 0.01 level holding other variables constant.

Lastly, a 1 unit increase in FDI has a positive effect on the change in REC by 0.6714 units with statistical significance at the 0.01 level holding other variables constant.

Energy is the key driver of economic growth. Thus, for the case of G7 countries, for the economic leaders of G7 countries to achieve the sustainable development goals, the urban population (URBAN) and the net FDI inflows (FDI) should be considered, as they have a more positive effect on the change in REC [12,13]. However, for the case of ASEAN-5 countries, for the economic leaders of ASEAN-5 countries to achieve the sustainable development goals, they should focus on the effects of the individuals using the internet (INT) and the net FDI inflows (FDI) on the REC [14,15].

4. Discussion

The empirical findings show the importance of macroeconomic, socioeconomic, and environmental variables to achieve REC in G7. For the case of ASEANN-5, it is macroeconomic and socioeconomic variables that affect REC. However, the environmental variable is not significant. This can be explained by the study in [16]. They found that carbon emissions per capita do not influence investments in renewable energy. This is because they are satisfied with the number of renewable energy initiatives in place to counter the perceived level of environmental degradation.

4.1. Factors Affecting REC in G7

For macroeconomic factors in the case of G7 countries, similar results have been found in previous studies [12,17,18]. Real GDP per capita affects India, and there is a long-term relationship between economic growth and renewable energy consumption [18]. While for the urban population, in the study of [12], the results show that urbanization is an important factor that affects REC. This can be explained by the increased intensity of energy demand in highly urbanized areas, which translates to an increased demand for alternative sources of energy. Based on the study of [17], the results show that trade openness exhibits a significant effect on hydropower energy, wind energy production, and solar and geothermal energy sources, which are renewable alternatives of energy supplied by the government based on domestic demand. While for FDI net inflows, the results of [13] show an effect between FDI and REC, i.e., a long-run U-shaped relationship is present for renewable sources and FDI.

For the socioeconomic variable, i.e., individuals using the internet, the results of [12] show internet usage is a significant factor affecting REC. This is because internet access enables a more efficient dissemination of information on environmental issues that help to increase consumer awareness towards environmental issues. While for the environmental variable, i.e., carbon dioxide emissions, the results are consistent with those of a previous study [19], which shows that energy utilization has a significant impact on carbon dioxide emissions.

4.2. Factors Affecting REC in ASEAN-5

For macroeconomic factors in the case of ASEAN-5 countries, similar results have been found in previous studies [15,20–22]. For the macroeconomic factors, the results of [22] show that economic growth (real GDP per capita) has a significant impact on renewable energy. This is especially during periods of high economic growth, and there is unidirectional causality from economic growth to renewable energy. Moreover, for urban population, the results of [21] show urbanization increases renewable energy significantly and there is bidirectional relationship between urbanization and renewable energy consumption. While for total trade, the results of [20] show a negative cointegration direction between trade openness and renewable energy, i.e., an increase in trade openness decreases renewable energy consumption. This is because, when countries' trade openness increases, they require more energy, which increases energy costs. However, countries can import advanced technologies at the same time, which reduces energy intensity and offsets the energy costs simultaneously. For FDI net inflows, [15] shows a strong relationship between FDI and renewable energy in South Asian nations, and it is suggested that policymakers incorporate FDI into policies to promote environmental sustainability.

For the socioeconomic variable, i.e., individuals using the internet, the results of [14], which used the system-generalized method of moments (GMM) estimator on Chinese provincial panel data from 2007 to 2016, show the popularity of the internet does improve the public's environmental protection awareness, which indicates a relationship between information and communication technology (ICT) and sustainable development.

5. Conclusions

In conclusion, this research study showed that REC is significantly affected by macroeconomic, socioeconomic, and environmental variables in G7 countries, while for the case of ASEANN-5, it is macroeconomic and socioeconomic variables that affect REC. This implies that policymakers should cater for the relevant indicators in policy formulation for the development of REC. Since ASEAN-5 countries are still in transition towards becoming developed countries, economic activities can become intensify in the mission to achieve sustained economic growth, which would subsequently lead to environmental degradation if REC is not improved. Thus, it is imperative for policymakers to incentivize business owners to transition towards greener alternatives to ensure that the future generations of the country are not left with a heavily polluted country.

Moreover, the significance of the socioeconomic variable is reflected in the importance of policymakers in G7 and ASEAN-5 countries utilizing internet platforms, such as social media, to raise environmental awareness effectively and efficiently. Since the youth is the most active user of social media and the future leaders of the nation, cultivating environmental awareness from a young age is important in ensuring the sustainable development of the country without compromising economic prosperity. Taking Singapore as an example, the Ministry of Sustainability and the Environment of Singapore runs social media accounts (i.e., Instagram and Facebook) and posts trendy pictures and videos that younger generations can relate to and "reshare" to their friends and families. Such initiatives are modern and up-to-date with the latest trends, which ensure that the youths are actively engaged with the work of policymakers.

The governments in G7 and ASEAN-5 countries need to set up special taskforces to track REC, which is crucial to identify the progress of the nation towards achieving the sustainable development goals. The governments can also invest in research and development projects (R&D) to improve public access to the required data, allowing researchers and even the general public to understand better the country's sustainable development progress.

Author Contributions: Conceptualization, A.A.K.; methodology, S.H.; validation, A.A.K.; writing—original draft preparation, K.M.T.; writing—review and editing, W.Y.W.; funding acquisition, K.M.T. All authors have read and agreed to the published version of the manuscript.

Funding: This research received no external funding.

Institutional Review Board Statement: Informed consent was obtained from all subjects involved in the study. Written informed consent was obtained from the University of Tunku Abdul Rahman to publish this paper.

Informed Consent Statement: Not applicable.

Data Availability Statement: Not applicable.

Acknowledgments: The authors are mainly thankful to the Universiti Tunku Abdul Rahman (UTAR), which provided the opportunity to conduct this research.

Conflicts of Interest: The authors declare no conflict of interest.

References

1. United States Energy Information Administration. What Is Renewable Energy? 2022. Available online: https://www.eia.gov/energyexplained/renewable-sources/ (accessed on 27 October 2022).
2. ASEAN. The Founding of ASEAN. 2022. Available online: https://asean.org/the-founding-of-asean/ (accessed on 23 October 2022).
3. Ministry of Investment Indonesia. Harnessing Renewable Energy Investment Sector in Indonesia. 2022. Available online: https://www5.bkpm.go.id/en/publication/detail/news/harnessing-renewable-energy-investment-sector-in-indonesia (accessed on 23 October 2022).
4. World Wildlife Fund (WWF). Renewables Best Way to Power Philippine Development. 2017. Available online: https://wwf.org.ph/what-we-do/climate/renewables/renewables-to-power-ph-dev/ (accessed on 25 October 2022).
5. International Trade Administration of United States. Thailand—Renewable Energy. 2022. Available online: https://www.trade.gov/energy-resource-guide-thailand-renewable-energy (accessed on 26 October 2022).
6. Energy Market Authority Singapore. Overview. 2022. Available online: https://www.ema.gov.sg/ourenergystory (accessed on 25 October 2022).
7. World Bank. Renewable Energy Consumption (% of Total Final Energy Consumption). 2022. Available online: https://data.worldbank.org/indicator/EG.FEC.RNEW.ZS (accessed on 27 October 2022).
8. Studenmund, A.H. *Using Econometrics: A Practical Guide*, 7th ed.; Pearson: Harlow, UK, 2017.
9. Gujarati, D.N.; Porter, D.C. *Basic Econometrics*, 5th ed.; McGraw-Hill: New York, NY, USA, 2009; ISBN 978-007-127625-2.
10. Levin, A.; Lin, C.F.; Chu, C.S.J. Unit root tests in panel data: Asymptotic and finite-sample properties. *J. Econom.* **2002**, *108*, 1–24. [CrossRef]
11. Stock, J.H.; Watson, M.W. *Introduction to Econometrics*, 3rd ed.; Pearson: Harlow, UK, 2015.
12. Baye, R.S.; Olper, A.; Ahenkan, A.; Musah-Surugu, I.J.; Anuga, S.W.; Darkwah, S. Renewable energy consumption in Africa: Evidence from a bias corrected dynamic panel. *Sci. Total Environ.* **2021**, *766*, 142583. [CrossRef] [PubMed]
13. Shahbaz, M.; Sinha, A.; Raghutla, C.; Vinh Vo, X. Decomposing scale and technique effects of financial development and foreign direct investment on renewable energy consumption. *Energy* **2022**, *238*, 121758. [CrossRef]
14. Zhang, J.; Zhang, H.; Gong, X. Government's environmental protection expenditure in China: The role of Internet penetration. *Environ. Impact Assess. Rev.* **2022**, *93*, 1318–1328. [CrossRef]
15. Kang, X.; Ullah Khan, F.; Ullah, R.; Arif, M.; Ur Rehman, S.; Ullah, F. Does Foreign Direct Investment Influence Renewable Energy Consumption? Empirical Evidence from South Asian Countries. *Energies* **2021**, *14*, 3470. [CrossRef]
16. Abban, A.R.; Hasan, M.Z. Revisiting the determinants of renewable energy investment—New evidence from political and government ideology. *Energy Policy* **2021**, *151*, 112184. [CrossRef]
17. Khezri, M.; Heshmati, A.; Khodaei, M. The role of R&D in the effectiveness of renewable energy determinants: A spatial econometric analysis. *Energy Econ.* **2021**, *99*, 105287. [CrossRef]
18. Das, N.; Bera, P.; Panda, D. Can economic development & environmental sustainability promote renewable energy consumption in India? Findings from novel dynamic ARDL simulations approach. *Renew. Energy* **2022**, *189*, 221–230. [CrossRef]
19. Ulucak, R.; Erdogan, S. The effect of nuclear energy on the environment in the context of globalization: Consumption vs. production-based CO_2 emissions. *Nucl. Eng. Technol.* **2022**, *54*, 1312–1320.
20. Zeren, F.; Akkus, H.T. The relationship between renewable energy consumption and trade openness: New evidence from emerging economies. *Renew. Energy* **2020**, *147*, 322–329. [CrossRef]
21. Hashemizadeh, A.; Bui, Q.; Kongbuamai, N. Unpacking the role of public debt in renewable energy consumption: New insights from the emerging countries. *Energy* **2021**, *224*, 120187. [CrossRef]
22. Chen, Y.; Mamon, R.; Spagnolo, F.; Spagnolo, N. Renewable energy and economic growth: A Markov-switching approach. *Energy* **2022**, *244*, 123089. [CrossRef]

engineering proceedings

MDPI

Proceeding Paper

Applying Data Mining and Machine Learning Techniques to Predict Powerlifting Results †

Jorge Medina-Romero [1], Antonio Miguel Mora [1,*], Juan Francisco Valenzuela-Valdés [1] and Pedro Ángel Castillo [2]

1 Departamento de Teoría de la Señal, Telemática y Comunicaciones, ETSIIT-CITIC, Universidad de Granada, 18071 Granada, Spain; medinajromero@correo.ugr.es (J.M.-R.); juanvalenzuela@ugr.es (J.F.V.-V.)
2 Departamento de Ingeniería de Computadores, Automática y Robótica, ETSIIT-CITIC, Universidad de Granada, 18071 Granada, Spain; pacv@ugr.es
* Correspondence: amorag@ugr.es
† Presented at the 9th International Conference on Time Series and Forecasting, Gran Canaria, Spain, 12–14 July 2023.

Abstract: This paper presents a study on the creation of a tool to help powerlifting athletes and coaches, as well as bodybuilders and other amateur gym athletes, to analyse their data and obtain useful information regarding the athlete's performance. The tool should also predict future personal records in lifting for both raw (non-equipped) and non-raw (equipped) attempts, and their various exercises. In order to achieve this, a dataset with entries of around 500 k lifters and more than 20 k official powerlifting competitions was used. Among those entries, biometric variables of the lifters and the weights they lift in each of the three movements of this sport discipline were included: squat, bench press, and deadlift. We applied data preprocessing and visualising as well as data splitting and scaling techniques in order to train the machine learning models that are used to make the predictions. Lastly, the best predictive models were used in the implemented tool.

Keywords: powerlifting; forecasting; prediction; dataset; data mining; data visualisation; machine learning

Citation: Medina-Romero, J.; Mora, A.M.; Valenzuela-Valdés, J.F.; Castillo, P.A. Applying Data Mining and Machine Learning Techniques to Predict Powerlifting Results. *Eng. Proc.* **2023**, *39*, 20. https://doi.org/10.3390/engproc2023039020

Academic Editors: Ignacio Rojas, Hector Pomares, Luis Javier Herrera, Fernando Rojas and Olga Valenzuela

Published: 29 June 2023

1. Introduction

Lifting exercises are classified as sports, and the category ranges from well-known sports, such as football (soccer) and basketball, to less popular sports, such as strongman, weightlifting, or powerlifting. Most of the popular sports also incorporate lifting exercises to increase performance. For instance, soccer players perform squats or deadlifts to enhance their sprint or jump performance.

Powerlifting is a sport that focuses on three movements: *squat*, *bench Press*, and *deadlift*. In this sport, the lifter has three attempts to perform one repetition in each movement with as much weight as possible. Competitions start with the squat and finish with the deadlift, where each lifter has three attempts in each exercise to lift the highest load they can. These competitions are known as SBD, which stands for the first letter of each movement. However, there are also competitions focused on only one of the lifts or a combination of them. In addition, attempts can also be either performed with or without equipment, such as squat suits, compression bands, and slingshots, among others. The equipped attempts are known as *non-raw*, whereas unequipped trials are called *raw* trials.

Progressive overload is the key to achieving an increase in performance over time. To improve in a particular movement, the lifter must increase the weight being lifted. However, knowing when to increase the weight is not straightforward, and can have negative consequences for the athlete, such as nervous fatigue, muscle fatigue, performance reduction, or demotivation. Performing such highly demanding lifts, for example, a *one-repetition max (1RM)*, also has additional risks associated with it.

Nowadays, some free online tools are available for athletes to estimate their 1RM based on their weight, age, height, and their results from less-demanding weight lifts, such as 5RM, 3RM, etc. However, there are no tools that can predict or estimate how much weight someone can lift in equipped powerlifting according to the same biometric inputs and their 1RM performance in non-equipped powerlifting, and vice versa.

Thus, the purpose of this work is to build a tool that can predict the future performance of lifters in both raw and non-raw competitions according to their current records in either discipline. It will be based on the application of data mining and machine learning methods on a complete real dataset, comprising data from more than 500 k athletes, both women and men, and belonging to different age and weightlifting category ranges. The application will allow lifters to predict how much a raw athlete will lift in a future non-raw competition, or how well a non-raw lifter can perform in a raw competition, based on their current progress in the raw/non-raw discipline.

To this end, an exploratory data analysis was conducted in order to better understand the data, and to extract some estimates that could be useful for the prediction models. Then, several regression algorithms were tuned and applied to the dataset in order to predict future weightlifting results from current variables, such as the athlete's gender, bodyweight, previous records, and equipment use.

2. Related Work

There are few studies in the literature related to powerlifting predictions. We have reviewed the main ones, which have served as inspiration and are the basis for comparison in the present study. In addition, these works have some weaknesses and flaws that have been addressed in our study.

- **Analysis of USA Powerlifting Federation data (2012–2016)** [1]: This work highlights the importance of considering equipped and raw powerlifting as different sports, and the substantial differences in performance between the two disciplines. Squats and bench presses are much more affected by equipment than deadlifts due to their eccentric and concentric phases. Lifters do not benefit as much from equipment when performing deadlifts, since there is no eccentric phase. Given this, in our study, we analysed raw and equipped lifting separately, and also considered gender differences. Women tend to reach their plateau sooner than men, but their performance declines earlier. Males reach their peak performance later, but tend to hold it for a much longer time period. The ideal age for performance ranges from 24 to 39 years, with the mean age established at 30 years. This work [1] has some limitations, such as the limited amount of data and the tedious process of manually searching through the International Powerlifting Federation (IPF) website. In our study, we processed an already well-developed database from OpenPowerlifting. Published statistics in [1] regarding ages, weights, genders, raw vs. equipped, and exercises were used as a reference for further analysis. We complement this information with additional graphic resources and analyses, in order to gain further insights.
- **Influence of Compressive Gear on Powerlifting Performance: Role of Blood Flow Restriction Training** [2]: In this paper, the blood occlusion induced by compressive gear is described. The summary proposes that the occlusion can be maximum while performing squats. That is, the gear used in non-raw competitions can totally occlude the blood flow in the legs, which resulted in an increase in performance. What is more, this increase is not as significant either in bench presses or deadlifts. This may be due to the nature of the movement itself (deadlift) and the fact that it lacks an eccentric phase, or that it is impossible to totally occlude the muscles involved in the exercise (pectoral muscles in bench press) to such an extent as is possible with squats. The fact of the matter is, wearing such gear will still increase the performance in both movements, especially when compared to raw users, but not to the extent as in squats. However, the deadlift is the least affected out of the three. For these reasons, the idea of splitting the dataset into two different disciplines (raw vs. equipped) was reinforced.
- **The Role of FFM Accumulation and Skeletal Muscle Architecture in Powerlifting Performance** [3]: The importance of distinguishing between muscle mass and fat mass

was explained in this study. Basically, it shows that more fat-free mass (FFM), i.e., muscle, is related to performance in powerlifting: the more the better. However, it is still limited by the muscle structure itself, as well as the muscle longitude and the degree in which those muscle fibres are distributed. All these variables are involved in determining how much strength each muscle section can produce. All in all, the more fat-free mass the lifter has, the stronger they will be, and stronger athletes tend to lift heavier weights, making them better-performing powerlifters.

- **The Effect of Experimental Alterations in Excess Mass on Pull-Up Performance in Fit Young Men** [4]: In this work, the authors explain how the differences in body fat percentage of a lean athlete (low body fat) and a heavier one (high body fat) can affect their pull-up and their bench press performance. To begin with, they show that there is a strong relationship between how heavy the lifter is (tall and/or muscular individual), or how high their body fat percentage is, and performance, since increasing body fat percentage by 10% will result in a significant reduction in performance (fewer repetitions). It is said that this could apply to other forms of exercise, such as the bench press. So, if an individual lifts 100 kg for eight repetitions, and this same individual adds 10 kg to the bar, they will then only be able to perform four repetitions, that is, half of what they achieved previously (assuming eight repetitions is close to the athlete's maximum performance). It is shown that pull-ups are not as good at measuring an athlete's performance as they are at measuring their bodyweight, either as body fat percentage or muscle mass. Essentially, this work showed us how important it is to take the athlete's weight into consideration when analysing our own dataset, since variations in weight do make a difference.
- **Adjusting Powerlifting Performances for Differences in Body Mass** [5]: In this work, there is a comparison between the formulas of Stiff and Sinclair as well as other aspects that must be taken into consideration when working with a dataset similar to the one we used. It is said that the Stiff formula is better, since it works regardless of weight, and this makes it more accurate than the Sinclair formula. Additionally, it is advised that one separates the data for males and females, as well as raw and non-raw, and thus avoids mixing everything together. It also mentions that the larger the number of instances the better. The key lesson we learned from this study is that it is very important to split the database into different groups, even though, supposedly, the fact that one group is raw and the other is non-raw does not influence the results of the Stiff formula as much as it does the Sinclair formula, since the former does not use the weight variable in its calculations, whereas the latter does.

In summary, the majority of the studies mentioned are in agreement that there are two disciplines (raw and non-raw), and they should be considered as two different sports altogether. For these reasons, in our study, we compared them as separate disciplines and sports from the beginning. Doing so provides further insights as to how these powerlifting variations compare to one another. More importantly, it will allow us to see how our results compare to those of [1], the most important work to date on this topic, given the fact that the considered database in our study has considerably more instances and it is up to date. Additionally, our study provides a very detailed visualisation of gender distribution, both at the very beginning of the sport and in current years, as well as how it fluctuates throughout the months and years. Finally, we showcase the record evolution for each movement according to gender and year.

3. Dataset Description

Data were obtained from a website named *OpenPowerlifting* [6]. It includes information about almost every international powerlifting competition since 1965 to present, and is constantly updated.

The dataset contains information about every lifter: name, age, gender, weight, age class, weight class, the three attempts of every movement (squat, bench, and deadlift), as well as a column with the best out of the three, and the sum of those best three. There is even information about whether or not the athletes were drug tested. It also contains data concerning the competition (powerlifting meet) itself: name, country, state, town, federation, parent federation, and date of the event.

Finally, it also provides the results of some formulas relevant to the sport, such as Goodlift (whether or not a certain lift was performed correctly), or Wilks/Glossbrenner/Dots (three different computations of the relative strength of the lifter according to their bodyweight category), as well as the lifter's place in each meet. However, these last variables are not considered in our model.

As described above, the dataset holds a great deal of information. In fact, the initial dataset had 41 columns (features), of which 22 were floats and the others were strings or dates. From these, we selected 19, these variables being the most relevant according to previous studies. They can be seen in Table 1.

Table 1. Selected variables for our study.

#	VARIABLE	TYPE	DESCRIPTION
1	Name	string	Full name of each athlete
2	Sex	string	Gender of the athlete ('F' for females or 'M' for males)
3	Event	string	Whether it was a SBD competition with the three movements or just one or a combination of them
4	Age	float	The athlete's age
5	AgeClass	string	The athlete's age class
6	BodyweightKg	float	The athlete's bodyweight
7	WeightClassKg	float	The athlete's weight class
8	Best3SquatKg	float	The heaviest and correct lift in kg for squats recorded in the previous attempts
9	Best3BenchKg	float	The heaviest and correct lift in kg for bench presses recorded in the previous attempts
10	Best3DeadliftKg	float	The heaviest and correct lift in kg for deadlifts recorded in the previous attempts
11	TotalKg	float	The sum of the three best marks (in the three previous variables)
12	Place	float	The final position of that athlete in the competition
13	Dots	float	The result of applying the Dots formula to these values in order to rate the performance
14	Wilks	float	The result of applying the Wilks formula to these values in order to rate the performance
15	Glossbrenner	float	The result of applying the Glossbrenner formula to these values in order to rate the performance
16	Goodlift	float	The result of applying the Goodlift formula to these values in order to rate the performance
17	Tested	boolean	Records whether the athlete was drug tested or not
18	Parent Federation	string	The parent federation where the competition took place
19	Date	date	The date of the competition. Format: yy-mm-dd

The dataset considered in the present study can be downloaded from https://openpowerlifting.gitlab.io/opl-csv/bulk-csv.html (accessed on 20 June 2023). There was a preprocessing step, in which the data were anonymised, unknown values were removed, and the patterns were separated by sex and equipment use, and grouped into age ranges.

After this, a preliminary exploratory analysis was conducted (described in the following section) in order to visualise and better understand the data.

4. Exploratory Data Analysis

This is an initial analysis of the considered dataset, which aims to show the magnitudes and distribution of the data as well as to present some facts about them to take into account.

It is important to remind the reader that raw means 'unequipped attempt', while non-raw corresponds to an 'equipped attempt'.

First of all, the number of samples can be seen in Table 2, grouped by gender and equipment use.

Table 2. Total number of samples per gender. Each one corresponds to the best weight lift attempt by each athlete in a competition for each of the three movements (squat, bench press, and deadlift).

	RAW	NON-RAW
Women	227,700	151,500
Men	631,000	521,200
TOTAL	858,700	672,700

As can be seen in the table, the amount of data is more than 1.5 million samples, whereas in the reference work [1], the data included 15,800 men and 6000 women patterns. Thus, we can push our study much further than the previous work, reaching stronger conclusions, i.e., better supported by the data.

Figure 1 plots the data in a density graph, grouped by gender, showing the distribution of samples per age range and per athlete weight class (weightlifting athletes compete in a different category or class according to their bodyweight as well as their gender).

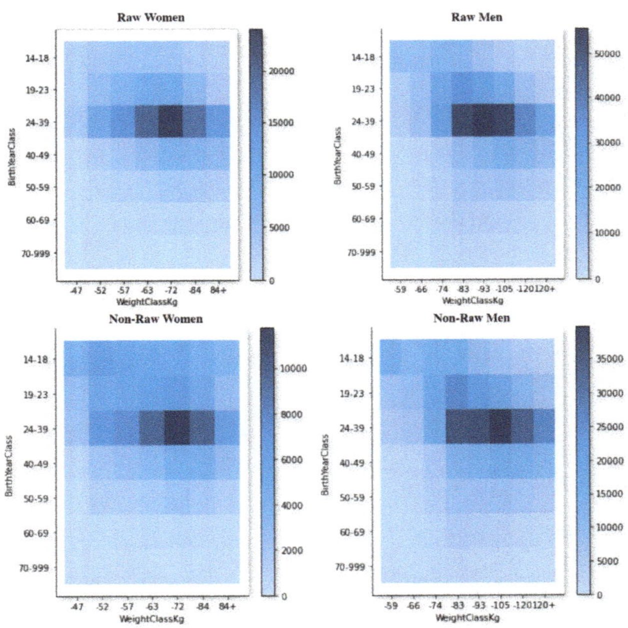

Figure 1. Women's and men's distributions by age range and weight categories according to the equipment used in their weight lift attempts. The colour of each cell represents the number of samples/instances in that specific age range and weight category (the darker the colour, the higher the number).

As can be seen in the density graphs, there are subtle distinctions that can be perceived. It seems that for non-raw male lifters, the weight category with the highest density is the under 105 kg category, whereas for raw lifters, the most common weight class is under the 93 kg category, followed very closely by the under 83 kg and under 105 kg categories. In spite of this, the age range in all of them is 24–39 years, as was also concluded from the previous study [1]. The graphs also show that, in the most popular weight classes for both men and women, there is a more balanced gender distribution in raw competitions, compared to equipped ones, in which there is a higher degree of male density (around 3.5 times more males than females).

With regard to the performance, we have analysed how it behaves according to each gender, equipment, and lift. Figures 2 and 3 show box plots of the weights lifted by the athletes according to their weight category.

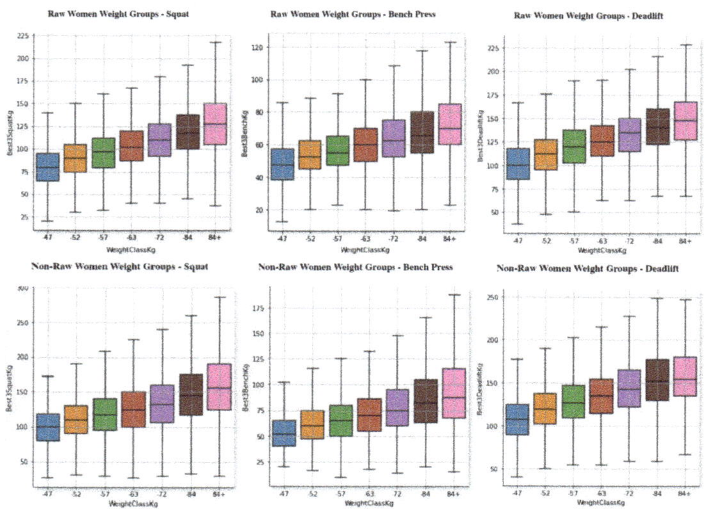

Figure 2. Powerlifting weight in the three movements for women, raw (**top**) and non-raw (**bottom**), separated by weight categories.

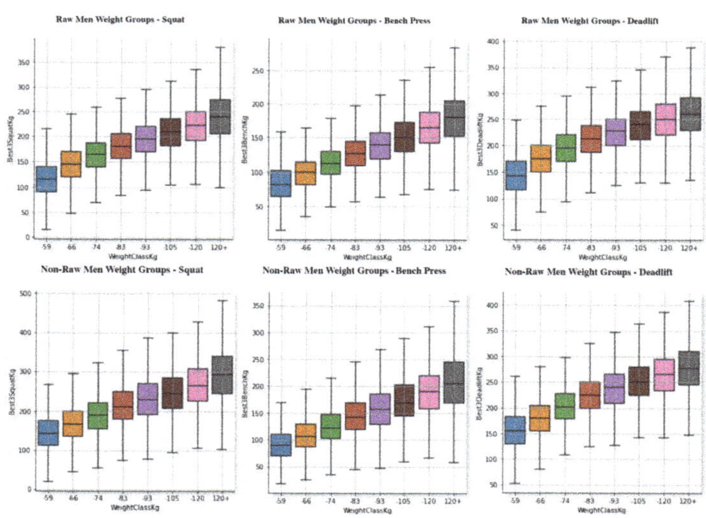

Figure 3. Powerlifting in the three movements for men, raw (**top**) and non-raw (**bottom**), separated by weight categories.

There are obvious trends, such as the correlation between the weight lifted and the weight class. Moreover, as expected, the main result is the positive impact of the equipment on the increase in performance in both genders and for all the weight classes. This affects men much more than women. Furthermore, the use of equipment does not help as much in deadlifts as in other movements.

The following analysis is similar to the previous one, but focused on the age range of the athletes. Thus, Figures 4 and 5 show box plots of the weights lifted by the athletes according to their age category.

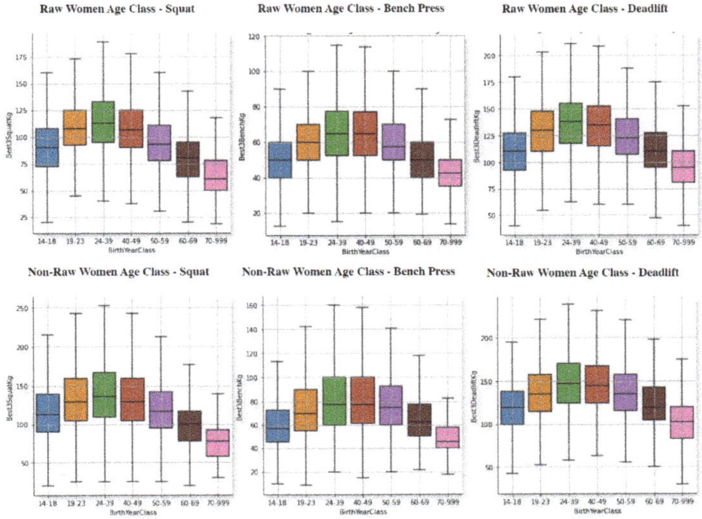

Figure 4. Powerlifting weight in the three movements for women, raw (**top**) and non-raw (**bottom**), separated by age ranges.

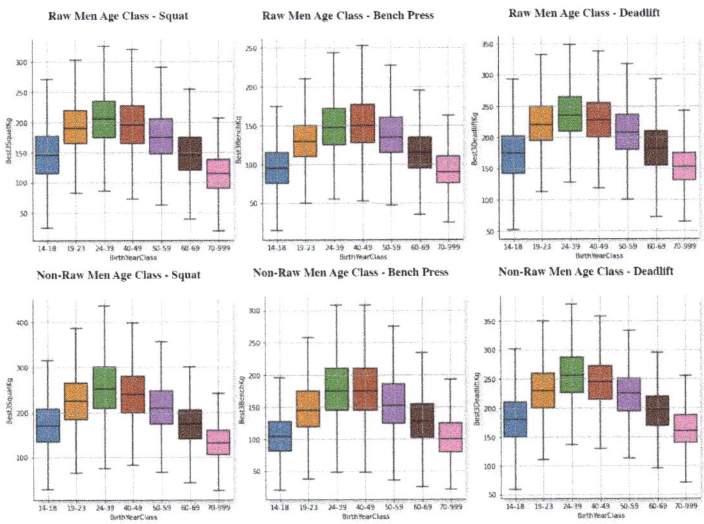

Figure 5. Powerlifting in the three movements for men, raw (**top**) and non-raw (**bottom**), separated by age ranges.

It is interesting to see, that the best performing age group is, in all of them 24 to 39 years of age. Which is the same age group, that has the highest amount of lifters as shown in Figure 1. Nonetheless, in the bench press exercise the differences between 40 to 49 yeas of age and this age group are almost non existent. So, it could be possible to say that in this particular movement, women and men tend to perform equally as good as the younger lifters or that athletes tend to maintain their bench press performance much better compared to the other two movements. In addition, with this data, it is save to say that younger male and female athletes perform better across all the movements except for bench press, where the differences are not as big and, in fact, can be surpassed by older athletes in the 40 to 49 age group. It is important to note, that the actual differences between 24–39 and 40–49 are not that notorious in any of the exercises to begin with, and neither in raw males or females. Additionally, it is confirmed, as pointed in [1] that males tend to reach their peak performance later than women and also hold it for longer, whereas women reach it faster but lose it sooner, as well. For instance, this can be perceived when looking at the performance in the youngest age group (blue) versus the 60–69 year group (brow box). For males, the performance is always either equal or even higher but for females is equal or lower.

Finally, we have analysed the evolution of the records per movement along the years. The data are plotted in Figure 6.

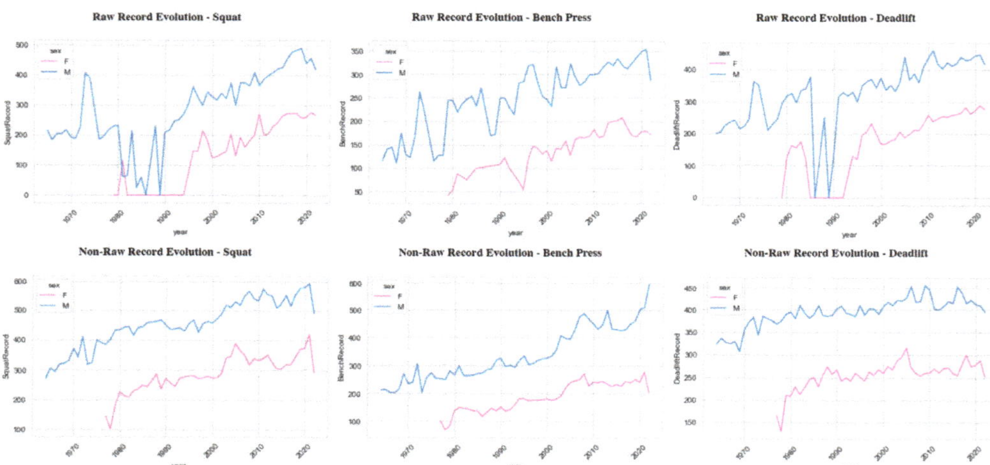

Figure 6. Records evolution along the years in the three movements for females (pink graphs) and males (blue graphs). Raw (**top**) and non-raw (**bottom**).

As can be seen, there are some missing pieces of information in a few periods, such as in the 1980s for squats and deadlifts, probably due to poor accuracy in the recollection of data. Moreover, there are no data for women before 1980. This is because women did not compete in the sport until this time. Furthermore, it is not until after 1995, more or less, that we start to see a somewhat steady increase in records, whereas before that date, the graph is rather unpredictable. The record improvement is a factor that everyone would logically consider important. Meaning that, in every sport, every year, new records are being set, due to improvements in techniques, coaching methods, and so on. Naturally, this applies for raw weightlifting as well. With regard to equipped powerlifting, there is a clear difference. Here, there are no years missing and the graphs show a very smooth increase in records year after year. As expected, the records for non-raw are much higher than for raw (since they wear performance-increasing gear).

In the following section, we will focus on the prediction of weightlifting for athletes.

5. Regression Models

In this section, we apply machine learning models to the data in order to predict results based on certain inputs. First of all, the dataset was preprocessed, removing the instances dating before 1996, since they could be deleterious during the learning phase of the model, as shown in the record evolution analysis in the previous section. Weight and age categories were also removed because they were redundant with regard to weight and age columns. Moreover, the Event column (describes the movement the athlete performed) was also removed, since it is information we decided not to address given the fact that we would create specific models for each movement, and that the vast majority of the instances were athletes that actually performed all three movements (squat, bench, and deadlift, also known as 'SBD'). We converted data in every column to integers. After this step, the 'Sex' column had only two values: '0' for females and '1' for males. Additionally, the 'Equipment' column followed the same strategy: '0' for raw and '1' for non-raw. We noticed that some athletes shared exactly the same name; in these instances, we removed inconsistent data. We performed a combination of all the instances for every athlete, with several comparisons between instances on a two-by-two basis (internally renamed as 'X' and 'Y'). This allowed us to compute a new variable 'DiffDays', where we stored the difference in days between the date of one instance and the other being compared. These actions provided us with the chance to select only those entries where the difference was no greater than 2 years and had a positive value (only predicting the future). By doing this, our model was capable of accurately predicting an athlete's performance up to 2 years into the future.

At this point, the dataset was ready to work, considering all the instances were correct and in numeric format, with only the essential and most relevant columns: sex; current equipment; current age; current bodyweight; current best squat, bench, and deadlift; future equipment; future age; future squat, bench, and deadlift; and the number of days until the next competition ('DiffDays'). Then, we separated the data into a different dataset per movement so that we could use a model devoted to each one of them. We split the datasets into 75% for training and 25% for testing.

Before applying the machine learning models, we analysed the relationship in terms of correlation between the dependent variable ('Y'), which is the weight lift per movement, and the rest of the variables ('X'), using Seaborn heat maps. The results can be seen in Figure 7.

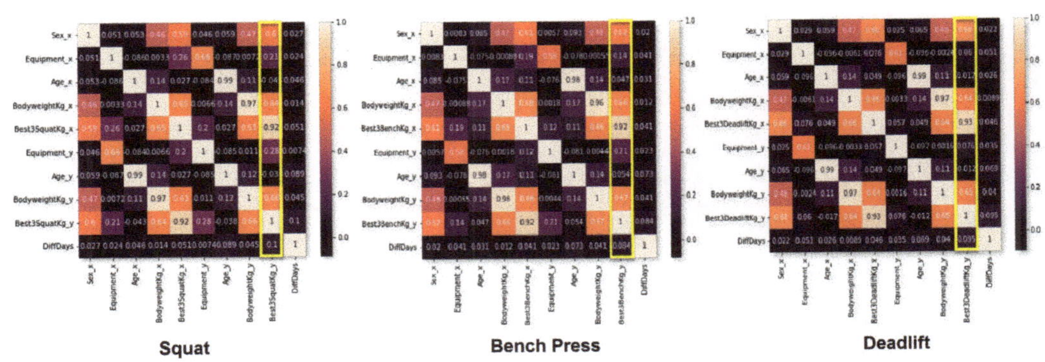

Figure 7. Correlation between features in each dataset (one per movement). Yellow rectangle highlights the target column.

As can be seen in the figure (squat subgraph), for the 'X' axis target column ("Best3SquatKg_y"), the "Best3SquatKg_x" feature is the the most influential towards the target (value 0.92), followed by future bodyweight (0.64), current bodyweight (0.62), sex (0.6), future equipment (0.28), and current equipment (0.2). The same results can be also seen in the case of bench press and deadlift, where the current '_x' best weight lifted, in that particular movement, is the feature with the strongest correlation towards the target best future weight lifted column '_y'.

We then used different regression models, since our goal was to predict a single numeric value depending on other numeric values. We applied linear, lasso, and ridge regression [7], as well as KNN [8], random forest [9], and decision tree [10] algorithms for regression. Additionally, we applied MARS (multivariate adaptive regression splines) [11] and ANN [12]. The configurations of all the methods were tuned after an exhaustive experimentation process, mainly in the case of the random forest (finally having 150 estimators, 5 minimum samples split, 1 minimum sample leaf, and no bootstrap) and ANN (3 layers: 9-4-1 neurons, RELU-TANH activation functions, MAE as loss metric, 128 batch size, 20 epochs). For all the methods, we computed, as performance metrics, the mean square error (MSE) [13], the mean absolute error (MAE) [14], and R-squared (R^2 Score), or the coefficient of determination [15], which indicates the percentage of the dependent variable interpreted from the independent variables, i.e., how well the regression model fits the data.

The obtained results can be seen in Tables 3–5, for each of the three movements.

All the tables show the best performance with the random forest method, followed very closely by ANN and KNN, with an R^2 between 91% and 92%, depending on the problem/movement, which is a relatively reliable result for a forecasting tool such as the one we aimed to build. However, the generated models in the case of the random forest required almost 20 GB of hard disk space, while ANN models only required 100 KB, which is a remarkable difference for just a 1% difference in R^2; so, in order to be used in a software tool, we chose to apply ANN models.

Table 3. Squat—regression model results.

Model	MSE	MAE	R^2 Score
Linear Regression	580.22	15.95	0.88
Lasso Regression	583.09	16.00	0.88
Ridge Regression	580.22	15.95	0.88
Random Forest	**433.59**	**13.56**	**0.91**
KNN	504.64	14.93	0.90
MARS	580.22	15.95	0.88
Decision Tree	849.55	18.47	0.84
ANN	489.37	14.75	0.90

Table 4. Bench press—regression model results.

Model	MSE	MAE	R^2 Score
Linear Regression	358.92	12.10	0.88
Lasso Regression	362.29	12.14	0.88
Ridge Regression	358.92	12.10	0.88
Random Forest	**193.86**	**8.53**	**0.94**
KNN	260.39	10.34	0.92
MARS	358.89	12.10	0.88
Decision Tree	407.55	11.86	0.88
ANN	243.12	9.88	0.92

Table 5. Deadlift—regression model results.

Model	MSE	MAE	R^2 Score
Linear Regression	413.61	13.10	0.88
Lasso Regression	416.12	13.13	0.88
Ridge Regression	412.27	13.07	0.88
Random Forest	**280.51**	**10.57**	**0.92**
KNN	351.06	12.38	0.90
MARS	418.07	13.21	0.88
Decision Tree	602.92	15.15	0.85
ANN	334.58	11.44	0.91

This tool has actually been implemented and deployed on the following website: https://medinajromero-webapptestfinal-webapplocal-6zlgiv.streamlitapp.com (accessed on 25 May 2023).

Figure 8 shows the webpage where the prediction application can be used.

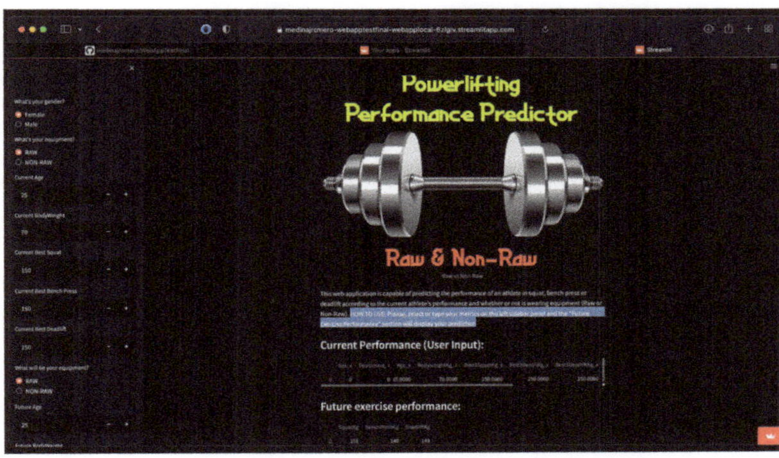

Figure 8. Webpage created with the implementation of the prediction tool.

The users can input their data, i.e., gender, current equipment (raw or non-raw), current age, current bodyweight, current best lift in each movement (squat, bench press, and deadlift), and their future age (could be the same), future bodyweight (could be the same), future equipment (can be raw or non-raw, in order to predict how much of a difference in performance the equipment will make), and the number of days until the desired competition takes place. The output will present the expected weight lift results for each of the movements.

6. Conclusions and Future Work

This paper presents a study on the creation of a tool based on machine learning (regression models) for the prediction of future results in powerlifting competitions for athletes. It takes into account gender, age range, weight category, and the possible use of equipment for the different attempts at three movements: squat, bench press, and deadlift.

To this end, a large dataset (more than 1.5 million instances) has been extracted and processed. After an exploratory analysis and a data preparation procedure, several regression techniques were applied to the data, obtaining R^2 (coefficient of determination of the regression models) results of around 91–92%. An artificial neural network was chosen as the predictor, and an actual web-based tool has been implemented.

This tool will be helpful for lifters and their trainers to assess whether they should compete in one competition or another without having to repeat lifts or wear uncomfortable equipment. This will help prevent unnecessary stress to their bodies, and save time associated with repeating lifts. The tool will also provide useful insights for trainers to monitor the progress of their athletes and help them prepare for future competitions.

As future work, we could try to improve the models' performances in certain ways, maybe using a meta-optimisation process on the parameters of each model by means of evolutionary algorithms.

Author Contributions: Conceptualization: J.M.-R. and A.M.M.; methodology: A.M.M. and J.M.-R.; software development: J.M.-R. and P.Á.C.; validation: J.F.V.-V., P.Á.C. and A.M.M.; formal analysis: all the authors; data preprocessing: J.F.V.-V. and P.Á.C.; data visualization, J.M.-R., P.Á.C. and A.M.M.; writing: J.M.-R. and A.M.M.; supervision: A.M.M.; project administration: J.F.V.-V. and A.M.M.; funding acquisition, J.F.V.-V. All authors have read and agreed to the published version of the manuscript.

Funding: This work was partially funded by projects PID2020-113462RB-I00 (ANIMALICOS), PID2020-115570GB-C22, and PID2020-115570GB-C21, granted by the Ministerio Español de Economía y Competitividad; project TED2021-129938B-I00, granted by the Ministerio Español de Ciencia e Innovación; projects P18-RT-4830 and A-TIC-608-UGR20, granted by Junta de Andalucía; and project B-TIC-402-UGR18 (FEDER and Junta de Andalucía).

Institutional Review Board Statement: Not applicable.

Informed Consent Statement: Not applicable.

Data Availability Statement: Data used in this study can be found at https://openpowerlifting.gitlab.io/opl-csv/bulk-csv.html (acceded on 20 June 2023)

Conflicts of Interest: The authors declare no conflict of interest.

References

1. Ball, R.; Weidman, D. Analysis of USA powerlifting federation data from January 1, 2012–June 11, 2016. *J. Strength Cond. Res.* **2018**, *32*, 1843–1851. [CrossRef] [PubMed]
2. Godawa, T.M.; Credeur, D.P.; Welsch, M.A. Influence of compressive gear on powerlifting performance: Role of blood flow restriction training. *J. Strength Cond. Res.* **2012**, *26*, 1274–1280. [CrossRef] [PubMed]
3. Brechue, W.F.; Abe, T. The role of FFM accumulation and skeletal muscle architecture in powerlifting performance. *Eur. J. Appl. Physiol.* **2002**, *86*, 327–336. [CrossRef] [PubMed]
4. Vanderburgh, P.M.; Edmonds, T. The effect of experimental alterations in excess mass on pull-up performance in fit young men. *J. Strength Cond. Res.* **1997**, *11*, 230–233.
5. Cleather, D.J. Adjusting powerlifting performances for differences in body mass. *J. Strength Cond. Res.* **2006**, *20*, 412–421. [PubMed]
6. OpenPowerlifting. Available online: https://www.openpowerlifting.org (accessed on 25 May 2023).
7. Freedman, D.A. *Statistical Models: Theory and Practice*; Cambridge University Press: Cambridge, UK, 2009.
8. Kramer, O.; Kramer, O. K-nearest neighbors. In *Dimensionality Reduction with Unsupervised Nearest Neighbors*; Springer: Berlin/Heidelberg, Germany, 2013; pp. 13–23.
9. Ho, T.K. Random decision forests. In Proceedings of the IEEE 3rd International Conference on Document Analysis and Recognition, Montreal, QC, Canada, 14–16 August 1995; Volume 1, pp. 278–282.
10. Edwards, W.; Von Winterfeldt, D.; Moody, D.L. Simplicity in decision analysis: An example and a discussion. In *Decision Making: Descriptive, Normative, and Prescriptive Interactions*; Cambridge University Press: Cambridge, UK, 1988; pp. 443–464.
11. Friedman, J.H. Multivariate adaptive regression splines. *Ann. Stat.* **1991**, *19*, 1–67. [CrossRef]
12. Haykin, S. *Neural Networks and Learning Machines, 3/E*; Pearson Education India: Uttar Pradesh, India, 2009.
13. Bickel, P.J.; Doksum, K.A. *Mathematical Statistics: Basic Ideas and Selected Topics, Volumes I-II Package*; CRC Press: Boca Raton, FL, USA, 2015.

14. Hyndman, R.J.; Koehler, A.B. Another look at measures of forecast accuracy. *Int. J. Forecast.* **2006**, *22*, 679–688. [CrossRef]
15. Draper, N.R.; Smith, H. *Applied Regression Analysis*; John Wiley & Sons: Hoboken, NJ, USA, 1998; Volume 326.

engineering proceedings

MDPI

Proceeding Paper

Modelling High-Dimensional Time Series with Nonlinear and Nonstationary Phenomena for Landslide Early Warning and Forecasting †

Hangfei Zheng *, Guoqi Qian and Antoinette Tordesillas

School of Mathematics and Statistics, University of Melbourne, Melbourne, Victoria 3010, Australia; qguoqi@unimelb.edu.au (G.Q.); atordesi@unimelb.edu.au (A.T.)
* Correspondence: hangfei.zheng@unimelb.edu.au
† Presented at the 9th International Conference on Time Series and Forecasting, Gran Canaria, Spain, 12–14 July 2023.

Abstract: Landslides are nonstationary and nonlinear phenomena, which are often recorded as high-dimensional vector time series manifesting spatiotemporal dependence. Contemporary econometric methods use error-correction cointegration (ECC) and vector autoregression (VAR) to handle the nonstationarity but ignore the nonlinear trend. Here, we improve the ECC-VAR methodology by inserting a nonlinear trend $\boldsymbol{c}(t)$ into the model and nonparametrically estimating it by penalised maximum likelihood, and name this method ECC-VAR-$\boldsymbol{c}(t)$. Assisted by the empirical dynamic quantiles (EDQ) dimension reduction technique, it is sufficient to apply ECC-VAR-$\boldsymbol{c}(t)$ to just a small number of representative EDQ series to surmise the whole dataset. The application of this ECC-VAR-$\boldsymbol{c}(t)$ is well fitted to the real-world slope dataset ($R^2 = 0.99$) that consists of 1803 time series, each having 5090 time states. In addition to the forecast values, we also provide three risk assessments to predict locations, time and risk of a future failure with quantified uncertainty for building an early-warning system (e.g., predicted time of failure (ToF), where the minimum error is 2.7 h before the actual ToF).

Keywords: vector time series; nonlinear and nonstationary; error-correction cointegration; dimension reduction

Citation: Zheng, H.; Qian, G.; Tordesillas, A. Modelling High-Dimensional Time Series with Nonlinear and Nonstationary Phenomena for Landslide Early Warning and Forecasting. *Eng. Proc.* **2023**, *39*, 21. https://doi.org/10.3390/engproc2023039021

Academic Editors: Ignacio Rojas, Hector Pomares, Luis Javier Herrera, Fernando Rojas and Olga Valenzuela

Published: 29 June 2023

1. Introduction

Recent advancements in modern technologies such as radars, satellites, sensors make it computationally feasible and accurate in monitoring the real-world complex systems [1–3]. Based on these advanced detection technologies, a move from a conventional *detection–diagnosis–mitigation* to a more proactive *prediction–prognosis–prevention* paradigm is becoming increasingly evident. A typical example is the prediction of geological hazards such as landslides. It is crucial to make reliable and timely predictions of an impending hazard for risk mitigation to protect lives, livelihoods and the environment [4,5]. However, observations of landslides are often recorded as high-dimensional, spatial–temporal-dependent vector time series with nonlinear and nonstationary phenomena. Time series forecasting of such complex systems are considered one of the emerging challenges of modern science [6].

The current existing methods for modelling and forecasting time series, for example [7–11], have some limitations in dealing with the diverse combinations of the nonlinear and nonstationary dynamic behaviours among the system and the computational infeasibility caused by high dimensions in the real-world dataset. The objective of this paper is to develop a statistical model used for high-dimensional, spatial–temporal-dependent, nonlinear and nonstationary time series–here, we focus on landslides—and provide reliable and timely prediction for early warnings.

We develop a data-driven model by combining several advanced techniques. First, we apply a dimension reduction technique called empirical dynamic quantiles (EDQ), proposed by Peña et al. [12] to present the high-dimension vector time series by a small number of EDQ series; then, we use error-correction and cointegration (ECC) form of the vector autoregession (VAR) model to deal with the nonstationarity in the time series and combine this with an empirical function $\boldsymbol{c}(t)$ used to capture the nonlinearity. To assess the performance of the proposed ECC-VAR-$\boldsymbol{c}(t)$-EDQ model, we apply it to real-world ground motion data, which have 1803 time series, with each having 5090 time states in total. Once we obtain the optimal model, we can calculate the forecast values and use them for further analysis to predict future failure.

The performance of a forecasting framework is not just about how accurately the forecast values can figure out the failure (i.e., true positive), it is also about how well the forecasting framework can confidently forecast a stable region (i.e., true negative). The studied slope data in this paper have both failure and stable regions, which is an ideal case for assessing our model. In addition to the forecast values that can be obtained from our proposed model, we also provide three risk assessments to predict the locations of failure, time of failure and risk of failure with quantified uncertainty, based on certain what-if-scenarios at each future time and location. From the forecast values and these three assessments, our developed forecasting framework can successfully tackle the high-dimension, nonstationary and nonlinearity among the spatial–temporal-dependent dynamic system, forecast the failure and stable in the slope data domain and provide a reliable prediction for early warnings, as shown later in the paper.

This paper is organized as follows. The slope data analysed in this paper are introduced in Section 2. The details of the method are described in Section 3 before we apply this method to the slope data in Section 4. The forecast results and three risk assessment discussions used for building an early warning system are presented in Section 5, and the conclusion about this forecasting framework is presented in Section 6.

2. Data Description

The studied slope data focus on a rock slope of an open-pit mine dominated by intact igneous rock that is heavily structured or faulted by many naturally occurring discontinuities [13]. Since the mine operation, location and year of the rock slides are confidential, we call this dataset Slope X data. The monitoring domain stretched to around 200 m in length and 40 m in height. Movements of the rock face were monitored over a 3-week period: 10:07 May 31 to 23:55 June 21. Displacement at each of the 1803 monitoring locations or pixels on the surface of the rock slope was updated every six minutes, with $T = 5090$ time states in total. This led to a vector time series data with dimensions 1803 and length 5090. A landslide occurred on the western side of the slope on June 15, with an arcuate back scar and a strike length of around 120 m. The time of failure (ToF) occurred at around 13:10 June 15 ($t = 3568$), close to when the global peak velocity of 33.61 mm/h was reached [13,14]. The observed displacement and locations are shown in Figure 1.

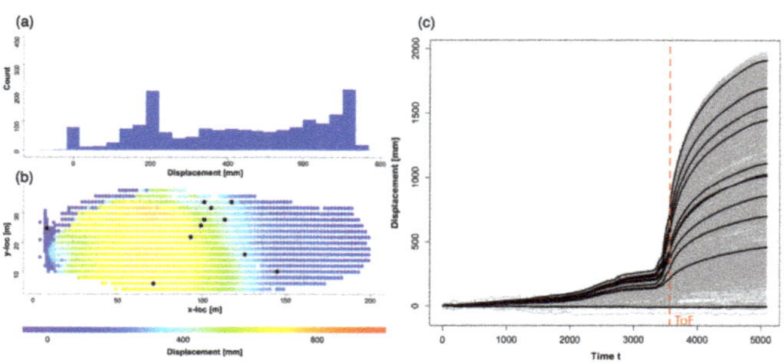

Figure 1. (**a**) Displacement histogram at time of failure (ToF), 13:10 June 15. (**b**) Displacement at all these 1803 monitoring locations. (**c**) The observed 1803 time series each has length 5090 in grey. The 11 EDQ series highlight in black in (**c**) and their corresponding locations showed as black dots in (**b**). The red dashed line is the ToF.

3. Method

Limited by computational infeasible for big spatial–temporal-dependent data, most of the existing models for geo-hazards [7,8,10] can only deal with univariate or low-dimension vector time series. To improve this limitation, our proposed data-driven model forecasts future behaviors for all time series in the study domain. This model accounts for the influence of past and present at location i and all nearby locations, as well as the changes in these interdependences across space and time. The forecasts from the existing geo-hazards forecasting model often lack quantification of the associated uncertainties in terms of probabilistic assessments. The method described in this paper provides new insights in the modeling of high-dimensional, spatial–temporal-dependent time series with nonlinear and nonstationary phenomena and provides a reliable prediction of where and when failure will occur, as well as the quantified uncertainty of a future failure. An overview of our forecasting framework is presented in Figure 2.

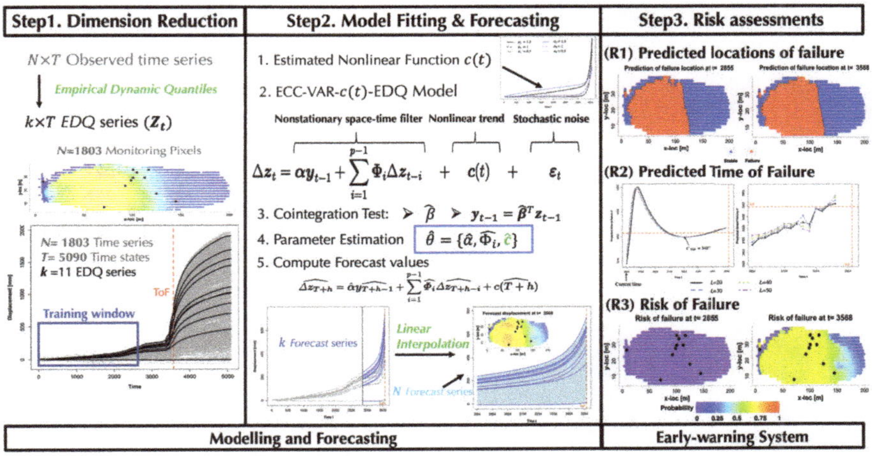

Figure 2. Overview of the forecasting framework. There are two parts of this forecasting framework. The first part is modelling and forecasting. The second part is building an early-warning system to predict future failure with three risk assessments (R1–R3).

In step 1, we first apply a dimension reduction technique to reduce the dimension of the large-scale dataset from $N = 1803$ to a small number k, which can make sure computation ia feasible for the further steps in our method, with a negligible loss of spatial and temporal dynamical information. We use the training sample to identify these k locations from the entire domain (i.e., $i = 1, \cdots, 1803; t = 1, \cdots, T$ with $T = 2854$ for Slope X data for purpose of illustration). To achieve this, we apply the idea of *empirical dynamic quantiles (EDQ)*, introduced by [12] and first used in a forecasting model for high-dimensional landslides data in [14]. The key idea behind this technique is that the dimension reduction to k EDQ series at k quantile levels $0 \leq q_1 < q_2 < \cdots < q_k \leq 1$ is achieved by selecting a small subset series from the original observed time series set, which is able to retain the dynamic dependence in the original dataset. This small set time series selected at k quantile levels optimally represents the whole time series set. Note that finding the k EDQ series from the original $N = 1803$ time series set is the same as finding the k locations from the 1803 monitoring locations. Statistical computations used to determine the k representative EDQ series essentially involve minimizing the sum of absolute differences between each observed time series and the prospective EDQ series at the given quantile level. This optimization procedure is computationally feasible and statistically consistent; more details ar eprovided in [12]. These k representative EDQ series enable us to perform various statistical inferences and forecasting in a manner that captures and retains the essential spatial–temporal dynamics that drive the slope's surface motion as damage spreads in the landslides. Similarly, the quantile level for each location and time series in the training data can also be determined by this technique.

In step 2, we develop a model using the training sample from the representative k EDQ series to capture the nonstationary, nonlinear trends and spatial–temporal dynamics in the ground motion system. It is important to have such a model to provide reliable and timely forecasts for this complex system. Unlike most existing time-series forecasting methods [7,8,11], our model considers both the nonstationary and nonlinear observed time series. To develop such a model, we use the concepts of vector autoregression (VAR) time series and error-correction cointegration (ECC) modelling [15,16] to deal with the nonstationarity and combine these with an empirical function $\boldsymbol{c}(t)$ in terms of time t to capture the nonlinear trends in the ground motion system. We call this method the ECC-VAR-$\boldsymbol{c}(t)$-EDQ model. Let $\{\boldsymbol{z}_t, t = 1, \cdots, T\}$ be the k dimension vector time series of displacement observed from k EDQ locations, and $\Delta \boldsymbol{z}_t = \boldsymbol{z}_t - \boldsymbol{z}_{t-1}$ be the velocity. The key equation for this ECC-VAR-$\boldsymbol{c}(t)$-EDQ model is of the following form

$$\Delta \boldsymbol{z}_t = \boldsymbol{\alpha} \boldsymbol{y}_{t-1} + \sum_{i=1}^{p-1} \boldsymbol{\Phi}_i \Delta \boldsymbol{z}_{t-i} + \boldsymbol{c}(t) + \boldsymbol{\epsilon}_t, \tag{1}$$

where $\boldsymbol{y}_{t-1} = \boldsymbol{\beta}^\top \boldsymbol{z}_{t-1}$, with $\boldsymbol{\beta}$ being the $k \times r$ co-integration matrix of rank r; $\boldsymbol{\alpha}$ is the $k \times r$ stationary adjustment matrix, which is also of rank r; $\boldsymbol{\Phi}_i$s are the $k \times k$ space–time association matrices; $\boldsymbol{c}(t)$ is the predefined function of time t used to capture the nonlinearity; $\{\boldsymbol{\epsilon}_t\}$ is a sequence of independent and identically distributed k random vectors with mean zero and covariance matrix Σ. Determining model (1) and estimating the parameters are the tasks undertaken in step 2 of Figure 2.

The first key property of Equation (1) contains a valid analysis of the nonstationary, spatial–temporal dynamics, where $\boldsymbol{z}_t$ is usually nonstationary in real-world dataset [8] but $\boldsymbol{y}_{t-1} = \boldsymbol{\beta}^\top \boldsymbol{z}_{t-1}$ must be stationary if $\Delta \boldsymbol{z}_t$ is stationary. Here, we apply the idea of cointegration [17–19]—a linear combination of these k unit-root nonstationary time series $\boldsymbol{z}_t$ that become r stationary series $\boldsymbol{y}_t$. The linear combination matrix $\boldsymbol{\beta}$ is called a cointegrating matrix with rank r. Cointegration implies a long-term stable relationship between variables in forecasting [15]. Another important feature of Equation (1) is its capacity for nonlinearity in the function $\boldsymbol{c}(t)$. A broad exploration and the literature show that empirical approaches are quite often used to overcome the difficulties encountered when using the nonlinear theoretical formulations [20,21]. Here, we use an empirical method to determine $\boldsymbol{c}(t)$-employ the flexible mathematical functions and represent the nonlinear trend by adapting parameters.

Specifically, to fit and estimate this model (1), we need to determine the cointegration rank r and the cointegration matrix $\boldsymbol{\beta}$. Consider the ECC-VAR-$\boldsymbol{c}(t)$-EDQ model (1) and replace $\boldsymbol{y}_{t-1} = \boldsymbol{\beta}^\top \boldsymbol{z}_{t-1}$; then, we have

$$\Delta \boldsymbol{z}_t = \boldsymbol{\Pi} \boldsymbol{z}_{t-1} + \sum_{i=1}^{p-1} \boldsymbol{\Phi}_i \Delta \boldsymbol{z}_{t-i} + \boldsymbol{c}(t) + \boldsymbol{\epsilon}_t, \tag{2}$$

where $\boldsymbol{\Pi} = \boldsymbol{\alpha}\boldsymbol{\beta}^\top$. The matrix $\boldsymbol{\Pi}$ plays an important role in the cointegration study: if $\text{rank}(\boldsymbol{\Pi}) = 0$, there is no cointegrating vector. In other words, the test for cointegration focuses on testing the rank of matrix $\boldsymbol{\Pi} = \hat{r}$. Johansen's method is the best-known cointegration test for VAR models; see more details in [22,23]. Impact matrix $\boldsymbol{\Pi}$ is the coefficient of the lagged levels in a nonlinear least squares regression of $\Delta \boldsymbol{z}_t$ on lagged differences and lagged levels $\sum_{i=1}^{p-1} \boldsymbol{\Phi}_i \Delta \boldsymbol{z}_{t-i}$ and function $\boldsymbol{c}(t)$ [15,17]. Three main steps are used to test the rank of the impact matrix $\boldsymbol{\Pi}$: since $\boldsymbol{\Pi}$ is related to the covariance matrix between $\Delta \boldsymbol{z}_t$ and $\boldsymbol{z}_{t-1}$, the mean will not influence the estimation of $\boldsymbol{\Pi}$. We can first simplify the Equation (2) by concentrating on the effect of the lagged differences $\sum_{i=1}^{p-1} \boldsymbol{\Phi}_i \Delta \boldsymbol{z}_{t-i}$ and function $\boldsymbol{c}(t)$. Let us achieve this by first regressing $\Delta \boldsymbol{z}_t$ on function $\boldsymbol{c}(t)$ and the lagged differences $\sum_{i=1}^{p-1} \boldsymbol{\Phi}_i \Delta \boldsymbol{z}_{t-i}$, providing the residuals $\hat{\boldsymbol{u}}_t$ and then regressing $\boldsymbol{z}_{t-1}$ on function

$\boldsymbol{c}(t)$ and the lagged differences $\sum_{i=1}^{p-1} \boldsymbol{\Phi}_i \Delta \boldsymbol{z}_{t-i}$, leading to the residuals $\hat{\boldsymbol{v}}_t$ as Equation (3). After performing these two regressions, we obtain the simplified model as Equation (4).

$$\Delta \boldsymbol{z}_t = \boldsymbol{c}(t) + \sum_{i=1}^{p-1} \boldsymbol{\Phi}_i \boldsymbol{\omega}_i \Delta \boldsymbol{z}_{t-1} + \boldsymbol{u}_t, \quad \boldsymbol{z}_{t-1} = \boldsymbol{c}(t) + \sum_{i=1}^{p-1} \boldsymbol{\Phi}_i \boldsymbol{\omega}_i{}^* \Delta \boldsymbol{z}_{t-1} + \boldsymbol{v}_t \tag{3}$$

where $\boldsymbol{c}(t)$ is our pre-defined function to capture the nonlinear trend. These two regressions in Equation (3) can be estimated by the least-squares (LS) method and $\hat{\boldsymbol{u}}_t$, $\hat{\boldsymbol{v}}_t$ are the residuals obtained from these two regressions. Then, we have the simplified model as follows,

$$\hat{\boldsymbol{u}}_t = \boldsymbol{\Pi} \hat{\boldsymbol{v}}_t + \boldsymbol{e}_t \tag{4}$$

The least-squares estimate of matrix $\boldsymbol{\Pi}$ is identical for Equation (2) and the simplified model (4). Hence, testing the rank of the covariance matrix between the $\boldsymbol{z}_t$ and $\Delta \boldsymbol{z}_{t-1}$ is equivalent to testing the rank of the covariance matrix between $\hat{\boldsymbol{u}}_t$ and $\hat{\boldsymbol{v}}_t$. We can focus on testing the rank of the matrix $\boldsymbol{\Pi}$ in Equation (4). We use the likelihood ratio test (LRT) to test the rank of $\boldsymbol{\Pi}$ and find the estimated cointegrating vector $\boldsymbol{\beta}$. Let

$$H(0) \subset H(1) \subset \cdots \subset H(r) \subset H(k)$$

be the nested models such that, under $H(r)$, there are r cointegrating vectors in z_t. In particular, under $H(0)$, we have $\boldsymbol{\Pi} = \boldsymbol{0}$. Under $H(r)$: $\text{rank}(\boldsymbol{\Pi}) = r$; that is, there are r linearly independent vectors among these k vectors and the maximum likelihood estimate (MLE) of Σ_r is

$$\hat{\Sigma}_r = \begin{bmatrix} \hat{\Sigma}_{r\times r} & \boldsymbol{0}_{r\times(k-r)} \\ \boldsymbol{0}_{(k-r)\times r} & \hat{\Sigma}_{(k-r)\times(k-r)} \end{bmatrix} \tag{5}$$

The maximized likelihood function is, therefore, approximate to

$$l_r \propto |\hat{\Sigma}_r|^{-\frac{T}{2}} = \left(|\hat{\Sigma}_{r\times r}||\hat{\Sigma}_{(k-r)\times(k-r)}|\right)^{-\frac{T}{2}} \tag{6}$$

Under $H(k)$: $\text{rank}(\boldsymbol{\Pi}) = k$, that is, the matrix $\boldsymbol{\Pi}$ is full rank, there is no constraint on the covariance matrix. Similarly, we can obtain the maximized likelihood function under $H(k)$ and the likelihood ratio is, therefore,

$$L = \left(\frac{|\hat{\Sigma}_k|}{|\hat{\Sigma}_r|}\right)^{\frac{T}{2}} = \left(|\boldsymbol{I}_{(k-r)\times(k-r)} - \hat{\Sigma}^{-1}_{(k-r)\times(k-r)}\hat{\Sigma}_{(k-r)\times r}\hat{\Sigma}^{-1}_{r\times r}\hat{\Sigma}_{r\times(k-r)}|\right)^{\frac{T}{2}} \tag{7}$$

We define

$$\hat{\boldsymbol{\Sigma}}_{uu} = \frac{1}{T}\sum_{t=1}^{T} \hat{\boldsymbol{u}}_t \hat{\boldsymbol{u}}_t^\top, \quad \hat{\boldsymbol{\Sigma}}_{vv} = \frac{1}{T}\sum_{t=1}^{T} \hat{\boldsymbol{v}}_t \hat{\boldsymbol{v}}_t^\top, \quad \hat{\boldsymbol{\Sigma}}_{uv} = \frac{1}{T}\sum_{t=1}^{T} \hat{\boldsymbol{u}}_t \hat{\boldsymbol{v}}_t^\top, \quad \hat{\boldsymbol{\Sigma}}_{vu} = \frac{1}{T}\sum_{t=1}^{T} \hat{\boldsymbol{v}}_t \hat{\boldsymbol{u}}_t^\top$$

Then, the sample matrix becomes $\boldsymbol{S} = \hat{\boldsymbol{\Sigma}}_{vv}^{-1}\hat{\boldsymbol{\Sigma}}_{vu}\hat{\boldsymbol{\Sigma}}_{uu}^{-1}\hat{\boldsymbol{\Sigma}}_{uv}$, and $\lambda_1 \geq \lambda_2 \geq \cdots \lambda_k \geq 0$ be the ordered eigenvalues of the sample matrix $\boldsymbol{S}$ and $\boldsymbol{g}_i$ be the eigenvector associated with eigenvalue λ_i. Here, $\lambda_{r+1} \geq \lambda_{r+2} \geq \cdots \lambda_k$ are the eigenvalues of $\hat{\Sigma}^{-1}_{(k-r)\times(k-r)}\hat{\Sigma}_{(k-r)\times r}\hat{\Sigma}^{-1}_{r\times r}\hat{\Sigma}_{r\times(k-r)}$. We can obtain the likelihood ratio test statistic as follows

$$LR = -2\log L = -T\sum_{i=r+1}^{k} \ln(1-\lambda_i). \tag{8}$$

Reject the null hypothesis if LR is larger than the critical value and the estimated cointegrating vector can be obtained from the r corresponding eigenvectors $\hat{\boldsymbol{\beta}} = [\boldsymbol{g}_1, \ldots, \boldsymbol{g}_r]$ [17,18].

Another aspect of fitting and estimating the ECC-VAR-$\boldsymbol{c}(t)$-EDQ model (1) is the determination of $\boldsymbol{c}(t)$. Johansen also pointed out that the function $\boldsymbol{c}(t)$ in Equation (2) has important implications in a cointegrated system [24]. Tavenas and Leroueil pointed out that the rationale for most time-of-failure predictions is that the slope displacement can be represented by a creep curve before rupture [25], which can be divided into three stages. According to the classic interpretation, the first stage is primary creep, with the strain rate decreasing logarithmically, followed by secondary (or steady-state) creep with a constant strain rate, and tertiary creep with an increasing creep rate, which leads to rupturing. Our aim was to capture the nonlinear trend by using a function $\boldsymbol{c}(t)$ that is equivalent to finding a method to represent the creep curves. Empirical methods have been used extensively to represent creep curves, which requires a user to define the functions to represent the curve models from the data trend [26,27]. Research and experiments have found that it is sufficient to obtain an accurate representation of the creep curve by adding the primary part and tertiary part together [21,28]. At present, these empirical approaches use the simple power function combined with exponential function to represent the creep curves [21,28,29]. Here, we apply the empirical method to determine the form of function $\boldsymbol{c}(t)$. Let $\boldsymbol{c}(t)$ have the form of $\boldsymbol{c}(t) = f_1(t) + f_2(t)$, where $f_1(t)$ is used to capture the trend in precursory failure regime and $f_2(t)$ is used to capture the nonlinear trend in tertiary creep. Then, the deterministic function $\boldsymbol{c}(t)$ has the following form:

$$\boldsymbol{c}(t) = \left(t^{a_1} + \exp\left[a_2(t - t_p)\right]\right)^a \tag{9}$$

where a_1, a_2, a are the pre-estimated parameters for the form of $\boldsymbol{c}(t)$ obtained by using penalized maximum likelihood; t_p is the time state, which can be determined by prior work. When $t < t_p, f_2(t) = \exp\left[a_2(t - t_p)\right] < 1$, then $\boldsymbol{c}(t)$ will mainly be influenced by $f_1(t)$; on the other hand, when $t > t_p$, $f_2(t)$ will have an exponential growth, and $\boldsymbol{c}(t)$ will mainly be influenced by $f_2(t)$, which can be used to represent the creep curve. Figure 3 provides several examples of $\boldsymbol{c}(t)$ with different values of a, a_1, a_2.

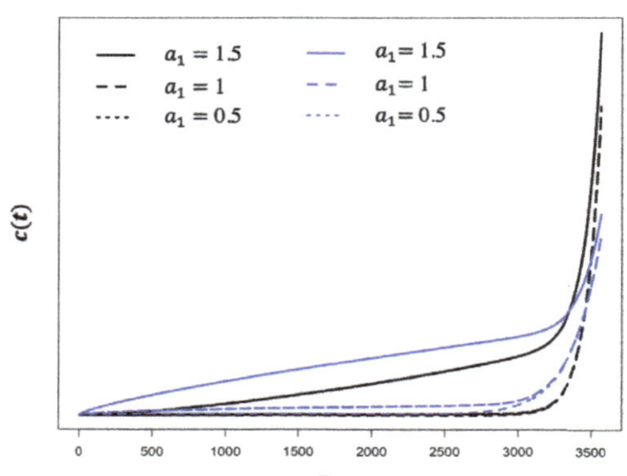

Figure 3. Curve of function $\boldsymbol{c}(t)$ with different values of a_1, a_2, a. The black lines are $a = 1, a_2 = 1$ with three different values of a_1. The blue lines are $a = 0.5, a_2 = 1$ with three different values of a_1.

We use the cointegration test with our pre-estimated function $\boldsymbol{c}(t)$ finding the estimated cointegrating matrix $\hat{\boldsymbol{\beta}}$. After that we can fit and estimate the ECC($\hat{r}$)-VAR(p)-$\boldsymbol{c}(t)$-EDQ model (1) with $\boldsymbol{y}_{t-1} = \hat{\boldsymbol{\beta}}^\top \boldsymbol{z}_{t-1}$. We use least square (LS) method estimate the unknown parameters $\{\boldsymbol{\alpha}, \boldsymbol{\Phi}_i, \boldsymbol{c}\}$,$\boldsymbol{c}$ is the parameters from $\boldsymbol{c}(t)$ function. Once we have the fitted model, we can calculate the forecast values of the k EDQ displacement series at times $t = T + 1$ and beyond. The forecast values $\Delta \boldsymbol{z}_t$ and $\boldsymbol{z}_t$ can be denoted as $\widehat{\Delta \boldsymbol{z}_t}$ and $\widehat{\boldsymbol{z}_t}$; at future time steps, $t = T + h, h = 1, 2, \cdots$, are

$$\widehat{\Delta \boldsymbol{z}_{T+h}} = \hat{\boldsymbol{\alpha}} \widehat{\boldsymbol{y}_{T+h-1}} + \sum_{i=1}^{p-1} \widehat{\boldsymbol{\Phi}}_i \widehat{\Delta \boldsymbol{z}_{T+h-i}} + \widehat{\boldsymbol{c}(T+h)}, \quad \widehat{\boldsymbol{z}_{T+h}} = \widehat{\Delta \boldsymbol{z}_{T+h}} + \widehat{\boldsymbol{z}_{T+h-1}} \tag{10}$$

where $\widehat{\Delta z_t} = \Delta z_t$ and $\widehat{z_t} = z_t$ if $t < T$ (i.e., if t is inside the training data), and $\{\hat{\boldsymbol{\alpha}}, \widehat{\boldsymbol{\Phi}}_i, \boldsymbol{c}(\widehat{\boldsymbol{T}+\boldsymbol{h}})\}$ are the estimated parameters in model (1). We then employ these forecast results to interpolate and approximate the displacement in the other $N - k$ locations at future time. For example, to forecast the displacement $z_{i,T+h}$ at location i with EDQ quantile level $q_i \in [0,1]$ at time $T + h$, we first figure out the two adjacent quantile levels q_j and q_{j+1} from the k EDQ series we found in part 1 with their forecast displacement $\widehat{z_{j,T+h}}$ and $\widehat{z_{j+1,T+h}}$, such that $q_j < q_i \leq q_{j+1}$. Then, $z_{i,T+h}$ can be calculated by

$$\widehat{z_{i,T+h}} = \frac{q_{j+1} - q_i}{q_{j+1} - q_j}\widehat{z_{j,T+h}} + \frac{q_i - q_j}{q_{j+1} - q_j}\widehat{z_{j+1,T+h}} \tag{11}$$

In step 3, once we obtain the forecast displacement across the entire domain, we can provide predictions for where and when a future failure will occur by using three risk assessments in parallel. The first uses the clustering method for the forecast values at all locations at a specific future time states to determine the likely locations of failure. The second one uses the adaptive Fukuzono method on the EDQ time series to estimate the earliest time of failure (ToF), which is more objective than traditional Fukuzono regression. The last one delivers a spatial map of the probability of risk of failure at each future time steps, which is calculated by the prescribed *what-if-scenario* (e.g., a scenario when the forecast velocity at a location exceeds a predefined threshold value). Details will be provided in Section 5 by analyzing the Slope X data described in Section 2.

4. Apply to Slope X Data

To assess the performance of our ECC(r)-VAR(p)-$\boldsymbol{c}(t)$-EDQ model, we applied it to real-world slope data. It is common to split the dataset into two parts when developing statistical and machine learning models [30,31]. Many researchers proposed a ratio of 80/20 for producing a training/testing dataset for landslide susceptibility problems [32,33]. The other principle for determining the training data is that they must not contain the actual failure but should contain some information on precursory failure. Therefore, we used the training set with length 80% of Slope X data, starting from $t = 1$ to $t = 2854$, and used the data starting from $t = 2855$ to $t = 3568$(the actual time of failure) as our test sample (20% of the observed data). There was no issue regarding which part was used as the training sample; to prove this, we also provide the results obtained when using moving training samples in Section 5. As Section 3 describes, first, we selected the $k = 11$ EDQ series to represent the entire $N = 1803$ time series based on the training sample. These 11 EDQ series at quantile levels- $\{0.0, 0.1, 0.2, \cdots, 0.8, 0.9, 1.0\}$ with the corresponding monitoring pixel ID are $\{33, 1451, 1130, 863, 1257, 930, 1059, 860, 825, 738, 534\}$. The displacement and locations for these 11 EDQ series are highlighted in black in Figure 1. We used the training sample to fit and estimate the ECC(r)-VAR(p)-$\boldsymbol{c}(t)$-EDQ model.

In Figure 3, the black solid is shown capture the shape of creep curve well; therefore, we chose the empirical function $\boldsymbol{c}(t) = t^{3/2} + exp(t - t_p)$ to represent the nonlinear trend. Based on the current knowledge from the data, when $t = 2200$, there is some acceleration trend; therefore, here, we took the current estimated time of failure as prior $t_p = 2200$. Next, we applied the cointegration test described in Section 3 for the 11 EDQ series $\{z_t\}$. The test results conclude that $r = 8$ for the Slope X data. A goodness of fit test using coefficient of determination shows that an auto-regression order of $p = 2$ is sufficient for modelling the Slope X data. Eventually, we determined the optimal values for all unknown parameters in ECC(8)-VAR(2)-$\boldsymbol{c}(t)$-EDQ model (1). The forecast displacements for the 11 selected EDQ locations at the test sample times $2855 \leq t \leq 3568$ are displayed in Figure 4.

These forecasts overall conform to the increasing trends of the observed displacement data (the grey curves in the plot). Such an observation is expected and was invariant to the selection of training data according to the established properties underlying the current vector time series forecasting methods. Clearly, the forecasts for each EDQ location are mostly greater than the actual observations, except when t is very close to the actual failure time ($t = 3568$). Apart from the 11 selected EDQ series, forecast displacement for the remaining 1792 time series over time $2855 \leq t \leq 3568$ was computed using Equation (11). The results for the displacement forecasts are displayed in the right plot of Figure 4. These displacement forecasts can also be depicted as a heat map for each future time t. Such a heat map, at $t = 2855$ and $t = 3568$, is displayed in Figure 5, along with its histogram plot.

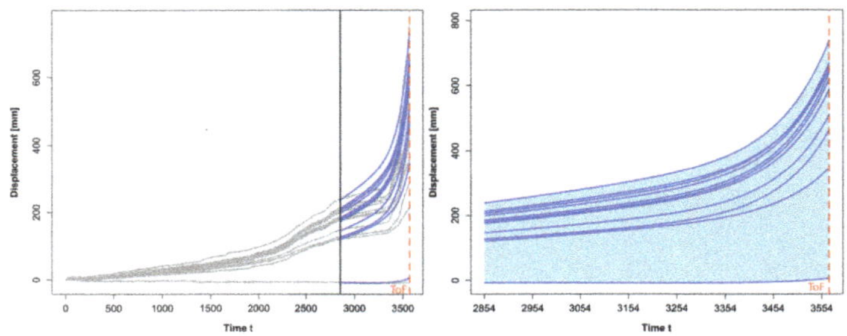

Figure 4. (**Left**) For the 11 selected EDQ series; (**Right**) for all 1803 time series. Forecast displacement at time $t = 2855$ to $t = 3568$ based on the training sample starting from $t = 1$ to $t = 2854$. The red dashed line is the actual time of failure at $t = 3568$.

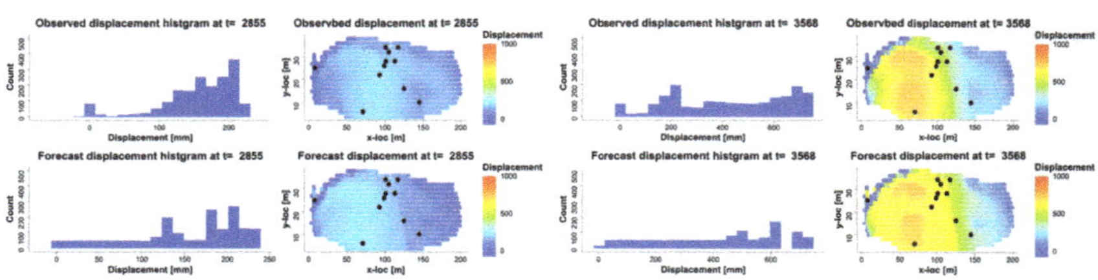

Figure 5. Forecast displacement at all these 1803 monitoring pixels. (**Left**) forecast displacement at $t = 2855$; (**Right**) forecast displacement at $t = 3568$, ToF.

5. Results and Discussion

Here, we built an early-warning system to predict the locations, time and risk of a future failure using three risk assessments. These risk assessments were based on the forecast values and provide more objective assessments. The results of these risk assessments apply to the forecast values for slope X data, and are shown in Figures 6–8.

The first risk assessment (R1) focuses on identifying the likely locations of the slope failure. This can be achieved by using the displacement forecasts to cluster all monitoring locations into *stable* and *failure* regions for each future time t. We used the simple K-means clustering to perform such a clustering for Slope X data [34,35] (e.g., K = 2 clusters, for stable and failure). From Figure 6, we can clearly identify the predicted failure pattern from $t = 2855$, the first time state at our forecast horizon (i.e., 2 days, 23:18 to ToF), and this failure geometry remains unchanged as the forecast time state moves forward until $t = 3568$ (ToF).

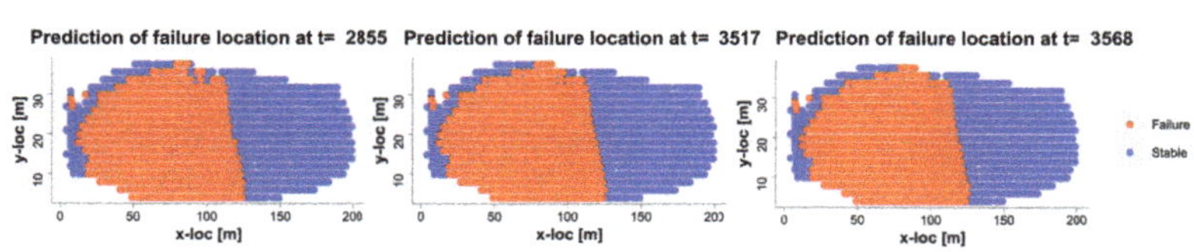

Figure 6. (R1) Predicted locations of failure using clustering at time $t = 2855, 3517, 3568$.

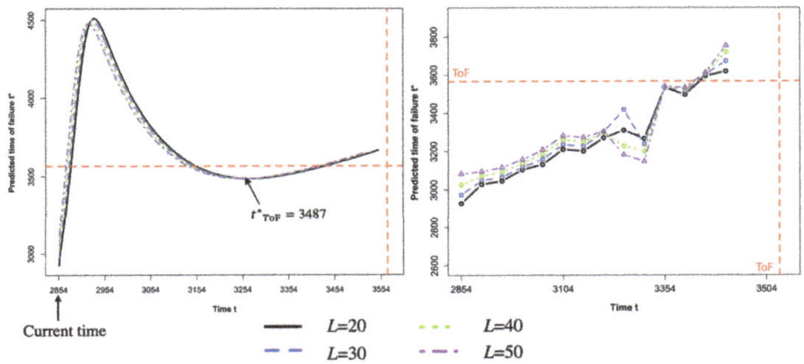

Figure 7. (R2) Predicted time of failure using adaptive Fukuzono analysis. The left plot is for fixed training data with rolling Fukuzono regression window. The right plot is for the rolling training sample with fixed Fukuzono regression window.

The second risk assessment (R2) focuses on predicting the time of failure (ToF) by using an adaptive Fukuzono method for the forecasts of each selected EDQ series. The classical Fukuzono method draws a regression line based on selected observed inverse velocities and extends the regression line forward until it intersects with the time axis, and the intersection is the Fukuzono estimated time of failure [36]. This method depends on (1) the start location and (2) the number of inverse velocities used for regression. Instead of determining these two things by personal choice or some prior work, here, we improved the Fukuzono method by using a more objective method. As the left panel in Figure 7 shows, we applied a moving Fukuzono regression window to the forecast timeline with different sizes: $L = 20, 30, 40, 50$. Each curve represents the predicted time of failure t^* using one Fukuzono regression window with size L at different time states. For example, the black lines is the t^* obtained from the moving Fukuzono regression window with length L towards the forecast horizon for the Max quantile EDQ series. The intersection of these four curves will be treated as the convergence point of the predicted time of failure. Figure 7 shows that our predicted time of failure is $t = 3487$, which is 8 h before the actual time of failure. We also assessed the performance of our model (1) using a moving training window. As the right plot in Figure 7 shows, there are 14 different training samples in total; we used the training sample starting at time interval $[1, 2854]$ and then moved forward by 50 with a fixed length 2854. Based on each training window, we used the first L forecasts to find the Fukuzono regression and obtain the predicted time of failure. Figure 7 shows that when the training window is closer to the actual ToF, our predicted ToF will be closer to the actual time of failure, with the minimum error of 2.7 h before the actual ToF.

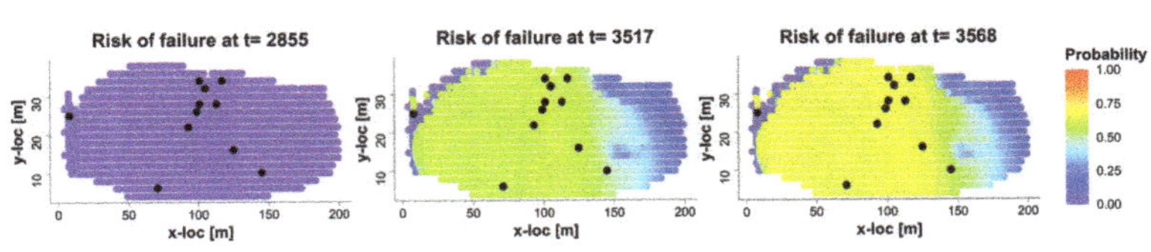

Figure 8. (R3) predicted risk of failure for what-if-scenario at time $t = 2855, 3517, 3568$.

The third risk assessment (R3) focuses on the risk of failure in terms of quantified uncertainty under a given *what-if-scenario*. In general, the risk of failure is a probability function of certain measured slope feature, aiming to reach a threshold hazard level at a future time state [37]. Our task is to generate a spatial map of the risk of failure for every monitoring location and every future time state until the actual failure occurs. For example, slope motion velocity data exceeding 10 mm/h = 1 mm/6 min are understood as dangerous, prompting an immediate red-alert warning [38]. Then, we can calculate the risk of failure for Slope X data $\Pr(\widehat{\Delta z_{i,t}} \geq 1)$ for location i from time $t = 2855$ to 3568. As

Figure 8 shows, we can identify the time for each location when the forecast risk of failure exceeds 60% probability. For example, $\Pr(\Delta z_{11,t} \geq 1) > 0.6$ for the level-1.0 EDQ location when $t \geq 3517$, 5.1 h before the actual time of failure $t = 3568$. It should be noted that these risk-of-failure results are obtained using the observations from $t = 1$ to $t = 2854$, about 71.4 h before the actual landslide. Therefore, we can conclude that our ECC-VAR-$\boldsymbol{c}(t)$-EDQ method is able to accurately predict an impending landslide, by providing a 60% or more risk of rock failure for each location, more than 71 h in advance. The spatial map of risks of failure at time $t = 2855, 3517, 3568$ and all monitoring locations are displayed in Figure 8. All the findings from Figures 6–8 show that this assessment is capable of providing timely and accurate ToF estimates.

6. Conclusions

We developed a statistical model and an early-warning system to forecast landslides. This ECC-VAR-$\boldsymbol{c}(t)$-EDQ model is computationally feasible and comprehensive to deal with high-dimensional, spatial–temporal-dependent time series with nonlinear and nonstationary phenomena. The results of the application of this ECC-VAR-$\boldsymbol{c}(t)$-EDQ model fitted the real-world Slope X data ($R^2 = 0.99$) and provided reliable early-warning predictions for decision-making and risk mitigation. Finally, there is still some further work that could improve our proposed ECC-VAR-$\boldsymbol{c}(t)$-EDQ model. For example, we used empirical functions to represent the nonlinear trend $\boldsymbol{c}(t)$, they are generally faster and straightforward to use, but make stronger assumptions about the parameters of the data distribution of time series; hence, non-parametric methods can be considered for further work.

Author Contributions: Conceptualization, G.Q. and A.T.; methodology, G.Q. and H.Z.; software, H.Z.; validation, H.Z., G.Q. and A.T.; formal analysis, H.Z. and G.Q.; investigation, H.Z.; data curation, A.T.; writing—original draft preparation, H.Z.; writing—review and editing, H.Z. and G.Q.; visualization, H.Z.; supervision, G.Q. and A.T.; funding acquisition, A.T. All authors have read and agreed to the published version of the manuscript.

Funding: This research was partly funded by the U.S. DoD High Performance Computing Modernization Program (HPCMP) and RDECOM International Technology Center-Pacific (ITC-PAC) contract number: FA5209-18-C-0002.

Institutional Review Board Statement: Not applicable.

Informed Consent Statement: Not applicable.

Data Availability Statement: The landslide data analyzed in this study is confidential and is subject to a non-disclosure agreement between the University of Melbourne and an undisclosed industry partner. Simulated data are available on request from the corresponding author.

Acknowledgments: The authors thank the ITISE 2023 Organization and the two anonymous reviewers for providing valuable comments and suggestions, leading to an improvement of the presentation of the paper.

Conflicts of Interest: The authors declare no conflict of interest.

References

1. Uhlemann, S.; Smith, A.; Chambers, J.; Dixon, N.; Dijkstra, T.; Haslam, E.; Meldrum, P.; Merritt, A.; Gunn, D.; Mackay, J. Assessment of Ground-based Monitoring Techniques Applied to Landslide Investigations. *Geomorphology* **2016**, *253*, 438–451. [CrossRef]
2. Zhao, C.; Lu, Z. Remote Sensing of Landslides—A Review. *Remote Sens.* **2018**, *10*, 279. [CrossRef]
3. Chuvieco, E. *Fundamentals of Satellite Remote Sensing: An Environmental Approach*; CRC Press: Boca Raton, FL, USA, 2016.
4. Riley, K.; Webley, P.; Thompson, M. *Natural Hazard Uncertainty Assessment: Modeling and Decision Support*; John Wiley & Sons: Hoboken, NJ, USA, 2016.
5. Dick, G.J.; Eberhardt, E.; Cabrejo-Liévano, A.G.; Stead, D.; Rose, N.D. Development of an Early-warning Time-of-failure Analysis Methodology for Open-pit Mine Slopes Utilizing Ground-based Slope Stability Radar Monitoring Data. *Can. Geotech. J.* **2015**, *52*, 515–529. [CrossRef]

6. Kumar, C.; Patel, N.; Jaudi, I. 21st Century Physics: Grand Challenges. *J. Fed. Am. Sci.* **2003**, *56*, 9–11.
7. Aggarwal, A.; Alshehri, M.; Kumar, M.; Alfarraj, O.; Sharma, P.; Pardasani, K.R. Landslide Data Analysis Using Various Time Series Forecasting Models. *Comput. Electr. Eng.* **2020**, *88*, 106858. [CrossRef]
8. Cheng, C.; Sa-Ngasoongsong, A.; Beyca, O.; Le, T.; Yang, H.; Kong, Z.; Bukkapatnam, S. Time Series Forecasting for Nonlinear and Non-stationary Processes: A Review and Comparative Study. *Iie Trans.* **2015**, *47*, 1053–1071. [CrossRef]
9. Tien Bui, D.; Tuan, T.A.; Klempe, H.; Pradhan, B.; Revhaug, I. Spatial Prediction Models for Shallow Landslide hazards: A comparative assessment of the efficacy of support vector machines, artificial neural networks, kernel logistic regression, and logistic model tree. *Landslides* **2016**, *13*, 361–378. [CrossRef]
10. Salles, R.; Belloze, K.; Porto, F.; Gonzalez, P.H.; Ogasawara, E. Nonstationary Time Series Transformation Methods: An Experimental Review. *Knowl.-Based Syst.* **2019**, *164*, 274–291. [CrossRef]
11. De, G.; Jan, G.; Hyndman, R.J. 25 Years of Time Series Forecasting. *Int. J. Forecast.* **2006**, *22*, 443–473.
12. Peña, D.; Tsay, R.S.; Zamar, R. Empirical Dynamic Quantiles for Visualization of High-dimensional Time Series. *Technometrics* **2019**, *61*, 429–444. [CrossRef]
13. Tordesillas, A.; Kahagalage, S.; Campbell, L.; Bellett, P.; Intrieri, E.; Batterham, R. Spatiotemporal Slope Stability Analytics for Failure Estimation (SSSAFE): Linking Radar Data to the Fundamental Dynamics of Granular Failure. *Sci. Rep.* **2021**, *11*, 1–18. [CrossRef]
14. Wang, H.; Qian, G.; Tordesillas, A. Modeling Big Spatio-temporal Geo-hazards Data for Forecasting by Error-correction Cointegration and Dimension-Reduction. *Spat. Stat.* **2020**, *36*, 100432. [CrossRef]
15. Tsay, R.S. *Multivariate Time Series Analysis: With R and Financial Applications*; John Wiley & Sons: Hoboken, NJ, USA, 2013.
16. Sa-Ngasoongsong, A.; Bukkapatnam, S.T.S.; Kim, J.; Iyer, P.S.; Suresh, R.P. Multi-step Sales Forecasting in Automotive Industry Based on Structural Relationship Identification. *Int. J. Prod. Econ.* **2012**, *140*, 875–887. [CrossRef]
17. Johansen, S. Statistical Analysis of Cointegration Vectors. *J. Econ. Dyn. Control* **1988**, *12*, 231–254. [CrossRef]
18. Reinsel, G.C.; Ahn, S.K. Vector Autoregressive Models with Unit-roots and Reduced Rank Structure: Estimation. Likelihood Ratio Test, and Forecasting. *J. Time Ser. Anal.* **1992**, *13*, 353–375. [CrossRef]
19. Pfaff, B. *Analysis of Integrated and Cointegrated Time Series with R*; Springer Science & Business Media: Berlin/Heidelberg, Germany, 2008.
20. Hadid, M.; Rechak, S.; Tati, A. Long-term Bending Creep Behavior Prediction of Injection Molded Composite Using Stress–Time Correspondence Principle. *Mater. Sci. Eng. A* **2004**, *385*, 54–58. [CrossRef]
21. Sandström, R. Basic Model for Primary and Secondary Creep in Copper. *Acta Mater.* **2012**, *60*, 314–322. [CrossRef]
22. Johansen, S.; Juselius, K. Maximum Likelihood Estimation and Inference on Cointegration—With Appucations to the Demand for Money. *Oxf. B Econ. Stat.* **1990**, *52*, 169–210. [CrossRef]
23. Johansen, S. Estimation and Hypothesis Testing of Cointegration Vectors in Gaussian Vector Autoregressive models. *Econ. J. Econ. Soc.* **1991**, *59*, 1551–1580. [CrossRef]
24. Johansen, S. *Likelihood-Based Inference in Cointegrated Vector Autoregressive Models*; Oxford University Press: Oxford, UK, 1995.
25. Tavenas, F.; Leroueil, S. Creep and Failure of Slopes in Clays. *Can. Geotech. J.* **1981**, *18*, 106–120. [CrossRef]
26. Abdallah, Z.; Gray, V.; Whittaker, M.; Perkins, K. A Critical Analysis of the Conventionally Employed Creep Lifing Methods. *Materials* **2014**, *7*, 3371–3398. [CrossRef] [PubMed]
27. Gray, V.; Whittaker, M. Development and Assessment of a New Empirical Model for Predicting Full Creep Curves. *Materials* **2015**, *8*, 4582–4592. [CrossRef] [PubMed]
28. Yang, L.; Li, Z. Nonlinear Variation Parameters Creep Model of Rock and Parametric Inversion. *Geotech. Geol. Eng.* **2018**, *36*, 2985–2993. [CrossRef]
29. Sattar, M.; Othman, A.R.; Kamaruddin, S.; Akhtar, M.; Khan, R. Limitations on the Computational Analysis of Creep Failure Models: A Review. *Eng. Fail. Anal.* **2022**, *134*, 105968. [CrossRef]
30. Nguyen, Q.H.; Ly, H.B.; Ho, L.S.; Al-Ansari, N.; Le, H.V.; Tran, V.Q.; Prakash, I.; Pham, B.T. Influence of Data Splitting on Performance of Machine Learning Models in Prediction of Shear Strength of Soil. *Math. Probl. Eng.* **2021**, *2021*, 1–15. [CrossRef]
31. Joseph, V.R.; Vakayil, A. Split: An Optimal Method for Data Splitting. *Technometrics* **2022**, *64*, 166–176. [CrossRef]
32. Chen, W.; Peng, J.; Hong, H.; Shahabi, H.; Pradhan, B.; Liu, J.; Zhu, A.; Pei, X.; Duan, Z. Landslide Susceptibility Modelling Using GIS-based Machine Learning Techniques for Chongren County, Jiangxi Province, China. *Sci. Total Environ.* **2018**, *626*, 1121–1135. [CrossRef]
33. Pham, B.T.; Tien Bui, D.; Pourghasemi, H.R.; Indra, P.; Dholakia, M.B. Landslide Susceptibility Assesssment in the Uttarakhand area (India) using GIS: A Comparison Study of Prediction Capability of Naïve Bayes, Multilayer Perceptron Neural Networks, and Functional Trees Methods. *Theor. Appl. Climatol.* **2017**, *128*, 255–273. [CrossRef]
34. Kodinariya, T.M.; Makwana, P.R. Review on Determining Number of Cluster in K-Means Clustering. *Int. J.* **2013**, *1*, 90–95.
35. Likas, A.; Vlassis, N.; Verbeek, J.J. The Global K-means Clustering Algorithm. *Pattern Recognit.* **2003**, *36*, 451–461. [CrossRef]
36. Fukuzono, T. A Method for Predicting the Failure Time of a Sandy Soil Slope using the Inverse Number of Velocity. In Proceedings of the 23rd Meeting of Japan Landslide Society, Tokyo, Japan, 2 July 1984; pp. 80–81.

37. Wang, R.; Li, L.; Simon, R. A Model for Describing and Predicting the Creep Strain of Rocks from the Primary to the Tertiary Stage. *Int. J. Rock Mech. Min. Sci.* **2019**, *123*, 104087. [CrossRef]
38. Mufundirwa, A.; Fujii, Y.; Kodama, J. A New Practical Method for Prediction of Geomechanical Failure-time. *Int. J. Rock Mech. Min. Sci.* **2010**, *47*, 1079–1090. [CrossRef]

engineering proceedings

MDPI

Proceeding Paper

Modeling Road Accessibility in a Flood-Prone Area in mboxRomania †

Cristian Popescu [1], Alina Bărbulescu [2,*] and Cristian Ștefan Dumitriu [3,*]

1 Doctoral School, Technical University of Civil Engineering of Bucharest, 020396 Bucharest, Romania; nicolae-cristian.popescu@phd.utcb.ro
2 Department of Civil Engineering, Transilvania University of Brasov, 900152 Brasov, Romania
3 Department of Machines and Advanced Technologies in Constructions, Faculty of Mechanical Engineering and Robotics for Constructions, Technical University of Civil Engineering of Bucharest, 020396 Bucharest, Romania
* Correspondence: alina.barbulescu@unitbv.ro (A.B.); cristian.dumitriu@utcb.ro (C.Ș.D.)
† Presented at the 9th International Conference on Time Series and Forecasting, Gran Canaria, Spain, 12–14 July 2023.

Abstract: Floods are repetitive and unpredictable phenomena. Interventions in these cases are carried out during or after a hazard's occurrence. The quality and density of the roads, the equipment of emergency intervention centers, and the proximity of these centers to risk areas influence accessibility and intervention capacity. An analysis of this issue is carried out in the present article for a small catchment in Romania in a predominantly hilly area. Areas with very high and very low road accessibility are identified. Proximity to emergency centers can reduce travel time, thus increasing the surfaces of well-served areas.

Keywords: road network; accessibility; emergency; catchment; intervention; flood

Citation: Popescu, C.; Bărbulescu, A.; Dumitriu, C.Ș. Modeling Road Accessibility in a Flood-Prone Area in Romania. *Eng. Proc.* **2023**, *39*, 22. https://doi.org/10.3390/engproc2023039022

Academic Editors: Ignacio Rojas, Hector Pomares, Luis Javier Herrera, Fernando Rojas and Olga Valenzuela

Published: 29 June 2023

1. Introduction

Areas subject to high hydrological pressure present a high risk of significant material and human damage [1]. At the international level, hydrological hazards occur in any season, whether during heavy snow, the rapid melting of snow in the spring that generates rapid floods, or heavy precipitation in May–July in areas of central and eastern Europe [2]. Romania is not an exception. In this country, over the last few decades, there have been situations when floods have caused great destruction [3]. Floods happen almost every year; they are unpredictable, and the countermeasures are complex and expensive, requiring a quick reaction time, specialized crews, and efficient intervention equipment. Intervention during the occurrence of a hazard is influenced, among other factors, by the road network (quality and density). The more modern the network, the shorter the travel time of special vehicles and crews [4].

Also, the number of intervention centers is essential, as is the level of equipment and training. In Romania, there are specialized detachments with sophisticated equipment and a high intervention capacity and small, local centers with a limited intervention capacity. The situation is complicated when the distance that the teams have to cover is significant as a result of the reduced number of centers; when the water covers the roads, making movement difficult or impossible; and the relief is rough (there are areas where special vehicles cannot enter) [5].

The authority that intervenes in the event of floods is called the Inspectorate for Emergency Situations, or the ISU [6]. Intervention plans should be developed based on the area's specific climate, hazard occurrence, equipment, preparation degree, and the road network.

Sekhar et al. [7] discuss mobility in transport and the importance of defining accessibility to services and facilities. They developed an accessibility index from different urban areas using transport resources.

Sahitya and Csrk [8] studied the decision factors required to increase intervention capacity and improve road accessibility. Such analyses are helpful to the authorities, who can implement and adopt specific measures in each case.

Considering this idea, the present article provides a case study of accessibility in the case of floods in the Vărbilău basin, Romania. The crews' travel times are analyzed considering the road network and the intervention units' resources (number of centers, and their importance and proximity).

2. Study Area

The study area, the Vărbilău catchment, Romania (Figure 1), is prone to flooding. It was studied for this analysis because of the repetition of hydrological phenomena, the production of material damage, and the presence of various road categories from county roads that have been modernized, where the travel speed is high, to residential or forest roads, where the speed (especially in the latter) is reduced.

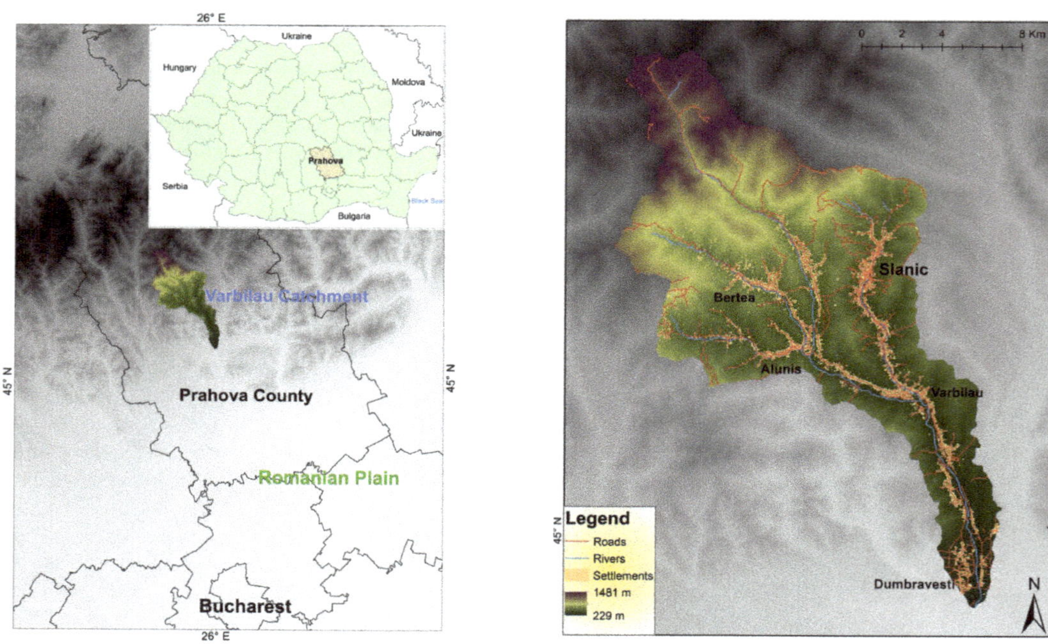

(a) (b)

Figure 1. The location of the Vărbilău catchment in Romania: (**a**) the location of the Vărbilău area in Prahova County; (**b**) road network and settlements inside the catchment.

The Vărbilău Basin, located in Prahova County (Figure 1a), has an area of 215 km^2, being a small hydrographic basin. It is located at altitudes between 229 m in the south, near the confluence of the Vărbilău River with the Teleajen collector river, and 1481 m in the north [9]. The basin is framed by the Romanian Plain in the south and the Carpathian Mountains in the north. Most of the basin's surface is represented by the alternation between hills and valleys, with gentle hills in the south and higher ones in the north [10]. The valleys are elongated, with an NW–SE arrangement in the north and center and N–S one in the basin's south. The elongated shape and flat surface of the riverbeds favor flooding, while the steep slopes and circular shape of the sub-basins in the north favor rapid runoff of water on the slope [11]. Precipitation, together with the previously mentioned elements, favors hydrological phenomena. Floods occur either in the spring after the rapid melting of the snow, or in the months with higher precipitation, namely in May and June [12]. Moreover, there have been recent floods in the localities of Aluniş, Bertea, Vărbilău, and Ştefeşti, but also landslides in the Lutu Roşu area, which have destroyed the roads and completely isolated the village.

The ISU intervention centers (emergency points) that serve this area are Ploieşti (two centers), Câmpina, Vălenii de Munte, and Slănic (a smaller intervention center, the only one located right in the hydrographic basin area). The population that lives here comprises about 30,000 people [13]. The roads show a higher density in the area of the Vărbilău commune, the town of Slănic, Dumbrăveşti, and in the Ostrovu–Aluniş area (Figure 1b). The quality of the roads is somewhat better in the central area of these places, while towards the periphery, it decreases, as does the density. In the mountainous north and the high hilly areas, the density of roads is very low; they are almost non-existent.

3. Materials and Methods

Field observations, GIS (Geographic Information System) databases, and data from the Emergency Situations Inspectorate were utilized. GIS maps were developed using different tools from ArcMap 10.2.2. Initially, the databases were added and prepared for processing. The road network, the essential component used in the elaboration of this study, is line-type data, and comprises segments that have set travel speeds depending on the road quality [14]. Based on their length and speed, it was possible to calculate travel speeds on a road section in GIS [15]. The four intervention centers mentioned in the previous "study area" were added as point data. The information related to them was obtained from the Prahova Emergency Unit website (https://www.isuprahova.ro/) (accessed on 20 May 2023).

Travel time maps that take into account the time away from an emergency point were built, taking into account the road network. The obtained results made up a polygon database. Five time intervals were set (<10 min, 10–20 min, 20–30 min, 30–40 min, and >40 min). They will be presented in the Results section. It should be mentioned that on three of the travel time maps (Vălenii de Munte, Ploieşti, and Câmpina), the points represent the entry points for the emergency teams in the hydrographic basin.

Further, the polygons were transformed into raster data (a raster is made of pixels of different sizes; the smaller the pixel, the higher the quality) [16]. A general accessibility map of the Vărbilău catchment for the emergency units depending on the road network and the intervention centers was built with the help of the following formula using Raster Calculator [16]:

$$R = MapS + MapC + 2 \times MapV + 4 \times MapP \quad (1)$$

where R is the sum of the rasters of the travel time maps for each emergency point taking into account the road network (general accessibility map), MapS is the raster for Slănic, MapC is the raster for Câmpina, MapV is the raster for Vălenii de Munte, and MapP is the raster for Ploieşti.

Marks were awarded to each of the five travel time intervals: 5 for <10 min, 4 for 10–20 min, 3 for 20–30 min, 4 for 30–40 min, and 5 for >40 min [17]. Depending on the size of the intervention center, a multiplier was added to the formula [18]. No multiplier was added to Slănic because this is a small intervention center. No multiplier was added for Câmpina because even if there is a detachment, it is a center situated at the longest distance from the Vărbilău basin compared to the others (20 km). Also, the distance the team has to cross to reach the Vărbilău catchment is in a hilly and winding area, so it is more difficult to pass. A multiplier of "2" was added to Vălenii de Munte center because it is closer to the basin than Câmpina (only 5 km from the Varbilau basin). In the case of Ploieşti, the multiplier "4" was added because there are two very important detachments there (Ploieşti is the county capital) located at a distance of 12 km from the hydrographic basin [19]. A general accessibility map was obtained after using formula values from 8 to 28.

A value of 8 means a small accessibility index, while 28 means a high index value. Using natural breaks in GIS (reclassification), five new intervals were set from 1 to 5 (Table 1).

Table 1. General accessibility value reclassification.

Intervals	Reclassification
8–9	1
9–11	2
11–16	3
16–22	4
22–28	5

Through a decision factor in which conditional statements were used, the best-served areas (the ones with under 10 min) were identified and a map was obtained [20]. Thus, the condition was defined using the four travel maps:

$$\begin{aligned} D_t = \mathrm{Con}(\mathrm{MapS} = \text{ under 10 min, 1}) \text{ and } \mathrm{Con}(\mathrm{MapC} = \text{ under 10min, 2}) \text{ and} \\ \mathrm{Con}(\mathrm{MapV} = \text{ under 10 min, 3}) \text{ and } \mathrm{Con}\ (\mathrm{MapP} = \text{ under 10 min, 4}) \end{aligned} \quad (2)$$

where D_t is the decision factor and Con is the condition.

The condition reads as follows: If MapS has an area equivalent to "under 10 min", then it is given a value of 1. In this way, only the area with the value "under 10 min" is displayed, which means time accessibility is very good. In order to include all four areas, "and" is added between conditions. To differentiate the accessibility of each of the areas, random values of 1, 2, 3, and 4 were added. The number of best-served settlements was determined based on a settlement database.

The obtained maps and other results will be presented in the next chapter.

4. Results

Figure 2a shows that, from the point of view of accessibility, the best-served area (under 10 min) by the Slănic intervention center is located in the northeast of the basin, around the point. Figure 2b shows the area that the Câmpina center best serves. Whole areas in the center, north, and south are poorly served due to the long distance (over 40 min). Figure 2c shows the area served by the center from Vălenii de Munte. It can be seen that a very small zone is very well served (under 10 min). Figure 2d indicates that the best-served area by the two intervention centers in Ploiesti is in the south. The shape of the areas was influenced by the times required to cross a road section and the distance from the intervention point.

The general accessibility map (Figure 3a) was obtained based on travel time maps. The best-served areas (4 and 5) are in the south in the area of Dumbravesti due to the high density of roads, their good quality, and the proximity to the centers of Ploieşti and Vălenii de Munte. The moderately served areas (3) are in the central and north-eastern parts and a few in the west as a result of the smaller importance of the intervention centers (Slănic), the greater distance of the Câmpina center from the basin, and the lower road density in the peripheral areas of certain localities (Slănic, Aluniş, and Vărbilău villages). The poorly served areas (2) include the communes of Bertea and Ştefeşti, and part of the commune of Vărbilău, the reason being the low density of the roads, the poor quality of the roads (travel speed is thus lower), and the greater distance from the intervention centers. The mentioned villages are located in hilly areas [21]. The least served areas, some inaccessible, include the mountainous area in the north with very steep slopes and certain hills bordering the Vărbilău commune to the west and east (the central area of the basin).

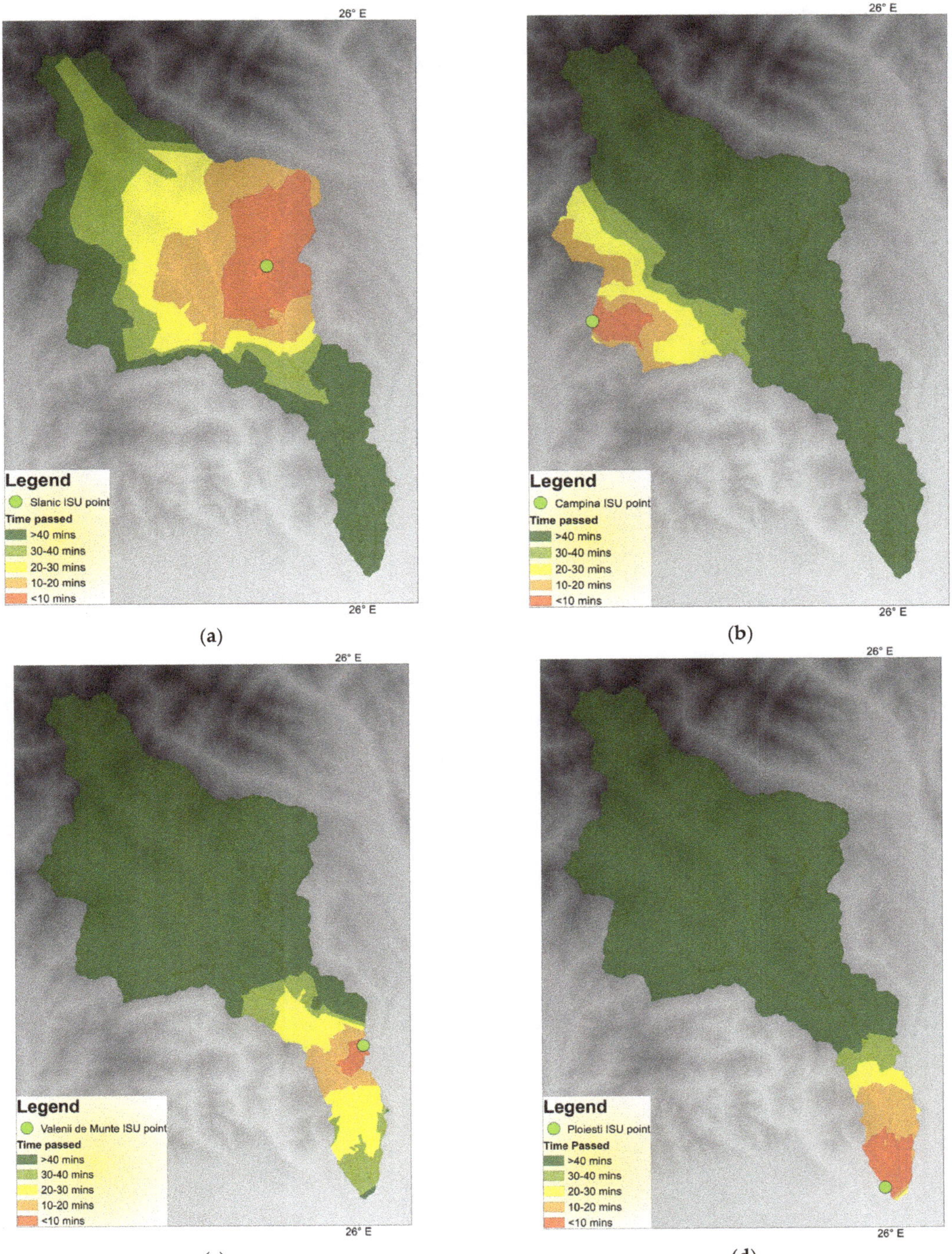

Figure 2. Travel time maps: (**a**) accessibility from Slănic; (**b**) accessibility from Câmpina; (**c**) accessibility from Vălenii de Munte; (**d**) accessibility from Ploieşti.

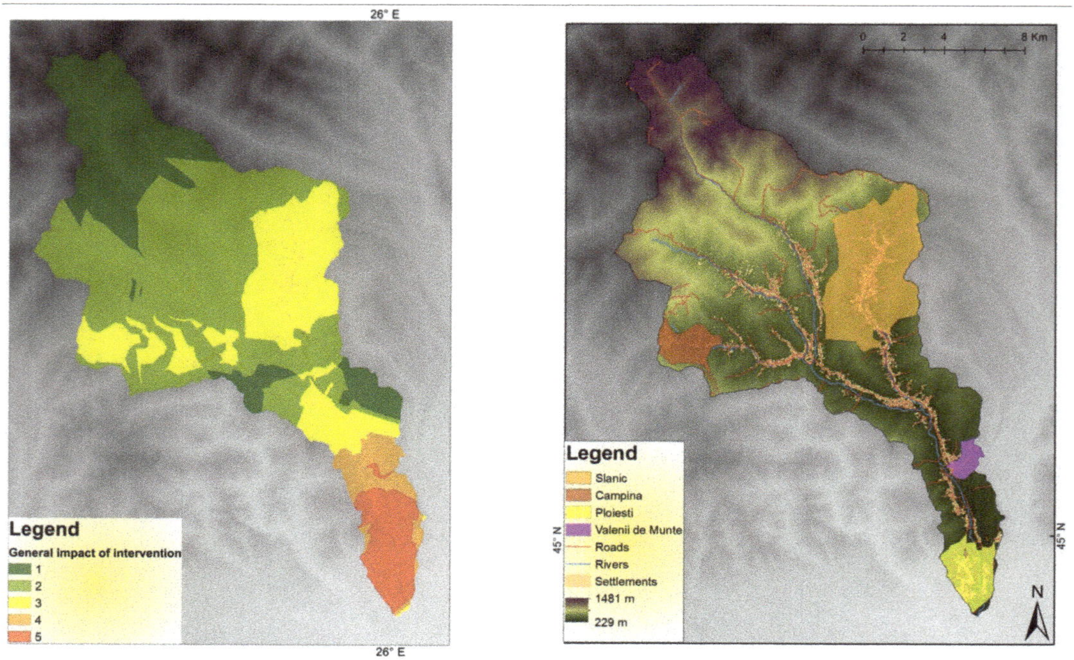

(**a**) (**b**)

Figure 3. Accessibility maps of Vărbilău catchment: (**a**) general accessibility index map; (**b**) best-served areas according to conditional statements.

In the situation of the best-served areas in terms of travel time, depending on the conditional statements used in the situation of the four travel time maps, Figure 3b shows that the best-served area is the Slănic area (in orange) served by the intervention center with the same name. This is the only town in the Vărbilău basin in which the density and quality of the roads are relatively good.

The smallest zone of the best-served areas is represented by the one under the influence of the center of Vălenii de Munte (in purple). This study used only time to determine the best-served areas (those with a journey of fewer than 10 min). Table 2 indicates that in the Slănic area, 3913 buildings are very well served, representing 18.5% of the total number of buildings, meaning that the intervention teams from Slănic arrive at that area in less than 10 min. The area served by the Vălenii de Munte center includes only 144 buildings—0.68% of the total number of buildings. It is important to note that the time starts from the entry point of the teams in the Vărbilau Basin to a settlement. Besides the time inside the basin, the crews need some extra minutes to travel from the center to the limits of the Vărbilau catchment [22]. The same applies to the Câmpina and Ploieşti intervention centers, but not Slănic, located inside the catchment.

Table 2. Number of buildings located in the best-served area (index 5).

Served by	Buildings Served
Vălenii de Munte	144
Ploieşti	1465
Slănic	3913
Câmpina	359
Buildings in Vărbilau basin	21,145

5. Conclusions

The capacity to intervene can be decisive in reducing the negative impact of floods and saving lives. It can be seen that from the viewpoint of time accessibility depending on the road network, not taking into account other factors, that the best-served areas are that of Slănic city, a small area in the west of the hydrographic basin, and two areas in the south.

Regarding general accessibility, factors such as the proximity of ISU intervention centers and their importance were included. Thus, it turned out that the area in the south (Dumbrăveşti area) is the best-served, while the extreme north is the least served due to the low density of roads and their poor quality (forest roads). In many areas, there is no way of communicating, with the slopes being steep and the mountains being mostly forested.

The situation can be improved by establishing new centers and equipping existing ones, such as Slănic, which is a small intervention point. Also, improving the quality of roads and building new routes in less accessible areas can increase travel times.

Author Contributions: Conceptualization, C.P. and C.Ş.D.; methodology, C.P. and A.B; software, C.P.; validation, A.B. and C.Ş.D.; formal analysis, A.B. and C.Ş.D.; investigation, C.P., A.B. and C.Ş.D.; resources, C.P.; data curation, C.P.; writing—original draft preparation, C.P. and A.B.; writing—review and editing, C.Ş.D.; visualization, C.P.; supervision, A.B.; project administration, A.B.; funding acquisition, C.P. All authors have read and agreed to the published version of the manuscript.

Funding: The research received fundings from the Doctoral School of Technical University of Civil Engineering, internal grants for Ph.D. students.

Institutional Review Board Statement: Not applicable.

Informed Consent Statement: Not applicable.

Data Availability Statement: Data will be available on request.

Conflicts of Interest: The authors declare no conflict of interest.

References

1. Toroimac, G.I.; Zaharia, L. Using Pressure and Alteration Indicators to Assess River Morphological Quality: Case Study of the Prahova River (Romania). *Water* **2015**, *7*, 2971–2989. [CrossRef]
2. Hamouda, M.F.; Pasquero, C. European extreme participation: The effects of spatio-temporal resolution of the data. *Weather Clim. Extrem.* **2021**, *33*, 100337. [CrossRef]
3. Croitoru, A.E.; Piticar, A.; Burada, D.C. Changes in precipitation extremes in Romania. *Quatern. Int.* **2016**, *415*, 325–335. [CrossRef]
4. Taniguachi, E.; Fereira, F.; Nicholson, A. A Conceptual Road network Emergency Model to Aid Emergency Preparedness and Response Decision-Making in the Context of Humanitarian Logistics. *Proced. Soc. Behav. Sci.* **2012**, *39*, 307–320. [CrossRef]
5. Rohr, A.; Priesmeier, P.; Tzavella, K.; Fekete, A. System Criticality of Road Network Areas for Emergency Management Services–Spatial Assessment Using a Tessellation Approach. *Infrastructures* **2020**, *5*, 99. [CrossRef]
6. Arrighi, C.; Pregnolato, M.; Dawson, R.J.; Castelli, F. Preparedness against mobility disruption by floods. *Sci. Total Environ.* **2019**, *654*, 1010–1022. [CrossRef]
7. Somenahalli, S.V. Road Network Accessibility and Socio-economic Disadvantage Across Adelaide Metropolitan Area. *Transp. Dev. Econ.* **2016**, *2*, 15. [CrossRef]
8. Sahytia, K.S.; Csrk, P. GIS-Based Urban Road Network Accessibility Modeling Using MLR, ANN and ANFIS Methods. *Transp. Telecommun.* **2021**, *22*, 15–28.
9. Ghinea, D. *The Geographical Encyclopedia of Romania*; Editura Enciclopedică: Bucharest, Romania, 2000. (In Romanian)
10. Velcea, V. *The Geography of Romanian Carpathians and Sub-Carpathians*; Editura Didactică şi Pedagogică: Bucharest, Romania, 1982. (In Romanian)
11. Roșu, A. *The Physical Geography of Romania*; Editura Didactică și Pedagogică: Bucharest, Romania, 1980. (In Romanian)
12. Drobot, R. *Lessons of Hydrology and Hydrogeology*; Editura Didactică și Pedagogică: București, Romania, 2020. (In Romanian)
13. Erdeli, G.; Dumitrache, L. *The Population Geography*; University Publishing: Bucharest, Romania, 2010. (In Romanian)
14. Toma, M.G.; Ungureanu, R.D.; Dicu, M. GIS, a tool for improving the road network management. *IOP Conf. Ser. Earth Environ. Sci.* **2021**, *664*, 012103. [CrossRef]
15. Safeer, M.; Anbuudayasankar, S.P.; Kartik, B.; Ganesh, K. Analyzing Transportation and Distribution in Emergency Humanitarian Logistics. *Procedia Eng.* **2014**, *97*, 2248–2258. [CrossRef]
16. Caniupan, M.; Torres-Aviles, R.; Gutierrez-Bunster, T.; Lepe, M. Efficient computation of map algebra over raster stored in the k^2-acc compact data structure. *GeoInformatica* **2022**, *26*, 95–123. [CrossRef]

17. Sedeghi-Niaraki, A. Real world representation of a road network for route planning in GIS. *Expert Syst. Appl.* **2011**, *38*, 11999–12008. [CrossRef]
18. Malczewski, J. On the Use of Weighted Linear Combination Method in GIS: Common and Best Practice approaches. *Trans. GIS* **2002**, *4*, 5–22. [CrossRef]
19. Slopek, J.; Netzel, P. Comparison of different implementations of a raster map calculator. *Comput. Geosci.* **2021**, *154*, 104824.
20. Eastman, J.R.; Kyen, P.A.K.; Toledano, J.; Jin, W. GIS and decision making. *Photogramm. Eng. Remote Sens.* **1995**, *61*, 539–547.
21. Shrestha, J. Rural Road Network Decision model for Hilly Regions of Nepal. *J. Adv. College Eng. Manag.* **2018**, *4*, 51–64. [CrossRef]
22. Graur, D.S.; Efros, V. GIS analysis or road network accessibility for emergencies managed by firefighters (case study–Suceava county, Romania). *Acta Geobalc.* **2017**, *3*, 65–70. [CrossRef]

engineering proceedings

MDPI

Proceeding Paper

Precipitation Time Series Analysis and Forecasting for Italian Regions †

Ebrahim Ghaderpour [1,2,*], Hanieh Dadkhah [1], Hamed Dabiri [1,*], Francesca Bozzano [1,2], Gabriele Scarascia Mugnozza [1,2] and Paolo Mazzanti [1,2]

1 Department of Earth Sciences & CERI Research Centre, Sapienza University of Rome, 00185 Rome, Italy; hanieh.dadkhah@uniroma1.it (H.D.); francesca.bozzano@uniroma1.it (F.B.); gabriele.scarasciamugnozza@uniroma1.it (G.S.M.); paolo.mazzanti@uniroma1.it (P.M.)

2 NHAZCA s.r.l., Via Vittorio Bachelet, 12, 00185 Rome, Italy

* Correspondence: ebrahim.ghaderpour@uniroma1.it (E.G.); hamed.dabiri@uniroma1.it (H.D.)

† Presented at the 9th International Conference on Time Series and Forecasting, Gran Canaria, Spain, 12–14 July 2023.

Abstract: In Italy, most of the destructive landslides are triggered by rainfall, particularly in central Italy. Therefore, effective monitoring of rainfall is crucial in hazard management and ecosystem assessment. Global precipitation measurement (GPM) is the next-generation satellite mission, which provides the precipitation measurements worldwide. In this research, we employed the available monthly GPM data to estimate the monthly precipitation for the twenty administrative regions of Italy from June 2000 to June 2021. For each region, we applied the non-parametric Mann–Kendall test and its associated Sen's slope to estimate the precipitation trend for each calendar month. In addition, for each region, we estimated a linear trend and the seasonal cycles of precipitation with the antileakage least-squares spectral analysis (ALLSSA) and showed the annual precipitation variations using box plots. Lastly, we compared machine-learning models based on the auto-regressive moving average for monthly precipitation forecasting and showed that ALLSSA outperformed them. The findings of this research provide a significant insight into processing climate data, both in terms of trend-season estimates and forecasting, and can potentially be used in landslide susceptibility analysis.

Keywords: ALLSSA; ARIMA; GPM; landslides; machine learning; Mann–Kendall; precipitation; remote sensing; time series forecasting; trend analysis

Citation: Ghaderpour, E.; Dadkhah, H.; Dabiri, H.; Bozzano, F.; Scarascia Mugnozza, G.; Mazzanti, P. Precipitation Time Series Analysis and Forecasting for Italian Regions. *Eng. Proc.* **2023**, *39*, 23. https://doi.org/10.3390/engproc2023039023

Academic Editors: Ignacio Rojas, Hector Pomares, Luis Javier Herrera, Fernando Rojas and Olga Valenzuela

Published: 29 June 2023

1. Introduction

Precipitation is one of the main components of water cycles, which plays a vital role in maintaining atmospheric balance, growing crops, and providing a fresh water supply. Heavy precipitation can damage crops and cause floods, landslides, soil erosion, etc., while the lack of precipitation can result in drought and forest fires [1–5]. Most parts of central Italy are ranked high or very high for landslide hazards, and most of the reported landslides were triggered by heavy precipitation [6,7]. For instance, a comprehensive description of rainfall-induced landslides in large cities was reported in [8]. Therefore, effective, and continuous monitoring of changes in precipitation is crucial for establishing/improving strategies to mitigate the socioeconomic damages from climate change.

Traditionally, ground precipitation gauges have been used to measure precipitation in high temporal resolution. However, in-situ precipitation gauges are limited in number and cannot sufficiently provide good coverage for all regions, particularly mountainous regions [9]. Recent advances in satellite remote sensing have made it possible to measure global precipitation remotely in a pleasant spatial and temporal resolution.

Global precipitation measurement (GPM) is an advanced satellite mission, which has been providing precipitation measurements since the beginning of the 21st century [10]. The integrated multi-satellite retrieval for GPM (IMERG) is a robust algorithm that considers all satellite microwave precipitation estimates as well as precipitation gauge measurements to provide reliable monthly precipitation estimates [11,12].

In this research, we analyze the monthly GPM measurements obtained for all twenty regions of Italy. We utilize several time series analysis methods to process these measurements. The main contributions of this research are summarized below.

(1) For each Italian region, we apply the non-parametric Mann–Kendall (MK) test and Sen's slope estimator to estimate the gradient of the precipitation time series corresponding to each calendar month since 2000.
(2) For each Italian region, we produce the monthly box plots to better visualize the annual variation of precipitation.
(3) We apply the antileakage least-squares spectral analysis (ALLSSA) to estimate the season and trend components of the 21-year-long monthly GPM time series for each Italian region.
(4) We compare the performance of time series forecasting methods on the monthly precipitation time series. These methods are auto-regressive integrated moving average (ARIMA), ARIMA with eXogenous factors (ARIMAX), ARIMAX with seasonal terms (SARIMAX), and ALLSSA.

To the best of the authors' knowledge, this is the first time that such analyses have been performed on the GPM measurements in Italy. The results presented in this paper can be used as a guide for a better understanding of the climate dynamics across Italy and for developing management strategies for landslide hazard management and a sustainable environment.

2. Materials and Methods

2.1. Study Region and Datasets

Italy, a country located in southwest Europe and the middle of the Mediterranean Sea, has 20 administrative regions (see Figure 1). Italy has four climate types: Alpine climate in northern Italy (high-elevation), Apennine climate in northern and central Italy (high-elevation), Peninsular climate in the central part of the peninsula, and Mediterranean climate in southern Italy and Sicily and Sardinia islands. The climate maps of Italy for different months are illustrated in [12].

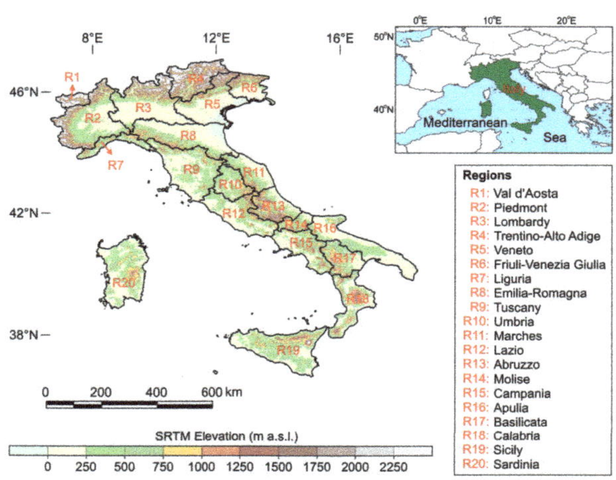

Figure 1. The study area: Italy and its 20 administrative regions. The background elevation in meters above the mean sea level is from the Shuttle Radar Topography Mission (SRTM) plus at ~30 m spatial resolution provided by Jet Propulsion Laboratory (JPL) available freely online at https://doi.org/10.5067/MEaSUREs/SRTM/SRTMGL1N.003 (accessed on 1 January 2023).

The monthly global precipitation measurement (GPM) level 3 data were obtained from all precipitation microwave and infrared satellite estimates and precipitation gauge analyses recommended to use for research [10]. The monthly GPM measurements at a spatial resolution of $0.1^\circ \times 0.1^\circ$ were averaged within each Italian region, and their associated standard errors were estimated with the covariance law of error propagation. Therefore, for each Italian region, an unequally weighted time series of size 252 was obtained spanning from June 2000 to June 2021. The weights are inversely proportional to the squares of the measurement errors used in the season-trend estimation with ALLSSA in this study.

2.2. Methods

2.2.1. The Mann–Kendall (MK) Method and Sen's Slope Estimator

The MK method is a non-parametric test for trend analysis widely used in climate studies [13,14]. The null hypothesis in MK is defined as no trend exists in the time series. Thus, if the null hypothesis is rejected, it means there is a statistically significant trend in the time series, the alternative hypothesis. Commonly used confidence levels in the MK test are 90%, 95%, and 99%. The slope of the trends can be obtained via its associated Sen's slope estimator, a non-parametric slope estimator that determines the magnitude and direction of the trend. The mathematical formulas for the MK test and Sen's slope can be found in [13,14]. Herein, to apply the MK test, the monthly precipitation measurements were categorized to obtain a time series of size 21 for each Italian region and each calendar month. Note that the new time series for each calendar month is yearly, i.e., one measurement per year, and so there is no seasonality or intra-annual component in the time series. Then, Sen's slope was estimated for each of the time series of size 21 along with its statistical significance using the MK test.

2.2.2. Boxplots

A boxplot is a graphical representation based on five numbers: the minimum value of the data excluding any outliers, the first quartile Q_1, the median, the third quartile Q_3, and the maximum value of the data excluding any outliers [15]. Values marked as ○ are classified as outliers. An outlier is a value that is either extremely large or extremely small when compared to the rest of the values in a sample. Specifically, an outlier is a value that is larger than the upper fence value or smaller than the lower fence value. These values are calculated as $Q_3 + 1.5(Q_3 - Q1)$ for the upper fence and $Q_1 - 1.5(Q_3 - Q1)$ for the lower fence. Note that the whiskers of the boxplot extend to the most extreme upper and lower points that are not outliers, not to the fence values. Herein, a boxplot for each time series of size 21 mentioned in Section 2.2.1 is also illustrated.

2.2.3. The Antileakage Least-Squares Spectral Analysis (ALLSSA)

ALLSSA is a season-trend fit model based on the least-squares principle [16,17]. In ALLSSA, an initial set of frequencies is first selected along with a type of trend (linear, quadratic, or cubic). Then, an iterative process is performed to simultaneously fit the trend and sinusoids to the time series at a certain confidence level, usually 99%. This process determines an optimal set of sinusoids that, along with the selected trend type, best fit the time series. The algorithm termination is based on the statistical significance of the peaks in the least-squares spectrum, i.e., the process is terminated if there are no more statistically significant spectral peaks in the least-squares spectrum, depending on the initial selected set of frequencies. The code of ALLSSA is described in [15] and is freely available online. ALLSSA can also be used for forecasting as it can prevent over/under-fitting issues. Furthermore, ALLSSA can consider the measurement errors for a more reliable season-trend estimate.

2.2.4. The Auto-Regressive Integrated Moving Average (ARIMA) and Its Extensions

ARIMA is a traditional time series forecasting model based on a time series initial values, lags, and lagged forecast errors [18,19]. If one considers exogenous variables (external data in a forecast), then the model is called ARIMAX [20]. If a time series has seasonal patterns, e.g., climate time series, seasonal terms need to be considered as well; therefore, in this case, the new model is called SARIMAX [21]. The mathematical formulas for ARIMA, ARIMAX, and SARIMAX are given in Equations (1)–(3), respectively.

$$y_t = c + \sum_{n=1}^{p} \alpha_n y_{t-n} + \sum_{n=1}^{q} \theta_n \epsilon_{t-n} + \epsilon_t \tag{1}$$

$$y_t = c + \sum_{n=1}^{p} \alpha_n y_{t-n} + \sum_{n=1}^{q} \theta_n \epsilon_{t-n} + \sum_{n=1}^{r} \beta_n x_{n_t} + \epsilon_t \tag{2}$$

$$y_t = c + \sum_{n=1}^{p} \alpha_n y_{t-n} + \sum_{n=1}^{q} \theta_n \epsilon_{t-n} + \sum_{n=1}^{P} \varphi_n y_{t-sn} + \sum_{n=1}^{Q} \eta_n \epsilon_{t-sn} + \sum_{n=1}^{r} \beta_n x_{n_t} + \epsilon_t \tag{3}$$

where y_t is the lag term; c is the intercept of the models; $\alpha_n, \theta_n, \varphi_n, \eta_n, \beta_n$ are coefficients; p is the trend autoregression order; q is the trend moving average order; P is the seasonal autoregressive order; Q is the seasonal moving average order; r is the number of exogenous variables; s stands for the seasonal effect; ϵ_t is the random error at time t; and x is the exogenous variable. The order of the models plays a crucial role on the result accuracy and could be obtained through a trial-and-error process by minimizing the autocorrelation function and partial autocorrelation function.

3. Results and Discussion

3.1. The MK and Sen's Slope Results

Table 1 lists the Sen's slope results and their MK statistical significance for the Italian regions and for the 12 calendar months. R16, R17, and R19 had significant declining trends in December, while R1 and R2 had significant decreasing trends in September since 2000. In addition, in April, R11, R15, R18, and R19 had significant declining trends since 2000. These results are also in agreement with the monthly gradient estimate maps presented in [12].

Table 1. Sen's slope (mm/year) and its MK significance (* = 90%) (** = 95%) (*** = 99%) for the Italian regions.

R	Jan	Feb	Mar	Apr	May	Jun	Jul	Aug	Sep	Oct	Nov	Dec
1	1.15	0.39	−0.67	−0.19	1.17	0.12	−0.87	−1.03	−2.14 **	0.80	1.75	2.85
2	0.34	−0.19	−0.11	0.48	0.30	0.20	−0.42	−0.76	−2.75 **	0.63	0.80	0.63
3	0.08	−0.22	−0.29	0.05	0.70	1.77	−2.07	−0.16	−1.01	1.72	−2.22	1.75
4	0.89	0.36	−0.95	−0.44	1.09	0.83	−1.56	0.94	0.142	0.68	−2.39	1.14
5	−0.48	1.30	−0.54	−1.34	1.30	0.54	−1.17	0.26	−1.23	0.56	−2.12	0.96
6	0.59	1.89	−1.16	−1.48	2.37	0.94	−1.37	0.46	−0.44	0.17	−1.97	1.34
7	−0.20	0.22	0.16	0.19	−0.35	0.86	0.07	−0.21	−1.08	−0.47	0.60	0.58
8	−0.34	0.92	−0.20	−1.14	0.70	0.50	−0.52	−0.33	−0.81	0.27	−0.29	0.42
9	0.35	1.49	−0.52	−0.91	0.47	0.19	−0.52	−0.25	−0.49	0.46	−0.44	−1.15
10	−0.35	1.17	1.24	−1.16	0.29	1.11	0.15	0.05	−0.04	0.31	0.84	−1.64
11	−0.08	2.06	1.47	−1.98 **	1.12	1.15	0.02	−0.69	−0.83	−0.40	0.24	−2.13
12	−0.75	0.72	0.93	−1.06	−0.03	0.53	0.20	−0.41	0.14	−0.58	0.55	−1.82
13	−0.04	−0.81	1.12	−1.23	−0.10	0.36	0.30	−0.62	−0.86	−0.71	−0.24	−2.89
14	−0.62	−0.31	0.35	−1.32	0.46	0.09	0.64	0.08	0.63	−0.49	2.95	−2.52
15	−0.37	0.22	0.33	−1.41 *	0.59	0.71	−0.02	0.19	0.38	0.52	2.07	−1.41
16	−0.27	0.42	0.50	−1.17	0.39	0.52	0.16	0.38	−0.68	0.42	2.52	−3.11 **
17	0.02	0.00	0.37	−1.60	−0.04	0.75	0.24	0.16	−0.87	0.03	1.96	−3.64 **
18	−0.47	0.27	0.30	−1.86 **	−0.06	0.32	0.44	0.08	−0.87	1.31	2.68	−2.83
19	−2.07	0.47	2.27	−1.98 **	−0.55	−0.32	0.12	0.14	−0.26	1.34	2.66	−4.12 ***
20	−0.23	−0.05	0.62	−1.25	−0.03	0.33	0.01	0.31	1.00	−1.32	0.36	−1.48

3.2. The Boxplots for Monthly Precipitation

The monthly boxplots corresponding to the Italian regions are illustrated in Figure 2. The mean values are shown by ×, while the outliers are shown by °. For example, the outlier observed in the boxplot in Lazio (R12) is for January 2021 when Lazio experienced heavy rainfall and flooding: https://world-weather.info/forecast/italy/rome/january-2021 (accessed on 1 January 2023).

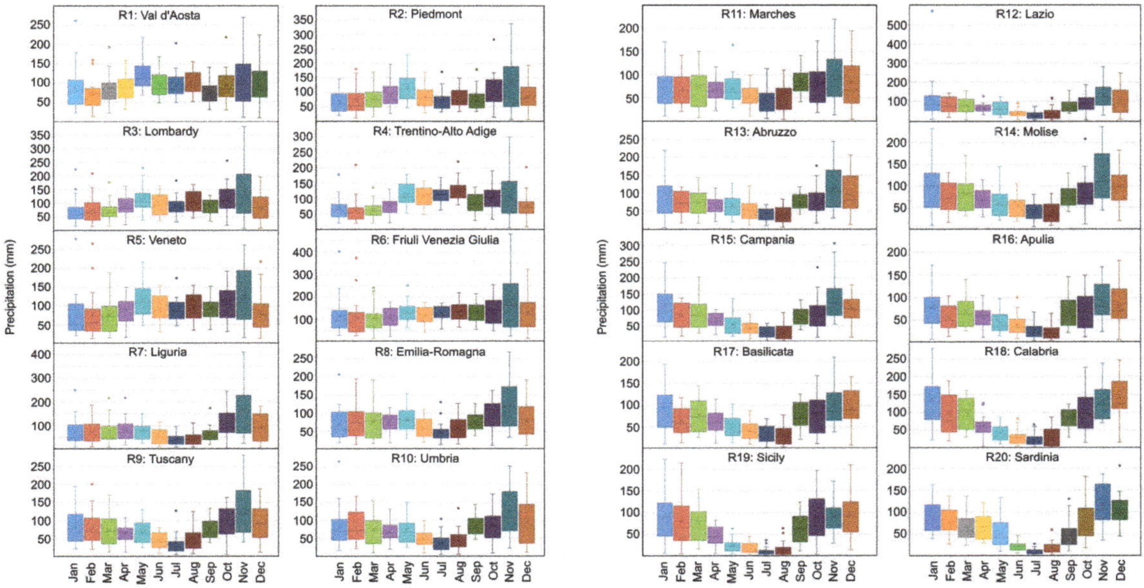

Figure 2. The boxplots for GPM monthly precipitation for the Italian regions. The mean is shown by ×, and the outlier is shown by °.

3.3. The ALLSSA Season-Trend Results

The monthly precipitation time series of size 252 corresponding to each Italian region with their simultaneously estimated linear trend and spectral components is illustrated in Figure 3. Except for Val d'Aosta (R1), the other regions had negative precipitation gradients. The monthly precipitation time series for Calabria (R18), Sicily (R19), and Sardinia (R20) had the largest amplitude for their annual cycles (>35 mm) as compared to other regions. Abruzzo (R13) had the most significant declining trend (0.61 mm/year) followed by Lazio (R12) that had a negative rate of 0.43 mm/year since 2000. The high peak observed in R12 in January 2021 and shown as an outlier in the boxplot (see Figure 2) is due to the heavy rainfall/flooding that is likely due to the gradual warming and declining precipitation trend over the past few decades. Note that the extreme values (very high precipitation) generally had higher errors, and therefore, they contributed less to the estimation of the season trend with ALLSSA. The estimated ALLSSA season trend may also be used for forecasting and near-real-time monitoring [22]. Herein, the ALLSSA forecasting performance results were also compared with the auto-regressive moving average-based models.

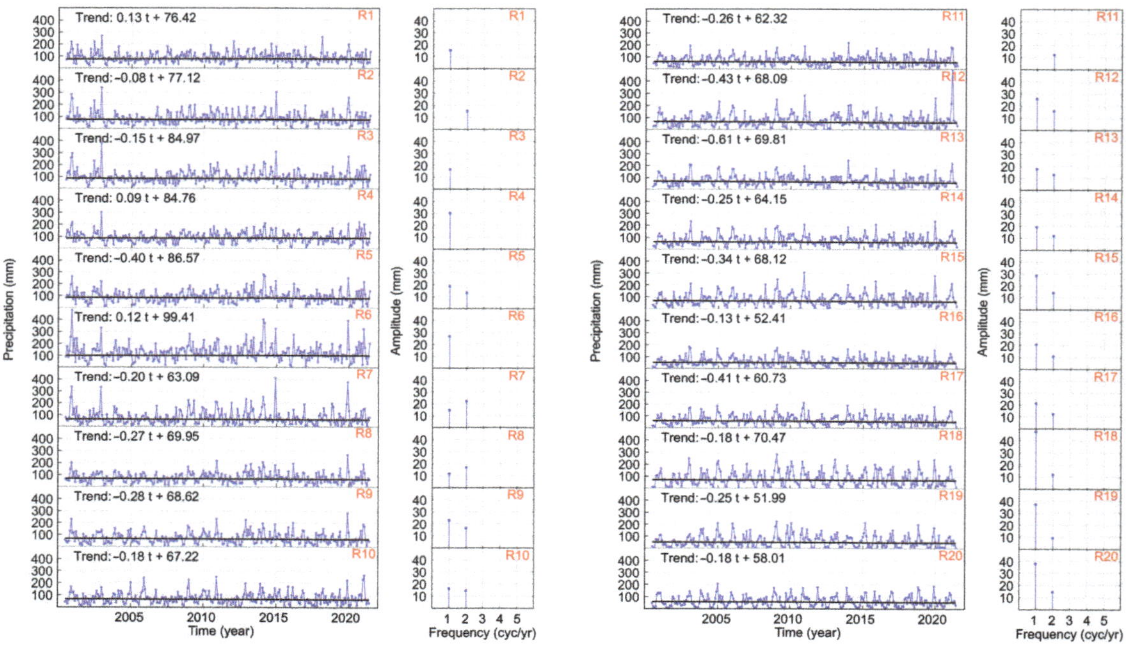

Figure 3. The simultaneous season-trend estimates with ALLSSA at a 99% confidence level. The region IDs shown in red correspond to the ones shown in Figure 1.

3.4. The ARIMA, ARIMAX, SARIMAX, and ALLSSA Results

After converting the time series of each region into a stationary time series by first-order differencing, the most optimized order for each model is proposed, as given in Table 2. The commonly used statistical metrics to evaluate the performance of each model are the mean absolute error (MAE), mean absolute percentage error (MAPE), and root mean square error (RMSE). If y_i and $\hat{y}_i$ are, respectively, the actual and predicted values with n observations, these metrics are mathematically defined by:

$$\text{MAE} = \frac{1}{n}\sum_{i=1}^{n}|\hat{y}_i - y_i| \tag{4}$$

$$\text{MAPE} = 100 \times \frac{1}{n}\sum_{i=1}^{n}\left|\frac{\hat{y}_i - y_i}{y_i}\right| \tag{5}$$

$$\text{RMSE} = \left(\frac{1}{n}\sum_{i=1}^{n}(\hat{y}_i - y_i)^2\right)^{0.5} \tag{6}$$

The values forecasted by the models are compared to the actual values in Figure 4 for Lazio and Calabria. The residual values $\hat{y}_i - y_i$ are also illustrated in Figure 4 for each region. To assess the models' reliability, MAE, MAPE, and RMSE are also calculated for the testing period (June 2020 to June 2021) and shown in Table 2. The period June 2000 to June 2020 was used for training the models (green transparent window in Figure 4), and the period June 2020 to June 2021 was used for testing (yellow transparent window in Figure 4) and forecasting 12 months afterward (blue transparent window in Figure 4).

Table 2. Performance metrics of the forecasting models. ALLSSA outperformed others in terms of MAE, MAPE, and RMSE. Parameters p, d, q are the lag order, the degree of differencing, and the order of moving average, respectively. Parameters P, D, Q, M are the autoregressive order for seasonal component, integration order of the seasonal process, moving average order of the seasonal component, and the number of periods in the season (e.g., 12 for annual data with 12 months in each year), respectively. Note that only the actual and forecasted values in the testing database shown within the yellow transparent window in Figure 4 were used to calculate MAE, MAPE, and RMSE.

	Lazio (R12)				Calabria (R18)			
	ARIMA (p,d,q) (1,1,1)	ARIMAX (p,d,q) (6,1,1)	SARIMAX $(p,d,q)(P,D,Q)[M]$ (3,1,6) (0,1,1) [12]	ALLSSA	ARIMA (p,d,q) (1,1,1)	ARIMAX (p,d,q) (6,1,1)	SARIMAX $(p,d,q)(P,D,Q)[M]$ (3,1,6) (0,1,1) [12]	ALLSSA
MAE (mm)	71.97	72.66	68.54	65.17	43.11	37.15	33.57	23.39
MAPE (%)	84.96	90.97	50.24	42.73	76.32	107.05	61.18	51.34
RMSE (mm)	153.39	153.23	150.25	145.71	57.49	50.20	39.10	27.75

From Figure 4 and the values provided in Table 2, one can observe that ARIMA and ARIMAX have approximately the same results. However, SARIMAX showed more accurate forecasting compared to ARIMA and ARIMAX, which highlights the importance of seasonality effect on the prediction reliability. The comparison between the different techniques is shown in Table 2, and Figure 4 clarifies that, for a time series with a seasonal pattern, SARIMAX can be used, whereas for a non-seasonal time series, both ARIMA or ARIMAX could lead to sufficiently acceptable prediction and forecasting results. For further comparison, ALLSSA was also applied to the monthly precipitation data from June 2000 to June 2020. The ALLSSA-estimated coefficients of harmonics and trend were used to forecast the monthly precipitation for June 2020 to June 2021. Using the actual measurements for June 2020 to June 2021, the statistical metrics were then calculated. From Table 2 obtained on the testing period, one can observe that ALLSSA outperformed ARIMA, ARIMAX, and SARIMAX for forecasting.

From the residual panel in Figure 4, all the models performed poorly for predicting the precipitation in January 2021 in Lazio. In fact, extreme precipitation events cannot be well predicted due to their abrupt behavior (e.g., precipitation value in January 2021 detected as an outlier in Figure 2). Figure 3 shows that Lazio historically had heavy precipitation in the months of November to February (e.g., November 2010 and 2019, December 2005 and 2009, January 2014 and 2021) when most of the destructive landslides occurred [7,8]. Due to factors such as gradual warming, declining annual precipitation, and changes in land cover [12], it is expected that the frequency of heavy rainfall in Lazio and other similar regions will increase in the upcoming years, posing a great risk for more landslides to occur. This also highlights the importance of conducting this research.

There are several other time series analysis methods for prediction and forecasting. For example, Song and Chissom [23] describe fuzzy time series and develop fuzzy time series models. The fundamental concept of this forecasting approach is replacing real values with fuzzy sets and could be performed through four main steps: (i) defining the universe of discourse and splitting it into intervals; (ii) determining the fuzzy sets on the universe of discourse and fuzzifying the time series; (iii) developing the model of the existing fuzzy-logic relationships in the fuzzified time series; and (iv) forecasting and eventually de-fuzzifying the forecasted values [24]. Novak et al. applied the fuzzy transform theory for time series analysis and forecasting for the first time [25]. Implementing such fuzzy models for processing precipitation time series could also be interesting and is subject to future studies.

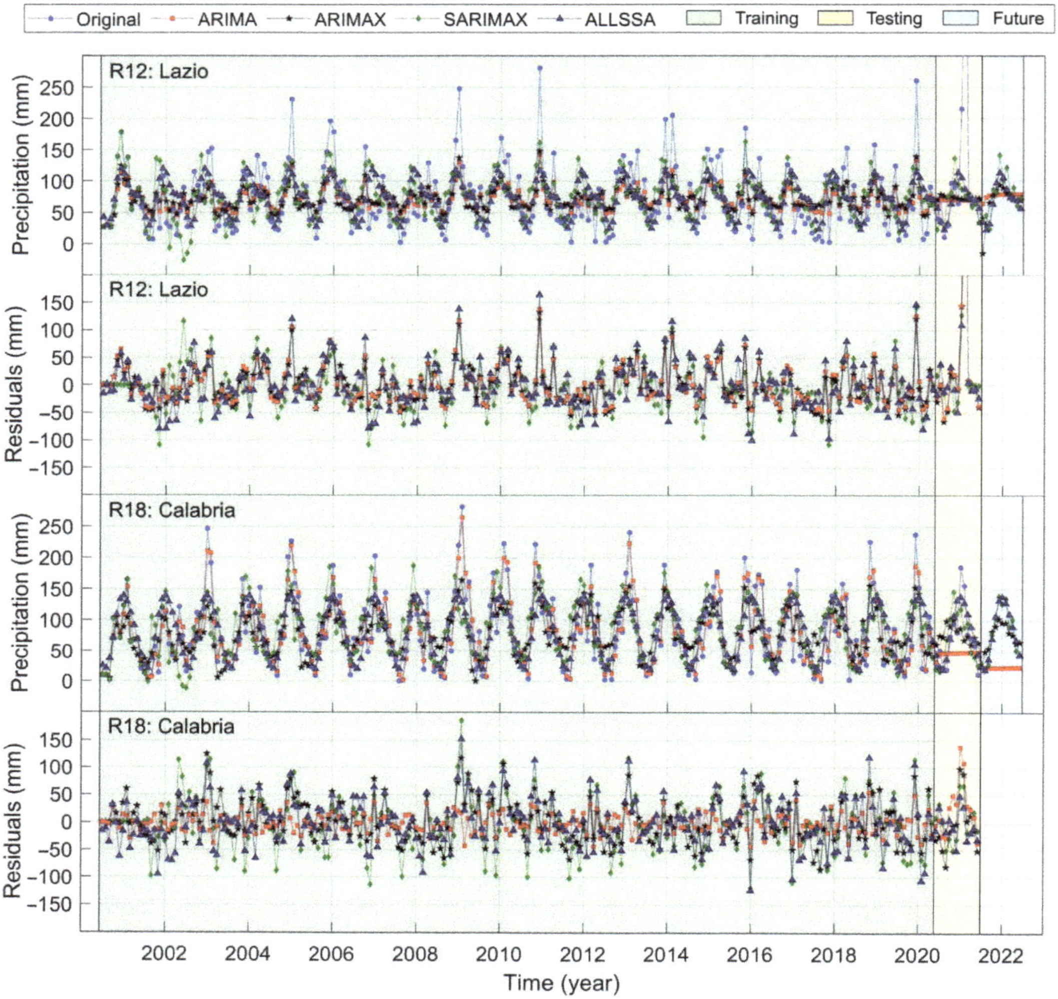

Figure 4. Comparison between the actual values and the predicted values with the ARIMA, ARIMAX, SARIMAX, and ALLSSA models for Lazio (R12) and Calabria (R18). The residual series are obtained by subtracting the predicted values from the actual values.

4. Conclusions

In this paper, we estimated the trend and seasonal components of the monthly precipitation time series for the twenty administrative regions of Italy. Using the Mann–Kendall test, we found that April, September, and December were the only months when precipitation had a declining trend at a 95% confidence level for northern and southern regions. The ALLSSA also showed negative declining trends for all the regions except Val d'Aosta, and annual cycles were most pronounced in the southern part of Italy: Calabria, Sicily, and Sardinia. We also compared the forecasting performances of ARIMA, ARIMAX, SARIMAX, and ALLSSA for Lazio that had extreme precipitation events during winter and Calabria that had the most significant seasonality. We found that ALLSSA had a better performance than ARIMA, ARIMAX, and SARIMAX for prediction and forecasting. However, all the models performed poorly for predicting extreme precipitation. Hybrid models that consider other parameters, such as temperature, soil moisture, and wind, may produce better and more reliable forecasting results. We hope our findings can help policymakers and stakeholders develop proper risk management strategies to maintain a sustainable environment.

Author Contributions: Conceptualization, E.G., H.D. (Hanieh Dadkhah), H.D. (Hamed Dabiri), F.B., G.S.M. and P.M.; methodology, E.G., H.D. (Hanieh Dadkhah) and H.D. (Hamed Dabiri); formal analysis, E.G., H.D. (Hanieh Dadkhah) and H.D. (Hamed Dabiri); writing—original draft preparation, E.G., H.D. (Hanieh Dadkhah) and H.D. (Hamed Dabiri); writing—review and editing, F.B., G.S.M. and P.M.; visualization, E.G., H.D. (Hanieh Dadkhah) and H.D. (Hamed Dabiri); supervision, E.G. All authors have read and agreed to the published version of the manuscript.

Funding: This research was funded by the CERI Research Centre at the Sapienza University of Rome, Italy.

Institutional Review Board Statement: Not applicable.

Informed Consent Statement: Not applicable.

Data Availability Statement: The monthly GMP dataset used in this research can be downloaded freely at https://doi.org/10.5067/GPM/IMERG/3B-MONTH/06 (accessed on 1 January 2023).

Acknowledgments: The authors thank the CERI Research Centre at the Sapienza University of Roma for supporting this research. The authors also thank NASA scientists and personnel for providing the datasets utilized in this research. The authors also expressed their sincere gratitude to the reviewers for their time and helpful suggestions.

Conflicts of Interest: The authors declare no conflict of interest.

References

1. Gentilucci, M.; Materazzi, M.; Pambianchi, G. Statistical Analysis of Landslide Susceptibility, Macerata Province (Central Italy). *Hydrology* **2021**, *8*, 5. [CrossRef]
2. Mazzanti, P.; Bozzano, F.; Cipriani, I.; Prestininzi, A. New insights into the temporal prediction of landslides by a terrestrial SAR interferometry monitoring case study. *Landslides* **2015**, *12*, 55–68. [CrossRef]
3. Koehler, J.; Dietz, A.J.; Zellner, P.; Baumhoer, C.A.; Dirscherl, M.; Cattani, L.; Vlahović, Ž.; Alasawedah, M.H.; Mayer, K.; Haslinger, K.; et al. Drought in Northern Italy: Long Earth Observation Time Series Reveal Snow Line Elevation to Be Several Hundred Meters Above Long-Term Average in 2022. *Remote Sens.* **2022**, *14*, 6091. [CrossRef]
4. Malandra, F.; Vitali, A.; Morresi, D.; Garbarino, M.; Foster, D.E.; Stephens, S.L.; Urbinati, C. Burn Severity Drivers in Italian Large Wildfires. *Fire* **2022**, *5*, 180. [CrossRef]
5. Marino, D.; Palmieri, M.; Marucci, A.; Soraci, M.; Barone, A.; Pili, S. Linking Flood Risk Mitigation and Food Security: An Analysis of Land-Use Change in the Metropolitan Area of Rome. *Land* **2023**, *12*, 366. [CrossRef]
6. Marchesini, I.; Ardizzone, F.; Alvioli, M.; Rossi, M.; Guzzetti, F. Non-susceptible landslide areas in Italy and in the Mediterranean region. *Nat. Hazards Earth Syst. Sci.* **2014**, *12*, 2215–2231. [CrossRef]
7. Lombardi, A.; Gallicchio, D.; Tomassetti, B.; Raparelli, E.; Tuccella, P.; Lidori, R.; Verdecchia, M.; Colaiuda, V. Evaluating the Response of Hydrological Stress Indices Using the CHyM Model over a Wide Area in Central Italy. *Hydrology* **2022**, *9*, 139. [CrossRef]
8. Alessi, D.; Bozzano, F.; Di Lisa, A.; Esposito, C.; Fantini, A.; Loffredo, A.; Martino, S.; Mele, F.; Moretto, S.; Noviello, A.; et al. Geological risks in large cities: The landslides triggered in the city of Rome (Italy) by the rainfall of 31 January–2 February 2014. *Ital. J. Eng. Geol. Environ.* **2014**, *1*, 15–34.
9. Shawky, M.; Moussa, A.; Hassan, Q.K.; El-Sheimy, N. Performance Assessment of Sub-Daily and Daily Precipitation Estimates Derived from GPM and GSMaP Products over an Arid Environment. *Remote Sens.* **2019**, *11*, 2840. [CrossRef]
10. Huffman, G.J.; Stocker, E.F.; Bolvin, D.T.; Nelkin, E.J.; Tan, J. GPM IMERG Final Precipitation L3 1 month 0.1 degree x 0.1 degree V06 (GPM_3IMERGM). 2019. [CrossRef]
11. Pradhan, R.K.; Markonis, Y.; Godoy, M.R.V.; Villalba-Pradas, A.; Andreadis, K.M.; Nikolopoulos, E.I.; Papalexiou, S.M.; Rahim, A.; Tapiador, F.J.; Hanel, M. Review of GPM IMERG performance: A global perspective. *Remote Sens. Environ.* **2022**, *268*, 112754. [CrossRef]
12. Ghaderpour, E.; Mazzanti, P.; Scarascia Mugnozza, G.; Bozzano, F. Coherency and phase delay analyses between land cover and climate across Italy via the least-squares wavelet software. *Int. J. Appl. Earth Obs. Geoinf.* **2023**, *118*, 103241. [CrossRef]
13. Shawky, M.; Ahmed, M.R.; Ghaderpour, E.; Gupta, A.; Dewan, A.; Hassan, Q.K. Remote sensing-derived land surface temperature trends over South Asia. *Ecol. Inform.* **2023**, *74*, 101969. [CrossRef]
14. Zaghloul, M.S.; Ghaderpour, E.; Dastour, H.; Farjad, B.; Gupta, A.; Eum, H.; Achari, G.; Hassan, Q.K. Long Term Trend Analysis of River Flow and Climate in Northern Canada. *Hydrology* **2022**, *9*, 197. [CrossRef]

15. McGill, R.; Tukey, J.; Larsen, W. Variations of box plots. *Am. Stat.* **1978**, *32*, 12–16.
16. Ghaderpour, E. JUST: MATLAB and Python software for change detection and time series analysis. *GPS Solut.* **2021**, *25*, 85. [CrossRef]
17. Ghaderpour, E.; Vujadinovic, T.; Hassan, Q.A. Application of the Least-Squares Wavelet software in hydrology: Athabasca River Basin. *J. Hydrol. Reg. Stud.* **2021**, *36*, 100847. [CrossRef]
18. McKenzie, E.D. General exponential smoothing and the equivalent ARMA process. *J. Forecast.* **1984**, *3*, 333–344. [CrossRef]
19. Barba, P.; Rosado, B.; Ramírez-Zelaya, J.; Berrocoso, M. Comparative Analysis of Statistical and Analytical Techniques for the Study of GNSS Geodetic Time Series. *Eng. Proc.* **2021**, *5*, 21.
20. Jalalkamali, A.; Moradi, M.; Moradi, N. Application of several artificial intelligence models and ARIMAX model for forecasting drought using the Standardized Precipitation Index. *Int. J. Environ. Sci. Technol.* **2015**, *12*, 1201–1210. [CrossRef]
21. Manigandan, P.; Alam, M.S.; Alharthi, M.; Khan, U.; Alagirisamy, K.; Pachiyappan, D.; Rehman, A. Forecasting Natural Gas Production and Consumption in United States-Evidence from SARIMA and SARIMAX Models. *Energies* **2021**, *14*, 6021. [CrossRef]
22. Ghaderpour, E.; Vujadinovic, T. The Potential of the Least-Squares Spectral and Cross-Wavelet Analyses for Near-Real-Time Disturbance Detection within Unequally Spaced Satellite Image Time Series. *Remote Sens.* **2020**, *12*, 2446. [CrossRef]
23. Song, Q.; Chissom, B.S. Fuzzy time series and its models. *Fuzzy Sets Syst.* **1993**, *54*, 269–277. [CrossRef]
24. Zhang, Y.; Qu, H.; Wang, W.; Zhao, J. A novel fuzzy time series forecasting model based on multiple linear regression and time series clustering. *Math. Probl. Eng.* **2020**, *2020*, 9546792. [CrossRef]
25. Novák, V.; Perfilieva, I.; Dvorak, A. *Insight into Fuzzy Modeling*; John Wiley & Sons: Hoboken, NJ, USA, 2016.

engineering proceedings

MDPI

Proceeding Paper

Using Risk Terrain Modeling for the Risk Assessment of Explosive ATM Attacks †

Katharina Schwarz * and Kai Seidensticker

Criminological Research Department, State Office of Criminal Investigation North Rhine-Westphalia, 40221 Düsseldorf, Germany
* Correspondence: katharina01.schwarz@polizei.nrw.de
† Presented at the 9th International Conference on Time Series and Forecasting, Gran Canaria, Spain, 12–14 July 2023.

Abstract: In this article, we present the use of risk terrain modeling for the risk assessment of explosive ATM attacks in North Rhine-Westphalia, Germany. We give a brief overview of three methods used for this purpose: risk terrain modeling, recapture rate index, and time series analysis. The results show that by using these methods, police can gain a deeper understanding of the patterns and trends associated with explosive ATM attacks and better allocate their resources by focusing on higher-risk ATMs.

Keywords: risk terrain modeling; time series analysis; recapture rate index; explosive ATM attacks; crime forecasting

Citation: Schwarz, K.; Seidensticker, K. Using Risk Terrain Modeling for the Risk Assessment of Explosive ATM Attacks. *Eng. Proc.* **2023**, *39*, 24. https://doi.org/10.3390/engproc2023039024

Academic Editors: Ignacio Rojas, Hector Pomares, Luis Javier Herrera, Fernando Rojas and Olga Valenzuela

Published: 29 June 2023

1. Introduction

Automated teller machines (ATM) have become ubiquitous in modern society, providing easy access to cash and other financial services. However, they have also become a prime target for criminals, particularly those seeking to carry out explosive attacks. These attacks not only pose a serious threat to public safety, but also cause significant financial losses to the banking industry.

In recent years, explosive ATM attacks have become a major issue in Germany. In 2022, more than a third of all explosive ATM attacks recorded throughout Germany were committed in North Rhine-Westphalia. In a nationwide comparison, North Rhine-Westphalia is, thus, the most severely affected federal state. This can be attributed in particular to the large number of opportunities to commit the crime: more than 10,000 ATMs, the good opportunities to escape due to the dense, well-developed road network, and the proximity to the border with the Netherlands—the perpetrators are predominantly members of Dutch criminal groups. Predicting the occurrence of these crimes is crucial for law enforcement agencies to allocate resources effectively, avoid losses for banks and customers, and prevent threats to citizens from explosive ATM attacks. To address this problem, law enforcement agencies need to be able to predict where and when these crimes are likely to occur. In recent years, various prediction models have been developed to tackle this issue, but their effectiveness remains to be proven.

Traditionally, law enforcement agencies and security experts have relied on historical crime data and expert knowledge to identify high-risk areas for ATM explosive attacks. While these methods can be effective, they are often limited by the availability and quality of data, as well as the subjective nature of expert opinion. In recent years, the field of risk terrain modeling (RTM) has emerged as a promising approach for identifying high-risk areas for various types of crime, including ATM explosive attacks. RTM is a spatial analysis technique that uses a combination of environmental and socio-demographic variables to predict the likelihood of crime occurring in a given area [1]. RTM is based on the idea that certain environmental and socio-demographic factors can create a "terrain" that is more conducive to criminal activity. For example, areas with poor lighting, high foot traffic, and limited natural surveillance may be more attractive to criminals looking to carry out an ATM explosive attack. RTM can identify these high-risk areas by analyzing a variety of spatial and non-spatial data, such as street networks, land use patterns, and demographic characteristics [2].

Thus, risk is understood here as a metric "to tie different parts of the crime problem together as it offers a probabilistic interpretation of crime analysis that allows us to suggest that certain things are likely to happen" [3] (p. 11). Following the idea of "risk metrics", Caplan and Kennedy [4] (p. 7) conceptualized risk as a "continuous dynamic value that increases or decreases intensity and clusters or dissipates in different places over time." In this understanding, crime risk is primarily tied to geography.

RTM methods are able to assess the risk of future crime occurrence from a spatial perspective. This approach is part of the SKALA (System for Crime Analysis and Anticipation) approach of the State Office of Criminal Investigation of North Rhine-Westphalia (LKA NRW). The intention of SKALA is to understand the main use of crime analysis and forecast algorithms to investigate crime patterns at different spatial and temporal scales to support crime prevention [5]. The findings concerning the risk assessment of explosive ATM attacks using RTM are presented and discussed below.

The purpose of this paper is to examine the effectiveness of RTM for predicting the risk of explosive attacks on ATMs. We will begin by providing a brief overview of the methodology and principles underlying RTM. We will then describe the data sources and variables used in our analysis, as well as the statistical techniques employed. Finally, we will present our findings and discuss the implications of our study for law enforcement and security professionals.

Overall, the use of RTM for identifying high-risk areas for ATM explosive attacks has the potential to greatly enhance the effectiveness of law enforcement and security efforts. By providing a more accurate and data-driven approach to risk assessment, RTM can help prevent and mitigate the impact of these dangerous and costly criminal activities.

2. Explosive ATM Attacks in North Rhine-Westphalia

In recent years, North Rhine-Westphalia (NRW) has become the most severely affected federal state in Germany in terms of explosive ATM attacks. In 2022, 496 explosive ATM attacks have been recorded by the police in Germany (Figure 1). The number of cases has, thus, reached its highest level to date. Over a third of all recorded explosive ATM attacks were committed in NRW (Figure 2). The amounts of loot and property damage resulting from these crimes are in the tens of millions of euros each year.

How can this high rate of explosive ATM attacks in a single federal state be theoretically explained? Environmental criminology provides two interrelated theories that explain why crime occurs in the places and at the times it does. Assuming the routine activity theory, crimes only occur when motivated offenders, suitable targets, and the absence of capable guardians against a violation converge in space and time [6]. The crime pattern theory [7] provides an explanation of where these convergences occur. Starting with the individual offender, the crime pattern theory postulates that the offender moves along fairly predictable paths between his nodes of routine activity, so-called awareness space. Crime can, therefore, only occur where the offender's awareness space intersects with opportunities for crime. In practice, to approximate the individual (and usually unknown) awareness space of the individual offender, the most likely locations for crimes are estimated based environmental factors that describe crime generators, crime attractors, and crime-neutral areas. Crime generators are areas that attract many people regardless of crime, e.g., shopping malls or train stations. They generate crime by their mere presence, as they can easily arouse desire in potential offenders, open up opportunities to commit crimes, and offer opportunities for escape (arrival/departure). Crime attractors are defined as areas that potential offenders find particularly attractive. The transitions between crime generators and crime attractors are sometimes fluid. Indicators are conceivable that, with a slightly different connotation, could be assigned equally to both (e.g., public transport stops). Crime neutral areas are areas that neither attract many people nor are particularly attractive to potential offenders [8].

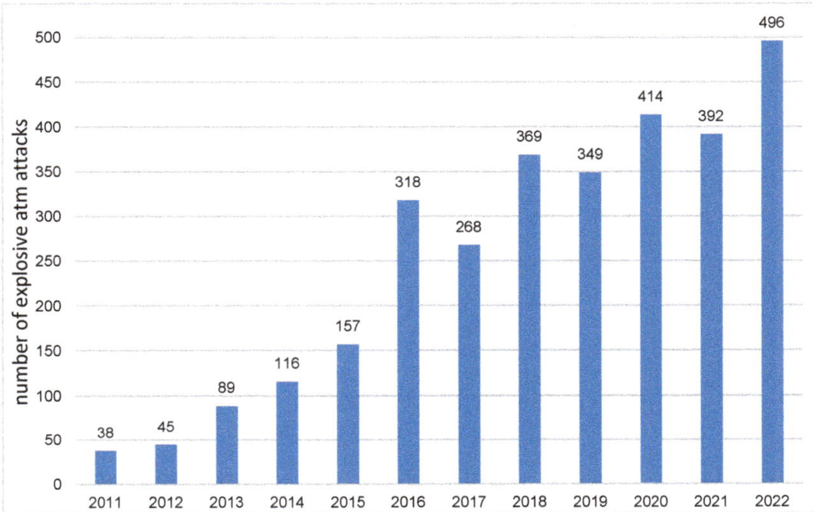

Figure 1. Number of explosive AMT attacks in Germany from 2011 to 2022.

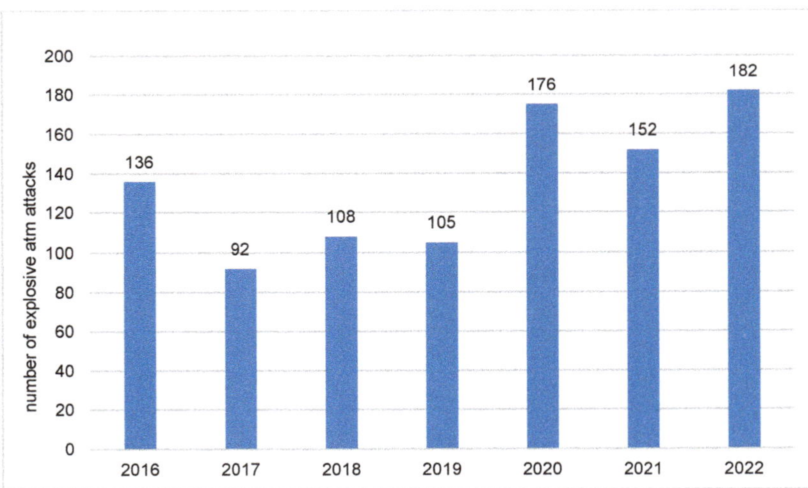

Figure 2. Number of explosive ATM attacks in NRW from 2016 to 2022.

Looking at the phenomenon of explosive ATM attacks, we see that the perpetrators are predominantly members of Dutch criminal groups; North Rhine-Westphalia borders the Netherlands [9]. With the dense, well-developed road network, and the preventive measures in the Netherlands, in particular the nightly closure or technical shutdown of ATMs, which have made the committing of such crimes in the Netherlands less attractive [10], these motivated offenders are more likely to act in areas "near" the border so that they are able to escape more quickly. Besides, the high rate of attacks can be attributed to the large number of suitable targets: there are more than 10,000 ATMs located in North Rhine-Westphalia, which have different installation locations and security measures. Due to the large number and varying security features of the ATMs, it is not possible to protect every ATM with police measures at all times. Besides, the attacks are often committed at night and in areas not heavily frequented, as there are fewer or no customers in the area of the ATMs, which can be regarded as informal guardians.

Given the framework of the Crime Pattern Theory, the occurrence of explosive ATM attacks is also closely linked to the spatial structure. It seems trivial that blasts only occur where ATMs are also installed. In context of explosive ATM attacks, ATM locations can, therefore, be understood as crime attractors themselves. In addition, North Rhine-Westphalia is characterized by its polycentric structure, i.e., it is shaped by many large cities that are connected by a dense road network. Many highways and expressways connect the cities with each other and also North Rhine-Westphalia with the Netherlands. ATMs are installed not only inside bank branches, but also on exterior walls, in shopping malls, at gas stations, and at train stations, and can be found both in inner cities and in rural areas. This spatial structure offers perpetrators diverse opportunities to expand their awareness space and locate potential targets. The road network contributes to the fact that the majority of ATM locations are a maximum of 10 min from the nearest highway access and can be reached within half an hour to an hour from the Netherlands. Looking at the distribution of explosive ATM attacks, this suggests that not all ATMs (as crime attractors within an awareness space) are equally at risk, but rather, certain site factors may contribute to a higher or lower risk of detonation.

3. Data and Methods

3.1. Methodological Approach

The risk assessment procedure comprises different steps. On the basis of scientific literature and discussions with internal and external police experts, a total of 20 hypotheses were formulated as to which factors may be decisive for perpetrators when selecting a crime object. A distinction was made between spatial and temporal factors (e.g., escape possibilities at the ATM location, informal social control at the ATM location), factors relating to characteristics of the ATMs and their locations (e.g., security measures at the building/at the ATM), police-related factors (e.g., distance to the nearest police station), and offender-related factors (e.g., degree of professionalization). In the next step, the established hypotheses were operationalized and tested.

3.2. Data and Preprocessing

For the risk assessment, data were collected from the police authorities of North Rhine-Westphalia and from the financial institutions. Police data consist of information on explosive ATM attacks, such as time and location of the attack, locality, building type, ATM equipment, and security measures. Not only the data on explosive ATM attacks be included in the analysis for a risk assessment, but also the risk assessment must be based on all ATMs; thus, further data were collected from the financial institutions in order to represent as many ATMs as possible in the dataset. In addition, sociostructural data, e.g., on building structure, population structure, transport accessibility, mobility, and pedestrian frequency, were included and some spatial distance variables were computed, e.g., the distance of every ATM to the nearest highway, to the nearest police station, and to the common border with the Netherlands.

Since the distance at which a particular risk factor affects crime is not generally known a priori, independent variables are calculated at several different distance bandwidths. Here, four bandwidth distances are used for each factor, such as distance the state borders or distance to the next highway junction. For each factor, content-specific distance categories were selected. Both proximity and density measures are converted to 0/1 indicators, with proximity effects calculated as 1 if the crime generator or attractor is within the particular distance bandwidth. This means that the continuous estimates were turned into dummy variables, e.g., if there is a highway access within 100 m, the dummy variable is coded 1; otherwise, it is coded 0. Since we also use the kernel density estimation (KDE), the respective dummy variable is coded 1, if the z-score of the KDE is above 2 [11]. Thus, in the end there is a collection of several dichotomous dummy variables created from the distance variables.

These separate datasets were then processed and merged. Based on the dataset generated in this way, the hypotheses were tested. Where hypotheses were empirically confirmed, the associated indicators were considered in the modeling.

3.3. Time Series Analysis

Time series analysis is a valuable tool for identifying patterns in temporal changes in crime data, which is critical for crime prediction and prevention. To achieve this, various methods, such as time series decomposition and auto-regressive integrated moving average (ARIMA) modeling, can be used [12].

ARIMA modeling is a powerful method for time series analysis and short-term forecasting and has been successfully applied in many fields including economics, marketing, industry production, and social issues, but also in crime forecasting [5]. In particular, ARIMA modeling has been used for forecasting property crime [12]. In this study, the ARIMA model was used to predict one year in advance from the observations of explosive ATM attacks in NRW.

The ARIMA modeling was performed according to the study by Seidensticker and Schwarz [5] and provides a forecast for one year for this offence with monthly aggregated data.

3.4. Recapture Rate Index

The recapture rate index (RRI) is a useful measure for assessing prediction precision. In this study, we utilized the RRI to examine spatial-temporal variations in police authority areas in NRW. The RRI is a ratio of recaptured crime hotspots during the predicted time period (period 2) to those in the historical time period (period 1), standardized for the change in crime density in the study region between the two periods [13].

Although the RRI is typically used to compare hotspot crime densities, we adapted it for our study by using police agency areas instead. Furthermore, we compared two past time periods to demonstrate the change in attack rates over time relative to space, which we then implemented in our risk model.

Interpreting the RRI values is straightforward: a value below 1.0 indicates a decrease in crime prediction from one period to another, while a value above 1.0 indicates an increase [14]. The RRI, therefore, provides valuable insights into temporal changes in crime patterns and can be a powerful tool for developing proactive law enforcement strategies.

3.5. Risk Terrain Modeling

In this paper, we propose using risk terrain modeling (RTM) to predict ATM attacks in Germany as a data-driven method for spatial risk assessment of crime. RTM utilizes crime and environmental data to identify areas at a higher risk for crime occurrence [1,4]. This raster-based approach examines the relationship between place-based factors, such as crime generators and attractors, and crime to generate a spatial risk profile and predict crime risk based on geographic attributes, rather than merely extrapolating past crime history [1]. Therefore, it is a genuinely predictive model [15]. In this study, we transferred the raster-based approach to the point level to evaluate each ATM separately.

RTM predicts the counts of crime aggregated to each ATM using count-based regression models, providing a consistent operationalization of the effects of various risk factors on crime. The effects of various crime generators are reduced to proximity effects, density effects, or both in RTM [16]. An estimated risk map is produced by RTM identifying potential causal factors which are contributing to the risk on crime, but also the potential areas of crime displacement. These are identified by analyzing similar criminogenic environmental characteristics. This model improves typical hotspot maps by not only predicting areas of future high risk but also identifying potential causal factors [2,16].

To create a better understandable risk assessment model, we applied a variable selection technique in RTM, which picks variables with a positive coefficient on the crime outcome and takes regularization into account [4,17]. A restricted regression strategy is used to select the best-fitting model with the lowest Bayesian information criterion (BIC) score [4,16].

In addition to the spatial approach, temporal factors were included, such as the recapture rate index (RRI) and auto-regressive integrated moving average (ARIMA) methods to account for the temporal variability of explosive ATM attacks.

The risk calculation results in a value between 0 and 1, with 0 representing no risk and 1 representing the highest possible risk. The risk of explosive ATM attacks is classified into three classes (class 1: high risk; class 2: medium risk; class 3: moderate risk) to assist in police force deployment and action planning. Based on these classifications, we prioritize ATMs with the highest risk of explosive attacks, on which police actions such as prevention measures can be planned. ATMs are assigned to the highest risk class with 5% of the highest risk. The lowest risk class is assigned to 60% of the ATMs with the lowest risk. The remaining 35% belong to the medium risk class. ATMs with a high number of missing data were not assigned a risk class.

4. Results and Discussion

4.1. Time Series Analysis

The results of the monthly aggregated offense time series ARIMA modeling demonstrate the usefulness of crime time series analysis. The ARIMA model fitting captures the fluctuating number of cases over the year and the continuous increasing trend (Figure 3). As there is no specific seasonality, as with other property crimes like residential burglary, ARIMA modeling should not represent seasonality. However, the low number of cases during the summer months in 2022, which was not previously evident, led to an overestimation of the numbers. Despite this, the predicted values fall within the confidence interval, making them suitable for further action planning.

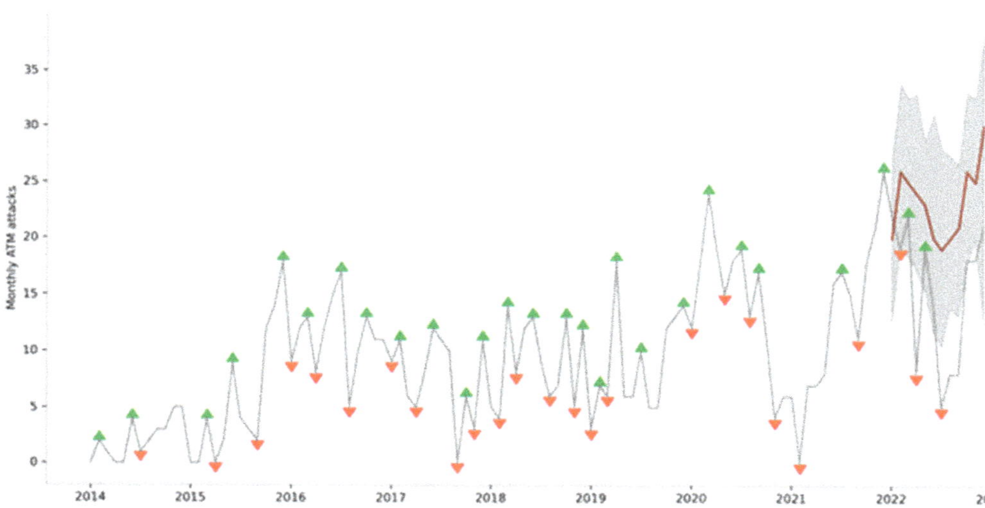

Figure 3. Time series of ATM attacks in NRW since 2014 (grey line) with the ARIMA modeled prediction for 2022 (red line) with confidence interval (grey area). Red and green triangles represent the local minima and maxima, respectively.

These findings suggest that ARIMA modeling is an effective tool for understanding and predicting temporal changes in crime patterns. It can provide valuable insights for proactive law enforcement strategies and resource allocation. However, caution should be taken in interpreting the results, especially when there are sudden changes in the data that may affect the accuracy of the predictions. In such cases, additional data sources and analytical techniques should be applied to improve the accuracy of such temporal predictions.

4.2. Recapture Rate Index

The importance of analyzing the temporal and spatial characteristics of explosive ATM attacks is apparent from the results of the adjusted RRI calculation (Figure 4). A RRI value above 1 indicates an increasing tendency, and the results reveal that large parts of NRW are experiencing a strong increase in explosive ATM attacks. In particular, the western regions near the Netherlands border have RRI values above 3, indicating a substantial increase in 2022 compared to 2021. Overall, half of the police authority areas exhibit increasing numbers of ATM attacks.

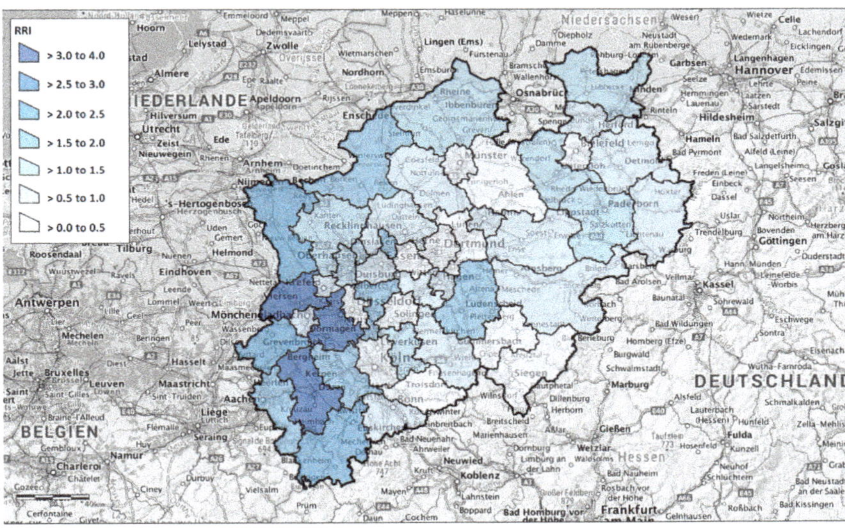

Figure 4. Adjusted RRI for each police authority in NRW based on ATM attacks from 2021 and 2022.

These findings were also integrated into the RTM analysis to account for temporal and spatial changes. The risk assessment considers the most recent developments by incorporating the past 365 days from the valuation date. This ensures that the latest trends and changes in explosive ATM attacks are taken into account, providing a more accurate risk prediction.

The results of this study emphasize the importance of continuous monitoring of crime patterns and trends, especially for offenses that are rapidly increasing. By integrating both temporal and spatial factors, law enforcement agencies can develop proactive strategies to prevent and reduce the incidence of such crimes.

4.3. Risk Terrain Modeling

The results of this study, presented in Figure 5, demonstrate that RTM is able to assess the risk of each ATM to be attacked and can be leveraged to improve resource allocation and prevention strategies. By identifying patterns in time series data, RTM methods can determine areas where explosive ATM attacks are more likely to occur. This enables law enforcement agencies to allocate resources effectively, prevent possible threats to citizens from explosive ATM attacks, and avoid losses for banks and customers.

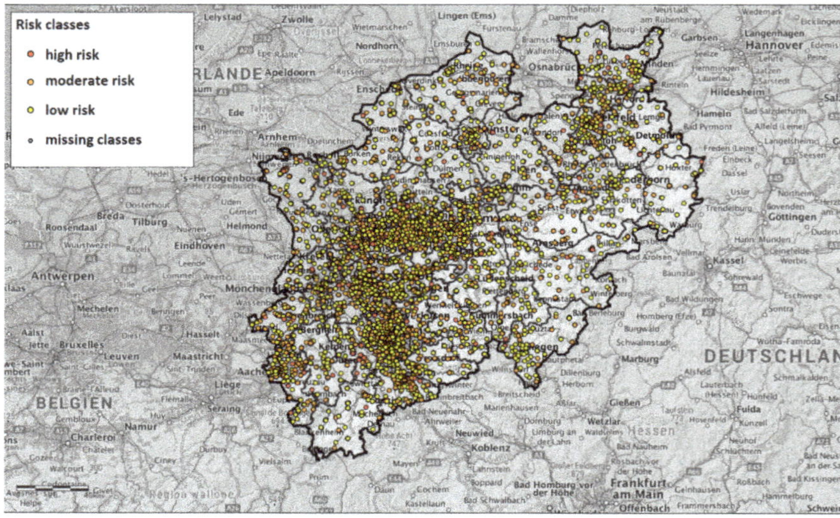

Figure 5. Risk classes of each ATM in NRW (in this figure, randomized classes have been assigned for security reasons).

The model accuracy was evaluated using retrograde analyses, which involved determining the risk for an earlier valuation date and evaluating it based on future transactions on the reporting date. The risk assessment for ATMs with at least an increased risk (risk classes 1 and 2) showed a model accuracy of 87%. However, if only ATMs with the highest risk (risk class 1) are considered, the model accuracy drops to 41%.

These findings suggest that RTM methods can be a powerful tool for predicting and preventing explosive ATM attacks in Germany. However, it is crucial to consider the limitations of the model and the accuracy of the risk assessments. Future studies should focus on improving the model accuracy and incorporating additional factors that may affect the incidence of explosive ATM attacks, such as socioeconomic factors and environmental conditions.

5. Potential and Limitations

The potential benefits of using RTM methods to predict explosive ATM attacks are clear. By using RTM, police can better allocate their resources by focusing their efforts on ATM of increased risk. Moreover, the use of RTM provides the police with a deeper understanding of the patterns and trends associated with explosive ATM attacks and empowers them to develop more effective prevention strategies. However, there are also limitations to the approach that need to be considered. For example, RTM methods may be limited by the quality and availability of data and may not capture all relevant factors that influence the occurrence of explosive ATM attacks. Moreover, such models can easily generate maps showing most of the populated area of a jurisdiction as a hotspot, because many geographic features can be mere proxies for populated areas (Perry et al., 2013). Thus, it is questionable whether a causal relationship actually exists between the particular risk factors and the crime.

In conclusion, our results demonstrate the potential of RTM methods to predict explosive ATM attacks in Germany and provide valuable insights for law enforcement agencies. By combining RTM methods with the SKALA approach, we can provide accurate predictions of explosive ATM attacks and support effective resource allocation and prevention strategies.

Overall, crime forecasting methods offer a wide range of possible applications that enable criminal expertise to be enriched by scientific findings, if these are understood as complementary tools and reflected upon in light of their respective limitations.

Author Contributions: Conceptualization, K.S. (Katharina Schwarz) and K.S. (Kai Seidensticker); methodology, K.S. (Katharina Schwarz); software, K.S. (Katharina Schwarz); validation, K.S. (Katharina Schwarz); formal analysis, K.S. (Katharina Schwarz) and K.S. (Kai Seidensticker); investigation, K.S. (Katharina Schwarz) and K.S. (Kai Seidensticker); resources, K.S. (Katharina Schwarz) and K.S. (Kai Seidensticker); data curation, K.S. (Katharina Schwarz); writing—original draft preparation, K.S. (Katharina Schwarz); writing—review and editing, K.S. (Kai Seidensticker); visualization, K.S. (Katharina Schwarz); supervision, K.S. (Kai Seidensticker); project administration, K.S. (Kai Seidensticker) and K.S. (Katharina Schwarz); funding acquisition, K.S. (Kai Seidensticker). All authors have read and agreed to the published version of the manuscript.

Funding: This research received no external funding.

Institutional Review Board Statement: Not applicable.

Informed Consent Statement: Not applicable.

Data Availability Statement: Data sharing not applicable.

Conflicts of Interest: The authors declare no conflict of interest.

References

1. Perry, W.; McInnis, B.; Price, C.; Smith, S.; Hollywood, J. *Predictive Policing. The Role of Crime Forecasting in Law Enforcement Operations*; RAND Corporation: Santa Monica, CA, USA, 2013.
2. Caplan, J.M.; Kennedy, L.W. *Risk Terrain Modeling: Crime Prediction and Risk Reduction*; University of California Press: Berkeley, CA, USA, 2016.
3. Kennedy, L.W.; Gibbs Van Brunschot, E. *The Risk in Crime*; Rowman & Littlefield: Lanham, MD, USA, 2009; p. 11.
4. Caplan, J.M.; Kennedy, L.W.; Miller, J. Risk terrain modeling: Brokering criminological theory and GIS methods for crime forecasting. *Justice Q.* **2010**, *28*, 360–381. [CrossRef]
5. Seidensticker, K.; Schwarz, K. Using Forecast Methods on Crime Data: The SKALA Approach of the State Office for Criminal Investigation of North Rhine-Westphalia. *Eng. Proc.* **2022**, *18*, 39.
6. Cohen, L.E.; Felson, M. Social change and crime rate trends: A routine activity approach. *Am. Sociol. Rev.* **1979**, *44*, 588–608. [CrossRef]
7. Brantingham, P.L.; Brantingham, P.J. Nodes, paths and edges: Considerations on the complexity of crime and the physical environment. *J. Environ. Psychol.* **1993**, *13*, 3–28. [CrossRef]
8. Brantingham, P.L.; Brantingham, P.J. The Theory of Target Search. In *The Oxford Handbook of Criminological Theory*; Cullin, F., Wilcox, P., Eds.; Oxford University Press: Oxford, UK, 2013; pp. 535–553.
9. Meyer, M.; Schwarz, K. Risikoanalyse zum Phänomen Geldautomatensprengungen. *Die Polizei* **2023**. *accepted*.
10. Bundeskriminalamt. *Angriffe auf Geldautomaten. Bundeslagebild 2021*; BKA: Wiesbaden, Germany, 2022.
11. Wheeler, A.P.; Steenbeek, W. Mapping the risk terrain for crime using machine learning. *J. Quant. Criminol.* **2021**, *37*, 445–480. [CrossRef]
12. Chen, P.; Yuan, H.; Shu, X. Forecasting crime using the ARIMA model. In Proceedings of the Fifth International Conference on Fuzzy Systems and Knowledge Discover, Jinan, China, 18–20 October 2008; IEEE Computer Society: Washington, DC, USA, 2008; pp. 627–630. [CrossRef]
13. Van Patten, I.T.; McKeldin-Coner, J.; Cox, D. A microspatial analysis of robbery: Prospective hot spotting in a small city. *Crime Mapp. A J. Res. Pract.* **2009**, *1*, 7–32.
14. Drawve, G. A metric comparison of predictive hot spot techniques and RTM. *Justice Q.* **2016**, *33*, 369–439. [CrossRef]
15. Belina, B.; Germes, M. Kriminalitätskartierung als Methode der kritischen Kriminologie? *Krim. J.* **2016**, *3*, 24–46.
16. Yoo, Y.; Wheeler, A.P. Using risk terrain modeling to predict homeless related crime in Los Angeles, California. *Appl. Geogr.* **2019**, *109*, 102039. [CrossRef]
17. Kennedy, L.W.; Caplan, J.M.; Piza, E.L.; Buccine-Schraeder, H. Vulnerability and exposure to crime: Applying risk terrain modeling to the study of assault in Chicago. *Appl. Spat. Anal. Policy* **2016**, *9*, 529–548. [CrossRef]

engineering proceedings

MDPI

Proceeding Paper

Analysis of GNSS Time Series Recorded on South Shetland Island and Antarctic Peninsula during the Geodynamic Activity in 2020 of the Orca Underwater Volcano (Brandfield Sea Rift, Antarctica) †

Belén Rosado [1,*], Alejandro Pérez-Peña [1], Paola Barba [1], Javier Ramírez-Zelaya [1], Enrique Carmona [2], Rosa Martín [2], Vanessa Jiménez [2], Jorge Gárate [1], Amós de Gil [1] and Manuel Berrocoso [1]

1 Laboratorio de Astronomía, Geodesia y Cartografía, Departamento de Matemáticas, Facultad de Ciencias, Campus de Puerto Real, Universidad de Cádiz, 11510 Puerto Real, Spain; alejandro.perezpena@uca.es (A.P.-P.); paola.barba@uca.es (P.B.); javierantonio.ramirez@uca.es (J.R.-Z.); jorge.garate@uca.es (J.G.); amos.degil@uca.es (A.d.G.); manuel.berrocoso@uca.es (M.B.)

2 Instituto Andaluz de Geofísica, Universidad de Granada, 18071 Granada, Spain; ecarmona@ugr.es (E.C.); rosa@ugr.es (R.M.); vanessa.jimenez@uca.es (V.J.)

* Correspondence: belen.rosado@uca.es

† Presented at the 9th International Conference on Time Series and Forecasting, Gran Canaria, Spain, 12–14 July 2023.

Abstract: The region defined by the South Shetland Islands, Bransfield Sea, and Antarctic Peninsula has complex geodynamic activity resulting from the active subduction process of the Phoenix Plate under the Antarctic Plate. This subduction produces a rift of expansion along the Bransfield Sea basin between the South Shetland Islands and the Antarctic Peninsula. There is also a chain of submarine (Orca, Three Sisters, and Building A) and emerged (Deception and Pinguin) volcanoes. In 2020, there was intense seismic activity around the Orca volcano with earthquakes of up to 6.9 Mw. This paper presents displacement models of this seismic activity produced in the region. The geodetic time series of the GNSS stations located in the region were analyzed: UYBA at the Uruguayan Artigas Antarctic Base (King George Island) and PAL2 at the U.S. Palmer Antarctic Base (Anvers Island). These data were taken from the Nevada Geodetic Laboratory.

Keywords: GNSS system; geodetic time series; geodynamic; volcano–tectonic deformation

Citation: Rosado, B.; Pérez-Peña, A.; Barba, P.; Ramírez-Zelaya, J.; Carmona, E.; Martín, R.; Jiménez, V.; Gárate, J.; de Gil, A.; Berrocoso, M. Analysis of GNSS Time Series Recorded on South Shetland Island and Antarctic Peninsula during the Geodynamic Activity in 2020 of the Orca Underwater Volcano (Brandfield Sea Rift, Antarctica). *Eng. Proc.* **2023**, *39*, 25. https://doi.org/10.3390/engproc2023039025

Academic Editors: Ignacio Rojas, Hector Pomares, Luis Javier Herrera, Fernando Rojas and Olga Valenzuela

Published: 29 June 2023

1. Introduction

Due to innovations and exponential advances over the past three decades, GNSS–GPS systems have become important tools in geodesy and geophysics, pushing forward the science and precise measurement of the Earth's various active processes on land, water, and ice, and in the atmosphere. Nowadays, GPS systems are essential components of the newest generation of Earth science and natural hazard assessment capabilities for monitoring and understanding earthquakes, volcanoes, and tsunamis, among others.

From the processing of the GPS observations, geodetic time series with subcentrimetric accuracy are obtained. GPS geodetic time-series analysis allows the study of the geodynamic behavior of a specific terrestrial area. These time series define the temporal evolution of the geocentric or topocentric coordinates obtained from geodetic stations, which are linear or nonlinear depending on the tectonic or volcanic–tectonic characteristics, respectively, of a region. Linear series are easily modeled, but for the study of nonlinear series, it is necessary to apply filtering techniques that provide a more detailed analysis of their behavior.

This work focused on the analysis of the geodynamic activity that occurred in 2020 in the region between the South Shetland Islands and the Antarctic Peninsula, close to the

ORCA volcano. For this, the time series of the stations located in the region were analyzed. These stations were UYBA at the Uruguayan Artigas Antarctic Base (King George Island) and PAL2 at the U.S. Palmer Antarctic Base (Anvers Island). For a better understanding of the geodynamic process, the preseismic, coseismic, and postseismic velocities were obtained. The analysis was extended by adding geodesic stations located in the region. All time series were provided by the Nevada Geodetic Laboratory (NGL).

2. Geodynamic Frame

The South Shetland Islands, the Bransfield Sea, and the Antarctic Peninsula define a geodynamically highly complex region. Two major tectonic plates converging in this region: the South American and Antarctic Plates. The boundary between both plates is even more complex due to the interaction of four minor tectonic plates: the Scotia, South Sandwich, Phoenix, and South Shetland plates; and two fracture zones, Shackethlon and Hero Figure 1 [1].

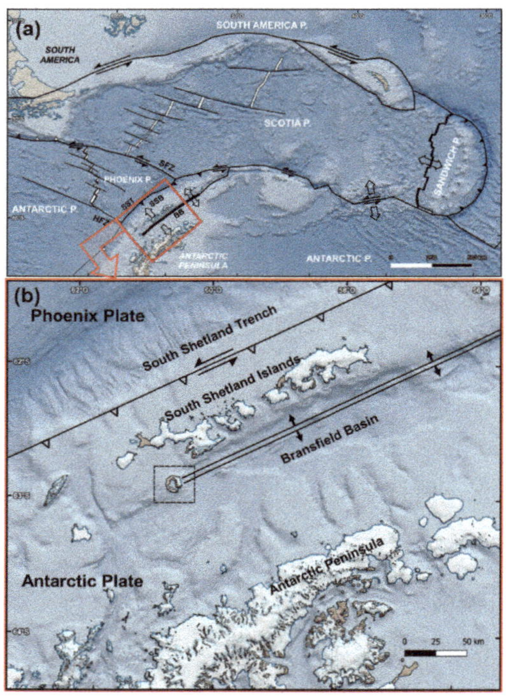

Figure 1. (**a**) Map of the Scotia microplate area and the most important plates boundaries. SST = South Shetland Trench; SSB = South Shetland Block; SFZ = Shackleton Fracture Zone; BB = Bransfield Basin; SOM = South Orkney Microcontinent. (**b**) Map of the Bransfield Strait and the South Shetland Islands showing the main faults in the region.

This region is a seismically active area, where more than 200 earthquakes of magnitude greater than 4 have occurred in the last 50 years (ISC catalogue). Most of these events have been shallow (<10 km) and, in some cases, have taken place at intermediate depths, which indicates the presence of active subduction of the Drake plate under the South Shetland continental block [2]. Seismic activity in the Bransfield Strait usually appears in the form of seismic swarms [3], which are usually located in the vicinity of volcanic edifices.

The largest recent series in Bransfield Strait took place in the Orca volcano area from from August 2020 to November 2021. The Orca volcano is a seafloor caldera shield volcano with a 900 m bathymetric height offshore the coast of King George Island. The epicentral locations (Figure 2) cluster in the region NE of the Orca volcano (USGS catalogue). The series has originated more than 80,000 earthquakes between King George Island and the Orca submarine volcano with maximum depths of up to approx. 10 km [4]. It started on 28 August 2020 and lasted approximately one year (September 2021). The greatest number of events in the series occurred in the first few months [5]. During this swarm, two cumulative seismic moments in the Bransfield Strait occurred: the first one in November 2020, when the greatest number of events with M > 4 occurred, along with the first largest earthquake (5.9 Mw) located near the Orca volcano; the second one in January 2021, with the biggest earthquake of the swarm (6.9 Mw), located more easterly, outside the Orca region (Figure 3). Most of moment tensor solutions were strike-slip and normal faults with a predominantly NWSE T-axis azimuth (GCMT catalogue, [4]).

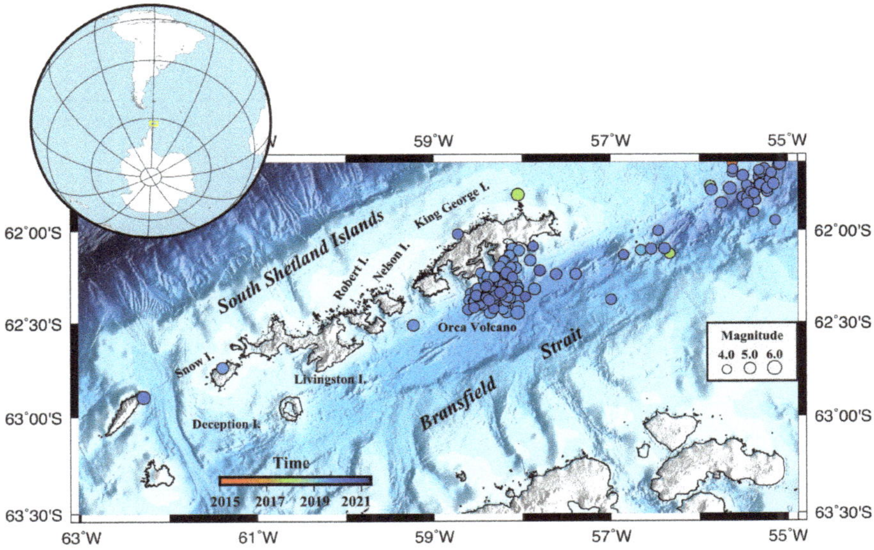

Figure 2. Epicentral map of seismic activity in the South Shetland Islands in the period 2015–2022 for earthquakes with magnitude greater than 4.0 (USGS catalogue).

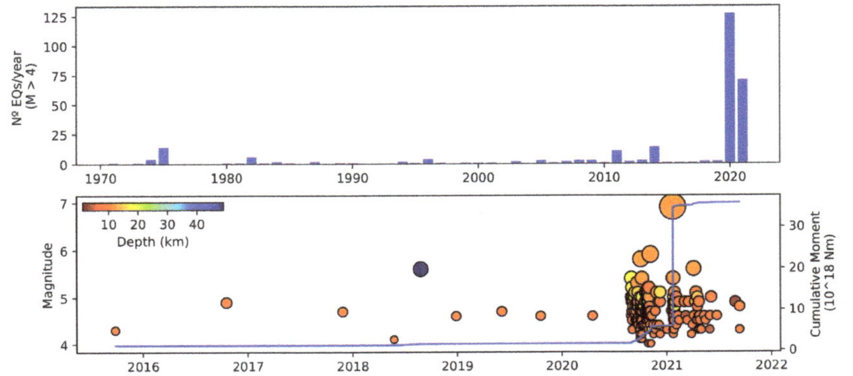

Figure 3. Histogram of number of events per year since 1971 (**top**). Histogram of number of events since 2015 by year, magnitude, and depth; plot of cumulative moment (**below**).

3. Data and Methods

3.1. Geodetic Stations and Data Processing

The time series analyzed in this study were from the geodetic stations provided by the Nevada Geodetic Laboratory (NGL). NGL collects and processes geodetic-quality GPS observations at more than 17,000 stations worldwide from many regional and commercial networks in addition to the commonly used International GNSS Service (IGS) network [6], (Figure 4). The system produces and makes publicly available data products such as position coordinates (latitude, longitude, and height) for geodetic-quality GPS stations around the globe with various data intervals and reference frames that might be useful to users [7]. NGL routinely processes the observations by using GipsyX version 1.0 software [8], released by the Jet Propulsion Laboratory (JPL), which has precise point positioning [9]. JPL's Repro3 final GPS orbits and clocks are also used. More details on the data-processing strategy are given in http://geodesy.unr.edu/gps/ngl.acn.txt. Data accessed on 1 June 2023.

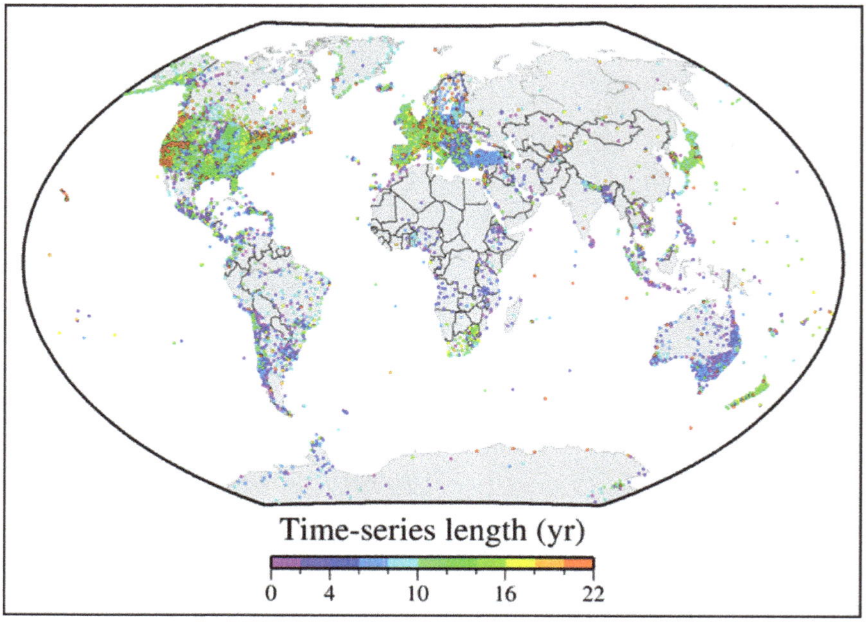

Figure 4. Geodetic GPS stations processed by the Nevada Geodetic Laboratory. Figure taken from [7].

NGL also routinely updates station velocities in the global reference frame, which can be used to capture the rates of deformation of the Earth's surface [7]. These velocities are robustly estimated using the Median Interannual Difference Adjusted for Skewness (MIDAS). This technique is explained in the next section.

3.2. Methodology

The methodology is summarized in Figure 5. First, the topocentric time series of the UYBA and PAL2 stations were obtain from NGL. A descriptive analysis of these time series was carried out, adding the seismic events that occurred from 2015 to 2023. Thus, an initial visualization of the data was carried out, correlating the jumps in the series with the seismic events.

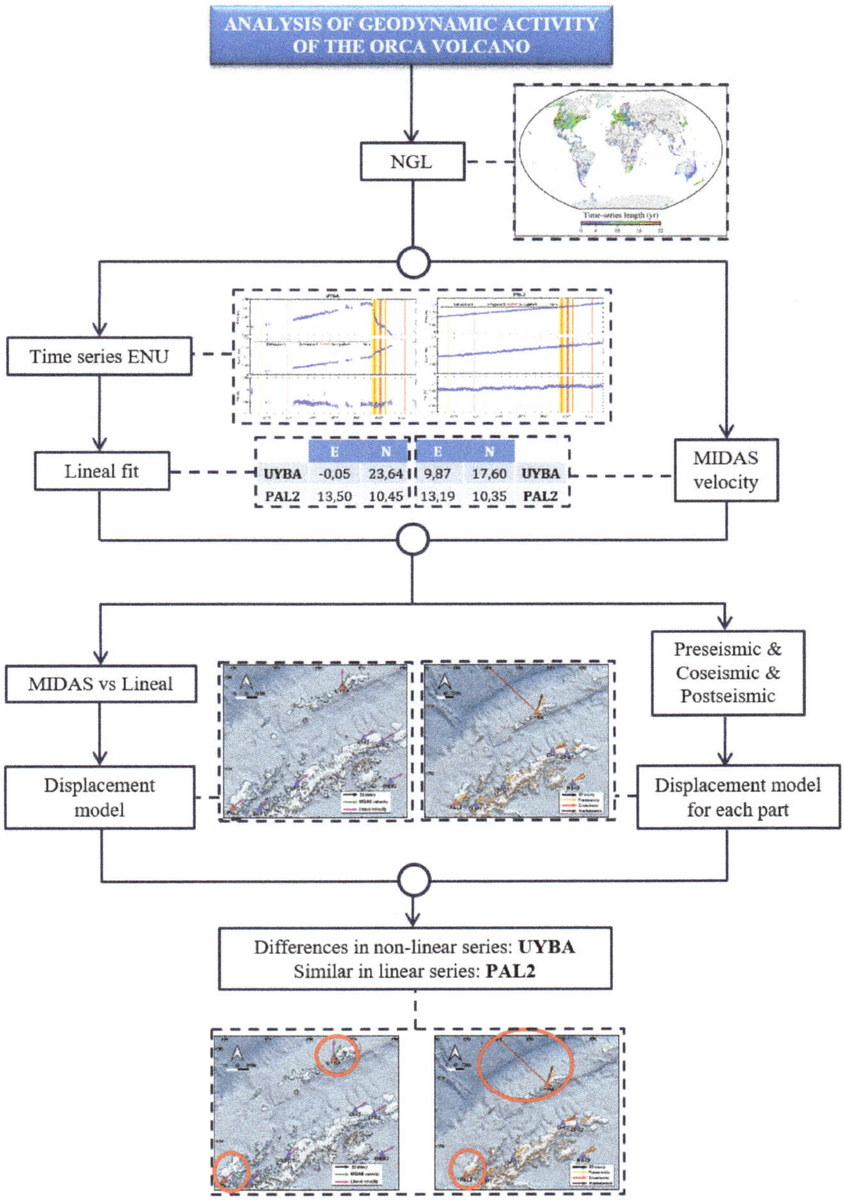

Figure 5. Scheme of the methodology.

Linear series are easily modeled, but for the study of nonlinear series, it is necessary to apply techniques that provide a more detailed analysis of their behavior [10,11]. For this reason, to estimate the trend in the series, two techniques were applied: a linear fit and the MIDAS technique. The trend obtained from the linear fit was valid if the series presented a linear behavior (PAL2 time series). For nonlinear series (UYBA time series), the MIDAS estimator was applied.

MIDAS is a median-based GPS station velocity estimator that is insensitive to outliers, seasonality, step functions (abrupt changes) arising from earthquakes or equipment changes, and statistical data variability [12]. MIDAS also provides velocity error bars based on subsampling the data that prove to be realistic, for example, when comparing the estimated velocities of stations that are close to each other or are separated by a tectonically stable region. MIDAS rates were estimated for all stations with at least one year of data and sufficiently dense time series. In cases where a large earthquake (Mw > 6.9) had occurred near enough to the station, an exponential decay function was solved: $A(1 - e^{-(t-t_0)/\tau})H(t - t_0)$, where t_0 is the time of the earthquake, τ is a relaxation time, A is the amplitude of the decay, and H is the Heaviside step function. In these cases, the background trends were solved after the exponential terms have been removed to obtain a self-consistent model for the time series [12].

Subsequently, displacement models of the region were obtained for each estimated velocity. On the other hand, the velocity of the preseismic, coseismic, and postseismic phases were estimated; the corresponding displacement models were obtained. The idea of the seismic cycle was developed by Harry Fielding Reid to explain his observations of the San Francisco earthquake of 1906. The earthquake-related deformation cycle consists mainly of four phases (Figure 6): the preseismic (nucleation), interseismic (long periods between large earthquakes during which elastic strain accumulation occurs in the broad region), coseismic (brief period during which the accumulated strain is released during earthquakes), and postseismic (the period immediately after an earthquake) phases, which exhibits relatively higher rates of deformation wherein the material deforms in response to the sudden coseismic release of strain [13].

Following this methodology, differences were detected between the behavior of linear and nonlinear time series.

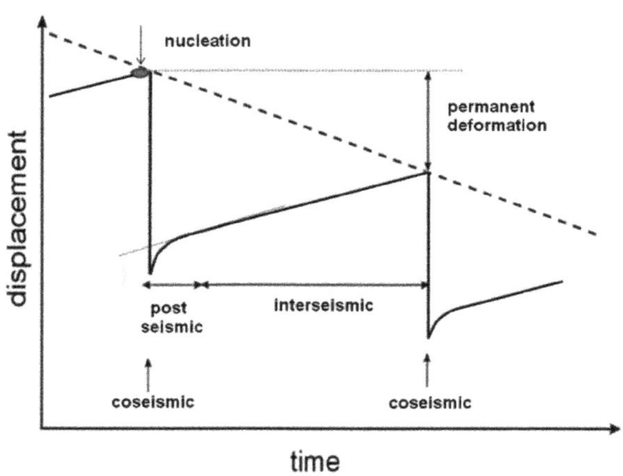

Figure 6. An idealized seismic cycle constituting four distinct phases: preseismic, interseismic, coseismic, and postseismic phases. Figure taken from [13].

4. Results

4.1. Time Series of the UYBA and PAL2 Stations

Figure 7 shows the topocentric time series of the UYBA (Figure 7a) and PAL2 (Figure 7b) stations from 2015 to 2023, represented in blue. These figures also include the seismic events that occurred in that period: earthquakes of magnitude between four and five in yellow, between five and six in orange, and earthquakes greater than six in red.

The PAL2 station is located at the U.S. Palmer Antarctic Base, on Anvers Island. These series are not affected by the geodynamic activity of the ORCA volcano. Therefore, its time series present a linear behavior in all components. The UYBA station is located at the Uruguayan Artigas Antarctic Base, on King George Island. This station is affected by the geodynamic process that occurred in 2020 at the ORCA volcano, so its time series are nonlinear.

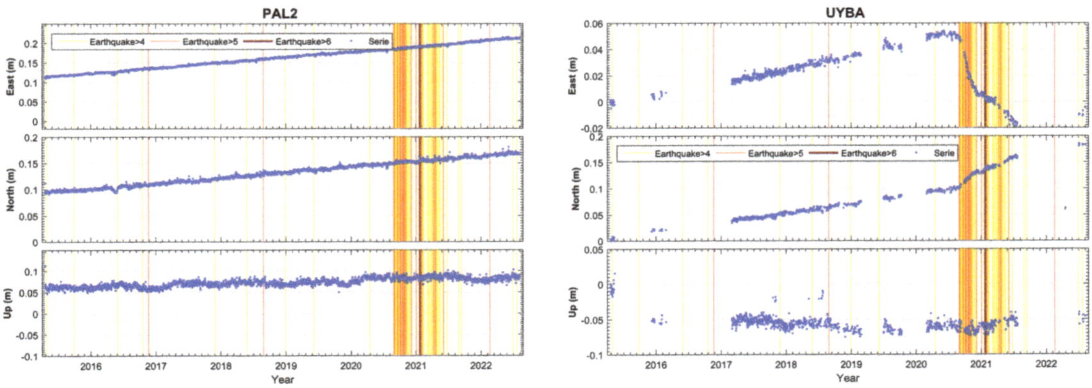

(**a**) (**b**)

Figure 7. Topocentric time series analyzed in this study. Seismic events were included: earthquakes of magnitude between 4 and 5 in yellow, between 5 and 6 in orange, and earthquakes greater than 6 in red. (**a**) Topocentric time series from PAL2 GNSS station located at the U.S. Palmer Antarctic Base, on Anvers Island. (**b**) Topocentric time series from UYBA GNSS station located at the Uruguayan Artigas Antarctic Base, on King George Island.

4.2. Evaluation of MIDAS Velocities versus Linear Velocities

Table 1 shows the velocities obtained from the linear fit and from the MIDAS estimator. For a better understanding, displacement models of the region were obtained for each estimated velocity (Figure 8).

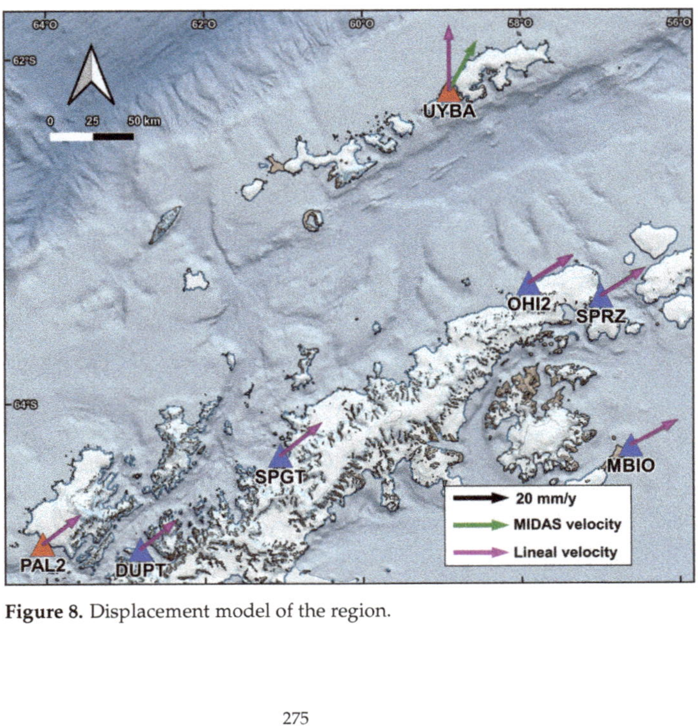

Figure 8. Displacement model of the region.

Table 1. Comparison of MIDAS velocities vs. linear fit, in mm/year.

	UYBA		PAL2	
Velocitt	**East**	**North**	**East**	**North**
MIDAS	9.87	17.60	13.19	10.35
Linear	−0.05	23.64	13.50	10.45

4.3. Preseismic, Coseismic, and Postseismic Displacement Models

The preseismic, coseismic, and postseismic phases were calculated for each component of the time series. The velocities of each phase are shown in Table 2. The different phases considered are shown in Figure 9, together with the velocity in millimeters per year (inset). The corresponding displacement model of each phase is shown in Figure 10.

Table 2. The preseismic, coseismic, and postseismic velocities in mm/year for UYBA and PAL2 stations.

	UYBA		PAL2	
Velocity	**East**	**North**	**East**	**North**
Preseismic	10.62	17.82	13.37	10.48
Coseismic	−64.78	61.38	13.92	9.55
Postseismic	7.79	24.70	14.94	11.09

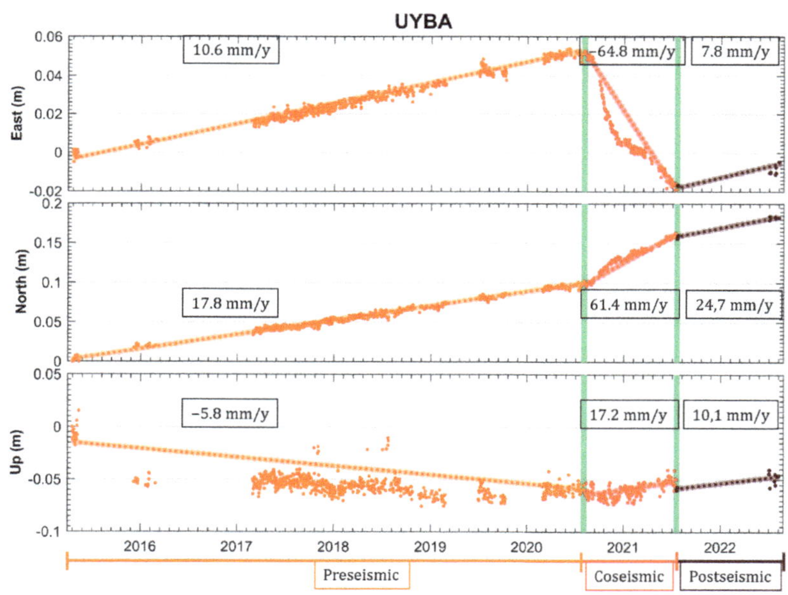

Figure 9. Topocentric time series of the UYBA station, separating the preseismic (orange), coseismic (red), and postseismic (brown) phases, and adding their linear fit. The corresponding velocities are shown in mm/year.

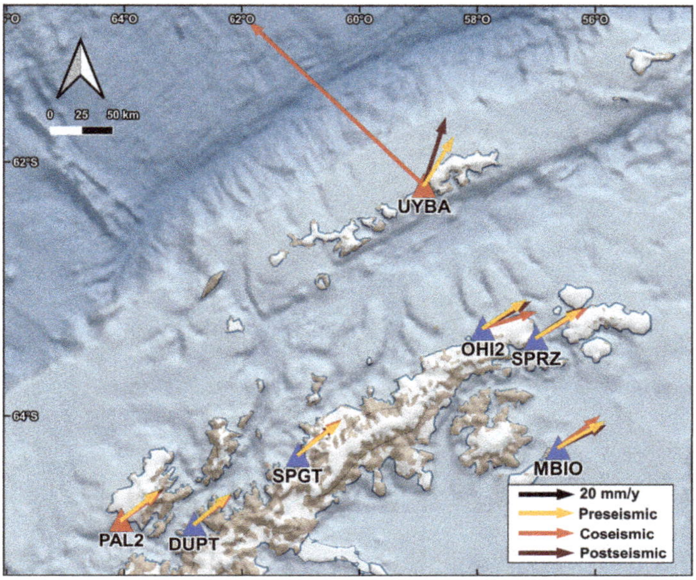

Figure 10. Preseismic, coseismic, and postseismic displacement model.

5. Conclusions

The region defined by the South Shetland Islands, the Bransfield Basin, and the Antarctic Peninsula is characterized by a highly complex tectonic environment. This region is also characterized by the presence of active volcanism, with emerged volcanoes, such as Deception, Penguin, and Bridgeman; and some underwater volcanic edifices, such as the Orca volcano. In 2020, intense seismic activity was recorded around the Orca volcano with earthquakes of up to 6.9 Mw.

This work focused on the analysis of the geodynamic activity that occurred in 2020 near the Orca volcano. The topocentric time series of the stations located in the region were analyzed: UYBA and PAL2, provided by the NGL. A descriptive analysis was carried out, adding the seismic events that occurred from 2015 to 2023. To estimate the trend in the series, two techniques were applied: a linear fit and MIDAS technique. The trend obtained from the linear fit was valid if the series presented a linear behavior (PAL2 time series). For nonlinear series (UYBA time series), the MIDAS estimator was applied. The displacement models of the region were obtained for each estimated velocity.

For a better understanding of the geodynamic process, the velocity of the preseismic, coseismic, and postseismic phases were estimated; the corresponding displacement models were obtained. These models showed that the preseismic and postseismic phases are similar in modulus and magnitude, while the coseismic phase presents different behavior.

Author Contributions: Conceptualization, B.R., M.B. and A.P.-P.; methodology, B.R., P.B., J.R.-Z. and M.B.; software, B.R. and P.B.; validation, M.B. and J.G.; formal analysis, B.R., J.R.-Z., V.J., J.G. and M.B.; investigation, B.R., V.J., R.M., E.C. and M.B.; resources, A.P.-P.; data curation, A.P.-P., A.d.G. and V.J.; writing—original draft preparation, B.R., M.B., A.P.-P., V.J., R.M. and E.C.; writing—review and editing, B.R and M.B.; visualization B.R. and J.R.-Z.; supervision, M.B. and J.G. All authors have read and agreed to the published version of the manuscript.

Funding: This research were possible thanks to the resources and funds of the astronomy, geodesy, and cartography laboratory of the University of Cadiz, directed by the principal researcher D. Manuel Berrocoso Domínguez.

Institutional Review Board Statement: Not applicable.

Informed Consent Statement: Not applicable.

Data Availability Statement: Data shown in this manuscript belong to the public catalog of the Nevada Geodetic Laboratory (NGL), http://geodesy.unr.edu/gps/ngl.acn.txt (accessed on 1 June 2023).

Acknowledgments: The contributions of both editor and of the reviewers are very much appreciated, which greatly helped to improve the manuscript.

Conflicts of Interest: The authors declare no conflict of interest.

References

1. Rosado, B.; Fernández-Ros, A.; Berrocoso, M.; Prates, G.; Gárate, J.; De Gil, A.; Geyer, A. Volcano-tectonic dynamics of Deception Island (Antarctica): 27 years of GPS observations (1991–2018). *J. Volcanol. Geotherm. Res.* **2019**, *381*, 57–82. [CrossRef]
2. Robertson-Maurice, S.D.; Wiens, D.A.; Shore, P.J.; Vera, E.; Dorman, L.M. Seismicity and tectonics of the South Shetland Islands and Bransfield Strait from a regional broadband seismograph deployment. *J. Geophys. Res.* **2003**, *108*, 2461. [CrossRef]
3. Ibáñez, J.M.; Carmona, E.; Almendros, J.; Saccorotti, G.; Del Pezzo, E.; Abril, M.; Ortiz, R. The 1998–1999 seismic series at Deception Island volcano, Antarctica. *J. Volcanol. Geotherm. Res.* **2003**, *128*, 65–88. [CrossRef]
4. Cesca, S.; Sugan, M.; Rudzinski, L.; Vajedian, S.; Niemz, P.; Plank, S.; Petersen, G.; Deng, Z.; Rivalta, E.; Vuan, A.; et al. Massive earthquake swarm driven by magmatic intrusion at the Bransfield Strait, Antarctica. *Commun. Earth Environ.* **2022**, *3*, 89. [CrossRef]
5. Olivet, J.L.; Bettucci, L.S.; Castro-Artola, O.A.; Castro, H.; Rodríguez, M.; Latorres, E. A seismic swarm at the Bransfield Rift, Antarctica. *J. S. Am. Earth Sci.* **2021**, *111*, 103412. [CrossRef]
6. Yuan, P.; Blewitt, G.; Kreemer, C.; Hammond, W.C.; Argus, D.; Yin, X.; Van Malderen, R.; Mayer, M.; Jiang, W.; Awange, J.; et al. An enhanced integrated water vapour dataset from more than 10000 global ground-based GPS stations in 2020. *Earth Syst. Sci. Data* **2023**, *15*, 723–743. [CrossRef]
7. Blewitt, G.; Hammond, W. C.; Kreemer, C. Harnessing the GPS data explosion for interdisciplinary science. *Eos* **2018**, *99*, 485. [CrossRef]
8. Bertiger, W.; Bar-Sever, Y.; Dorsey, A.; Haines, B.; Harvey, N.; Hemberger, D.; Heflin, M.; Lu, W.; Miller, M.; Moore, A.W.; et al. GipsyX/RTGx, a new tool set for space geodetic operations and research. *Adv. Space Res.* **2020**, *66*, 469–489. [CrossRef]
9. Zumberge, J.; Heflin, M.; Jefferson, D.; Watkins, M.; Webb, F. Precise point positioning for the efficient and robust analysis of GPS data from large networks. *J. Geophys. Res. Solid Earth* **1997**, *102*, 5005–5017. [CrossRef]
10. Barba, P.; Rosado, B.; Ramírez-Zelaya, J.; Berrocoso, M. Comparative Analysis of Sta-tistical and Analytical Techniques for the Study of GNSS Geodetic Time Series. *Eng. Proc.* **2021**, *5*, 21. [CrossRef]
11. Rosado, B.; Ramírez-Zelaya, J.; Barba, P.; De Gil, A.; Berrocoso, M. Comparative Analysis of Non-Linear GNSS Geodetic Time Series Filtering Techniques: El Hierro Volcanic Process (2010–2014). *Eng. Proc.* **2021**, *5*, 23. [CrossRef]
12. Blewitt, G.; Kreemer, C.; Hammond, W.C.; Gazeaux, J. MIDAS robust trend estimator for accurate GPS station velocities without step detection. *J. Geophys. Res. Solid Earth.* **2016**, *121*, 2054–2068. [CrossRef] [PubMed]
13. Reddy, C.D.; Arora, S.K.; Sunil, P.S.; Prajapati, S.K. Earthquake Related Deformation Cycle: Perspectives from 2004 Sumatra and 2010 Chile Mega-Earthquakes. *Disaster Adv.* **2011**, *4*, 13–21.

engineering proceedings

Proceeding Paper

BERT (Bidirectional Encoder Representations from Transformers) for Missing Data Imputation in Solar Irradiance Time Series †

Llinet Benavides Cesar *, Miguel-Ángel Manso-Callejo and Calimanut-Ionut Cira

Departamento de Ingeniería Topográfica y Cartográfica, E.T.S.I. en Topografía Geodesia y Cartografía, Universidad Politécnica de Madrid, C/Mercator 2, 28031 Madrid, Spain; m.manso@upm.es (M.-Á.M.-C.); ionut.cira@upm.es (C.-I.C.)

* Correspondence: llinet.bcesar@upm.es

† Presented at the 9th International Conference on Time Series and Forecasting, Gran Canaria, Spain, 12–14 July 2023.

Abstract: The availability of solar irradiance time series without missing data is an ideal scenario for researchers in the field. However, it is not achievable for a variety of reasons, such as measurement errors, sampling gaps, or other factors. Time series imputation methods can be a solution to the lack of data and, in this paper, we study the applicability of Bidirectional Encoder Representations from Transformers (BERT) as an irradiance time series imputation solution. In this regard, a BERT model was trained from scratch for the masked language modelling (MLM) task, and the quality of the imputation was evaluated according to the number of missing values and the position within the series. The experiments were conducted over a dataset of 165 stations, captured by meteorological stations distributed over the Spanish regions of Galicia, Castile, and León. In the evaluation process, an average coefficient of determination (R^2 score) of 0.89% was obtained, the maximum result being 0.95%.

Keywords: time series; imputation; solar irradiance; transformers; BERT

Citation: Cesar, L.B.; Manso-Callejo, M.-Á.; Cira, C.-I. BERT (Bidirectional Encoder Representations from Transformers) for Missing Data Imputation in Solar Irradiance Time Series. *Eng. Proc.* **2023**, *39*, 26. https://doi.org/10.3390/engproc2023039026

Academic Editors: Ignacio Rojas, Hector Pomares, Luis Javier Herrera, Fernando Rojas and Olga Valenzuela

Published: 30 June 2023

1. Introduction

Time series are sequences of data that are recorded at regular intervals of time, and are ordered according to the time in which they are recorded [1]. In many fields where it is important to have complete and accurate data for analysis and decision making, such as meteorology, healthcare, and solar energy, missing data imputation is a common challenge.

The imputation method is an important time series analysis, in which missing values in the series are filled in using available observed values [2]. In [2], Fang et al. defined nine types of the missing value imputation methods, based on the methodology used for filling the values. The methods included were: (1) deletion methods, (2) neighbour-based methods, (3) constraint-based methods, (4) regression-based methods, (5) statistical-based methods, (6) matrix factorization/based methods, (7) expectation-maximization-based methods, (8) multi-layer perceptron-based methods, and (9) methods based on deep learning (DL). For each of the established types, examples were given, although the study focused on imputation methods based on DL.

In this regard, imputation of solar energy time series [3] is a topic that has been explored with traditional statistical methods [4,5] and with more modern machine learning (ML) methods [6,7]. Demirhan et al. [5] evaluated 36 imputation methods for solar irradiance series with a dataset collected in Australia—the considered methods of imputation were variants of the methods listed hereafter, namely: (1) interpolation (such as linear, spline, or Stineman), (2) Kalman filters, (3) persistence, (4) weighted moving average, and (5) random sample. The authors defined sixteen experimental scenarios, and concluded that the most accurate methods for minutely and hourly series were the linear and the Stineman interpolations (based on a function that runs through a set of points in the *xy*-plane and returns the estimates of the calculated slopes); for daily or weekly series, the weighted moving average delivered the best result.

De-Paz-Centeno et al. [7] proposed a neural network to impute values for series with missing values in ranges of 30% to 70% of the total number of values, and recommended its use for scenarios with 50% of lost vales. The proposed neural network was a convolutional neural network following the encoder–decoder structure, and the experiments were carried out on a private and a public dataset, each containing two years of samples. The application of the proposed architecture resulted in coefficients of determination (R^2 score) ranging range between 0.81 and 0.98, considerably higher when compared to the other models evaluated.

Due to the sequential nature of the time series, it is possible to use models developed for natural language processing (NLP), such as transformers, in their processing. Transformers are based on attention mechanisms (they relate different positions of the same sequence to compute a representation of the sequence; also known as self-attention) and proved successful at solving sequential tasks, while easily handling long-range dependencies [8]. Transformer-based models have been applied and achieved good results in the imputation of time series [9,10].

Bidirectional Encoder Representations from Transformers (BERT) [11] is a DL model based on transformers that uses bidirectional self-attention by jointly conditioning the left and right context, being is one of the most popular DL-based linguistic models [12]. BERT can be pre-trained using two unsupervised tasks, the masked language model (MLM) and next sentence prediction (NSP). The MLM randomly masks a part of the tokens in the input sequence, and the goal is to predict the masked words based solely on its context. For the NSP task, BERT model is pre-trained on representations of pairs of texts to predict a sequence from the previous sequence. BERT has also been pre-trained for other areas of knowledge, such as vision [13,14] bioinformatics and computational biology [15–17], or geospatial representation learning based on a point of interest [18].

In this study, the BERT's performance in irradiance time series imputation will be assessed by training the model from scratch for the MLM task. To the authors' knowledge, this is the first time the model has been trained on irradiance data. We hypothesized that training directly on a specialized corpus and using a specialized vocabulary could lead to more adapted embeddings and, thus, help performance.

The main contributions of this paper are as follows.

(1) To the best of the authors' knowledge, the first BERT model trained from scratch with solar irradiance data is introduced;
(2) The implementation is evaluated for time series imputation in two scenarios, namely (1) the imputation of a single missing value at a specific position and (2) imputed a missing value where all values were missing after this position in the sequence.

The remaining part of the document is organised as follows. Section 2 describes the model, the data and the methodology used. Section 3 the experiments, and presents the analysis performed. The works ends with Section 4.

2. Methodology

2.1. Studied Model (BERT)

In NLP tasks, the first step in the processing pipeline is the tokenization (the process of dividing the text into small units, called tokens; tokens can be the words of a sentence or a sequence of characters).

BERT is a complex and advanced linguistic model, where the sentence is parsed as a token chain—each token in the chain is compared against all other tokens to gather information and learn the dynamics of the context. This information is stored in the form of embeddings (a numerical representation of the information). Figure 1 shows the representation of the BERT input for a sequence of irradiance values.

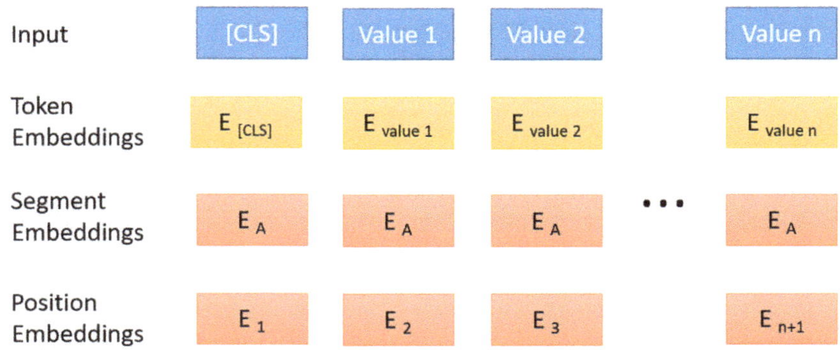

Figure 1. Representation of the BERT input for sequences of irradiances value—BERT's input embedding are the sum of token embedding, segmentation embedding and position embedding (sum of the column). Notes: (1) Token embedding is represented in yellow and represent the value in another dimension space. (2) Segmentation embedding indicate which sentence it belongs to A or B, while (3) the position embedding represents the position in the sequence. Abbreviations: CLS is used to identify the beginning of the sequence, E_A is used to code the segment embedding, and $E_1, E_2 \ldots E_{n+1}$ represent positions of the embedding.

In the field of NLP, the MLM task (mentioned in Section 1) enforces bi-directional learning from text by masking (or covering) a word in a sentence. In the process of tokenization, BERT uses special tokens, such as "[MASK]", to cover the word to be predicted. This way, BERT is forced to use neighboring words of the masked word to predict it. In the process, the model will generate the most likely substitution for any input containing one or more "[MASK]" tokens. For example, if BERT's input would be the following sequence of irradiance values, $(5, 56, 76, 89, 112, [\mathbf{MASK}], 145, 172, 189)$, a probable output would be $(45, 56, 76, 89, 112, \mathbf{123}, 145, 172, 189)$. The model assigned the masked token a value based on the learning. It is important to note that BERT uses other special tokens, such as "[CLS]", to identify the beginning of the sequence; "[UNK]", to signal an unknown word; "[PAD]", when sentences are not of the same length to fill in missing spaces; or "[SEP]", a sentence separator token used for input/output in the NSP task described in Section 1.

2.2. Data Description

In this study, the experiments were carried out on two solar irradiance datasets. The first one is composed of records from 112 meteorological stations in Galicia, stored in a tabular .csv file format, containing two years of information (from February 2017 to February 2019), with a time resolution of 10 min. The variables observed at the stations were temperature, atmospheric pressure, precipitation, wind speed, wind direction, and solar irradiance. The second dataset, CyL-GHI [19], contains information from the 53 stations located in Castile and León, and continuously covers 21 years (the period from January 2001 to December 2021), with a temporal resolution of 30 min. The spatial distribution of the stations is presented in Figure 2.

The variables observed at the stations were temperature, relative humidity, precipitation, wind speed, wind direction and solar irradiance. In addition, there is an identifier for each of the stations, as well as their geographical coordinates and height. However, only the solar irradiance time series of both datasets were used for the study. The data is grouped by stations and, using one time series for each station, implies that 165 GHI time series were used in this study.

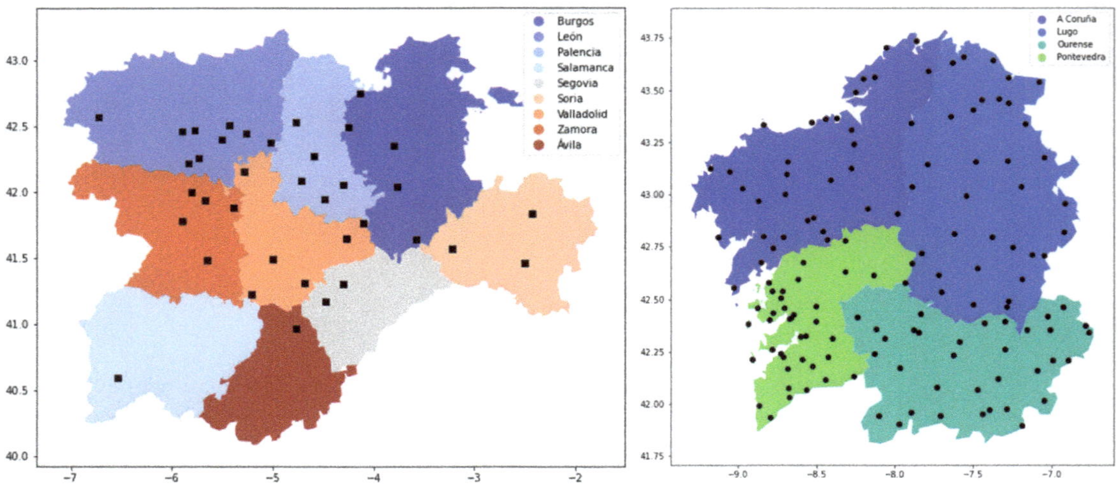

(**a**) (**b**)

Figure 2. Spatial distribution of stations capturing the data used in the experiments located in the Spanish regions of (**a**) Castile and León (the CyL-GHI dataset), and (**b**) Galicia.

2.3. Methodology

The BERT model was trained from scratch for the MLM task. Experiments were conducted to evaluate two scenarios: (1) Scenario 1, where a single missing value was imputed at a specific position in the sequence; and (2) Scenario 2, imputing a missing value where all values were missing after a position in the sequence.

In our study, the sentence (or sequence) will be the GHI time series corresponding to a day, each of the time step values shall represent one token for the model. The masked value to be completed can be found in any position into the time series of the day. In this regard, the data pre-processing step should ensure that all sequences to be used in training the model are complete, and ensure that test data are not part of the input.

The first operation applied was data preparation. In this regard, the datasets featured different temporal resolutions for the irradiance values—data from Galicia had a temporal resolution of 10 min, while data from Castile and León featured a temporal resolution of 30 min. To unify the temporal resolutions for all the data, the frequency of data from Galicia was shifted to 30 min.

Next, the last year of each dataset was separated for the test set. Night hours were eliminated because during night-time, solar irradiance is zero. Sequences of values were created with daytime data only, ensuring also that only days with no missing values were used.

To prepare the input data for the BERT model, it was necessary to reformat the data to plain text and save the time series of each station in a separate file, where each row contains the irradiance values of one day. WordPiece embeddings [20] were used in this study—WordPiece splits each irradiance value for a time step into a token. In our case, a vocabulary of 1600 tokens was selected, to include all values present in the training data (the irradiance can take values from 0 to 1600). The MLM task involves training the model by randomly placing the special token "[MASK]" at different positions in the chain, so that the model learns how to predict it. The special classification token "[CLS]" is always the first token in each sequence, and the "[UNK]" was also used to indicate that there are unknown values in the sequence (their capabilities are indicated in Section 2.1).

The following search space was considered: (1) the number of training epochs ("num_train_epochs"), (2) the training batch size ("per_device_train_batch_size"), (3) the number of gradients to accumulate ("gradient_accumulation_steps") before updating the weights (between the values (2, 6, 8, 10, 12, and (4) the batch size (with 32, 64, and 128 samples). The hyperparameter configuration that achieved the best results featured ten training epochs, a batch training size of ten, a validation batch size of training of sixty-four, eight accumulable gradients, twelve attention heads, and twelve hidden layers. The model saved checkpoints every 500 steps.

The selected performance metric for evaluation is the coefficient of determination, or R^2 score, a statistical measure that indicates how well a model fits the observed data. The R^2 score is calculated using the actual value (y_i) and the predicted value ($\hat{y}_i$) with Equation (1), where $\bar{y}_i = \frac{1}{n}\sum_{i=1}^{n} y_i$.

$$R^2 \text{ score} = 1 - \frac{\sum_{i=1}^{n}(y_i - \hat{y}_i)^2}{\sum_{i=1}^{n}\left(y_i - \bar{y}_i\right)^2} \quad (1)$$

3. Results and Discussion

The analysis of the results will be carried out according to the two established scenarios outlined in Section 2.3. The BERT model was trained once and evaluated for the two considered scenarios.

3.1. Scenario 1: Imputation of a Single Missing Value at a Specific Position

In Scenario 1, the impact of the imputation of a missing value was assessed according to its position in the sequence. The values for sunrise and sunset were left out, due to the discontinuous nature of solar radiation, where forecasts in the immediate vicinity of sunrise and sunset are problematic.

The experiments were carried out by moving the mask from the position corresponding to 10 a.m. to the position corresponding to 5 p.m., with all positions within that interval evaluated. Masking was performed in a separate experiment for each field position. The R^2 score in the set varied within the range of 0.83 to 0.95, as shown in Figure 3, with a mean of 0.89 and a variance of 0.13.

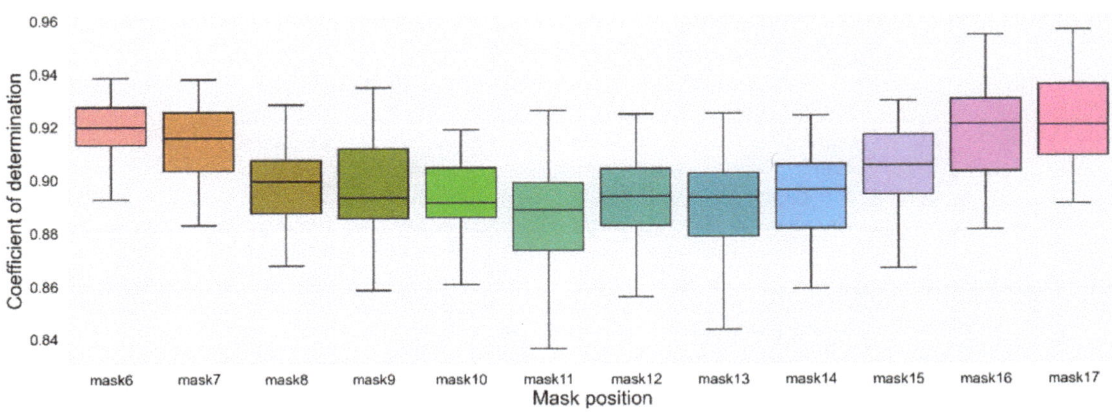

Figure 3. Analysis of the coefficient of determination for the position of the MASK token within the sequence.

In Figure 3 it can be found that seen that the best predictions are corresponding to the mask at the start and end values of the sequence. If the analysis is transposed to the solar data, it is observed that the morning and afternoon values feature smaller differences when compared to the values of the central hours of the day. It is expected that these variations are more difficult to model, considering that the differences are also highlighted depending on the month of the year.

3.2. Scenario 2: Imputation of a Missing Value after Several Unknown Values at a Random Position

In Scenario 2, all values were missing from a specific position in the sequence. We assessed the quality of the model to impute one value without the rest of the sequence. As in Scenario 1, the experiments were made by moving the mask position corresponding to 10 a.m. to 5 p.m. In the sequence, the mask number is the position where the "[MASK]" token was set; from that position onwards, the values are replaced by the "[UNK]" token. As expected, the R^2 score within the sequence increases as the number of unknown values decreases (as shown in Figure 4). The R^2 score varied within the range of 0.08 to 0.93, with a mean of 0.59 and a variance of 0.25.

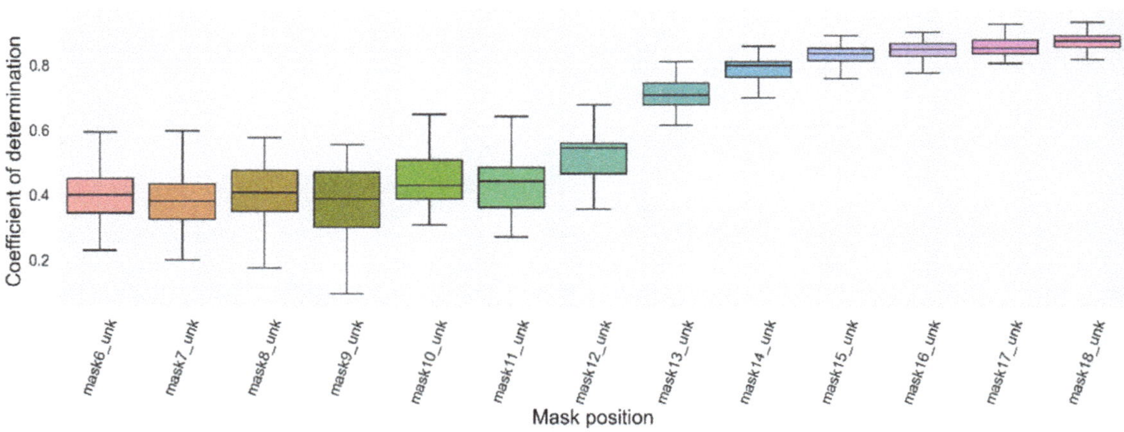

Figure 4. Analysis of the coefficient of determination for the position of the MASK token within the sequence with unknown values.

4. Conclusions and Future Work

In this study, the application of BERT as an imputation technique for missing values in solar radiation time series data was analyzed. The BERT model was trained from scratch with historical data of two Spanish regions, and was evaluated for two scenarios, a first scenario where a single missing value was imputed at a specific position in the sequence, and a second scenario where all values were missing from a specific position in the sequence. The metric evaluated was the R^2 score, and the average performance was 0.89%, the best result being 0.95% for imputation on the final values of the sequence.

The results achieved show how DL models can be used to impute missing data in time series. The work can be considered as a first step in the introduction of this model in the field of renewable energies, and raises new questions on how the addition of spatial (such as latitude and longitude) and temporal data (such as day of the year and year) affects the quality of the imputation.

In future studies, the BERT model could be evaluated from a spatio-temporal perspective, to analyze whether the model can model the spatial location of the weather station and is able to improve the imputation operation with the introduction of these new features. The model could also retrained on the next sentence prediction (NSP) task, to predict the subsequent day of irradiance values from the previous day. In addition, the exploration of the automation of the variables pre-processing to the format expected by BERT, or the conversion of the output back to the time series format, is recommended.

Author Contributions: Conceptualization, L.B.C. and M.-Á.M.-C.; methodology, L.B.C. and M.-Á.M.-C.; software, L.B.C.; validation, L.B.C., M.-Á.M.-C. and C.-I.C.; investigation, L.B.C.; resources, M.-Á.M.-C. and C.-I.C.; data curation, L.B.C.; writing—original draft preparation, L.B.C.; writing—review and editing, L.B.C., M.-Á.M.-C. and C.-I.C.; visualization, L.B.C.; supervision, M.-Á.M.-C. and C.-I.C. All authors have read and agreed to the published version of the manuscript.

Funding: This research received no external funding.

Institutional Review Board Statement: Not applicable.

Informed Consent Statement: Not applicable.

Data Availability Statement: The CyL-GHI dataset in this study are openly available in Zenodo repository (https://doi.org/10.5281/zenodo.7404167) and the Galicia dataset was compiled from an online service (https://www.meteogalicia.gal/web/RSS/rssIndex.action?request_locale=es, accessed on 1 March 2019).

Conflicts of Interest: The authors declare no conflict of interest.

References

1. Chatfield, C. *The Analysis of Time Series*, 6th ed.; Chapman and Hall/CRC: New York, NY, USA, 2003; ISBN 9780203491683.
2. Fang, C.; Wang, C. Time Series Data Imputation: A Survey on Deep Learning Approaches. *arXiv* **2020**, arXiv:2011.11347.
3. Glasbey, C.A. Imputation of Missing Values in Spatio-Temporal Solar Radiation Data. *Environmetrics* **1995**, *6*, 363–371. [CrossRef]
4. Layanun, V.; Suksamosorn, S.; Songsiri, J. Missing-Data Imputation for Solar Irradiance Forecasting in Thailand. In Proceedings of the 2017 56th Annual Conference of the Society of Instrument and Control Engineers of Japan (SICE), Kanazawa, Japan, 19–22 September 2017; pp. 1234–1239.
5. Demirhan, H.; Renwick, Z. Missing Value Imputation for Short to Mid-Term Horizontal Solar Irradiance Data. *Appl. Energy* **2018**, *225*, 998–1012. [CrossRef]
6. Zhang, W.; Luo, Y.; Zhang, Y.; Srinivasan, D. SolarGAN: Multivariate Solar Data Imputation Using Generative Adversarial Network. *IEEE Trans. Sustain. Energy* **2021**, *12*, 743–746. [CrossRef]
7. de-Paz-Centeno, I.; García-Ordás, M.T.; García-Olalla, Ó.; Alaiz-Moretón, H. Imputation of Missing Measurements in PV Production Data within Constrained Environments. *Expert Syst. Appl.* **2023**, *217*, 119510. [CrossRef]
8. Vaswani, A.; Shazeer, N.; Parmar, N.; Uszkoreit, J.; Jones, L.; Gomez, A.N.; Kaiser, Ł.; Polosukhin, I. Attention Is All You Need. *Adv. Neural Inf. Process. Syst.* **2017**, *30*, 5999–6009.
9. Yldz, A.Y.; Koc, E.; Koc, A. Multivariate Time Series Imputation With Transformers. *IEEE Signal Process. Lett.* **2022**, *29*, 2517–2521. [CrossRef]
10. Bansal, P.; Deshpande, P.; Sarawagi, S. Missing Value Imputation on Multidimensional Time Series. *Proc. VLDB Endow.* **2021**, *14*, 2533–2545. [CrossRef]
11. Devlin, J.; Chang, M.-W.; Lee, K.; Toutanova, K. BERT: Pre-Training of Deep Bidirectional Transformers for Language Understanding. *arXiv* **2018**, arXiv:1810.04805.
12. Koroteev, M.V. BERT: A Review of Applications in Natural Language Processing and Understanding. *arXiv* **2021**, arXiv:2103.11943.
13. Dong, X.; Bao, J.; Zhang, T.; Chen, D.; Zhang, W.; Yuan, L.; Chen, D.; Wen, F.; Yu, N. Bootstrapped Masked Autoencoders for Vision BERT Pretraining. In Proceedings of the 17th European Conference on Computer Vision (ECCV), Tel Aviv, Israel, 23–27 October 2022; pp. 247–264.
14. Wang, R.; Chen, D.; Wu, Z.; Chen, Y.; Dai, X.; Liu, M.; Jiang, Y.-G.; Zhou, L.; Yuan, L. BEVT: BERT Pretraining of Video Transformers. In Proceedings of the 2022 IEEE/CVF Conference on Computer Vision and Pattern Recognition (CVPR), New Orleans, LA, USA, 18–24 June 2022; pp. 14713–14723.
15. Lee, H.; Lee, S.; Lee, I.; Nam, H. AMP-BERT: Prediction of Antimicrobial Peptide Function Based on a BERT Model. *Protein Sci.* **2023**, *32*, e4529. [CrossRef] [PubMed]
16. Ghazikhani, H.; Butler, G. TooT-BERT-M: Discriminating Membrane Proteins from Non-Membrane Proteins Using a BERT Representation of Protein Primary Sequences. In Proceedings of the 2022 IEEE Conference on Computational Intelligence in Bioinformatics and Computational Biology (CIBCB), Ottawa, ON, Canada, 15 August 2022; pp. 1–8.
17. Wen, N.; Liu, G.; Zhang, J.; Zhang, R.; Fu, Y.; Han, X. A Fingerprints Based Molecular Property Prediction Method Using the BERT Model. *J. Cheminform.* **2022**, *14*, 71. [CrossRef] [PubMed]
18. Gao, Y.; Xiong, Y.; Wang, S.; Wang, H. GeoBERT: Pre-Training Geospatial Representation Learning on Point-of-Interest. *Appl. Sci.* **2022**, *12*, 12942. [CrossRef]

19. Benavides Cesar, L.; Manso Callejo, M.Á.; Cira, C.-I.; Alcarria, R. CyL-GHI: Global Horizontal Irradiance Dataset Containing 18 Years of Refined Data at 30-Min Granularity from 37 Stations Located in Castile and León (Spain). *Data* **2023**, *8*, 65. [CrossRef]
20. Wu, Y.; Schuster, M.; Chen, Z.; Le, Q.V.; Norouzi, M.; Macherey, W.; Krikun, M.; Cao, Y.; Gao, Q.; Macherey, K.; et al. Google's Neural Machine Translation System: Bridging the Gap between Human and Machine Translation. *arXiv* **2016**, arXiv:1609.08144.

engineering proceedings

MDPI

Proceeding Paper

A Machine Learning Approach for Bitcoin Forecasting †

Stefano Sossi-Rojas [1,*], Gissel Velarde [1,2,*] and Damian Zieba [3,*]

1 Computational Systems Engineering, Universidad Privada Boliviana, Cochabamba 3967, Bolivia
2 Vodafone GmbH., 40549 Düsseldorf, Germany
3 Faculty of Economic Sciences, University of Warsaw, 00927 Warsaw, Poland
* Correspondence: tenosossi2012@gmail.com (S.S.-R.); gissel.velarde@vodafone.com (G.V.); dzieba@wne.uw.edu.pl (D.Z.)

† Presented at the 9th International Conference on Time Series and Forecasting, Gran Canaria, Spain, 12–14 July 2023.

Abstract: Bitcoin is one of the cryptocurrencies that has gained popularity in recent years. Previous studies have shown that closing price alone is not enough to forecast its future level, and other price-related features are necessary to improve forecast accuracy. We introduce a new set of time series and demonstrate that a subset is necessary to improve directional accuracy based on a machine learning ensemble. In our experiments, we study which time series and machine learning algorithms deliver the best results. We found that the most relevant time series that contribute to improving directional accuracy are open, high, and low, with the largest contribution of low in combination with an ensemble of a gated recurrent unit network and a baseline forecast. The relevance of other Bitcoin-related features that are not price-related is negligible. The proposed method delivers similar performance to the state of the art when observing directional accuracy.

Keywords: Bitcoin; forecasting; time series; machine learning

Citation: Sossi-Rojas, S.; Velarde, G.; Zieba, D. A Machine Learning Approach for Bitcoin Forecasting. *Eng. Proc.* **2023**, *39*, 27. https://doi.org/10.3390/engproc2023039027

Academic Editors: Ignacio Rojas, Hector Pomares, Luis Javier Herrera, Fernando Rojas and Olga Valenzuela

Published: 29 June 2023

1. Introduction

Bitcoin price forecasting has been the focus of many studies in the literature over the years. Nevertheless, unexpected price movements, smaller and larger bubbles, and different short- and long-term trends mean this task is an ongoing topic for research. In one of the recent studies exploring this area of research [1], the authors verify the performance of different machine learning algorithms and mention the current state of the knowledge in Bitcoin forecasting. One aspect refers to [2] the price of Bitcoin being mainly driven by the spot market rather than the futures market. Another aspect is the division of approaches to forecasting into the "Blockchain approach" and "Financial market's approach". The former is based on technical variables such as hash rates or mining difficulty, while the latter uses standard econometric variables such as stocks, bonds, or gold price. We can add to this another approach, which involves taking variables related to investor sentiment such as Google trends data [3], uncertainty indices (VIX, UCRY, see [4]), or the Fear and Greed index, which has not been explored much in the literature. In our study, we used a mix of features from these three approaches and the open, high, low, and close prices to predict the Bitcoin price using a novel approach. Based on this, we were able to verify which variables contribute to the forecast accuracy the most. In fact, the features with the largest contribution are only the price-related ones. The other variables, whether coming from the "Blockchain approach", the "Financial markets approach", or the "Sentiment approach" have a negligible impact in terms of improving Bitcoin's price performance.

We used a method that explores features and machine learning algorithms for Bitcoin's closing price prediction. In recent forecasting competitions, it was observed that machine learning models and hybrid approaches demonstrated superiority over alternative methods [5]. Therefore, in this study, we evaluated the following machine learning algorithms:

Long Short-Term Memory (LSTM), Bidirectional Long-Term Memory (BiLSTM), Gated Recurrent Unit (GRU), Bidirectional Gated Recurrent Unit (BiGRU), and Light Gradient Boosting Machine (LightGBM). In addition we used ensembling. The results were evaluated observing the predicted and actual Bitcoin closing price measuring Root Mean Squared Error (RMSE), Mean Squared Error (MSE), Mean Absolute Error (MAE), and Directional Accuracy (DA).

Although previous studies show that the financial time series closing price, either in the case of stocks [6,7], commodities [8], or cryptocurrencies [9], is not enough for prediction when training deep learning models, we found that ensembling does help to improve the forecasting accuracy. In this work, we demonstrate that the Bitcoin closing price is not sufficient for forecasting and additional features are necessary when using machine learning algorithms. Evaluation on return shows that the method developed in this work presents one of the highest scores, with a directional accuracy score equal to 0.7645, exceeding a baseline by 58.24 percent.

We compare our work to related studies for Bitcoin prediction. In [10], LSTM was used in combination with the Empirical Wavelet Transform (EWT) decomposition technique. The authors used the Intrinsic Mode Function (IMF) to optimize and estimate outputs with Cuckoo Search (CS) [10]. In [11], Linear Regression (LR) techniques and particle swarm optimization were used to train and forecast data from beginning of 2012 to the end of March 2020. The best setup for the model was obtained with 42 days plus 1 standard deviation [11]. In [12], Autoregressive Integrated Moving Average (ARIMA) was used for data from 1 May 2013 to 7 June 2019. This model works best for short-term predictions and can be used to predict Bitcoin for one to seven days ahead [12]. Finally, in [13], a BiLSTM with Low-Middle-High features (LMH-BiLSTM) was tested with two primary steps: data decomposition and bidirectional deep learning. The results demonstrate that the proposed model outperforms other benchmark models and achieved high investment returns in the buy-and-hold strategy in a trading simulation [13].

In this work, we tested the previously mentioned machine learning algorithms one by one, and also in ensemble. We fed the algorithms with a new set of 13 time series (see Section 2.1). In addition, we included 11 signals that come from Variational Mode Decomposition (MMD) as proposed in [13]. However, in our experiments, data decomposition did not provide significant improvements. Next, we explain our proposed method.

2. Materials and Methods

An overview of the method is presented in Figure 1. The input data is prepared in three steps. First, it is normalized between 0 and 1. Then, the train and test set partitions are created, where the train set is used for training several machine learning algorithms. The next phase considers algorithm training. The details regarding hyperparameter selection are described in Section 2.4. Next, the evaluation phase is performed as a rolling forecast for 1-step ahead prediction over the test set.

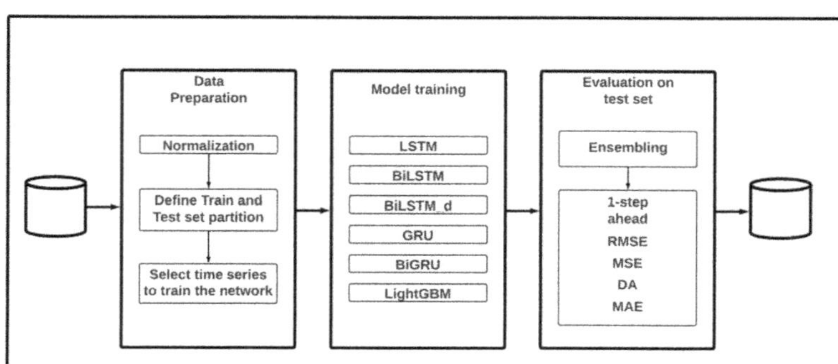

Figure 1. Visual summary of the method.

2.1. Data Collection

We collected the daily Bitcoin closing price from 7 October 2013 to 6 November 2022. We used a public API from the Kraken page [14]; the data collected were the values of close, open, high, low, volume, and date. With the Nasdaq-Data-Link library for Python [15], the following values were obtained: transaction fee, estimated Bitcoin USD transaction volume, Bitcoin USD exchange trade volume, and Bitcoin hash rate. Bitcoin Google trends were obtained with the pytrends library [16]. The gold to USD exchange rate was obtained from the Investing.com page. The Fear and Greed Index was obtained from the Kaggle page [17]. The moving average of the closing value was added, taking the last 30 days. Table 1 presents all features used in this study. A similar set of features was used in [13]: Bitcoin price, Bitcoin transaction fees as Bitcoin miner's revenue divided by transactions, USD trade volume from the top Bitcoin exchanges, Bitcoin transaction volume, USD exchange or trade volume from the top Bitcoin exchanges, gold exchange rate to US dollar, hash rate, and Google trends of Bitcoin.

Table 1. Series, short name, description, and count. Originally some signals had more samples than others; therefore, the dates where no information was recorded for all signals were removed to obtain the same number of samples per signal, down sampling to 3812 in the range 19 November 2013 to 4 November 2022.

Series	Features	Description	Count
Series 1	Close	Daily Bitcoin close price	4379
Series 2	Low	Daily Bitcoin low price	4379
Series 3	High	Daily Bitcoin high price	4379
Series 4	Open	Daily Bitcoin open price	4379
Series 5	Trans_Volume	Bitcoin transaction volume in dollars	3812
Series 6	Volume	Daily quantity of Bitcoins bought or sold	4379
Series 7	Hash_Rate	Number of giga hashes Bitcoin network performed	3812
Series 8	Trans_Fees	Bitcoin miner's revenue divided by transactions	3812
Series 9	XAU_USD	Gold (XAU) Exchange rate to US dollar (USD)	3812
Series 10	Trade_Volume	Bitcoin trade volume in dollars	3812
Series 11	Google_Trend	Bitcoin's Google Trend	3812
Series 12	Fear_Greed	Fear and Greed Index. It is a way to gauge stock market movements and whether stocks are fairly priced	3812
Series 13	Moving_Avg_30	Moving average of Bitcoin's closing price for the last 30 days.	3783

2.2. Variational Mode Decomposition (VMD)

Bitcoin' closing price was decomposed using the VMD method as proposed in [13]. Each decomposed mode was labeled M0 through M10, where M0 has the lowest frequency and M10 has the highest. We can observe the graph of the decomposition in Figure 2.

Variational Mode Decomposition (VMD) is a completely nonrecursive signal decomposition technique proposed by [18]. VMD is a problem of variational optimization that aims to minimize the total bandwidth of each mode. This work used the vmdpy python library [19], with parameters by default using Bitcoin's close price as the input and a bandwidth of 5000.

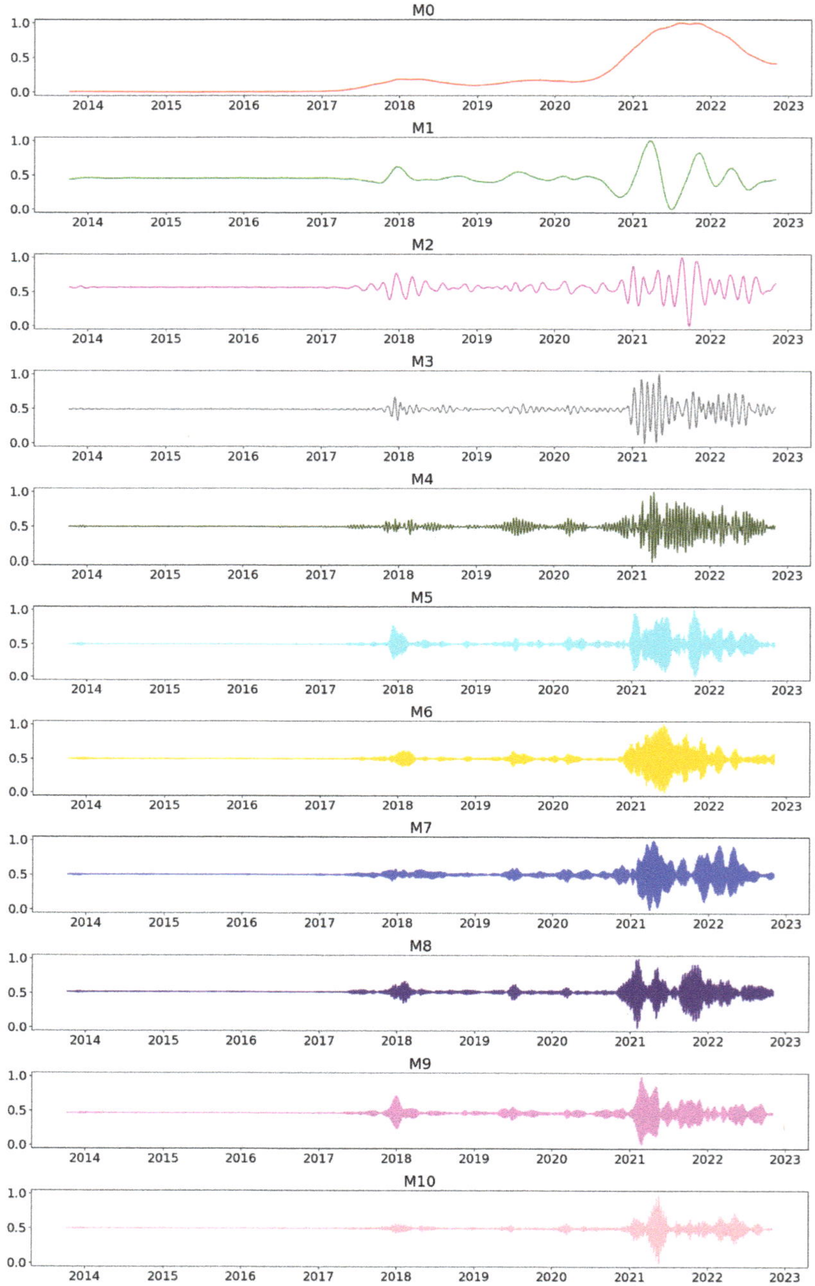

Figure 2. Variational Mode Decomposition (VMD) decomposition of Bitcoin's closing price, normalized between 0 and 1, from 7 October 2013 to 6 November 2022.

2.3. Data Preparation

For the LSTM, BiLSTM, GRU, and BiGRU models, the data were scaled between 0 and 1 before training, except the LightGBM model for which data scaling was performed for evaluation only. Next, the data was divided into 25-day windows, converting the data table into 3D lists (arrays), where the first dimension corresponds to the batch size, the second to the number of time-steps, and finally, the third dimension to the number of units of one input sequence [20]. Next to each list, the expected future value was saved. This last value is taken from the closing values of the previous day. For the LightGBM model, the complete data Table 1 was used without modifications. Data partitioning was performed as follows: the set was trained from 7 October 2013 to 8 August 2022, and the set was tested from 9 August 2022 to 6 November 2022, where 90 days were used for testing, and the rest of data were used for training.

2.4. Model Training

We tested five deep learning architectures and one tree boosting method. All deep learning networks present an input layer of 90 units, a set value of 500 epochs, and a batch size of 64 without early stopping.

- **Long Short-Term Memory (LSTM)**: The network trains with five layers, an input layer with the activation function seen in [6], the bias initializer glorot uniform, kernel regulator l1, l2, kernel constraint unit norm, and the time_major activated, followed by a dense layer of 90 units and linear activation. Then, an output layer and another dense layer is used. The model uses the Adam optimizer with a learning rate of 0.002;
- **Gated Recurrent Unit (GRU)**: The network trains with five layers, an input layer followed by a dropout layer set to 0.3, an output layer followed by another dropout layer, and a dense layer of one unit. The model uses the Adam optimizer with a learning rate of 0.0001. A similar network was used in [21];
- **Bidirectional Long Short-Term Memory (BiLSTM)**: The network consists of an input layer with the *tanh* activation function, followed by a backward learning layer and a dense layer. The model uses the Adam optimizer with a learning rate of 0.01. A similar network was used in [13];
- **Bidirectional Long Short-Term Memory with dropout (BiLSTM_d)**: The network consists of an input layer with *tanh* activation function, followed by a dropout layer, followed by a backward learning layer, followed by a dropout layer and a dense layer. The model uses the Adam optimizer with a learning rate of 0.01 and the dropout set to 0.3;
- **Bidirectional Gated Recurrent Unit (BiGRU)**: The network trains with five layers, an input bidirectional layer followed by a dropout layer set to 0.3, an output bidirectional layer followed by another dropout layer and a dense layer. The model uses the Adam optimizer with a learning rate of 0.0001;
- **Light Gradient Boosting Machine (LighGBM)**: This presents an early stopping round set to 50, and verbose evaluation set to 30 with 3600 number of boost rounds. The model trains with a gradient booting decision tree, with the objective set to *tweedie* and a variance power of 1.1, and uses an RMSE metric with n-jobs set to -1. In addition, it uses 42 seeds with a learning rate of 0.2, the bagging fraction is set to 0.85 and the bagging frequency is set to 7. Moreover, colsample by tree and colsample by node are set to 0.85 with a min data per leaf of 30, and the number of leaves is 200 with lambda *l1* and *l2* set to 0.5. A similar network was used in [5].

2.5. Evaluation

The predicted results were normalized between 0 and 1 before evaluation measurements were made.

We measure the Mean Squared Error (MSE), Root Mean Squared Error (RMSE), Mean Absolute Error (MAE), and Directional Accuracy (DA) between the predicted and actual closing price as described in [22], such that n is the number of samples, and y_t and x_t are the predicted and actual closing price at time t:

$$MSE = n^{-1} \sum_{t=1}^{n} (x_t - y_t)^2. \tag{1}$$

$$RMSE = \sqrt{MSE}. \tag{2}$$

$$MAE = \frac{\sum_{t=1}^{n} |y_t - x_t|}{n}. \tag{3}$$

$$DA = \frac{100}{n} \sum_{t=1}^{n} d_t, \tag{4}$$

where

$$d_t = \begin{cases} 1 & (x_t - x_{t-1})(y_t - y_{t-1}) \geq 0 \\ 0 & otherwise. \end{cases}$$

2.6. Return

The return is a financial measure used to assess the efficiency of an asset investment. It is an growth indicator of the value of an investment during a certain period of time. Return On Investment (ROI) is one of the main financial measures used both in the traditional stock market and in the world of cryptocurrencies [23]. The formula can be expressed in terms of the final Value of Inversion (FVI) and the Initial Value of Inversion (IVI):

$$ROI = (\frac{FVI - IVI}{IVI})100\%. \tag{5}$$

3. Results

Experiments were conducted to see how different time series influence Bitcoin closing price prediction and how different models perform. The experiments were carried out analyzing the results with different factors. Tables 2–11 show the results obtained in the different experiments. An RMSE, MSE, MAE close to zero and a DA close to one are preferred. The best results are highlighted. The values were compared with values obtained with a Baseline prediction. Baseline prediction means that the predicted value is the last observed value. Series importance for the prediction was obtained according to the LightGBM model in order to classify them later and continue with the experiments, as can be seen in Figure 3.

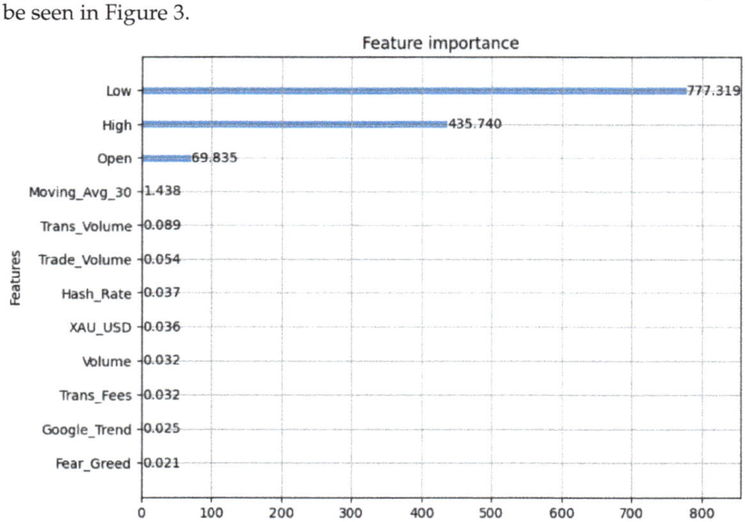

Figure 3. Feature importance found by LightGBM. The most important features are low, high, and open.

We performed the following experiments:

- **Experiment 1.** The first experiment sent a subset of all tested time series, as we assumed these features represent price action over a set period of time, and in combination could be used to predict price movements. The time series used were close, open, high, low, and volume of Bitcoin. These were obtained from the Kraken API [14]. See Table 2 for the results of the prediction. Table 2 shows that BiLSTM was best DA = 0.5056 but BiGRU produced the lowest MAE = 0.0467;
- **Experiment 2.** In the second experiment, we tested all time series presented in Table 1: close, open, high, low, volume, transaction fee, estimated Bitcoin USD transaction volume, Bitcoin USD exchange trade volume, rate of Bitcoin hash, Bitcoin Google trends, and gold to USD exchange rate, Fear and Greed index, and the moving average of the closing value. The main idea was to enrich the input data to help improve the prediction. The results can be seen in Table 3. LightGBM had the best performance and there was an improvement in comparison with Experiment 1 whereby less input data were used;
- **Experiments 3.** For the following experiments, series importance for the prediction was analyzed according to the LightGBM model. The order of importance can be seen in Figure 3. The experiments were performed indicating different combinations of the four most important series: open, high, low, and close values. These time series exhibited a significant correlation with the closing price of Bitcoin and exerted an influence over its behavior. For this experiment, open, high, and low values were used. Table 4 shows that BiLSTM had the best DA = 0.4832, but GRU had the best MAE = 0.0496. The results show a decrease in performance;
- **Experiment 4.** For this experiment, high and low values were used. Table 5 shows that BiLSTM had the best performance with DA = 0.5169. There was an improvement in comparison with Experiment 1, but it did not reach the performance of Experiment 2;
- **Experiment 5.** For this experiment, low values were used. Table 6 shows that LSTM had the best performance with DA = 0.5169. The results show a similar performance in comparison with Experiment 4;
- **Experiment 6.** For this experiment, open and low values were used. Table 7 shows that LSTM had the best DA = 0.5393 and BiLSTM delivered the lowest errors. The results show an improvement in performance compared with previous experiments, but it was not better than Experiment 2;
- **Experiment 7.** Since Experiment 2 had the best performance until this point, we tested all time series used in Experiment 2 and added 11 VMD modes to assess their impact on prediction. That is, the input data included the values of close, open, high, low, volume, transaction fee, estimated Bitcoin USD transaction volume, Bitcoin USD exchange trade volume, rate of Bitcoin hash, Bitcoin Google trends, and gold to USD exchange rate, Fear and Greed index, and the moving average of the closing value and 11 VMD modes. The results are presented in Table 8. This time, BiGRU had the best DA = 0.5730 but LSTM had the lowest errors. The results show a big improvement in performance compared with the previous experiments;
- **Experiment 8.** This experiment consisted of sending the values of all 11 VMD modes as input data, that is, only the 11 VMD modes were added to assess the impact these had on the prediction (see Table 9). BiGRU had the best DA = 0.5618, BiLSTM the best MAE, and LSTM the best MSE and RMSE. Results show an improvement in performance compared with previous experiments, but it was not better than Experiment 7.

In all the experiments, we observed a noticeable improvement in the BiLSTM model when it did not present the dropout layer.

Table 2. Prediction measured values with Bitcoin close, open, high, low, and volume values.

Network	Measured Value MAE	MSE	RMSE	DA
GRU	0.0541	0.0111	0.1053	0.4494
BiGRU	**0.0467**	0.0107	0.1032	0.4494
LSTM	0.0660	0.0125	0.1120	0.4719
BiLSTM	0.0954	0.0163	0.1277	**0.5056**
BiLSTM_d	0.1610	0.0667	0.2583	0.4832
LightGBM	0.0513	**0.0106**	**0.1028**	0.4157

Table 3. Prediction measured values with all factors.

Network	Measured Value MAE	MSE	RMSE	DA
GRU	0.1426	0.0496	0.2228	0.3371
BiGRU	0.1501	0.0660	0.2569	0.3596
LSTM	0.1731	0.0666	0.2580	0.3708
BiLSTM	0.2019	0.0864	0.2940	0.4382
BiLSTM_d	0.2685	0.1109	0.3330	0.3371
LightGBM	**0.0566**	**0.0154**	**0.1240**	**0.5618**

Table 4. Prediction measured values with Bitcoin open, high, and low values.

Network	Measured Value MAE	MSE	RMSE	DA
GRU	**0.0496**	**0.0103**	**0.1014**	0.4494
BiGRU	0.0508	0.0110	0.1050	0.4719
LSTM	0.1029	0.0352	0.1875	0.4382
BiLSTM	0.0812	0.0178	0.1334	**0.4832**
BiLSTM_d	0.1317	0.0571	0.2390	0.4607
LightGBM	0.0864	0.0157	0.1254	0.3483

Table 5. Prediction measured values with high and low values of Bitcoin.

Network	Measured Value MAE	MSE	RMSE	DA
GRU	0.0572	0.0112	0.1059	0.4157
BiGRU	0.0487	0.0112	0.1060	0.4607
LSTM	0.0508	0.0117	0.1083	0.4832
BiLSTM	**0.0402**	**0.0092**	**0.0961**	**0.5169**
BiLSTM_d	0.1709	0.0599	0.2447	0.4832
LightGBM	0.0796	0.0217	0.1472	0.2472

Table 6. Prediction measured values with low value of Bitcoin.

Network	Measured Value MAE	MSE	RMSE	DA
GRU	0.0558	0.0116	0.1076	0.4719
BiGRU	0.5616	0.0117	0.1080	0.4607
LSTM	**0.0378**	**0.0098**	**0.0988**	**0.5169**
BiLSTM	0.0460	0.0104	0.1019	0.4719
BiLSTM_d	0.1155	0.0523	0.2286	0.4944
LightGBM	0.1213	0.5991	0.2448	0.1348

Table 7. Prediction measured values with Bitcoin open and low values.

Network	Measured Value MAE	MSE	RMSE	DA
GRU	0.5125	0.0102	0.1011	0.4607
BiGRU	0.0518	0.0104	0.1022	0.4494
LSTM	0.0613	0.0151	0.1228	**0.5393**
BiLSTM	**0.0344**	**0.0092**	**0.0958**	0.4719
BiLSTM_d	0.1221	0.0623	0.2496	0.5056
LightGBM	0.0801	0.0227	0.1505	0.2809

Table 8. Prediction measured values with all factors plus VMD modes.

Network	Measured Value MAE	MSE	RMSE	DA
GRU	0.1002	0.0245	0.1565	0.5506
BiGRU	0.0751	0.0230	0.1517	**0.5730**
LSTM	**0.0705**	**0.0206**	**0.1423**	0.5169
BiLSTM	0.0963	0.0222	0.1489	0.4832
BiLSTM_d	0.1081	0.0322	0.1793	0.4832
LightGBM	0.0740	0.0222	0.1490	0.4719

Table 9. Prediction measured values with 11 VMD modes.

Network	Measured Value MAE	MSE	RMSE	DA
GRU	0.0806	0.0179	0.1338	0.5169
BiGRU	0.0762	0.0176	0.1327	**0.5618**
LSTM	0.0591	**0.0167**	**0.1290**	0.4607
BiLSTM	**0.0521**	0.0176	0.1329	0.4607
BiLSTM_d	0.0837	0.0209	0.1446	0.5169
LightGBM	0.1299	0.0486	0.2204	0.4494

Ensembling and Return Performance

Since we are predicting one day in the future (rolling forecast), the value of IVI seen in Equation (5) is the same value as the baseline, and FVI is the prediction of the models. Ensembling was obtained with the simple arithmetic average, using Equation (5), combining each model with the baseline. We tested each model individually. The results show that the best combination was GRU and the baseline, with the model being trained using the open, high, and low factors.

- **Experiment 9.** In this experiment, we tested ensembling. Return values were calculated over the prediction of all previous experiments. The best results were obtained using open, high, and low values as inputs. See Table 10 for the results. GRU had the best performance with DA = 0.7645. The results show a big improvement in DA compared to the previous experiments;
- **Experiment 10.** The last experiment was executed in order to make a close comparison with LMH-BiLTSM [13]. Therefore, we used approximately the same date range as used in [13]. Return values were calculated over the prediction period. See Table 11 for the results. This time, GRU had the best DA = 0.7865 but BiGRU produced the lowest errors. The DA obtained here is comparable to that reported by LMH-BiLTSM [13]. Indeed, in this experiment, we observed the highest DA prediction out of all the experiments we performed. However, the lowest errors were measured using the BiLSTM network with the open and low series: MAE = 0.0344, MSE = 0.0092, and RMSE = 0.0958, see Table 7.

Table 12 shows a comparison between previous studies and the best performing model presented in this paper. The model with the best performance consists of a GRU trained with open, high, and low values of Bitcoin. The GRU ensemble achieved 0.7865 ± 0.2113 DA, matching the performance obtained by LMH-BiLSTM [13] in a similar time range. Notice that for the period from October 2013 to 6 November 2022, the DA was slightly lower but the standard deviation was also lower. Making a comparison of the GRU network (Table 10) and the baseline, we see an improvement of 58.14 percent in directional accuracy; this was the model with the highest score.

Table 10. Return measured values for prediction with Bitcoin open, high, and low values. Range between 7 October 2013 and 6 November 2022.

Network	Measured Value			
	MAE	**MSE**	**RMSE**	**DA**
Baseline	0.0371	0.0091	0.0952	0.4831
GRU	**0.0843**	**0.0296**	**0.1716**	**0.7645**
BiGRU	0.2299	0.0735	0.2711	0.7191
LSTM	0.2629	0.1113	0.3337	0.5169
BiLSTM	0.1687	0.0553	0.2352	0.5169
BiLSTM_d	0.3082	0.1072	0.3274	0.7303
LightGBM	0.1730	0.0488	0.2208	0.6547

Table 11. Return measured values for prediction with all factors plus VMD modes from 7 October 2013 to 1 January 2021, which is approximately same range as that used in [13]. Baseline DA was 0.5281.

Network	Measured Value			
	MAE	**MSE**	**RMSE**	**DA**
Baseline	0.0496	0.0138	0.1176	0.5281
GRU	0.1745	0.0548	0.2341	**0.7865**
BiGRU	**0.1137**	**0.0400**	**0.2001**	0.7303
LSTM	0.2039	0.0860	0.2933	0.5955
BiLSTM	0.2772	0.1044	0.3232	0.6067
BiLSTM_d	0.3134	0.1332	0.3649	0.5056
LightGBM	0.2806	0.1078	0.3283	0.5730

Table 12. Performance comparisons. The methods presented in this table were trained using different datasets in different date ranges. Therefore, this comparison is relative.

Method	DA
GRU (This work Table 10, 7 October 2013 to 6 November 2022)	0.7645 ± 0.1299
GRU (This work Table 11, 7 October 2013 to 1 January 2021)	0.7865 ± 0.2113
LMH-BiLSTM [13] (29 April 2013 to 1 January 2021)	0.8170 ± ?
ARIMA [12] (1 May 2013 to 7 June 2019)	0.5719 ± ?
LSTM [10] (18 July 2010 to 28 March 2019)	0.5409 ± ?
LR [11] (Beginning of 2012 until the end of March 2020)	0.5155 ± ?

4. Conclusions

We confirm the hypothesis that Bitcoin is difficult to predict with the closing price alone, that is, the closing price does not contain enough information to predict Bitcoin, and a set of price-related time series are necessary to improve prediction. We tested 13 series as shown in Table 1, plus 11 modes decomposed using Variational Mode Decomposition (VMD). In addition, we tested various machine learning algorithms and found that a selected set of time series consisting of open, high, and low values and an ensemble based on a GRU network combined with the value of return, or a baseline prediction, demonstrates a great improvement in the results of the experiments. Our method delivers a comparable DA when compared to the state of the art, which in contrast uses a BiLSTM with Low-Middle-High features (LMH-BiLSTM) [13].

Author Contributions: S.S.-R. Development, implementation, and writing. G.V. Conceptualization, supervision, and writing. D.Z. Advise on the project, and writing. All authors have read and agreed to the published version of the manuscript.

Funding: This research received no external funding.

Institutional Review Board Statement: Not applicable.

Informed Consent Statement: Not applicable.

Data Availability Statement: Not applicable.

Acknowledgments: We thank the anonymous reviewers for their feedback.

Conflicts of Interest: The authors declare no conflict of interest.

References

1. Chevallier, J.; Guégan, D.; Goutte, S. Is it possible to forecast the price of bitcoin? *Forecasting* **2021**, 3, 377–420. [CrossRef]
2. Baur, D.G.; Dimpfl, T. Price discovery in bitcoin spot or futures? *J. Futur. Mark.* **2019**, 39, 803–817. [CrossRef]
3. Kristoufek, L. BitCoin meets Google Trends and Wikipedia: Quantifying the relationship between phenomena of the Internet era. *Sci. Rep.* **2013**, 3, 3415. [CrossRef] [PubMed]
4. Lucey, B.M.; Vigne, S.A.; Yarovaya, L.; Wang, Y. The cryptocurrency uncertainty index. *Financ. Res. Lett.* **2022**, 45, 102147. [CrossRef]
5. Makridakis, S.; Spiliotis, E.; Assimakopoulos, V. M5 accuracy competition: Results, findings, and conclusions. *Int. J. Forecast.* **2022**, 38, 1346–1364. [CrossRef]
6. Velarde, G.; Brañez, P.; Bueno, A.; Heredia, R.; Lopez-Ledezma, M. An Open Source and Reproducible Implementation of LSTM and GRU Networks for Time Series Forecasting. *Eng. Proc.* **2022**, 18, 30.
7. Velarde, G. *Forecasting with Deep Learning*; White Paper; Technical Report 2(8); Vodafone, The Data Digest: Duesseldorf, Germany, 2022.
8. Ben Ameur, H.; Boubaker, S.; Ftiti, Z.; Louhichi, W.; Tissaoui, K. Forecasting commodity prices: empirical evidence using deep learning tools. *Ann. Oper. Res.* **2023**, 1–19. [CrossRef] [PubMed]
9. Lamothe-Fernández, P.; Alaminos, D.; Lamothe-López, P.; Fernández-Gámez, M.A. Deep learning methods for modeling bitcoin price. *Mathematics* **2020**, 8, 1245. [CrossRef]
10. Altan, A.; Karasu, S.; Bekiros, S. Digital currency forecasting with chaotic meta-heuristic bio-inspired signal processing techniques. *Chaos Solitons Fractals* **2019**, 126, 325–336. [CrossRef]
11. Cohen, G. Forecasting Bitcoin trends using algorithmic learning systems. *Entropy* **2020**, 22, 838. [CrossRef] [PubMed]
12. Wirawan, I.M.; Widiyaningtyas, T.; Hasan, M.M. Short term prediction on bitcoin price using ARIMA method. In Proceedings of the 2019 International Seminar on Application for Technology of Information and Communication (iSemantic), Semarang, Indonesia, 21–22 September 2019; pp. 260–265.
13. Li, Y.; Jiang, S.; Li, X.; Wang, S. Hybrid data decomposition-based deep learning for Bitcoin prediction and algorithm trading. *Financ. Innov.* **2022**, 8, 1–24. [CrossRef]
14. Kraken. 2022. Available online: https://docs.kraken.com/rest/ (accessed on 6 November 2022).
15. Nasdaq. Nasdaq Data Link. 2021. Available online: https://data.nasdaq.com/tools/python (accessed on 6 November 2022).
16. Pytrends. 2015. Available online: https://pypi.org/project/pytrends/ (accessed on 6 November 2022).
17. de Araujo, A. Crypto Fear and Greed Index. Available online: https://www.kaggle.com/datasets/adelsondias/crypto-fear-and-greed-index/code (accessed on 6 November 2022).
18. Dragomiretskiy, K.; Zosso, D. Variational Mode Decomposition. 2013. Available online:. (accessed on 28 January 2023). [CrossRef]
19. Carvalho, V.R.; Moraes, M.F.; Braga, A.P.; Mendes, E.M. Evaluating Five Different Adaptive Decomposition Methods for EEG Signal Seizure Detection and Classification. 2020. Available online:. (accessed on 6 February 2023). [CrossRef]
20. Verma, S. Input and Output Shape in LSTM (Keras). 2019. Available online: https://www.kaggle.com/code/shivajbd/input-and-output-shape-in-lstm-keras (accessed on 28 January 2023).
21. Rasifaghihi, N. LSTM-GRU-BiLSTM-in-TensorFlow-for-Predictive-Analytics. 2020. Available online: https://github.com/NioushaR/LSTM-GRU-BiLSTM-in-TensorFlow-for-predictive-analytics (accessed on 6 January 2023).

22. Wang, J.J.; Wang, J.Z.; Zhang, Z.G.; Guo, S.P. Stock index forecasting based on a hybrid model. *Omega* **2012**, *40*, 758–766. [CrossRef]
23. Phemex. Cómo Calcular el Retorno de la Inversión (ROI) de las Criptomonedas? 2021. Available online: https://phemex.com/es/academy/como-calcular-el-roi-de-las-criptomonedas (accessed on 6 February 2023).

engineering proceedings

MDPI

Proceeding Paper

Dynamic Tikhonov State Forecasting Based on Large-Scale Deep Neural Network Constraints †

Cristhian Molina [1], Juan Martinez [2] and Eduardo Giraldo [3,*]

1 Department of Systems, Instituto Tecnológico Metropolitano, Medellín 050012, Colombia; cristhianmolina@itm.edu.co

2 School Applied Sciences and Engineering, Universidad EAFIT, Medellín 050021, Colombia; jdmartinezv@eafit.edu.co

3 Department of Electrical Engineering, Research Group in Automatic Control, Universidad Tecnológica de Pereira, Pereira 660003, Colombia

* Correspondence: egiraldos@utp.edu.co

† Presented at the 9th International Conference on Time Series and Forecasting, Gran Canaria, Spain, 12–14 July 2023.

Abstract: This work presents dynamic Tikhonov state forecasting based on large-scale deep neural network constraint for the solution to a dynamic inverse problem of electroencephalographic brain mapping. The dynamic constraint is obtained by using a large-scale deep neural network to approximate the dynamics of the state evolution in a discrete large-scale state-space model. An evaluation by using neural networks with several hidden layer configurations is performed to obtain the adequate structure for large-scale system dynamic tracking. The proposed approach is evaluated over two models of 2004 and 10,016 states in discrete time. The models are related to an electroencephalographic problem for EEG generation. A comparison analysis is performed by using static and dynamic Tikhonov approaches with simplified dynamic constraints. By considering the obtained results it can be concluded that the deep neural networks adequately approximate large-scale state dynamics by improving the dynamic inverse problem solutions.

Keywords: dynamic state forecasting; deep neural network; large scale

Citation: Molina, C.; Martinez, J.; Giraldo, E. Dynamic Tikhonov State Forecasting Based on Large-Scale Deep Neural Network Constraints. *Eng. Proc.* **2023**, *39*, 28. https://doi.org/10.3390/engproc2023039028

Academic Editors: Ignacio Rojas, Hector Pomares, Luis Javier Herrera, Fernando Rojas and Olga Valenzuela

Published: 29 June 2023

1. Introduction

Deep neural networks (DNNs) have emerged as promising tools for state estimation. DNNs can learn complex non-linear relationships between inputs and outputs, making them well-suited for estimating dynamic systems. Additionally, DNNs can handle high-dimensional data and can be adapted to handle various state estimation problems. For example, Zhang et al. [1] and Li et al. [2] discussed the impact of deep learning on the field of inverse problems, with the former proposing a residual learning-based deep convolutional neural network (CNN) approach for image denoising. They reviewed existing work in this area and covered the basics of deep learning and its application for inverse problems. However, Zhang et al. [3] and Chien et al. [4] focused on the use of Tikhonov regularization for training DNNs. Zhang et al. [1] used the convergent block coordinate descent (CBCD) algorithm for training with Tikhonov regularization, showing its effectiveness through experimental results on various datasets. Chien et al. [4] explored the use of Tikhonov regularization in acoustic modelling, showing that adding Tikhonov regularization can improve the generalization performance of DNNs.

In order to solve inverse problems using DNNs and regularization, in Fkham, et al. [5] the authors developed a DNN-based method for automatically learning the regularization parameters in inverse problems, resulting in improved accuracy and robustness. Alternatively, Nguyen et al. [6] incorporated prior knowledge about the inverse problem into the network architecture, resulting in improved accuracy and robustness compared to traditional DNN-based methods. Furthermore, Romano et al. [7] developed a method called regularization by denoising (RED) for inverse problems, using a deep denoising neural network to regularize the solution of an inverse problem. In contrast, Mao et al. [8] introduced a deep learning-based approach for image restoration using a profound convolutional encoder–decoder network architecture with symmetric skip connections to handle the inverse problem and the experimental results that demonstrate the effectiveness of the proposed approach were tested on several benchmark datasets.

According to Kolowrocki et al. [9], large-scale complex systems need to be modelled in order to identify the cross-correlation of variables through their inherent complex dynamics. In many cases, their dynamics are hard to describe using non-linear equations due to their inherent couplings and complexity. As Sockeel et al. [10] reported, electroencephalographic signals are a clear example of large-scale systems where the discrete state-space model is represented by a large-scale non-linear state evolution equation and a measurement equation. The state estimation in EEG is an ill-conditioned, ill-posed inverse problem that requires additional constraints to be solved adequately as Sanchez-Bornot [11] mentioned. In many cases, as Wang et al. [12] reported, the number of states to be estimated is large and requires high-performance computing.

This paper proposes a dynamic Tikhonov state forecasting method based on large-scale DNNs as dynamic constraints, evaluated for the solution to an EEG inverse problem for neural activity estimation and compared with the static version of the method. The non-linear dynamic of state evolution is approximated by using the non-linear structure of the DNNs for two brain models with 2004 and 10,016 sources in which the EEG dynamics are simulated and approximated using the state measurements. Qualitative and quantitative analysis is performed for the state estimation for dynamic tracking, where the quantitative analysis is measured in terms of the least-squared estimation error. As a result, an improvement in the dynamic tracking of the EEG is achieved for the proposed dynamic Tikhonov based on the DNN approach obtained in comparison with the static and dynamic Tikhonov approaches.

The paper is organized as follows: in Section 2 the dynamic Tikhonov structure based on DNN constraints is presented; in Section 3, the state forecasting results for the proposed approach and the static and dynamic Tikhonov approaches are shown. Finally, in Section 4, the conclusions and final remarks are presented.

2. Materials and Methods

2.1. Forward Dynamic Problem

The measurements equation for the state-space representation of the EEG dynamics can be described as follows:

$$y_k = Ax_k + \epsilon_k \tag{1}$$

where y_k are the vector time series measurements at time k, x_k is the state vector, and A is the lead-field matrix. In addition, the intrinsic dynamic evolution of the states x_k is defined through a non-linear differences equation, as follows:

$$x_k = f(x_{k-1}, x_{k-2}, \ldots) + \eta_k \tag{2}$$

where f is a non-linear dynamic difference equation that describes the state evolution. The structure of $f(.)$ can be defined as a non-linear physically motivated model, as used in [13].

2.2. Dynamic Tikhonov Based on DNN

Consider a cost function defined by

$$J_k = \|y_k - Ax_k\|_2^2 + \lambda^2\|x_k - x_k^-\|_2^2 \tag{3}$$

where x_k^- the a priori state estimation, and where the solution can be computed as

$$\hat{x}_k = (A^T A + \lambda^2 I)^{-1}(A^T y_k + \lambda^2 x_k^-) \tag{4}$$

In this work, the a priori estimation is performed by using a DNN in order to consider the dynamic evolution of the states, therefore approximating the function described in (2) as follows

$$x_k^- = \Phi(\hat{x}_{k-1}) \tag{5}$$

where Φ is the DNN. Therefore,

$$J_k = \|y_k - Ax_k\|_2^2 + \lambda^2 \|x_k - \Phi(\hat{x}_{k-1})\|_2^2 \tag{6}$$

where the solution for state forecasting can be computed as

$$\hat{x}_k = (A^T A + \lambda^2 I)^{-1}(A^T y_k + \lambda^2 \Phi(\hat{x}_{k-1})) \tag{7}$$

The DNN in Figure 1 shows the structure of the DNN Φ of (5), used to approximate f in (5). It has an input layer x_{k-1}, three hidden layers which are all fully connected, and an output layer x_k with two outputs.

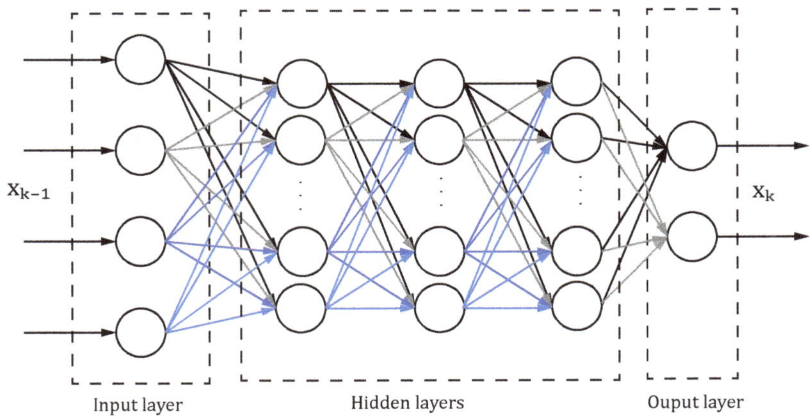

Figure 1. DNN used to consider the dynamic evolution of the states in the Tikhonov method.

3. Results

3.1. Experimental Setup

In order to evaluate the performance of the proposed dynamic Tikhonov approach based on DNN dynamic constraints, a simulation of the time series corresponding to the EEG was performed using Equations (1) and (2). To this end, the lead-field matrix corresponding to the New York head model was used [14].

The model considered two different source configurations, a detailed representation of which can be seen in Figure 2.

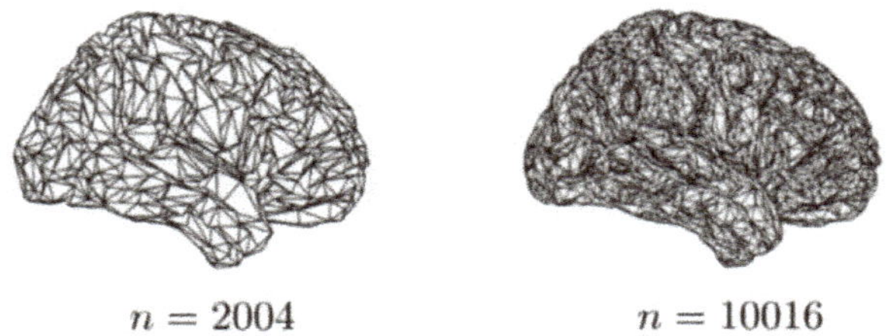

Figure 2. New York head model was selected for two source configurations: a model with $n = 2004$ sources, and a model with $n =$ 10,016 sources.

The non-linear function for states evolution $f(.)$ was simulated considering the structures proposed in [13] as follows:

$$x_k = A_1 x_{k-1} + A_2 x_{k-2} + A_3 x_{k-1}^{\circ 2} + A_4 x_{k-2}^{\circ 3} + A_5 x_{k-\eta} \tag{8}$$

where $\circ$ is the Hadamard product and η is the delayed state. The simulation of the EEG was developed considering $A_1 = 0.8I$, and A_2, A_3, A_4, and A_5 equal to zero. The approximation of the function f was performed by the DNN Φ by using a structure with three hidden layers, ReLU activation functions and an L2 regularization parameter to avoid overfitting and reduce model complexity to improve performance on the test data. A comparative analysis in terms of the least-squared error was performed considering the static and dynamic Tikhonov approaches. The implementation of the proposed DNN approach was performed in Python using TensorFlow.

Figures 3 and 4 show the simulated EEG using a brain model with 2004 and 10,016 states of four EEG time series measurements, respectively.

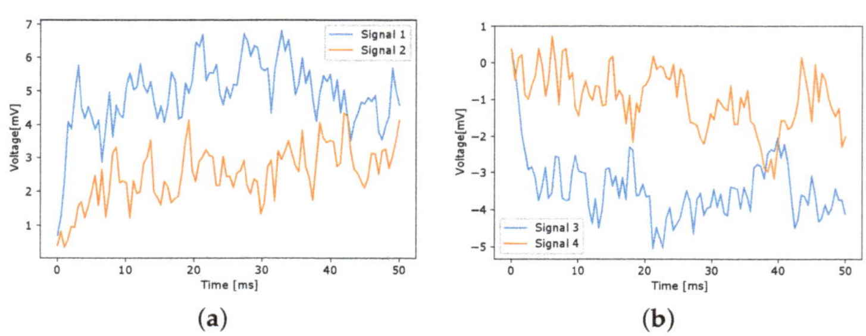

Figure 3. Simulated EEG for a head model with 2004 states. (**a**) Two channels of simulated EEG: channels 1 and 2. (**b**) Two channels of simulated EEG: channels 3 and 4.

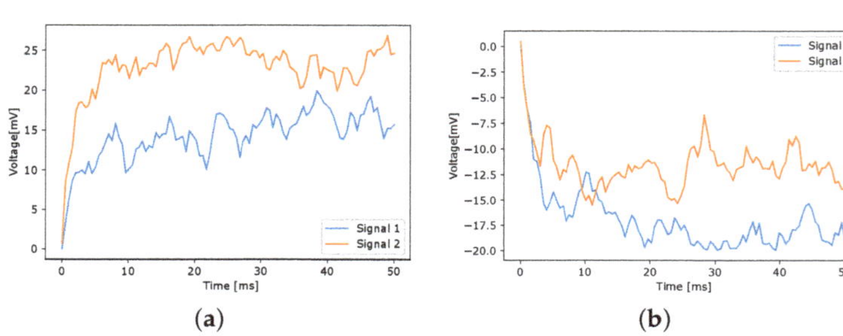

Figure 4. Simulated EEG for a head model with 10,016 states. (**a**) Two channels of simulated EEG: channels 1 and 2. (**b**) Two channels of simulated EEG: channels 3 and 4.

3.2. State Forecasting Results

The state forecasting results show that the proposed dynamic Tikhonov based on the DNN approach achieves the closest fit to the trustworthy sources, as evidenced by the lower estimation error compared to the static and dynamic Tikhonov approaches, as shown in the following graphs for the two models and the estimation with different hidden layers. Results were obtained by testing the New York model for 2004 and 10,016 states, comparing the behaviour of the data estimates with different architectures, making variations from zero layers, for a simple linear regression model, to three layers, obtaining the best performance with the dynamic Tikhonov based on the DNN approach.

Real and estimated states using the Tikhonov and dynamic Tikhonov based on the DNN are shown in Figures 5 and 6 using one and three hidden layers, respectively. Four states (10, 600, 1200, and 2000) were used for the model to adequate estimate the behaviuor of the source in different states. In all four cases, it can be seen that the behaviour of the dynamic Tikhonov based on the DNN approach overcomes the Tikhonov approach, and that the three-layer architecture achieves a better performance compared to the one-layer architecture. In order to observe the behaviour of the implemented methodology, a test with a 2004-state model (considered small) was made.

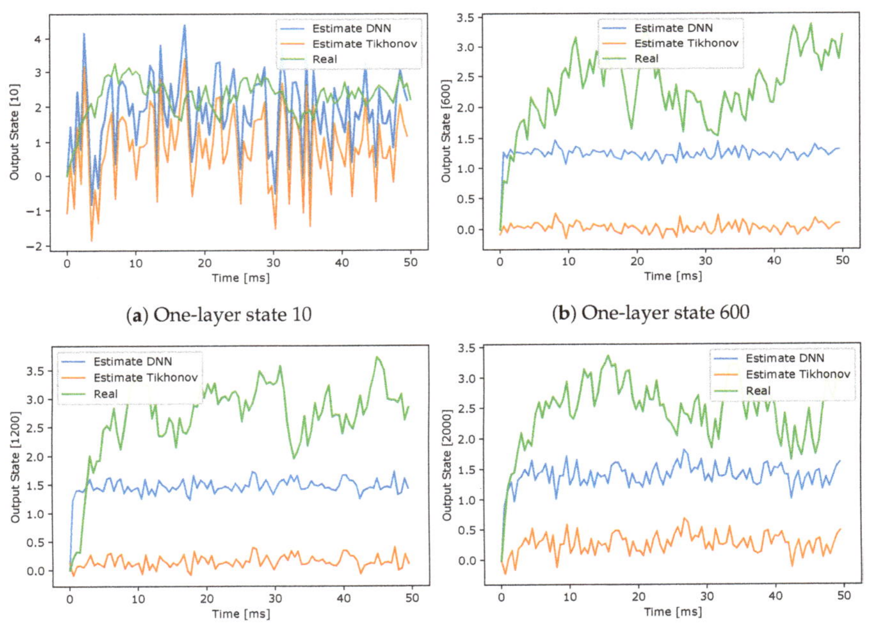

(**a**) One-layer state 10 (**b**) One-layer state 600

(**c**) One-layer state 1200 (**d**) One-layer state 2000

Figure 5. Output estimation for the one-layer architecture of the Tikhonov and dynamic Tikhonov based on the DNN models for 2004 states.

Once several architectures for the model with 2004 states were tested, they were extrapolated to a model with 10,016 state to observe the dynamic behaviour of the system with Tikhonov and dynamic Tikhonov based on DNN approaches. This comparison for the architectures was also made for the model with 10,016 states. Figures 7 and 8 show the results obtained by using one and three hidden layers, respectively. Four states (states 100, 2500, 5000, and 10,000) were used to observe the behaviour of the approaches with respect to the source in different states for a bigger model.

Overall, the models with 2004 and 10,016 sources and three hidden layers demonstrated the efficacy of the proposed dynamic Tikhonov based on the DNN approach for state estimation in the context of the EEG inverse problem. The approach achieved lower estimation errors, was closer to sources, and had faster convergence rates than the static and dynamic Tikhonov approaches.

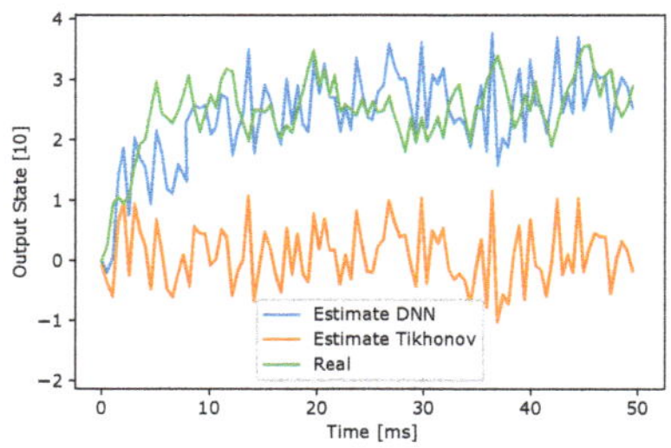

(**a**) Three-layer state 10

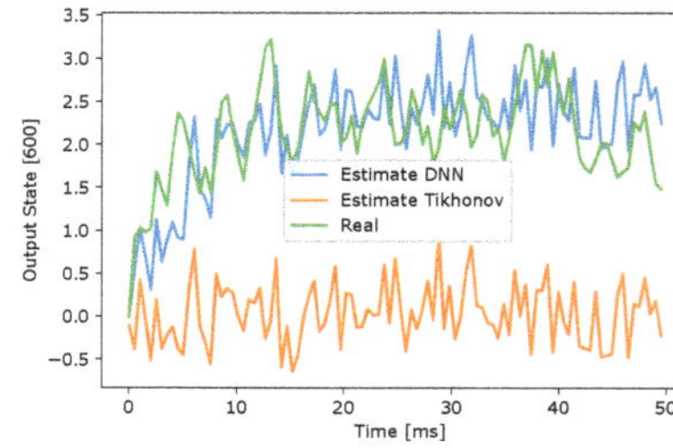

(**b**) Three-layer state 600

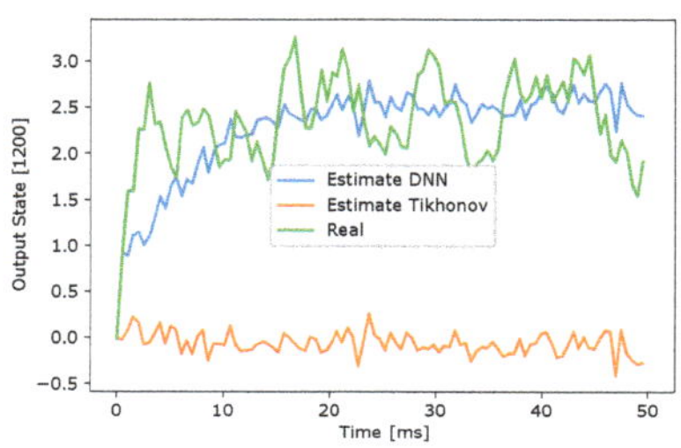

(**c**) Three-layer state 1200

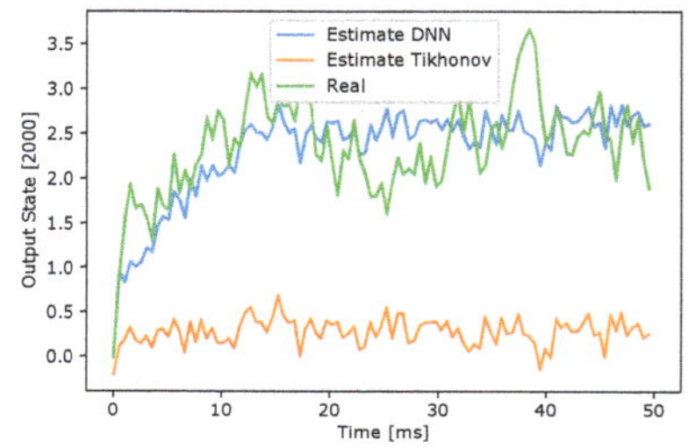

(**d**) Three-layer state 2000

Figure 6. Output estimation for the three-layer architecture of the Tikhonov and dynamic Tikhonov based on the DNN models for 2004 states.

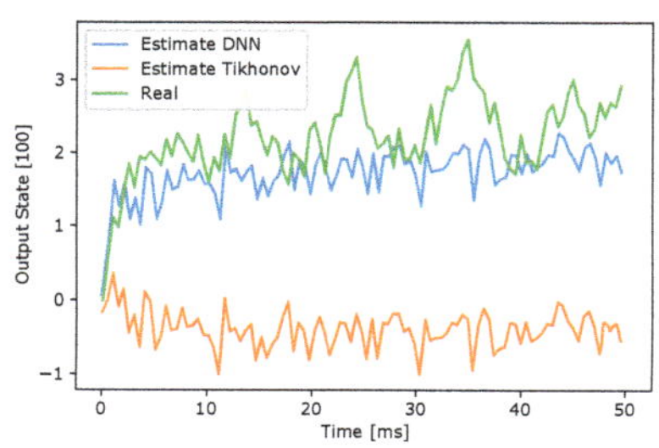

(**a**) One-layer state 100

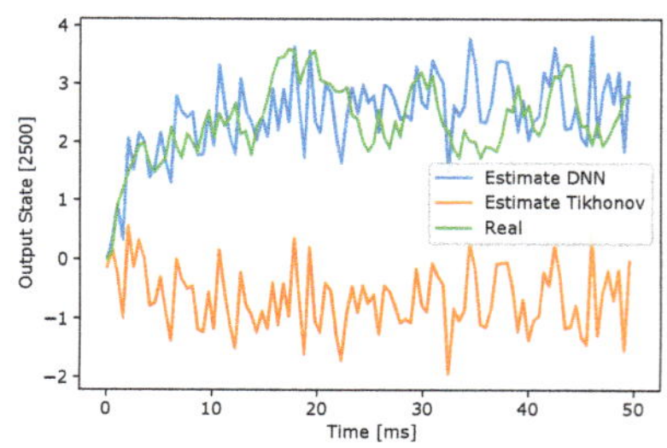

(**b**) One-layer state 2500

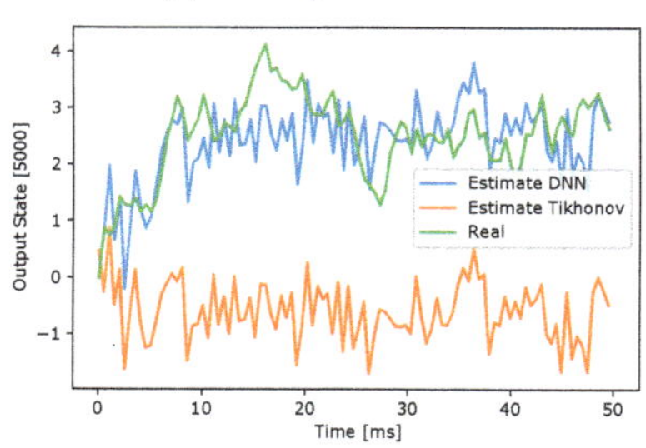

(**c**) One-layer state 5000

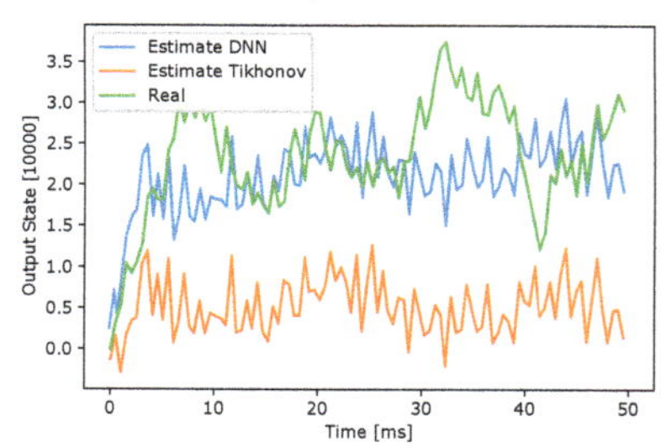

(**d**) One-layer state 10,000

Figure 7. Output estimation for the one-layer architecture of the Tikhonov and dynamic Tikhonov based on the DNN models for 10,016 states.

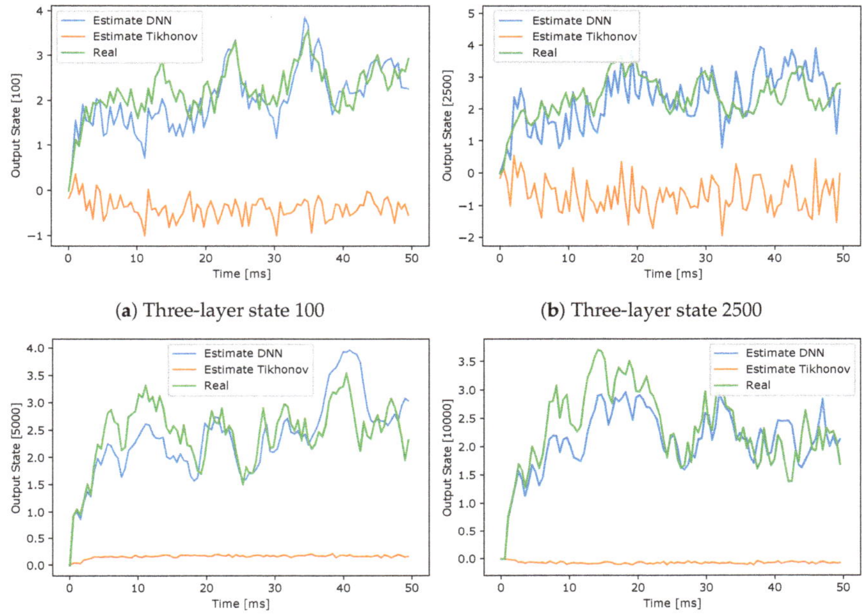

(**a**) Three-layer state 100 (**b**) Three-layer state 2500

(**c**) Three-layer state 5000 (**d**) Three-layer state 10,000

Figure 8. Output estimation for the three-layer architecture of the Tikhonov and dynamic Tikhonov based on the DNN models for 10,016 states.

Figure 9 shows the estimation results from the forecasting, demonstrating that the DNN approach continues to achieve the closest fit to the actual sources and exhibits the lowest prediction error compared to the static and dynamic Tikhonov approaches. This suggests that the DNN approach is more effective in capturing the complex non-linear dynamics of the EEG sources.

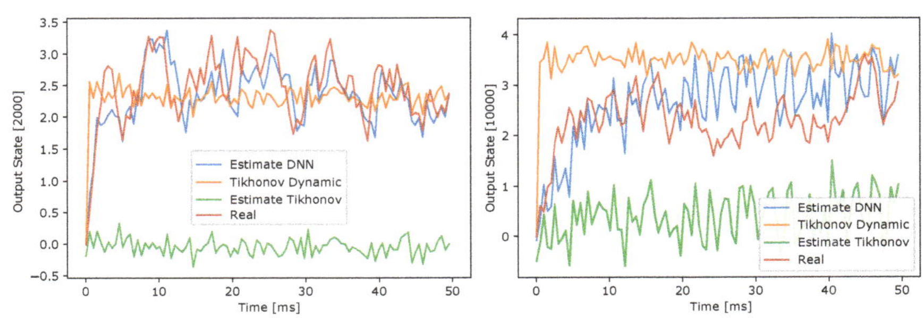

(**a**) Model 2004 states (**b**) Model 10,016 states

Figure 9. Output estimation of the Tikhonov and dynamic Tikhonov based on the DNN models for 2004 and 10,016 states.

Table 1 shows the estimation results for the static and dynamic Tikhonov approach in terms of the least-squared error for a model with 2004 and 10,016 states.

Table 1. Mean-squared estimation error.

Model	Regularized LS Model 2 K	Regularized LS Model 10 K
Estatic	639.016	602.9669
Dynamic	127.4491	95.4469

The least-square error for the dynamic Tikhonov approach is significantly lower than for the static Tikhonov approach for both models. This suggests that incorporating the temporal dynamics in the state estimation problem can significantly improve the accuracy of the estimation. In addition, Table 2 shows the estimation errors by using the proposed dynamic Tikhonov with DNN constraints approach using several configurations of hidden layers in terms of the least-squared error.

Table 2. Mean-squared estimation error.

Hidden Layers	Regularized LS-DNN Model 2 K	Regularized LS-DNN Model 10 K
0	439.4857	897.7086
1	348.5712	60.4926
2	177.608	42.6174
3	104.813	34.2116

In terms of the least-squared estimation error, the dynamic Tikhonov approach with DNN constraints outperformed both the static and dynamic Tikhonov approaches, as well as achieved better results with an increasing number of hidden layers.

4. Discussion and Conclusions

The proposed dynamic Tikhonov based on the DNN constraints approach was evaluated to solve an EEG inverse problem for neural activity estimation. By approximating the non-linear dynamic of state evolution using DNNs for two brain models with 2004 and 10,016 sources, the method improves the dynamic tracking of EEG. Qualitative and quantitative analyses were conducted for state forecasting for dynamic tracking, and the proposed method showed better results than the static and dynamic Tikhonov approaches. The findings indicate that using a dynamic approach that considers changing dynamics over time can improve the estimation accuracy, especially when combined with DNN constraints and multiple hidden layers. The proposed method also has the potential for future behaviour prediction and could be valuable in solving inverse problems in various applications with dynamic behaviour and non-linear dynamics.

Author Contributions: Conceptualization, E.G., J.M.; methodology, E.G., C.M., J.M.; software, E.G., C.M., J.M.; validation, E.G., C.M., J.M.; formal analysis, E.G.; investigation, C.M.; resources, E.G., C.M., J.M.; data curation, E.G., C.M., J.M.; writing—original draft preparation, E.G., C.M., J.M.; writing—review and editing, E.G., C.M., J.M.; visualization, E.G., C.M., J.M.; supervision, E.G., J.M.; project administration, E.G.; funding acquisition, E.G., C.M., J.M. All authors have read and agreed to the published version of the manuscript.

Funding: This work is funded by Project No. 6-22-8 entitled "Identificación y control de sistemas multivariables interconectados a gran escala" by Universidad Tecnológica de Pereira, Pereira, Colombia.

Institutional Review Board Statement: Not applicable.

Informed Consent Statement: Not applicable.

Data Availability Statement: Not applicable.

Conflicts of Interest: The authors declare no conflict of interest.

References

1. Zhang, K.; Zuo, W.; Chen, Y.; Meng, D.; Zhang, L. Beyond a gaussian denoiser: Residual learning of deep CNN for image denoising. *IEEE Trans. Image Process.* **2017**, *26*, 3142–3155. [CrossRef] [PubMed]
2. Li, H.; Schwab, J.; Antholzer, S.; Haltmeier, M. Nett: Solving inverse problems with deep neural networks. *Inverse Probl.* **2020**, *36*, 065005. [CrossRef]
3. Zhang, Z.; Brand, M. Convergent block coordinate descent for training tikhonov regularized deep neural networks. In Proceedings of the Advances in Neural Information Processing Systems, Long Beach, CA, USA, 4–9 December 2017; Volume 30.
4. Chien, J.T.; Lu, T.W. Tikhonov regularization for deep neural network acoustic modeling. In Proceedings of the 2014 IEEE Spoken Language Technology Workshop (SLT), South Lake Tahoe, NV, USA, 7–10 December 2014; pp. 147–152.
5. Afkham, B.M.; Chung, J.; Chung, M. Learning regularization parameters of inverse problems via deep neural networks. *Inverse Probl.* **2021**, *37*, 105017. [CrossRef]
6. Nguyen, H.V.; Bui-Thanh, T. Tnet: A model-constrained tikhonov network approach for inverse problems. *arXiv* **2021**, arXiv:2105.12033.
7. Romano, Y.; Elad, M. The little engine that could: Regularization by denoising (red). In Proceedings of the IEEE Conference on Computer Vision and Pattern Recognition, Honolulu, HI, USA, 21–26 July 2017.
8. Mao, X.J.; Shen, C.; Yang, Y.B. Image restoration using very deep convolutional encoder-decoder networks with symmetric skip connections. In Proceedings of the European Conference on Computer Vision (ECCV), Amsterdam, The Netherlands, 11–14 October 2016.
9. Kolowrocki, K. 10—Large complex systems. In *Reliability of Large and Complex Systems;* Elsevier: Amsterdam, The Netherlands, 2014; Volume 2.
10. Sockeel, S.; Schwartz, D.; Pélégrini-Issac, M.; Benali, H. Large-scale functional networks identified from resting-state EEG using spatial ICA. *PLoS ONE* **2016**, *11*, e0146845. [CrossRef] [PubMed]
11. Sanchez-Bornot, J.M.; Sotero, R.C.; Kelso, S.; Coyle, D. Solving large-scale meg/EEG source localization and functional connectivity problems simultaneously using state-space models. *arXiv* **2022**, arXiv:2208.12854.
12. Wang, Q.; Loh, J.M.; He, X.; Wang, Y. A latent state space model for estimating brain dynamics from electroencephalogram (EEG) data. *Biometrics* 2022, *Early View*. [CrossRef]
13. Giraldo-Suarez, E.; Martinez-Vargas, J.D.; Castellanos-Dominguez, G. Reconstruction of neural activity from EEG data using dynamic spatiotemporal constraints. *Int. J. Neural Syst.* **2016**, *26*, 1650026. [CrossRef] [PubMed]
14. Huang, Y.; Parra, L.C.; Haufe, S. The new york head—A precise standardized volume conductor model for EEG source localization and tes targeting. *NeuroImage* **2016**, *140*, 150–162. [CrossRef] [PubMed]

engineering proceedings

MDPI

Proceeding Paper

Statistical Haplotypes Based on Functional Sequence Data Analysis for Genome-Wide Association Studies †

Pei-Yun Sun and Guoqi Qian *

School of Mathematics and Statistics, University of Melbourne, Parkville, VIC 3010, Australia; peiyuns@student.unimelb.edu.au
* Correspondence: qguoqi@unimelb.edu.au; Tel.: +61-3-8344-4899
† Presented at the 9th International Conference on Time Series and Forecasting, Gran Canaria, Spain, 12–14 July 2023.

Abstract: Functional data analysis has demonstrated significant success in time series analysis. In recent biomedical research, it has also been used to analyze sequence variations in genome-wide association studies (GWAS). The observations of genetic variants, called single-nucleotide polymorphisms (SNPs), of an individual are distributed over the loci of a DNA sequence. Thus, it can be regarded as a realization of a stochastic process, which is no different from a time series. However, SNPs are usually coded as the number of minor alleles, which are categorical. The usual least-square smoothing in FDA only works well when the data is continuous and normally distributed. The normality assumption will be violated for categorical SNP data. In this work, we propose a two-step method for smoothing categorical SNPs using a novel method and constructing haplotypes having strong associations with the disease using functional generalized linear models. We show its effectiveness through a real-world PennCATH dataset.

Keywords: stochastic process; functional data analysis; genome-wide association study; epistasis; haplotype; variable selection

1. Introduction

Functional data analysis (FDA) is a common tool for analyzing complex datasets that are collected over a sequence, such as time and location. It is based on the assumption that there is an underlying stochastic process represented by a continuous function $x_i(t)$, but the data can only be observed at discrete time points or locations. To overcome this challenge, FDA represents the datasets as functions or curves by smoothing the observations. FDA has found extensive application in many fields related to time series analysis. Examples include predicting stock prices, discovering geophysical and meteorological patterns, and forecasting traffic volumes. Recently, there has been increasing interest in using FDA in biomedical research, such as genome-wide association studies (GWAS).

GWAS aim to find associations between single-nucleotide polymorphisms (SNPs) and diseases or traits by case–control studies, providing insights into the genetic risk factors for diseases and offering opportunities for preventive measures and treatments [1]. As a result, GWAS plays a critical role in personal genomics, and thousands of GWAS have been conducted over the past few decades. Fan et al. [2] have highlighted that genetic variant data can be viewed as a collection of random variables forming a stochastic process, and can be treated as functional data. One of the key advantages of FDA in GWAS is its ability to naturally incorporate the correlation, linkage, and linkage disequilibrium (LD) information of the genetic variants into the association tests. FDA can capture the complex dependency structure and higher order LD among the genetic variants, which is often missed by other methods such as the sequence kernel association test (SKAT) [3] and its optimal unified test (SKAT-O) [4]. Another advantage of the functional data representation is its suitability for large-scale genomic data, providing a computationally efficient way to test the association between multiple variants and the phenotype.

Citation: Sun, P.-Y.; Qian, G. Statistical Haplotypes Based on Functional Sequence Data Analysis for Genome-Wide Association Studies. *Eng. Proc.* **2023**, *39*, 29. https://doi.org/10.3390/engproc2023039029

Academic Editors: Ignacio Rojas, Hector Pomares, Luis Javier Herrera, Fernando Rojas and Olga Valenzuela

Published: 29 June 2023

In their work, Fan et al. [2] proposed two methods for fitting the discrete SNP values: either using ordinary least-square smoothers, or approximating SNPs with K functional principal components. The effectiveness of FDA in GWAS, as demonstrated by Fan et al., has also been confirmed in the subsequent studies [5–7]. Nevertheless, SNPs are typically coded as the number of minor alleles, resulting in categorical data that violates the normality assumption required for the least-square smoothing. The objective of this paper is to propose a more robust approach to represent genetic variants using functional data analysis, which addresses the problem of normality assumption violation in the least-square smoothing method. In addition, we introduce a two-step method that integrates our novel smoothing method with functional generalized linear models to identify haplotypes, i.e., blocks of SNPs, that are strongly associated with the phenotype.

Our proposed approach is described in detail in the following sections. In Section 2, we describe the PennCATH dataset used in our experiments and explain our two-step method, which includes curve smoothing and variable selection by regression analysis, for identifying haplotypes associated with the phenotype. Section 3 presents the performance of our method, compared with single SNP test results and haplotypes constructed based on genetic information. In Section 4, we discuss the strengths and limitations of our approach, as well as possible future directions. We conclude the work in Section 5 by summarizing our findings and emphasizing the potential impact of our approach on genetics and genomics research.

2. Materials and Methods

2.1. GWAS Data

In GWAS, SNPs are often coded as the number of minor alleles observed at a given locus. For a particular locus, there are two alleles on a homologous chromosome pair, and the SNPs can be coded as 0, 1, or 2. As an example, if 90% of the genomes have nucleotide A at the locus, and 10% have T, this locus would be an SNP with two alleles. The nucleotide A, the more common allele, is referred to as the major allele or reference allele, while T, the less common allele, will be the minor allele or non-reference allele. Consequently, a genotype of AA is coded as 2, AT or TA as 1, and TT as 0. Section 3 of this paper will present a GWAS example using the PennCATH dataset. PennCath [8] is a GWAS of coronary artery disease (CAD) conducted by the University of Pennsylvania Medical Centre. The study has recruited 3850 patients undergoing cardiac catheterization and coronary angiography between 1 July 1998 and 31 March 2003. All of the patients have provided written informed consent. Their age, gender, ethnicity, medical history, physical exams, and other clinical data have been extracted from their medical records. The samples in PennCath are genotyped by the calling algorithm "Birdseed", offered by the Affymetrix Genome-Wide Human SNP Array 6.0 platform. The dataset comprises 656,890 SNPs from 1401 patients after a quality control procedure [9]: 933 patients with coronary artery disease (CAD) and 468 with no or minimal CAD. In this work, 3758 SNPs on Chromosome 9p21 are extracted and divided into 4 chunks of equal window size. It is expected to find strong associations between some of the SNPs at around 9p21.3 (position 19.9Mb to 25.6 Mb) and CAD according to previous studies [10–12].

2.2. Curve Smoothing

The characteristics of FDA make it natural to apply FDA to numerical data, but it may not seem to be applicable to categorical data such as SNPs. To address this issue, we make a different assumption about the underlying stochastic curves. Instead of assuming that all the categorical values $k = 0, \dots, K-1$ came from a single underlying curve $x_i(t)$, we assume there are K probability curves $p_{ik}(t)$ associated with the continuum t. Hence, an observation y_{ij} can be seen as a sample drawn from a categorical distribution:

$$y_{ij} \sim \text{Categorical}(p_{i0}(t_{ij}), ..., p_{iK-1}(t_{ij})) \quad (1)$$

The probability curves for any sample i at any position t need to follow the law of total probability and lie between 0 and 1:

$$\sum_{k=0}^{K-1} p_{ik}(t) = 1$$
$$0 < p_{i0}(t), \ldots, p_{iK-1}(t) < 1$$

In the case where the categorical observations $k = 0, \ldots K-1$ can be regarded as the number of "success" events in $K-1$ trials, y_{ij} can be treated as a sample from the binomial distribution:

$$y_{ij} \sim \text{Binomial}(n = K-1, p = p_i(t_{ij})) \quad (2)$$

in which the probability of "success" depends on the value of a single probability curve $p_i(t_{ij})$. The bounds of 0 and 1 for probabilities still apply in this setting. In both (1) and (2), the rationale behind smooth probability curves assumption naturally accommodated the correlations among adjacent SNPs: for any two SNPs j and j' of an individual i, if the SNPs have close positions t_{ij} and $t_{ij'}$, they should have similar probabilities for minor alleles. Therefore, the probabilities are assumed to change smoothly as the position varies.

In minor-allele SNP coding, SNPs are coded as categorical variables 0, 1, and 2. There are at least two ways to represent SNPs with functional data. The first method is to consider them as 3 unordered categories with probabilities

$$P(\text{SNP}_{ij} = 0) = p_{ij0}$$
$$P(\text{SNP}_{ij} = 1) = p_{ij1}$$
$$P(\text{SNP}_{ij} = 2) = p_{ij2}$$

which has weak assumptions on interdependence among the categories. The observations of SNPs can then be adapted into the FDA framework based on Assumption (1) with $K = 3$. However, each sample will be determined by two probability curves, which may give too much flexibility in fitting the data. In Assumption (2), SNPs are regarded as the outcome of 2 binomial trials with probability p, which is also a reasonable interpretation since each SNP is equivalent to the number of occurrences of the less common allele in two alleles at a position. With the Hardy–Weinberg Equilibrium (HWE) principle, which states that "genotype frequencies in a population remain constant between generations in the absence of disturbance by outside factors" [13], the occurrences of minor alleles in the two alleles are independent of one another, and the probability of each outcome will be as follows:

$$P(\text{SNP}_{ij} = 0) = (1 - p_{ij})^2$$
$$P(\text{SNP}_{ij} = 1) = 2p_{ij}(1 - p_{ij})$$
$$P(\text{SNP}_{ij} = 2) = p_{ij}^2$$

In GWAS, the check of violations of HWE is usually part of the quality control. The departures from HWE can indicate potential genotyping errors [14,15], and it is a common practice to remove those SNPs from the studies. Therefore, it is reasonable to make HWE assumptions about the probabilities for minor alleles, and the functional representation can also be built on (2). In either setting, the probabilities vary between individuals and over SNP positions. Therefore, we can find the associations between the differences in their probability curves and the target disease. In this paper, we will be converting the observed SNP into functional data based on Assumption (2).

Smoothing for probability curves of SNP data is an example of curve smoothing with constraints. One solution to this is to transform them into unconstrained curves. For SNP data, fitting the probability curves can be achieved by fitting the following unconstrained log-odds curves:

$$g_i(t) = \log \frac{p_i(t) + \alpha}{(1 - p_i(t)) + \alpha} \tag{3}$$

where α is a Laplace smoothing parameter set to 0.01 to avoid zero probabilities. Once the log-odds curves have been smoothed, they can easily be transformed back to probability curves $p_i(t)$ by the logistic function. To smooth the log-odds curves, we can use maximum likelihood estimation (MLE) to solve for the coefficients iteratively with gradient descent. Alternatively, we can smooth the empirical log-odds values (−4.62, 0, 4.62) that are transformed from the empirical probabilities (0, 0.5, 1) for categorical SNP observations of (0, 1, 2), respectively. This transformation brings the values closer to a normal distribution, enabling us to obtain a better approximation via least-square smoothing.

2.3. Variable Selection through Regression Analysis

Finite-dimensional regression models such as GLMs and MLMs are commonly used in GWAS. These models treat the phenotype or disease as the response variable, and the SNPs as explanatory variables. By examining the coefficients and p-values of the SNPs, we can determine which SNPs are strongly associated with the phenotype/disease. Similar to the finite-dimensional regression models, coefficients and p-values in functional regression models can also indicate the relationship between phenotype and covariates through the "scalar-on-function" model. Depending on the response data type, the model can be a functional linear regression or a functional logistic regression. When the phenotype is continuous V_i, such as height or blood pressure, a functional linear model can be employed, where β_0 is a scalar intercept term and $\beta(t)$ is the functional coefficient of the probability curve:

$$V_i = \beta_0 + \int \beta(t) p_i(t) dt + \epsilon_i \tag{4}$$

On the other hand, when the phenotype V_i is binary, such as a disease indicator, the probability of the individual having the phenotype, $\pi_i = Pr(V_i = 1)$, can be modeled by:

$$\log(\frac{\pi_i}{1 - \pi_i}) = \beta_0 + \int \beta(t) p_i(t) dt \tag{5}$$

Due to the complication of GWAS data, there may be some confounding factors making it difficult to find the true relationships between probability curves and the target disease. These factors can be accounted for by including them in the regression model. Therefore, the final disease-SNP model that accounts for age, sex, and population substructure can be formulated as:

$$\log(\frac{\pi_i}{1 - \pi_i}) = \beta_0 + \beta_1 \text{age}_i + \beta_2 \text{sex}_i + \beta_3 \text{pc}_{i1} + \ldots + \beta_{12} \text{pc}_{i,10} + \int \beta(t) p_i(t) dt \tag{6}$$

The number of basis functions $q^{(\beta)}$ for $\beta(t)$ needs to be carefully selected. If $q^{(\beta)}$ is too large, $\beta(t)$ will have more curvature and may be capturing noise instead of signals in $p_i(t)$. A common method to determine $q^{(\beta)}$ is to select the value that minimizes $\text{AIC} = -2l + 2q^{(\beta)}$ such that each additional basis function in $\beta(t)$ should significantly increase the log-likelihood in order to justify a lower AIC value.

In regression analysis, determining if an explanatory variable has a statistically significant association with the response variable requires taking into account not only the coefficient but also the standard errors of the estimated coefficients. The same principle applies to functional regression analysis. The statistically significant SNPs can be found by examining the confidence band of $\beta(t)$, which connects the point-wise confidence intervals of $\beta(t)$ at each position t:

$$\hat{\mathbf{b}}^T\boldsymbol{\phi}(t) \pm z_{\alpha/2}\sqrt{\sum_{k=1}^{q^{(\beta)}}\sum_{l=1}^{q^{(\beta)}}\phi_k(t)\phi_l(t)Cov(b_k, b_l)}$$

where $\boldsymbol{\phi}(t)$ is a set of basis functions for $\beta(t)$, and $\mathbf{b}$ is a vector of coefficients corresponding to the basis functions.

3. Results

In our experiments, we aim to evaluate the performance of our two-step method in identifying important SNPs by comparing it with single SNP tests and genetic-based haplotypes using real-world data from the PennCATH study of coronary artery disease (CAD). We will use two different approaches to create genetic-based haplotypes: a fixed number of SNPs and a fixed window size. For these haplotypes, we will conduct GWAS by fitting logistic models. In contrast, for our FDA-based haplotypes, we will use functional logistic models to analyze all SNPs at once and identify haplotypes with strong associations with the target disease.

3.1. Single SNP Tests

The single SNP test is an association analysis that regresses the phenotype on each SNP separately. While it can identify SNPs with large marginal effects on the phenotype, it does not consider epistatic effects among the SNPs. The test results shown in Figure 1 (left) are based on logistic models fitted on each SNP using CAD as the response, adjusted for confounding factors. By setting the threshold for p-value to 10^{-3}, we identify 9 significant SNPs between position 22064465 and 22125503. Additional information about the significant SNPs and their test results is provided in Table 1.

Table 1. Single SNP tests found 9 SNPs strongly associated with CAD in chromosome 9p21 with p-value $< 10^{-3}$.

SNP	Position	Estimate	Std. Error	t-Value	p-Value
rs6475606	22081850	−0.445673	0.092841	−4.800395	1.583531×10^{-6}
rs1333049	22125503	−0.442759	0.094169	−4.701767	2.579192×10^{-6}
rs4977574	22098574	−0.436480	0.094321	−4.627610	3.699095×10^{-6}
rs2891168	22098619	−0.432042	0.094221	−4.585422	4.530707×10^{-6}
rs1333042	22103813	−0.431915	0.094976	−4.547645	5.424963×10^{-6}
rs10757272	22088260	−0.421598	0.094232	−4.474050	7.675192×10^{-6}
rs1333048	22125347	−0.404207	0.094842	−4.261892	2.027037×10^{-5}
rs10757270	22072719	0.389317	0.094121	4.136343	3.528847×10^{-5}
rs8181047	22064465	−0.425314	0.109852	−3.871706	1.080763×10^{-4}

3.2. Genetic Haplotypes

We used two common approaches to construct the genetic-based haplotypes: (1) blocks containing a fixed number of SNPs and (2) fixed genomics window of size. In these two approaches, we have constructed blocks containing 5, 10, 20, and 50 SNPs and blocks with window sizes 10 kb, 20 kb, 50 kb, and 100 kb, respectively. It ends up with 1392 blocks generated for (1) and 2394 blocks generated for (2).

To compute the p-value of each haplotype, we fit a GLM for each haplotype while adjusting for confounding factors and compare it to a base model with only an intercept and confounding factors. However, due to missing values in our genotype data, samples containing missing values for any SNP in a haplotype were excluded when fitting the models. To ensure sufficient data for fitting GLMs, we imputed missing values using the most frequent value (mode) for each SNP before computing the p-values.

Applying the same p-value threshold of 10^{-3} as we did for single SNP tests, we found 9 significant blocks associated with CAD using Approach (1) and 11 significant blocks using Approach (2). Figure 1 (middle) displays the positions of the significant blocks, which are primarily located around 22 Mb with a few at around 24 Mb. These findings are consistent with our results from single SNP tests, which identified significant single SNPs near 22 Mb. The haplotypes at around 24 Mb have larger p-values ($>10^{-4}$) as compared with the haplotypes at 22 Mb. Therefore, they could be considered as noise instead of true associations. Table 2 provides further details on the significant blocks, including their p-values.

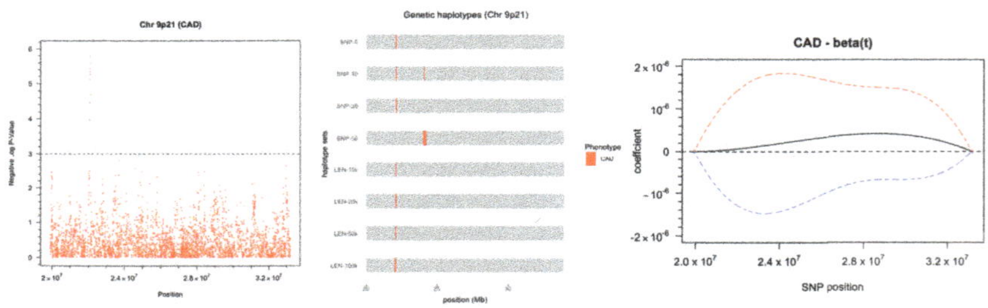

Figure 1. (**Left**) Manhattan plot displaying the distribution of p-values for 3758 SNPs located on chromosome 9p21 in relation to their genomic position. (**Middle**) Genetic-based haplotypes constructed using a fixed number of SNPs and fixed window size that showed significant associations with CAD. (**Right**) Estimate and 95% confidence bands of $\beta(t)$ in functional logistic model using 3758 SNPs.

Table 2. Significant genetic haplotypes found by fixed SNP and fixed window size methods.

Method	Haplotype No.	Position	Number of SNPs	p-Value
SNP-5	1	22073334–22088260	5	1.522134×10^{-5}
	2	22098574–22125503	5	1.941005×10^{-4}
	3	22128709–22133251	5	9.663982×10^{-5}
	4	24095607–24112974	5	1.712341×10^{-4}
SNP-10	5	22073334–22125503	10	2.645748×10^{-4}
	6	22128709–22136489	10	6.936598×10^{-5}
	7	24075082–24112974	10	9.016075×10^{-4}
SNP-20	8	22073334–22136489	20	1.599449×10^{-4}
SNP-50	9	24018470–24233746	50	8.881007×10^{-4}
LEN-10k	10	22070008–22080008	2	1.605103×10^{-4}
	11	22080008–22090008	4	5.748821×10^{-6}
	12	22090008–22100008	2	1.189989×10^{-5}
	13	22100008–22110008	1	3.743386×10^{-6}
	14	22120008–22130008	3	1.541748×10^{-5}
LEN-20k	15	22060008–22080008	5	1.955793×10^{-4}
	16	22080008–22100008	6	1.896116×10^{-5}
	17	22100008–22120008	1	3.743386×10^{-6}
	18	22120008–22140008	15	4.555725×10^{-4}
LEN-50k	19	22100008–22150008	17	2.301992×10^{-4}
LEN-100k	20	22100008–22200008	30	9.809913×10^{-4}

3.3. Two-Step FDA Approach

In the two-step approach, we construct the FDA-based haplotypes based on the functional GLM coefficients, as explained in Section 2. This involved smoothing the log-odds curves $g(t)$ using 64 cubic spline basis functions to approximate the genotype, with the positions of the SNPs in the genome treated as the continuum t for the functions. Then, we transform the log-odds curves back to probability curves and fit the functional logistic model as described in Equation (6). In regression analysis, we used AIC to select the appropriate number of basis functions for $\beta(t)$, resulting in $q^{(\beta)} = 4$ cubic spline basis functions with AIC = 1440.4. The confidence bands of the estimated coefficient curve are displayed in Figure 1 (right). However, in a large number of SNPs, small signals in the SNPs may be overlooked. In fact, the confidence bands in Figure 1 barely found any signals in SNPs. As a result, we divided the SNPs into four chunks with an equal window size of approximately 3.3 Mb per chunk and fit the curves independently. For each chunk, the number of basis functions for SNPs $q^{(\text{SNP})}$ was set to 64. The number of bases for $\beta(t)$ in the functional logistic models of the four chunks were chosen as $q^{(\beta)} = 18$ (AIC = 1442.4), $q^{(\beta)} = 5$ (AIC = 1444.3), $q^{(\beta)} = 4$ (AIC = 1442.2), and $q^{(\beta)} = 4$ (AIC = 1442.0) in the model selection. The final estimated coefficient curves and their confidence bands for all four chunks are presented in Figure 2.

The plots suggest that SNPs located around 22 Mb exhibit strong associations with the phenotype, CAD. The confidence bands indicate strong positive associations of 68 SNPs in the position range of 21986218 to 22219365 at the 5% significance level. When the significance level is 1%, the number of SNPs with significant associations is reduced to 62 in the position range of 21988896 to 22176961. It is worth noting that there are some weak associations between CAD and a small number of SNPs located at 19914792 to 19915135, 20028452 to 20057787, and 21747672 to 21869079. These associations are significant at 5% significance level but not at 1%, and therefore, are considered noises rather than signals.

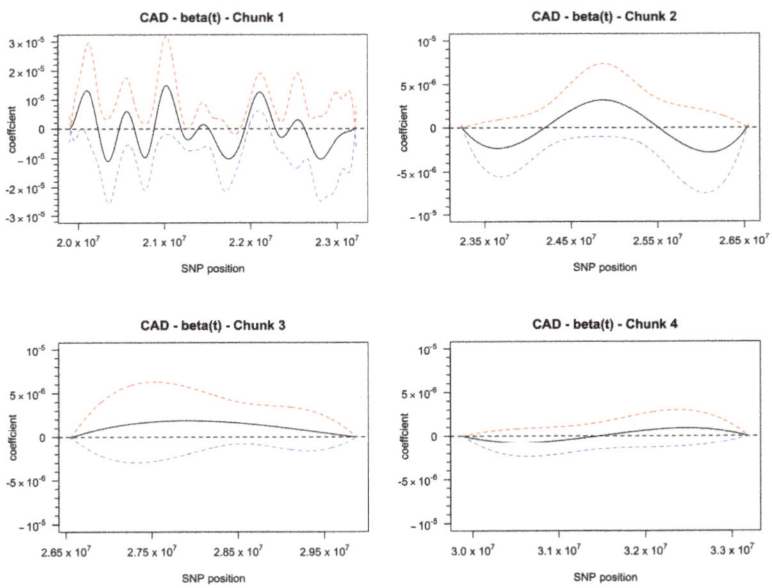

Figure 2. Coefficient curve $\beta(t)$ with its 95% confidence band. It indicates significant associations between CAD and SNPs in Chunk 1 at around position 22.1 Mb only.

4. Discussion

As stated in Section 2.1, we anticipated finding associations between SNPs in 9p21.3 and CAD. In our experiments, all three methods identified significant SNPs and haplotypes within this region (19.9 Mb to 25.6 Mb). Specifically, the SNPs with the smallest p-values in single SNP tests were located between 22.06 Mb and 22.13 Mb, while most of the strongly associated haplotypes in genetic haplotypes were found at 22.06 Mb to 22.20 Mb. Similarly, in FDA haplotypes, the most significant haplotype was found around 21.98 Mb to 22.22 Mb. Notably, in the region where these associations were detected, two genes, cyclin-dependent kinase 2A and 2B: *CDKN2A* (21.97 Mb to 21.99 Mb) and *CDKN2B* (22.00 Mb to 22.01 Mb), have been implicated in conferring risk for CAD in multiple studies [16–18], thus validating our findings.

5. Conclusions

In this paper, we presented a novel method for smoothing the categorical SNP data in GWAS and demonstrated its effectiveness in identifying haplotypes strongly associated with disease using the PennCATH dataset. The FDA haplotypes successfully retrieved all significant regions in the SNPs while maintaining low false-positive rates at low significance levels. Both functional and non-functional methods were successful, but the main advantage of the functional method is its ability to efficiently identify important SNPs without an exhaustive search while naturally incorporating spatial information and correlation among SNPs into the regression models. Additionally, in the presence of missing values, it implicitly imputes the missing SNP values when smoothing the curves, eliminating the need to exclude or impute missing values in SNPs. This prevents it from getting biased results when missing values are not handled correctly. There is not a large discrepancy between the curves smoothed directly by the least-square method and the curves smoothed by our novel approach. However, our approach, which considers the data as binomial probability curves, has better interpretability and resolves the issue of violating the normality assumption.

Author Contributions: Conceptualization, G.Q.; methodology, P.-Y.S. and G.Q.; software, P.-Y.S.; validation, P.-Y.S. and G.Q.; formal analysis, P.-Y.S. and G.Q.; investigation, P.-Y.S. and G.Q.; resources, P.-Y.S. and G.Q.; data curation, P.-Y.S.; writing—original draft preparation, P.-Y.S.; writing—review and editing, G.Q.; visualization, P.-Y.S.; supervision, G.Q.; project administration, P.-Y.S. and G.Q.; funding acquisition, G.Q. All authors have read and agreed to the published version of the manuscript.

Funding: This research received no external funding.

Institutional Review Board Statement: Not applicable.

Informed Consent Statement: Not applicable.

Data Availability Statement: No new data were created or analyzed in this study. Data sharing is not applicable to this article.

Acknowledgments: Pei-Yun Sun's research in this paper was supported by the University of Melbourne Graduate Research Training Scholarship.

Conflicts of Interest: The authors declare no conflict of interest.

References

1. Uffelmann, E.; Huang, Q.Q.; Munung, N.S.; de Vries, J.; Okada, Y.; Martin, A.R.; Martin, H.C.; Lappalainen, T.; Posthuma, D. Genome-wide association studies. *Nat. Rev. Methods Prim.* **2021**, *1*, 59. [CrossRef]
2. Fan, R.; Wang, Y.; Mills, J.L.; Wilson, A.F.; Bailey-Wilson, J.E.; Xiong, M. Functional linear models for association analysis of quantitative traits. *Genet. Epidemiol.* **2013**, *37*, 726–742. [CrossRef] [PubMed]
3. Wu, M.C.; Lee, S.; Cai, T.; Li, Y.; Boehnke, M.; Lin, X. Rare-variant association testing for sequencing data with the sequence kernel association test. *Am. J. Hum. Genet.* **2011**, *89*, 82–93. [CrossRef] [PubMed]
4. Lee, S.; Emond, M.J.; Bamshad, M.J.; Barnes, K.C.; Rieder, M.J.; Nickerson, D.A.; Christiani, D.C.; Wurfel, M.M.; Lin, X.; Project, N.G.E.S.; et al. Optimal unified approach for rare-variant association testing with application to small-sample case-control whole-exome sequencing studies. *Am. J. Hum. Genet.* **2012**, *91*, 224–237. [CrossRef] [PubMed]
5. Li, Y.; Wang, F.; Wu, M.; Ma, S. Integrative functional linear model for genome-wide association studies with multiple traits. *Biostatistics* **2022**, *23*, 574–590. [CrossRef] [PubMed]
6. Jadhav, S.; Tong, X.; Lu, Q. A functional U-statistic method for association analysis of sequencing data. *Genet. Epidemiol.* **2017**, *41*, 636–643. [CrossRef] [PubMed]

7. Chiu, C.y.; Zhang, B.; Wang, S.; Shao, J.; Lakhal-Chaieb, M.L.; Cook, R.J.; Wilson, A.F.; Bailey-Wilson, J.E.; Xiong, M.; Fan, R. Gene-based association analysis of survival traits via functional regression-based mixed effect cox models for related samples. *Genet. Epidemiol.* **2019**, *43*, 952–965. [CrossRef] [PubMed]
8. Reilly, M.; Li, M.; He, J.; Ferguson, J.; Stylianou, I.; Mehta, N.; Burnett, M.; Devaney, J.; Knouff, C.; Thompson, J.; et al. Identification of ADAMTS7 as a novel locus for coronary atherosclerosis and association of ABO with myocardial infarction in the presence of coronary atherosclerosis: Two genome-wide association studies. *Lancet* **2011**, *377*, 383–392. [CrossRef] [PubMed]
9. Reed, E.; Nunez, S.; Kulp, D.; Qian, J.; Reilly, M.P.; Foulkes, A.S. A guide to genome-wide association analysis and post-analytic interrogation. *Stat. Med.* **2015**, *34*, 3769–3792. [CrossRef] [PubMed]
10. Jarinova, O.; Stewart, A.F.; Roberts, R.; Wells, G.; Lau, P.; Naing, T.; Buerki, C.; McLean, B.W.; Cook, R.C.; Parker, J.S.; et al. Functional analysis of the chromosome 9p21. 3 coronary artery disease risk locus. *Arterioscler. Thromb. Vasc. Biol.* **2009**, *29*, 1671–1677. [CrossRef] [PubMed]
11. Shen, G.Q.; Li, L.; Rao, S.; Abdullah, K.G.; Ban, J.M.; Lee, B.S.; Park, J.E.; Wang, Q.K. Four SNPs on chromosome 9p21 in a South Korean population implicate a genetic locus that confers high cross-race risk for development of coronary artery disease. *Arterioscler. Thromb. Vasc. Biol.* **2008**, *28*, 360–365. [CrossRef] [PubMed]
12. Chen, Z.; Qian, Q.; Ma, G.; Wang, J.; Zhang, X.; Feng, Y.; Shen, C.; Yao, Y. A common variant on chromosome 9p21 affects the risk of early-onset coronary artery disease. *Mol. Biol. Rep.* **2009**, *36*, 889. [CrossRef] [PubMed]
13. Edwards, A. Anecdotal, Historical and Critical Commentaries on Genetics: GH Hardy (1908) and Hardy–Weinberg Equilibrium. *Genetics* **2008**, *179*, 1143. [CrossRef] [PubMed]
14. Turner, S.; Armstrong, L.L.; Bradford, Y.; Carlson, C.S.; Crawford, D.C.; Crenshaw, A.T.; De Andrade, M.; Doheny, K.F.; Haines, J.L.; Hayes, G.; et al. Quality control procedures for genome-wide association studies. *Curr. Protoc. Hum. Genet.* **2011**, *68*, 1–19. [CrossRef] [PubMed]
15. Marees, A.T.; de Kluiver, H.; Stringer, S.; Vorspan, F.; Curis, E.; Marie-Claire, C.; Derks, E.M. A tutorial on conducting genome-wide association studies: Quality control and statistical analysis. *Int. J. Methods Psychiatr. Res.* **2018**, *27*, e1608. [CrossRef] [PubMed]
16. Almontashiri, N.A. The 9p21. 3 risk locus for coronary artery disease: A 10-year search for its mechanism. *J. Taibah Univ. Med. Sci.* **2017**, *12*, 199–204. [PubMed]
17. McPherson, R.; Pertsemlidis, A.; Kavaslar, N.; Stewart, A.; Roberts, R.; Cox, D.R.; Hinds, D.A.; Pennacchio, L.A.; Tybjaerg-Hansen, A.; Folsom, A.R.; et al. A common allele on chromosome 9 associated with coronary heart disease. *Science* **2007**, *316*, 1488–1491. [CrossRef] [PubMed]
18. Zhong, J.; Chen, X.; Ye, H.; Wu, N.; Chen, X.; Duan, S. CDKN2A and CDKN2B methylation in coronary heart disease cases and controls. *Exp. Ther. Med.* **2017**, *14*, 6093–6098. [CrossRef]

engineering proceedings

MDPI

Proceeding Paper

Stock Embeddings: Representation Learning for Financial Time Series †

Rian Dolphin [1,*], Barry Smyth [1,2] and Ruihai Dong [1,2]

1 School of Computer Science, University College Dublin, D04 V1W8 Dublin, Ireland; barry.smyth@ucd.ie (B.S.); ruihai.dong@ucd.ie (R.D.)
2 Insight Centre for Data Analytics, University College Dublin, D04 V1W8 Dublin, Ireland
* Correspondence: rian.dolphin@ucdconnect.ie

† Presented at the 9th International Conference on Time Series and Forecasting, Gran Canaria, Spain, 12–14 July 2023.

Abstract: Identifying meaningful and actionable relationships between the price movements of financial assets is a challenging but important problem for many financial tasks, from portfolio optimization to sector classification. However, recent machine learning research often focuses on price forecasting, neglecting the understanding and modelling of asset relationships. To address this, we propose a neural model for training stock embeddings that harnesses the dynamics of historical returns data to reveal the nuanced correlations between financial assets. We describe our approach in detail and discuss several practical ways it can be used in the financial domain. Specifically, we present evaluation results to demonstrate the utility of this approach, compared to several benchmarks, in both portfolio optimization and industry classification.

Keywords: time series; representation learning; asset relationships; risk management; industry classification

Citation: Dolphin, R.; Smyth, B.; Dong, R. Stock Embeddings: Representation Learning for Financial Time Series. *Eng. Proc.* **2023**, *39*, 30. https://doi.org/10.3390/engproc2023039030

Academic Editors: Ignacio Rojas, Hector Pomares, Luis Javier Herrera, Fernando Rojas and Olga Valenzuela

Published: 29 June 2023

1. Introduction

The stock market is a challenging but appealing target for time series analysis [1,2], and it has long attracted the attention of researchers. While state-of-the-art time series techniques have demonstrated an excellent performance in the financial domain [3], much of the recent literature applying time series analysis to financial markets overlooks relational information. Instead, assets are analysed in isolation while valuable relational information is overlooked [4]. Understanding asset relationships is essential for several important financial tasks like portfolio optimization, hedging, and sector classification [5]. The conventional measure of asset similarity is correlation, popularised by Markowitz's seminal paper on modern portfolio theory [6]. However, there has been criticism of the application of correlation to financial returns [7] and recently proposed alternative similarity measures include geometric [8] and adjusted correlation-based approaches [9].

In recent years, learning embedding representations have led to breakthroughs in capturing semantic relationships in natural language processing [10]. However, the applications of embeddings in finance are mainly limited to applying pre-trained large language models to textual data, with very limited work on learning embeddings directly from non-textual financial data such as historical returns [11]. For example, the authors in [12–14] use event embeddings from financial news for stock return forecasting, [15] employ BERT in annual report texts, and [16] uses word embeddings for stock selection.

In this paper, we outline a novel methodology that allows for rich relational information to be extracted from financial returns time series and encoded using embedding representations. After a detailed description of the approach in Section 2, we present two evaluations to showcase how the learned representations are useful in tackling the real-world financial problems of sector classification and portfolio optimization. The main contributions can be summarized as follows:

- A novel approach to learning embedding representations of time series is proposed and applied in the context of financial assets.
- In contrast to existing sector classification schemes, which are highly subjective, we showcase how the learned representations can be used to objectively segment stocks into industry sectors.
- The learned embeddings are used within a novel approach for portfolio construction that results in portfolios with statistically significantly lower out-of-sample volatility.

2. Architecture and Approach

Inspired by the distributional semantics area of natural language processing, the model described in this section uses the idea of *context stocks* to learn the embeddings of target stocks. In linguistics, the distributional hypothesis, which underpins a number of popular language models [10], encodes the idea that words commonly occurring in the same contexts tend to have similar meanings.

In the financial domain, a similar hypothesis also holds: companies with similar characteristics—such as those operating in the same business sectors—tend to exhibit similar stock price fluctuations [17]. By engineering the selection of context stocks to reflect this hypothesis, and adding noise reduction strategies, our proposed framework generates embeddings that capture nuanced relationships between financial assets purely based on historical pricing time series.

2.1. Generating Training Data

Consider a universe of stocks $U = \{a_1, \ldots, a_n\}$. For each stock a_i, there is a time series $\mathbf{p}_{a_i} = \{p_0^{a_i}, \ldots, p_T^{a_i}\}$ representing its price at discrete points in time $t \in \{0, 1, \ldots, T\}$ (daily or weekly, for example). From these pricing data, we can compute a *returns time series* $\mathbf{r}_{a_i} = \{r_1^{a_i}, \ldots, r_T^{a_i}\}$ using Equation (1).

$$r_t^{a_i} = \frac{p_t^{a_i} - p_{t-1}^{a_i}}{p_{t-1}^{a_i}} \tag{1}$$

Using these returns time series, we can generate sets of stocks called *target:context sets* made up of a target stock and its set of context stocks. More concretely, for a context size C (a hyperparameter), the context stocks for the target asset a_i at time t are the C stocks, which have the closest return at that point in time. The closest return is defined by the lowest absolute value difference in return between candidate stock a_j and the target asset a_i, formulated as $\left|r_t^{a_i} - r_t^{a_j}\right|$. An example of this process is outlined in Figure 1, with AAPL as the context stock and t is 3 January 2000. We compute the absolute value difference between the return of AAPL at that point in time with the return of each other stock at the same point in time. Then, we choose the C stocks with the lowest values (most similar) as the context stocks, excluding AAPL itself. In this case, IBM and MSFT have the smallest difference with AAPL and so are chosen as context stocks. We generate a target:context set for every stock at each point in time, which results in a total of $|U| \times T$ sets for training.

An example of a target:context set for $C = 3$ is $\mathcal{S}(a_1, t) = [a_1 : a_{270}, a_{359}, a_{410}]$, which corresponds to *[Apple: IBM, Microsoft, Oracle]* since, for example, 270 is the index value for IBM in the dataset and so a_{270} corresponds to IBM. This tells us that, at a certain point in time t, the three stocks with the closest returns to Apple Inc. were IBM, Microsoft and Oracle.

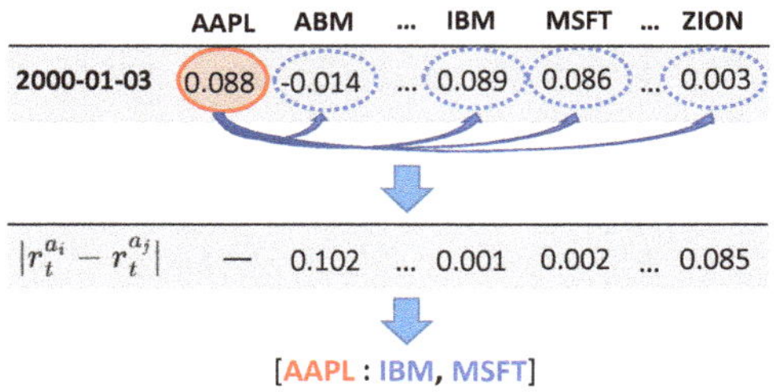

Figure 1. Generating training data, i.e., target:context stock sets.

2.2. Base Model Architecture

The proposed model architecture is illustrated in Figure 2 with *the stock embeddings as the model parameters*. Thus, each row in the weight matrix $\mathbf{W} \in \mathbb{R}^{|U| \times N}$ is a stock embedding, all of which are randomly initialized. Here, N is the embedding size (a hyperparameter), and $|U|$ is the number of stocks/assets in the dataset. The architecture is unusual because the goal of the model is not to make predictions in a downstream task; rather, its sole purpose is to learn parameters such that the resulting embeddings encapsulate the relationships present in the underlying returns time series. In the remainder of this section, the model architecture is described in detail from *input* (the context stocks $a_{j_1}, a_{j_2}, \ldots, a_{j_C}$, where $j_1, j_2, \ldots, j_C \in \{1, 2, \ldots, |U|\}$ are the index values of context stocks) to *output* (the probability of each stock being the associated target).

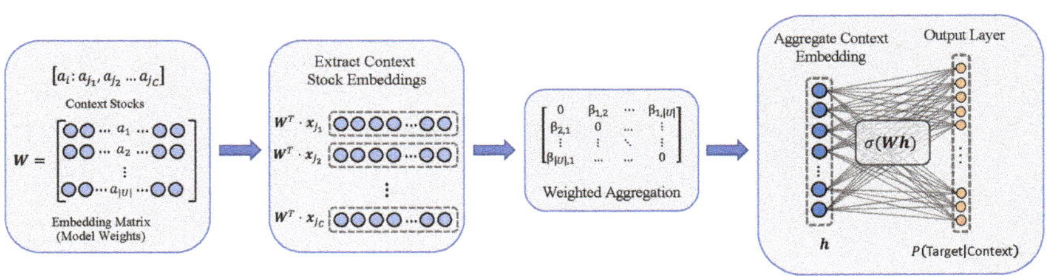

Figure 2. Model Architecture

The first step is to compute the hidden layer $\mathbf{h} \in \mathbb{R}^N$, which can be thought of as an aggregation of the context stocks' embeddings. To do this, each of the context stocks is one-hot encoded, which allows for us to easily extract the relevant embedding (row) from $\mathbf{W}$. For example, the first context stock a_{j_1} is encoded as $\mathbf{x}_{j_1} \in \mathbb{R}^{|U|}$, a one-hot vector of all zeros except for the element in position j_1. As a result, computing $\mathbf{W}^T \cdot \mathbf{x}_{j_1}$ will extract a single row from $\mathbf{W}$— the embedding corresponding to the first context stock a_{j_1}.

In this way, C one-hot vectors are obtained, $\{\mathbf{x}_{j_1}, \mathbf{x}_{j_2}, \ldots, \mathbf{x}_{j_C}\}$, one for each context stock. The embeddings corresponding to the C context stocks are then extracted by pre-multiplying each one-hot vector by $\mathbf{W}^T$ as previously described. Finally, to compute the hidden layer $\mathbf{h}$, the context stock embeddings are aggregated. The most basic form of aggregation proposed is an element-wise average of the extracted embeddings, which is formalized in Equation (2).

$$\mathbf{h} = \frac{1}{C}\mathbf{W}^T(\mathbf{x}_{j_1} + \mathbf{x}_{j_2} + \cdots + \mathbf{x}_{j_C}) \quad (2)$$

Thus, the hidden layer, $\mathbf{h}$, is an N-dimensional vector and can be thought of as an aggregate embedding representation of the context stocks. In Equation (2), each embedding receives an equal weighting of C^{-1}; however, in Section 2.3, we describe a more complex aggregation approach intended to reduce noise.

The next step is to estimate the conditional probability of a particular stock a_i being the target stock *given* the context stocks, which have aggregate embedding $\mathbf{h}$. This is computed as shown in Equation (3), where we define $\mathbf{v}_{a_k}^{\mathbf{T}} := \mathbf{W}^{\mathbf{T}} \cdot \mathbf{x}_{a_k}$, interpreted as the embedding corresponding to asset a_k.

$$\mathbb{P}(a_T = a_i \mid a_{j_1}, \ldots, a_{j_C}) = \frac{\exp(\mathbf{v}_{a_i}^{\mathbf{T}} \cdot \mathbf{h})}{\sum_{k=1}^{|U|} \exp(\mathbf{v}_{a_k}^{\mathbf{T}} \cdot \mathbf{h})} \tag{3}$$

Ensured by the softmax activation, the output is a conditional probability expressing the probability stock a_i is the target stock, given the context stocks that were observed. The goal is to learn the weight/embedding matrix that maximizes the conditional probability for the correct target stock a_T. As a result, we frame this as an optimization problem where the goal is minimizing the loss function shown in Equation (4) with respect to $\mathbf{W}$, which can be achieved using stochastic gradient descent.

$$\mathcal{L} = -\mathbf{v}_{a_T}^{\mathbf{T}} \cdot \mathbf{h} + \log \sum_{k=1}^{|U|} \exp(\mathbf{v}_{a_k}^{\mathbf{T}} \cdot \mathbf{h}) \tag{4}$$

In this way, after training, stocks that commonly co-occur in target–context sets will have high similarity representations in the latent space, as desired.

2.3. Noise Reduction Strategies

Financial returns data are notoriously noisy [18], and so, in addition to the base model architecture, we propose two amelioration strategies to improve performance. Firstly, a weighting strategy based on overall distributional co-occurrence is introduced, and included in Figure 2. With this, the hidden layer $\mathbf{h}$ is computed via a weighted average, and is implemented by scaling each $\mathbf{x}_j$ using a weight, which is proportional to the rate at which the given context stock a_j appears in the context of the target stock a_i over every time point in the training dataset ($t = 1, 2, \ldots, T$). This is outlined in Equation (5), where $\mathbb{1}$ denotes the indicator function.

$$w_{i,j,t} \propto \beta_{i,j} = \frac{1}{T} \sum_{t=1}^{T} \mathbb{1}\left(a_j \in \mathcal{S}(a_i, t)\right) \tag{5}$$

The constant of proportionality here, $k_{i,t}$, is computed such that the weightings over all context stocks sum to one.

$$w_{i,j,t} = k_{i,t} \cdot \beta_{i,j} \quad : \quad k_{i,t} = \left(\sum_{j:a_j \in \mathcal{S}(a_i,t)} \beta_{i,j} \right)^{-1} \tag{6}$$

Secondly, the distribution of returns over a short time period, such as daily, contains a large proportion of values close to 0, which indicates little movement in stock price. In an effort to isolate meaningful cases and reduce noise, a context set $\mathcal{S}(a_i, t)$ was deleted from the training data if the target stock return, $r_t^{a_i}$, was within the interquartile range (IQR) of returns on that day. As a result, only sets where the target stock had a movement outside the IQR of the market average on a given day were included in training.

3. Evaluation Dataset

In the following two sections, we present the results of initial evaluations of the proposed stock embeddings approach. We describe two experiments to evaluate different aspects of the learned distributed representations. In Section 4, we evaluate the ability of the embeddings to capture sectoral similarity by: visualizing the embeddings in 2D space, examining nearest-neighbour stocks, and finally quantifying the results through sector classification. In Section 5, we describe an evaluation within the portfolio construction setting and show that the proposed approach results in out-of-sample portfolios with statistically significantly lower volatility than conventional hedging strategies.

In both experiments, we used a publicly available dataset of daily pricing data for 611 US stocks during the period 2000–2018. In addition to daily returns, each stock is also associated with a *sector* and *industry* classification label from the Global Industry Classification Scheme (GICS). The former corresponds to the business sector in which the company operates—there are eleven sectors in total, including Finance, Health Care and Technology, for example—while the latter represents a finer-grained classification so that a stock in the Technology sector may have Computer Software as its industry label, for example, to contrast it with another Technology stock in the Electronic Components industry. The embeddings used in both experiments were generated using a context size of 3 and an embedding dimension of 20. Results are reported both with and without noise reduction techniques.

4. Evaluation 1: Sector Classification

Stock prices are influenced by a myriad of hidden factors and unpredictable events, making investing a risky venture. Individual stocks expose investors to both market risk (systematic) and asset-specific risk (idiosyncratic). Exchange-traded funds (ETFs) have been growing in popularity because they allow investors access to mitigate idiosyncratic risk, through diversification, at very low costs. They are securities traded on public exchanges, that provide partial ownership in large portfolios of stocks. These portfolios often track a specific market sector or geographical region, allowing for investors diversified exposure to desired market segments. By 2016, ETFs' market share surpassed 10% of total US market capitalization, accounting for over 30% of trading volume [19].

However, ETF providers must decide on portfolio constituents, which is currently a very subjective process, particularly for ETFs tracking particular market sectors. For instance, a strong case could be made for Amazon, classified as consumer discretionary by the GICS, to be included in consumer discretionary, technology, or consumer staples ETFs.

Aside from deciding on ETF constituents, segmenting stocks into market sectors is also crucial for many other types of financial and economic analysis—measuring economic activity, identifying peers and competitors, quantifying market share and bench-marking company performance—none of which would be possible without industry classifications [5]. A well-defined sector classification system also facilitates relative valuation and sector-specific return and risk estimates [20].

We demonstrate the utility of stock embeddings in classifying stocks into business sectors and identifying inconsistencies in existing classification schemes. We first visualize the latent space of learned embeddings to examine stock clustering and relationships. We then present a nearest neighbors analysis, and finally train a classification model using the embeddings to assign sector labels to companies.

4.1. Clustering Stocks Using Embeddings

Visualizing latent embeddings in a lower dimensional space can often be useful to identify relationships and clustering behavior. Figure 3 shows a graphical representation of the embeddings for stocks in four of the largest business sectors: Energy, Finance, Public Utilities and Technology. Each node represents a stock, colored by sector, and an edge indicates that the two nodes it connects have greater than 0.7 cosine similarity between their embeddings. The plot is generated using a force-directed graph-drawing algorithm.

The clustering of stocks into business sectors is clearly evident in Figure 3. We can also see that nearly all the edges in the graph are between nodes from the same sector. From this, we conclude that the proposed model architecture and training procedure result in embeddings that successfully capture relationships between stocks from their time series.

When we consider that the training data used are purely derived from historical returns data, this is a very positive result because it suggests that it is possible to reconstruct important sectoral information from the embeddings, and indeed this can likely be achieved a way that is more nuanced than might be possible using simple sectoral labels.

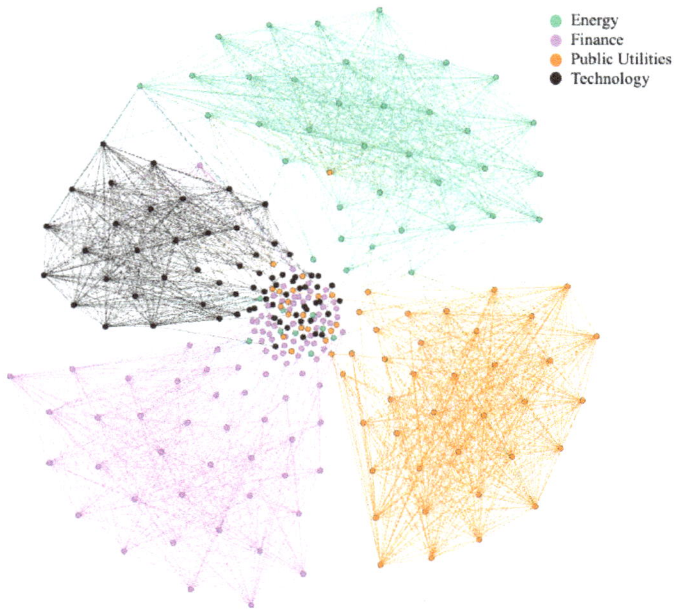

Figure 3. Graphical Visualization of Stock Embeddings Colored by Business Sector.

4.2. *Identifying Nearest Neighbor Assets*

Ideally, we should expect the embeddings of related stocks to be 'semantically' closer, by some suitable similarity metric, than dissimilar stocks, and the ability to identify similar stocks is an important tool for effective portfolio design. Here, we use cosine similarity as our similarity metric to provide examples of the k nearest neighbors for a sample of example stocks; similar results can be obtained when using alternative metrics such as Euclidean distance.

Table 1 shows the top-3 nearest neighbors for JP Morgan Chase, Analog Devices, and Exxon Mobil Corp, three well-known companies in very different sectors. In each case, the nearest neighbors pass the "sanity test" in that they belong to similar sectors and industries. For example, the three nearest neighbors of JP Morgan, a major bank, are also all major banks. Remember, no sectoral or industry information was used to determine these nearest neighbors, and only daily returns time series were used to generate the embeddings used for similarity assessments.

Table 1. Examples of Top-3 Nearest neighbors for Given Query Stocks.

Query Stock Sector-Industry	3 Nearest Neighbors-Sector-Industry	Similarity
JP Morgan Chase Finance Major Bank	Bank of America Corp-Finance-Major Bank State Street Corp-Finance-Major Bank Wells Fargo & Company-Finance-Major Bank	0.88 0.82 0.81
Analog Devices Technology Semiconductors	Maxim Integrated-Technology-Semiconductors Texas Instruments-Technology-Semiconductors Xilinx, Inc.-Technology-Semiconductors	0.93 0.91 0.90
Chevron Corporation Energy Oil & Gas	Exxon Mobil-Energy-Oil & Gas BP P.L.C.- Energy-Oil & Gas Occidental Petroleum-Energy-Oil & Gas	0.89 0.82 0.78

Nearest neighbor stocks can be used in a variety of ways by investors. By focusing on the *k* nearest neighbor stocks, we can develop a basic stock recommendation system which, when given a target stock—a novel stock for the investor or one already in their portfolio—can generate a ranked list of similar stocks based on their historical returns data. Conversely, the ability to identify maximally dissimilar stock is an important way to improve portfolio diversity in order to provide an investor with the ability to guard against volatility and *hedge* against sudden sectoral shocks.

4.3. Sector Classification Performance

This section uses the generated embeddings to objectively segment companies into industry sectors. To do this, the learned embeddings serve as the input to a classification model and their corresponding GICS sector labels represent the ground truth. However, classification accuracy is limited by the unpredictable factors inherent in historical returns data and inconsistencies in current subjective approaches to stock labeling [5]. Therefore, sector classification in finance is a challenging task, particularly when relying solely on returns data.

A considerable class imbalance exists among the sectors within the dataset, which can introduce algorithmic bias and adversely affect the results. To mitigate this issue, we implemented the Synthetic Minority Oversampling Technique (SMOTE) [21] on the training data.

Table 2 shows the performance of the proposed methodology in the sector classification task. We also include alternative several time series classification models as baselines. The embedding model with both noise reduction techniques achieves the highest accuracy of 60%. In addition, our method provides the added benefit of producing learned representations that can serve as features in other asset-related tasks.

Table 2. Sector Classification Results.

Model	Precision	Recall	F1	Accuracy
Catch22	0.31	0.35	0.31	35%
Contractable BOSS	0.47	0.39	0.37	39%
RBOSS	0.57	0.42	0.45	42%
Shapelet	0.54	0.42	0.45	42%
Shapelet Transform	0.39	0.46	0.40	46%
WEASEL	0.50	0.47	0.47	47%
MUSE	0.54	0.54	0.51	54%
Time Series Forest Classifier	0.55	0.55	0.53	55%
Canonical Interval Forest	0.57	0.56	0.52	56%
Arsenal	**0.64**	0.58	0.53	58%
Embedding	0.57	0.54	0.55	54%
Embedding + IQR	0.59	0.56	0.56	56%
Embedding + Weight	0.59	0.57	0.57	57%
Embedding + Weight + IQR	0.62	**0.60**	**0.60**	**60%**

This is an impressive result considering the aforementioned limitations and the fact that there are 11 sector classes (Basic Industries, Capital Goods, Consumer Durables, Consumer Non-Durables, Consumer Services, Energy, Finance, Health Care, Public Utilities, Technology, Transportation). A more in-depth analysis shows a variation in accuracy across sectors, with the more populated sectors being very accurately classified (F1 > 0.9 in some cases) while other minority sectors have relatively low accuracy. With this in mind, we believe the accuracy could be improved with a larger dataset.

The objective sector classification approach presented has considerable potential for addressing and mitigating inconsistent company segmentation—a well-documented issue [22]—in practical applications.

5. Evaluation 2: Hedging/Diversification

As previously mentioned, the existing literature on computational methods for stock markets mainly targets returns forecasting. However, not all investors prioritize maximizing returns; for some, portfolio protection is more important. For instance, a defined benefit pension fund portfolio manager primarily focuses on covering agreed benefits, making risk management more crucial than maximizing returns. To ensure protection, investors and portfolio managers use diversification and hedging, measuring effectiveness in terms of volatility reduction. As a result, identifying dissimilar stocks that behave oppositely to similar ones is essential for traders to hedge their target stocks and limit overall risk.

Typically, hedging involves negatively correlated assets and various correlation metrics. We propose an alternative: using generated embeddings to find maximally dissimilar stocks and inform hedging strategies. We evaluate a scenario where an investor holds a position in a stock (query stock) and seeks a single stock (hedge stock) to reduce risk, measured as volatility, as much as possible.

To do this, we create a hedged two-asset long-only portfolio for each stock in the dataset, resulting in 611 portfolios, each containing a query stock and its lowest similarity hedge stock. The similarity is based on cosine similarity between embeddings or baselines from Table 3. We then simulate each portfolio's out-of-sample performance using different similarity metrics, recording realized volatility values. See Algorithm 1.

Algorithm 1 Finding Hedged Portfolios and Simulating Realized Volatility

```
for target_stock in stocks do
    # Initialize variables
    min_similarity ← ∞
    selected_hedge ← None
    # Iterate over all stocks to find the least similar stock as a hedge
    for candidate_hedge in stocks do
        if similarity(target_stock, candidate_hedge) < min_similarity then
            # Update the minimum similarity and the selected hedge stock
            min_similarity ← similarity(target_stock, candidate_hedge)
            selected_hedge ← candidate_hedge
        end
    end
    # Simulate the performance of the hedged portfolio
end
```

Table 3. Portfolio hedging experiment results along with Tukey HSD test indicating significantly lower volatility than Pearson baseline at $\alpha = 0.01$.

Method	Avg Volatility	Significant
Pearson	23.8%	-
Spearman	24.0%	✗
Geometric	23.9%	✗
Embedding	22.9%	✓
Embedding + Weight	22.8%	✓
Embedding + IQR	**21.3%**	✓
Embedding + Weight + IQR	21.9%	✓

A train-test split is crucial to resemble real-world out-of-sample trading applications. Due to financial data's time sensitivity, we use the first 70% of data (2000–2013) for training embeddings and computing baseline similarity metrics, and the remaining 30% (2014–2018) for simulating portfolios and computing realized volatility.

In modern portfolio theory [6], similarity is defined using Pearson correlation, a common method in industry and academia. We suggest that cosine similarity between proposed embeddings may be superior. The evaluation includes Pearson correlation, Spearman rank-order correlation coefficient, and a recent geometric shape similarity [8] as baselines.

Table 3 displays the average volatility results. To ensure the robustness of results, we reran the experiment 100 times; instead of choosing the single most dissimilar stock as the hedge stock, we randomly chose one of the 25 most dissimilar results for each target stock on each iteration. Overall, the proposed embeddings approach with IQR noise reduction results in portfolios with the lowest average volatility, at 21.3%. Post-hoc Tukey HSD tests indicate that the volatility in all of the embedding-based methods is statistically significantly lower than the Pearson baseline at $\alpha = 0.01$; none of the other baseline approaches generate a significantly lower mean volatility compared to the Pearson approach.

Thus, the proposed embedding methodology can be used to inform a hedging strategy that is superior to a number of baselines, at least within the simplified setting used for this experiment. Obviously, real-world settings are based on more complex portfolios with many different stocks that need to be collectively hedged, and it is a matter for future work to further evaluate our embeddings approach in these more realistic settings. That being said, the approach used here still serves as an important indication of success: had the embeddings approach not been able to demonstrate improved volatility in these simple two-stock portfolios, this would cast doubt on its likely future success in more complex portfolios. Similarly, there is more work to be carried out when it comes to understanding the dynamics of the distributed representations and how they are learned: for example, how changing the context size or embedding dimension impacts these findings.

6. Conclusions

In this paper, we presented a novel methodology for quantifying the relationships between financial assets by learning embedding representations derived solely from historical returns. The effectiveness of the approach is demonstrated through a nearest neighbour case study and benchmark comparisons in two key financial tasks: (1) accurately classifying stocks into their respective industry sectors, and (2) constructing portfolios that exhibit statistically significant reductions in volatility compared to traditional baseline methods.

Moving forward, we aim to further assess the potential of this methodology by examining more intricate portfolio management scenarios and incorporating additional datasets. The proposed technique is versatile and applicable to any group of financial assets with accessible pricing information, allowing us to extend our analysis to encompass multiple asset classes beyond equities. Furthermore, we plan to conduct a thorough exploration of the model parameter space used for learning the embeddings, as well as investigating alternative approaches for generating context stocks.

Author Contributions: Conceptualization, R.D. (Rian Dolphin), R.D. (Ruihai Dong) and B.S.; methodology, R.D. (Rian Dolphin); software, R.D. (Rian Dolphin); validation, R.D. (Rian Dolphin); formal analysis, R.D. (Rian Dolphin); investigation, R.D. (Rian Dolphin); resources, R.D. (Rian Dolphin); data curation, R.D. (Rian Dolphin); writing—original draft preparation, R.D. (Rian Dolphin), and B.S.; writing—review and editing, R.D. (Rian Dolphin), R.D. (Ruihai Dong) and B.S.; visualization, R.D. (Rian Dolphin); supervision, R.D. (Ruihai Dong) and B.S.; project administration, R.D. (Ruihai Dong) and B.S.; funding acquisition, R.D. (Ruihai Dong) and B.S. All authors have read and agreed to the published version of the manuscript.

Funding: This publication has emanated from research conducted with the financial support of Science Foundation Ireland under Grant number 18/CRT/6183.

Institutional Review Board Statement: Not applicable.

Informed Consent Statement: Not applicable.

Data Availability Statement: All code and data is available at github.com/rian-dolphin/stock-embeddings (accessed on 15 June 2023).

Conflicts of Interest: The authors declare no conflict of interest.

References

1. Bachelier, L. Théorie de la spéculation. *Ann. Sci. l'École Norm. Supérieure* **1900**, *17*, 21–86. [CrossRef]
2. Fama, E.F. The behavior of stock-market prices. *J. Bus.* **1965**, *38*, 34–105. [CrossRef]
3. Sezer, O.B.; Gudelek, M.U.; Ozbayoglu, A.M. Financial time series forecasting with deep learning: A systematic literature review: 2005–2019. *Appl. Soft Comput.* **2020**, *90*, 106181. [CrossRef]
4. Sharma, A.; Bhuriya, D.; Singh, U. Survey of stock market prediction using machine learning approach. In Proceedings of the 2017 International Conference of Electronics, Communication and Aerospace Technology (ICECA), Coimbator, India, 20–22 April 2017; IEEE: Piscataway, NJ, USA, 2017; Volume 2, pp. 506–509.
5. Phillips, R.L.; Ormsby, R. Industry classification schemes: An analysis and review. *J. Bus. Financ. Librariansh.* **2016**, *21*, 1–25. [CrossRef]
6. Markowitz, H. *Portfolio Selection*; Wiley: Hoboken, NJ, USA, 1952.
7. Lhabitant, F.S. *Correlation vs. Trends: A Common Misinterpretation*; EDHEC: Roubaix, France, 2020.
8. Chun, S.H.; Ko, Y.W. Geometric Case Based Reasoning for Stock Market Prediction. *Sustainability* **2020**, *12*, 7124. [CrossRef]
9. Dolphin, R.; Smyth, B.; Xu, Y.; Dong, R. Measuring Financial Time Series Similarity with a View to Identifying Profitable Stock Market Opportunities. In Proceedings of the International Conference on Case-Based Reasoning, Salamanca, Spain, 13–16 September 2021; Springer: Berlin/Heidelberg, Germany, 2021; pp. 64–78.
10. Mikolov, T.; Chen, K.; Corrado, G.; Dean, J. Efficient estimation of word representations in vector space. *arXiv* **2013**, arXiv:1301.3781.
11. Dolphin, R.; Smyth, B.; Dong, R. Industry Classification Using a Novel Financial Time-Series Case Representation. *arXiv* **2023**, arXiv:2305.00245.
12. Ding, X.; Zhang, Y.; Liu, T.; Duan, J. Deep learning for event-driven stock prediction. In Proceedings of the Twenty-Fourth International Joint Conference on Artificial Intelligence, Buenos Aires, Argentina, 25–31 July 2015.
13. Ding, X.; Zhang, Y.; Liu, T.; Duan, J. Knowledge-driven event embedding for stock prediction. In Proceedings of the Coling 2016, the 26th International Conference on Computational Linguistics: Technical Papers, Osaka, Japan, 11–16 December 2016; pp. 2133–2142.
14. Cheng, D.; Yang, F.; Wang, X.; Zhang, Y.; Zhang, L. Knowledge graph-based event embedding framework for financial quantitative investments. In Proceedings of the 43rd International ACM SIGIR Conference on Research and Development in Information Retrieval, Virtual Event, 25–30 July 2020; pp. 2221–2230.
15. Ito, T.; Camacho Collados, J.; Sakaji, H.; Schockaert, S. Learning Company Embeddings from Annual Reports for Fine-Grained Industry Characterization. In Proceedings of the Second Workshop on Financial Technology and Natural Language Processing, Kyoto, Japan, 5 January 2020.
16. Hirano, M.; Sakaji, H.; Kimura, S.; Izumi, K.; Matsushima, H.; Nagao, S.; Kato, A. Selection of related stocks using financial text mining. In Proceedings of the 2018 IEEE International Conference on Data Mining Workshops (ICDMW), Singapore, 17–20 November 2018; IEEE: Piscataway, NJ, USA, 2018; pp. 191–198.
17. Gopikrishnan, P.; Rosenow, B.; Plerou, V.; Stanley, H.E. Identifying business sectors from stock price fluctuations. *arXiv* **2000**, arXiv:cond-mat/0011145.
18. De Long, J.B.; Shleifer, A.; Summers, L.H.; Waldmann, R.J. Noise trader risk in financial markets. *J. Political Econ.* **1990**, *98*, 703–738. [CrossRef]
19. Ben-David, I.; Franzoni, F.; Moussawi, R. *Exchange Traded Funds (ETFs)*; Technical Report; National Bureau of Economic Research: Cambridge, MA, USA, 2016.
20. S&P.; MSCI. *Global Industry Classification Standard (GICS) Methodology*; MSCI: New York, NY, USA, 2020.
21. Chawla, N.V.; Bowyer, K.W.; Hall, L.O.; Kegelmeyer, W.P. SMOTE: Synthetic minority over-sampling technique. *J. Artif. Intell. Res.* **2002**, *16*, 321–357. [CrossRef]
22. Chan, L.K.; Lakonishok, J.; Swaminathan, B. Industry classifications and return comovement. *Financ. Anal. J.* **2007**, *63*, 56–70. [CrossRef]

engineering proceedings

MDPI

Proceeding Paper

Efficient Forecasting of Large-Scale Hierarchical Time Series via Multilevel Clustering †

Xing Han [1,*], Tongzheng Ren [2], Jing Hu [3], Joydeep Ghosh [1] and Nhat Ho [4]

1 Department of Electrical and Computer Engineering, University of Texas at Austin, Austin, TX 78712, USA; jghosh@utexas.edu
2 Department of Computer Science, University of Texas at Austin, Austin, TX 78712, USA; tongzheng@utexas.edu
3 Intuit, Mountain View, CA 94043, USA; jing_hu@intuit.com
4 Department of Statistics and Data Science, University of Texas at Austin, Austin, TX 78712, USA; minhnhat@utexas.edu
* Correspondence: aaronhan223@utexas.edu

† Presented at the 9th International Conference on Time Series and Forecasting, Gran Canaria, Spain, 12–14 July 2023.

Abstract: We propose a novel approach to cluster hierarchical time series (HTS) for efficient forecasting and data analysis. Inspired by a practically important but unstudied problem, we found that leveraging local information when clustering HTS leads to a better performance. The clustering procedure we proposed can cope with massive HTS with arbitrary lengths and structures. In addition to providing better insights, this method can also speed up the forecasting process for a large number of HTS. Each time series is first assigned the forecast from its cluster representative, which can be considered as "prior shrinkage" for the set of time series it represents. Then, the base forecast can be efficiently adjusted to accommodate the specific attributes of the time series. We empirically show that our method substantially improves performance for large-scale clustering and forecasting tasks involving HTS.

Keywords: hierarchical time series; clustering; Wasserstein distance

Citation: Han, X.; Ren, T.; Hu, J.; Ghosh, J.; Ho, N. Efficient Forecasting of Large-Scale Hierarchical Time Series via Multilevel Clustering. *Eng. Proc.* **2023**, *39*, 31. https://doi.org/10.3390/engproc2023039031

Academic Editors: Ignacio Rojas, Hector Pomares, Luis Javier Herrera, Fernando Rojas and Olga Valenzuela

Published: 29 June 2023

1. Introduction

Time series with hierarchical aggregation constraints are commonly seen in many practical scenarios [1]. In applications such as finance or e-commerce, an HTS normally represents historical records from one user (e.g., the cash flow example in Figure 1). Normally, separately building a predictive model for each user is inefficient, particularly when the number of users is quite large, or the length of user records varies significantly. To address this problem, we design a novel clustering procedure. It effectively finds the cluster representatives of a large group of HTS, followed by fine-tuning forecasts on these representatives to obtain user-specific forecasts.

Clustering time series is an important tool for discovering patterns over sequential data when categorical information is not available. Most clustering approaches fall into discriminative and generative categories. Discriminative approaches normally define a proper distance measure [2] or construct features [3] that capture temporal information. Generative approaches [4] specify the model type (e.g., HMM) a priori and estimate the parameters using maximum likelihood algorithms. Deep learning has also been applied to time series clustering. Most state-of-the-art discriminative approaches first extract useful temporal representations followed by clustering in the embedding space [5]. However, there is no prior work on clustering HTS data. This problem is more challenging since data at different level of HTS have distinct properties. Regular clustering methods for time series lead to inferior performance, particularly when the hierarchy is complex. When clustering HTS, we need to leverage level-wise information, but it is difficult to completely respect the hierarchy since the data are not easy to partition given the imposed constraints.

There is little prior work on the clustering of multilevel-structured data. A pioneering effort [6] proposed to simultaneously partition data at both local and global levels and discover latent multilevel structures. This work proposed an optimization objective for two-level clustering based on Wasserstein metrics. Its core idea was to perform global clustering based on a set of local clusters. However, this work mainly applies to discrete and semi-structured data such as annotated images and documents. It cannot be applied to HTS involving continuous or structured data, which has more constraints. A follow-up work [7] extended this to continuous data by assuming that the data at the local level is generated by predefined exponential family distributions. The authors then performed model-based clustering at both levels. However, model-based clustering for time series is computationally expensive and crucially depends on the modelling assumptions. Moreover, both these works were limited to two-level structures, whereas for several HTS applications, given a set of pre-specified features as aggregation variables, it is possible to have a multilevel hierarchy. Note that, our problem is different from hierarchical clustering [8]: the hierarchy comes from the time series data instead of the method that builds a hierarchy of clusters.

In this paper, we propose `HTS-Cluster`, an efficient model-free clustering method that can handle HTS with various types of individual components and hierarchies. `HTS-Cluster` employs a combined objective that involves clustering terms from each aggregated level. This formulation uses Wasserstein distance metrics coupled with Soft-DTW divergence [9] to cater to variable length series that are grouped together. In addition to providing superior clustering results for multilevel hierarchies, `HTS-Cluster` significantly improves the efficiency of forecasts when applied to large HTS datasets containing hundreds of thousands of time series.

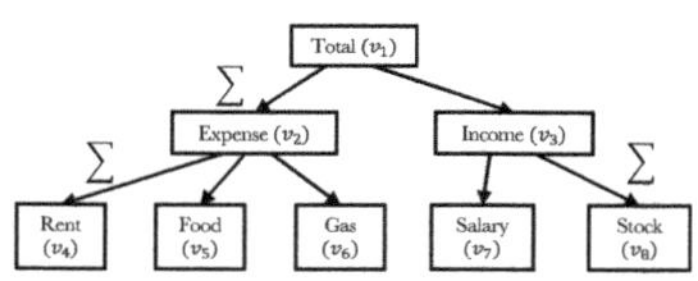

$$S = \begin{bmatrix} S_{\text{aggregated}} \\ I_5 \end{bmatrix} = \left[\begin{array}{ccccc} 1 & 1 & 1 & 1 & 1 \\ 1 & 1 & 1 & 0 & 0 \\ 0 & 0 & 0 & 1 & 1 \\ \hline & & I_5 & & \end{array}\right]$$

Figure 1. Left: an example of hierarchical time series (HTS) with five bottom-level time series and three-level hierarchical structure. Each vertex ($V = \{v_i\}_{i=1}^{8}$) represents time series aggregated on different variables related through a domain-specific conceptual hierarchy (e.g., product categories, locations, etc.). Right: the summation matrix $S \in \{0,1\}^{8\times 5}$ used to denote the given hierarchy.

2. Backgrounds

Hierarchical time series: Given the time stamps $t = 1, \ldots, T$, let $\mathbf{x}_t \in \mathbb{R}^n$ be the value of HTS at time t, where $x_{t,i} \in \mathbb{R}$ is the value of the ith (out of n) univariate time series. Figure 1 shows an example of HTS with a three-level structure. We refer to the time series at the leaf nodes of the hierarchy as bottom-level time series and the remaining nodes as aggregated-level time series. We split the vector of $\mathbf{x}_t$ into m-bottom time series and l-aggregated time series such that $\mathbf{x}_t = [\mathbf{a}_t \ \mathbf{b}_t]^\top$ where $\mathbf{a}_t \in \mathbb{R}^l$ and $\mathbf{b}_t \in \mathbb{R}^m$ with $n = l + m$. The summation matrix $S \in \{0,1\}^{n\times m}$ satisfies $\mathbf{x}_t = S \cdot \mathbf{b}_t$, which can later be used to calibrate forecasting results to be aligned with a given hierarchical structure. For notational simplicity, we omit the time stamp of each series in the following discussion.

Dynamic Time Warping (DTW) [10]: DTW is a popular method for computing the optimal alignment between two time series with arbitrary lengths. Given $\mathbf{X}$ and $\mathbf{Y}$ of length T_1 and T_2, respectively, DTW computes the $T_1 \times T_2$ pairwise distance matrix between each time stamp and solves a dynamic program (DP) using Bellman's recursion in $\mathcal{O}(T_1 \cdot T_2)$ time. DTW discrepancy can be used to describe the average similarity within a set of time series [2]. However, DTW is not a differentiable metric given its DP recursion nature. To address this issue, the authors of [11] proposed Soft-DTW by smoothing the `min` operation using the `log-sum-exp` trick. Specifically, we assume $\mathbf{A} \in \{0,1\}^{T_1 \times T_2}$ is the alignment matrix between two time series and $\mathbf{C} \in \mathbb{R}^{T_1 \times T_2}$ is the cost matrix, the formulation of the Soft-DTW can be written as

$$\mathrm{SDTW}_\gamma(\mathbf{C}(\mathbf{X}, \mathbf{Y})) = \min_{\mathbf{A} \in \mathcal{A}(T_1, T_2)} {}^{\gamma} \langle \mathbf{A}, \mathbf{C} \rangle = -\gamma \log \sum_{\mathbf{A} \in \mathcal{A}(T_1, T_2)} \exp(-\langle \mathbf{A}, \mathbf{C} \rangle / \gamma), \tag{1}$$

where $\gamma > 0$ is a parameter that controls the trade-off between the approximation and smoothness, and $\mathcal{A}(T_1, T_2)$ is the collection of all possible alignments between two time series. Soft-DTW is differentiable with respect to all of its variables and can be used for a variety of tasks such as averaging, clustering, and prediction of time series. However, Soft-DTW also has several drawbacks. Ref. [9] recently showed that Soft-DTW is not a valid divergence given its minimum is not achieved when two time series are equal; furthermore, the value of Soft-DTW is not always non-negative. Ref. [9] proposed Soft-DTW divergence, which can address these issues and achieves a better performance. This divergence $\mathcal{D}$ can be written as

$$\mathcal{D}(\mathbf{X}, \mathbf{Y}) := \mathrm{SDTW}_\gamma(C(\mathbf{X}, \mathbf{Y})) - \frac{1}{2}\mathrm{SDTW}_\gamma(C(\mathbf{X}, \mathbf{X})) - \frac{1}{2}\mathrm{SDTW}_\gamma(C(\mathbf{Y}, \mathbf{Y})). \tag{2}$$

Our method incorporates the Soft-DTW divergence as a base distance measure for variable length sequences, and use it as a differentiable loss during the clustering procedure.

Wasserstein distance: For any given subset $\Theta \subset \mathbb{R}^d$, let $\mathcal{P}(\Theta)$ denote the space of Borel probability measures on Θ. The Wasserstein space of order r of probability measures on Θ is defined as $\mathcal{P}_r(\Theta) = \{G \in \mathcal{P}(\Theta) : \int \|x\|^r dG(x) < \infty\}$, where $\|\cdot\|$ denotes the Euclidean metric in $\mathbb{R}^d$. For any P or Q in $\mathcal{P}_r(\Theta)$, the r-Wasserstein distance W_r between P and Q is

$$W_r(P, Q) = \left(\inf_{\pi \in \Pi(P,Q)} \int_{\Theta^2} \|x - y\|^r \, d\pi(x, y) \right)^{1/r}, \tag{3}$$

where $\Pi(P, Q)$ contains all the joint (coupling) distributions whose margins are P and Q, and the π coupling that achieves the minimum of Equation (3) is called the transportation plan. In other words, $W_r(P, Q)$ is the optimal cost of moving mass from P to Q, which is proportional to the r-power of the Euclidean distance in Θ. Furthermore, by recursion of concepts, we define $\mathcal{P}_r(\mathcal{P}_r(\Theta))$ as the space of Borel measures on $\mathcal{P}_r(\Theta)$, then $\forall P', Q' \in \mathcal{P}_r(\mathcal{P}_r(\Theta))$,

$$W_r^{(2)}(P', Q') = \left(\inf_{\pi \in \Pi(P',Q')} \int_{\mathcal{P}_r(\Theta)^2} W_r^r(P, Q) \, d\pi(P, Q) \right)^{1/r}.$$

Similarly, the cost of moving unit mass in its space of support $\mathcal{P}_r(\Theta)$ is proportional to the r-power of the W_r distance in $\mathcal{P}_r(\Theta)$. The Wasserstein distance can be thought of as a special case of the Wasserstein barycenter problem. Computation of the Wasserstein distance and Wasserstein barycenter has been studied by many prior works, where [12] proposed an efficient algorithm to find its local solutions. The well-known K-means clustering algorithm can also be viewed as a method to solve the Wasserstein means problem [6].

3. Hierarchical Time Series Clustering

In this section, we present the `HTS-Cluster` for clustering time series with both two-level and multilevel hierarchical structures. We use $x_{j,i}$ to denote the ith univariate time series of the jth HTS, where $1 \leq j \leq N$ and $1 \leq i \leq n_j$. We assume the index i of each series is given by the level-order traversal of the hierarchical tree from left to right at each level. We will use $a_{j,i}$ and $b_{j,i}$ for the corresponding aggregated and bottom-level series, respectively.

3.1. Two-Level Time Series Clustering

We define a new Wasserstein distance measure W_{sdtw} as

$$W_{\text{sdtw}}(P,Q) = \inf_{\pi \in \Pi(P,Q)} \int_{\Theta^2} \mathcal{D}(x,y)\, d\pi(x,y). \tag{4}$$

For any $j = 1, \ldots, N$, we denote the empirical measure of all bottom-level series as $P_{N'} = \frac{1}{N'} \sum_{j=1}^{N} \sum_{i=1}^{n_j} \delta_{b_{j,i}}$, where $N' \geq N$ given that each HTS has at least one bottom-level series. For local (bottom-level) clustering, we assume that at most k_2 clusters can be obtained, we perform K-means that can be viewed as finding a finite discrete measure $G = \sum_{k=1}^{k_2} u_k \delta_{\mu_k} \in \mathcal{O}_{k_2}(\Theta)$ that minimizes $W_{\text{sdtw}}(G, P_{N'})$, where $\mu_k \in \mathbb{R}^T$ is the "cluster mean" time series to be optimized in support of the finite discrete measure G and $u \in \Delta_{k_2}$, where $\Delta_k = \left\{ w \in \mathbb{R}^k : w_i \geq 0, \sum_{i=1}^{k} w_i = 1 \right\}$ is the probability simplex for any $k \geq 1$.

Although this approach can be extended to any aggregated level, such a method cannot leverage the connections with adjacent levels. As Figure 2 shows, aggregation of data will cause the loss of information: it is less likely to obtain reasonable results by simply clustering data at the aggregated level. Therefore, we believe that with the help of bottom-level information, clustering at the aggregated level can be further improved.

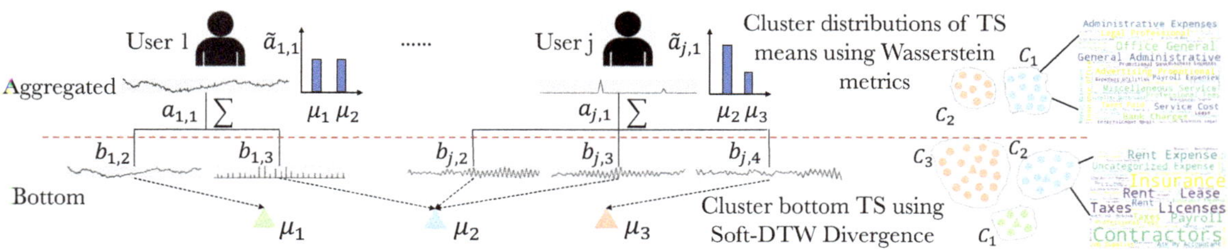

Figure 2. Leveraging local clustering results for HTS clustering. We improve the clustering performance at the aggregated level by clustering empirical distributions over cluster representatives obtained from the bottom-level.

Problem formulation: A direct solution is to replace each top-level series with a large feature vector obtained by concatenating all bottom-level series, but this will introduce redundancy and require large training datasets due to the induced high dimensionality. Instead, we propose to leverage local information by utilizing local clustering results. For the jth HTS, we denote $\mathcal{F}_j(i)$ as the set that contains all descendant indices of its ith series, and assume each top-level series $a_{j,1}$ is aggregated from the bottom-level series $\{b_{j,i}\}_{i \in \mathcal{F}_j(1)}$. First, we cluster all bottom-level series $\{b_{j,i}\}_{j \in [N], i \in \mathcal{F}_j(1)}$ into k_2 clusters $\{C_k\}_{k \in [k_2]}$ with C_k centred at μ_k. We then assign the following probability measure to each $a_{j,1}$:

$$\tilde{a}_{j,1} = \frac{1}{|\mathcal{F}_j(1)|} \sum_{i \in \mathcal{F}_j(1)} \sum_{k \in [k_2]} \mathbf{1}_{\mu_k} \mathbf{1}_{b_{j,i} \in C_k}, \tag{5}$$

that is, we represent the top-level time series as an empirical distribution of $\{\mu_k\}_{k \in [k_2]}$, where the weight of each μ_k is determined by the number of $b_{j,i}$ that belongs to cluster C_k. Note that, we distinguish $\tilde{a}_{j,1}$ from $a_{j,1}$, where $\tilde{a}_{j,1}$ is its corresponding probability measure of $a_{j,1}$. This formulation represents the top-level time series using a finite number of bottom-level clusters, reducing computation time from concatenating bottom-level time series while simultaneously leveraging local information. We now define the objective function to jointly optimize both local and global clusters as follows

$$\inf_{\substack{G \in \mathcal{O}_{k_2}(\Theta), \\ \mathcal{H} \in \mathcal{O}_{k_1}(\mathcal{P}_r(\Theta))}} W_{\text{sdtw}}(G, P_{N'}) + W^{(2)}_{\text{sdtw}}(\mathcal{H}, \frac{1}{N}\sum_{j=1}^{N} \delta_{\tilde{a}_{j,1}}), \tag{6}$$

where k_1 is the number of clusters at the top-level, and $\mathcal{H}$ is the distribution over top-level cluster centroids. Similar to each top-level time series $\tilde{a}_{j,1}$, the supports of $\mathcal{H}$ are also finite discrete measures. Specifically, $\mathcal{H} = \sum_{k=1}^{k_1} v_k \delta_{\nu_k} \in \mathcal{P}_r(\mathcal{P}_r(\Theta))$, where $v \in \Delta_{k_1}$, $\nu_k \in \mathcal{O}_{\bar{n}}(\Theta)$ and $\bar{n} = \frac{1}{N}\sum_{j=1}^{N} \mathcal{F}_j(1)$. Equation (6) is our formulation for the two-level HTS clustering problem, where the first term $W_{\text{sdtw}}(G, P_{N'})$ is the Wasserstein distance defined in the space of measures $\mathcal{P}_r(\Theta)$ and the second term is defined in $\mathcal{P}_r(\mathcal{P}_r(\Theta))$.

3.2. Multilevel Time Series Clustering

For HTS with multiple levels, we employ a "bottom-up" clustering procedure that recursively uses lower-level information for higher-level clustering, till the root is reached. We assume we are given N HTS with L levels; we denote by $x^l_{1:N}$ a collection of the lth-level time series from the first N HTS and $\mathbf{x}^l_{1:N}$ as the replacement of the corresponding time series represented by lower-level clusters. We formulate the objective function to cluster multilevel time series as

$$\inf_{\substack{G \in \mathcal{O}_{\mathsf{k}_\mathsf{L}}(\Theta), \\ \mathcal{H}^l \in \mathcal{O}_{\mathsf{k}_l}(\mathcal{P}_r(\Theta))}} W_{\text{sdtw}}(G, P_{N'}) + \sum_{l=1}^{\mathsf{L}-1} W^{(2)}_{\text{sdtw}}(\mathcal{H}^l, \frac{1}{\mathcal{G}(l)}\sum_{j=1}^{\mathcal{G}(l)} \delta_{\mathbf{x}^l_{1:N}[j]}), \tag{7}$$

where $\mathcal{G}(l)$ is the total number of time series at level l among N HTS. Similarly, we have $\mathcal{H}^l = \sum_{k=1}^{\mathsf{k}_l} v_k \delta_{\nu^l_k}$, $v \in \Delta_{\mathsf{k}_l}$ and $\nu^l_k \in \mathcal{O}_{\bar{n}_l}(\Theta)$. Algorithm 1 shows the full procedure of the bottom-up clustering for HTS with arbitrary levels, while its core steps also apply to two-level HTS clustering. Specifically, steps 4 and 5 are the centring and cluster assignment steps for clustering time series at the Lth (bottom) level, which uses Soft-DTW divergence as a distance measure between each input pair. After obtaining the clustering result at the Lth-level, its cluster indices and means are used to construct probability measures for each time series at the $\mathsf{L} - 1$ level (step 8). Meanwhile, we still have access to the original time series in the aggregated levels. Steps 10 and 11 perform clustering in the space of probability measures. To efficiently compute the Wasserstein barycenter, we optimize the support of the barycenter ν^l_k, featured as the free-support method studied in [12]. For cluster assignment, we compute the Wasserstein distance between pairs of probability measures in the Soft-DTW divergence space. Both the assignment and centring steps utilize information from lower aggregation levels, where these steps are repeated until the cluster assignments are stable. We then use the results of the cluster assignment to compute the cluster means of the original time series at that level, used as supports of the probability measure to represent time series in the next aggregation level, until the clustering procedure for all levels is finished.

Computational efficiency: Compared with model-based clustering, `HTS-Cluster` waives the extensive computation of HMM parameters and the data assumptions. Note that, computing the Wasserstein barycenter at step 10 is very efficient since one just needs to compute the Soft-DTW divergence between the supports of each distribution, which can be obtained beforehand. The process of finding the optimal barycenter (steps 4 and 10) is differentiable. Therefore, the clustering time is progressively reduced as we proceed to higher levels.

Algorithm 1 `HTS-Cluster`.

```
Input: L, total aggregation level; x^l_{1:N}, collection of the l^th level series from #1 to #N HTS
Initialize: cluster assignment: {C^l_k}_{k∈[k_l], l∈[L]}
while not converged do
    μ^L_k = arg min_{μ∈R^T} 1/|C^L_k| Σ_{i∈C^L_k} D(x^L_{1:N}[i], μ)
    C^L_k = {i : D(x^L_{1:N}[i], μ^L_k) = min_{k∈[k_L]} D(x^L_{1:N}[i], μ^L_k)}
end while
for l = [L − 1, L − 2, ..., 1] do
    x^l_j[n] = 1/|F_j(n)| Σ_{i∈F_j(n)} Σ_{k∈[k_{l+1}]} 1_{ν^{l+1}_k} 1_{x^{l+1}_j[i]∈C^{l+1}_k}
    while not converged do
        ν^l_k = arg min_ν Σ_{i∈C^l_k} λ_i W_sdtw(x^l_{1:N}[i], ν)
        C^l_k = {i : W_sdtw(x^l_{1:N}[i], ν^l_k) = min_{k∈[k_l]} W_sdtw(x^l_{1:N}[i], ν^l_k)}
    end while
    ν^l_k = arg min_{ν∈R^T} 1/|C^l_k| Σ_{i∈C^l_k} D(x^l_{1:N}[i], ν)
end for
```

HTS forecasting; Forecasts for individual HTS can "borrow strength" from the forecasts of the nearest cluster means at each level. Specifically, we first perform forecasts for the bottom- and aggregated-level cluster-mean time series $\{\mu_k^{\mathsf{L}}\}_{k\in[k_{\mathsf{L}}]}$ and $\{\nu_k^l\}_{k\in[k_l],l\in[\mathsf{L}-1]}$, respectively. The forecast for each time series can be represented as the weighted combination of forecasts of the corresponding cluster means at that level. We define the weight between time series i and cluster mean j at level l as

$$w_{i,j}^l = \frac{1}{\sum_{k=1}^{k_l}\left(\frac{\mathcal{D}(x_i^l,\nu_j^l)}{\mathcal{D}(x_i^l,\nu_k^l)}\right)^{\frac{2}{m-1}}}, \quad m \in (1,\infty), \tag{8}$$

where the closer x_i^l is to a certain cluster mean, the higher its weight is. Equation (8) is well known in fuzzy clustering, where a data point can belong to more than one cluster, and m is the parameter that controls how fuzzy the cluster assignments are. One can use post-reconciliation methods, such as in [1], to calibrate the results for individual forecasts.

4. Experiments

We evaluate `HTS-Cluster` in multiple applications. Overall, our experiments include (1) clustering time series with multilevel structures (Section 4.1); (2) facilitating time series forecasting with the help of clusters (Section 4.2).

4.1. HTS Clustering

Two-level HTS: We first conduct experiments on synthetic data using ARMA simulations, to provide a feel for the setting and the results attainable. We generate a simple HTS with two levels: one parent node with four children vertices, i.e., for the jth hierarchy $X_j = \{x_i\}_{i=1}^5, x_1 = \sum_{i=2}^5 x_i$. The length of each X is different, ranging from 80 to 300. We use the following simulation function for each time series $x_{1:T}$

$$x_t = 0.75x_{t-1} - 0.25x_{t-2} + 0.65\varepsilon_{t-1} + 0.35\varepsilon_{t-2} + \varepsilon_t + c,$$

where ε_t is a white noise error term at time t, and c is an offset that is used to separate different clusters. We simulate four clusters, each having 30 HTS as members. Additionally, the evaluation is performed on a real-world HTS dataset containing financial records from multiple users for tax purposes. This dataset contains 12,000 users' electronic records of expenses in different categories. The bottom-level time series are summed across all categories to obtain the total expenses. Each user owns an HTS but the length of records varies from user to user.

Multilevel HTS: We also test our method on HTS with multiple aggregated levels. It is simple to extend simulated two-level HTS to multiple levels by modifying the summation matrix S. The evaluation is also performed on a large, real-world financial dataset that contains HTS with ≥ 3 aggregated levels. Each HTS represents the expense records of a small business, where the bottom level (or the lowest two levels) time series are user-defined accounts (or sub-accounts), which are then aggregated by different tax purposes to obtain the middle-level time series. The top-level time series are the total expenses aggregated from the middle level, including the overall information of the business. The dataset contains 18,568 HTS with 222,989 bottom-level time series in total.

Experiment baselines: Our baselines for evaluating `HTS-Cluster` include the recent state-of-the-art method DTCR [5], which employs an encoder–decoder structure integrated with a fake sample generation strategy. The authors of [5] showed that DTCR can learn better temporal representations for time series data that improve the clustering performance. Here, we implement DTCR to treat HTS as regular multivariate time series data. In addition, we implemented independent level-wise clustering using Soft-DTW divergence (`Soft-DTW`), i.e., without local information and clustering aggregated-level data via simply concatenating lower-level time series (`concat`). We used three prevalent methods for clustering evaluation: normalized mutual information (NMI) [13], adjusted mutual information (AMI) [14], and adjusted rand index (ARI) [15].

Clustering results: We conducted 10 experiments, with different random seeds, on both simulated and real datasets. As shown in Table 1 (upper), for the synthetic two-level HTS, our method is superior to the baseline methods in both clustering performance and computational efficiency. Specifically, in terms of clustering performance, level-wise clustering approaches are better than DTCR, at both global (aggregated) and local (bottom) levels, since separating information from different granularities can improve the partitioning of data. As for computation time, DTCR training consists of two stages: it first learns temporal representations and then performs K-means clustering. This results in a longer computation time compared with `HTS-Cluster`. For level-wise approaches, clustering using Soft-DTW divergence and simple concatenation yield the same results at the bottom level, but concatenating bottom-level data provides better results at the top level since aggregation causes the loss of information. Finally, the alternating updates using the global and local cluster formulations of Equation (6), lead to improved performance due to leveraging both local and global information. Specifically, the top-level time series are represented by empirical distributions over bottom-level cluster means, and the cluster means at the top level can be obtained more efficiently via fast computation of the Wasserstein barycenter. Based on user-specified domain knowledge or constraints, we utilize the global cluster assignment to calibrate local time series that are far from the nearest cluster centre. This procedure improves both the local and global clustering results while simultaneously reducing the total computation time.

`HTS-Cluster` also demonstrates improved performance over baseline methods on multilevel HTS. As shown in Table 2, all methods are evaluated on HTS datasets with four aggregation levels, where level one is the top level and four is the bottom level. Here, `HTS-Cluster` employs the bottom-up procedure of Algorithm 1, where the clustering results from the lower level are leveraged for upper-level clustering until the root is reached. Therefore, the level-wise clustering methods (`Soft-DTW`, `Concat`, and `HTS-Cluster`) share the same results at the bottom level. At aggregated levels, `HTS-Cluster` consistently outperforms `DTCR` and `Soft-DTW` with the help of local information and achieves a competitive performance with `Concat` at a much smaller computational cost.

For the financial data, there are no cluster labels. Therefore, we use the "business type", included in the metadata of each HTS, as a weak "ground truth" label for clustering. Unsurprisingly, the results metrics for all the methods are low (Table 1 bottom), and the utility of `HTS-Cluster` really emerges when we examine the downstream forecasting results later on. For now, to show that the clusters are still meaningful, we visualize the HTS metadata at the tax code level using our method (Figure 3). We see that `HTS-Cluster` does create meaningful partitions for HTS by accounting for features from a local time series. Finally, we monitor the level-wise clustering time of each method. The compared baselines include (1) level-wise clustering using Soft-DTW divergence without leveraging local clustering results; (2) simple concatenation of lower-level time series for higher-level clustering. All three methods are conducted in a bottom-up fashion, with the same bottom-level clustering procedure. As shown in Table 3 (left), `HTS-Cluster` provides the most efficient method for clustering aggregated-level time series. This is because (1) computing the Wasserstein barycenter at aggregated levels based on [12] is more efficient than obtaining the barycenters using Soft-DTW divergence; (2) `HTS-Cluster` only leverages lower-level clustering result instead of the entire set of time series at that level.

Table 1. Level-wise clustering results on HTS with two aggregated levels. The upper part shows the results of simulated data. The lower part gives the results on real-world financial record data using a weak proxy for the cluster labels.

Method\Metric	**Time** (s)	**Global**			**Local**		
		NMI	**AMI**	**ARI**	**NMI**	**AMI**	**ARI**
`DTCR`	132	0.325 ± 0.012	0.257 ± 0.023	0.21 ± 0.011	0.392 ± 0.014	0.313 ± 0.006	0.284 ± 0.009
`Soft-DTW`	67	0.412 ± 0.009	0.326 ± 0.019	0.277 ± 0.008	0.411 ± 0.022	0.342 ± 0.009	0.304 ± 0.014
`Concat`	186	0.436 ± 0.015	0.342 ± 0.014	**0.314** ± 0.016	0.411 ± 0.022	0.342 ± 0.009	0.304 ± 0.014
`HTS-Cluster`	**37**	**0.455** ± 0.018	**0.354** ± 0.015	0.302 ± 0.013	**0.424** ± 0.018	**0.366** ± 0.013	**0.321** ± 0.018
`DTCR`	72	0.065 ± 0.002	0.015 ± 0.001	0.008 ± 0.002	0.105 ± 0.011	0.059 ± 0.002	0.054 ± 0.003
`Soft-DTW`	49	0.119 ± 0.005	0.043 ± 0.003	0.027 ± 0.003	0.126 ± 0.008	**0.082** ± 0.006	0.061 ± 0.005
`Concat`	174	**0.135** ± 0.004	0.073 ± 0.007	**0.045** ± 0.006	0.126 ± 0.008	**0.082** ± 0.006	0.061 ± 0.005
`HTS-Cluster`	**34**	0.134 ± 0.005	**0.075** ± 0.005	0.041 ± 0.004	**0.128** ± 0.014	0.064 ± 0.005	**0.065** ± 0.002

Table 2. Level-wise clustering results on HTS with multiple aggregated levels. On the left are the results on simulated data while the right shows the results on real-world user financial record data. Since cluster labels are not available for the financial data, scores obtained from a weak proxy are lower than expected.

Level	**Metric**	**Simulation**				**Financial Record**			
		`DTCR`	`Soft-DTW`	`Concat`	`HTS-Cluster`	`DTCR`	`Soft-DTW`	`Concat`	`HTS-Cluster`
1	NMI	0.28	0.313	0.342	**0.356**	0.037	0.124	**0.156**	0.154
	AMI	0.243	0.277	0.301	**0.322**	0.021	0.079	**0.112**	0.106
	ARI	0.221	0.265	0.285	**0.304**	0.009	0.056	**0.094**	0.092
2	NMI	0.298	0.317	0.357	**0.375**	0.056	0.116	0.147	**0.152**
	AMI	0.271	0.282	0.314	**0.346**	0.034	0.087	0.115	**0.121**
	ARI	0.236	0.259	0.302	**0.317**	0.016	0.034	0.083	**0.092**
3	NMI	0.272	0.324	0.364	**0.372**	0.055	0.134	0.163	**0.172**
	AMI	0.234	0.295	0.322	**0.33**	0.028	0.098	0.132	**0.141**
	ARI	0.217	0.268	0.307	**0.309**	0.012	0.057	0.106	**0.113**
4	NMI	0.303	**0.369**	**0.369**	**0.369**	0.076	**0.136**	**0.136**	**0.136**
	AMI	0.275	**0.341**	**0.341**	**0.341**	0.043	**0.102**	**0.102**	**0.102**
	ARI	0.264	**0.316**	**0.316**	**0.316**	0.026	**0.061**	**0.061**	**0.061**

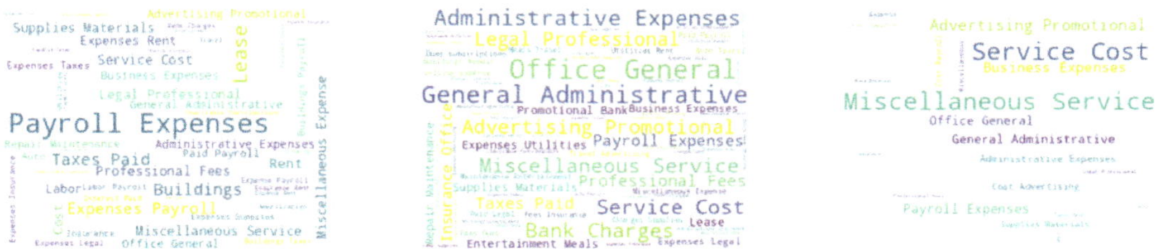

Figure 3. Word cloud visualization of time series metadata from massive financial records. The keywords represent different types of expenses for tax purpose. The results show keywords from three representative clusters at the expense level, by leveraging the local information of clustering results obtained at the user level. The clusters provide meaningful partitions, such as "payroll expenses", "administrative expenses", and "service cost", which are from distinct categories.

4.2. HTS Forecasting

We propose two forecasting applications that can utilize our proposed method. We use the mean absolute scaled error (MASE) [16] to evaluate the forecasting accuracy.

Case 1: Forecast single-HTS with complex structure: Many public datasets comprise a single hierarchy that includes a large number of time series; this is common when HTS has many categorical variables to be aggregated. Forecasting a large number of correlated time series requires extensive computation for global models or parameter tuning for local models. `HTS-Cluster` provides an efficient way of modelling such HTS. For the bottom k-levels that have large numbers of time series, one just needs to forecast their cluster means obtained from clustering "sub-trees" at these levels. The forecasts of each time series at the bottom k-levels can be "reconstructed" using a soft combination of cluster means in Equation (8). We test our method using two popular models: DeepAR [17] and LSTNet [18] and two public HTS: Wiki and M5 [19]. In Table 4, this strategy achieves competitive results with less computation compared with the original methods. This could also improve aggregated levels without applying clustering.

Table 3. Left: level-wise computation time of different clustering approaches, DTCR is excluded since it is not a level-wise approach and requires a much longer time. Right: forecasting massive HTS with the help of clustering, results are measured by MASE and relative computing time. Results are averaged across ten runs on four-level simulated HTS.

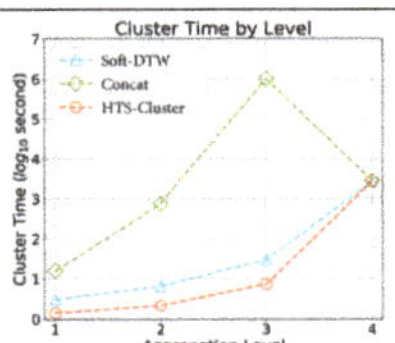

Method \ Level	1	2	3	4	Total Time
Without cluster	62.39	76.26	78.25	84.14	1
DTCR	82.35	96.09	104.85	104.33	0.39
Soft-DTW	78.61	93.04	93.12	96.76	0.27
Concat	74.24	84.65	83.73	96.76	0.57
`HTS-Cluster`	72.99	80.07	85.29	96.76	0.16

Table 4. `HTS-Cluster` can be used to improve HTS forecasting when a large number of forecasts are required. Results are measured by the mean absolute scaled error (MASE, the lower the better) using two multivariate time series models. Both Wiki and M5 possess a single hierarchy with many time series; we cluster "sub-trees" at the bottom two levels (out of five) of Wiki and the bottom three levels (out of twelve; we only show levels eight to twelve) of M5 to reduce the total number of time series to be modelled.

Dataset	Wiki					M5				
Levels	1	2	3	4	5	8	9	10	11	12
LSTNet	76.36	76.89	79.65	81.13	**86.22**	63.74	69.43	73.35	**76.46**	**82.36**
LSTNet-Cluster	**76.33**	**76.56**	**77.68**	**78.07**	95.16	**62.48**	**69.14**	**71.11**	76.52	98.78
DeepAR	**73.98**	74.54	77.42	79.12	**84.77**	59.36	67.18	**72.04**	76.41	**80.24**
DeepAR-Cluster	74.21	**74.37**	**77.36**	**77.56**	89.67	**58.74**	**65.46**	74.39	**75.04**	90.49

Case 2: Forecast massive HTS with simple structures: Similarly, we forecast cluster means obtained from each level of HTS, and then use Equation (8) to obtain a prediction for each HTS. To ensure forecasts are consistent with respect to the hierarchy, we apply reconciliation from [1] to the forecasts of each HTS. Table 3 shows the effectiveness of the clustering, where the total time is normalized by the method without cluster. From the results, `HTS-Cluster` can greatly reduce the overall computation time without compromising the forecasting accuracy.

5. Conclusions

In this paper, we addressed an important but understudied problem for clustering time series with hierarchical structures. Given that time series at different aggregated levels possess distinct properties, regular clustering methods are not ideal. We introduced a new clustering procedure for HTS such that when clustering is conducted at the same aggregated level it simultaneously utilizes clustering results from an adjacent level. In each clustering iteration, both local and global information are leveraged. Our proposed method shows improved clustering performance in both simulated and real-world HTS and proves to be an effective solution when a large number of HTS forecasting is required as a downstream task. For future work, we plan to extend this framework to model-based clustering for HTS with some known statistical properties.

Author Contributions: Conceptualization, X.H., T.R., and N.H.; methodology, X.H. and T.R.; software, X.H.; validation, X.H., and J.H.; formal analysis, X.H., T.R., and N.H.; investigation, X.H.; resources, X.H.; data curation, X.H.; writing—original draft preparation, X.H.; writing—review and editing, X.H., T.R., J.G., and N.H.; visualization, X.H.; supervision, J.G., and N.H. All authors have read and agreed to the published version of the manuscript.

Funding: This work is funded by Intuit AI.

Data Availability Statement: This study applied a simulation dataset and open access datasets (M5 [19], Wiki https://www.kaggle.com/code/muonneutrino/wikipedia-traffic-data-exploration (accessed on 1 April 2023)), which is referenced accordingly. The financial record dataset cannot be publicly accessed due to privacy concerns.

Conflicts of Interest: The authors declare no conflict of interest.

References

1. Wickramasuriya, S.L.; Athanasopoulos, G.; Hyndman, R.J. Optimal forecast reconciliation for hierarchical and grouped time series through trace minimization. *J. Am. Stat. Assoc.* **2019**, *114*, 804–819. [CrossRef]
2. Petitjean, F.; Ketterlin, A.; Gançarski, P. A global averaging method for dynamic time warping, with applications to clustering. *Pattern Recognit.* **2011**, *44*, 678–693. [CrossRef]
3. Aghabozorgi, S.; Shirkhorshidi, A.S.; Wah, T.Y. Time-series clustering–a decade review. *Inf. Syst.* **2015**, *53*, 16–38. [CrossRef]

4. Zhong, S.; Ghosh, J. A unified framework for model-based clustering. *J. Mach. Learn. Res.* **2003**, *4*, 1001–1037.
5. Ma, Q.; Zheng, J.; Li, S.; Cottrell, G.W. Learning representations for time series clustering. *Adv. Neural Inf. Process. Syst.* **2019**, *32*, 3781–3791. [CrossRef]
6. Ho, N.; Nguyen, X.; Yurochkin, M.; Bui, H.H.; Huynh, V.; Phung, D. Multilevel clustering via Wasserstein means. In Proceedings of the International Conference on Machine Learning, PMLR, Sydney, Australia, 6–11 August 2017; pp. 1501–1509.
7. Ho, N.; Huynh, V.; Phung, D.; Jordan, M. Probabilistic multilevel clustering via composite transportation distance. In Proceedings of the 22nd International Conference on Artificial Intelligence and Statistics, PMLR, Okinawa, Japan, 16–18 April 2019; pp. 3149–3157.
8. Rodrigues, P.P.; Gama, J.; Pedroso, J. Hierarchical clustering of time-series data streams. *IEEE Trans. Knowl. Data Eng.* **2008**, *20*, 615–627. [CrossRef]
9. Blondel, M.; Mensch, A.; Vert, J.P. Differentiable divergences between time series. In Proceedings of the International Conference on Artificial Intelligence and Statistics, PMLR, Virtual, 13–15 April 2021; pp. 3853–3861.
10. Müller, M. Dynamic time warping. *Inf. Retr. Music. Motion* **2007**, 69–84.
11. Cuturi, M.; Blondel, M. Soft-dtw: A differentiable loss function for time-series. In Proceedings of the International Conference on Machine Learning, PMLR, Sydney, Australia, 6–11 August 2017; pp. 894–903.
12. Cuturi, M.; Doucet, A. Fast computation of Wasserstein barycenters. In Proceedings of the International Conference on Machine Learning, PMLR, Beijing, China, 21—26 June 2014; pp. 685–693.
13. Schütze, H.; Manning, C.D.; Raghavan, P. *Introduction to Information Retrieval;* Cambridge University Press: Cambridge, UK, 2008; Volume 39,
14. Hubert, L.; Arabie, P. Comparing partitions. *J. Classif.* **1985**, *2*, 193–218. [CrossRef]
15. Vinh, N.X.; Epps, J.; Bailey, J. Information theoretic measures for clusterings comparison: Variants, properties, normalization and correction for chance. *J. Mach. Learn. Res.* **2010**, *11*, 2837–2854.
16. Hyndman, R.J.; Koehler, A.B. Another look at measures of forecast accuracy. *Int. J. Forecast.* **2006**, *22*, 679–688. [CrossRef]
17. Salinas, D.; Flunkert, V.; Gasthaus, J.; Januschowski, T. DeepAR: Probabilistic forecasting with autoregressive recurrent networks. *Int. J. Forecast.* **2020**, *36*, 1181–1191. [CrossRef]
18. Lai, G.; Chang, W.C.; Yang, Y.; Liu, H. Modeling long-and short-term temporal patterns with deep neural networks. In Proceedings of the 41st International ACM SIGIR Conference on Research & Development in Information Retrieval, Ann Arbor, MI, USA, 8–12 July 2018; pp. 95–104.
19. Makridakis, S.; Spiliotis, E.; Assimakopoulos, V. The M5 accuracy competition: Results, findings and conclusions. *Int. J. Forecast.* **2022**, *38*, 1346–1364. [CrossRef]

engineering proceedings

MDPI

Proceeding Paper

Genetic Algorithm Approach for Modeling the Structural Global Stiffness †

Cristian Ștefan Dumitriu, Ștefan Mocanu *, Radu Panaitescu, Anca Ruxandra Sasu * and Oana Tonciu

Department of Machines and Advanced Technologies in Construction, Faculty of Mechanical Engineering and Robotics in Construction, Technical University of Civil Engineering of Bucharest, 122-124 Lacul Tei Bd., 020396 Bucharest, Romania; cristian.dumitriu@utcb.ro (C.Ș.D.); liess.panaitescu@utcb.ro (R.P.); oana.tonciu@utcb.ro (O.T.)

* Correspondence: stefan.mocanu@utcb.ro (Ș.M.); anca-ruxandra.sasu@utcb.ro (A.R.S.)

† Presented at the 9th International Conference on Time Series and Forecasting, Gran Canaria, Spain, 12–14 July 2023.

Abstract: In recent decades, Artificial Intelligence (AI) has become an essential tool for modeling and forecasting in different research fields. Mechanical engineering is no exception because practical problems that classical methods can hardly solve can receive more efficient solutions using AI. Given a support scheme of a structural system, the article aims to determine the maximum stiffness of the system based on the series of moments' variation for a variable dimensional parameter of the support. The series represents the input for a Gene Expression Programming (GEP) aiming to determine the model for a specific geometric parameter in mechanical structures, namely, deflection.

Keywords: Gene Expression Programming (GEP); stiffness; structure deflection; modeling

Citation: Dumitriu, C.Ș.; Mocanu, Ș.; Panaitescu, R.; Sasu, A.R.; Tonciu, O. Genetic Algorithm Approach for Modeling the Structural Global Stiffness. *Eng. Proc.* **2023**, *39*, 32. https://doi.org/10.3390/engproc2023039032

Academic Editors: Ignacio Rojas, Hector Pomares, Luis Javier Herrera, Fernando Rojas and Olga Valenzuela

Published: 30 June 2023

1. Introduction

Structural design, material selection, and optimizing fabrication costs are the main subjects in the machinery industry. The goals of structural design are rational exploitation, economical maintenance, and safety operation. All of these demand a good structural design confirmed by optimal structural stiffness.

Engineering optimization problems are often very complex and present non-linear behavior, so finding their solution represents a topic of interest in the scientific literature. Classical approaches might fail in finding the best solution, so other techniques are necessary. It was shown that Artificial Intelligence (AI) methods could provide alternative solutions without restrictions on the data series. Methods like Gene Expression Programming and Generalized Regression Neural Networks (GRNN) have been successfully used in optimization problems in material science [1], mechanics [2], signal processing [3], meteorology and hydrology [4–7], natural sciences [8], and economics [9].

This paper proposes to find the solution to a mechanical engineering problem using GEP. The proposed approach can determine a better solution in a shorter time [2].

2. The Study Problem

The application presented in the paper [10] requires finding a value of a geometric parameter "*a*" of a particular support scheme to determine the maximum rigidity of the system (Figure 1, top-left representation). The structure is designed as a circular frame, laterally supported through two beams by fixed support and slide support. The frame is loaded in its plane with two forces P in the highest and lowest points (in circle quadrants). The frame has the R radius, and the supporting beams are equal in length ("a" value).

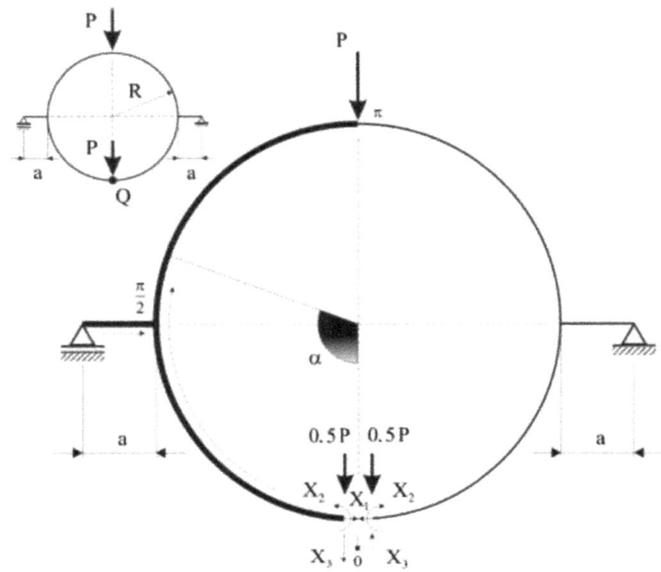

Figure 1. The initially indeterminate static system (top-left representation) and the basic form used in removing indeterminacy.

Figure 1 presents a possible basic form of the proposed statically three times indeterminate system. Considering the symmetry of the support schemes, respectively loading, the value of the unknown antisymmetric shear force type is zero [11]. By using the bending moment terms in Maxwell–Mohr [11] exclusively, eliminating the system indeterminacy involves solving the system of two equations with two unknowns (1) of the form:

$$\begin{cases} \delta_{11}X_1 + \delta_{12}X_2 + \Delta_{10} = 0 \\ \delta_{21}X_1 + \delta_{22}X_2 + \Delta_{20} = 0 \end{cases} \quad (1)$$

Using the moment diagram, the system's solution has the following form [10]:

$$\begin{cases} X_1 = \frac{P}{\pi}\left(1 + 2\frac{a}{R}\right) \\ X_2 = \frac{PR}{\pi}\left[\left(\frac{\pi}{2} - 1\right) + \left(\frac{\pi}{2} - 2\right)\frac{a}{R}\right] \end{cases} \quad (2)$$

The performance criterion chosen for quantifying the global stiffness of system (2) is the projection of the displacement of point Q on the vertical plane given by the relation:

$$v_Q = \frac{1}{EI}\int m_Q^0 \cdot M \cdot ds \quad (3)$$

where m_Q^0, M are the polynomial forms of the corresponding bending moment variations [10].

Considering the working variable $q = a/R$, the vertical projection displacement of the point Q will depend on q, that is $v_Q = f(q)$. The goal is to determine a GEP model for the deflection v_Q.

In the following, we shall solve the problem for a sample case in which the values for the force, frame radius, and material properties are $P = 150$ daN, $R = 5$ m, $E = 2.1 \times 10^5$ MPa, and $I = 4.93 \times 10^6$ mm^4, respectively.

3. Methodology

The first step is to generate the deflection values for different q values. To reduce the computation effort, the AxisVM13LT (https://axisvm.eu/ (accessed on 3 February 2023)) and Solidworks (https://www.solidworks.com/ (accessed on 3 February 2023)) software were utilized. The software was chosen because it is designed for structure calculations using Finite Element Analysis (FAE) [12].

Since the aim of the article is not the FEA computation, we shall not present these details here. For more information on the topic, the reader may see [10].

In the second step, GEP was employed to model the deflection.

GEP has futures from Genetic Algorithms (GA) and Genetic Programming (GP), working with function and terminal sets, evaluating the individual using a fitness function, and evolving the generation until a termination condition is met [13,14]. In GEP, a candidate solution to a problem represents a formula formed by a composition of constants, variables, and functions structured as a mathematical expression parse tree. GEP individuals are strings of symbols encoding non-linear expressions formed by genes with a fixed length. The chromosomes are formed by genes in a constant number for all the population's members. The structure of a gene is formed by a head (composed of any symbol) and a tail (composed of constants and variables). The operators used in GEP are transposition, mutation, and crossover. The candidate solutions are evaluated using the fitness function. For details on GEP, the reader may see [1,13].

The input was formed by the data series generated at the previous step and the corresponding a values.

To perform the modeling data series was divided into two parts: 70% for training and 30% for validation. The following settings of the GEP parameters were used: population size: 50, maximum tries for the initial population: 10,000, number of genes per chromosome: 4, the length of the gene's head: 8, the maximum number of generations: 2000, and the number of generations without improvement: 1000. The fitness function was the variance explained by the model (R^2), and the hit tolerance was 0.01. The addition was the linking function. The inversion, IS, and RIS transposition rates were set to 0.1, the one- and two-point recombination to 0.3, the gene transposition and recombination rates to 0.1, and the mutation rate was 0.03. The following operators and functions were used to obtain the GEP model: +, −, ×, /, sqrt. Leave-one-out method was used to check the model's quality.

The modeling was carried out in DTREG.

4. Results and Discussion

Figure 2 shows the system deformation resulting from the AxisVM simulation in the studied case.

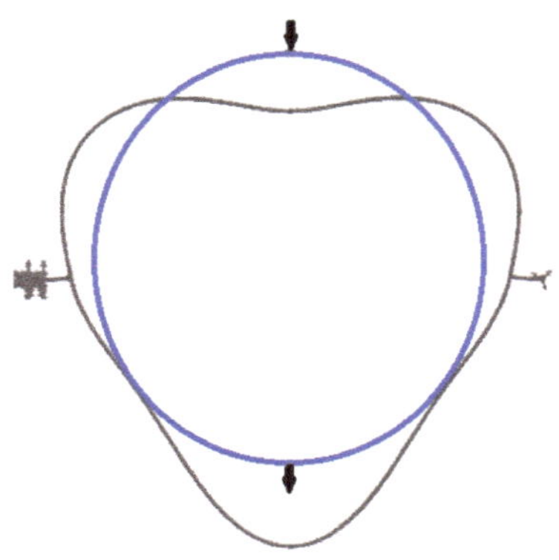

Figure 2. The system deformation: q—on the abscissa, and v_q on the ordinate.

After running the GEP algorithm, the following expression was generated for the deflection as a function of a:

$$v_q = \sqrt[4]{a - 0.47697} - 1.63476 + 1.35321/a + (a + 1.04081) \times \sqrt{a}, \tag{4}$$

or

$$v_q = \sqrt[4]{5q - 0.47697} - 1.63476 + 0.27064/q + (11.18034q + 2.3273221) \times \sqrt{q}. \quad (5)$$

Figure 3 shows the chart of the actual vs. predicted error in the model. Most values are situated at a small distance (measured on the parallel of the Ox axis) from the first bisectrix of the axes of coordinates, indicating good model quality. To confirm this last assertion, the goodness of fit indicators (R^2), mean square error (MSE), mean absolute error (MAE), mean absolute percentage error (MAPE), and the correlation between the actual and predicted values (r_{ap}) are displayed in Table 1.

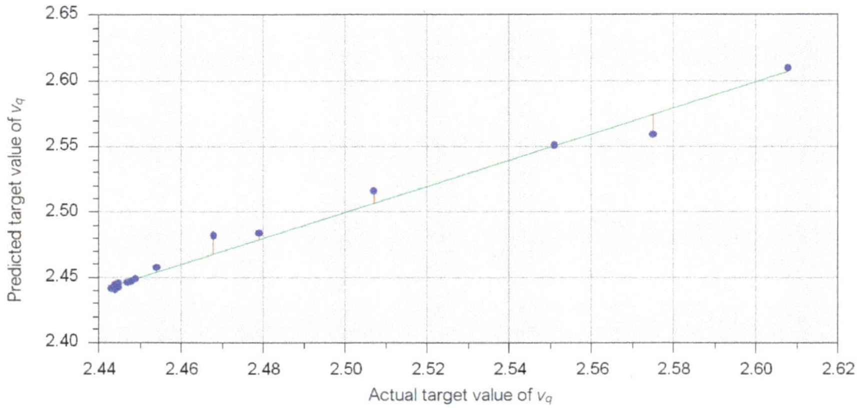

Figure 3. Predicted target values vs. actual target values in the GEP model.

Table 1. The goodness of fit indicators in the GEP model.

		Training					Test		
R^2	MSE	MAE	r_{ap}	MAPE	R^2	MSE	MAE	r_{ap}	MAPE
98.705%	0.00003	0.0035	0.9935	0.1333	97.299%	0.00009	0.0073	0.9936	0.2935

The R^2 values are over 0.97, showing that 98.705% (97.299%, respectively) of the deflection variation is explained by the variation of the explicative variable. Moreover, there is a very good correlation between the actual and target values, with both r_{ap} values being above 0.99. In terms of MAE and MSE, the model performs better on the Training set than on the Test data set, but all these values are very low (MAEs < 0.036 and MSE < 0.00009), indicating that the model fits very well the data series.

An analogous study has been conducted using the SolidWorks output. The results are similar in terms of goodness of fit indicators on the Training set, while on the Test set, they are worse. The explanation is that the accuracy mainly depends on the type of discretization used in FEA computation by each application (beam meshing is a simpler discretization than solid meshing) and on the structure supporting schema.

5. Conclusions

Based on a set of deflection values obtained through FEA software or manual calculation, an Artificial Intelligence algorithm was built using GEP. This algorithm finds the expression of deflection variation as a function of parameter a, where a/R quantifies the structure stiffness in the studied problem. For a practical case, using this model, the deflection's extreme values can be determined to analyze the structural behavior in different loading cases for this supporting schema. Similar modeling expressions for deflection variation using Artificial Intelligence must be generated for other supporting schema.

Author Contributions: Conceptualization, C.Ș.D. and Ș.M.; methodology, Ș.M. and R.P.; software, R.P.; validation, C.Ș.D. and O.T.; formal analysis, O.T. and A.R.S.; investigation, Ș.M. and A.R.S.; resources, Ș.M.; writing—original draft preparation, C.Ș.D. and ȘM.; writing—review and editing, C.Ș.D.; visualization, Ș.M.; supervision, C.Ș.D. All authors have read and agreed to the published version of the manuscript.

Funding: This research received no external funding.

Institutional Review Board Statement: Not Applicable.

Informed Consent Statement: Not applicable.

Data Availability Statement: Not applicable.

Conflicts of Interest: The authors declare no conflict of interest.

References

1. Dumitriu, C.Ș.; Bărbulescu, A. Artificial intelligence models for the mass loss of copper-based alloys under the cavitation. *Materials* **2022**, *15*, 6695. [CrossRef] [PubMed]
2. Șerban, C.; Dumitriu, C.Ș.; Bărbulescu, A. Solving Single Nesting Problem Using a Genetic Algorithm. *An. Ştiinţifice Univ. Ovidius Constanţa Ser. Mat.* **2020**, *30*, 259–272.
3. Dumitriu, C.Ș.; Dragomir, F.-L. Modeling the signals collected in cavitation field by stochastic and Artificial intelligence methods. In Proceedings of the 2021 13th International Conference on Electronics, Computers and Artificial Intelligence (ECAI), Pitesti, Romania, 1–3 July 2021; pp. 1–4. [CrossRef]
4. Bărbulescu, A. *Time Series with Applications*; Junimea: Iasi, Romania, 2002. (In Romanian)
5. Bărbulescu, A.; Băutu, E. Time Series Modeling Using an Adaptive Gene Expression Programming. *Int. J. Math. Models Methods Appl. Sci.* **2009**, *2*, 85–93.
6. Băutu, E.; Bărbulescu, A. Forecasting meteorological time series using soft computing methods: An empirical study. *Appl. Math. Inf. Sci.* **2013**, *7*, 1297–1306. [CrossRef]
7. Bărbulescu, A.; Șerban, C.; Indrecan, M.-L. Computing the Beta Parameter in IDW Interpolation by Using a Genetic Algorithm. *Water* **2021**, *13*, 863. [CrossRef]
8. Bărbulescu, A.; Șerban, C.; Caramihai, S. Assessing the soil pollution using a genetic algorithm. *Rom. J. Phys.* **2021**, *66*, 806.
9. Bărbulescu, A. Do the time series statistical properties influence the goodness of fit of GRNN models? Study on financial series. *Appl. Stoch. Models Bus. Ind.* **2018**, *34*, 586–596. [CrossRef]
10. Mocanu, Ș. The stiffness as an optimization parameter of a structure. *Sinteze de Mecanică Teoretică și Aplicată* **2016**, *7*, 335–342. (In Romanian)
11. Popov, E.P. *Mechanics of Materials*; Prentice-Hall: Englewood Cliffs, NJ, USA, 1976.
12. Akin, J. *Finite Element Analysis Concepts via SolidWorks*; World Scientific: Singapore; Rice University: Houston, TX, USA, 2009.
13. Ferreira, C. *Gene Expression Programming: Mathematical Modeling by an Artificial Intelligence*; Springer: Berlin/Heidelberg, Germany, 2006.
14. Sharifi, Y.; Moghbeli, A. New predictive models via gene expression programming and multiple nonlinear regression for SFRC beams. *J. Mater. Res. Technol.* **2020**, *9*, 14294–14306. [CrossRef]

engineering proceedings

MDPI

Proceeding Paper

A CNN–BiLSTM Architecture for Macroeconomic Time Series Forecasting †

Alessio Staffini

Department of Economics and Finance, Catholic University of Milan, 20123 Milan, Italy; alessio.staffini@bocconialumni.it

† Presented at the 9th International Conference on Time Series and Forecasting, Gran Canaria, Spain, 12–14 July 2023.

Abstract: In forecasting socio-economic processes, it is essential to have tools that are highly performing, with results as close to reality as possible. Forecasting plays an important role in shaping the decisions of governments and central banks about macroeconomic planning, and it is an essential analytical tool in defining economic strategies of countries. The most common forecasting methods used in the analysis of macroeconomic processes are based on extrapolation, i.e., extending the trend observed in the past (and present) to the future. However, the presence of non-linearity in the socio-economic systems under uncertainty, as well as the partial observability of the processes, has contributed to make researchers and practitioners consider other methodologies, too. In this paper, we analyze 18 time series of macroeconomic variables of the United States of America. We compare the benchmark results obtained with "classic" forecasting techniques with those obtained with our proposed architecture. The model we construct can be defined as "hybrid" since it combines a Convolutional Neural Network (CNN) with a Bidirectional Long Short-Term Memory Network (BiLSTM) backend. We show that, for what concerns minimizing the forecast error, our model competes with and often improves the results obtained with the benchmark techniques. The goal of this work is to highlight that, due to the recent advances in computing power, new techniques can be added to the set of tools available to a policymaker for forecasting macroeconomic data.

Keywords: time series forecasting; economic forecasting; macroeconometric forecasting; deep learning; CNN-BiLSTM

Citation: Staffini, A. A CNN–BiLSTM Architecture for Macroeconomic Time Series Forecasting. *Eng. Proc.* **2023**, *39*, 33. https://doi.org/10.3390/engproc2023039033

Academic Editors: Ignacio Rojas, Hector Pomares, Luis Javier Herrera, Fernando Rojas and Olga Valenzuela

Published: 30 June 2023

1. Introduction

Machine Learning is now deeply entwined with many aspects of modern society, from optimizing web searches and filtering content on social networks to personalized recommendations on e-commerce sites. Increasingly, these applications make use of a class of techniques called Deep Learning, which forms a subfield of Machine Learning. Deep Learning techniques are based on Artificial Neural Networks (ANNs), which are vaguely inspired by the structure and functioning of the brain. ANNs are made by artificial neurons, and these neurons are organized in many different layers: a given layer computes the values to be used as input for the next layer in order to process the data in an increasingly complex and complete way.

With a sufficient amount of data, such networks are able to learn the correct representation and to outperform Machine Learning algorithms [1,2]; in other words, Deep Learning can be thought of as a learning technique in which ANNs are exposed to very huge amounts of data so that they can efficiently learn to perform tasks [3].

Deep Learning is one of the main sources of success for the recent advances in Artificial Intelligence: because of ANNs and their variations, which can efficiently analyze images, video, audio and sequential data, this area is developing rapidly.

For some years now, researchers have been trying to apply Deep Learning for forecasting analysis on time series data. Excellent results have been achieved in some fields (among all, that of weather forecasts [4,5]), but in the economic–financial field, the use of these new tools has not yet matured except for some applications in portfolio selection [6,7] and in stock price forecasting [8–10]. Even so, Machine Learning techniques and Deep Learning algorithms have shed light on new approaches to dealing with prediction problems, in which the relationships between input and output variables are modeled in a deep and layered hierarchy. Machine Learning techniques such as Support Vector Machines (SVMs) and Random Forests, as well as Deep Learning techniques such as Convolutional Neural Networks (CNNs), Recurrent Neural Networks (RNNs) and Long Short-Term Memory (LSTMs) have gained a lot of attention recently, and they are increasingly being applied in many disciplines [11–13]. Unlike simpler classification and regression problems, time series forecasting adds a further order of complexity, caused by the time dependence between observations. Traditionally, time series forecasting has been dominated by linear methods, such as ARIMA, because they have a good theoretical framework and are effective on many problems [11,14]. However, these methods also suffer from some limitations, such as being sensible to the presence of missing data, excluding complex non-linear mappings between inputs and outputs, and requiring a careful data preparation to find the optimal number of delays.

Forecasting time series data in the macroeconomic domain is an even more challenging task, mainly due to the possible unprecedent changes in economic trends on the one hand, and incomplete information on the other [15–17]. The market volatility in recent years has created serious concerns over the precision of economic and financial forecasts; therefore, the accuracy of the predictions is an extremely important parameter to consider when selecting a forecasting method [18,19].

One of the goals of this paper is to study which techniques offer the best forecasts, starting from a dataset of US macroeconomic data. In this regard, we first employed three econometric models (ARIMA, ARIMAX, and VAR), as well as Machine Learning models (Elastic Net, XGBoost, and BiLSTM), to measure their prediction accuracy on the examined variables. We have chosen to use these models because they are the most popular forecasting tools in the econometric field [20,21] or because they have been deemed as optimal for forecasting certain macroeconomic variables [15] and because, in a comparison with other techniques (naive forecasting, exponential smoothing, etc.) they demonstrated better performances. Next, we measured the prediction accuracy reported by our Deep Learning architecture.

Deep Learning methods add non-linearity and complexity in time series forecasting, and therefore they can provide good performance [3]. Furthermore, neural networks should be able to exploit all the capabilities of classical linear prediction methods, since they basically try to find the best mapping from inputs to outputs [2]. However, when it comes to time series forecasting, ANNs and complex methods should be evaluated and applied carefully, as depending on the data at hand they do not always deliver better results than more traditional tools [22,23]. It is therefore important to include and evaluate the performance of multiple methods to justify the application of neural networks.

CNNs are feedforward neural networks with shared weights and sparse interactions, often used for processing data that has a known grid-like topology (such as images or time series data). In CNN architectures, it is custom to process the data with a certain number of convolutional layers, followed by a max-pooling layer. Informally, the convolutional layers filter the data by removing noise and extracting features of the input, while the max-pooling layer acts as an information contractor, keeping only a summary of those filtered features. RNNs are a family of neural networks obtained by adding feedback connections to the feedforward architecture, and they are again used mainly for processing sequential data (such as text and time series data); LSTMs are a particular kind of them, which introduced the idea of gates to better control the flow of information and learn long-term dependencies.

The main goal of this paper is to show that, due to the recent advances in computing power, there are additional methods of analysis that are available to policymakers, especially when they are planning some interventions based on the forecasts of economic variables. We suggest that these methods should be taken into consideration.

2. The Model

Our proposed architecture is a hybrid CNN-BiLSTM. First, we stacked two one-dimensional convolutional layers (64 filters each, with a filter size of 3) to filter the input signal. The results of these convolutions were then passed to a one-dimensional MaxPooling layer (pooling size of 2, no strides and "valid" padding), in order to down-sample the feature maps, reduce the dimensionality, extract only the relevant information and in general make the network more robust. Notice that we only halved the dimensionality, applying a modest pooling size of 2. Applying a pooling layer after (usually two or more) convolutional layers in order to filter the relevant information while also making the network easier to train is a standard procedure in CNNs, even though recently (see, for example, [24]) some researchers proposed efficient architectures made only of convolutional layers that often reached state-of-the-art performance in a series of object recognition tasks. We then flattened the down-sampled results into a vector of neurons, and we provided it as a sequence to three-layer stacked Bidirectional LSTMs (BiLSTMs), each one, respectively, with 200, 100 and 50 units. A BiLSTM cell [25] can be simply thought of as made of two LSTMs: one reading the input sequence forward and the other reading it backwards, before concatenating both interpretations. Doing so increases the amount of information available to the model because it provides more context and, while so far it has been typically applied for Natural Language Processing tasks, it recently proved to be very useful also for time series forecasting [13].

Finally, the output of the BiLSTM submodel is followed by a fully connected (FC) layer of 25 neurons to interpret its outcomes. We then stacked the output layer, which was again a fully connected one, this time made by 12 neurons (when we wanted to forecast the next 12 months) or just a single neuron (when we performed single-step forecasting).

When training complex Deep Learning models, overfitting the train set is a very common phenomenon. To deal with this problem and provide a good generalization, we applied extensive dropout. Dropout [26] is an algorithm for which, at every training step, each neuron in the considered layer has a certain defined probability p of being "dropped". Dropout offers regularization because it approximates training in parallel many neural networks with different architectures: since at each training step a neuron can be active or not, we are effectively training 2^N different networks (where N represents the number of neurons to which dropout is applied). It is also a very suitable technique when we do not have a high amount of training data [27,28].

We applied dropout after the CNN submodel, after each BiLSTM layer, and after the FC layer (before the output layer), for a total of five dropout layers. We applied stronger dropout ($p = 0.5$) after each BiLSTM layer, while after the CNN submodel and the FC layer we applied a milder one ($p = 0.1$).

As highlighted by [26], dropout works even better when combined with other regularization techniques, such as weight and bias constraints. Large weight size can lead to the problem of exploding gradients [29] which in turn makes the network unstable. To deal with this problem, we can force the norm of all the weights in a layer to be below a certain threshold c. Following again [26], we combined dropout with max-norm regularization, bounding both the weights and the biases of each hidden layer to have their norms be below three.

The activation function of a neuron/node is used to map the weighted sum of the inputs to the output. For the entire network, we applied a ReLU (Rectified Linear Unit) activation function, which is a piecewise linear function that outputs the summed weighted input if positive, otherwise it outputs zero:

$$ReLU(x) = \max(0, x). \tag{1}$$

Non-linearity is needed in activation functions to learn complex enough (non-linear) mappings from the inputs to the output. ReLU and its variants have become the standard choice as activation functions for training deep neural networks because of their many desirable properties, one among which is the fact that they deal well with the problem of vanishing gradients [30]. Traditionally, for the LSTM cells, the hyperbolic tangent (*tanh*) function is used as activation. However, empirically, we found that passing the BiLSTM output as argument to a ReLU function performed slightly better, which is in line with other results in the literature [31]. We initialized the weights of each layer applying the Glorot uniform initializer [30], which in practice works well for deep networks [32].

Another important choice is the optimization algorithm. We selected Adam [33] because it combines both the ideas of momentum estimation and RMSProp, and it is fast and efficient. It is typically the default choice [34], and indeed in our experiments we found it to perform the best.

Finally, we trained our network using mini-batch gradient descent, which works as a compromise between gradient descent and stochastic gradient descent (SGD). Using mini-batches, we obtained a more stable convergence than using SGD while still injecting enough noise in the model, which is required for non-convex optimization [35]. We chose mini-batches of size $m = 32$, which usually works well in practice [36,37].

Figure 1 shows a stylized picture of our architecture.

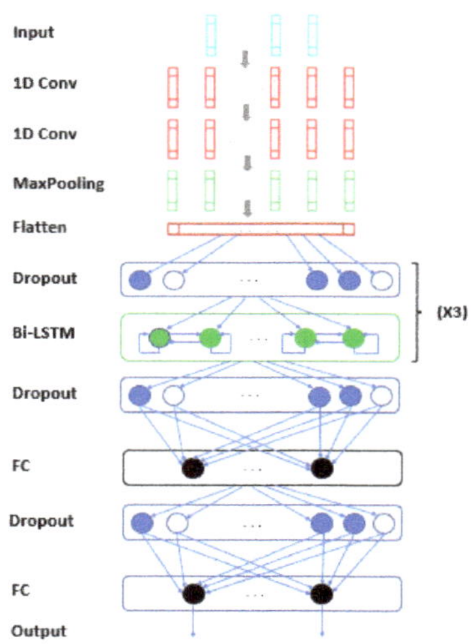

Figure 1. The network structure of our CNN-BiLSTM architecture.

Consider a normalized time series X_n, where n denotes the number of input variables ($n = 1$ for the univariate case) and $y \in X_n$ represents our variable of interest. After reframing the data as a supervised learning problem (where a given segment of X_n is used to forecast the next t steps of y—which we denote here as y_t), our described model first processes X_n by extracting different features with different filters, down-sampling the resulting feature maps via MaxPooling. The resulting outcome F is then flattened and provided as input to the BiLSTM backend, which further processes the data and makes the final prediction, $\hat{y}_t$. In short,

$$F = \text{CNN}(X_n), \tag{2}$$

$$\hat{y}_t = \text{BiLSTM}(F). \tag{3}$$

3. Empirical Analysis

We conducted our empirical analysis using data collected by the Bureau of Economic Analysis (BEA) and the Bureau of Labor Statistics (BLS) on different economic indicators of the United States of America. Among the many variables collected by the two agencies, we chose those that presented a low correlation (see Figure 2) in order to capture different aspects of the economy as a whole. The 18 chosen variables have already been seasonally adjusted, and they range from June 1947 to December 2019. They were measured monthly (we linearly interpolated those that were recorded on a quarterly basis).

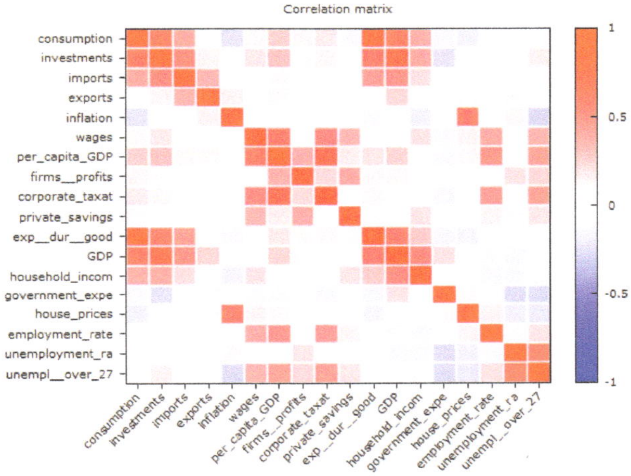

Figure 2. Correlation matrix of the considered variables.

The collected variables are consumption of goods, expenditure on durable goods, private investments, import, export, inflation, GDP, available household income, government expenditure, house prices, employment rate, and unemployment rate (all expressed as growth rate over the same period of the previous year); average wages, Per Capita GDP (in dollars); corporate profits, corporate taxation, private savings (in billions of dollars); long-term unemployment (as a percentage of total unemployment).

Each variable is numerical, and we had no missing values in the original dataset. We had a total of 871 observations for each variable. We split each time series into a train set and a test set, using an 80–20 split ratio (i.e., the first 697 values were used for training our models while the last 174 values were used to evaluate their performance). The reason behind this split is because it is one of the of the most common ones and has been shown to work well in practice [38]. We ran both multi-step forecasting (where the models tried to forecast at once the next 12 values) and one-step forecasting (where the models iteratively forecasted the next time step).

We first implemented three "classic" econometric models to perform forecasting on the test set: an ARIMA model, an ARIMAX model, and a VAR model. We also considered three Machine Learning models: a regularized linear regression model (Elastic Net), a gradient boosting algorithm (XGBoost), and a baseline Deep Learning architecture (BiLSTM). The forecast errors obtained from these models were the benchmark results to which to compare our proposed architecture.

First, we needed to verify that the considered processes were stationary. We applied the Augmented Dickey Fuller (ADF) test to check for the existence of unit roots. For only 5 of the 18 variables (average wages, Per Capita GDP, corporate profits, corporate taxation, and private savings) we did not reject the null hypothesis of having a unit root at a significance level of 0.01. We set for these variables the differencing term to $d = 1$ in the ARIMA and ARIMAX models, and a further test on the differentiated data of these variables confirmed the absence of unit roots. In addition, the autocorrelation function (ACF) and the partial autocorrelation function (PACF) of these five variables further suggested the idea that $d = 1$. We employed ACF and PACF also for finding the appropriate order of the autoregressive term p and the moving average term q. We further compared the model specification suggested by ACF and PACF with the one suggested by information criteria, such as the Akaike Information Criterion (AIC). The specification suggested by ACF/PACF and AIC often matched; when it did not, we preferred the one that empirically performed better. This selection criterion led to the definition of different models depending on the variable. In regard to the specifications for ARIMAX and VAR models, we selected the additional covariates relying on the results of the Granger causality test, i.e., we included only further explanatory variables that "Granger-cause" the dependent one; this can be understood as including only the variables that improve the forecast results of the dependent one.

The choices of the values of p and q in the ARIMAX models were conducted as for the ARIMA case. Naturally, since we were also considering further covariates, the implemented specifications were different from those of ARIMA.

The number of lags p in the VAR models were selected using the multivariate version of AIC. Since for each VAR model we included different covariates in the system (recall that we selected only those that passed the Granger causality test for the considered dependent variable of interest), we obtained different specifications for each model. We also considered the possible presence of cointegration but, given the few processes that were $I(1)$ in first place, we never encountered such a case. We ran both the Engle–Granger cointegration test and the Johansen cointegration test, and indeed they excluded such a possibility. Therefore, when we had one or more $I(1)$ processes in the system, we implemented a VAR in differences, without the risk of it being mis-specified.

For what concerns Elastic Net and XGBoost, we used a grid search approach on the train set of each variable to optimize the values of the hyperparameters as well as the number of inputs, testing the performance of multiple different configurations and choosing the most performing one. The BiLSTM architecture is the same three-layer stacked configuration chosen for the BiLSTM backend in our proposed CNN-BiLSTM model.

4. Results

After having implemented the econometric models, we ran diagnostic checks to verify the homoskedasticity, normality and absence of autocorrelation in the residuals: all the models passed these tests, confirming the correctness of the selected specifications.

In order to compare the performance of the results obtained with the baseline models with that obtained from our CNN-BiLSTM architecture, we needed to select an error metric. We chose two popular metrics regularly employed in evaluation studies: the Mean Absolute Error (MAE) and the Root Mean Squared Error (RMSE), which is the square root of MSE. MAE is considered a natural measure of average error [39], while RMSE has the property of penalizing large errors more, and it is often the preferred measure for regression tasks in absence of many outliers. Considering N observations, their equations are given by

$$MAE = \frac{1}{n}\sum_{i=1}^{n}|y_i - \hat{y}_i|, \quad (4)$$

$$RMSE = \sqrt{\frac{1}{N}\sum_{i=1}^{N}(y_i - \hat{y}_i)^2}, \quad (5)$$

where y_i is the true value and $\hat{y}_i$ is the value predicted by the model.

The econometric models have been implemented using the statistical package gretl (Version 2020b), while the Machine Learning models and our proposed architecture were built with Python (Version 3.7.3). We collected all the results in Tables 1–4.

Table 1. Multi-step forecast errors for the univariate input models. Text in bold denotes the best performance (95% confidence level).

Variable	ARIMA	Elastic Net	XGBoost	BiLSTM	Our CNN-BiLSTM
	RMSE MAE	RMSE MAE	RMSE MAE	RMSE (std. dev.) MAE (std. dev.)	RMSE (std. dev.) MAE (std. dev.)
consumption	4.0965 2.8597	2.9343 1.9814	3.0921 2.0763	2.9725 (0.0686) 2.0737 (0.0217)	**1.9736 (0.0463)** **1.5941 (0.0459)**
investments	8.8201 6.0830	6.0735 4.0524	6.1441 3.9671	6.2068 (0.7248) 4.1441 (0.4170)	**3.7446 (0.1710)** **2.9190 (0.1372)**
imports	11.088 7.1216	8.3575 6.1214	7.7900 5.2379	7.8201 (0.4894) 5.1850 (0.4264)	**4.3390 (0.1758)** **3.3621 (0.1416)**
exports	10.599 6.6301	7.1607 5.0656	7.2648 4.9356	7.1996 (0.2561) 4.6305 (0.4166)	**5.1139 (0.1191)** **3.3972 (0.0941)**
inflation	2.2526 1.4850	1.5051 0.9833	1.5849 0.9830	1.5760 (0.0418) 1.0704 (0.0357)	**0.8927 (0.0305)** **0.6801 (0.0404)**
wages	206.60 143.02	148.31 101.07	166.21 127.92	164.75 (8.7017) 130.71 (13.638)	**141.20 (2.4414)** **95.986 (3.4415)**
per capita GDP	145.61 107.06	102.61 76.707	114.19 87.994	148.54 (9.9937) 125.79 (11.249)	**86.209 (2.9799)** **68.618 (3.3470)**
firms' profits	39.849 28.648	28.078 19.137	29.166 19.690	29.548 (1.0818) 19.660 (0.7318)	**23.911 (0.9131)** **18.481 (0.5890)**
corporate taxation	4.0462 3.1330	3.2166 2.4978	3.7096 2.9433	4.5836 (0.6401) 3.6909 (0.2091)	**3.0049 (0.0761)** **2.3460 (0.0533)**
private savings	47.670 35.448	37.938 27.982	38.543 28.764	39.569 (1.5244) 29.538 (1.2444)	**36.840 (03155)** **27.392 (0.3602)**
expen. durable goods	8.5890 5.9085	6.0217 3.8684	6.3203 4.2891	6.5361 (0.1523) 4.4667 (0.1151)	**4.2486 (0.0985)** **3.2753 (0.0737)**
GDP	2.9405 2.0696	2.2276 1.6256	2.2654 1.5530	2.1998 (0.1925) 1.4258 (0.0305)	**1.3007 (0.0669)** **0.9965 (0.0445)**
household income	4.4913 2.9752	3.0415 2.0485	3.3690 2.1789	3.0825 (0.0092) 1.9547 (0.0143)	**2.9496 (0.0172)** **1.8623 (0.0161)**
government expenditure	2.4130 1.9668	2.4078 1.8971	2.6883 2.0271	2.4379 (0.0322) 2.0667 (0.0413)	**1.9379 (0.0894)** **1.5853 (0.0892)**
house prices	5.5399 4.2488	2.8767 1.9062	3.4029 2.1275	3.6669 (0.1034) 2.5121 (0.1170)	**2.3246 (0.0597)** **1.8021 (0.0506)**
employment rate	0.9396 0.5943	**0.4462** **0.2688**	0.6097 0.3803	1.0915 (0.1840) 0.6800 (0.2310)	0.8562 (0.0315) 0.7342 (0.0032)
unemployment rate	1.2203 0.8644	0.5638 0.4000	0.7087 0.4847	0.6782 (0.0911) 0.4163 (0.0714)	**0.4307 (0.0561)** **0.3565 (0.0417)**
unemp. over 27 weeks	5.8469 3.8736	3.9508 2.9983	11.856 8.7233	5.4433 (0.8422) 4.0504 (0.6085)	**1.9146 (0.1641)** **1.4947 (0.1287)**

Table 2. Single-step forecast errors for the univariate input models. Text in bold denotes the best performance (95% confidence level).

Variable	ARIMA	Elastic Net	XGBoost	BiLSTM	Our CNN-BiLSTM
	RMSE MAE	RMSE MAE	RMSE MAE	RMSE (std. dev.) MAE (std. dev.)	RMSE (std. dev.) MAE (std. dev.)
consumption	0.9643 0.7120	0.9553 0.6583	1.3197 0.9289	0.9081 (0.0280) 0.6078 (0.0211)	**0.8653 (0.0238)** **0.5385 (0.0168)**
investments	1.4326 0.9711	1.4369 0.9419	2.3368 1.4167	1.4349 (0.0392) 0.9849 (0.0383)	**1.3400 (0.0365)** **0.9246 (0.0147)**
imports	1.7607 1.1129	2.3095 1.6583	2.5519 1.7599	1.7439 (0.1267) 1.1268 (0.0544)	**1.5588 (0.0478)** **1.0276 (0.0273)**
exports	2.283 1.6165	2.2169 1.5940	2.7194 2.0555	2.3930 (0.1076) 1.5815 (0.0847)	**2.0841 (0.0452)** **1.3116 (0.0749)**
inflation	**0.4846** **0.2919**	0.9553 0.6593	0.6456 0.4136	0.6115 (0.0376) 0.3695 (0.0223)	0.5554 (0.0149) 0.3225 (0.0204)
wages	117.50 50.580	**98.920** **44.557**	127.54 78.866	126.99 (4.5553) 81.072 (9.1984)	121.86 (1.1490) 71.469 (2.7451)
per capita GDP	65.879 37.380	65.311 **32.470**	78.466 46.829	80.204 (5.5873) 56.759 (6.7406)	**64.167 (0.4711)** 40.493 (1.7013)
firms' profits	20.872 10.754	19.399 9.1343	23.328 13.805	20.771 (1.5599) 12.611 (0.8903)	**17.165 (0.0945)** **9.4561 (0.1147)**
corporate taxation	2.2362 1.3484	2.2831 1.2905	2.3848 1.4029	2.5278 (0.1010) 1.7611 (0.1016)	**2.0766 (0.0557)** **1.1868 (0.0732)**
private savings	26.963 18.998	28.001 20.012	34.706 23.230	31.628 (2.3708) 21.659 (1.8836)	**25.765 (0.3857)** **15.805 (08.280)**
expen. durable goods	2.1026 1.3557	2.0493 1.4233	3.2429 2.1295	1.9780 (0.1184) 1.4542 (0.0796)	**1.7694 (0.0524)** **1.2527 (0.0757)**
GDP	0.6006 0.4002	0.6097 0.3924	0.7673 0.5539	0.5439 (0.0094) 0.3475 (0.0222)	**0.5086 (0.0291)** **0.3165 (0.0240)**
household income	0.9904 0.5434	1.1643 0.6927	1.7724 0.9226	1.1304 (0.0906) 0.6783 (0.0442)	**0.8914 (0.0281)** 0.6085 (0.0277)
government expenditure	0.5634 0.4001	0.6093 0.4405	0.7855 0.6336	0.5975 (0.0394) 0.4525 (0.0304)	**0.5191 (0.0098)** **0.3923 (0.0062)**
house prices	0.7943 0.5159	**0.7671** **0.4363**	1.2116 0.7466	0.9762 (0.0797) 0.6241 (0.0592)	0.8293 (0.0291) 0.5085 (0.0313)
employment rate	**0.1316** **0.1029**	0.1373 0.1059	0.2263 0.1703	0.3539 (0.0297) 0.2360 (0.0404)	0.3716 (0.0058) 0.2422 (0.0141)
unemployment rate	**0.1547** **0.1238**	0.1592 0.1279	0.2037 0.1526	**0.1552 (0.0022)** **0.1230 (0.0027)**	0.1569 (0.0007) 0.1247 (0.0011)
unemp. over 27 weeks	**1.2845** **0.9715**	1.2896 0.9756	9.7659 6.6189	1.3294 (0.0988) 1.0333 (0.0757)	1.5099 (0.0656) 1.1661 (0.0587)

Table 3. Multi-step forecast errors for the multivariate input models. Text in bold denotes the best performance (95% confidence level).

Variable	ARIMAX	VAR	Elastic Net	XGBoost	BiLSTM	Our CNN-BiLSTM
	RMSE MAE	RMSE MAE	RMSE MAE	RMSE MAE	RMSE (std. dev.) MAE (std. dev.)	RMSE (std. dev.) MAE (std. dev.)
consumption	4.0460 2.7761	4.1935 3.0393	3.6708 2.6287	3.4773 2.3817	3.4177 (0.1048) 2.5099 (0.0734)	**3.2332 (0.0232)** **2.2925 (0.0376)**
investments	8.3121 5.4483	11.113 8.8338	6.9493 4.5236	6.4072 4.2031	6.2005 (0.3294) 4.2836 (0.2893)	**5.6495 (0.0631)** **3.8215 (0.1032)**
imports	9.5076 7.1720	8.7729 6.8316	7.6435 5.4115	7.8722 5.5894	6.9976 (0.1286) 4.3225 (0.1295)	**6.5982 (0.0757)** **4.0813 (0.0615)**
exports	8.6620 6.4796	8.7729 6.8316	7.6435 5.4115	7.3792 5.3136	7.3524 (0.1209) 5.2083 (0.2321)	**6.9591 (0.1499)** **4.7479 (0.1894)**
inflation	1.9595 1.5155	1.7622 1.1454	1.6580 1.2344	**1.5018** **0.9855**	**1.5093 (0.0164)** 1.0497 (0.0339)	**1.5165 (0.0353)** 1.0930 (0.0473)
wages	173.51 135.74	164.74 123.20	142.60 94.343	147.79 99.890	154.29 (7.3877) 108.36 (11.398)	**141.65 (0.4800)** **91.314 (1.1180)**
per capita GDP	131.39 103.27	127.81 99.720	108.66 80.566	109.46 84.470	104.87 (2.6097) 79.508 (2.7268)	**97.836 (0.5310)** **72.670 (0.4794)**
firms' profits	29.515 20.162	29.606 20.127	29.169 20.061	28.808 19.758	28.881 (0.1190) 19.517 (0.1231)	**28.029 (0.0803)** **18.927 (0.1183)**
corporate taxation	4.1696 3.0078	3.9330 2.8608	3.0939 2.3600	3.5095 2.7044	3.2635 (0.0528) 2.5187 (0.0696)	**2.9983 (0.0300)** **2.2922 (0.0125)**
private savings	39.156 29.315	40.942 30.340	37.644 27.215	37.657 27.197	36.509 (0.3576) 26.957 (0.1213)	**35.862 (0.2169)** **26.416 (0.0249)**
expen. durable goods	7.5644 5.0644	6.9866 5.3048	6.7296 4.3630	6.6235 4.4179	6.7366 (0.1282) 4.7585 (0.2162)	**6.3273 (0.1463)** **4.2023 (0.0660)**
GDP	3.2227 2.4332	3.3951 2.3968	2.5851 1.8965	2.3209 1.6542	2.2353 (0.1251) 1.6550 (0.2241)	**2.0479 (0.0587)** **1.3924 (0.0620)**
household income	3.5447 2.5716	3.0540 2.0809	3.1000 2.1340	3.4116 2.1781	3.0019 (0.0269) 2.0218 (0.0438)	**2.8639 (0.1174)** **1.9530 (0.0519)**
government expen.	3.5540 2.7296	4.8023 3.4904	2.0151 1.5048	2.3825 1.8464	2.3299 (0.0241) 1.8869 (0.0480)	**1.5517 (0.0455)** **1.2469 (0.0295)**
house prices	4.1156 2.7895	5.2339 4.0954	3.4730 2.5971	3.4337 2.4875	3.6274 (0.0694) 2.7029 (0.3359)	**3.2234 (0.0437)** **2.3406 (0.0359)**
employment rate	2.0070 1.5946	4.2618 3.9164	0.4970 0.3281	0.5008 0.3830	0.5733 (0.1733) 0.4789 (0.1207)	**0.3754 (0.0168)** **0.2988 (0.0182)**
unemployment rate	2.0894 1.6533	2.3794 1.7583	0.5858 0.4443	0.5554 0.4153	0.6595 (0.0508) 0.4667 (0.0342)	**0.4636 (0.0306)** **0.3147 (0.0314)**
unemp. over 27 w.	17.772 15.120	17.696 17.218	2.9366 2.2811	9.5743 6.9341	3.7248 (0.6547) 3.1515 (0.5486)	**2.8321 (0.0330)** **2.2403 (0.0075)**

Table 4. Single-step forecast errors for the multivariate input models. Text in bold denotes the best performance (95% confidence level).

Variable	ARIMAX	VAR	Elastic Net	XGBoost	BiLSTM	Our CNN-BiLSTM
	RMSE MAE	RMSE MAE	RMSE MAE	RMSE MAE	RMSE (std. dev.) MAE (std. dev.)	RMSE (std. dev.) MAE (std. dev.)
consumption	1.7566 1.2117	2.9969 2.4505	**1.1707** **0.8992**	1.6084 1.2032	1.9307 (0.1019) 1.4358 (0.1144)	2.1592 (0.1323) 1.6629 (0.1393)
investments	5.3641 3.6766	10.213 9.2144	2.7433 2.2516	2.8168 1.8298	2.5233 (0.1986) **1.8002 (0.1657)**	**2.2397 (0.2277)** **1.6682 (0.1700)**
imports	5.3600 4.3036	6.0002 4.7285	2.8470 2.1585	3.4139 2.0686	3.2036 (0.2811) 2.4632 (0.1612)	**2.3958 (0.0599)** **1.7573 (0.0653)**
exports	7.5044 5.1107	7.2302 5.4215	3.9126 2.8886	3.9077 **2.7276**	4.4484 (0.3598) 3.0564 (0.1857)	**3.6799 (0.1089)** **2.7379 (0.1062)**

Table 4. *Cont.*

Variable	ARIMAX	VAR	Elastic Net	XGBoost	BiLSTM	Our CNN-BiLSTM
inflation	2.4004 1.9318	5.8533 4.6454	**0.6923** **0.4677**	0.7326 0.4734	1.3577 (0.1675) 1.0098 (0.1148)	1.1152 (0.2807) 0.8316 (0.2151)
wages	144.23 107.37	138.73 92.062	117.72 68.958	124.43 75.590	118.26 (2.1950) 68.691 (2.4276)	**114.22 (2.0562)** **63.500 (1.9796)**
per capita GDP	87.899 62.521	78.678 59.400	64.985 45.917	83.200 54.800	71.752 (2.4801) 49.911 (0.9571)	**61.601 (1.3644)** **42.292 (2.1096)**
firms' profits	23.038 16.984	24.186 17.132	22.729 12.229	28.622 18.731	21.102 (0.6201) 13.122 (0.3800)	**18.358 (0.3645)** **11.501 (0.6107)**
corporate taxation	3.5300 2.6559	3.2948 2.6556	2.3399 1.5964	2.5744 1.5975	2.4300 (0.1239) 1.6751 (0.0506)	**2.2880 (0.0093)** **1.5749 (0.0094)**
private savings	29.915 21.136	32.496 22.814	28.248 16.479	37.090 23.940	33.859 (1.1122) 22.989 (1.0929)	**25.749 (0.4926)** **15.895 (0.2147)**
expen. durable goods	7.6315 4.9930	15.597 12.497	5.2789 3.4234	5.0983 3.2737	7.0292 (0.3995) 5.0255 (0.7138)	**4.2094 (0.1529)** **3.1987 (0.1189)**
GDP	1.3641 0.9897	2.2758 1.9223	0.9309 0.7082	0.9349 0.7232	1.1034 (0.0647) 0.7880 (0.0315)	**0.8727 (0.0372)** **0.6515 (0.0343)**
household income	3.1828 2.2418	7.3236 5.7770	2.1474 1.4936	**1.9454** **1.0294**	2.2263 (0.1348) 1.5581 (0.1717)	2.2520 (0.1713) 1.6404 (0.1257)
government expen.	3.8283 3.1686	2.8563 2.2076	1.2604 1.0073	**1.0056** **0.8085**	1.1806 (0.1617) **0.8747 (0.1080)**	1.0664 (0.0420) **0.8037 (0.0320)**
house prices	4.0064 2.7954	3.2791 2.7633	**0.7930** **0.6006**	1.3561 0.9131	1.3317 (0.1338) 1.0029 (0.1015)	1.2537 (0.0430) 0.9268 (0.0668)
employment rate	2.0802 1.6528	0.6177 0.4981	**0.1319** **0.1027**	0.2055 0.1618	0.2540 (0.0274) 0.1981 (0.0160)	0.2182 (0.0099) 0.1700 (0.0067)
unemployment rate	1.8229 1.3537	2.2742 1.8059	**0.1414** **0.1287**	0.3147 0.2085	0.2357 (0.0259) 0.1915 (0.0240)	0.2013 (0.0022) 0.1566 (0.0021)
unemp. over 27 w.	19.020 16.049	12.429 11.726	**1.1769** **0.9036**	10.796 7.5981	1.9132 (0.2058) 1.4810 (0.2187)	1.7467 (0.0925) 1.3394 (0.0701)

For the sake of equal comparisons, we compared the MAE/RMSE of the ARIMA models and univariate Elastic Net/XGBoost/BiLSTM with the MAE/RMSE obtained from a univariate input CNN-BiLSTM (that used only lagged values of the dependent variable), and we compared the MAE/RMSE of the ARIMAX, VAR, and multivariate Elastic Net/XGBoost/BiLSTM models with the ones obtained from a multivariate input CNN-BiLSTM (that used, in addition to the lagged values of the dependent variable, also lagged values from additional covariates). As mentioned in Section 3, we selected as additional covariates to be added to the multivariate input models only those variables that Granger-caused the dependent variable. We reported the selected variables in the Appendix A (Table A1).

In the above tables, for each variable, the text in bold denotes the best performance for the considered metric. From Tables 1 and 2, we can see that for the univariate time series analysis, the forecast errors of our architecture were in general much lower than the competitor models, in particular for the multi-step forecast. Indeed, when trying to forecast the next 12 values, the CNN-BiLSTM architecture significatively performed better for 17 variables out of 18. For the one-step forecast, there were only six variables where the results of the competitor models were significantly better, and even in those cases the results of our architecture were often close.

We obtained the same excellent results for multi-step forecasting when we operated in a multivariate time series context (see Table 3). The multivariate one-step forecast results (Table 4) were also satisfactory: our model performed better than the multivariate competitor models for 10 variables out of 18.

It should be noted that, for all the models, the multivariate results were in general significantly worse than the univariate case. In particular, for what concerns our proposed model, we guess that one reason behind this worsening is related to the higher amount of information (coming from multiple time series) that was processed by the network: to determine the best configuration of the parameters, the network tries to minimize a non-convex cost function that lies in a very high dimensional space, and non-relevant information could make the optimization process to become more easily stuck in poor local minima. We think that a longer training combined with injecting more noise (for instance, by reducing the mini-batch size) could solve the problem, and we plan to study this issue in greater detail in future studies.

It should also be mentioned that we cannot evaluate the quality of a Deep Learning architecture from a single run. Indeed, as a consequence of the random weight initialization and the optimization process, there is an inherent stochastic component in the obtained results. The results reported in Tables 1–4 are the average mean of 10 different evaluations of the Deep Learning models, where in brackets we reported the standard deviation. For each variable, to confirm statistically significant differences in the results, we ran independent t-tests with $p < 0.05$.

We can also visualize the predictions of the competitor models with those of our architecture. As an example, let us consider the variable "Investments". We compared the one-step forecasts (Figure 3) and the multi-step forecasts (Figure 4) obtained with ARIMA, Elastic Net, and the CNN-BiLSTM architecture.

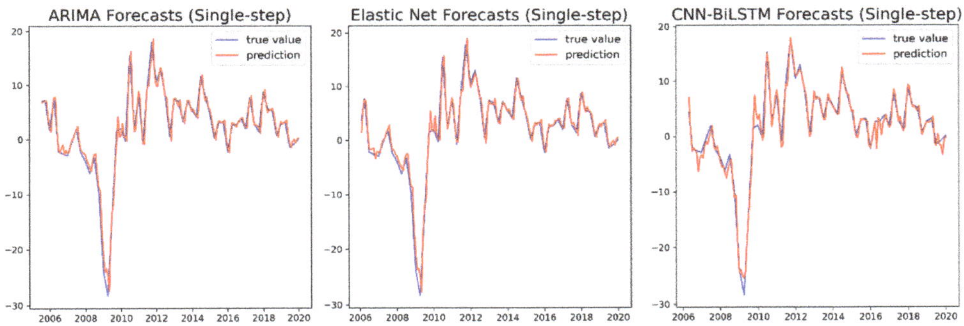

Figure 3. A comparison between the one-step forecast results for ARIMA (**left**), Elastic Net (**middle**), and our CNN-BiLSTM architecture (**right**) for the variable "Investments".

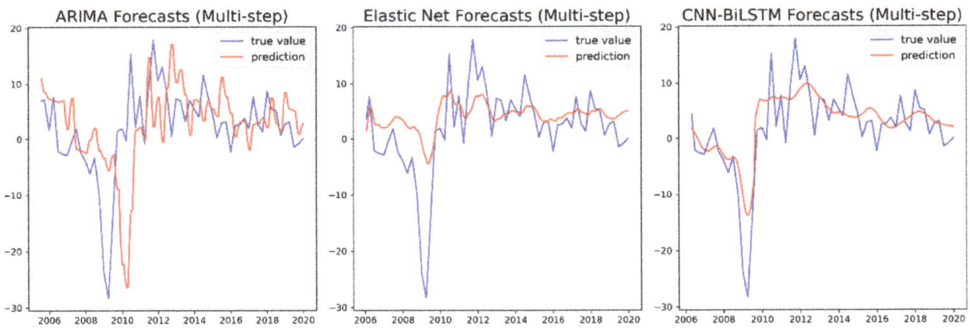

Figure 4. A comparison between the multi-step forecast results for ARIMA (**left**), Elastic Net (**middle**), and our CNN-BiLSTM architecture (**right**) for the variable "Investments".

5. Conclusions

This paper addresses the problem of obtaining increasingly accurate forecasts for different macroeconomic indicators in order to offer a valid support for policymakers to better plan economic policies. With the recent advances in Machine Learning (and in particular Deep Learning), new investigation techniques are emerging, and they are gaining popularity among researchers from different disciplines. However, in the context of macroeconomic forecasting, these techniques are not yet widely used (except for some financial applications), and other more "traditional" econometric models are often preferred.

The question we asked ourselves was to understand how much a new model implemented with these techniques was able to compete with the traditional tools; to this purpose, we selected two error metrics (MAE and RMSE) and we compared the forecast errors of the considered models. In constructing our architecture, we implemented a "hybrid" model, combining a CNN to extract the salient features of the time series and a Bidirectional LSTM backend to learn the timing relationships and perform the forecasts. While econometric models are certainly easier to fit and are more interpretable, for what purely concerns the forecasting component the results obtained with the proposed Deep Learning architecture are very promising.

The main problem when applying these techniques to time series (in particular macroeconomic time series) is that to obtain good forecast results, one in general needs a huge number of observations. Deep Learning is not suitable in contexts where the time series are too short, or where the measurements are not frequent enough. However, we can conclude that, with the data availability that increases more and more over time, Deep Learning techniques should be taken into serious consideration to make predictions about the evolution of the economy.

Our model can be further improved: first, we selected the model hyperparameters with a "trial and error" approach, and an accurate grid search is likely to lead to better results. However, the focus of this work was on producing better (or on par) results than "traditional" methods, and not on finding the best possible architecture configuration. Effective Deep Learning applications to time series analysis are being developed only recently and there is certainly room for further improvement. We believe that new more performing algorithms and architectures will be available shortly.

Funding: This research received no external funding.

Institutional Review Board Statement: Not applicable.

Informed Consent Statement: Not applicable.

Data Availability Statement: The datasets used for the empirical analyses have been provided by the U.S. Bureau of Economic Analysis (BEA) and the U.S. Bureau of Labor Statistics (BLS) and are publicly available. The datasets used in the empirical experiments, as well as the code of our model, are available at: https://github.com/staale92/cnn-bilstm-macroeconomic-forecasting.

Conflicts of Interest: The author declares no conflict of interest.

Appendix A

Appendix A.1

Table A1. Additional covariates added in the multivariate input models for each dependent variable.

consumption	inflation, per capita GDP, corporate taxation, private savings, expen. durable goods, household income, unemployment rate, unempl. over 27 weeks
investments	consumption, imports, exports, per capita GDP, firms' profits, expen. durable goods, GDP, house prices, employment rate, unemployment rate, unempl. over 27 weeks

Table A1. *Cont.*

imports	consumption, exports, inflation, GDP, expen. durable goods, household income, government expenditure, house prices
exports	consumption, investments, inflation, GDP, expen. durable goods, government expenditure, unemployment rate
inflation	corporate taxation, private savings, GDP, government expenditure, unempl. over 27 weeks
wages	per capita GDP, firms' profits, corporate taxation, household income
per capita GDP	investments, wages, firms' profits, corporate taxation, private savings, GDP, household income, employment rate
firms' profits	wages, per capita GDP, Private savings, GDP
corporate taxation	wages, firms' profits, private savings, unemployment rate
private savings	wages, per capita GDP, firms' profits, household income
expen. durable good	inflation, wages, firms' profits, private savings, employment rate, unemployment rate, unempl. over 27 weeks
GDP	consumption, investments, per capita GDP, firms' profits, expen. durable goods, household income, government expenditure, house prices, unemployment rate
household income	consumption, imports, inflation, wages, firms' profits, private savings, expen. durable goods, house prices, employment rate, unemployment rate, unempl. over 27 weeks
government expenditure	investments, exports, inflation, expen. durable goods, household income, unemployment rate
house prices	imports, exports, inflation, corporate taxation, expen. durable goods, GDP, government expenditure, employment rate, unemployment rate
employment rate	consumption, inflation, per capita GDP, GDP, household income, government expenditure
unemployment rate	consumption, inflation, expen. durable goods, GDP, employment rate
unempl. over 27 weeks	private savings, imports, exports, GDP, house prices, unemployment rate

Appendix A.2. Specifications of the ARIMA, ARIMAX, and VAR Models

For each variable we report, in order, (p,d,q) specification of multi-step ARIMA; (p,d,q) specification of single-step ARIMA; (p,d,q) specification of ARIMAX; (l) lags for VAR.

Consumption (3,0,4), (4,0,1), (4,0,4), (8); investments (2,0,2), (4,0,1), (4,0,4), (5); imports (2,0,4), (4,0,2), (0,0,4), (8); exports (6,0,4), (4,0,1), (0,0,4), (8); inflation (8,0,2), (4,0,1), (4,0,2), (8); wages (3,1,2), (2,1,1), (4,1,2), (7); per capita GDP (8,1,4), (2,1,1), (5,1,4), (7); firms' profits (2,1,2), (3,1,1), (4,1,4), (7); corporate taxation (8,1,1), (3,1,1), (7,1,4), (7); private savings (3,1,4), (3,1,1), (4,1,4), (7); expen. durable goods (3,0,4), (4,0,1), (0,0,4), (7); GDP (8,0,2), (4,0,1), (8,0,4), (5); household income (3,0,4), (3,0,2), (0,0,4), (5); government expenditure (4,0,4), (4,0,1), (4,0,4), (8); house prices (6,0,4), (4,0,1), (8,0,4), (8); employment rate (8,0,3), (4,0,1), (4,0,2), (8); unemployment rate (8,0,4), (4,0,1), (7,0,3), (8); unemp. over 27 weeks (8,0,4), (4,0,2), (4,0,4), (7).

References

1. Fan, J.; Ma, C.; Zhong, Y. A selective overview of deep learning. *Stat Sci.* **2021**, *36*, 264–290. [CrossRef] [PubMed]
2. Mathew, A.; Amudha, P.; Sivakumari, S. Deep Learning Techniques: An Overview. In *Advanced Machine Learning Technologies and Applications, Proceedings of the International Conference on Advances in Intelligent Systems and Computing (AMLTA 2020), Jaipur, India, 13–15 February 2020*; Hassanien, A., Bhatnagar, R., Darwish, A., Eds.; Springer: Singapore, 2020; Volume 1141, p. 1141. [CrossRef]
3. Chakraborty, C.; Joseph, A. *Machine Learning at Central Banks*; Working Paper 674; Bank of England: London, UK, 2017.
4. Espeholt, L.; Agrawal, S.; Sonderby, C.; Kumar, M.; Heek, J.; Bromberg, C.; Gazen, C.; Carver, R.; Andrychowicz, M.; Hickey, J.; et al. Skillful twelve hour precipitation forecasts using large context neural networks. *arXiv* **2021**, arXiv:2111.07470.
5. Chen, G.; Liu, S.; Jiang, F. Daily Weather Forecasting Based on Deep Learning Model: A Case Study of Shenzhen City, China. *Atmosphere* **2022**, *13*, 1208. [CrossRef]

6. Corsaro, S.; De Simone, V.; Marino, Z.; Scognamiglio, S. l1-Regularization in portfolio selection with machine learning. *Mathematics* **2022**, *10*, 540. [CrossRef]
7. Asawa, Y.S. Modern Machine Learning Solutions for Portfolio Selection. *IEEE Eng. Manag. Rev.* **2022**, *50*, 94–112. [CrossRef]
8. Kumbure, M.M.; Lohrmann, C.; Luukka, P.; Porras, J. Machine learning techniques and data for stock market forecasting: A literature review. *Expert Syst. Appl* **2022**, *197*, 116659. [CrossRef]
9. Khalid, A.; Huthaifa, K.; Hamzah, A.A.; Anas, R.A.; Laith, A. A New Stock Price Forecasting Method Using Active Deep Learning Approach. *J. Open Innov. Technol. Mark. Complex.* **2022**, *8*, 96, ISSN 2199-8531. [CrossRef]
10. Staffini, A. Stock Price Forecasting by a Deep Convolutional Generative Adversarial Network. *Front. Artif. Intell.* **2022**, *5*, 837596. [CrossRef]
11. Namini, S.S.; Tavakoli, N.; Namin, A.S. A Comparison of ARIMA and LSTM in Forecasting Time Series. In Proceedings of the 17th IEEE International Conference on Machine Learning and Applications (ICMLA), Orlando, FL, USA, 17–20 December 2018; pp. 1394–1401. [CrossRef]
12. Zulfany, E.R.; Reina, S.; Andy, E. A Comparison: Prediction of Death and Infected COVID-19 Cases in Indonesia Using Time Series Smoothing and LSTM Neural Network. *Procedia Comput. Sci.* **2021**, *179*, 982–988, ISSN 1877-0509. [CrossRef]
13. Jin, X.; Yu, X.; Wang, X.; Bai, Y.; Su, T.; Kong, J. Prediction for Time Series with CNN and LSTM. In *Proceedings of the 11th International Conference on Modelling, Identification and Control (ICMIC2019)*; Lecture Notes in Electrical Engineering; Wang, R., Chen, Z., Zhang, W., Zhu, Q., Eds.; Springer: Singapore, 2020; Volume 582. [CrossRef]
14. Jama, M. Time Series Modeling and Forecasting of Somaliland Consumer Price Index: A Comparison of ARIMA and Regression with ARIMA Errors. *Am. J. Theor. Appl. Stat.* **2020**, *9*, 143–153. [CrossRef]
15. Hall, A.S. Machine Learning Approaches to Macroeconomic Forecasting. *Econ. Rev. Fed. Reserve Bank Kans. City* **2018**, *103*, 63–81. [CrossRef]
16. Khan, M.A.; Abbas, K.; Su'ud, M.M.; Salameh, A.A.; Alam, M.M.; Aman, N.; Mehreen, M.; Jan, A.; Hashim, N.A.A.B.N.; Aziz, R.C. Application of Machine Learning Algorithms for Sustainable Business Management Based on Macro-Economic Data: Supervised Learning Techniques Approach. *Sustainability* **2022**, *14*, 9964. [CrossRef]
17. Coulombe, P.G.; Leroux, M.; Stevanovic, D.; Surprenant, S. How is machine learning useful for macroeconomic forecasting? *J. Appl. Econom.* **2022**, *37*, 920–964. [CrossRef]
18. Nosratabadi, S.; Mosavi, A.; Duan, P.; Ghamisi, P.; Filip, F.; Band, S.S.; Reuter, U.; Gama, J.; Gandomi, A.H. Data Science in Economics: Comprehensive Review of Advanced Machine Learning and Deep Learning Methods. *Mathematics* **2020**, *8*, 1799. [CrossRef]
19. Yoon, J. Forecasting of Real GDP Growth Using Machine Learning Models: Gradient Boosting and Random Forest Approach. *Comput. Econ.* **2021**, *57*, 247–265. [CrossRef]
20. Vafin, A. Forecasting macroeconomic indicators for seven major economies using the ARIMA model. *Sage Sci. Econ. Rev.* **2020**, *3*, 1–16. [CrossRef]
21. Shijun, C.; Xiaoli, H.; Yunbin, S.; Chong, Y. Application of Improved LSTM Algorithm in Macroeconomic Forecasting. *Comput. Intell. Neurosci.* **2021**, *2021*, 4471044. [CrossRef]
22. Makridakis, S.; Spiliotis, E.; Assimakopoulos, V. Statistical and Machine Learning forecasting methods: Concerns and ways forward. *PLoS ONE* **2018**, *13*, e0194889. [CrossRef]
23. Staffini, A.; Svensson, T.; Chung, U.-I.; Svensson, A.K. Heart Rate Modeling and Prediction Using Autoregressive Models and Deep Learning. *Sensors* **2021**, *22*, 34. [CrossRef]
24. Springenberg, J.; Dosovitskiy, A.; Brox, T.; Riedmiller, M. Striving for Simplicity: The All Convolutional Net. *arXiv* **2015**, arXiv:1412.6806.
25. Graves, A.; Schmidhuber, J. Framewise phoneme classification with bidirectional LSTM networks. In Proceedings of the IEEE International Joint Conference on Neural Networks 2005, Montreal, QC, Canada, 31 July–4 August 2005; Volume 4, pp. 2047–2052. [CrossRef]
26. Srivastava, N.; Hinton, G.; Krizhevsky, A.; Sutskever, I. Dropout: A Simple Way to Prevent Neural Networks from Overfitting. *J. Mach. Learn. Res.* **2014**, *15*, 1929–1958.
27. Yuanyuan, C.; Zhang, Y. Adaptive sparse dropout: Learning the certainty and uncertainty in deep neural networks. *Neurocomputing* **2021**, *450*, 354–361, ISSN 0925-2312. [CrossRef]
28. Bikash, S.; Angshuman, P.; Dipti, P.M. Deterministic dropout for deep neural networks using composite random forest. *Pattern Recognit. Lett.* **2020**, *131*, 205–212, ISSN 0167-8655. [CrossRef]
29. Bengio, Y.; Simard, P.; Frasconi, P. Learning long-term dependencies with gradient descent is difficult. *IEEE Trans. Neural Netw.* **1994**, *5*, 157–166. [CrossRef] [PubMed]
30. Glorot, X.; Bengio, Y. Understanding the difficulty of training deep feedforward neural networks. *J. Mach. Learn. Res.* **2010**, *9*, 249–256.
31. Le, Q.; Jaitly, N.; Hinton, G. A Simple Way to Initialize Recurrent Networks of Rectified Linear Units. *arXiv* **2015**, arXiv:1504.00941.
32. Calin, O. *Deep Learning Architectures: A Mathematical Approach*; Springer Series in the Data Sciences; Springer International Publishing: New York, NY, USA, 2020; ISBN 978-3-030-36721-3.
33. Kingma, D.; Ba, J. Adam: A Method for Stochastic Optimization. *arXiv* **2014**, arXiv:1412.6980v9.
34. Ruder, S. An Overview of Multi-Task Learning in Deep Neural Networks. *arXiv* **2017**, arXiv:1706.05098.

35. Ge, R.; Huang, F.; Jin, C.; Yuan, Y. Escaping from Saddle Points---Online Stochastic Gradient for Tensor Decomposition. *arXiv* **2015**, arXiv:1503.02101.
36. Bengio, Y. Practical recommendations for gradient-based training of deep architectures. In *Tricks of the Trade*, 2nd ed.; Volume 7700 of Theoretical Computer Science and General Issues; Springer: Berlin/Heidelberg, Germany, 2012; pp. 437–478.
37. Masters, D.; Luschi, C. Revisiting Small Batch Training for Deep Neural Networks. *arXiv* **2018**, arXiv:1804.07612.
38. Gholamy, A.; Kreinovich, V.; Kosheleva, O. *Why 70/30 or 80/20 Relation Between Training and Testing Sets: A Pedagogical Explanation*; University of Texas at El Paso Departmental Technical Reports (CS): El Paso, TX, USA, 2018.
39. Willmott, C.J.; Matsuura, K. Advantages of the Mean Absolute Error (MAE) over the Root Mean Square Error (RMSE) in Assessing Average Model Performance. *Clim. Res.* **2005**, *30*, 79. [CrossRef]

engineering proceedings

MDPI

Proceeding Paper

Forecasting Tangency Portfolios and Investing in the Minimum Euclidean Distance Portfolio to Maximize Out-of-Sample Sharpe Ratios †

Nolan Alexander * and William Scherer

Department of Systems Engineering, University of Virginia, Charlottesville, VA 22903, USA; wts@virginia.edu
* Correspondence: nolan_alex2018@yahoo.com

† Presented at the 9th International Conference on Time Series and Forecasting, Gran Canaria, Spain, 12–14 July 2023.

Abstract: We propose a novel model to achieve superior out-of-sample Sharpe ratios. While most research in asset allocation focuses on estimating the return vector and covariance matrix, the first component of our novel model instead forecasts the future tangency portfolio, and the second component then determines the optimal investment portfolio. First, to forecast the tangency portfolio, we forecast the efficient frontier by decomposing its functional form, a square root second-order polynomial, into three interpretable coefficients, which can then be used to calculate a forecasted tangency portfolio. These coefficients can be forecasted using vector autoregressions. Second, the model invests in the portfolio on the efficient frontier that is the minimum Euclidean distance from this forecasted tangency portfolio. A motivation for our approach is to address the limitation that the tangency portfolio only maximizes the Sharpe ratio when future returns and covariances are stationary, and can be directly estimated with historical data, which often does not hold in out-of-sample data. Our approach addresses this shortcoming in a novel way by forecasting the tangency portfolio, rather than estimating return and covariance. For empirical testing, we employ two sets of assets that span the market to demonstrate and validate the performance of this novel method.

Keywords: modern portfolio theory; mean–variance optimization; tangency portfolio; forecasting; efficient frontier; asset allocation

Citation: Alexander, N.; Scherer, W. Forecasting Tangency Portfolios and Investing in the Minimum Euclidean Distance Portfolio to Maximize Out-of-Sample Sharpe Ratios. *Eng. Proc.* **2023**, *39*, 34. https://doi.org/10.3390/engproc2023039034

Academic Editors: Ignacio Rojas, Hector Pomares, Luis Javier Herrera, Fernando Rojas and Olga Valenzuela

Published: 30 June 2023

1. Introduction

Mean–variance optimization, developed by Markowitz [1], is the foundation to Modern Portfolio Theory (MPT). The model uses quadratic programming to select portfolios that minimize risk, as measured by volatility, while generating a fixed expected return. The model constructs a pareto frontier that consists of these optimal portfolios, known as an efficient frontier. This efficient frontier allows investors to visualize tradeoffs between portfolio risk and return. A natural question that arises is which portfolio on the efficient frontier is best to invest in. In 1964, William Sharpe developed a portfolio metric known as the Sharpe ratio, defined as the ratio of the portfolio return exceeding the risk-free rate and the portfolio's volatility [2]. The portfolio that maximizes the Sharpe ratio when there is a risk-free rate is the one that corresponds with the market. This portfolio intersects with the capital market line (CML) under the Capital Asset Pricing Model (CAPM), which has an intercept at the risk-free rate and a slope defining the Sharpe ratio. This Sharpe-maximizing portfolio is known as the tangency portfolio because the CML is tangent to the efficient frontier as the feasible region is convex.

The tangency portfolio model makes several assumptions that researchers have attempted to relax, such as transaction costs [3]. This paper, however, will only focus on relaxing one: that the future expected asset returns and covariances will be identical to the asset's historical data. This paper will focus on relaxing this assumption, as there often exist significant estimation errors in practice because of nonstationary market behavior. This limitation is significant, as Dickinson [4] found that the misestimation of these parameters causes the portfolio weights to be unstable, and Kan and Zhou [5] found that estimating the population parameters with samples leads to poor performance out-of-sample. There exist multiple models that attempt to relax this assumption already. Some of the most significant are the Black–Litterman model, Covariance Matrix Shrinkage, and Dynamic Conditional Correlation Multivariate GARCH (DCC MV-GARCH).

In 1992, Black and Litterman [6] proposed a closed-form model to provide better posterior estimates of the return vector and covariance matrix using a Bayesian approach. The Black–Litterman model uses two priors: investors' beliefs in the assets and the market capitalized equilibrium view, which weights assets by their market capitalization. The model requires investors to provide several inputs: a matrix of their views on the assets, a vector of the returns in each view, the level-of-unconfidence in each view, and the hyperparameter weight-on-views, which determines how much weight should be put on each prior.

Engle and Sheppard [7] proposed a method to forecast covariance matrices using Generalized Autoregressive Conditional Heteroskedasticity (GARCH), which is a time-series method to estimate volatility (Engle and Sheppard, 2001). DCC MV-GARCH transforms residuals from GARCH models applied to each asset to obtain conditional correlation estimators, which can be used to estimate the covariance matrix.

Ledoit and Wolf [8] proposed a model to obtain better estimates of the covariance matrix through shrinkage (Ledoit and Wolf, 2003). Shrinkage reduces estimation error by reducing the sample covariance matrix towards a highly structured estimator, which helps ensure that estimates of the covariance matrix are stable when the number of columns is greater than the number of rows.

While these models improve estimation of the MPT parameters, and therefore provide more accurate tangency portfolios, they still have limitations. Mankert [9] explains that, in practice, the Black–Litterman model can be difficult to implement because it requires that the investor input a large number of estimates including matrices and vectors. Covariance matrix shrinkage and DCC MV-GARCH may provide better estimates of covariance matrices, but are unable to provide estimates of the return vector, which is the other parameter required for mean–variance estimation. Finally, the complexity of the Black–Litterman model and DCC MV-GARCH relative to the Markowitz model makes them difficult for investors to interpret and trust.

We provide a novel approach to relax this assumption. While multiple researchers have attempted to relax this assumption to provide more accurate tangency portfolios by providing better parameter estimates, we propose an alternative approach: forecasting the tangency portfolio through the dimensionality reduction of the efficient frontier's square root second-order polynomial coefficients and selecting the portfolio that is the minimum Euclidean distance from this forecasted tangency portfolio.

This novel approach is related to two previous works. The standard approach to creating the efficient frontier is to use mean–variance optimization to find several efficient portfolios and extrapolate between these, but Merton derived a closed-form solution of the efficient frontier as a square root second-order polynomial function with three coefficients [10]. Alexander et al. [11] previously developed a model to forecast these coefficients to provide better parameter estimates. However, this previous method was more prone to overfit than our proposed method, since the coefficients were calculated using yearly data rather than rolling windows, and used a less-robust weights extraction method that did not use the forecasted tangency portfolio. In addition, the previous paper forecasted the same coefficients that Merton derived, rather than a set of interpretable coefficients. The set of assets used in the previous empirical analysis was also not as diverse as the two sets used in this paper.

This paper provides three main contributions: (1) This paper develops three novel, interpretable coefficients: r_{MVP}, σ_{MVP}, and u. (2) This paper provides a method to forecast the coefficients with time-series regression. (3) With these forecasted coefficients, we propose a method of selecting the portfolio that is the minimum Euclidean distance from the forecasted tangency portfolio, because it is the most robust to market movement. We provide empirical results demonstrating that this novel approach outperforms four benchmark portfolios across metrics like Sharpe ratio and max drawdown relative to annual return. We also provide a visual showing that this novel approach consistently outperforms the benchmarks over time.

1.1. Assets Data

To test our proposed models, we used two asset universes: one set corresponding to market capitalization and growth/value, and the other set corresponding to the market sectors. The first set included six mutual funds representing the French–Fama three-factor model's [12] classification of portfolios: growth/value stocks and large/small market capitalization (we refer to this set as GVMC).

The mutual funds in the first set were:

- Fidelity OTC Portfolio—large-cap growth (FDGRX)
- Fidelity Growth and Income Portfolio—large-cap value (FGRIX)
- Fidelity Growth Company Fund—mid-cap growth (FLPSX)
- Fidelity Low-Priced Stock Fund—mid-cap value (FOCPX)
- Invesco Oppenheimer Discovery Fund—small-cap growth (HRTVX)
- Heartland Value Fund Investor class—small cap value (OPOCX)

The second set included all nine original S&P Sector ETFs:

- Materials (XLB)
- Energy (XLE)
- Finance (XLF)
- Industrial (XLI)
- Technology (XLK)
- Consumer Staples (XLP)
- Utilities (XLU)
- Health Care (XLV)
- Consumer Discretionary (XLY)

Both also included a mutual fund to represent an investment in bonds: FPA New Income Fund (FPNIX). These asset datasets were collected using the Yahoo Finance API. In addition, we used the French–Fama three-factor data from Dr. French's website [13] for the risk-free rate of our tangency portfolios.

1.2. Formulation of Markowitz Efficient Frontiers

The MPT model performs mean–variance optimization to select portfolios and creates an efficient frontier as a visual. The following is the formal notation used in MPT: r_i is the expected ln return of an asset in $\boldsymbol{r}$. w_i is the weight of an asset in $\boldsymbol{w}$, satisfying $\sum_{i=1}^{n} w_i = 1$. r_{target} is the target expected return of the portfolio. $\boldsymbol{V}$ is the covariance matrix. $\boldsymbol{e}$ is the ones vector. The formulation of the mean–variance optimization is:

$$\min_{\boldsymbol{w}} \boldsymbol{w}^T \boldsymbol{V} \boldsymbol{w}$$
$$\text{s.t. } \boldsymbol{w}^T \boldsymbol{r} = r_{target} \text{ and } \boldsymbol{e}^T \boldsymbol{w} = 1$$

By adding a constraint that $w_i \geq 0$, we can forbid shorting; this paper uses models that allow shorting to better simulate a breadth of investment strategies.

We used both the CVXPY optimizer and the Sequential Least-Squares Quadratic Programming algorithm (SLSQP) in the SciPy package of Python to perform mean–variance optimization for the 33 years of assets data.

1.3. The Tangency Portfolio

In MPT, there is often a question of which portfolio to select on the efficient frontier, i.e., what target return to select. Tobin's Separation Theorem [14] states that rational investors would choose a combination of the risk-free asset and an optimal portfolio known as the tangency portfolio. The tangency portfolio is the portfolio that intersects the Capital Market Line (CML) under the Capital Asset Pricing Model (CAPM). The tangency portfolio maximizes a standard portfolio metric, the Sharpe ratio [15], which is defined by $\frac{r_p - r_f}{\sigma_p}$. The standard approach to finding the tangency portfolio [16] poses it as an optimization problem. Further below, we provide a method to find the tangency portfolio from efficient frontier coefficients.

1.4. Efficient Frontiers Coefficients

Merton [10] used Lagrange multipliers to derive a functional representation of the efficient frontier as a square root second-order polynomial with three coefficients [17]:

$$\begin{aligned} A &= \boldsymbol{e}^T \boldsymbol{V}^{-1} \boldsymbol{e} > 0 \\ B &= \boldsymbol{r}^T \boldsymbol{V}^{-1} \boldsymbol{e} \\ C &= \boldsymbol{r}^T \boldsymbol{V}^{-1} \boldsymbol{r} > 0 \end{aligned} \tag{1}$$

The equation for the efficient frontier is

$$\sigma^2(r) = \frac{Ar^2 - 2Br + C}{AC - B^2} \tag{2}$$

2. A Novel Set of Interpretable Efficient Frontier Coefficients

We derive an equation with the same form as Equation (2), but with more interpretable coefficients, as

$$\sigma^2(r) = \left(u^{-1}(r - r_{MVP})\right)^2 + \sigma^2_{MVP} \tag{3}$$

r_{MVP} and σ_{MVP} are the return and standard deviation associated with the minimum variance point (MVP). u is the rate of curvature of the efficient frontier utility function, and represents the usefulness of the set of assets returns for mean–variance optimization. An efficient frontier with a higher u diminishes more slowly, and therefore has better tradeoffs of risk and return at every efficient portfolio except for the minimum variance point. We can rearrange terms to calculate each of these more interpretable coefficients as functions of A, B, C as shown in Equation (4).

$$r_{MVP} = \frac{B}{A}, \; \sigma_{MVP} = \frac{1}{\sqrt{A}}, \; u = \sqrt{\frac{AC - B^2}{A}} \tag{4}$$

These more interpretable coefficients each control one graphical efficient frontier component: each coordinate of the vertex, and the rate of curvature. An increase in r_{MVP} implies the market demanding greater returns for all levels of risk. An increase in σ_{MVP} implies greater risk for all returns. An increase in u signals a better market for risk-seeking investors.

The u coefficient can be reduced to a more interpretable form in Equation (5).

$$u = \sqrt{\boldsymbol{r}^T \boldsymbol{V}^{-1} \boldsymbol{r}} \cdot \sqrt{1 - S_c(\boldsymbol{r}, \boldsymbol{e})^2} \tag{5}$$

We can observe that u is a product of the Mahalanobis distance of the return vector to the zero vector and a function of the cosine similarity between the return vector and the vector of ones.

The Tangency Portfolio Defined by Coefficients

When using a forecasted efficient frontier, we cannot find the tangency portfolio by solving the standard optimization problem to maximize the Sharpe ratio, because it requires knowledge of the future return vector and covariance matrix. Alternatively, with the efficient frontier coefficients, we can solve for the return and standard deviation of the tangent portfolio. We will start with Equation (3). The equation for the line that intersects the efficient frontier and goes through (σ, r) is $\sigma - \sigma_P = \sigma'(r)(r - r_P)$, where σ_P and r_P are portfolio volatility and return, respectively. The CML must be tangent to the efficient frontier, so we take the derivate of $\sigma(r)$:

$$\sigma'(r) = \frac{r - r_{MVP}}{u^2\sqrt{\left(u^{-1}(r - r_{MVP})\right)^2 + \sigma_{MVP}^2}}$$

The tangent portfolio has an intercept at the risk-free rate, so let $r = r_f$ and $\sigma = 0$, then:

$$0 - \sigma_{TP} = \sigma'(r_{TP})\left(r_f - r_{TP}\right)$$

$$-\sqrt{\left(u^{-1}(r_{TP} - r_{MVP})\right)^2 + \sigma_{MVP}^2} = \frac{r_{TP} - r_{MVP}}{u^2\sqrt{\left(u^{-1}(r_{TP} - r_{MVP})\right)^2 + \sigma_{MVP}^2}}\left(r_f - r_{TP}\right)$$

Solving for return of the tangency portfolio, r_{TP}, yields

$$r_{TP} = \frac{r_{MVP}^2 + u^2\sigma_{MVP}^2 - r_{MVP}r_f}{r_{MVP} - r_f} \text{ and } \sigma_{TP} = \sigma(r_{TP}) \quad (6)$$

Therefore, to invest in the tangent portfolio of a forecasted efficient frontier, we extract the weights on the efficient frontier at this return.

3. Efficient Frontier Forecasting

The model uses an online vector autoregression with exogenous variables (VARX) with lag order 1 to forecast the average efficient frontier coefficients for the forward time period that is the same length as the lookback period used in the mean–variance optimization. In 1980, Sims [18] proposed the use of vector autoregressions in macroeconomic forecasting as a theory-free model, which is still a particularly relevant model to this day. To demonstrate our approach, we elected to limit our feature set to only the lag of the coefficients, and the historical equal-weighted return moving average, as a smaller feature set is less likely to overfit. For each model, we only selected up to two predictors to create more parsimonious models that have greater interpretability. This regularization was incorporated into the model by forcing coefficient parameters of the VARX model in vector notation to 0 as the associating features were not significantly predictive in-sample.

We selected a VARX model to forecast the coefficients so that the results would be interpretable while being able to model autocorrelation. We forecasted the 1 month (21 business days) forward coefficients. The forecasts are online, so after each daily rolling forecast, the current day's coefficients are observed and then included. We elected to use online models as static models do not update as new information arrives. Table 1 shows the out-of-sample R^2s for the three proposed online VARX models. The subscript denotes the time the variable is measured, and the superscript denotes the window length used for the calculation of the rolling variable. The predictors all used a 1-year (252 business days) window, demonstrating that long-term efficient frontier coefficients can provide predictive power to forecast the shorter-term future efficient frontier coefficients.

Table 1. Out-of-Sample R^2s of the online VARX Efficient Frontier Coefficients forecasting models.

VARX Model	GVMC OoS R^2 (%)	Sectors OoS R^2 (%)
$r_{MVP}{}_{t+21}^{(21)} = \beta_{1,t} r_{MVP}{}_t^{(252)} + \beta_{0,t}$	13	2
$\sigma_{MVP}{}_{t+21}^{(21)} = \beta_{1,t} \sigma_{MVP}{}_t^{(252)} + \beta_{0,t}$	34	9
$u_{t+21}^{(21)} = \beta_{2,t} \sigma_{MVP}{}_t^{(252)} + \beta_{1,t} \bar{r}_{EW}{}_t^{(252)} + \beta_{0,t}$	1	2

While these R^2s are low, these are measured out-of-sample to ensure the model is not overfitted, and measuring R^2 out-of-sample is significantly harsher than standard R^2.

4. The Minimum Distance Portfolio to the Forecasted Tangency Portfolio

With the forecasted coefficients, we can find the forecasted tangency portfolio using Equation (6). Because a forecasted efficient frontier does not provide its return vector and covariance matrix, we cannot directly solve for the forecasted tangency portfolio weights. In addition, the forecasted tangency portfolio likely does not exist on the current efficient frontier given the nonstationary behavior of the market. Instead, we are able to solve for the portfolio on the current efficient frontier that is the minimum Euclidean distance from the current efficient frontier to the forecasted tangency portfolio, formulated as

$$\min_r \sqrt{(\hat{r}_{TP} - r)^2 + \left(\hat{\sigma}_{TP} - \sqrt{(u^{-1}(r - r_{MVP}))^2 + \sigma_{MVP}^2}\right)^2}$$

To solve this optimization problem, we can solve for the square of this objective function because the distance function is convex and cannot be negative. This optimization problem can be solved by taking the derivative of the square of the objective function, and solving for the root using Newton's Method.

$$0 = 2(\hat{r}_{TP} - r) + \frac{2(r - r_{MVP})\left(\hat{\sigma}_{TP} u - \sqrt{(r - r_{MVP})^2 + \sigma_{MVP}^2 u^2}\right)}{u^2 \sqrt{(r - r_{MVP})^2 + \sigma_{MVP}^2 u^2}} \tag{7}$$

Now, with the return of the minimum distance portfolio to the forecasted tangency portfolio, we can find the weights by performing mean–variance optimization at this return. Figure 1 provides a visual of the minimum distance portfolio.

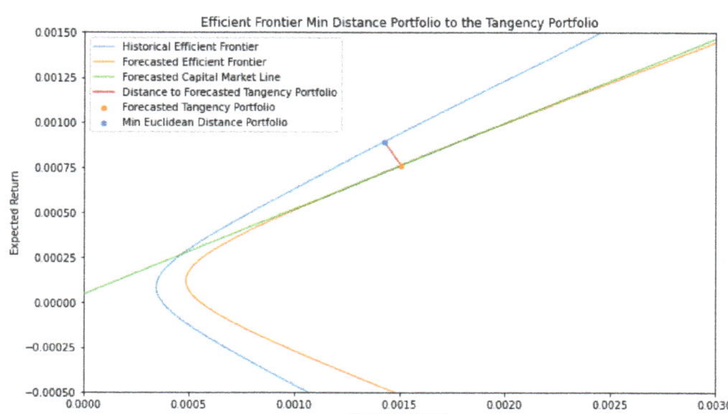

Figure 1. The Minimum Distance Portfolio to the Forecasted Tangency Portfolio on 1 February 2018 with a 1-month lookback.

5. Empirical Analysis

5.1. Benchmarks and Ensuring Realistic Portfolios

To select the best benchmark for our proposed model, we must determine the best version of the tangency portfolio model. The standard tangency portfolio model is sensitive to the update frequency and the lookback period of the historical data. More frequent updates allow the model to use the most relevant data, so we used daily updates. A shorter lookback period uses only the most relevant data, so we used a 1-month window.

To ensure the portfolios do not transfer all of their risk to leverage, we limited the leverage to 1.5× with the following scaling:

$$\widetilde{w}_{<0} = w_{<0}\frac{\frac{l-1}{2}}{\sum(|w_{<0}|)} \text{ and } \widetilde{w}_{>0} = w_{>0}\frac{\frac{l-1}{2}+1}{\sum(|w_{>0}|)}$$

where l is the maximum leverage allowed, and in our case, $l = 1.5$. While this leverage restriction could have been added as a constraint to the optimization, we chose not to do so, so that our empirical results can be generalized to standard tangency portfolios.

Finally, we added 1% daily transaction costs calculated by subtracting the 1% of the sum of absolute changes in weights.

5.2. Empirical Results and Discussion

For the first set of assets, the online VARX was initially trained on 1990–1999 data, and the portfolio returns were measured out-of-sample 2000–2022. The second set was trained on 1999–2007 data, and the portfolio returns were measured out-of-sample 2008–2022. We measured the performance of our proposed model against four benchmarks: The Tangency Portfolio with a 1-month rolling window, the equal-weighed portfolio, the S&P 500 total return, and the 60/40 stock and bonds portfolio (60% S&P 500 and 40% FPNIX). Figures 2 and 3 show that the proposed model (black line) outperforms all the benchmarks in terms of return and volatility. Across all years, the proposed model is able to provide consistent returns, while suffering low drawdown during crashes, like in 2008 and 2020. The proposed model in Figure 3 is 2×-levered to provide a better visual comparison, as the proposed model had higher Sharpe than the benchmarks, but lower return.

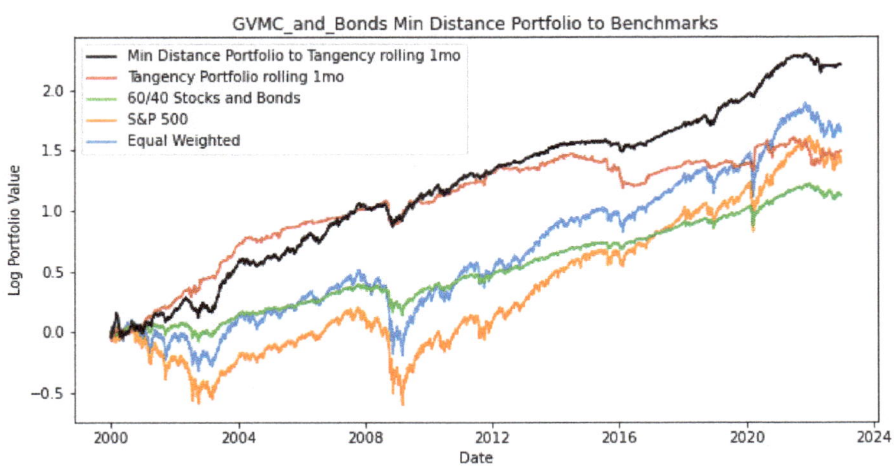

Figure 2. The performance of the proposed portfolio model as compared to four benchmarks with the GVMC data.

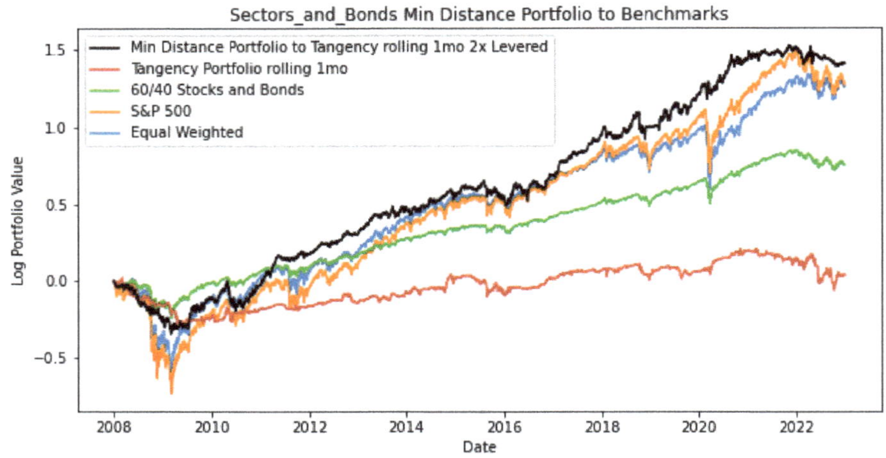

Figure 3. The performance of the proposed portfolio model as compared to four benchmarks with the S&P Sector ETFs data.

The proposed model has a greater Sharpe ratio compared to all of the benchmark portfolios, as shown in Tables 2 and 3, for each set of assets, with high annual return relative to max drawdown.

Table 2. Portfolio metrics for the market cap and growth/value assets from 2000 to 2022.

Portfolio	Sharpe Ratio	Sortino Ratio	Annual Return	Max Drawdown
Minimum Distance Portfolio to Tangency rolling 1 mo	1.00	1.31	10.7%	−18.7%
Tangency Portfolio rolling 1 mo	0.67	0.71	7.0%	−25.0%
Equal Weighted	0.41	0.53	9.2%	−50.6%
S&P 500 Total Return	0.33	0.42	8.3%	−55.3%
60/40 Stocks and Bonds	0.47	0.60	5.3%	−22.6%

Table 3. Portfolio metrics for the S&P sector assets from 2008 to 2022.

Portfolio	Sharpe Ratio	Sortino Ratio	Annual Return	Max Drawdown
Minimum Distance Portfolio to Tangency rolling 1 mo 2× Levered	0.76	0.99	10.7%	−29.3%
Tangency Portfolio rolling 1 mo	0.01	0.01	6.3%	−25.8%
Equal Weighted	0.52	0.63	10.6%	−46.7%
S&P 500 Total Return	0.48	0.58	11.2%	−51.8%
60/40 Stocks and Bonds	0.57	0.69	5.5%	−22.0%

We conducted alpha regression to demonstrate that the proposed model statistically significantly outperforms the benchmarks, as shown in Table 4.

Table 4. Alpha regressions against baseline portfolios for each universe.

Universe	GVMC and Bonds		Sectors and Bonds	
Baseline Portfolio	**Alpha**	***p*-Value**	**Alpha**	***p*-Value**
Tangency Portfolio rolling 1 mo	0.06	$<1 \times 10^{-4}$	0.10	0.0003
Equal Weighted	0.07	$<1 \times 10^{-4}$	0.06	0.01
S&P 500	0.06	$<1 \times 10^{-4}$	0.06	0.02
60/40 Stocks and Bonds	0.06	$<1 \times 10^{-4}$	0.05	0.02

6. Conclusions and Future Research

This paper presents a novel approach to improve out-of-sample Sharpe ratios in comparison to investing in the tangency portfolio. This method is interpretable and does not require input estimates like the Black–Litterman model. This approach decomposes efficient frontiers into three interpretable coefficients: r_{MVP}, σ_{MVP}, and u. The utility coefficient, u, controls the rate of curvature of the efficient frontier. u measures the value of using the set of assets in mean–variance optimization, as we show that it can be decomposed into the Mahalanobis distance of the return vector to the zero vector and a function of the cosine similarity between the return vector and the ones vector. This method forecasts these coefficients using an online VARX, and determines a forecasted tangency portfolio. This approach invests in the portfolio on the current efficient frontier that is the minimum Euclidean distance from the forecasted tangency portfolio. Using this method out-of-sample from 2000–2022 and 2008–2022 for the two universes, this model outperformed four benchmark portfolios: the tangency portfolio, the equal-weighted, the S&P 500 total return, and the 60/40. The proposed model achieved a higher Sharpe ratio compared to these benchmarks.

While this analysis yielded significant results, there exist certain limitations that we will investigate in future research. The results may have been dependent on the selected assets, so we will test the model on other universes that span the market than the two we tested. The results may also be dependent on the up to 1.5× leverage allowed, so we will perform further analysis on the portfolio risk of using leverage. Another limitation is that the forecasting model used a VARX with only one or two features, so for future research we will explore whether technical indicators such as stochastic oscillators or economic data can improve the regression. Additionally, with a larger feature set, we will employ automated approaches such as Least Absolute Shrinkage and Selection Operator (LASSO) and nonlinear models including decision trees.

Author Contributions: Conceptualization, N.A. and W.S.; methodology, N.A. and W.S.; software, N.A.; validation, N.A. and W.S.; formal analysis, N.A. and W.S.; investigation, N.A. and W.S.; resources, N.A. and W.S.; data curation, N.A.; writing—original draft preparation, N.A.; writing—review and editing, W.S.; visualization, N.A.; supervision, W.S.; project administration, W.S.; funding acquisition, W.S. All authors have read and agreed to the published version of the manuscript.

Funding: This research received no external funding.

Institutional Review Board Statement: Not applicable.

Informed Consent Statement: Not applicable.

Data Availability Statement: Publicly available datasets were analyzed in this study. This data can be found on Yahoo Finance here: https://finance.yahoo.com/.

Conflicts of Interest: The authors declare no conflict of interest.

References

1. Markowitz, H. Portfolio selection. *J. Financ.* **1952**, *7*, 77. [CrossRef]
2. Sharpe, W.F. Capital asset prices: A theory of market equilibrium under conditions of risk. *J. Financ.* **1964**, *19*, 425–442. [CrossRef]

3. Davis, M.H.A.; Norman, A.R. Portfolio Selection with Transaction Costs. *Math. Oper. Res.* **1990**, *15*, 676–713. [CrossRef]
4. Dickinson, J.P. The Reliability of Estimation Procedures in Portfolio Analysis. *J. Financ. Quant. Anal.* **1974**, *9*, 447–462. [CrossRef]
5. Kan, R.; Zhou, G. Optimal Portfolio Choice with Parameter Uncertainty. *J. Financ. Quant. Anal.* **2007**, *42*, 621–656. [CrossRef]
6. Black, F.; Litterman, R. Global portfolio optimization. *Financ. Anal. J.* **1992**, *48*, 28–43. [CrossRef]
7. Engle, R.F.; Sheppard, K. *Theoretical and Empirical Properties of Dynamic Conditional Correlation Multivariate GARCH*; Working Paper No. 8554; National Bureau of Economic Research: Cambridge, MA, USA, 2001. [CrossRef]
8. Ledoit, O.; Wolf, M. Improved estimation of the covariance matrix of stock returns with an application to portfolio selection. *J. Empir. Financ.* **2003**, *10*, 603–621. [CrossRef]
9. Mankert, C. The Black-Litterman Model: Towards Its Use in Practice. Ph.D. Thesis, KTH Royal Institute of Technology, Stockholm, Sweden, 2010. Available online: http://urn.kb.se/resolve?urn=urn:nbn:se:kth:diva-26798 (accessed on 1 November 2019).
10. Merton, R.C. An Analytic Derivation of the Efficient Portfolio Frontier. *J. Financ. Quant. Anal.* **1972**, *7*, 1851. [CrossRef]
11. Alexander, N.; Scherer, W.; Burkett, M. Extending the Markowitz model with dimensionality reduction: Forecasting efficient frontiers. In Proceedings of the 2021 Systems and Information Engineering Design Symposium (SIEDS), Charlottesville, VA, USA, 29–30 April 2021; pp. 1–6. [CrossRef]
12. Fama, E.F.; French, K.R. Common Risk Factors in the Returns on Stocks and Bonds. In *The Fama Portfolio: Selected Papers of Eugene F. Fama*; University of Chicago Press: Chicago, IL, USA, 1993; pp. 392–449.
13. Kenneth, r. French—Data Library. Available online: https://mba.tuck.dartmouth.edu/pages/faculty/ken.french/data_library.html (accessed on 28 March 2023).
14. Tobin, J. Liquidity Preference as Behavior Towards Risk. *Rev. Econ. Stud.* **1958**, *25*, 65. [CrossRef]
15. Sharpe, W.F. Mutual fund performance. *J. Bus.* **1966**, *39*, 119–138. Available online: https://www.jstor.org/stable/2351741 (accessed on 1 November 2019). [CrossRef]
16. Luenberger, D.G. *Investment Science*, 1st ed.; Oxford University Press: Oxford, UK, 1997.
17. Petters, A.O.; Dong, X. *An Introduction to Mathematical Finance with Applications: Understanding and Building Financial Intuition*, 1st ed.; Springer: Cham, Switzerland, 2016.
18. Sims, C. Macroeconomics and Reality. *Econometrica* **1980**, *48*, 1–48. [CrossRef]

engineering proceedings

MDPI

Proceeding Paper

Forecasting Agricultural Area Using Nerlovian Model in Côte d'Ivoire †

Gueï Cyrille Okou [1,*], Kolé Keita [1], Yao Aubin N'Dri [1] and Auguste K. Kouakou [2]

1 Laboratoire des Sciences et Technologies de l'Environnement (LSTE), Université Jean Lorougnon GUEDE, Daloa P.O. Box 150, Côte d'Ivoire
2 Laboratoire de Recherche en Economie et Gestion (LAREG), Université Jean Lorougnon GUEDE, Daloa P.O. Box 150, Côte d'Ivoire
* Correspondence: okou.guei.cyrille@gmail.com
† Presented at the 9th International Conference on Time Series and Forecasting, Gran Canaria, Spain, 12–14 July 2023.

Abstract: In this article, we develop the Nerlove models that give the area of cacao and cashew nuts in terms of the area, the price and the rainfall. These models are estimated using the methods of ordinary least square and likelihood maximum and are used to analyze the link in a short time between the agricultural determinants. The results showed that the anticipation elasticity had an effect on the price practiced and forecast model. In a short time, the price delayed by one year, and the area delayed by one and two years, had decreasing returns to scale for the current area.

Keywords: price elasticity; Nerlove model; econometric modeling; econometric forecasting

Citation: Okou, G.C.; Keita, K.; N'Dri, Y.A.; Kouakou, A.K. Forecasting Agricultural Area Using Nerlovian Model in Côte d'Ivoire. *Eng. Proc.* **2023**, *39*, 35. https://doi.org/10.3390/engproc2023039035

Academic Editors: Ignacio Rojas, Hector Pomares, Luis Javier Herrera, Fernando Rojas and Olga Valenzuela

Published: 29 June 2023

1. Introduction

The agricultural sector has always been one of the pillars in the Ivorian economy. As the world's leading producer of cocoa beans and the cashew nuts, Côte d'Ivoire is a key country in this sector. Export agriculture (coffee and cocoa) has long been the mainstay of the economy. Despite the emergence of a growing industrial fabric and the government's commitment to invest in other sectors such as education, Ivorian agriculture continues to contribute significantly to government revenues. Thus, in 2021, the agricultural sector represented 20% of Côte d'Ivoire's GDP and 60% of the country's exports in 2018. The agricultural sector employs 46% of the workforce and provides a living for two-thirds of the population.

After independence, Côte d'Ivoire inherited cocoa. Its production, which was 1.5 million tons in 1964, reached the mark of 4.162 million tons in 2013–2014 [1]. It has now become one of the most profitable crops in the world, generating 7.4 billion USD in 2008 among small producers [2]. In contrast, cashew nuts were introduced in northern Côte d'Ivoire in the late 1950s for reforestation and soil protection. Progressively, from a purely ecological aspect, the establishment of the cashew tree met a socio-economic need, since this tree can produce marketable nuts. Thus, cashew became a real speculation from the 1990s, due to the increasing demand for cashew nuts on the international market. The cashew sector has thus experienced spectacular development, with national production of raw cashew nuts increasing from 19,000 tons in 1990 to about 750,000 tons in 2018.

In Cote d'Ivoire, the development of the agricultural sector has had a significant and considerable impact on the economic and social well-being of the population. The economy of Côte d'Ivoire is still based on the exploitation and export of raw materials, mainly agricultural materials [3]. This economy is much more oriented towards the analysis of agricultural supply. This consists of analyzing the supply response to product prices and to the prices of production factors and intermediate consumption. This analysis also concerns the agricultural income of suppliers (producers). Producers are rational economic agents who offer the total quantity of a good on the market in exchange for a price.

Several econometric models have based their studies on the relationship between agricultural determinants such as supply and price. We can cite the work of Nerlove, who was the first to develop a theory known as "the Nerlovian models of supply response" in 1956 and 1958. This article is part of the econometric modeling and forecasting of agricultural determinants based on the Nerlove model. The first part aims at estimating the price elasticities of the agricultural products cocoa and cashew nuts and analyzing their effects on forecasting using the Nerlove model. In the second part, we develop an estimation technique for the parameters of the Nerlove model, based on the maximum likelihood method. This study is applied to data on rainfall, agricultural areas and prices paid to farmers in Côte d'Ivoire over the period 1980–2022.

2. Materials and Methods

According to Nerlove [4,5] the most robust starting point to determine what is likely is to assume that the expectation price of agricultural determinants depend on past price values. Nerlove defined the adjustment coefficient as a ratio between the actual price variation and the expected price change. This assumption is similar to that of Cagan's [6] theory of adaptive expectations, which states that "The expected rate of price change is revised in each period in proportion to the deviation of the observed rate of price change from the previously expected rate of price change".

2.1. Specification of the Nerlove Model of the Cultivated Area

The basic Nerlove model formed by the three hypotheses is, thus, finally written

$$\begin{cases} SF_t^e = a_0 + a_1 P_t^a + a_2 Z_t + u_t \\ P_t^a = P_{t-1}^a + \beta\left(P_{t-1} - P_{t-1}^a\right) \\ SF_t = SF_{t-1} + \delta\left(SF_t^e - SF_{t-1}\right) \end{cases} \tag{1}$$

where SF_t^e is the area desired; SF_t is the area practiced; P_t^a is the expected price deemed "normal" by producers in period t; P_t is the actual price in period t; β is the "expectation coefficient", or "anticipation elasticity", it is the correction factor assumed constant and such that $0 \leq \beta \leq 1$; and a_0, a_1, a_3 and δ are adjustment coefficients.

Nerlove said that the actual change in area is a proportion of the difference between the equilibrium level of area (noted δ) and the area actually cultivated in the previous period. Eliminating the unobservable variable (SF_t^e) using the Koyck transformation [7], we consider that the random variable u_t is a white Gaussian noise if V_t is a moving average process of order 1 (MA(1) or ARMA(0, 1)).

Model (1) can still be written as follows:

$$\begin{aligned} \left[SF_t - (1-\beta)^2 SF_{t-2}\right] = b_1 + b_2 P_{t-1} + b_3\left[SF_{t-1} - (1-\beta)SF_{t-2}\right] \\ + b_5\left[Z_t - (1-\beta)Z_{t-1}\right] + V_t \end{aligned} \tag{2}$$

2.2. Estimation of the Cultivated Area Model

We estimated the parameter β from the second equation of Model (1) with the method of ordinary least squares (OLS). The estimated parameter was denoted $\hat{\beta}$.

The $\hat{\beta}$ statistic was injected into Model (2) and we also estimated the other parameters with the OLS method.

The estimation errors of the OLS method, denoted V_t, must be independent and identically distributed (normal distribution). In some cases, we had assumptions that were not verified; then, the models were adjusted with the maximum likelihood method. In this case, the estimation errors must be a moving average process of order 1.

(a) Estimation of the β parameter

From Equation (1), we derive the following linear fitting model:

$$P_t^a = P_{t-1}^a + \beta(P_{t-1} - P_{t-1}^a) + u_t,$$

where u_t is the assumed independent and identically distributed normal distribution error term. Using the OLS method, we obtained the expression for the estimator $\hat{\beta}$ of the parameter β defined by

$$\hat{\beta} = \frac{\sum_{t=1}^{T}(P_{t-1} - P_{t-1}^a)(P_t^a - P_{t-1}^a)}{\sum_{t=1}^{T}(P_{t-1} - P_{t-1}^a)^2}$$

(b) Ordinary least squares method (OLS)

In Model (2), we denote

$$\lambda = 1 - \hat{\beta} \text{ and } y_t = SF_t - \lambda 2SF_{t-2}$$

Then, the model is written

$$y_t = b_1 + b_2 P_{t-1} + b_3(SF_{t-1} - \lambda SF_{t-2}) + b_5(Z_t - \lambda Z_{t-1}) + V_t. \quad (3)$$

In matrix form, the multiple regression model is written (the matrix for Model (3)) is

$$Y = XB + V,$$

with

$$Y = \begin{pmatrix} y_1 \\ y_2 \\ y_3 \\ y_4 \\ \vdots \\ y_T \end{pmatrix}, X = \begin{pmatrix} 1 & P_0 & SF_1 - \lambda^2 SF_{-1} & Z_1 - \lambda Z_0 \\ 1 & P_1 & SF_1 - \lambda SF_0 & Z_2 - \lambda Z_1 \\ 1 & P_2 & SF_2 - \lambda SF_1 & Z_3 - \lambda Z_2 \\ 1 & P_3 & SF_3 - \lambda SF_2 & Z_4 - \lambda Z_3 \\ \vdots & \vdots & \vdots & \vdots \\ 1 & P_{T-1} & SF_{T-1} - \lambda SF_{T-2} & Z_T - \lambda Z_{T-1} \end{pmatrix}$$

$$B = \begin{pmatrix} b_1 \\ b_2 \\ b_3 \\ b_5 \end{pmatrix} \text{ and } V = \begin{pmatrix} V_1 \\ V_2 \\ V_3 \\ V_4 \\ \vdots \\ V_T \end{pmatrix}.$$

We obtained the estimator $\hat{B}$ of B defined by

$$\hat{B} = \begin{pmatrix} \hat{b}_1 \\ \hat{b}_2 \\ \hat{b}_3 \\ \hat{b}_5 \end{pmatrix} = (X'X)^{-1}X'Y.$$

(c) Maximum likelihood method

For this method, we used Model (2) with the error term $V_t = \delta[u_t - (1-\beta)u_{t-1}]$, which is a moving average of order. Model (2) is written

$$y_t = b_1 + b_2 P_{t-1} + b_3(SF_{t-1} - \lambda SF_{t-2}) + b_5(Z_t - \lambda Z_{t-1}) + \delta(u_t - \lambda u_{t-1})$$

The model is then rewritten

$$Y = XB + AU,$$

with

$$A = \begin{pmatrix} \delta & 0 & & 0 & 0 & \cdots & \cdots & 0 & \\ -\lambda\delta & \delta & & 0 & 0 & \cdots & \cdots & 0 & \\ & & & & & & \cdots & & 0 \\ 0 & -\lambda\delta & \delta & 0 & \cdots & & & & \\ & & & & & & \cdots & & 0 \\ 0 & 0 & -\lambda\delta & \delta & \cdots & & & & \\ \vdots & \vdots & \vdots & \vdots & \vdots & & \vdots & & \vdots \\ 0 & 0 & 0 & 0 & 0 & & \delta & & \vdots \\ & & & & & & -\lambda\delta & & \delta \end{pmatrix} \text{ and } U = \begin{pmatrix} u_1 \\ u_2 \\ u_3 \\ u_4 \\ \vdots \\ u_{T-1} \\ u_T \end{pmatrix}.$$

We have

$$Y = XB + AU \Leftrightarrow AU = Y - XB \Leftrightarrow U = A^{-1}(Y - XB)$$

The vector U is a vector whose components are Gaussian noise. Therefore,

$$U = A^{-1}(Y - XB) \sim \mathcal{N}\left(\begin{pmatrix} 0 \\ 0 \\ 0 \\ \vdots \\ \vdots \\ 0 \end{pmatrix}; \sigma^2 I_T\right)$$

where I_T is the identity matrix of order T.

The density function of the vector U is given by

$$f(u_1, \ldots, u_T) = \frac{1}{(2\pi)^{T/2}} \frac{1}{(\sigma^2)^{T/2}} exp\left(-\frac{1}{2}\left[A^{-1}(Y - XB)\right]' \sum{}^{-1}[(Y - XB)]\right),$$

where $\sum = \sigma^2 I_t$ and $\sum^{-1} = \frac{1}{\sigma^2} I_t$.

The likelihood can be written as follows

$$\mathcal{L}(b_1, b_2, b_3, b_5, \sigma) = \prod_{t=1}^{T}\left(\frac{1}{(2\pi)^{1/2}} \frac{1}{(\sigma^2)^{1/2}} exp\left(-\frac{1}{2\sigma^2}\left[A^{-1}(Y - XB)\right]'\left[A^{-1}(Y - XB)\right]\right)\right).$$

We show by recurrence that:

$$A^{-1} = \frac{1}{\delta}\begin{pmatrix} \lambda^0 & 0 & 0 & 0 & \cdots & \cdots & 0 \\ \lambda^1 & \lambda^0 & 0 & 0 & \cdots & \cdots & 0 \\ \lambda^2 & \lambda^1 & \lambda^0 & 0 & \cdots & \cdots & 0 \\ \lambda^3 & \lambda^2 & \lambda^1 & \lambda^0 & \vdots & \cdots & 0 \\ \vdots & \vdots & \vdots & \vdots & \vdots & \vdots & \cdots \\ \vdots & \vdots & \vdots & \vdots & \vdots & \lambda^0 & \cdots \\ \lambda^{T-1} & \lambda^{T-2} & \lambda^{T-3} & \cdots & \cdots & \lambda^1 & \lambda^0 \end{pmatrix}$$

We can easily define the likelihood

$$\mathcal{L}(b_1, b_2, b_3, b_5, \sigma) = \left(\frac{1}{2\pi}\right)^{T/2}\left(\frac{1}{\sigma^2}\right)^{T/2} exp\left[-\frac{1}{2\sigma^2\delta^2}\sum_{t=1}^{T}\left(\sum_{j=1}^{t}\lambda^{t-j}\left(y_{t-j+1} - \sum_{k=1}^{4} X_{t-j+1;k}B_k\right)\right)^2\right]$$

The log-likelihood is given by

$$log(\mathcal{L}(b_1, b_2, b_3, b_5, \sigma)) = -\frac{T}{2}log\left(2\pi\sigma^2\right) - \frac{1}{2\sigma^2\delta^2}\sum_{t=1}^{T}\left(\sum_{j=1}^{t}\lambda^{t-j}\left(y_{t-j+1} - \sum_{k=1}^{4}X_{t-j+1;k}B_k\right)\right)^2.$$

We obtained the optimal values of the parameters by maximizing the log-likelihood function. To do this, we have

$$\nabla log(\mathcal{L}(b_1, b_2, b_3, b_5, \sigma)) = 0. \tag{4}$$

We used the first four equations. We obtained

$$(S)\begin{cases} \sum_{t=1}^{T}\left(\sum_{j=1}^{t}X_{j';1}\lambda^{t-j}\left(y_{j'} - X_{j';1}b_1 - X_{j';2}b_2 - X_{j';3}b_3 - X_{j';4}b_5\right)\right) = 0 \\ \sum_{t=1}^{T}\left(\sum_{j=1}^{t}X_{j';2}\lambda^{t-j}\left(y_{j'} - X_{j';1}b_1 - X_{j';2}b_2 - X_{j';3}b_3 - X_{j';4}b_5\right)\right) = 0 \\ \sum_{t=1}^{T}\left(\sum_{j=1}^{t}X_{j';3}\lambda^{t-j}\left(y_{j'} - X_{j';1}b_1 - X_{j';2}b_2 - X_{j';3}b_3 - X_{j';4}b_5\right)\right) = 0 \\ \sum_{t=1}^{T}\left(\sum_{j=1}^{t}X_{j';4}\lambda^{t-j}\left(y_{j'} - X_{j';1}b_1 - X_{j';2}b_2 - X_{j';3}b_3 - X_{j';4}b_5\right)\right) = 0 \end{cases}.$$

We developed each equation of the system (S) and applied the double sum on each term to write our system (S) in matrix form. We obtained the following matrix equation

$$N = MB \Longleftrightarrow B = M^{-1}N$$

where M is a symmetric matrix given by

$$M = \begin{pmatrix} M_{11} & M_{12} & M_{13} & M_{14} \\ M_{21} & M_{22} & M_{23} & M_{24} \\ M_{31} & M_{32} & M_{33} & M_{34} \\ M_{41} & M_{42} & M_{43} & M_{44} \end{pmatrix} N = \begin{pmatrix} \sum_{t=1}^{T}\left(\sum_{j=1}^{t}X_{j';1}\lambda^{t-j}y_{j'}\right) \\ \sum_{t=1}^{T}\left(\sum_{j=1}^{t}X_{j';2}\lambda^{t-j}y_{j'}\right) \\ \sum_{t=1}^{T}\left(\sum_{j=1}^{t}X_{j';3}\lambda^{t-j}y_{j'}\right) \\ \sum_{t=1}^{T}\left(\sum_{j=1}^{t}X_{j';4}\lambda^{t-j}y_{j'}\right) \end{pmatrix} \text{ and } B = \begin{pmatrix} b_1 \\ b_2 \\ b_3 \\ b_5 \end{pmatrix}.$$

We denoted

$$\mathrm{t} - j + 1 = j', M_{ik} = \sum_{t=1}^{T}\left(\sum_{j=1}^{t}X_{j';i}X_{j';k}\lambda^{t-j}\right) \text{ and } i, k\epsilon\{1,2,3,4\}.$$

We define Equation (5) from the gradient of the log-likelihood function given by Equation (4).

$$\frac{1}{\sigma^3\delta^2}\sum_{t=1}^{T}\left(\sum_{j=1}^{t}\lambda^{t-j}\left(y_{j'} - X_{j';1}b_1 - X_{j';2}b_2 - X_{j';3}b_3 - X_{j';4}b_5\right)\right)^2 = T \tag{5}$$

After the computation the estimator $\hat{B}$ of the parameter matrix B, we injected the estimated values $\hat{b}_1, \hat{b}_2, \hat{b}_3$ and $\hat{b}_5$ into Equation (5) of our system (S) to calculate the estimated variance $\hat{\sigma}^2$ of the errors.

2.3. Data

This study was carried out on two types of agricultural products most practiced in Côte d'Ivoire. It was, in particular, about cocoa and cashew nuts, where Cote d'Ivoire occupies the first world rank. These two agricultural products were chosen because they contribute significantly to reducing poverty in Côte d'Ivoire. The determinants of cocoa and cashew used in our study were the cultivated area (hectare), the effective price (field price), the average annual rainfall in the Savane and Denguélé regions where the cashews are produced and the average annual rainfall (in millimeters) in the southeast, east, central-east, central-west, central and west, where cocoa is produced. The cultivated area was assumed to be the harvested area that is relatively available. The effective price was the edge-of-field price, which is the price at which a kilogram of cocoa or cashew nuts is purchased from the producer (farmer). Our data for cashew nuts were collected from the databases of the Cotton and Cashew Council (CCA), the Food and Agriculture Organization (FAO) and SODEXAM. These data cover the period from 1990 to 2019 (sample of 30 observations). For cocoa, the data for our study came from the databases of the Coffee-Cacao Council, the World Bank and the FAO (Food and Agriculture Organization). These data cover the period from 1980 to 2020 (sample of 41 observations). We worked with 26 observations for the cashew nut variables and 38 observations for the cocoa variables. In any case, the four remaining observations were used to test the predictive quality of our models.

3. Results and Discussions

3.1. Estimation of the Elasticity Coefficient

The elasticity estimation for the cashew nuts was calculated by assuming the expected price to be within the range of prices announced by the board. We had $P_t^a \in [P_{min}, P_{max}]$, where $[P_{t,min}, P_{tmax}]$ is the range of prices charged to buyers of the cashew nuts given by Cotton and Cashew Council. The effective price (P_t) was the average annual price practiced. We assumed the expected price of cocoa (P_t) as the average annual market price. The actual price was the field price of cocoa. The results are given in the following.

The price elasticity coefficients in Table 1 are included in [0, 1].

Table 1. Estimation of the elasticity coefficients.

Variables	Anticipation Elasticity	Estimated Values
Cacao	β	0.012
	β_{min}	0.37
	β_{q1}	0.71
Cashew nuts	β_{med}	0.65
	β_{q3}	0.31
	β_{max}	0.23

- For the cashew nuts: we remark that the elasticities β_{min}, β_{q3} and β_{max} of the minimum, third quartile and maximum prices, respectively, were moderately sensitive to the prices actually practiced. On the other hand, the elasticities β_{q1} and β_{med}, respectively, of the first quartile and median prices were highly sensitive to the price practiced.
- In the case of cacao, we also noted a low sensitivity of the stock market price in relation to the field price of cocoa.

These estimated values of the elasticities were then fixed in the Nerlove model to estimate the parameters of the model.

3.2. Estimation of the Nerlove Model Parameters of the Cultivated Area

<u>**Cashew nuts:**</u>

We estimated the parameters using the OLS method. These estimates were made to calculate the elasticity values because they presented the best conditions of estimation.

The results of the estimates are given in Table 2.

Table 2. Parameters estimated using the OLS method.

Models	Anticipation Elasticity	$\hat{b}_1$	$\hat{b}_0$	$\hat{b}_2$	$\hat{b}_3$	$\hat{b}_5$	R^2	*p* Value DW
1	β_{min}	1.716 **	0.694 ***	−0.00114 *	0.268 **	−0.0027	0.9665	0.8602
2	β_{q1}	2.092 **	0.514 ***	−0.00112	0.459 **	−0.0016	0.9647	0.8018
3	β_{med}	1.992 **	0.545 ***	−0.00113.	0.427 **	−0.00173	0.9649	0.8143
4	β_{q3}	1.696 **	0.7469 ***	−0.00112 *	0.209 **	−0.0031	0.9668	0.8543
5	β_{max}	-1.752×10^6 **	1.708×10^5 ***	4.546×10	2.046×10^0 ***	-1.381×10^3	0.9221	0.01856

*** significance to 1%, ** to 5% and * to 10%, R^2, the coefficient of determination. *p* value test, Durbin–Watson.

Table 2 presents the estimated parameters of the Nerlove model of the cultivated area of cashew nuts using OLS. The results show that the good quality given by the R^2 was close to one. The residuals of these models were independent and identically Gaussian. The estimated Models 1, 2, 3 and 4 are given by the following equation:

$$\log(SF_t) = \hat{b}_1 + \hat{b}_0 * \log\left(\lambda^2 SF_{t-2}\right) + \hat{b}_2 * P_{t-1}$$
$$+\hat{b}_3 \log(SF_{t-1} - \lambda SF_{t-2}) + \hat{b}_5 * (Z_t - \lambda Z_{t-1}) + V_t.$$

The last model, 5, is given by

$$SF_t = \hat{b}_1 + \hat{b}_0 * \lambda^2 SF_{t-2} + \hat{b}_2 * P_{t-1} + \hat{b}_3 (SF_{t-1} - \lambda SF_{t-2}) + \hat{b}_5 * (Z_t - \lambda Z_{t-1}) + V_t,$$

where $V_t; t = 1, 2, \ldots; n$ is the residual of the model, independent and normally distributed; and $\lambda = 1 - \beta$.

<u>Cacao:</u>

The estimation of the Nerlove model with the cocoa data was done using the maximum likelihood method. The residuals of this model did not respect the assumptions required by the OLS method. Table 3 presents the results of the estimation using the maximum likelihood method.

Table 3. Parameters estimated using the maximum likelihood method.

Model	Elasticity	$\hat{b}_1$	$\hat{b}_2$	$\hat{b}_3$	$\hat{b}_5$
6	β_{cacao}	0.375	0.010	0.002	−0.001

The results show that all the variables, namely the delayed cocoa price, the delayed area of one and two periods, rainfall and the delayed rainfall of one period, statistically and significantly permit explanations for the variations in the area actually practiced and, therefore, in the supply of cocoa in Côte d'Ivoire. Nervole model of the cocoa did not give the good prediction qualities. For this reason, we used models 1, 2, 3 and 4 to predict the area practiced for the cashew nuts.

3.3. Forecast of Cashew Area

In this section, we present the prediction results of our estimated cashew model. Figure 1 shows the graphs of the cultivated area and the predictions of Models 1, 2, 3 and 4. For the choice of the best model, we compared the results of the Root of the Sum of the Mean Squares (RSMS) of the bias. We calculated the bias between the practiced area and the cultivated area estimated by the model. We define the bias by:

$$Biais_t = \frac{Surf_t - Surf_{t,estim}}{Surf_{t,estim}} \text{ and RSMS} = \sqrt{\frac{1}{N-1}\sum_t^N Biais_t^2}$$

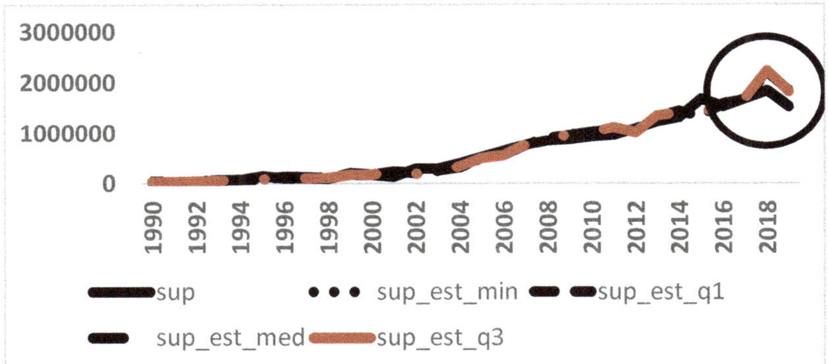

Figure 1. Graph of Nerlove model predictions of cashew nut.

The results in Table 4 show that the RSMS of the practiced areas and those given by the different models are quite close. The bias was calculated with 26 observations used in our model estimation. Table 5 shows the RSMS between the four observations that were conserved and those given by the estimated models. We can see that the RSMS are also quite close.

Table 4. RSMS results of the estimated values.

Model	1	2	3	4
RSMS	0.02841696	0.02842018	0.02841973	0.02841385

Table 5. RSMS results of the predicted values.

Model	1	2	3	4
RSMS	0.0109437	**0.01070902**	**0.01072822**	0.01119934

We noticed that Models 2 and 3 gave smaller RSMS results. The expectation coefficients (anticipation elasticity) calculated from the median and first quartile expected prices contributed to the better fit of the Nerlove model of the practiced area.

The circled portion of Figure 1 shows the predictions of the estimated Models 1, 2, 3, 4.

3.4. Discussion

To apply the logarithm, the estimated parameters were considered as elasticities at short times in Models 1, 2, 3 and 4 [8]. The practiced area delayed by one year and two years had positive coefficients. This means that they contributed positively to the variation of the practiced area. On the other hand, the coefficient of the variation price ($\hat{b}_2$) of the cashew nuts was not positive. This variable contributed negatively to the variation in the practiced area. The sum of the elasticities of the practiced area delayed by one year, by two years and the price variation of the cashew nuts were positive and less than one. This is decreasing returns to scale. This means that when all factors of production are increased by one unit (1%), supply will decrease by one unit (1%). These determinants of production contributed significantly and decreasingly to explain the area variation in cashew nuts.

The coefficient of the rainfall factor was not significant and positive. The effect of rainfall was negligible (short time) in the supply variation of cashew nuts. The significance of the parameters and the RSMS results (Table 5) indicated that Models 2 and 3 were better fits for the areas of the cashew nuts. They give good predictions for the practiced area. When the sum of the elasticity of cocoa was equal to one, then we had constant returns to scale. Indeed, if we increase or decrease the variables (factors of production) by 1%, then the supply will also increase or decrease by 1%. The estimated parameters represent the elasticities in a short period. The estimated value of β, which corresponded to the elasticity of anticipation, was 0.012 or 1.2%, which means that Ivorian producers assume that the price in the current period is equal the sum of 98.8% of the price anticipated in the previous period and 1.2% of the actual price.

4. Conclusions

This study considers the relationships between certain factors of production in Côte d'Ivoire. Using the Nerlove model, the econometric results of the study confirm that even in Côte d'Ivoire, the price variation has a negative effect on supply. Nerlove's econometric model, which is a pioneering model of farmer behavior and has been used in a multitude of studies around the world, with quite variable results [9] allowed us to show in this study that price (delayed), rainfall (current and delayed) and area (delayed by one and two periods) are key variables in production decisions for farmers. Producers are positively sensitive in the short run to the variation in delayed price, area and all other supply determinants except current rainfall. Thus, over the study period, changes in price, area delayed by one and two years and rainfall delayed by one period contributed differently to increasing cocoa and cashew supplies in Côte d'Ivoire.

Author Contributions: G.C.O.: Conceptualization, methodology, software, formal analysis, visualization, investigation and writing—original draft preparation. K.K.: Conceptualization, validation, methodology, software and writing—original draft preparation. Y.A.N.: Methodology, validation and review. A.K.K.: Conceptualization, validation and review. All authors have read and agreed to the published version of the manuscript.

Funding: This research received no external funding.

Institutional Review Board Statement: Not applicable.

Informed Consent Statement: Not applicable.

Data Availability Statement: Databases of the Coffee Cacao Council (ICCO), the Cotton Cashew council in Côte d'Ivoire and SODEXAM.

Acknowledgments: The authors would like to thank the anonymous referees for their constructive comments that improved the quality of this paper.

Conflicts of Interest: The authors declare no conflict of interest.

References

1. ICCO. *Quarterly Bulletin of Cocoa Statistics*; Cocoa Year 2013/2014; ICCO: London, UK, 2014.
2. ICCO. *Quarterly Bulletin of Cocoa Statistics*; Cocoa Year 2017/18; ICCO: London, UK, 2008.
3. PND 2016–2020, Groupe Consultatif, Plan National De Developpement, Côte d'Ivoire. Available online: http://www.caidp.ci/uploads/506b0bce6be504b64d5359c6407cd7df.pdf (accessed on 1 January 2016).
4. Nerlove, M. Estimates of the Elasticities of Supply of Selected of Agricultural Commodities. *J. Farm Econ.* **1956**, *38*, 496–509.
5. Nerlove, M. *The Dynamics of Supply: Estimation of Farmers Response to Price*; John Hopkins University Press: Baltimore, MD, USA, 1958.
6. Cagan, P. The Monetary Dynamics of Hyperinflation. In *Studies in the Quantity Theory of Money*; Friedman, M., Ed.; University of Chicago Press: Chicago, IL, USA, 1956; pp. 25–117.
7. Koyck, L.M. *Distributed Lags and Investment Analysis*; North-Holland Publishing Company: Amsterdam, The Netherlands, 1954.
8. Gujarati, D.N. *Basic Econometrics*, 4th ed.; McGraw-Hill Companies: Singapore, 2004.
9. Askari, H.; John, T. *Cumming, Agricultural Supply Response: A Survey of the Econometric Evidence*; Praeger Publishers: Westport, CT, USA, 1977.

engineering proceedings

Proceeding Paper

Improving Predictive Accuracy in the Context of Dynamic Modelling of Non-Stationary Time Series with Outliers †

Fernanda Catarina Pereira [1,*], Arminda Manuela Gonçalves [2,‡] and Marco Costa [3,‡]

1 Centre of Mathematics, University of Minho, 4710-057 Braga, Portugal
2 Department of Mathematics and Centre of Mathematics, University of Minho, 4710-057 Braga, Portugal
3 Centre for Research and Development in Mathematics and Applications, Águeda School of Technology and Management, University of Aveiro, 3810-193 Aveiro, Portugal
* Correspondence: id9976@alunos.uminho.pt
† Presented at the 9th International Conference on Time Series and Forecasting, Gran Canaria, Spain, 12–14 July 2023.
‡ These authors contributed equally to this work.

Abstract: Most real time series exhibit certain characteristics that make the choice of model and its specification difficult. The objective of this study is to address the problem of parameter estimation and the accuracy of forecasts *k*-steps ahead in non-stationary time series with outliers in the context of state-space models. In this paper, three methods for detecting and treating outliers are proposed. We also present a comparative study of the proposed methods using data simulated from a local level model with sample sizes of 50 and 500 and with various combinations of parameters, with a 5% contamination error rate of the observation equation. The results were evaluated in terms of the accuracy of model parameters and the forecasts *k*-steps ahead, as well as the detection rate of true outliers. These methodologies are applied to three real examples. This study shows that the local level model is sufficiently robust even for non-stationary contaminated series, in the sense that they are able to handle non-stationary time series and outliers in a satisfactory way.

Keywords: outliers; contaminated data; non-stationary time series; state-space models; Kalman filter; simulation study

Citation: Pereira, F.C.; Gonçalves, A.M.; Costa, M. Improving Predictive Accuracy in the Context of Dynamic Modelling of Non-Stationary Time Series with Outliers. *Eng. Proc.* **2023**, *39*, 36. https://doi.org/10.3390/engproc2023039036

Academic Editors: Ignacio Rojas, Hector Pomares, Luis Javier Herrera, Fernando Rojas and Olga Valenzuela

Published: 29 June 2023

1. Introduction

State-space models were originally developed in aerospace engineering in the early 1960s for the purpose of monitoring and correcting the trajectory of a spacecraft headed to the moon. Today, these models have wide applicability in many areas, such as finances [1], ecology [2], machine learning [3], and time series analysis and forecasting [4–7]. These models, associated with the Kalman filter algorithm [8], are a very powerful tool given their ability to update predictions both in real time and in a recursive procedure as new observations of the time series become available, thus improving the accuracy of predictions. In addition, state-space models are very flexible due to their ability to incorporate fixed effects and stochastic components that can represent the different unobserved components, such as periodic structures, trends, seasonality, and temporal correlation. These components describe the structural variation of the time series under study. Furthermore, potential covariates can be added because they are important to explain the process and complement the information introduced by the different stochastic components of the model. These models include two sources of variability: one corresponding to measurement errors and the other to process variations. In this way, it becomes simpler to interpret both errors separately. One advantage of these models is that they do not require the assumption of stationarity and can handle time series with missing values in a particularly simple way [4,9]. However, the existence of outliers in real data can influence the estimation and prediction accuracy of both the parameters.

Outliers can be a problem for model specification and prediction accuracy, since the Kalman filter is not generally robust to the presence of outliers. An incorrectly specified model can lead to incorrect covariance matrices of predictions given by the Kalman filter, and thus there is no way to describe the actual quality of the filter [10]. According to [11], the presence of outliers in a time series can induce non-Gaussian heavy-tailed noise, leading to misspecified models, biased estimates, and inaccurate forecasts. The authors of [12] showed that simple linear Gaussian state-space models can present estimation problems. Therefore, in this paper, several methods of detecting and treating outliers are discussed. These methods will be compared and illustrated with a simulation study that considers a simple Gaussian stationary state-space model with 5% data contamination. To create the non-stationarity scenario, the local level model, which is a particular case of the state-space model, will be considered for the sake of simplicity. Detection and treatment of the methods' performance is evaluated by the root-mean-square error (RMSE) and the mean absolute error (MAE) of the Gaussian likelihood of the parameters' estimates and the one-step ahead predictions of the time-series variable. Several scenarios are considered accounting for different combinations of parameters and times series sizes, n in this specific case, (n = 50,500). Time series simulations are generated until 1000 time series have a state-space model with valid estimates, i.e., estimates within the space parameter.

2. Methodologies

The univariate state-space model can be represented by the observation and state equations, respectively, given by

$$Y_t = W_t\beta_t + e_t \tag{1}$$

$$\beta_t = \mu + \phi(\beta_{t-1} - \mu) + \varepsilon_t \tag{2}$$

where t represents the time, Y_t is the observed data, W_t is a factor assumed to be known that relates the observation Y_t to the latent variable β_t at time t. The disturbances e_t and ε_t are independent and identically distributed, with Gaussian distribution of zero mean and variances σ_e^2 and σ_ε^2, respectively, and are uncorrelated with each other.

The state β_t is a latent variable and therefore must be estimated. The Kalman filter algorithm ([8]) provides optimal unbiased linear one-step ahead and update estimators of the unobservable state β_t. Let $\Theta = \{\phi, \sigma_e^2, \sigma_\varepsilon^2\}$ be the vector of the model's unknown parameters, let $\hat{\beta}_{t|t-1}$ denote the predictor of β_t based on the observations $Y_1, Y_2, \ldots, Y_{t-1}$ and $P_{t|t-1}$ be its mean square error, i.e., $E[(\hat{\beta}_{t|t-1} - \beta_t)^2]$. The one-step ahead forecast for the observable vector Y_t is given by $\hat{Y}_{t|t-1} = W_t\hat{\beta}_{t|t-1}$. When, at time t, Y_t is available, the prediction error or innovation, $\eta_t = Y_t - \hat{Y}_{t|t-1}$, is used to update the estimate of β_t (filtering) through the equation

$$\hat{\beta}_{t|t} = \hat{\beta}_{t|t-1} + K_t\eta_t,$$

where K_t is called the Kalman gain and is given by $K_t = P_{t|t-1}W_t(W_t^2P_{t|t-1} + \sigma_e^2)^{-1}$. The mean square error of the updated estimator $\hat{\beta}_{t|t}$, represented by $P_{t|t}$, verifies the relationship $P_{t|t} = P_{t|t-1} - K_tW_tP_{t|t-1}$. Furthermore, the predictor of β_{t+k} at time t is given by

$$\hat{\beta}_{t+k|t} = \mu + \phi^k\left(\hat{\beta}_{t|t} - \mu\right),$$

and its mean square error is $P_{t+k|t} = \phi^{2k}P_{t|t} + \sum_{i=0}^{k-1}\phi^{2i}\sigma_\varepsilon^2$.

Outlier Detection and Treatment Procedures

Three approaches to outlier detection and treatment are presented. The first approach is based on linear interpolation, which represents the naive method. The other two approaches are based on iterative processes from the robust Kalman filter and from the Kalman filter in the missing values perspective.

1. Linear interpolation (LI)
    - Outlier detection: Observations are considered outliers if they are less than $Q_1 - 1.5IQR$ or greater than $Q_3 + 1.5IQR$, where Q_1 and Q_3 denote the first and third quartiles, respectively, and IQR (interquartile range) is the difference between the third and first quartiles (IQR rule).
    - Outlier treatment: Any outliers that are identified are replaced by LI using the neighbouring observations [13].
2. Iterative method based on the robust Kalman filter (RKF)
    - Outlier detection: Outlier detection is performed by applying the IQR rule on the standardized residuals after fitting a state-space model to the data.
    - Outlier treatment: An alternative to the state estimator $\hat{\beta}_{t|t}$, inspired by the work by [14] and subsequently by [15], is proposed. In this approach, the state prediction $\hat{\beta}_{t|t}$ is replaced by

    $$\hat{\beta}^*_{t|t} = \underset{\beta}{\operatorname{argmin}}\left\{ \left(\hat{\beta}_{t|t-1} - \beta\right)^2 P^{-1}_{t|t-1} + \left(Y_t^{\text{out}} - W_t\beta\right)^2 \sigma_e^{-2} \right\} \tag{3}$$

    where Y_t^{out} is an identified outlier that is replaced by $\hat{Y}_t^* = W_t\hat{\beta}^*_{t|t}$. This proposal considers the robust version of the Kalman filter only at moments at which outliers are detected, as opposed to the original work, in which it is applied at all moments. In the end, the model is iteratively fitted j times to the corrected time series until $\|\hat{\Theta}^{(j)}_{ML} - \hat{\Theta}^{(j-1)}_{ML}\| < \delta, j \in \mathbb{N}$, or for some value j.
3. Iterative method based on the Kalman filter for time series with missing values (naKF)
    - Outlier detection: Outlier detection is performed by applying the IQR rule to the standardized residuals after fitting a state-space model to the data.
    - Outlier treatment: Outlier observations Y_t^{out} are assumed to be missing values and the state estimator $\hat{\beta}_{t|t}$ and its mean square error $P^*_{t|t}$ are replaced by $\hat{\beta}^*_{t|t} = \hat{\beta}_{t|t-1}$ and $P^*_{t|t} = P_{t|t-1}$, respectively. The missing observations Y_t^{out} are replaced by $\hat{Y}_t^* = W_t\hat{\beta}^*_{t|t}$ and the state-space model is fitted j times to the corrected time series until $\|\hat{\Theta}^{(j)}_{ML} - \hat{\Theta}^{(j-1)}_{ML}\| < \delta, j \in \mathbb{N}$, or for some value j.

The aim of this paper is to investigate under which conditions the presence of outliers affects the estimation of parameters and states in the state-space model and to propose competitive approaches for outlier detection and treatment. Thus, we simulate time series of size n ($n = 50{,}500$), considering for all simulation studies the local level model, which is a simple and particular case of the state-space model (2)–(4), where $W_t = 1$, $\forall t$ and $\phi = 1$, which will be used to illustrate the non-stationary case. The local level model is given by:

$$\begin{aligned} Y_t &= \beta_t + e_t \\ \beta_t &= \beta_{t-1} + \varepsilon_t \end{aligned} \tag{4}$$

In the literature, some approaches have been proposed for the initialization of the Kalman filter for non-stationary stochastic processes. Perhaps the best known is the diffuse initialization ([16]). In this paper, we will use the approximate diffuse initialization, assuming a zero mean and a very large variance of the state ($\sigma_e^2 \times 10^7$).

This study examines two distinct situations: one characterized by non-contaminated data, i.e., the clean data where $e_t \sim N(0, \sigma_e^2)$; $\varepsilon_t \sim N(0, \sigma_\varepsilon^2)$, and the other involving data that has been contaminated at a rate of $p = 0.05$, i.e., $e_t \sim (1-p)N(0, \sigma_e^2) + pN(10\sigma_e, \sigma_e^2)$; $\varepsilon_t \sim N(0, \sigma_\varepsilon^2)$.

For each of the scenarios, the simulation design was formulated with a sample sizes of $n = (50{,}500)$, and σ_ε^2 and σ_e^2 (0.10, 1.00, 0.05). For each parameter combination, 1000 replicates with valid estimates were considered, i.e., $\sigma_\varepsilon > 0$, and $\sigma_e > 0$; It was considered as convergence criteria $\|\hat{\Theta}_{ML}^{(j)} - \hat{\Theta}_{ML}^{(j-1)}\| < 10^{-4}$ or until $j = 100$. To initialize the Kalman filter, $\mu_1 = 0$ and $P_1 = \sigma_e^2 \times 10^7$ was taken.

To evaluate the quality of the parameter estimates and the k-steps ahead forecasts, it was considered that

- $\text{RMSE}(\Theta) = \sqrt{\frac{1}{n}\sum_{i=1}^{n}\left(\Theta_i - \widehat{\Theta}_i\right)^2}$;
- $\text{MAE}(\Theta) = \frac{1}{n}\sum_{i=1}^{n}\left|\Theta_i - \widehat{\Theta}_i\right|$.

To evaluate the rate of true outliers detected, two rates were used rate $1 = \text{A}/\text{B}$; rate $2 = \text{A}/\text{C}$, where A is the number of true outliers detected, B is the total number of outliers detected by the method (total number of true and false outliers), and C is the total number of true outliers.

3. Results

In this section, the results obtained from the proposed methodologies are presented. The results of the simulation study are represented in the first subsection. In the second subsection, the application of outlier detection and treatment methodologies are demonstrated via three illustrative examples.

3.1. Simulation Results

Tables 1 and 2 show the RMSE and MAE of the local level model parameters and the one-step ahead forecasts for sample sizes $n = 50$ and $n = 500$ for the simulation study, respectively. In most scenarios, the methodologies improved the accuracy of model parameters and one-step ahead forecasts. However, this improvement was minimal. In fact, there are scenarios where the RMSE and MAE evaluation measures are lower in the non-treated case compared to when outliers are treated; for example, for the scenario $n = 500$, $\sigma_\varepsilon^2 = 0.10$ and $\sigma_e^2 = 0.05$. In particular, LI performed least favourably in comparison to RKF and naKF, especially to estimate the variance of the observation error σ_e^2. For example, for $n = 500$, $\sigma_\varepsilon^2 = 0.10$ and $\sigma_e^2 = 1.00$, in the case of treating outliers by LI, the RMSE of σ_e^2 was 2.0559, while for RKF it was 0.2428 and for naKF it was 0.1168. Overall, it can also be seen that naKF was the method that showed the better performance to improve the accuracy of the parameters and one-step ahead forecasts, especially for $n = 500$. The proposed methodologies had problems in improving the accuracy of the estimates of the level variance σ_ε^2. Finally, regarding the detection of outliers, it is clearly seen the advantage of identifying outliers over standardized residuals, whose means of rate 1 and 2 were higher.

Table 1. Root-mean-square error (RMSE), mean absolute error (MAE), rate 1, and rate 2 of Θ with 1000 simulations of non-stationary time series of sample sizes $n = 50$, considering Gaussian errors (NC = non-contaminated; C = contaminated; RKF = robust Kalman filter; naKF = Kalman filter for time series with missing values).

Parameters			RMSE			MAE			Outlier	Mean	Mean
σ_ε^2	σ_e^2		σ_ε^2	σ_e^2	$\hat{Y}_{t\|t-1}$ vs. Y_t	σ_ε^2	σ_e^2	$\hat{Y}_{t\|t-1}$ vs. Y_t	Detection	Rate 1	Rate 2
		NC	**0.0416**	**0.0276**	**0.4271**	**0.0335**	**0.0217**	**0.3399**		-	-
		C	0.0621	0.2614	0.5243	0.0475	0.2214	0.4033		-	-
0.10	0.05	LI	0.0584	0.1772	0.4910	0.0438	0.1286	0.3781	Time series	84%	42%
		RKF	0.0665	0.0910	0.4910	0.0456	0.0718	0.3781	Standardized residuals	74%	88%
		naKF	0.0536	0.0556	0.4667	0.0393	0.0337	0.3607			
		NC	**0.3114**	**0.1453**	**1.0734**	**0.2488**	**0.1088**	**0.8539**		-	-
		C	0.4638	0.6275	1.2216	0.3644	0.4951	0.9507		-	-
1.00	0.10	LI	0.4255	0.5723	1.2127	0.3432	0.4499	0.9421	Time series	45%	8%
		RKF	0.4216	0.4347	1.2048	0.3384	0.3422	0.9387	Standardized residuals	61%	42%
		naKF	0.4285	0.3821	1.2210	0.3422	0.2706	0.9383			
		NC	**0.0840**	**0.2456**	1.1675	**0.0618**	**0.1977**	0.9326		-	-
		C	14.5332	468.2479	1.4690	1.3638	77.8606	1.1298		-	-
0.10	1.00	LI	0.1025	0.3266	**1.1653**	0.0719	0.2373	**0.9250**	Time series	91%	99%
		RKF	0.3768	0.5958	1.2860	0.1245	0.3587	0.9876	Standardized residuals	78%	98%
		naKF	0.4510	0.3155	1.2844	0.1582	0.2525	0.9620			
		NC	**0.0275**	**0.0329**	**0.4413**	**0.0212**	**0.0260**	**0.3517**		-	-
		C	0.0564	0.4242	0.5416	0.0333	0.3516	0.4180		-	-
0.05	0.10	LI	0.0343	0.1501	0.4663	0.0237	0.0830	0.3652	Time series	91%	83%
		RKF	0.0586	0.0710	0.4914	0.0327	0.0557	0.3798	Standardized residuals	75%	97%
		naKF	0.0476	0.0391	0.4714	0.0279	0.0294	0.3635			

3.2. Illustrative Examples

In this subsection, a comparative analysis of the proposed outlier detection and treatment methods using the local level model is presented based on three illustrative examples. The aim is to evaluate the performance of the methodologies from a practical point of view, in terms of outlier detection and treatment and validation of the assumptions (normality and independence of residuals). The three time series that present outliers and are used for illustrative purposes are the following:

- TS1: Number of earthquakes per year of magnitude 7.0 or greater, between 1900 and 1998 (Figure 1);
- TS2: Kiewa River at Kiewa, Victoria, Australia, between 1885 and 1954 (Figure 2);
- TS3: Tree: Beyond Burn, Australia. Pencil Pine, between 1028 and 1975 (Figure 3).

The data is available on GitHub (https://github.com/FinYang/tsdl (accessed on 27 June 2023)) in the Time Series Data Library (TSDL), created by Professor Rob Hyndman.

The data was divided into a training sample (80%) and a test sample (20%). TS1 presents one outlier in the training sample corresponding to the year 1943; TS2 presents one outlier in the training sample (1916) and one in the test sample (1955). TS3 presents 18 outliers in the training sample (16 outliers before 1335 and two outliers corresponding to the years 1770 and 1777, respectively) and three outliers in the test sample, namely 1972, 1973 and 1975.

The results of the local level model fit to the three time series are shown in Table 3.

Table 2. Root-mean-square error (RMSE), mean absolute error (MAE), rate 1, and rate 2 of Θ with 1000 simulations of non-stationary time series of sample sizes $n = 500$, considering Gaussian errors (NC = non-contaminated; C = contaminated; RKF = robust Kalman filter; naKF = Kalman filter for time series with missing values).

Parameters			**RMSE**			**MAE**			**Outlier**	**Mean**	**Mean**
σ_ε^2	σ_e^2		σ_ε^2	σ_e^2	$\hat{Y}_{t\|t-1}$ **vs.** Y_t	σ_ε^2	σ_e^2	$\hat{Y}_{t\|t-1}$ **vs.** Y_t	**Detection**	**Rate 1**	**Rate 2**
		NC	**0.0138**	**0.0086**	**0.4315**	**0.0109**	**0.0068**	**0.3443**		-	-
		C	0.0170	0.2228	0.5303	0.0137	0.2187	0.4115		-	-
0.10	0.05	LI	0.0184	0.2156	0.5561	0.0147	0.2112	0.4193	Time series	52%	4%
		RKF	0.0189	0.0696	0.4913	0.0146	0.0684	0.3822	Standardized residuals	77%	91%
		naKF	0.0181	0.0133	0.4656	0.0137	0.0103	0.3613			
		NC	**0.1156**	**0.0524**	**1.0891**	**0.0934**	**0.0419**	**0.8685**		-	-
		C	0.1376	0.4955	1.2374	0.1112	0.4775	0.9679		-	-
1.00	0.10	LI	0.1454	0.4962	1.3117	0.1165	0.4788	0.9915	Time series	19%	1%
		RKF	0.1366	0.3261	1.2226	0.1102	0.3114	0.9550	Standardized residuals	65%	41%
		naKF	0.1643	0.2065	1.2561	0.1272	0.1803	0.9634			
		NC	**0.0235**	**0.0771**	**1.1685**	**0.0188**	**0.0610**	**0.9324**		-	-
		C	0.0351	4.7013	1.4334	0.0275	4.6231	1.1320		-	-
0.10	1.00	LI	0.0341	2.0559	1.2754	0.0242	1.3978	1.0019	Time series	94%	68%
		RKF	0.0299	0.2428	1.2191	0.0227	0.2255	0.9664	Standardized residuals	89%	100%
		naKF	0.0423	0.1168	1.1950	0.0254	0.0976	0.9436			
		NC	**0.0086**	**0.0100**	**0.4466**	**0.0068**	**0.0079**	**0.3561**		-	-
		C	0.0125	0.4614	0.5647	0.0098	0.4517	0.4417		-	-
0.05	0.10	LI	0.0119	0.3605	0.5628	0.0094	0.3348	0.4290	Time series	81%	23%
		RKF	0.0116	0.0722	0.4893	0.0088	0.0702	0.3854	Standardized residuals	84%	99%
		naKF	0.0104	0.0129	0.4617	0.0077	0.0106	0.3644			

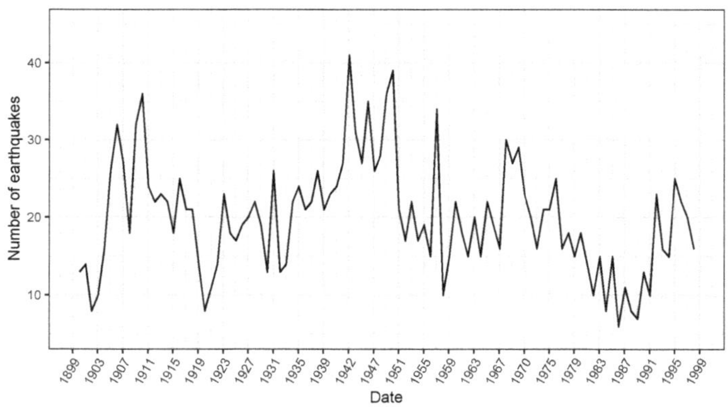

Figure 1. Number of earthquakes per year of magnitude 7.0 or greater, between 1900 and 1998 (TS1).

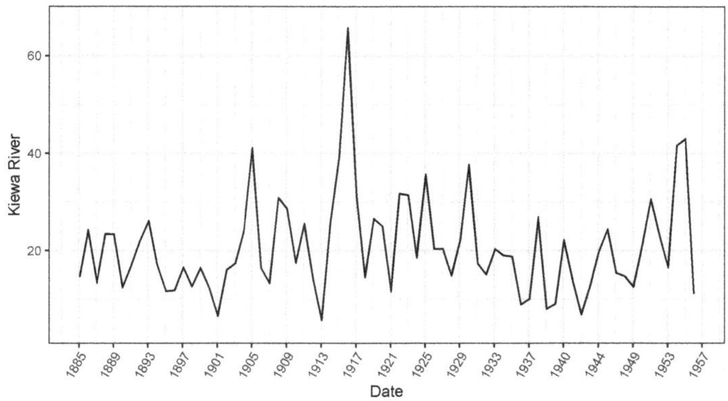

Figure 2. Kiewa River at Kiewa, Victoria, Australia, between 1885 and 1954 (TS2).

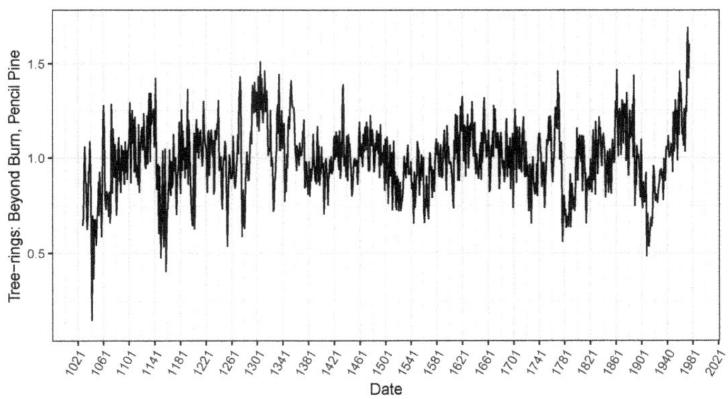

Figure 3. Tree: Beyond Burn, Australia. Pencil Pine, between 1028 and 1975 (TS3).

Table 3. Parameter estimates and respective standard errors of the non-stationary state-space model (local level model); LI—linear interpolation; RKF—robustified Kalman filter; naKF—Kalman filter for time series with missing values.

		σ_ε		σ_e		**log** L
		Estimate	**(SE)**	**Estimate**	**(SE)**	
TS1	Non-treated	2.7103	(0.6932)	4.8341	(0.5760)	−192.8515
	LI	2.6438	(0.6735)	4.6330	(0.5578)	−190.1958
	RKF	2.9174	(0.6983)	4.0890	(0.5653)	−185.7237
	naKF	3.0671	(0.7237)	3.8387	(0.5844)	−183.6041
TS2	Non-treated	1.6446	(0.8822)	9.3662	(0.9774)	−170.7793
	LI	1.2913	(0.7006)	7.8502	(0.8092)	−161.2743
	RKF	1.1704	(0.6859)	7.7522	(0.7959)	−160.3136
	naKF	1.0999	(0.6905)	7.7692	(0.7967)	−158.2455
TS3	Non-treated	0.0623	(0.0058)	0.1054	(0.0046)	1096.8770
	LI	0.0597	(0.0057)	0.0971	(0.0045)	1149.8800
	RKF	0.0614	(0.0055)	0.1000	(0.0044)	1129.9350
	naKF	0.0601	(0.0055)	0.1020	(0.0044)	1124.1500

After fitting the model to the non-treated data, outliers were detected in the standardized residuals, and these outliers were treated in the two iterative methods, RKF and naKF.

In TS1, two outliers were detected (1943 and 1957). In example TS2, the detected outlier initially remained (1916). Finally, in TS3, where eighteen outliers were initially detected, after the adjustment the residuals showed eight outliers, of which three (1042, 1158 and 1777) were initially detected in the time series.

Table 4 shows the observed evaluation measures and predicted values in the test sample. This table highlights the lowest RMSE and MAE values, with the naKF method performing best. However, the difference between these values is minimal, especially in the case of TS3; therefore, these results are in line with those obtained in the simulation study.

Table 4. Root-mean-square error (RMSE) and mean absolute error (MAE) between the observed and forecasted values via the local level model in the test sample.

		Non-Treated		LI		RKF		naKF	
		RMSE	MAE	RMSE	MAE	RMSE	MAE	RMSE	MAE
TS1	Y_t vs. $\widehat{Y}_{t+k\|t}$	7.0245	6.0496	7.0087	6.0353	6.8205	5.8609	**6.7342**	**5.7788**
	Percentage reduction	-	-	0.22%	4.14%	2.90%	3.12%	**4.13%**	**4.48%**
TS2	Y_t vs. $\widehat{Y}_{t+k\|t}$	11.4091	8.1459	11.3624	8.1456	11.2833	**8.1455**	**11.2249**	**8.1455**
	Percentage reduction	-	-	0.41%	0.004%	1.10%	**0.01%**	**1.61%**	**0.01%**
TS3	Y_t vs. $\widehat{Y}_{t+k\|t}$	0.3759	0.3231	0.3757	0.3229	0.3756	0.3228	**0.3742**	**0.3213**
	Percentage reduction	-	-	0.05%	0.06%	0.08%	0.09%	**0.45%**	**0.56%**

Figures 4–6 show TS1, TS2 and TS3 in black, respectively, the forecasts in red, and the 95% prediction intervals using naKF for the treatment of outliers. The amplitude of the prediction intervals for TS1 (Figure 4) and TS3 (Figure 6) show a considerable increase over time, whereas for TS2 (Figure 5) this increase is minimal, and the interval does not cover all the observations in the test sample.

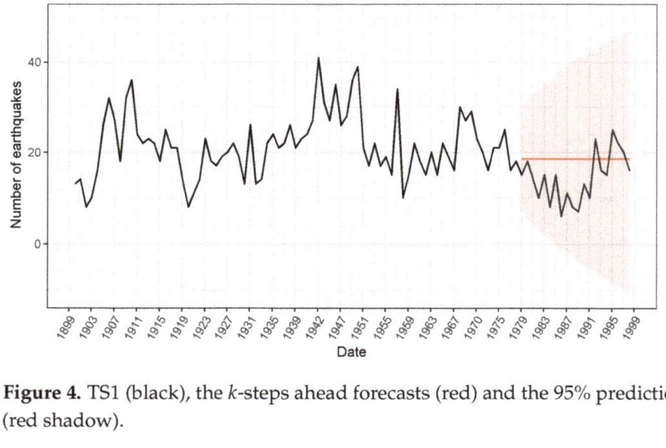

Figure 4. TS1 (black), the k-steps ahead forecasts (red) and the 95% prediction intervals using naKF (red shadow).

Regarding the analysis of the model assumptions, the residuals should behave similarly to white noise. Normality was verified for all models and for all time series: Kolmogorov–Smirnov p values between 0.398 (RKF and TS2) and 0.967 (RKF and TS1). The models for TS1 and TS2 verified the independence assumption: p values ranging between 0.314 (non-treated and TS1) and 0.574 (NA and TS1) from the Ljung–Box test. However, this assumption was not verified for TS3 (all p values of the Ljung–Box test were less than 0.003).

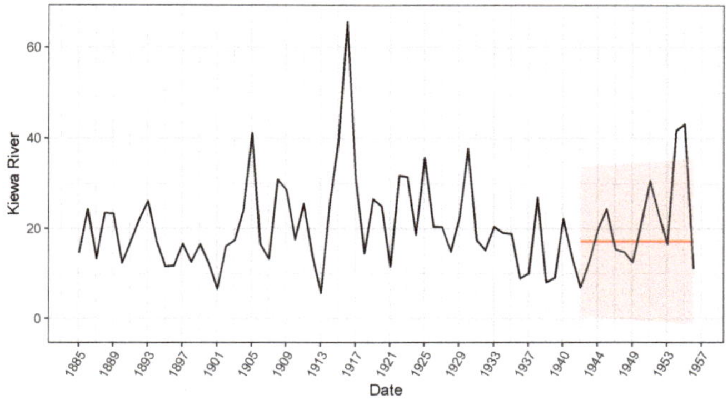

Figure 5. TS2 (black), the k-steps ahead forecasts (red) and 95% prediction intervals using naKF (red shadow).

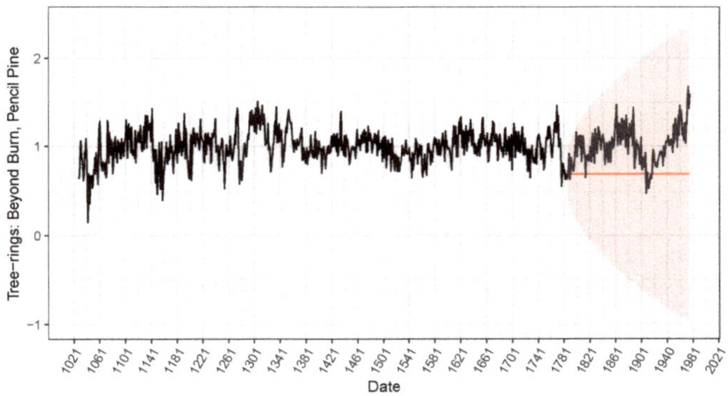

Figure 6. TS3 (black), the k-steps ahead forecasts (red) and 95% prediction intervals using naKF (red shadow).

4. Discussion

In this work, three methods for detecting and treating outliers in time series were proposed. This study highlighted the problem of contaminated non-stationary time series from a state-space modelling perspective. To study the impact of outliers on parameter estimates and the observation forecasts, and to make a comparative analysis of the proposed methods, a simulation study was conducted with sample sizes of 50 and 500 with various combinations of parameters, generated using a non-stationary local level model. The data were contaminated at a 5% error rate of the observations. It was found that the proposed methods overall improved the accuracy of the parameters and forecasts; however, this improvement was minimal compared to the contaminated data. The treatment of outliers by naKF and RKF were found to be the most favourable, therefore highlighting the performance of naKF. LI was overall performed the worse. These proposed methodologies were applied to three real time series, where the same conclusion was drawn. In other words, in view of the study's results, the state-space models are generally sufficiently robust, given that they are able to handle non-stationary time series and outliers in a satisfactory way.

Author Contributions: F.C.P., A.M.G. and M.C. contributed to this work. All authors have read and agreed to the published version of the manuscript.

Funding: This research received no external funding.

Institutional Review Board Statement: Not applicable.

Informed Consent Statement: Not applicable.

Data Availability Statement: The data that support the findings of this study are available in GitHub (https://github.com/FinYang/tsdl (accessed on 27 June 2023)).

Acknowledgments: F. Catarina Pereira was funded by national funds through FCT (Fundação para a Ciência e a Tecnologia) through the individual PhD research grant UI/BD/150967/2021 of CMAT-UM. A. Manuela Gonçalves was partially financed by Portuguese Funds through FCT within the Projects UIDB/00013/2020 and UIDP/00013/2020 of CMAT-UM. Marco Costa was partially supported by The Center for Research and Development in Mathematics and Applications (CIDMA-UA) through the Portuguese Foundation for Science and Technology—FCT, references UIDB/04106/2020 and UIDP/04106/2020.

Conflicts of Interest: The authors declare no conflict of interest.

References

1. Triantafyllopoulos, K. The State Space Model in Finance. In *Bayesian Inference of State Space Models*; Springer Texts in Statistics; Springer: Cham, Switzerland, 2021. [CrossRef]
2. Auger-Methe, M.; Newman, K.; Cole, D.; Empacher, F.; Gryba, R.; King, A.A.; Leos-Barajas, V.; Flemming, J.M.; Nielsen, A.; Petris, G.; et al.A guide to state–space modeling of ecological time series. *Ecol. Monogr.* **2021**, *91*, 1–38. [CrossRef]
3. Wu, H.; Matteson, D.; Wells, M. Interpretable Latent Variables in Deep State Space Models. *arXiv* **2022**, arXiv:2203.02057
4. Matsuura, K. Time Series Data Analysis with State Space Model. In *Bayesian Statistical Modeling with Stan, R, and Python*; Springer: Singapore, 2022. [CrossRef]
5. Monteiro, M.; Costa, M. Change Point Detection by State Space Modeling of Long-Term Air Temperature Series in Europe. *Stats* **2023**, *6*, 113–130. [CrossRef]
6. Pereira, F.C.; Gonçalves, A.M.; Costa, M. Short-term forecast improvement of maximum temperature by state-space model approach: The study case of the TO CHAIR project. *Stoch. Environ. Res. Risk Assess.* **2023**, *37*, 219–231. [CrossRef]
7. Shumway, R.H.; Stoffer, D.S. *Time Series Analysis and its Applications: With R Examples*; Springer: New York, NY, USA, 2017.
8. Kalman, R. A New Approach to Linear Filtering and Prediction Problems. *ASME J. Basic Eng.* **1960**, *82*, 35–45. [CrossRef]
9. Harvey, A. *Forecasting, Structural Time Series Models and the Kalman Filter*; Cambridge University Press: Cambridge, UK, 1990. [CrossRef]
10. Teunissen, P.J.G.; Khodab, A.; Psychas, D. A generalized Kalman filter with its precision in recursive form when the stochastic model is misspecified. *J. Geod.* **2021**, *95*, 108. [CrossRef]
11. Huang, Y.; Zhang, Y.; Zhao, Y.; Shi, P.; Chambers, J.A. A Novel Outlier-Robust Kalman Filtering Framework Based on Statistical Similarity Measure. *IEEE Trans. Autom. Control* **2021**, *66*, 2677–2692. [CrossRef]
12. Auger-Méthé, M.; Field, C.; Albertsen, C.M.; Derocher, A.E.; Lewis, M.A.; Jonsen, I.D.; Flemming, J.M. State-space models' dirty little secrets: Even simple linear Gaussian models can have estimation problems. *Sci. Rep.* **2016**, *6*, 26677. [CrossRef] [PubMed]
13. Hyndman, R.J.; Athanasopoulos, G. *Forecasting: Principles and Practice*, 2nd ed.; OTexts: Melbourne, Australia, 2018.
14. Cipra, T.; Romera, R. Kalman filter with outliers and missing observations. *Test* **1997**, *6*, 379–395. [CrossRef]
15. Crevits, R.; Croux, C. Robust estimation of linear state space models. *Commun. Stat.- Simul. Comput.* **2019**, *48*, 1694–1705. [CrossRef]
16. Durbin, J.; Koopman, S.J. *Time Series Analysis by State Space Methods*, 2nd ed.; Oxford Statistical Science Series; Oxford University Press: Oxford, UK, 2013. [CrossRef]

engineering proceedings

MDPI

Proceeding Paper

Gaining Flexibility by Rethinking Offshore Outsourcing for Managing Complexity and Disruption †

Michela Pellicelli

Department of Economics and Management, University of Pavia, 27100 Pavia, Italy; michela.pellicelli@unipv.it

† Presented at the 9th International Conference on Time Series and Forecasting, Gran Canaria, Spain, 12–14 July 2023.

Abstract: The challenges confronting management in making decisions on offshore outsourcing have inevitably changed dramatically over the last decade. Offshore outsourcing today is characterized by a higher level of complexity and disruption brought about mainly factors of change, as: (1) the new waves of globalization; (2) the rapidly shifting conditions in the marketplace, which have made offshoring a vital part of global strategies, leading the way to new business models; (3) the increasing variety of models in the global supply chain; (4) the inroads of previous providers that have emerged as world-class competitors; and (5) the irresistible march of technological disruption. Threatening events, as terrorist attacks and wars, had disrupted the world economy since the 2000s. In these cases, supply chains were disrupted although soon restored. After these critical events, globalization was severely weakened by the arrival of the COVID-19 pandemic, on the threshold of 2020, and Russia's invasion of Ukraine dealt a blow to global supply chains, by 2022. Due to these complexities, the trend to outsource production to other countries might increasingly push companies to transform themselves into virtual organizations. Outsourcing strategies could therefore evolve moving towards a digital era.

Keywords: digital technologies; flexibility; innovation; offshore outsourcing; supply chain

Citation: Pellicelli, M. Gaining Flexibility by Rethinking Offshore Outsourcing for Managing Complexity and Disruption. *Eng. Proc.* **2023**, *39*, 37. https://doi.org/10.3390/engproc2023039037

Academic Editors: Ignacio Rojas, Hector Pomares, Luis Javier Herrera, Fernando Rojas and Olga Valenzuela

Published: 3 July 2023

1. Introduction

In the globalized economy, multinational companies have led to the emergence in many countries of local enterprises capable of producing at low costs and at an acceptable level of quality. They are entrusted with all kinds of production and processing by an increasing number of companies that, with these decisions, not only reduce their production costs but also streamline their organizational structure.

In recent years, outsourcing strategies have undergone a profound evolution from simple forms of production contracts entrusted to third parties to arrangements encompassing more strategically relevant functions: from technological innovation to logistics, from customer relations to after-sales services.

The decision to outsource is not limited to production of modest technological and marginal content, but is increasingly extended to activities that-requiring 'core competencies', or being part of the 'core business' were considered inseparable from the company and not outsourceable [1–3].

In the course of this evolution, outsourcing has shifted the center of gravity from a 'tactical' choice-tending to have production outsourced in order to reduce production costs-to a 'strategic' choice, to redefine the very boundaries of the enterprise.

Quinn and Hilmer [1] have stated that management have to consider two strategic processes: (1) concentrating the own resources on the distinct and inimitable core capacities that generate value for the customer; (2) entrusting to third parties the activities considered non-core and for which the company has no special competencies. The concept of core competence was introduced into the economic literature by Prahalad and Hamel in 1990 [2], giving rise to many other definitions.

The initial interest in outsourcing rapidly gave way to serious concerns. The threat that cheap jobs in China, India and other countries pose to employment in more advanced economies has been a central theme in economic policy choices and election campaigns in recent years [4].

Larger supplier companies have moved out of the borders of their countries and into those of their customers in order to provide the best mix of cost, risk and quality. While companies from Western countries moved their production to India, Indian companies invested in Western countries.

Proximity, both geographically and culturally, is important to build customer relationships, to provide certain types of services and to give guarantees regarding data security and confidentiality.

The outsourcing sector is also gradually moving towards maturity. This suggests that the flow of jobs from rich to poor countries may be slowing down [4].

After a strong development in the 1990s and early 2000s, outsourcing has started to show signs of slowing down (measured in terms of number of contracts). As the development trend slowed down, the margins for supplier companies also began to fall.

Nowadays, global markets are subject to intense phenomena, which are sometimes even difficult to predict. Offshore outsourcing is characterized by a higher level of complexity and disruption caused mainly by change factors such as: (1) the new waves of globalization; (2) the rapidly shifting conditions in the marketplace, which have made offshoring a vital part of global strategies, leading the way to new business models; (3) the increasing variety of models in the global supply chain; (4) the inroads of previous providers that have emerged as world-class competitors; and (5) the irresistible march of technological disruption.

Threatening events, as terrorist attacks and wars, had disrupted the world economy since the 2000s. In these cases, supply chains were disrupted although soon restored. After these critical events, globalization was severely weakened by the arrival of the COVID-19 pandemic, on the threshold of 2020, and Russia's invasion of Ukraine dealt a blow to global supply chains, by 2022.

2. The Main Drivers of Change

Offshore outsourcing today is characterized by a higher level of complexity and disruption brought about mainly factors of change, as: (1) the new waves of globalization; (2) the rapidly shifting conditions in the marketplace, which have made offshoring a vital part of global strategies, leading the way to new business models; (3) the increasing variety of models in the global supply chain; (4) the inroads of previous providers that have emerged as world-class competitors; and (5) the irresistible march of technological disruption.

(1) Since their inception, the forces of globalization have pushed companies to develop new capabilities to enter international markets. The rules for competition have changed. Competing globally requires supply chains that seek the most efficient sources of supply, which means developing more complex supply chains, adapting to larger operational scales, and establishing relationships with more providers covering more functions and processes, and in more countries.

In recent times, an abrupt change in the degree of internalization of production and in the scope of offshore transactions has led to what we define as "the new waves of globalization". Leading Western economies are seeking alternative models. Three main consequences for offshore outsourcing have resulted from these changes: (1) an increase in the number of "participants" in international trade, and thus in the demand for products and services and in their supply; i.e., low-wage countries are now able to produce high-quality manufactured goods; (2) a revival of protectionism in Western countries; and (3) an "unbeatable" uncertainty that is forcing outsourcers to frequently revise their offshoring strategies. These and other consequences force firms to often restructure the supply chain.

(2) The recent wave of changes in the economic, technological and regulatory environment worldwide, as well as shifts in corporate thinking and digital disruption, have

made offshoring a vital part of any search for sustainable competitive advantage, and consequently for global strategies. Being competitive on a global scale has become increasingly important in most industries. Companies are exploring alternative ways that can raise the level of efficiency in the use of resources at all stages of the value chain, from R&D to production, logistics, sales and marketing. This move is leading the way to new business models, on the one hand by reducing costs, thereby improving profitability, and on the other by reducing the investment needs of outsourcers to increase shareholder value. Constantly reviewing core competencies is the basic premise for defining activities that a firm should retain for competitive advantage. The "core" is a shifting target, as competition changes and new technologies emerge, new activities become the core, and other activities that do not add competitive value are shifted to other organizations that can perform them better, faster, cheaper, or at least as efficiently.

(3) The complexity of business functions being moved offshore has increased the range of possible relations between the lead firms and their suppliers. As a consequence, organizations using the offshore approach have developed a variety of models to manage the relationships with their providers and to build international networks of production. We distinguish between the type of outsourcing agreement and the way in which the firm manages the supply chain by means of the distribution of value added and the sharing of technology along the various phases. As for the type of agreement with foreign vendors or with allied partners, solutions are far from being homogeneous. The characteristics of providers exert a profound effect as well: size, experience, skills, financial resources and organization. By managing the share of value added and of technology assigned to the various elements in the supply chain, outsourcers have strengthened their power asymmetry, thereby achieving the opportunity to pursue different strategies in their relationships with providers.

(4) Over the past two decades, some 'old' suppliers have become leaders in their industry. They are now able to compete with traditional multinationals on the same level and they have the advantage to produce in low-cost countries. Their entrance into the outsourcing arena has forced many players to restructure the organization of the global supply chain.

(5) In recent years, technology has profoundly impacted offshoring decisions as well as their implementation. There are no obstacles to the disruption of the previous equilibria. Modern methods have continued to uproot traditional practices, with many supply chains facing challenging times, necessitating in many instances a reorganization of the supplier network. Digital technologies are reshaping supply chains. The Internet of Things is set to transform the supply chain of many industries, from transport to manufacturing, banking to retailing. Acquiring digital skills has become a must for those firms that want to be competitive and in the vanguard in the near future. In this challenging and crowded market, new competitors are emerging all the time. New technologies can suddenly make room for new competitors from other industries, competitors who are often cash-rich and looking for new investments (i.e., Apple and Google in driverless cars). In such a context, flexibility is more important than ever for companies. An important reason for lead firms to establish a global supply chain is the flexibility this provides in allowing for rapid adjustments to changes in market demand and, to a greater extent, a shift in the risk to supplier firms from possible declines in demand and excessive inventory.

3. The Recent Tensions on Globalization and the Effects of COVID-19

The supply chain has been at the center of globalization over the past two decades. However, there has been no shortage of stresses on globalization, such as the financial crisis of 2007–2009, the slowdown in global economic growth, uncertainties about the future, Brexit, and the trade wars between the US and China.

On the threshold of 2020, with the arrival of the COVID-19 pandemic, globalization has been considerably weakened. Previously, many threatening events had occurred, from the 9/11 terrorist attack to the SARS-CoV-1 virus and the tsunami in Japan. However, the consequences of these events had only affected the affected geographical areas to a limited extent. Supply chains had been disrupted, but nevertheless restored within a short time.

The consequences caused by COVID-19 were traumatic and in this case the shock affected the entire global supply chain. Major economies went into recession within a few months and the effects are likely to remain in the economy and society for a long time.

The pandemic caused the largest and fastest decline in international flows in modern history [5], even giving some governments the excuse to withdraw from multilateral and free trade agreements.

The management strategies of companies with supply chains located in different countries, and far away from the market of origin, have been very complex and debated. The pandemic has reduced demand for many products in end markets. The need to provide flexibility and resilience to supply chains has forced companies to sacrifice short-term profit margins [6].

However, it is widely believed that globalization can contribute strongly to growth. Indeed, countries with higher scores on the DHL Global Connectedness Index tend to experience faster economic growth.

4. The Effects of Ukraine War on Globalization

Even after the COVID-19 pandemic, big companies showed less interest in true globalization. Non-economic facts, such as activist pressure against sourcing from countries known for human rights violations, further contributed to this trend. After two years of pandemic, Western governments were grappling with rising energy prices that threatened to slow economic recovery.

Russia's invasion of Ukraine in late February 2022 began to disrupt supply chains and the export of critical raw materials, from food and mining to car and aircraft production.

Energy prices increased precipitously and thereby inflation as well. "From crude oil to diesel to natural gas, the fossil fuels that power the global economy are trading at or towards record levels, threatening to redraw geopolitical relations between producers and consumers, drive up inflation and potentially even disrupt the fight against climate change" [7].

The war in Ukraine seems even to have debilitated the economy's ability to overcome its difficulties and regain its balance towards development, which it had previously demonstrated.

The globalization of supply chains has been weakened by the collision between the world's largest economies, on the one hand, and their major raw material suppliers, on the other.

It is widely deemed that the long-term impact will be to accelerate the division of the world into economic blocs and Russia will move towards trade and financial ties with China. "The invasion of Ukraine might not cause a global economic crisis today but it will change how the world economy operates for decades to come" [8].

5. The Evolution of Offshore Outsourcing towards Greater Flexibility for Managing Complexities and Disruptions

Globalization has driven the dispersion of markets. The reaction of companies has been to distribute core assets and create clusters of skills around the world. In this perspective, *"the innovative firm of the future will restructure each individual cell, the basic building block of the firm consisting of an occupation devoted to a product, and redeploy and relocate them globally, where it is most advantageous"* [9].

As early as 2000, Quinn remarked that: "Strategically outsourcing innovation–using the most current technologies an management techniques–can put a company in a sustainable leadership position [10].

It is widely recognized that many companies have difficulty to inject flexibility into the global value chain in the short term because they are anchored to old tools and unable to adapt to a changing environment. "The growing trend towards globalization and outsourcing is in fact leading many industries to outsource important parts of their business to suppliers often located in developing countries" [11].

Technological progress in the last two decades has strongly impacted offshoring decisions and implementation. As novel methods continue to overturn conventional practices, supply chains are being challenged and companies are forced to overhaul the existing supplier networks. In addition, new market segments may open up to competitors, who are often cash rich and in search of new investments, from sectors far removed from those the firms had to deal with in the past. These changes have led to the re-emergence of reshoring and to frequent power shifts among firms in the supply chain. Most significantly, there now exists an "unbeatable" uncertainty that can be dealt with only through increased flexibility in the design and management of the supply chain. For these reasons, offshore outsourcing is transforming the functions of traditional firms into a network of competencies [12].

Some authors argue that business competition in the future will lead the traditional company towards supply chain networks [13–15] and this evolution could lead towards virtual organizations. A virtual organization is defined as: *"a set of geographically distributed, functionally and/or culturally diverse entities linked by electronic forms of communication and relying on lateral and dynamic relationships for coordination"* ([16], p. 695).

In particular, Tuma ([17], p. 642) defines a virtual enterprise as a "production system with mainly independent enterprises as single elements, which can be dynamically insourced or outsourced depending on the market demands".

"The recent proliferation of cloud-based technologies will further boost *'virtualization' of firms. The progression of the corporation from a national to a multinational entity will evolve to the next generation, possibly into a seamless web of best-sourced people and processes around the globe bound together by a common vision and mission and managed through collaborative tools leading to a plethora of innovations"* ([18], p. 165).

The evolution of supply chains, prompted by the Fourth Industrial Revolution, allows an increasing use of digital technologies. Supply chain management has made the most from going digital.

As defined by Kinnett ([19], p. 5), digital supply chain is *"an intelligent, value-driven network that leverages new approaches with technology and analytics to create new forms of revenue and business value, through a centric platform that captures and maximizes the utilization of real-time information emerging from a variety of sources"*.

Digital supply chain can *"lower operational costs by more than 30 percent, reduce lost sales opportunities by more than 60 percent, and even reduce inventory requirements by more than 70 percent, all while making companies faster, more agile, granular, accurate, and efficient"* ([20], p. 1).

"The effectiveness of the development of modern business entities of various scales and profiles in a dynamically developing digital economy largely depends on the quality of the use of virtual outsourcing opportunities" ([21], p. 63).

In this perspective, the trend to outsource most of the production and economic activities and processes could transform companies into virtual organizations, improving the formation of flexible networks and agile organizational structures. This evolution will even push companies towards open innovation [22] and greater flexibility.

6. Conclusions

Nowadays, global markets are subject to intense phenomena, which are sometimes even difficult to predict. Management is challenged with a higher level of complexity and disruptions brought about in particular by the new waves of globalization and the irresistible march of technological changes. In addition, globalization has been severely weakened by the arrival of the COVID-19 pandemic, on the threshold of 2020, and Russia's invasion of Ukraine dealt a blow to global supply chains, by 2022.

There is now 'unbeatable' uncertainty that can only be addressed through greater flexibility in supply chain design and management. For these reasons, offshore outsourcing is transforming the functions of traditional companies into a network of competencies.

The evolution of supply chains, prompted by the Fourth Industrial Revolution, allows an increasing use of digital technologies. In this perspective, the trend to outsource most of the production and economic activities and processes could transform companies into virtual organizations, improving the formation of flexible networks and agile organizational structures.

Outsourcing strategies could therefore evolve by exploiting digital technologies, moving towards a digital era.

Funding: This research received no external funding.

Institutional Review Board Statement: Not applicable.

Informed Consent Statement: Not applicable.

Data Availability Statement: Not applicable.

Conflicts of Interest: The authors declare no conflict of interest.

References

1. Quinn, J.B.; Hilmer, F.G. Strategic outsourcing. *Sloan Manag. Rev.* **1994**, *35*, 43–55.
2. Hamel, G.; Prahalad, C.K. Strategic intent. *Harv. Bus. Rev.* **1990**, *68*, 18–38.
3. Thompson, A.; Strickland, A. *Strategic Management*; John Wiley & Sons: New York, NY, USA, 2005.
4. Pellicelli, M. *L'outsourcing e L'offshoring Nell'economia Dell'impresa*; Giappichelli: Torino, Italy, 2009.
5. Altman, M. Smart thinking, lockdown and Covid-19: Implications for public policy. *J. Behav. Econ. Policy* **2020**, *4*, 23–33.
6. Pellicelli, M. *The Digital Transformation on Supply Chain Management*; Elsevier: Amsterdam, The Netherlands, 2023.
7. Fera, M.; Fruggiero, F.; Lambiase, A.; Macchiaroli, R.; Miranda, S. The role of uncertainty in supply chains under dynamic modeling. *Int. J. Ind. Eng. Comput.* **2017**, *8*, 119–140. [CrossRef]
8. Van Agtmael, A. *The Emerging Markets Century*; Simon & Schuste: London, UK, 2007.
9. Bardhan, A.D. Managing globalization of R&D: Organizing for offshoring innovation. *Hum. Syst. Manag.* **2006**, *25*, 103–114.
10. Quinn, J.B. Outsourcing innovation: The new engine of growth. *MIT Sloan Manag. Rev.* **2000**, *41*, 13.
11. Martino, G.; Fera, M.; Iannone, R.; Miranda, S. Proposal of a multi-method decision support system for the fashion retail industry. In *Workshop on Business Models and ICT Technologies for the Fashion Supply Chain*; Springer: Berlin, Germany, 2016; pp. 187–199.
12. Pellicelli, M. Gaining flexibility and innovation through offshore outsourcing. *Sustainability* **2018**, *10*, 1672. [CrossRef]
13. Wang, W.Y.; Chan, H.K. Virtual organization for supply chain integration: Two cases in the textile and fashion retailing industry. *Int. J. Prod. Econ.* **2010**, *127*, 333–342. [CrossRef]
14. De Souza, R.; Song, Z.; Liu, C. Supply chain dynamics and optimization. *Integr. Manuf. Syst.* **2000**, *11*, 348–364. [CrossRef]
15. Min, H.; Zhou, G. Supply chain modeling: Past, present, and future. *Comput. Ind. Eng.* **2002**, *43*, 231–249. [CrossRef]
16. DeSanctis, G.; Monge, P. Introduction to the special issue: Communication processes for virtual organizations. *Organ. Sci.* **1999**, *10*, 693–703. [CrossRef]
17. Tuma, A. Configuration and coordination of virtual production networks. *Int. J. Prod. Econ.* **1998**, *56–57*, 641–648. [CrossRef]
18. Pingali, S.R.; Rovenpor, J.; Shah, G. From Outsourcing to Best-Sourcing? The Global Search for Talent and Innovation. In *Human Capital and Innovation*; Palgrave Macmillan: London, UK, 2017; pp. 161–191.
19. Kinnett, J. Creating a digital supply chain: Monsanto's Journey. In Proceedings of the 7th AnnualBCTIM Industry Conference, Washington, DC, USA, 13 October 2015.
20. GEP. Impact of Industry 4.0 on Supply Chains, All You Need to Know. March 08, Digital Supply Chain Transformation, Supply Chain & Procurement. 8 March 2019. Available online: https://www.gep.com/blog/strategy/impact-of-industry-4-on-supply-chain (accessed on 31 May 2023).
21. Yeo, C.; Saboori-Deilami, V. Strategic challenges of outsourcing innovation in global market. *Asia Pac. J. Innov. Entrep.* **2017**, *11*, 5–16. [CrossRef]
22. Bogoviz, A.V.; Suglobov, A.E.; Maloletko, A.N.; Kaurova, O.V. *Cooperation and Sustainable Development*; Springer International Publishing: Berlin, Germany, 2022.

engineering proceedings

MDPI

Proceeding Paper

Modelling of Leishmaniasis Infection Dynamics: A Comparative Time Series Analysis with VAR, VECM, Generalized Linear and Markov Switching Models †

Fadoua Badaoui [1], Souad Bouhout [2], Amine Amar [3,*] and Kenza Khomsi [4]

[1] GEAS3D Laboratory, National Institute of Statistics and Applied Economics (INSEA), Rabat 10100, Morocco; fbadaoui@insea.ac.ma
[2] Direction of Epidemiology and Disease Control (DELM), Ministry of Health, Rabat 10100, Morocco; souabouhout@gmail.com
[3] School of Science and Engineering, Al Akhawyn University, Ifrane 53000, Morocco
[4] Independent Researcher, Casablanca 20230, Morocco; k.khomsi@gmail.com
* Correspondence: a.amar@aui.ma
† Presented at the 9th International Conference on Time Series and Forecasting, Gran Canaria, Spain, 12–14 July 2023.

Abstract: In this paper, we are interested in modeling the dynamics of cutaneous leishmaniasis (CL) in Errachidia province (Morocco), using epidemiologic data and the most notable climatic factors associated with leishmaniasis, namely humidity, wind speed, rainfall, and temperature. To achieve our objective, we compare the performance of three statistical models, namely the Vector Auto-Regressive (VAR) model, the Vector Error Correction model (VECM), and the Generalized Linear model (GLM), using different metrics. The modeling framework will be compared with the Markov Switching (MSM) approach.

Keywords: leishmaniasis dynamics; generalized linear model (GLM); Markov switching model (MSM); meteorological data; vector auto-regressive model (VAR); vector error correction model (VECM)

Citation: Badaoui, F.; Bouhout, S.; Amar, A.; Khomsi, K. Modelling of Leishmaniasis Infection Dynamics: A Comparative Time Series Analysis with VAR, VECM, Generalized Linear and Markov Switching Models. *Eng. Proc.* **2023**, *39*, 38. https://doi.org/10.3390/engproc2023039038

Academic Editors: Ignacio Rojas, Hector Pomares, Luis Javier Herrera, Fernando Rojas and Olga Valenzuela

Published: 3 July 2023

1. Introduction

Despite new developments in disease control and advanced treatment methods, leishmaniasis is still one of the most prevalent tropical diseases in the world. The World Health Organization (WHO) defines leishmaniasis as an infectious disease caused by protozoan parasites in the genus Leishmania. The transmission of the disease occurs through the bite of a sandfly infected with Leishmania parasites. Infection may be restricted to the skin in cutaneous leishmaniasis (CL), to the mucous membranes in mucosal leishmaniasis (MCL), or spread internally in visceral leishmaniasis (VL). This disease is in fact a vector-borne disease whose transmission is highly influenced by climatic factors, whereas the nature and magnitude differ between geographical regions. Further, it is known that spatial heterogeneity influences shifting patterns of vector parasite interactions, vector–host contact, and susceptibility of the population [1].

Many recent studies and research papers [2–4] suggest that the incidence of leishmaniasis is influenced by climatic variables. Therefore, prediction approaches are needed to achieve a better outcome in disease forecasting. In this context, the predictions made through time series analysis are extremely important in light of recent developments. This will undoubtedly help identify trends and possible disease outbreaks, which may ultimately facilitate the smooth and timely implementation of control programs through appropriate precautionary interventions.

In this paper, we compare the performance of three statistical models, namely the Vector Auto-Regressive (VAR) model, the Vector Error Correction Model (VECM), and the Generalized Linear Model (GLM), using different metrics. These models are used to measure the impact of climate change on the epidemiology of leishmaniases in Errachidia province, Morocco.

These models were selected based on a benchmarking study that showed their usefulness in explaining the dynamics in different fields. Furthermore, the time series modeling framework will be compared with the Markov Switching approach.

2. Materials and Methods

The main contribution of this study is to model cutaneous leishmaniasis (CL) dynamics in Errachidia province (Morocco) using epidemiologic data and the most notable climatic factors associated with leishmaniosis, namely humidity, wind speed, rainfall, and temperature. To achieve our objective, we use three statistical models, namely the Vector Auto-Regressive (VAR) model, the Vector Error Correction model (VECM), and the Generalized Linear Model (GLM). The modeling framework will be compared with the Markov Switching approach.

2.1. A Brief Overview of Generalized Linear Models (GLM)

Generalized Linear Models include several types of models, such as linear regression, logistic regression, Poisson regression, and Negative Binomial Regression [5]. In these models, the response variable Y_i is assumed to follow an exponential family distribution. The mean μ_i of the response variable is often assumed to be a nonlinear function of $x_i^T\beta$. The model is given by:

$$g(E(Y)) = \beta_0 + \beta_1 X_1 + \beta_2 X_2 + \cdots + \beta_n X_n + \varepsilon, \quad (1)$$

where $g(E(Y))$ specifies the link between the mean $E(Y)$ and the linear combination of predictors $X_1, X_2, \ldots, X_n$. GLMs are generally fitted using a Newton-type method, such that an iteratively re-weighted least square (IWLS) algorithm is also referred to as a Fisher Scoring algorithm [6].

Various predictor selection methods are used to compromise the stability of a final model from several nested models. Based on the significance of the predictors and their correlation values with the outcome Y_i we can reduce the number of predictors. Using forward-selection or backward-elimination variable-selection algorithms, the deviance, and the AIC criterion, we can determine the best-fitting model [7]. In order to assess the relevance of GLMs, we use the residual deviance test, defined as:

$$D = 2\{l(saturated) - l(\mathrm{M})\}, \quad (2)$$

where $l(\mathrm{M})$ is the log-likelihood of the model M, and $l(saturated)$ is the log-likelihood of a saturated model. A saturated model is where the number of parameters is equal to the number of data points. Thus, models with high likelihoods will have low deviances, and vice versa. If the model M is correct and has $p+1$ parameters, including the intercept, then the deviance will generally approach a chi-square distribution with degrees of freedom equal to $n-(p+1)$.

Poisson and negative binomial regression models (Figure 1) are part of the family of Generalized Linear Models that are commonly used in epidemiological studies [8]. These models are widely used to model disease incidence data with a non-negative integer, no upper limit, and highly skewed distribution. Otherwise, a zero-inflated Poisson (ZIP) model, a zero-inflated negative binomial (ZINB), or a negative binomial need to be used [9].

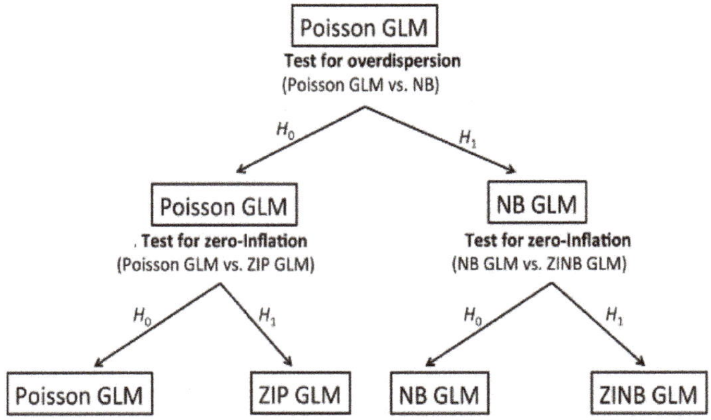

Figure 1. Checking for zero-inflation and overdispersion in the analysis of count data.

2.2. A Brief Overview of Multivariate Time Series Models

The Multivariate Time Series approach (Figure 2) is used to model and explain the interactions among a group of time series. In this framework, the strength of associations among different variables is expressed across time lags.

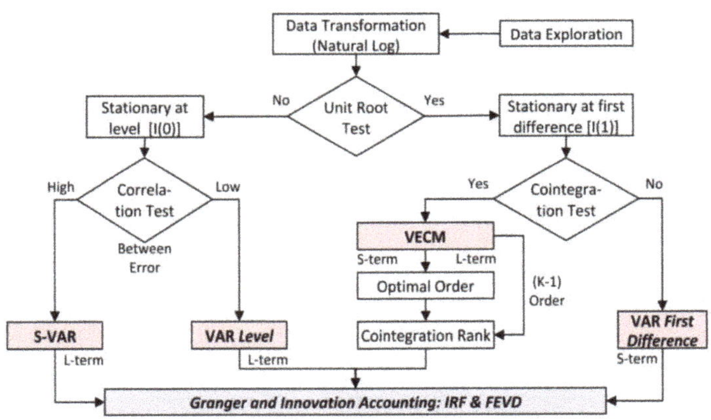

Figure 2. VAR and VECM Analysis model.

The Vector Auto-Regressive (VAR) model is one of the most commonly used techniques. It is considered as an extension of the auto-regressive model. The VAR model involves multiple independent variables and therefore has more than one equation, explaining the behavior of the relationship between endogenous variables as well as between endogenous and exogenous variables. Each equation uses as its explanatory variables the lags of all the variables and likely a deterministic trend.

If all of the original variables have unit roots and are not cointegrated, then they should be differenced and the resulting stationary variables should be used in the VAR. Let $x(t) = (x_1(t), \ldots, x_m(t))'$ be an m-dimensional stationary process admitting the following $VAR(p)$ representation:

$$x(t) = A_1 x_{t-1} + \cdots + A_p x_{t-p} + \varepsilon_t, \quad t \in \mathbb{Z} \tag{3}$$

where $A_1, \cdots, A_p$ are $(m \times m)$ coefficient matrices, p is the model order, and $\varepsilon_t = (\varepsilon_{1t}, \cdots, \varepsilon_{mt})'$ is a $(m \times 1)$ vector of white noises with $E[\varepsilon_t \varepsilon_s'] = 0$ for $t \neq s$ and $\varepsilon_t \sim N(0, \sum_\varepsilon)$. The coefficient matrices $A_1, \cdots, A_p$ describe the temporal relationships within the m time series in the system.

If the cointegration exists then a vector error correction model (VECM), which combines levels and differences, can be estimated instead of a VAR in levels. The VECM regression equation is given by:

$$\Delta y_t = \alpha_1 + \rho_1 e_1 + \sum_{i=0}^{n} \beta_i \Delta y_{t-i} + \sum_{i=0}^{n} \delta_i \Delta x_{t-i} + \sum_{i=0}^{n} \gamma_i \Delta z_{t-i} \quad (4)$$

$$\Delta x_t = \alpha_2 + \rho_2 e_2 + \sum_{i=0}^{n} \beta_i \Delta y_{t-i} + \sum_{i=0}^{n} \delta_i \Delta x_{t-i} + \sum_{i=0}^{n} \gamma_i \Delta z_{t-i} \quad (5)$$

2.3. *A Brief Overview of the Markov Switching Model (MSM)*

Although ARMA models are quite successful in numerous applications, they are unable to represent many nonlinear dynamic patterns such as asymmetry, amplitude dependence, and volatility clustering [10]. In addition, nonlinear time series models are not a panacea and have their own limitations. First, the nonlinear optimization algorithms easily get stuck at a local optimum in the parameter space. Second, most nonlinear models are designed to describe certain nonlinear patterns of data and hence may not be so flexible.

The Markov Switching model of Hamilton [11], also known as the regime-switching model, is one of the most popular nonlinear time series. This model involves multiple equations that can characterize the time series behavior in different regimes. By permitting switching between these structures, this model is able to capture more complex dynamic patterns. This model and its variants have been widely applied [12,13]. The following network mapping [14] summarized the co-occurrence of author's keywords when using the Markov Switching model (Figure 3).

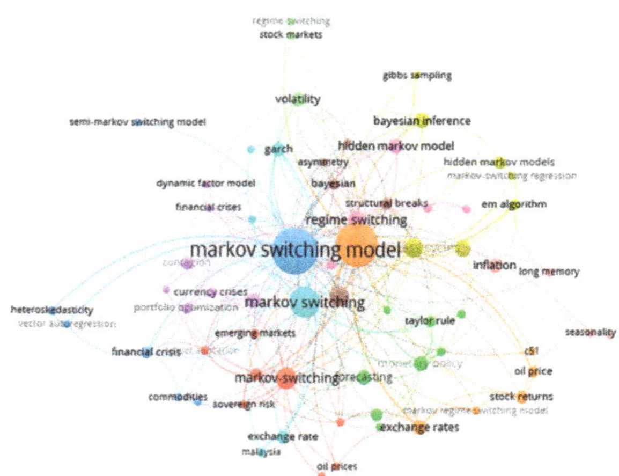

Figure 3. A network analysis of keyword co-occurrence.

In this network mapping, the sizes reflect the frequency of the author's keywords in the Markov Switching literature, while the colors represent the number of clusters.

Let s_t denote an unobservable state variable assuming the value of one or zero. As mentioned by Kuan (2002), a simple switching model for the variable z_t involves two AR specifications:

$$z_t = \begin{cases} \alpha_0 + \beta z_{t-1} + \varepsilon_t, & s_t = 0 \\ \alpha_0 + \alpha_1 + \beta z_{t-1} + \varepsilon_t, & s_t = 1 \end{cases} \quad (6)$$

where $|\beta| < 1$ and ε_t are i.i.d random variables with mean zero and variance σ_ε^2. This is a stationary AR(1) process with mean $\alpha_0/(1-\beta)$ when $s_t = 0$, and it switches to another stationary AR(1) process with mean $(\alpha_0 + \alpha_1)/(1-\beta)$ when s_t changes from 0 to 1. This model admits two dynamic structures at different levels, depending on the value of the state variable s_t. In this case, z_t are governed by two distributions with distinct means, and s_t determines the switching between these two distributions (regimes).

To implement our methodology and test all aforementioned modeling approaches, in this research we used monthly CL incidence data from Errachidia province and climatic variables. Data covers the period from January 2010 to December 2019.

Errachidia province is located in the Ziz Ghris Valley in the south-east of Morocco (Figure 4), including the Saharan areas, plains, and highlands at an altitude above 1900 m and covering a surface of 46,000 km^2. Errachidia has an arid climate with temperatures between −4 °C and 48 °C, with large daily and seasonal temperature variations. The annual mean temperature is 21 °C. Rainfall is scarce and usually occurs between February and March. The annual total precipitation is 134.64 mm [15].

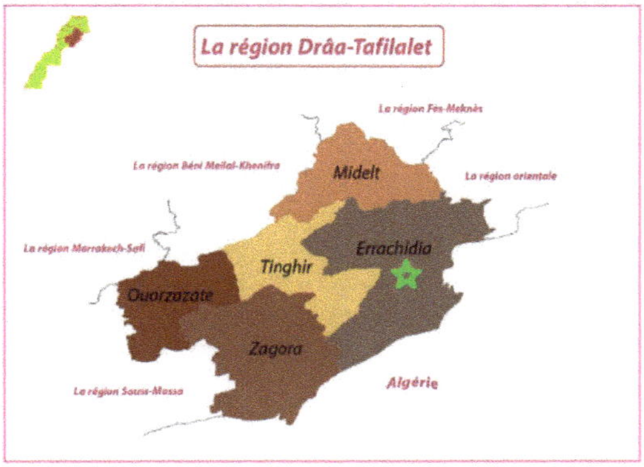

Figure 4. Location of Errachidia province (HCP).

3. Results and Discussions

The analysis of collected data shows that Errachidia province recorded 8487 cases of cutaneous leishmaniasis between 2010 and 2019 (Figure 5). It can be observed from Figure 5 that the monthly CL incidence peaked between 2010 and 2011, declined between 2011 and 2016, and rose around the end of 2016 to 2018. In addition, the number of CL cases had seasonal fluctuations. Most cases were recorded in the months of November, December, and January. The trend starts to increase in October, with high peaks in December and January, and declines until February. A steady-state period is observed until September.

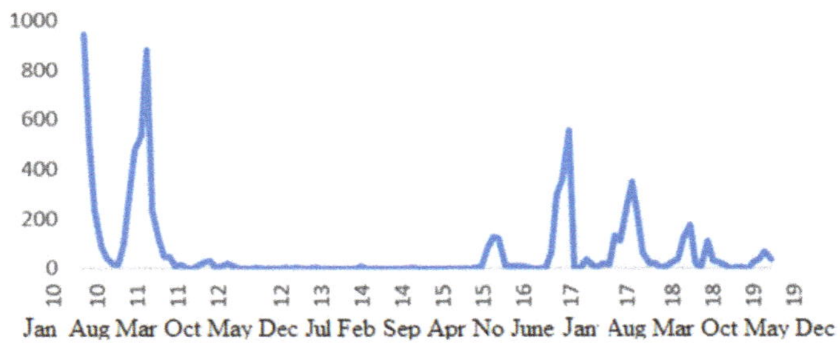

Figure 5. The evolution of LC incidence in Errachidia province from 2010 to 2019.

To study the impact of climate change on the CL incidence in this region, we selected seven predictors, namely the monthly average humidity (Hmoy), precipitation (Prec), average wind speed (Vmoy), minimum temperature (Tmin), mean temperature (Tmoy), and maximum temperature (Tmax), as well as evaporation (Evap).

A linear correlation between the covariables was assessed using the Pearson coefficient, and predictors that present a strong correlation with CL cases were retained. The Spearman coefficient and the cross-correlation between the CL cases and climatic data were estimated to determine the predictors and the adequate lags to be included in the model.

Our first fitted model is Poisson regression. However, the dispersion coefficient showed severe over-dispersion ($\Phi_{\text{Poisson}} = 145.18$). Thus, negative binomial regression was used as an alternative approach to model over-dispersion in the data.

The use of the stepwise method for 3- and 6-month gaps between CL cases and predictors leads to the retention of the model given by (Table 1):

$$\text{CL} \sim \text{Vmoy}_6 + \text{Tmin}_6 \tag{7}$$

Table 1. Results of the stepwise selection algorithm for the 3- and 6-month gaps between CL and predictors.

GLM: Negative Binomial Models	AIC
Log(E(CL))~9.0747 Hmoy − 28.3713 Prec + 1.0572	1030.9
Log(E(CL))~6.90924 Hmoy_3 + 0.18451 Tmoy_3 − 2.26628	980.74
Log(E(CL))~1.20795886 Vmoy_6 + 0.07317 Tmin_6 − 2.34280875	**932.59**

Note: The bold means that this is the optimal value.

To examine the adequacy of the fitted model, a residual diagnosis as well as the deviance test were performed.

The multivariate time-series analysis shows that data can be considered as stationary time series. The optimal VAR model identified is given by:

$$\begin{cases} \text{CL}_t = 0.629\ \text{CL}_{t-1} + 3.326\ \text{Tmoy}_{t-1} - 45.573 \\ \text{Tmoy}_t = -0.010\ \text{CL}_{t-1} + 0.894\ \text{Tmoy}_{t-1} + 2.844 \end{cases} \tag{8}$$

To assess the performance of this model a post analysis is required. The test of the presence of serial correlation in the residual, the Portmanteau Test, shows a p-value equal to $0.0874 > 0.05$, meaning that we can accept the null hypothesis of nonexistence of serial correlations.

Concerning the test of heteroskedasticity, the p-value of the arch test is equal to $0.009107 < 0.05$, meaning that we reject the null hypothesis and thus the absence of heteroskedasticity.

The Jarque-Bera test was then used to check if the residuals fit the normal distribution. Because the p-value is higher than 0.05, the normal distribution is accepted. The following figure (Figure 6) shows no structural break.

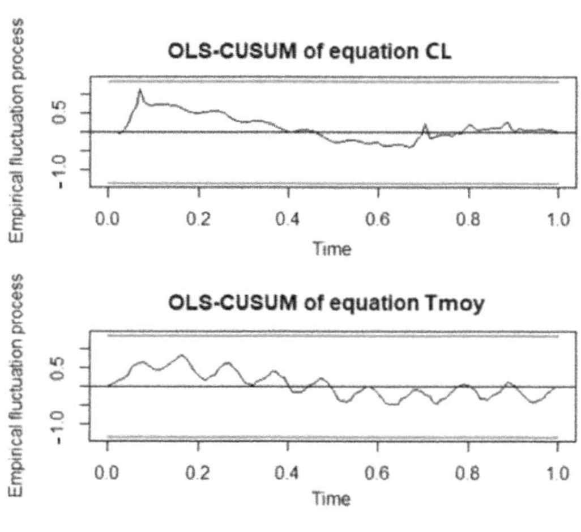

Figure 6. Stability Test of the VAR model adjusted.

One of the most useful tools to characterize the dependence among time series when running the VAR model is the causality test. In our case, the objective of this test is to check if the variable "Tmoy" contributes to forecasting the "CL incidence" variable.

The *p*-value of the Granger causality test is equal to $0.0023 < 5\%$, meaning that we can reject the null hypothesis. Hence, "Tmoy" is considered a pertinent feature (risk factor) when predicting the "CL" variable. As displayed in Figure 7, a good performance is shown, in terms of predictions obtained from VAR(1).

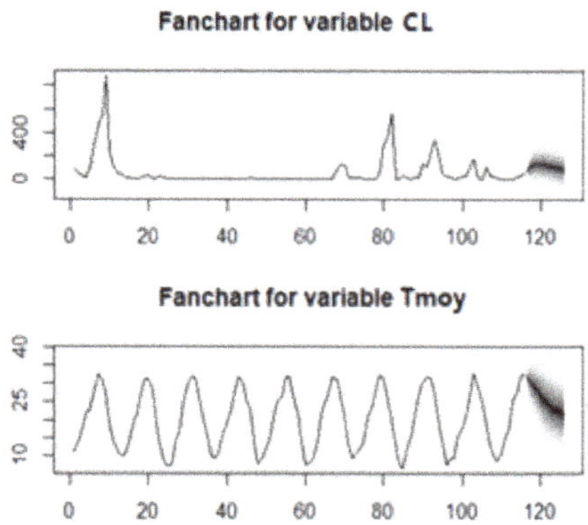

Figure 7. Forecasts for CL cases and the mean temperature from VAR(1).

The following figure (Figure 8) provides results about the Markov Switching model fitting with the estimation of regime 1 and regime 2. However, based on the AIC and residuals diagnosis, the VAR(1) performs better than the Markov Switching model.

```
Markov Switching Model

Call: msmFit(object = Model, k = 2, sw = rep(TRUE, 3))

       AIC      BIC    logLik
  1117.522 1147.619 -554.7608

Coefficients:
          (Intercept)(S) temp_decal(S)     Std(S)
Model 1         7.169842    -0.0435326   8.441853
Model 2        -8.854126     8.0269161 179.400494

Transition probabilities:
          Regime 1  Regime 2
Regime 1 0.91532615 0.2019992
Regime 2 0.08467385 0.7980008
```

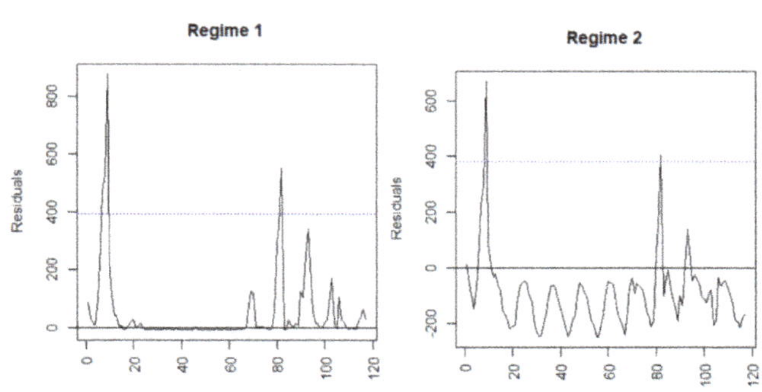

Figure 8. Results about Markov Switching model fitting.

4. Conclusions

In this study, we are interested in modeling the spread of cutaneous leishmaniasis (CL) using the Vector Auto-Regressive (VAR) model, the Vector Error Correction model, (VECM) and the Markov Switching approach (MSM). Based on our findings, we can mention that the CL time series are characterized by extreme values; therefore, it was cumbersome to explain them using meteorological data. It is worth noting that among all candidate models, the VAR model performs well in terms of underlying hypotheses such as the stationary series, so there is no need to use the VECM. In addition, the VAR model provides good results in terms of prediction. In our case, the implication of the adequacy of the VAR is that the CL variable can be considered as a function of its own past values.

It is worth noting that the VAR model is one of the most successful, flexible, and easy to use models for the analysis of multivariate time series. This model often provides superior forecasts to those from univariate time series models. In addition to data description and the underlying theory based on simultaneous equations and forecasting, the VAR is also used for structural inference.

Author Contributions: F.B.: Conceptualization, methodology, software, formal analysis, investigation, writing—original draft preparation, visualization; S.B.: Resources, data curation; A.A.: Conceptualization, validation, investigation, writing—review and editing; K.K.: Resources, data curation. All authors have read and agreed to the published version of the manuscript.

Funding: This research received no external funding.

Institutional Review Board Statement: Not applicable (Only aggregate number of CL incidence are included).

Informed Consent Statement: Not applicable.

Data Availability Statement: Data supporting reported results can be found: Direction of Epidemiology and Disease Control (DELM), Ministry of Health, Rabat, Morocco; The Climate Data Store of the Copernicus Service (https://cds.climate.copernicus.eu/cdsapp#!/home (accessed on 1 February 2023)).

Acknowledgments: We are grateful to the Souad Bouhout and Kenza Khomsi for their contributions and for providing support for this research. We would like to thank all the participants in this study for their time and willingness to share their experiences.

Conflicts of Interest: The authors declare no conflict of interest.

References

1. Wijerathna, T.; Gunathilaka, N.; Gunawardena, K.; Rodrigo, W. Population dynamics of phlebotomine sand flies (diptera: Psychodidae) in cutaneous leishmaniasis endemic areas of Kurunegala district, Sri Lanka. *Acta Trop.* **2022**, *230*, 106406. [CrossRef] [PubMed]
2. Medenica, S.; Miladinović-Tasić, N.; Stojanović, N.M.; Lakićević, N.; Rakočević, B. Climate Variables Related to the Incidence of Human Leishmaniosis in Montenegro in Southeastern Europe during Seven Decades (1945–2014). *Int. J. Environ. Res. Public Health* **2023**, *20*, 1656. [CrossRef] [PubMed]
3. Vieira, T.M.; de Oliveira Silva, S.; Lima, L.; Sabino-Santos, G.; Duarte, E.R.; Lima, S.M.; Pereira, A.A.S.; Ferreira, F.C.; de Araújo, W.S.; Teixeira, M.M.G.; et al. Leishmania diversity in bats from an endemic area for visceral and cutaneous leishmaniasis in Southeastern Brazil. *Acta Trop.* **2022**, *228*, 106327. [CrossRef] [PubMed]
4. Hakkour, M.; Hmamouch, A.; Mahmoud El Alem, M.; Bouyahya, A.; Balahbib, A.; El Khazraji, A.; Fellah, H.; Sadak, A.; Sebti, F. Risk factors associated with leishmaniasis in the most affected provinces by leishmania infantum in Morocco. *Interdiscip. Perspect. Infect. Dis.* **2020**, *2020*, 6948650. [CrossRef] [PubMed]
5. McCullagh, P.; Nelder, J.A. *Generalized Linear Models*; Monographs on Statistics and Applied Probability; Chapman & Hall: London, UK, 1989; Volume 37.
6. Heinze, G.; Wallisch, C.; Dunkler, D. Variable selection—A review and recommendations for the practicing statistician. *Biom. J.* **2018**, *60*, 431–449. [CrossRef] [PubMed]
7. Kianifard, F.; Gallo, P.P. Poisson regression analysis in clinical research. *J. Biopharm. Stat.* **1995**, *5*, 115–129. [CrossRef] [PubMed]
8. Fekedulegn, D.; Andrew, M.; Violanti, J.; Hartley, T.; Charles, L.; Burchfiel, C. Comparison of statistical approaches to evaluate factors associated with metabolic syndrome. *J. Clin. Hypertens.* **2010**, *12*, 365–373. [CrossRef] [PubMed]
9. Kuan, C.-M. *The Markov Switching Model*; Institute of Economics Academia Sinica: Taipei, Taiwan, 2002; Available online: http://homepage.ntu.edu.tw/~ckuan/pdf/Lec-Markov_note.pdf (accessed on 1 February 2023).

10. Hamilton, J.D. A new approach to the economic analysis of nonstationary time series and the business cycle. *Econom. J. Econom. Soc.* **1989**, *57*, 357–384. [CrossRef]
11. Lu, H.M.; Zeng, D.; Chen, H. Prospective infectious disease outbreak detection using Markov switching models. *IEEE Trans. Knowl. Data Eng.* **2009**, *22*, 565–577. [CrossRef]
12. Van Os, B.; van Dijk, D. Accelerating peak dating in a dynamic factor Markov-switching model. *Int. J. Forecast.* 2023; *in press*.
13. Bouteska, A.; Sharif, T.; Abedin, M.Z. COVID-19 and stock returns: Evidence from the Markov switching dependence approach. *Res. Int. Bus. Financ.* **2023**, *64*, 101882. [CrossRef] [PubMed]
14. Phoong, S.W.; Phoong, S.Y.; Khek, S.L. Systematic literature review with bibliometric analysis on Markov switching model: Methods and applications. *SAGE Open* **2022**, *12*, 21582440221093062. [CrossRef]
15. Bounoua, L.; Kahime, K.; Houti, L.; Blakey, T.; Ebi, K.L.; Zhang, P.; Imhoff, M.L.; Thome, K.J.; Dudek, C.; Sahabi, S.A.; et al. Linking climate to incidence of zoonotic cutaneous leishmaniasis (*L. major*) in pre-Saharan North Africa. *Int. J. Environ. Res. Public Health* **2013**, *10*, 3172–3191. [CrossRef] [PubMed]

engineering proceedings

Proceeding Paper

Time Series Regression Modelling: Replication, Estimation and Aggregation through Maximum Entropy †

Jorge Duarte [1,2], Maria Costa [2] and Pedro Macedo [2,*]

1 Faculty of Exact and Technological Sciences, University Púnguè, Bairro Heróis Moçambicanos, Chimoio 323, Mozambique; jorgeduarte17@ua.pt
2 CIDMA—Center for Research and Development in Mathematics and Applications, Department of Mathematics, University of Aveiro, 3810-193 Aveiro, Portugal; lopescosta@ua.pt
* Correspondence: pmacedo@ua.pt; Tel.: +351-234-372-545
† Presented at the 9th International Conference on Time Series and Forecasting, Gran Canaria, Spain, 12–14 July 2023.

Abstract: In today's world of large volumes of data, where the usual statistical estimation methods are commonly inefficient or, more often, impossible to use, aggregation methodologies have emerged as a solution for statistical inference. This work proposes a novel procedure for time series regression modelling, in which maximum entropy and information theory play central roles in the replication of time series, estimation of parameters, and aggregation of estimates. The preliminary results reveal that this three-stage maximum entropy approach is a promising procedure for time series regression modelling in big data contexts.

Keywords: big data; bootstrap; maximum entropy; neagging; regression modelling; time series

Citation: Duarte, J.; Costa, M.; Macedo, P. Time Series Regression Modelling: Replication, Estimation and Aggregation through Maximum Entropy. *Eng. Proc.* **2023**, *39*, 39. https://doi.org/10.3390/engproc2023039039

Academic Editors: Ignacio Rojas, Hector Pomares, Luis Javier Herrera, Fernando Rojas and Olga Valenzuela

Published: 3 July 2023

1. Introduction

Understanding, predicting, and interpreting information collected over different periods of time (days, weeks, years, etc.) has been of major concern in recent decades. Time series analysis, concerning longer or shorter periods, has been of major importance in various areas, such as economics and finance, agriculture, meteorology, astronomy, medicine, and health sciences, just to name a few. Current reality is characterized by large volumes of data, generally complex, unstructured and/or non-homogeneous, which are usually referred to as big data to emphasize the challenges they present to traditional computational and statistical techniques. In many situations, it is not possible to treat all data as a whole, either because of the inadequacy of the methodology or the insufficient computational capacity available. In such situations, traditional methods of estimation become inefficient and new statistical approaches are needed. The methodology adopted in this study was contextualized in the reality described above and is presented here as an alternative tool to deal with large volumes of data.

The strategy that is followed in this work is based on a three-stage maximum entropy approach for time series analysis. We started by considering a given data set to which the linear regression model could be applied, but represented a high computational burden when considering estimation. Instead of trying to tackle this issue using the whole data set, we started by considering only a few, say G, small samples or groups (with a few observations, say *Obs*) that were extracted from the original data using a random sampling procedure. The first step, after collecting the G small dimension random samples, was to replicate each of them following the maximum entropy bootstrap methodology proposed by H. D. Vinod [1,2]. Each replica could then be easily estimated using a proper estimation procedure. In the second stage, the generalized maximum entropy (GME) estimator was used to obtain estimates for the linear regression model parameters. Other estimators could have been considered in this stage, but due to the stability of the GME estimator in ill-posed problems (which are common in a big data context) this estimator was chosen for this work. Having an estimate for each parameter of the model and for each replica obtained of each original small sample, it was time to aggregate this information and obtain an overall estimate for each parameter of the model. Considering the total number of replicas, the percentile method can be used to obtain a confidence interval with any given confidence level for each of the model parameters (for a particular small sample of G groups collected). Alternatively, an overall estimate for each of the G group can be obtained by calculating the median of the estimates. As the normalized entropy is also calculated for each replica, the median normalized entropy can be obtained for each of the G samples. The third and last stage of the procedure consisted of aggregating the G estimates into one single estimate for each model parameter. To accomplish this last step, an entropy-based procedure was used once again, as the normalized entropy was used as the weighting factor for the aggregated overall model parameter estimate. Details about this aggregation technique, called neagging, can be found in [3,4]. This work reports the first time these three maximum entropy-based procedures have been combined into one single three-stage framework for regression analysis of time series. There are still many details that need further exploration, such as the adequate number of small samples or groups, G, and the proper number of observations per group, *Obs*, to consider; however, the preliminary results presented here are promising and certainly deserve to be brought to light.

2. Materials and Methods

This study considered daily available prices (USD; United States Dollar) for carbon dioxide (CO_2) emissions, coal, natural gas, and Brent oil. Data from 2 July 2018 to 29 June 2022 were collected from Investing.com [5], considering only days with simultaneous information for all variables. The collected data for 994 days were considered a "big data set" that could only be analyzed using an aggregation methodology. Naturally, given the purpose of this work, the dimension was carefully chosen so that, although of high dimension, it was still computationally tractable, and the results of the model with 994 observations could be compared with those from the aggregation methodology. It is also important to note that the purpose of this work was not to provide an economic interpretation of the regression model defined below, but rather to illustrate how the combined methods may work in time series regression modelling scenarios. The methods considered in this work for replication of time series, estimation of parameters, and aggregation of estimates are briefly presented next.

2.1. Maximum Entropy Bootstrap for Time Series

The maximum entropy bootstrap for time series was proposed by H. D. Vinod [1,2]; a package (meboot) in the R software was developed by H. D. Vinod and J. López-de-Lacalle [6]. The methodology creates a large number of replicates of the time series for inference purposes (1000 replications were considered in this study). Those generated elements of the ensemble retain the shape of the original time series using maximum entropy density, as well as the time dependence structure of the autocorrelation and the partial autocorrelation functions. Moreover, and probably most importantly, the methodology avoids all structural changes and unit root type testing, and all the usual shape-destroying transformations, such as detrending or differencing, to achieve stationarity. Details of the algorithm can be found in [6].

2.2. Generalized Maximum Entropy Estimation

The GME estimator was proposed by A. Golan, G. Judge, and D. Miller [7] based on the maximum entropy (ME) principle of E. Jaynes [8,9] and the entropy measure of C. E. Shannon [10]. Recently, A. Golan [11] presented a new area of research at the intersection of statistics, computer science, and decision theory, entitled info-metrics, in which the GME estimator and other information–theoretic estimation methods are included. The

GME estimator represents a stable estimation procedure in ill-posed models, namely in ill-conditioned models, in underdetermined models, and when only samples of small size are available for inference purposes (the micronumerosity problem).

Under GME estimation, the linear regression model $\boldsymbol{y} = \boldsymbol{X\beta} + \boldsymbol{e}$, where $\boldsymbol{y}$ denotes a $(N \times 1)$ vector of noisy observations, $\boldsymbol{\beta}$ is a $(K \times 1)$ vector of parameters to be estimated, $\boldsymbol{X}$ is a $(N \times K)$ matrix of explanatory variables, and $\boldsymbol{e}$ is a $(N \times 1)$ vector of random errors, needs to be reformulated as $\boldsymbol{y} = \boldsymbol{XZp} + \boldsymbol{Vw}$, where $\boldsymbol{Z}$ is a $(K \times KM)$ matrix of support spaces, $\boldsymbol{V}$ is a $(N \times NJ)$ matrix of support spaces, $\boldsymbol{p}$ is a $(KM \times 1)$ vector of probabilities to be estimated, $\boldsymbol{w}$ is a $(NJ \times 1)$ vector of probabilities to be estimated, and M and J are the number of points in the support spaces. The GME estimator can then be stated as

$$\mathrm{argmax}_{\boldsymbol{p},\boldsymbol{w}}\{-\boldsymbol{p}' \ln \boldsymbol{p} - \boldsymbol{w}' \ln \boldsymbol{w}\}, \tag{1}$$

subject to the model and the additivity constraints $\boldsymbol{y} = \boldsymbol{XZp} + \boldsymbol{Vw}$, $\boldsymbol{1}_K = (\boldsymbol{I}_K \otimes \boldsymbol{1}'_M)\boldsymbol{p}$ and $\boldsymbol{1}_N = \left(\boldsymbol{I}_N \otimes \boldsymbol{1}'_J\right)\boldsymbol{w}$, where $\otimes$ represents the Kronecker product. Using numerical optimization techniques, because the statistical problem has no closed form solution, the estimates of the parameters are given by $\widehat{\boldsymbol{\beta}} = \boldsymbol{Z}\widehat{\boldsymbol{p}}$. Additional details regarding GME estimation can be found in [7,11–14].

2.3. Normalized Entropy Aggregating (Neagging)

Aggregation methods become useful in big data contexts, where a significant computational effort is required and traditional statistical estimation methods become inefficient or, more often, impossible to apply. Normalized entropy aggregating (neagging) is based on identifying the information content of a given (randomly selected) group of observations ($g = 1, 2, \ldots, G$) through the calculation of the normalized entropy; [3,4]. The aggregated (global) estimate is given by

$$\widehat{\boldsymbol{\beta}} := \sum_{g=1}^{G} w_g \widehat{\boldsymbol{\beta}}_g, \tag{2}$$

where $\widehat{\boldsymbol{\beta}}_g$ is the estimate in group g, $w_g \propto 1 - S(\widehat{\boldsymbol{p}})_g$, with $S(\widehat{\boldsymbol{p}})_g = (-\widehat{\boldsymbol{p}}' \ln \widehat{\boldsymbol{p}})/(K \ln M)$, and $\sum_{g=1}^{G} w_g = 1$. Thus, neagging consists of weighting the estimates obtained from the GME estimator for each group g according to the information content of the group.

2.4. Statistical Model and Estimation Configuration

The regression model is defined as

$$CO2_t = b_0 + b_1 COAL_t + b_2 GAS_t + b_3 OIL_t + e_t, \tag{3}$$

where $CO2$ represents the price of CO_2 emissions, $COAL$ represents the price of coal, GAS represents the price of natural gas, OIL represents the price of Brent oil, t represents the day, and e is the error component.

First, maximum entropy bootstrap was used to create 1000 replications of the four series under study, using the package (meboot) in the R software [6]. Next, the four parameters of the 1000 models were estimated using the GME estimator (implemented in MATLAB by the authors) considering symmetric supports about zero in matrix $\boldsymbol{Z}$, with five ($M = 5$) equally spaced points, and the following lower and upper bounds: $[-2, 2]$, $[-5, 5]$, $[-10, 10]$, $[-100, 100]$, and $[-1000, 1000]$. The use of five supports was intended to illustrate different a priori information scenarios, and also to evaluate the performance of the estimation process. Regarding the matrix $\boldsymbol{V}$, the supports were defined symmetrically and centered on zero with three ($J = 3$) equally spaced points, using the three-sigma rule with the empirical standard deviation of the noisy observations, namely $[-3\widehat{\sigma}_y, 3\widehat{\sigma}_y]$. With 1000 estimates for each parameter of the model in (3), the percentile method was used to construct confidence intervals for each one at 90%, 95%, and 99%. The twelve confidence intervals obtained from the 1000 models with 994 observations were then used to evaluate the performance of neagging.

Finally, to implement neagging in different scenarios, random sampling with replacement was performed considering two different number of groups, $G = 5$ and $G = 10$, where the numbers of observations per group were 20 ($Obs = 20$), 50 ($Obs = 50$), and 100 ($Obs = 100$). The parameters of the different models obtained by random sampling were estimated using the GME estimator with the same configuration mentioned above. As there were 1000 estimates and 1000 values of normalized entropy for each scenario, the median of the 1000 estimates was considered the estimate of each parameter, and the value of w_g was obtained using the median (and the mean for comparison purposes; see next section) of the 1000 values of normalized entropy.

3. Results and Discussion

The results obtained in this work are summarized in Table 1, in which the rows contain the five different supports used: [−2, 2], [−5, 5], [−10, 10], [−100, 100], and [−1000, 1000]. For each support, two different numbers of groups were tested, $G = 5$ and $G = 10$. The number of observations per group is represented in the columns of Table 1, where the confidence intervals obtained by the percentile method for the confidence levels of 90%, 95%, and 99% are also listed. As these confidence intervals were calculated for each model parameter, we had twelve confidence intervals per support that related to the model with 994 observations. To evaluate results from the aggregation methodology, we checked if the overall aggregated confidence intervals, obtained for each of the random sampling scenarios listed in Table 1, had the same sign of the corresponding confidence interval. The best case scenario, evaluated for each of the 90 different scenarios tested, was four correct identifications of the corresponding confidence intervals signs (one for each model parameter), which we called four hits out of four. The worst case scenario would have been zero hits, which did not occur. Table 1 presents the number of hits for each scenario tested.

Table 1. Number of hits in the comparison between confidence intervals for original data and overall confidence intervals for aggregated data.

Supports	G	*Obs* = 20			*Obs* = 50			*Obs* = 100		
		CI90	CI95	CI99	CI90	CI95	CI99	CI90	CI95	CI99
[−2, 2]	5	4	4	3	4	4	4	4	4	4
	10	4	4	3	3	3	4	4	4	4
[−5, 5]	5	4	3	3	4	4	4	4	4	4
	10	3	3	3	3	3	4	4	4	4
[−10, 10]	5	4	4	3	3	4	4	4	4	4
	10	3	3	3	3	3	3	4	4	4
[−100, 100]	5	2	2	2	4	2	2	4	4	3
	10	2	2	2	2	2	2	4	4	3
[−1000, 1000]	5	2	2	2	4	2	2	4	4	3
	10	2	2	2	3	2	2	4	4	2

Note that the number of possible hits for each scenario was four, given that there were four model parameters to estimate. Each confidence level listed in Table 1 is represented in the HDR graphs with the following color correspondence: 99% CI—red, 95% CI—green, and 90% CI—blue.

The weighting of the estimates obtained from the GME estimator for each group according to the information content of the group—the neagging procedure—was done considering two alternative forms: using the median and the mean of the normalized entropy. The results are presented in a single table because they were exactly the same in terms of the number of hits. This equality of the median and the mean can be explained by the fact that the w_g of the median and the mean were very close.

The discussion of the results in Table 1 has two focuses of analysis. The first one, looking at the lower amplitude supports [−2, 2], [−5, 5] and [−10, 10], where it is expected that this lower amplitude comes from the fact that the user has a lot of information about the model (relating to a situation of a high a priori information), and the second one, looking at the supports [−100, 100] and [−1000, 1000], with greater amplitude, where it is expected that the user has little information about the model (low a priori information scenarios).

In the first case, where the supports have smaller amplitude, it can be seen that with less observations per group (20 and 50), as the amplitude of the supports decreases, the number of hits increases and the results improve in cases where the aggregation is done with a smaller number of groups. However, the results are better for the scenario with 50 observations, when compared to the scenario with 20 observations. Finally, in the scenario with samples of 100 observations, the normalized entropy aggregation procedure presents 100% of hits in all supports [−2, 2], [−5, 5] and [−10, 10], regardless of the number of groups.

For the second case, where we have little information about the model and the supports are of greater amplitude, whatever the support [−100, 100] or [−1000, 1000], it can be seen that as the number of observations increases, the number of hits improves significantly, and for 50 and 100 observations it improves even more when the number of groups decreases from 10 to 5.

Therefore, in general, we can say that the results improve as the number of observations increase and the amplitude of the supports decrease, as well as the number of groups.

Looking at the highest density region (HDR) graphs below, which represent the model with 994 observations, we can see that the constant (b_0) in Figure 1 and the coefficient b_2 in Figure 2 are not statistically significant, as all confidence intervals (at confidence levels of 90%, 95%, and 99%) include zero. On the other hand, coefficients b_1 and b_3 can be considered statistically significant, as the corresponding confidence intervals do not include zero at any confidence level tested.

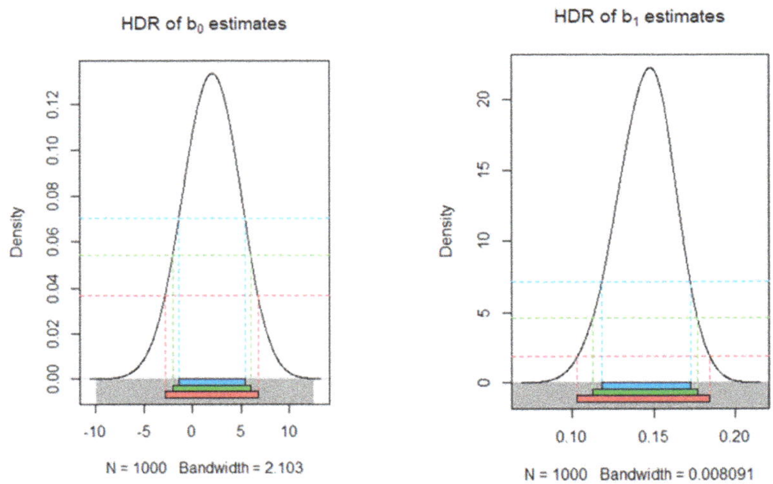

Figure 1. HDRs for the sampling distribution of estimates of b_0 and b_1.

Figure 3 shows the trend in the price of CO_2 emissions. Figure 4 (on the left) shows trends in the price of coal, the price of natural gas, and the price of Brent oil in the period under study. Figure 4 (on the right) highlights only the price of natural gas, because it is not clear in the figure on the left.

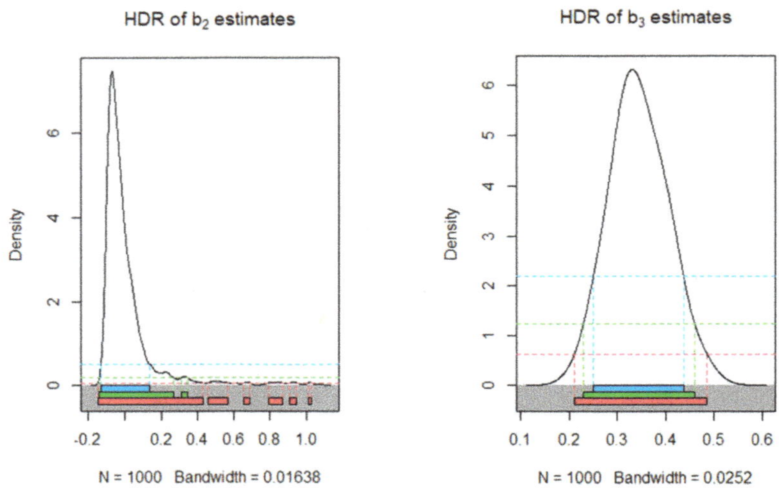

Figure 2. HDRs for the sampling distribution of estimates of b_2 and b_3.

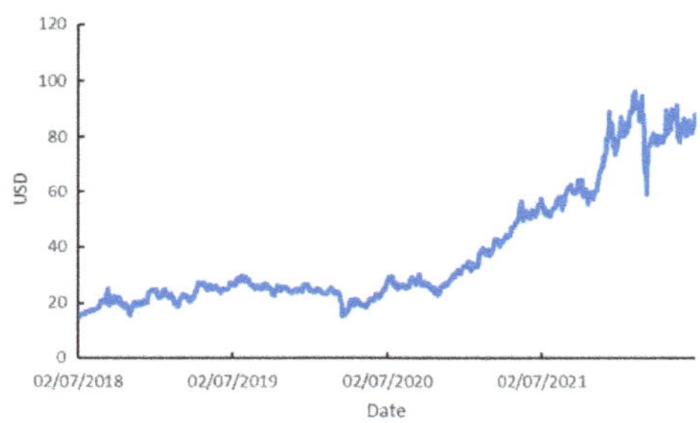

Figure 3. Price of CO_2 emissions.

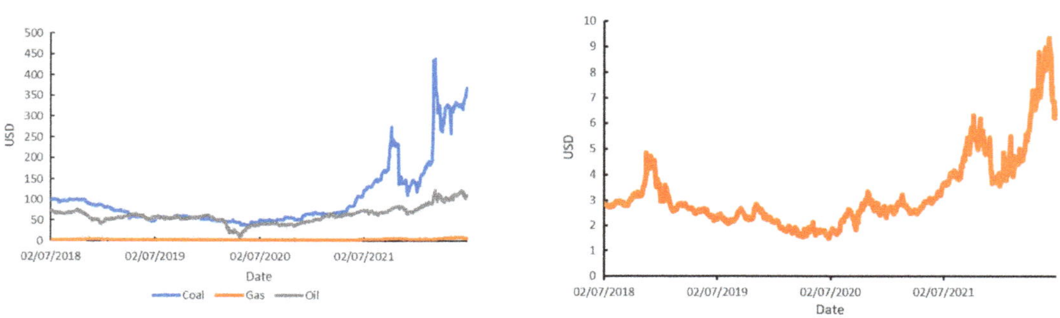

Figure 4. Prices of coal, natural gas, and Brent oil (**left**), and the price of natural gas (**right**).

The values for all observed variables had an increasing trend, including the one represented separately in the graph on the right.

4. Conclusions

Recently, a two-stage maximum entropy approach for time series regression modelling was proposed in the literature [14]. This work extends that idea to big data contexts by adding a third stage with the neagging procedure. Although further research is needed, the preliminary results reveal that this three-stage approach, in which maximum entropy is used in replication of the time series, estimation of parameters, and aggregation of estimates, is a promising procedure in big data contexts. As future work, an exhaustive simulation study is being planned to investigate the adequate number of groups and the adequate number of observations per group. Additionally, comparisons with other aggregation methods, such as bagging, are expected to be analyzed, in order to explore the advantages and disadvantages of this three-stage approach.

Author Contributions: Conceptualization, methodology, software, investigation, data curation, writing—original draft preparation, and writing—review and editing: J.D., M.C. and P.M. All authors have read and agreed to the published version of the manuscript.

Funding: This work was supported by the Center for Research and Development in Mathematics and Applications (CIDMA) through the Portuguese Foundation for Science and Technology (FCT—Fundação para a Ciência e a Tecnologia), reference UIDB/04106/2020.

Institutional Review Board Statement: Not applicable.

Informed Consent Statement: Not applicable.

Data Availability Statement: Data collected from Investing.com (https://www.investing.com/, accessed on 5 August 2022).

Acknowledgments: We would like to express our gratitude to the Editor and two anonymous reviewers for all helpful comments.

Conflicts of Interest: The authors declare no conflict of interest.

References

1. Vinod, H.D. Ranking mutual funds using unconventional utility theory and stochastic dominance. *J. Emp. Financ.* **2004**, *11*, 353–377. [CrossRef]
2. Vinod, H.D. Maximum entropy ensembles for time series inference in economics. *J. Asian Econ.* **2006**, *17*, 955–978. [CrossRef]
3. Costa, M.; Macedo, P. Normalized entropy aggregation for inhomogeneous large-scale data. In *Theory and Applications of Time Series Analysis*; Valenzuela, O., Rojas, F., Pomares, H., Rojas, I., Eds.; Springer: Cham, Switzerland, 2019; pp. 19–29. [CrossRef]
4. Costa, M.; Macedo, P.; Cruz, J.P. Neagging: An aggregation procedure based on normalized entropy. In Proceedings of the International Conference on Numerical Analysis and Applied Mathematics, Rhodes, Greece, 17–23 September 2020; Simos, T.E., Tsitouras, C., Eds.; AIP Conference Proceedings: New York, NY, USA, 2022; Volume 2425, p. 190003. [CrossRef]
5. Energy Prices 2022. Available online: https://www.investing.com (accessed on 5 August 2022).
6. Vinod, H.D.; Lopez-de-Lacalle, J. Maximum entropy bootstrap for time series: The meboot R package. *J. Stat. Softw.* **2009**, *29*, 1–19. [CrossRef]
7. Golan, A.; Judge, G.; Miller, D. *Maximum Entropy Econometrics: Robust Estimation with Limited Data*; Wiley: Chichester, UK, 1996.
8. Jaynes, E.T. Information theory and statistical mechanics. *Phys. Rev.* **1957**, *106*, 620–630. [CrossRef]
9. Jaynes, E.T. Information theory and statistical mechanics. II. *Phys. Rev.* **1957**, *108*, 171–190. [CrossRef]
10. Shannon, C.E. A mathematical theory of communication. *Bell Syst. Tech. J.* **1948**, *27*, 379–423. [CrossRef]
11. Golan, A. *Foundations of Info-Metrics: Modeling, Inference, and Imperfect Information*; Oxford University Press: New York, NY, USA, 2018. [CrossRef]
12. Jaynes, E.T. *Probability Theory—The Logic of Science*; Cambridge University Press: Cambridge, UK, 2003. [CrossRef]
13. Mittelhammer, R.; Cardell, N.S.; Marsh, T.L. The data-constrained generalized maximum entropy estimator of the GLM: Asymptotic theory and inference. *Entropy* **2013**, *15*, 1756–1775. [CrossRef]
14. Macedo, P. A Two-Stage Maximum Entropy Approach for Time Series Regression. *Commun. Stat. Simul. Comput.* 2022, *in press*. [CrossRef]

engineering proceedings

MDPI

Proceeding Paper

The Dutch Disease in Angola: An Empirical Analysis †

Zsuzsanna Biedermann [1], Tamás Barczikay [2] and László Szalai [3,*]

1 Centre for Economic and Regional Studies, Institute of World Economics, Hungarian Academy of Sciences Centre of Research Excellence, 4. Tóth Kálmán Utca, 1097 Budapest, Hungary
2 Centre for Social Sciences, Hungarian Academy of Sciences Centre of Research Excellence, 4. Tóth Kálmán Utca, 1097 Budapest, Hungary; barczikay.tamas@tk.hu
3 Department of Economics, Budapest University of Technology and Economics, 2. Magyar Tudósok Körútja, 1117 Budapest, Hungary
* Correspondence: szalai.laszlo@gtk.bme.hu

† Presented at the 9th International Conference on Time Series and Forecasting, Gran Canaria, Spain, 12–14 July 2023.

Abstract: Despite being the second largest oil exporter in Africa, Angola continues to lag behind in most macroeconomic and institutional indicators. At least partially, this is a consequence of the Dutch disease, a phenomenon that establishes a clear link between high resource endowments and lack of economic diversity through the loss of international competitiveness in non-resource sectors. In this paper, we use a nonlinear autoregressive distributed lag (NARDL) model to identify the cointegrated relationship between international oil prices and the real effective exchange rate of the kwanza, which is a striking sign of the presence of the Dutch disease.

Keywords: resource curse; spending effect; real exchange rate; cointegration

1. Introduction

A significant body of empirical literature shows that the economic performance of resource-rich countries is typically inferior to that of their resource-deficient counterparts. Auty [1], for example, found that resource-poor economies grew four times faster than big oil exporters between 1970 and 1993. In case of OPEC members, Gylfason [2] calculated that the real GDP per capita fell by an average of 1.3% between 1965 and 1998. Furthermore, examining 51 resource-rich economies during the 1990s, Weber-Fahr [3] reported that in countries where more than 50% of export revenues came from mineral exports, the real GDP per capita declined by an average of 2.3%. This phenomenon, commonly known as the resource curse, has been explained by several theories of neoclassical economics and political economy [4]. This study predominantly focuses on the former, as it uses statistical techniques to identify the presence of the Dutch disease in Angola.

The Dutch disease refers to the deindustrialization of a national economy as real appreciation of the domestic currency induces a loss of international competitiveness in non-resource sectors. First developed by Corden and Neary [5], the theoretical model deals with the effects of a resource boom in a three-sector, small, open economy. As windfalls from commodity exports drive domestic demand, the relative price of non-tradables increases, causing real appreciation through the Balassa–Samuelson mechanism. Referred to as the spending effect, this change in terms of trade induces indirect deindustrialization as domestic manufactures are crowded-out by cheaper imports. Furthermore, the Dutch disease also undermines the competitiveness of agricultural tradables. Although Angola was once the fourth largest coffee exporter in the world, currently no more than 8–14% of the arable land is being utilized [6]. Agricultural exports in 2013 counted only for 8.5% of the peak figure in 1974, while the related revenues constituted a negligible 0.4% share in the national budget [7].

Citation: Biedermann, Z.; Barczikay, T.; Szalai, L. The Dutch Disease in Angola: An Empirical Analysis. *Eng. Proc.* **2023**, *39*, 40. https://doi.org/10.3390/engproc2023039040

Academic Editors: Ignacio Rojas, Hector Pomares, Luis Javier Herrera, Fernando Rojas and Olga Valenzuela

Published: 3 July 2023

Until the end of the 2014 commodity supercycle, these negative effects did not have a direct impact on growth. However, the complete loss of the agricultural sector and dominance of the oil industry suggest that Angola did suffer and keeps suffering from the Dutch disease. Our aim is to address the involvement of this problem in Africa's second largest oil exporter by examining the cointegration—i.e., long-term statistical co-movement, free of confounding factors—between oil prices and the real effective exchange rate (REER) of the national currency (the Angolan kwanza)—in the framework of a nonlinear autoregressive distributed lag (NARDL) model. In other words, we are looking for signs of the spending effect to explain the lack of economic diversity. More specifically, we seek to answer the following two questions:

Can cointegration—a long-term statistical link—be detected between the real price of oil on the world market and the REER of the kwanza? We expect to find a cointegrated nexus as a remarkable symptom of the Dutch disease in Angola.

Furthermore, if this link exists, is it asymmetrical? That is, does a unit change in the price of oil affect the appreciation or depreciation of the exchange rate to a different extent? This asymmetry would suggest a long-term deterioration in terms of trade, which commonly characterizes the trade relationships of developing nations. We assume that the cointegration is asymmetrical, and that the amplitude of appreciation is higher than that of depreciation.

2. Data and Methods

In the related empirical literature, the technique of cointegration is typically used to detect a long-term statistical relationship between commodity prices and different real effective exchange rates [8,9]. This procedure filters out the intermittent local co-movements caused by accidental fluctuations, which otherwise would result in biased and inconsistent estimations [10,11]. Based on Shin et al. [12], we formulated our error correction model using an asymmetrical form so that we could address both research questions outlined in the introduction.

To develop the asymmetric distributed lag model, first we broke down the logarithm of the real price of oil on the world market (p_t) into partial sum processes by separating the positive (p_t^+) and negative (p_t^-) changes:

$$\log p_t^+ = \sum_{j=1}^{t} \Delta \log p_j^+ = \sum_{j=1}^{t} \max\left(\Delta \log p_j; 0\right) \tag{1}$$

$$\log p_t^- = \sum_{j=1}^{t} \Delta \log p_j^- = \sum_{j=1}^{t} \min\left(\Delta \log p_j; 0\right) \tag{2}$$

If we add up the time series produced by Equations (1) and (2), we obtain our original $\log p_t$ series. The ξ_t error term representing the deviation from the long-term stochastic equilibrium can be calculated as the difference between e_t, the real affective exchange rate, and the logarithms of the decomposed oil price [12]:

$$\xi_t = \log e_t - \beta^+ \log p_t^+ - \beta^- \log p_t^- \tag{3}$$

To estimate both the long- and short-term effects, we used the following equation:

$$\begin{aligned}\Delta \log e_t = {} & \rho \log e_{t-1} + \theta^+ \log p_{t-1}^+ + \theta^- \log p_{t-1}^- + \sum_{j=1}^{n-1} \gamma_j \Delta \log e_{t-j} \\ & + \sum_{j=0}^{m-1} \left(\phi^+ \Delta \log p_{t-j}^+ + \phi^- \Delta \log p_{t-j}^-\right) + \varepsilon_t\end{aligned} \tag{4}$$

$$\Delta \log e_t = \rho \xi_{t-1} + \sum_{j=1}^{n-1} \gamma_j \Delta \log e_{t-j} + \sum_{j=0}^{m-1} \left(\phi^+ \Delta \log p_{t-j}^+ + \phi^- \Delta \log p_{t-j}^-\right) + \varepsilon_t \tag{5}$$

Equation (4) was used to estimate the long-term asymmetry parameters of Equation (5), β^+ and β^-, which are based on $-\theta^+/\rho$ and $-\theta^-/\rho$, respectively. Parameter ρ in Equation (5) measures the convergence rate of return to the common stochastic equilibrium, a phenomenon that occurs when this parameter is negative and statistically significant.

In summary, the relationship between the real effective price of the kwanza and the real price of oil—and thus the presence of the Dutch disease—is confirmed if parameter ρ is significant and negative. Furthermore, if the estimates for parameters β^+ and β^- are significantly different, the model verifies the presence of asymmetry.

For the exchange rate, we used the Bruegel Institute's REER time series, while nominal oil prices were derived from the World Bank's pink sheet databank and converted to real terms using the consumer price index from the IMF's IFS statistics database. Thus, we obtained monthly time series data for the period between January 2005 and December 2018.

3. Results

The upper section of Figure 1 demonstrates the quasi-permanent appreciation of the kwanza, while the lower section shows the effects of the financial crisis on the oil price, as well as the end of the 2014 supercycle. In order to establish a link between the variables, we examined the time series in terms of stationarity and cointegration.

Figure 1. Time series of real oil prices and kwanza REER.

Based on our model, the ADF and Phillips–Perron postestimation tests augmented with an intercept were unable to reject the null hypothesis that the time series contained a unit root at a 5% significance level ($e_t : p_{ADF} = 0.14$, $p_{PP} = 0.45$, $p_t : p_{ADF} = 0.08$, $p_{PP} = 0.07$). The 5.73 statistic of the Pesaran–Shin–Smith (PSS) test for asymmetric cointegration exceeded the upper critical value of 3.23 and thus indicated hidden cointegration [13].

The fit of the estimated model shown in Table 1 was satisfied with an adjusted R^2 of 15%. In addition, all postestimation tests were passed and there were no signs of autocorrelation ($\chi^2_{SC} = 0.54$) nor ARCH effect ($\chi^2_H = 0.88$). The value of the ρ convergence parameter was, as the theory suggested, significant and negative with a value of -0.05, indicating cointegration and a slow return to the joint equilibrium. Furthermore, the model was symmetrical, that is, we were not able to reject the null hypothesis of the long-term asymmetry test at any conventional significance level ($W_{LR} = 0.18$).

Table 1. Regression results of the NARDL model.

Parameters	Coeff.	S.E.	*t*-Stat	*p*-Value
EC Parameters				
Const.	0.23 **	0.10	2.28	0.02
ρ	−0.05 **	0.02	−2.21	0.03
θ^+	0.01	0.01	1.28	0.20
θ^-	0.01	0.01	0.77	0.44
γ_1	0.35 ***	0.08	4.41	0.01
ϕ^+	−0.02	0.05	−0.39	0.70
ϕ^-	−0.03	0.04	−0.92	0.36
Asymmetric LR parameters				
β^+	0.20 **	0.09	2.22	0.03
β^-	0.11 **	0.05	2.21	0.03
Diagnostics				
R^2		0.18		
adj. R^2		0.15		
χ^2_{SC}		0.54		
χ^2_{H}		0.88		
W_{LR}		0.18		
Observations		168		

** Significant at 5% level; *** Significant at 1% level

4. Discussion

The oil sector in Angola counts for one-third of the GDP and more than 90% of export revenues. This extreme oil-dependency raises serious doubts since reserves are expected to be depleted by 2032 at current production levels [7]. At the same time, economic growth is clearly linked to the dynamics of international oil prices and has been fuelled by extraction since oil production doubled between 2002 and 2008. The 2014 oil price slump, however, revealed the unsustainability of the oil-based growth strategy. While Angola enjoyed an average annual growth rate of more than 8% between 2002 and 2014, the GDP declined by 0.98% yearly between 2015 and 2018. Currently, macro-stability is threatened by indebtedness, a decline in production, and a lack of foreign investments due to corruption.

The model shown in Table 1 supports our first hypothesis about the cointegration of oil prices and the kwanza REER, indicating a long-term statistical relationship. This result suggests that the Dutch disease is present and evolves through the spending effect. The subsequent deindustrialization exposes the country to the volatility of international commodity prices and divests the economy from the positive spillovers associated with manufacturing. Ultimately, Angola becomes even more dependent on the natural resource, which poses substantial risks to long-term economic growth. On the other hand, we did not find statistically significant evidence of asymmetry, that is, the effects of oil price shocks on the exchange rate are symmetrical. This result suggests that there is no long-term deterioration in terms of trade and oil price plunges still provide an opportunity to restore the competitiveness of non-resource sectors. Conclusively, policies aimed at diversifying the economy of Angola have the highest potential when international oil prices are low.

Author Contributions: Conceptualization, Z.B. and L.S.; Methodology, T.B.; Formal analysis, T.B.; Investigation, Z.B. and L.S.; Writing–original draft preparation, L.S. and T.B.; Writing–review and editing, L.S. All authors have read and agreed to the published version of the manuscript.

Funding: This research received no external funding.

Institutional Review Board Statement: Not applicable.

Informed Consent Statement: Not applicable.

Data Availability Statement: Not applicable.

Conflicts of Interest: The authors declare no conflict of interest.

References

1. Auty, R.M. Natural Resource Endowment, the State and Development Strategy. *J. Int. Dev.* **1997**, *9*, 651–663. [CrossRef]
2. Gylfason, T. Natural Resources and Economic Growth: From Dependence to Diversification. In *Economic Liberalization and Integration Policy*; Broadman, H.G., Paas, T., Welfens, P.J.J., Eds.; Springer: Berlin/Heidelberg, Germany, 2006; pp. 201–231. [CrossRef]
3. Weber-Fahr, M. *Treasure or Trouble? Mining in Developing Countries*; World Bank: Washington, DC, USA, 2002.
4. Szalai, L. A Review on the Resource Curse. *Period. Polytech. Soc. Manag. Sci.* **2018**, *26*, 179–189. [CrossRef]
5. Corden, M.W.; Neary, P.J. Booming Sector and De-Industrialisation in a Small Open Economy. *Econ. J.* **1982**, *92*, 825–848. [CrossRef]
6. Kyle, S. *Angola's Macroeconomy and Agricultural Growth*; (No. 2010–04); Cornell University: Ithaca, NY, USA, 2010.
7. Fiess, N.; Alnafeesi, A.; Barroso, R.; Calderon, C.; Didier, T.; Engelke, W.; Trost, M. *Angola: Country Economic Memorandum*; World Bank: Washington, DC, USA, 2018.
8. Barczikay, T.; Biedermann, Z.; Szalai, L. An investigation of a partial Dutch disease in Botswana. *Resour. Policy* **2020**, *67*, 101665. [CrossRef]
9. Jahan-Parvar, M.R.; Mohammadi, H. Oil Prices and Real Exchange Rates in Oil-Exporting Countries: A Bounds Testing Approach. *J. Dev. Areas* **2011**, *45*, 313–322. [CrossRef]
10. Granger, C.; Newbold, P. Spurious regressions in econometrics. *J. Econom.* **1974**, *2*, 111–120. [CrossRef]
11. Hendry, D. Econometrics-Alchemy or Science? *Economica* **1980**, *47*, 387–406. [CrossRef]
12. Shin, Y.; Yu, B.; Greenwood-Nimmo, M. Modelling Asymmetric Cointegration and Dynamic Multipliers in a Nonlinear ARDL Framework. In *Festschrift in Honor of Peter Schmidt*; Springer: New York, NY, USA, 2014. [CrossRef]
13. Pesaran Shin, Y.; Smith, R.J. Bounds testing approaches to the analysis of level relationships. *J. Appl. Econom.* **2001**, *16*, 289–326. [CrossRef]

engineering proceedings

MDPI

Proceeding Paper

Intrinsic Explainable Self-Enforcing Networks Using the ICON-D2-Ensemble Prediction System for Runway Configurations †

Dirk Zinkhan [1], Anneliesa Greisbach [2], Björn Zurmaar [3], Christina Klüver [4,*] and Jürgen Klüver [4]

1 Deutscher Wetterdienst, 63067 Offenbach, Germany; dirk.zinkhan@dwd.de
2 Master's Program Digital Business Innovation and Transformation, University of Duisburg-Essen, 45141 Essen, Germany; anneliesa.greisbach@stud.uni-due.de
3 Centre for Information and Media Services, University of Duisburg-Essen, 45127 Essen, Germany; bjoern.zurmaar@uni-due.de
4 CoBASC Research Group, 45130 Essen, Germany; cobasc@rebask.de
* Correspondence: kluever@rebask.de
† Presented at the 9th International Conference on Time Series and Forecasting, Gran Canaria, Spain, 12–14 July 2023.

Abstract: Weather forecasts are indispensable for the decision on the direction of operation of a runway system. Since the forecasts contain uncertainties, additional challenges arise for runway configuration management (RCM). With developments in machine learning, numerous models have been developed to improve forecasts and assist management. In this contribution, an intrinsically explainable Self-Enforcing Network (SEN) is presented as a decision support system for the RCM at Frankfurt Airport.

Keywords: Self-Enforcing Networks (SEN); Explainable AI (xAI); ICON-D2-EPS; runway configurations

Citation: Zinkhan, D.; Greisbach, A.; Zurmaar, B.; Klüver, C.; Klüver, J. Intrinsic Explainable Self-Enforcing Networks Using the ICON-D2-Ensemble Prediction System for Runway Configurations. *Eng. Proc.* **2023**, *39*, 41. https://doi.org/10.3390/engproc2023039041

Academic Editors: Ignacio Rojas, Hector Pomares, Luis Javier Herrera, Fernando Rojas and Olga Valenzuela

Published: 3 July 2023

1. Introduction

Weather forecasts are fundamental to flight safety and management. The functioning of airport-related meteorological services supports decision-making regarding flight routes and planning [1].

Despite the greatest efforts in recent years to improve weather forecasts, the uncertainties that still exist must be considered [2,3]. In addition, new challenges arise from climatic changes, extreme weather conditions, strong wind shear at low altitudes [4,5], lateral boundary perturbations [6], or general hazardous meteorological conditions [7] that require variation or extension of previous models [8–10].

The digitalization and increase in, e.g., sensors and satellite imagery, create a large amount of data that is analyzed with various tools (for an overview, see [8]). For example, Key Performance Indicators (KPIs) are recommended to be used as a propensity metric in the preparation of Terminal Aerodrome Forecasts (TAFs) for future weather conditions [1]. Parameters for traffic management initiatives (TMI) under uncertain weather conditions are proposed using an epsilon greedy approach and a Softmax algorithm [2]. For postprocessing models, natural gradient boosting (NGB), quantile random forests (QRF), distributional regression forests (DRF), or Support Vector Quantile Regression (SVQR) are used [11,12].

Advances in machine learning and deep learning have led to many methods being developed in recent years to improve weather forecasts and support air traffic management (ATM) [13]. These methods are correspondingly diverse, such as an encoder-decoder U-net neural network to forecast convective storms and lightning [10], offline model-free reinforcement learning, or eXtreme Gradient Boosting (XGBoost), to support runway configuration management (RCM) [14,15], detection of adverse weather with EEG-enabled Bayesian neural networks [7], or anomaly detection and hierarchical clustering to spot anomalous data and group similar forecasts [1]. Additional methods are used, e.g., Transductive Long Short Term Memory (T-LSTM) for weather forecasting [16] or Spatio-Temporal Graph Convolutional Network to estimate the arrival and departure capacity under weather impact [17].

Hybrid systems combine different methods and learning concepts, such as supervised, unsupervised, and self-attention [18,19], or decomposition methods with adaptive learning strategies [20].

In comparison, few articles were found in which self-organizing neural networks are used for weather prediction, e.g., structural self-organizing maps (S-SOMs) for weather typing [21], combining SOM with ensemble climate forecasts to investigate the predictability of European summer climate in relation to the North Atlantic jet stream [22], to predict ocean currents [23], or to analyze radar data for nowcasting [24].

A big challenge remains regarding the explainability of the results in the context of ATM [5,25–27]. According to Degas et al. [28], the problem for responsible end users in ATM is that the results are difficult to understand or are not transparent for safety-critical areas such as air traffic. Explainability is necessary to (a) describe the algorithms or the results (descriptive xAI), (b) predict the behavior of an algorithm (predictive xAI), or (c) detect errors or undesirable behavior of an AI method (prescriptive xAI).

Explainable AI algorithms are distinguished by their scope. The scope describes the area to which the explanation refers. It can refer to the logic of the entire model (global xAI) or to the explanation of the background of a particular decision (local xAI) [29].

A further differentiation is made between post-hoc and ante-hoc explanation methods. In a post-hoc method, an explanation is given after training, in an ante-hoc method, it is given during training and is already available through the design of the algorithm. This method is also called intrinsic xAI, which is model-specific and explainable due to its internal structure. The learning process is transparent (algorithmic transparency), the technical operation is understandable (simulatability), or the algorithm can be decomposed into its individual parts (decomposability) [30].

We address the problem by showing how decision support for runway selection based on weather forecasts is accomplished with a Self-Enforcing Network (SEN), a self-organized learning neural network [31]. To explain the recommendations, Shapley Values are used to indicate which wind speeds and directions are determinants for the recommendation. Due to the way the overall system works, we are referring to an intrinsically explainable SEN.

The contribution is organized as follows: the next section provides a brief introduction to the methods used. Subsequently, the model as well as the results are presented. Finally, the challenges posed by the new, more complex, and detailed data structure are pointed out, and solution concepts are proposed.

2. Methods

The ensemble forecast system COSMO-DE-EPS, the Self-Enforcing Network (SEN), and Feature Importance using Shapley Values (SV) are briefly described below.

2.1. ICON-D2-EPS

Air traffic management at Frankfurt Airport requires not only weather data for a specific point in time but also forecast data for the next period. The forecast data comes from the ICON-D2 ensemble prediction forecast system (ICON-D2-EPS), which is operated by the German Weather Service (DWD). The use of an ensemble forecast system makes it possible to quantify uncertainties or probabilities for the predicted wind situations.

The main purpose of running an ensemble system is to enable an estimation of the forecast uncertainty, in our case, the development of the wind situation, by running a number of physically consistent scenarios of future development. The different scenarios arise from uncertainties in the initial conditions and model errors. An additional source of forecast uncertainty arises from the border conditions of a limited area model [32].

ICON-D2-EPS is a limited area model and currently consists of 20 ensemble members. It runs eight times a day with 48 h of forecasts for the 00, 03, 06, 09, 12, 15, 18, and 21 UTC runs.

In the case of ICON-D2-EPS, the 20 ensemble members were created through perturbations of the initial state and perturbations of a number of parametrizations of the model physics (e.g., turbulence, microphysics, or shallow convection). In addition, the boundary conditions were received from the global ICON-EPS forecast model, which consists of 40 ensemble members. From the total 40 ensemble members, the first 20 (arbitrarily chosen) were used to provide the boundary conditions for the 20 ensemble members of the limited area model ICON-D2-EPS [32].

On the one hand, this is an advantage over the use of traditional deterministic forecast models, but on the other hand, it poses an additional challenge for management as uncertainty must be included in the decision-making process [33].

2.2. Self-Enforcing Network (SEN)

The Self-Enforcing Network (SEN) is a deterministic, two-layered, and self-organized learning neural network that acquires and orders knowledge according to cognitive theory learning models [34,35].

The characteristic features of the SEN algorithm and SEN tool are: (a) a semantic matrix; (b) the so-called cue validity factor (cvf) for highlighting essential attributes; (c) the transformation of the data from the semantic matrix into the weight matrix by the specific learning rule; and (d) various visualizations.

(a) The semantic matrix is the basis of the learning process. It contains the essential attributes (features) and their degree of membership in an object [35]. The data is normalized in the interval $(-1, 1)$.

(b) For each feature, a cue validity factor (cvf) is set when building the model. The cue validity factor influences the strength of an attribute's effect on activation by the network. In the model, the cvf influences how much of the wind from the eleven measurement points is used for the decision. The measurement points that are farthest away have the lowest cvf; beyond that, the middle quantile (median) has the highest cvf (see below, Section 3).

(c) The Self-Enforcing Rule (SER) is the learning rule used in SEN, which transforms the values of the semantic matrix v_{sm} into a weight value between attribute and object w_{ao} with the learning rate c, a parameter responsible for adjusting weights in a neural network, and the cue validity factor (cvf_a):

$$w_{ao} = c \times v_{sm} \times cvf_a. \quad (1)$$

The peculiarity of SEN is that the weight matrix is not randomly generated.

There are several activation functions available that return the activation value a_j of a neuron; for this problem, the enforcing activation function (EAF) is the most suitable [33]. In the general representation of the functions in neural networks, w_{ij} (the weight w between the sending neuron i and the receiving neuron j) is the equivalent of w_{ao}:

$$a_j = \sum_{i=1}^{n} \frac{w_{ij} \times a_i}{1 + |w_{ij} \times a_i|}. \quad (2)$$

(d) The following visualizations are used to display the results: The so-called map visualization places the reference types on a two-dimensional map. The greater the Euclidean distance between the objects, the further apart they are shown on the map; correspondingly, similar objects are close together.

In the SEN-visualization, the activation values indicate the degree of similarity between the new input data (weather forecasts) and the reference types, i.e., the runway directions. The higher the final activation values are, the more the input data resembles the reference types. The computed Euclidean distances are also displayed, where similarity is indicated by the smallest distance between the input data and the reference types (see below, Section 3).

2.3. Shapley Values

The calculation of the Feature Importance is based on the concept of Shapley Values (SV) from the cooperative game theory of Shapley [36], according to which the influence of a player can be computed considering the effects of cooperation and individual performance on the game outcome.

The value a player contributes to the payoff is called the SV, with four defining properties: "zero player property", "efficiency", "symmetry", and "linearity" [37].

The zero player property means that an actor receives no share of the payoff if that actor contributes nothing to the final outcome. The efficiency property refers to the actual influence of an actor on the outcome and allows inferences to be made about that actor. Symmetry states that two actors who have the same influence on the outcome have the same SV. Apart from that, the actor who contributes the most has the highest SV. Linearity ensures that the sum of all SVs accounts for the total influence on the outcome.

Only if these four properties are true can a value be called a Shapley Value. The SV Φ of a player i on the game outcome v can be calculated with the following formula [38]:

$$\Phi_i(v) = \sum_{S \subseteq N \setminus \{i\}} \frac{|S|!(n-1-|S|)!}{n!}(v(S \cup \{i\}) - v(S)). \tag{3}$$

From the set of players participating in the game, all possible player sets S are formed, each containing a subset n of the entire number of players N. The influence Φ of player i on the outcome of the game v is calculated for each player set, considering the effects of cooperation between the players. The total contribution of the player to the game outcome $\Phi_i(v)$ is the sum of all partial influences resulting from coalitions in the respective player sets.

It turns out that the features in SEN have the properties of dummy players [39]. Dummy players are SV whose constant contribution to the game result is independent of the other players.

This property results from the fact that the weight matrix is not initialized with random values but with the values from the semantic matrix. This leads to the effect that the weight values directly reflect the importance of the individual attributes in the reference types, and thus the SV has corresponding effects on the output. Therefore, the formula for calculating the SV is shortened to [39]:

$$\Phi_i(v) = v(i). \tag{4}$$

The four properties of Shapley Values described above are met since, on the one hand, the feature's individual impact on the total activation is extracted, and on the other hand, the added values constitute the total activation value.

Because of the structure and operation of SEN and the immediate identification of its Feature Importance, we can consider SEN to be an intrinsically Explainable AI.

3. Model and Results

The operating direction of the runway system has to be selected according to the wind conditions between the ground and up to approximately 5000 m. The two possible operating directions at Frankfurt airport are "Direction 07", used during generally easterly winds, and "Direction 25", used during westerly winds.

The weather data necessary for the appropriate decisions are derived from the COSMO-DE-EPS ensemble prediction System of the German Weather Service (Deutscher Wetterdienst (DWD)).

The required weather data are ensemble forecasts for 11 reference positions located at different distances along the glide path to the airport and on the airport itself. For each position, five quantiles (i.e., a statistical measure reflecting the dispersion of the ensemble forecasts) are computed for the parallel wind component (i.e., the headwind or tailwind component of a departing or approaching aircraft). The ensemble forecast data of the DWD, processed over years, are the basis for our experiments.

For the development of the reference types, representing the operation directions 07 and 25 at the airport in Frankfurt, the decisions of air traffic controllers based on the weather forecasts were considered. The knowledge of several experts (including meteorologists, air traffic controllers, and SEN developers) was incorporated into the development of the model shown in Figure 1.

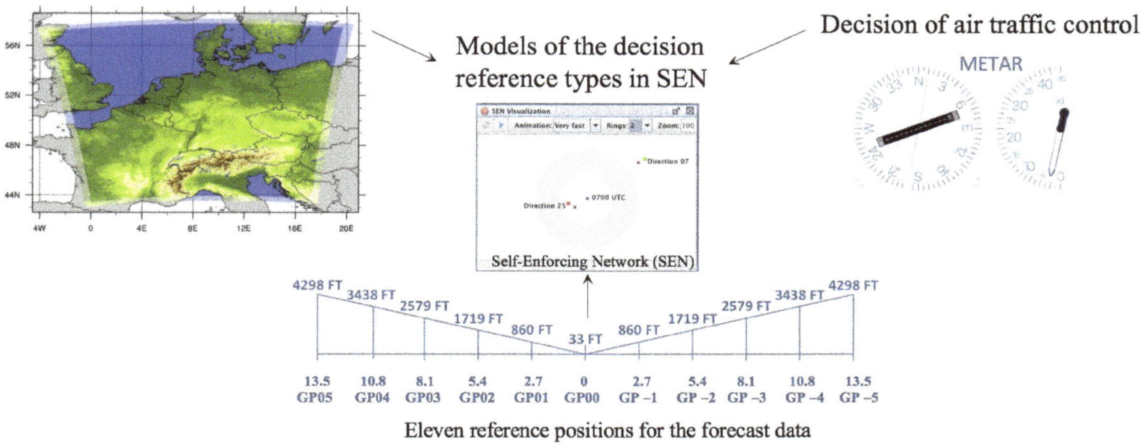

Figure 1. Model of the decision reference types (Figure on the left ([32], p. 9); Figure on the right METAR: https://metar-taf.com/de/EDDF accessed on 27 June 2023).

The reference types for directions 07 and 25 are defined in the semantic matrix, which contains 55 attributes (5 wind components reflecting the 5 quantiles for each of the 11 measurement points of the forecasted wind direction at the reference positions along the glide path—GP). The normalized data are transformed into the weight matrix using the learning rule and are learned by SEN as "representative or ideal wind conditions" with a learning rate c of 0.5 and three iterations using the Enforcing Activation Function (EAF). In Figure 2, there is a section of the reference types in which the west direction always contains negative values [35].

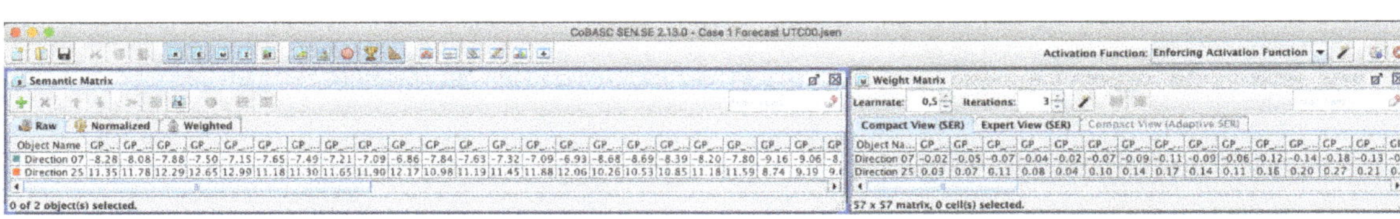

Figure 2. Excerpt of the semantic matrix (on the left) and the weight matrix.

After the learning process and given new input data (weather forecasts), different visualizations are available, as shown in Figure 3 with a recent weather forecast.

The input vectors include the forecasted data at 0000 UTC for the next fifteen hours. The two visualizations next to the input vectors show the similarity of the forecasts to the reference types for the corresponding operation direction. In the SEN visualization, only one selected prediction is focused, and the appropriate reference type is attracted according to the similarity of the wind conditions. In the map visualization, all predictions are classified into the reference types Direction 07 (at the bottom) and Direction 25 (at the top). The calculated activation values are visualized in tabular form (colored green for high positive activation, white for low activation and red for high negative activation) as well as by bars sorted according to highest activation (Ranking) and smallest Euclidean distance (Distance). In this case, the unambiguous recommendation is runway 07.

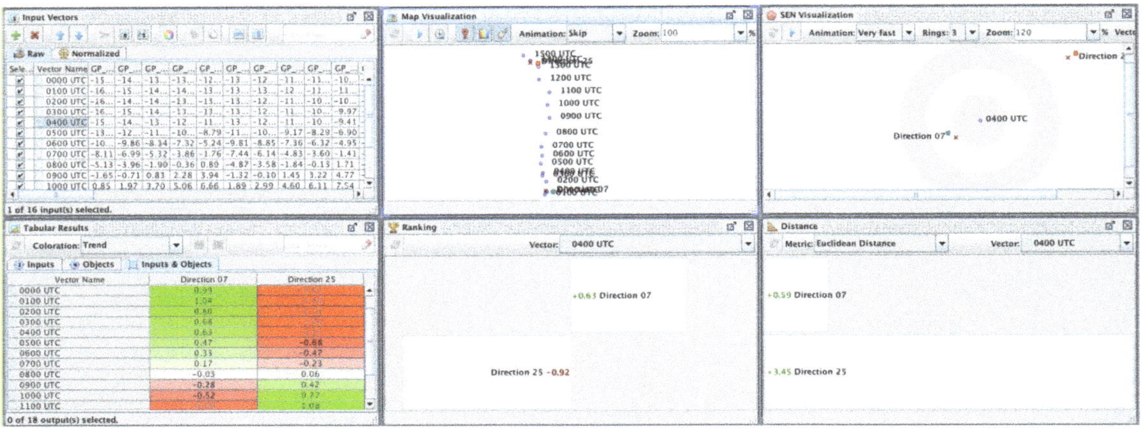

Figure 3. Results and visualizations of the SEN recommendation at 0000 UTC.

The intrinsic explainability of SEN allows the straightforward determination and visualization of the feature's importance for the final activation of a forecast. For a better interpretation of the results, it was decided to calculate the feature importance for the 11 measurement points along the glide path and to adjust the visualization by rotating the visualization by 90 degrees and sorting the values according to the location on the glide path. Figure 4 shows the feature importance of the two reference types, Directions 07 (colored green) and Direction 25 (colored blue).

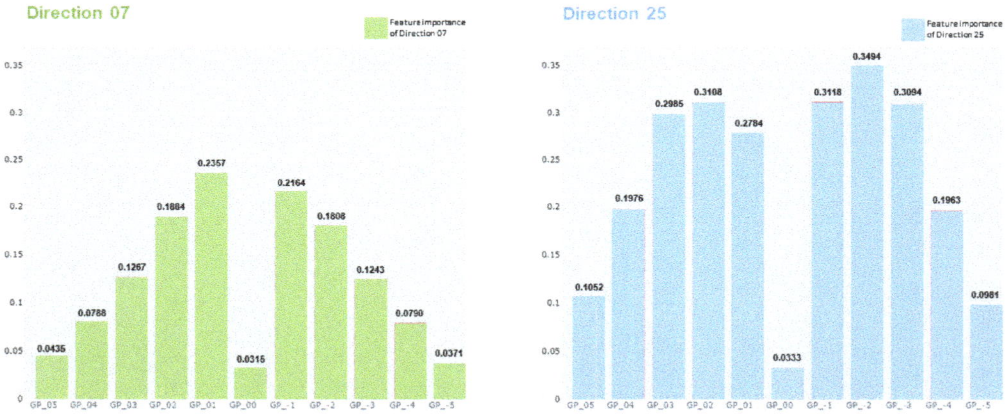

Figure 4. Illustration of the Feature Importance of Directions 07 and 25.

The forecast at 0000 UTC for 0400 UTC is classified by SEN as Direction 07 with a final activation of 0.63 (Figure 3). An analysis of the feature importance of this prediction in Figure 5 shows the influence of the conditions at the measurement points on the final activation compared to the classified direction. For 0400 UTC, the high-altitude winds (GP_5, GP_4) have a stronger influence on the final result than for Direction 07, while the medium-altitude winds at GP_3, GP_2, and GP_1 have a weaker influence. The winds directly on the runway (GP_00) have a negative Feature Importance. Here, wind conditions oppose the use of Direction 07 and thus support the use of Direction 25 by reducing the activation of Direction 07.

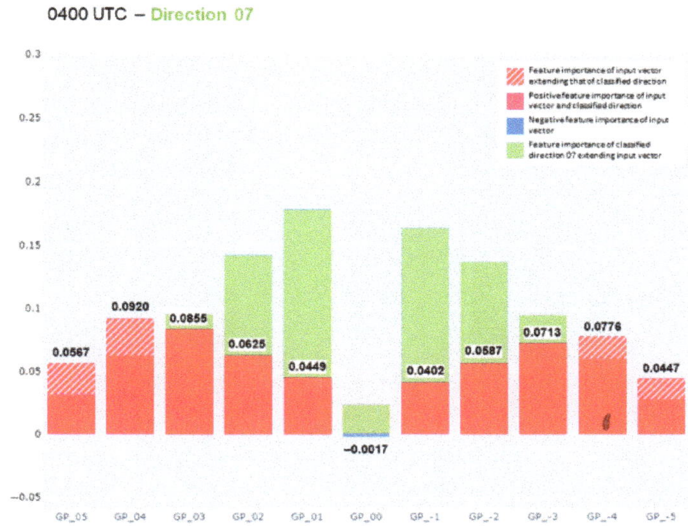

Figure 5. Forecast for 0400 UTC at 0000 UTC.

The prediction at 0000 UTC (Figure 3) can be further analyzed by examining the change in feature importance at the fifteen predicted times (Figure 6).

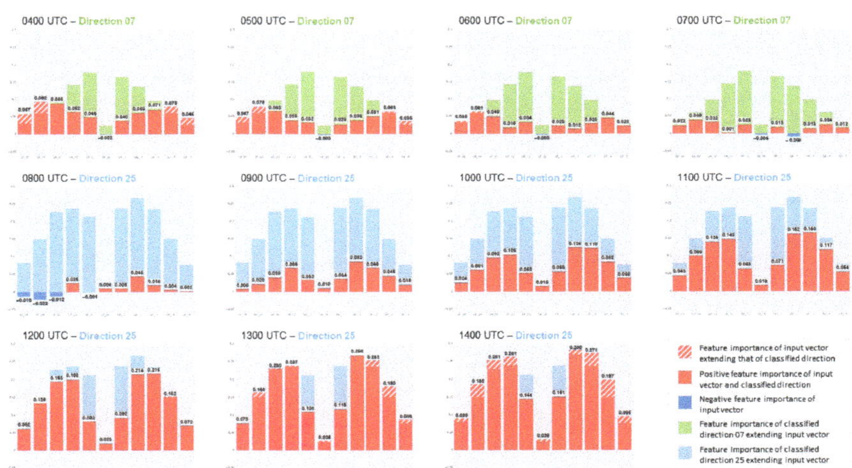

Figure 6. Feature Importance between 0400 UTC and 1400 UTC at 0000 UTC.

The influence of the high-altitude winds decreases in the course of time until 0700 UTC, and the ground-level wind has less and less influence on the final activation, too. At 0800 UTC, the winds are classified by SEN as being in direction 25, and thus a change of the operating direction is recommended. Here the correspondence between ranking and distances is not correct, an indication of the upcoming change (Figure 7). Currently, the winds close to the ground are crucial for the decision. Until 1400 UTC, the feature importance increases at all measurement points, and finally, even the high-altitude winds have a stronger activation than the classified direction 25.

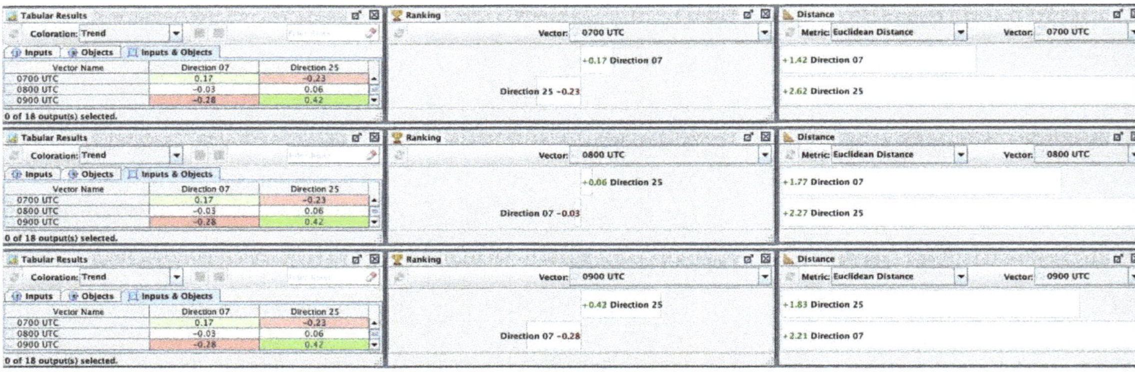

Figure 7. Prediction of SEN at 00 UTC between 07 UTC and 09 UTC for Case.

The forecast at 0000 UTC predicts a change of direction from Direction 07 to Direction 25 at 0800 UTC. Examining the forecasts between 1800 UTC of the previous day and 0900 UTC of the following day, the change was already predicted at that time. Figure 8 shows the map visualization of SEN for the different forecasts, which was created in Python. This visualization was not generated by attraction and repulsion as in the SEN tool but by calculating the concrete coordinates; nevertheless, it gives the same representation. The line between the forecasts allows the user to see the progression of the forecast. The corridor immediately indicates the time at which a change in wind direction occurs.

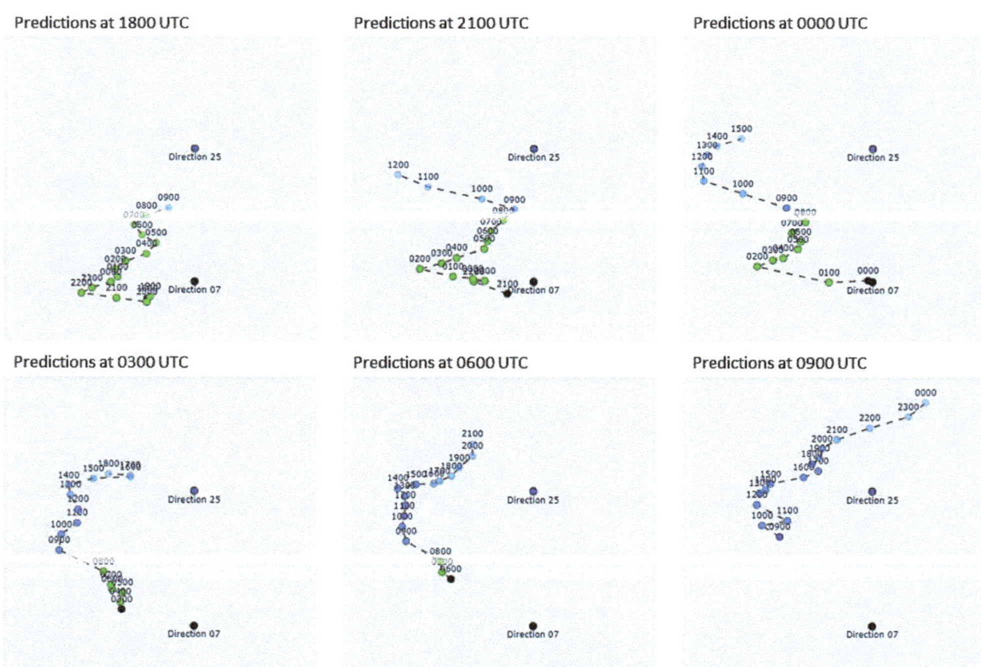

Figure 8. Predictions of forecasts between 1800 UTC and 0900 UTC.

Upon analyzing the Feature Importance at the time of the change, it can be seen how the recommendation for a change at 0800 UTC strengthened over time as the time of the change got closer (Figure 9).

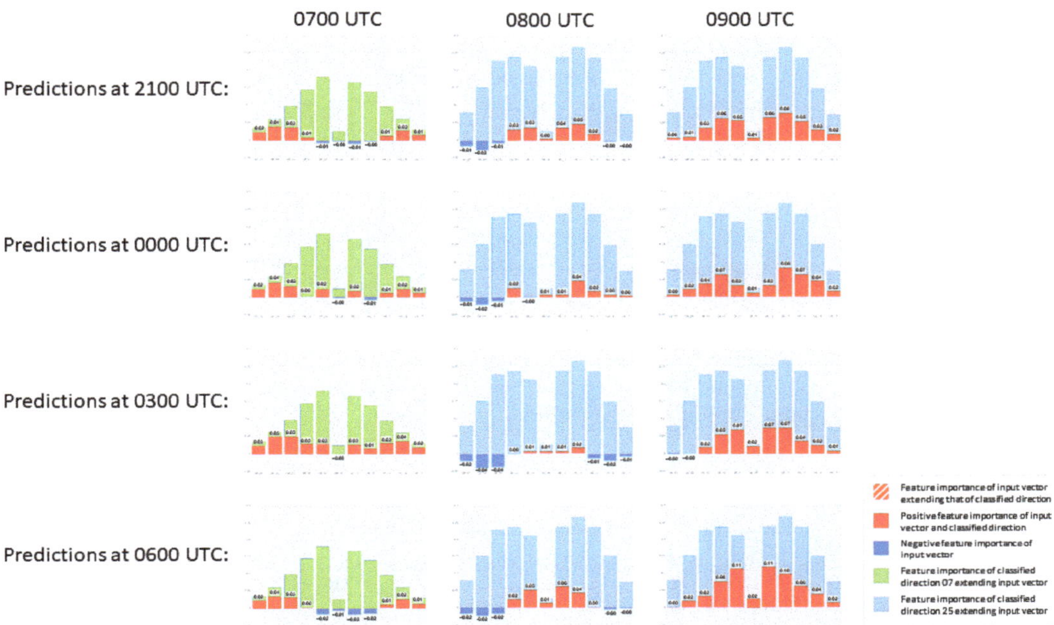

Figure 9. Feature Importance of 0700, 0800, and 0900 UTC at 2100, 0000, 0300, and 0600 UTC.

In this case, at 0249 UTC, it was decided to start air traffic with Direction 25 at the start of flight operations at 0400 UTC. This was justifiable because, until 0800 UTC, the winds were very weak, and in this case, flight operations with a tailwind were possible. This saved the costly change during the day.

4. Conclusions and Recent Work

For complex systems such as weather forecasting and runway configuration management, we have shown how Feature Importance can be used to help understand the results of the recommendation. The architecture and learning procedures of SEN fulfill the properties of Shapley Values with no loss of performance in determining Feature Importance. Feature Importance can be read directly by decomposing a vector into its individual components; thus, SEN can be classified as intrinsically Explainable AI.

While the first version used for predicting the best runway to approach used 11 selected measurement points in the planes' glide path, a currently enhanced model utilizes wind data from around the airport in an area of 2.500 km^2. When taking into account the area, the grid's density, and the number of ensembles, this increases the amount of input data to be processed for a single prediction by a factor of more than 168.000. With thousands of available training data entries, the amount of data to be processed is staggering. The fact that weight matrices grow exponentially with increasing object numbers complicates matters even further.

The amount of data to deal with has two consequences: firstly, raw data can no longer be inspected reasonably because the human mind is unable to cope with the given amount. Secondly, the required computational power to run experiments is not available on current consumer desktop systems. This applies especially to the number of CPU cores available.

The above considerations lead to the following consequences:

1. It is necessary to build tools to view raw data in an aggregated and easily perceivable manner.
2. It is necessary to move the experimentation software away from a desktop application and towards a command-line tool that can be executed on a HPC. Techniques and optimizations have to be introduced to deal with the new dimension of data.
3. The modeling process can no longer be accomplished by a human alone but instead needs to be computer-aided in terms of data selection for the training process and validation of the resulting model.
4. When discussing the results with domain experts, i.e., air traffic controllers, the decisions taken by the system must be retraceable and must be presented to them in a manner that makes it clear how the predicted result came about.

We are confident that the usage of Explainable AI, as presented in this contribution, will support the professional exchange with flight controllers by enabling us to quickly comprehend why the system came to a specific prediction and which edge cases can be ignored or must be considered.

Author Contributions: D.Z., A.G., B.Z., C.K. and J.K. enumerated in this paper participated equally. All authors have read and agreed to the published version of the manuscript.

Funding: This research received no external funding.

Institutional Review Board Statement: Not applicable.

Informed Consent Statement: Not applicable.

Data Availability Statement: Not applicable.

Conflicts of Interest: The authors declare no conflict of interest.

References

1. Patriarca, R.; Simone, F.; Di Gravio, G. Supporting weather forecasting performance management at aerodromes through anomaly detection and hierarchical clustering. *Expert Syst. Appl.* **2023**, *213*, 119210. [CrossRef]
2. Jones, J.C.; Ellenbogen, Z.; Glina, Y. Recommending strategic air traffic management initiatives in convective weather. *J. Air Transp.* **2023**, *31*, 45–56. [CrossRef]
3. Scala, P.; Mota, M.M.; Ma, J.; Delahaye, D. Tackling uncertainty for the development of efficient decision support system in air traffic management. *IEEE Trans. Intell. Transp. Syst.* **2019**, *21*, 3233–3246. [CrossRef]
4. Bombelli, A.; Sallan, J.M. Analysis of the effect of extreme weather on the US domestic air network. A delay and cancellation propagation network approach. *J. Transp. Geogr.* **2023**, *107*, 103541. [CrossRef]
5. Khattak, A.; Chan, P.-W.; Chen, F.; Peng, H. Prediction of a Pilot's Invisible Foe: The Severe Low-Level Wind Shear. *Atmosphere* **2023**, *14*, 37. [CrossRef]
6. Zhang, L.; Min, J.; Zhuang, X.; Wang, S.; Qiao, X. The Lateral Boundary Perturbations Growth and Their Dependence on the Forcing Types of Severe Convection in Convection-Allowing Ensemble Forecasts. *Atmosphere* **2023**, *14*, 176. [CrossRef]
7. Yiu, C.Y.; Ng, K.K.H.; Li, X.; Zhang, X.; Li, Q.; Lam, H.S.; Chong, M.H. Towards safe and collaborative aerodrome operations: Assessing shared situational awareness for adverse weather detection with EEG-enabled Bayesian neural networks. *Adv. Eng. Inform.* **2022**, *53*, 101698. [CrossRef]
8. Fathi, M.; Haghi Kashani, M.; Jameii, S.M.; Mahdipour, E. Big Data Analytics in Weather Forecasting: A Systematic Review. *Arch. Comput. Methods Eng.* **2022**, *29*, 1247–1275. [CrossRef]
9. Gonzalo, J.; Domínguez, D.; López, D.; García-Gutiérrez, A. An analysis and enhanced proposal of atmospheric boundary layer wind modelling techniques for automation of air traffic management. *Chin. J. Aeronaut.* **2021**, *34*, 129–144. [CrossRef]
10. Li, Y.; Liu, Y.; Sun, R.; Guo, F.; Xu, X.; Xu, H. Convective Storm VIL and Lightning Nowcasting Using Satellite and Weather Radar Measurements Based on Multi-Task Learning Models. *Adv. Atmos. Sci.* **2023**, *40*, 887–899. [CrossRef]
11. Benáček, P.; Farda, A.; Štěpánek, P. Postprocessing of Ensemble Weather Forecast Using Decision Tree–Based Probabilistic Forecasting Methods. *Weather Forecast.* **2023**, *38*, 69–82. [CrossRef]
12. Takamatsu, T.; Ohtake, H.; Oozeki, T. Support Vector Quantile Regression for the Post-Processing of Meso-Scale Ensemble Prediction System Data in the Kanto Region: Solar Power Forecast Reducing Overestimation. *Energies* **2022**, *15*, 1330. [CrossRef]
13. Tang, J.; Liu, G.; Pan, Q. Review on artificial intelligence techniques for improving representative air traffic management capability. *J. Syst. Eng. Electron.* **2022**, *33*, 1123–1134. [CrossRef]
14. Churchill, A.; Coupe, W.J.; Jung, Y.C. Predicting Arrival and Departure Runway Assignments with Machine Learning. In Proceedings of the 2021 AIAA AVIATION FORUM, Virtual, 2–6 August 2021. [CrossRef]

15. Memarzadeh, M.; Puranik, T.G.; Battistini, J.; Kalyanam, K.M.; Ryan, W. Airport Runway Configuration Management with Offline Model-Free Reinforcement Learning. In Proceedings of the 2023 AIAA SCITECH Forum, National Harbor, MD, USA, 23–27 January 2023. [CrossRef]
16. Venkatachalam, K.; Trojovský, P.; Pamucar, D.; Bacanin, N.; Simic, V. DWFH. An improved data-driven deep weather forecasting hybrid model using Transductive Long Short Term Memory (T-LSTM). *Expert Syst. Appl.* **2023**, *213*, 119270. [CrossRef]
17. Tang, S.; Fang, Q.; Yang, Y.; Chen, J.; Cai, K. A Learning Estimation Approach for Arrival and Departure Capacity considering Weather Impact. In Proceedings of the 2022 IEEE/AIAA 41st Digital Avionics Systems Conference (DASC), Portsmouth, VA, USA, 18–22 September 2022. [CrossRef]
18. Ren, X.; Li, X.; Ren, K.; Song, J.; Xu, Z.; Deng, K.; Wang, X. Deep Learning-Based Weather Prediction: A Survey. *Big Data Res.* **2021**, *23*, 100178. [CrossRef]
19. Schultz, M.G.; Betancourt, C.; Gong, B.; Kleinert, F.; Langguth, M.; Leufen, L.H.; Mozaffari, A.; Stadtler, S. Can deep learning beat numerical weather prediction? *Philos. Trans. R. Soc. A* **2021**, *379*, 20200097. [CrossRef]
20. Zouaidia, K.; Rais, M.S.; Ghanemi, S. Weather forecasting based on hybrid decomposition methods and adaptive deep learning strategy. *Neural Comput. Appl.* **2023**, *35*, 11109–11124. [CrossRef]
21. Doan, Q.-V.; Kusaka, H.; Sato, T.; Chen, F. S-SOM v1. 0: A structural self-organizing map algorithm for weather typing. *Geosci. Model Dev.* **2021**, *14*, 2097–2111. [CrossRef]
22. Carvalho-Oliveira, J.; Borchert, L.F.; Zorita, E.; Baehr, J. Self-organizing maps identify windows of opportunity for seasonal European summer predictions. *Front. Clim.* **2022**, *4*, 844634. [CrossRef]
23. Vilibić, I.; Šepić, J.; Mihanović, H.; Kalinić, H.; Cosoli, S.; Janeković, I.; Žagar, N.; Jesenko, B.; Tudor, M.; Dadić, V.; et al. Self-Organizing Maps-based ocean currents forecasting system. *Sci. Rep.* **2016**, *6*, 22924. [CrossRef]
24. Czibula, G.; Mihai, A.; Mihuleţ, E.; Teodorovici, D. Using self-organizing maps for unsupervised analysis of radar data for nowcasting purposes. *Procedia Comput. Sci.* **2019**, *159*, 48–57. [CrossRef]
25. Midtfjord, A.D.; De Bin, R.; Huseby, A.B. A decision support system for safer airplane landings: Predicting runway conditions using XGBoost and explainable AI. *Cold Reg. Sci. Technol.* **2022**, *199*, 103556. [CrossRef]
26. Rudd, K.; Eshow, M.; Gibbs, M. Method for Generating Explainable Deep Learning Models in the Context of Air Traffic Management. In Proceedings of the Machine Learning, Optimization, and Data Science: 7th International Conference, LOD 2021, Grasmere, UK, 4–8 October 2021; Revised Selected Papers, Part I; 2022. [CrossRef]
27. Xie, Y.; Pongsakornsathien, N.; Gardi, A.; Sabatini, R. Explanation of machine-learning solutions in air-traffic management. *Aerospace* **2021**, *8*, 224. [CrossRef]
28. Degas, A.; Islam, M.R.; Hurter, C.; Barua, S.; Rahman, H.; Poudel, M.; Ruscio, D.; Ahmed, M.U.; Begum, S.; Rahman, A.; et al. A survey on artificial intelligence (ai) and explainable ai in air traffic management: Current trends and development with future research trajectory. *Appl. Sci.* **2022**, *12*, 1295. [CrossRef]
29. Schwalbe, G.; Finzel, B. A comprehensive taxonomy for explainable artificial intelligence: A systematic survey of surveys on methods and concepts. *Data Min. Knowl. Discov.* **2023**. [CrossRef]
30. Arrieta, A.B.; Díaz-Rodríguez, N.; Del Ser, J.; Bennetot, A.; Tabik, S.; Barbado, A.; García, S.; Gil-Lopez, S.; Molina, D.; Benjamins, R.; et al. Explainable Artificial Intelligence (XAI): Concepts, taxonomies, opportunities and challenges toward responsible AI. *Inf. Fusion* **2020**, *58*, 82–115. [CrossRef]
31. Greisbach, A.; Klüver, C. Determining Feature Importance in Self-Enforcing Networks to achieve Explainable AI (xAI). In *Proceedings 32. Workshop Computational Intelligence*; Schulte, H., Hoffmann, F., Mikut, R., Eds.; KIT Scientific Publishing: Karlsruhe, Germany, 2022; pp. 237–256.
32. Reinert, D.; Prill, F.; Frank, H.; Denhard, M.; Baldauf, M.; Schraff, C.; Gebhardt, C.; Marsigli, C.; Zängl, G. DWD Database Reference for the Global and Regional ICON and ICON-EPS Forecasting System. 2023. Available online: https://www.dwd.de/SharedDocs/downloads/DE/modelldokumentationen/nwv/icon_d2/icon_d2_dbbeschr_aktuell.pdf?view=nasPublication&nn=13934 (accessed on 27 June 2023).
33. Klüver, C.; Klüver, J.; Zinkhan, D. A self-enforcing neural network as decision support system for air traffic control based on probabilistic weather forecasts. In Proceedings of the IEEE International Joint Conference on Neural Networks (IJCNN), Anchorage, AK, USA, 14–19 May 2017; pp. 729–736. [CrossRef]
34. Klüver, C.; Klüver, J. Self-organized learning by self-enforcing networks. In *IWANN 2013, LNCS 7902*; Part I; Rojas, I., Joya, G., Cabestany, J., Eds.; Springer: Berlin/Heidelberg, Germany, 2013; pp. 518–529. [CrossRef]
35. Zinkhan, D.; Eiermann, S.; Klüver, C.; Klüver, J. Decision Support Systems for Air Traffic Control with Self-enforcing Networks Based on Weather Forecast and Reference Types for the Direction of Operation. In *Advances in Computational Intelligence*; IWANN 2021. Lecture Notes in Computer Science; Rojas, I., Joya, G., Catala, A., Eds.; Springer: Cham, Switzerland, 2021; Volume 12862, pp. 404–415. [CrossRef]
36. Shapley, L.S. A Value for n-Person Games. In *Contributions to the Theory of Games (AM-28)*; Kuhn, H.W., Tucker, A.W., Eds.; Princeton University Press: Princeton, NJ, USA, 1953; Volume 2, pp. 307–318.
37. Rozemberczki, B.; Watson, L.; Bayer, P.; Yang, H.-T.; Kiss, O.; Nilsson, S.; Sarkar, R. The shapley value in machine learning. *arXiv* **2022**, arXiv:2202.05594.

38. Bokati, L.; Kosheleva, O.; Kreinovich, V.; Thach, N.N. Why Shapley Value and Its Variants Are Useful in Machine Learning (and in Other Applications). Proc., 15. Workshop Computational Intelligence. Departmental Technical Reports (CS). 1729. 2022. Available online: https://scholarworks.utep.edu/cs_techrep/1729/ (accessed on 27 June 2023).
39. Casajus, A.; Huettner, F. Null, nullifying, or dummifying players: The difference between the Shapley value, the equal division value, and the equal surplus division value. *Econ. Lett.* **2014**, *122*, 167–169. [CrossRef]

engineering proceedings

MDPI

Proceeding Paper

Hints of Earlier and Other Creation: Unsupervised Machine Learning in Financial Time-Series Analysis †

James Ming Chen [1,*] and Charalampos Agiropoulos [2]

1 College of Law, Michigan State University, East Lansing, MI 48824, USA
2 School of Economics, Business, and International Studies, University of Piraeus, 18534 Piraeus, Greece; hagyropoylos@gmail.com
* Correspondence: chenjame@law.msu.edu

† Presented at the 9th International Conference on Time Series and Forecasting, Gran Canaria, Spain, 12–14 July 2023.

Abstract: This study extends previous work applying unsupervised machine learning to commodity markets. The first article in this sequence examined returns and volatility in commodity markets. The clustering of these time series supported the conventional ontology of commodity markets for precious metals, base metals, agricultural commodities, and crude oil and refined fuels. A second article used temporal clustering to identify critical periods in the trading of crude oil, gasoline, and diesel. This study combines the ontological clustering of financial time series with the temporal clustering of the matrix transpose. Ontological clustering, contingent upon the identification of structural breaks and other critical periods within financial time series, is this study's distinctive contribution. Conditional, time-variant ontological clustering should be applicable to any set of related time series, in finance and beyond.

Keywords: unsupervised machine learning; clustering; financial time series; commodities; energy; fossil fuels

> The sea is the land's edge also, the granite
> Into which it reaches, the beaches where it tosses
> Its hints of earlier and other creation:
> The starfish, the horseshoe crab, the whale's backbone;
> The pools where it offers to our curiosity
> The more delicate algae and the sea anemone.
>
> T.S. Eliot, "The Dry Salvages", *Four Quartets* (1943) [1] (p. 36)

Citation: Chen, J.M.; Agiropoulos, C. Hints of Earlier and Other Creation: Unsupervised Machine Learning in Financial Time-Series Analysis. *Eng. Proc.* **2023**, *39*, 42. https://doi.org/10.3390/engproc2023039042

Academic Editors: Ignacio Rojas, Hector Pomares, Luis Javier Herrera, Fernando Rojas and Olga Valenzuela

Published: 3 July 2023

1. Introduction

This paper extends previous work using unsupervised machine learning to evaluate commodity markets. "Clustering Commodity Markets in Space and Time" examined returns and volatility in commodity markets [2]. That paper supported the conventional ontology of commodity markets for precious metals, base metals, agricultural commodities, and crude oil and refined fuels. These groupings emerged from the application of clustering methods and a nonlinear manifold to the matrices formed by the concatenation of daily logarithmic returns for individual commodities or conditional volatility forecasts based on a GARCH(1, 1, 1) process.

A sequel to [2], "A Pattern New in Every Moment," used temporal clustering to identify critical periods in energy-related commodity markets [3]. That article applied a suite of clustering methods to the transpose of the time-series matrix evaluated in [2]. The temporal clustering of financial markets reveals market events that can be readily interpreted as shifts in volatility, cumulative logarithmic returns, or both. As applied to energy-related commodities trading during the first two decades of the twenty-first century, temporal clustering isolated critical periods associated with wars, terrorist attacks, comprehensive economic crises, and other disruptions in energy supply or demand (Figure 1).

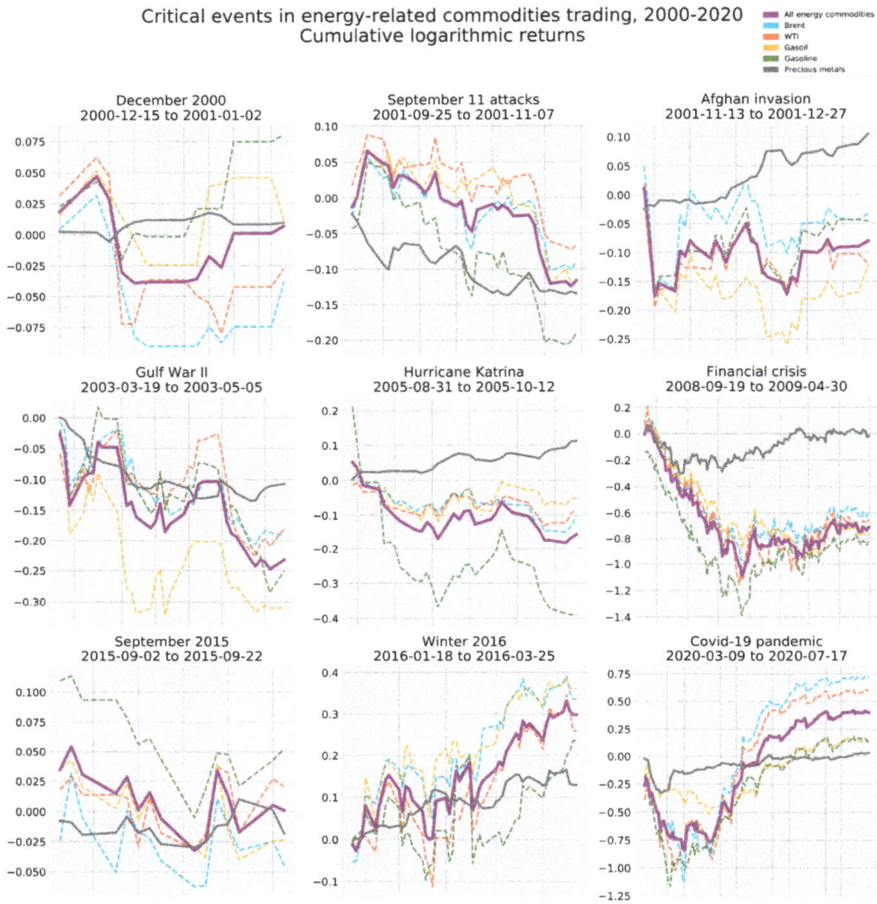

Figure 1. Cumulative logarithmic returns during critical periods for four oil and fuel commodities (dotted colored lines), with cumulative log returns on precious metals (solid gray line) as a benchmark [3] (p. 41).

This study proposes further elaborations in unsupervised machine learning. By combining the ontological (or spatial) approach to clustering of [2] with the temporal approach of [3], this study makes its own distinctive contribution to the application of unsupervised machine learning to financial time series. Ontological clustering, conditioned upon the identification of structural breaks and other critical periods, should reveal information about co-movement among asset classes and discrete assets as markets shift between normal and extraordinary states. By focusing on critical periods identified by temporal clustering, this study's novel hybrid clustering method reveals the extent, if any, to which the unconditional spatial ontology among financial assets changes under economic stress.

Together with its predecessor articles, this study expands the machine-learning toolkit for time-series analysis to three methods: (1) ontological clustering, (2) temporal clustering, and (3) the new hybrid method of conditional, time-variant ontological clustering. In principle, all three methods can be applied to any time series, in finance and beyond. Indeed, future research applying unsupervised machine learning may address forecasting tasks involving meteorology, air quality, and other environmental sciences.

This study contemplates the application of time-variant ontological clustering to markets for crude and refined fossil fuels as well as related commodities. Energy-related commodity markets are affected not only by background macroeconomic events, but also by events specific to energy markets. Shocks to supply should be distinguished from shocks to demand. Co-movement in prices for crude oil and refined fuel is asymmetrical insofar as these markets respond differently to rising and falling prices. Finally, biofuel demand affects markets for agricultural commodities that can provide human food or animal feed in addition to serving as biofuel feedstocks.

2. Materials and Methods

2.1. Materials

This study marshals market data across a variety of commodity markets, ranging from markets for crude oil, natural gas, and refined fuels to agricultural commodities and precious and base metals. Agricultural commodities are both substitutes for and complements to fossil fuels. Precious and base metals track inflation, consumer demand, and other macroeconomic conditions.

- Raw energy commodities: crude oil (Cushing and Brent) and natural gas;
- Refined energy commodities: gasoline, diesel/gasoil, and heating oil;
- Agricultural commodities: corn, soybeans, and sugar;
- Precious metals: gold, silver, platinum, and palladium;
- Base metals: copper, tin, nickel, and aluminum.

These five categories comprise 17 distinct commodity markets. Those 17 markets, in turn, are divided between six energy-specific markets and 11 commodity markets not typically regarded as energy-related. Data used in this study were drawn from Datastream and the United States Energy Information Administration. The data cover the period from 2000 to 2022.

Different combinations among commodity markets advance different research objectives. For instance, their traditional contribution to portfolio hedging and diversification enables precious metals to serve as a control variable for certain macroeconomic conditions, such as inflation or flights to safe havens [4–6]. Perhaps surprisingly, oil itself serves as a hedge, safe haven, and diversifier relative to conventional currencies during periods of turbulence [7]. Base metals indicate industrial demand, especially during declines in demand attributable to events exogenous to the business cycle, such as the COVID-19 pandemic [8].

Hypotheses anticipating asymmetric relationships between crude oil and refined fuels may be evaluated with as few as two individual commodity series, such as Brent or WTI and gasoline. Divergence between crude oil and natural gas prices may prove especially revealing. Finally, agricultural commodity markets may reveal the impact, if any, of renewable energy policies prescribing ethanol additives, E85 fuel, or biodiesel. Soft food commodities, such as corn, cotton, and cocoa, demonstrated safe-haven properties during the COVID-19 pandemic [9]. These crops are intriguingly diverse: whereas corn serves as human food, animal feed, and a biofuel feedstock, cotton is both food and fiber. As a component of sweets and a stimulant in its own right, cocoa serves more as a complement to staple grains and oilseeds than as a substitute for those crops [10].

2.2. Methods

This study applies a broad variety of clustering methods to different subsets of these commodity market time series. The clustering methods deployed in this study include spectral clustering [11–13], mean-shift clustering [14], affinity propagation [15–17], hierarchical agglomerative clustering [18], and *k*-means clustering [19]. A single method of manifold learning—*t*-distributed stochastic neighbor embedding (*t*-SNE)—facilitates the visualization of clusters among commodity markets [20–22]. A comprehensive discussion of those five clustering methods and *t*-SNE appears in [3] (pp. 12–14).

3. Anticipated Results

The application of conditional, time-variant ontological clustering to energy-related commodities should provide deeper insights into a wide range of contestable and controversial propositions about the behavior of these markets. Oil price shocks comprise two distinct components: effects endogenous to the global business cycle and effects specific to markets for energy-related commodities [23]. The endogenous component of oil price volatility reflects cyclical differences in the performance of equity markets in the United States and other advanced economies [24,25].

Industry-specific effects may be usefully divided according to a dichotomy between supply-side and demand-side effects [26,27]. Supply-side disruptions are conventionally ascribed to geographic events, such as storms [28,29], or to or geopolitical events, such as wars or acts of terrorism [30]. Supply-side disruptions have their most pronounced, enduring impacts on poorer countries [31,32]. Disruptions in oil supply leave an especially deep footprint on oil-exporting countries [33–36].

By contrast, many disruptions in demand arise from broader macroeconomic phenomena, such as the financial crisis of 2008–2009 and the ensuing Great Recession. Unsupervised machine learning has suggested that the COVID-19 pandemic should be evaluated as a stochastic "black swan" event in commodity markets rather than an artifact of the business cycle [3]. Demand-side effects are more pronounced in the wealthy, industrialized countries that account for most of the world's consumption of exhaustible and renewable fuels [37–39].

Refined fuel markets, particularly for gasoline, move asymmetrically vis-à-vis crude oil markets [40–43]. According to the "rockets and feathers" hypothesis, increases in crude oil prices are transmitted more quickly to gasoline than decreases [44–46]. Other sources identify Edgeworth price cycles, which are characterized by sawtooth-shaped time series consisting of many price decreases punctuated by occasional upward jumps [47,48]. Since they identify periodicity within otherwise stochastic phenomena, Edgeworth price cycles may be regarded as a special instance of "rockets and feathers" [49]. Edgeworth cycles may arise from consumers who are extremely loyal to a brand and therefore unaware of lower retail gasoline prices, or at least are unwilling to search for bargains [50,51].

Other sources contest the alleged asymmetry of oil and refined fuel markets [52]. Recent crises in energy-related commodity markets have neither exhibited "rocket and feathers" behavior nor followed Edgeworth cycles. The "rockets and feathers" hypothesis may partially explain oil–gasoline asymmetry, but not completely. When oil prices are falling, gasoline prices follow a contrary "boulders and balloons" dynamic by which gasoline more swiftly tracks oil price declines than increases [53]. Reversals in oil–gasoline asymmetry strongly suggest that volatility transmission between crude oil and refined fuels varies over time [54].

Finally, energy commodities move in tandem with agricultural commodities that supply fuel as well as food, feed, or fiber [55–59]. Fuel feedstock crops, such as corn and soybeans, either compete directly against crude oil as renewable substitutes or provide complements to fuels refined from petroleum [60–62]. Although biofuel policies in wealthy countries are suspected of affecting volatility transmission between energy-related and agricultural commodities [63], firm evidence supporting such hypotheses has not emerged [64–68].

4. Discussion

This study advances the understanding of energy-related commodity markets. This study also expands the toolkit for unsupervised machine learning in time-series forecasting. This section discusses each of these contributions.

4.1. Commodity-Specific Insights

Energy-related commodity markets have a disproportionate impact on developmental economics, international trade, and environmental policy. Factors affecting oil prices include wars and other political disturbances, shifts in global supply and demand, and technological and regulatory changes promoting demand for renewable energy. OPEC production decisions and extreme weather events must also be taken into account.

Interactions between fossil fuels and renewable fuel feedstocks attract especially intense security. Crude oil, gasoline, and diesel affect not only energy policy but also demand for agricultural feedstocks for ethanol and biodiesel. Biofuel feedstock demand may be swayed by domestic and international policies responding to global climate change.

Asymmetry, persistence, and cyclicality in volatility must be understood in the context of other financial markets and the macroeconomy. Beyond its impact on public policy, co-movement among all commodity markets and between commodities and other asset classes influences the leadership of energy companies and other forms of private risk management, including portfolio allocation. Fuel taxes, renewable energy policy, and the impact of energy prices on the behavior of industries and households hang in the balance.

4.2. Prelude and Performance: Unsupervised Machine Learning and Time-Series Forecasting

Unsupervised machine learning coexists comfortably alongside conventional methods for time-series forecasting. Indeed, more complete integration of unsupervised machine learning with forecasting forms the basis for future work. Each of the energy market propositions raised in this study—(1) macroeconomic versus industry-specific effects, (2) supply-side versus demand-side shocks, (3) upside versus downside asymmetry in oil and refined fuel prices, and (4) interactions with agricultural commodities—can be described and visualized through clustering and manifold learning.

The unsupervised machine-learning methods in this article exhibit strengths as well as limitations. Unlike traditional forecasting methods or even generalized linear methods for panel data, unsupervised machine learning does not rely on the formal apparatus of null hypothesis significance testing [69]. The statistical community has raised particularly sharp concerns over the rampant misunderstanding and misuse of *p*-values [70].

Willingness to use unsupervised machine learning does not hinge on a researcher's position on *p*-values or the growing movement seeking alternatives to statistical conventions based on them [71,72]. Nor should it. Unsupervised machine learning reveals mathematical properties and relationships within the data in ways not restricted by the rigid conventions of null hypothesis significance testing. In the absence of *p*-values and other conventional indicators of statistical significance, unsupervised learning converts raw data into mathematical outputs that can, in turn, enable more fruitful applications of economic domain knowledge and expert judgment. Other applications of machine learning and artificial intelligence, particularly in natural language processing, readily accommodate a blend of formal mathematics and subjective but mathematically informed analyst judgment [73–75].

The application of unsupervised machine learning to time series provides both prelude and performance in this branch of financial economics. Although unsupervised learning cannot directly forecast time series, unsupervised learning does generate insights beyond those available through descriptive statistics or exploratory data analysis. As the poet T.S. Eliot rendered the sentiment, unsupervised machine learning scours "the beaches where [the sea] tosses / Its hints of earlier and other creation" [1] (p. 36).

Time-variant ontological clustering represents a methodological innovation in its own right. The time-series data of greatest interest to the research questions in this study consist of the transpose of a more conventional matrix whose rows designate trading days and whose columns represent distinct commodity markets. Temporal clustering identifies mathematically distinct periods within the historical record. Transitions between clusters may indicate structural breaks in a trading regime [76]. Successfully locating such events addresses a known weakness of conventional forecasting methods. As the most labor- and data-intensive form of unsupervised machine learning for time series, time-variant ontological clustering can report changes in co-movement among market components—dynamic shifts, as it were, punctuating longer episodes of evanescent equilibrium among assets within a financial ecosystem [77].

Further applications of this study's methods include the investigation of changes in co-movement among asset classes or, more narrowly, among equity or bond market subsectors during shifts in macroeconomic conditions. This study's application of unsupervised machine learning may shed light on jump-diffusion models [78–81], a venerable economic tradition well represented in the literature on commodity markets [82–85]. If jump-diffusion processes can generate random sampling algorithms in pattern theory, computer vision, and medical imaging [86], unsupervised machine learning might facilitate the extraction of previously undetected mixtures within time series.

This study's methods also invite the extension of unsupervised machine learning to forecasting tasks that supervised machine learning methods have begun to tackle. For instance, all three methods of clustering presented in [2,3] and this study can be applied to the immensely popular task of forecasting air pollution in Beijing [87–89]. Unsupervised machine learning holds promise for addressing similar problems in meteorology, pollution control, and other environmental sciences [90].

The application of unsupervised machine learning to time series in finance, meteorology, and ecology should follow a Hegelian dialectic [91,92]. Unconditional ontological clustering as *thesis* stands beside its *antithesis*, temporal clustering of the matrix transpose. As the *synthesis* of space and time, time-variant ontological clustering reveals shifts within these stylized ecosystems during critical periods. What financial economics calls jumps or structural shifts, upon closer inspection, may display the mathematical properties that distinguish recessions from ordinary macroeconomic equilibria [93–95]. Given the common origins of economics and ecology [96], the similarities between these phenomena and punctuated equilibria in biology [77] should come as no surprise. The only difference is the frequency of ticks on the clocks measuring financial and geological time.

5. Conclusions

This study anticipates the completion of a toolkit for applying unsupervised machine learning to financial time series. An initial application of clustering and manifold learning to logarithmic returns and forecast volatility garnered quantitative support for the traditional ontology of commodity markets [2]. Temporal clustering then identified critical periods within energy-related commodity markets [3]. This study combines these methods into a novel hybrid called time-variant ontological clustering.

At a higher level of theoretical abstraction, this study unveils the distinct contributions of machine-learning methodology and data-gathering to time-series analysis. In machine learning, there is no such thing as a free lunch [97,98]. No single method or family of algorithms should be expected to outperform others, with respect to any dataset or even a broad category of problems [99]. The absence of a free lunch counsels the deployment of all plausible methodologies.

Unsupervised machine learning offers a meaningful combination of predictive success and explanatory power. The intrinsic parsimony of mathematics [100], remarkably effective in the natural sciences [101], notoriously fails when applied to economics [102]. By relying on all available data and the mathematical relationships lurking therein, unsupervised machine learning captures the most probable source of "unreasonable effectiveness" in the otherwise dismal science of economics: the data [103]. In that spirit, this study reveals the mechanics underlying energy-related commodity markets through machine-learning methods that neither require nor request human judgment.

Author Contributions: Conceptualization, J.M.C.; methodology, J.M.C.; software, J.M.C.; formal analysis, J.M.C.; investigation, C.A.; data curation, C.A.; writing—original draft preparation, J.M.C.; writing—review and editing, J.M.C. and C.A. All authors have read and agreed to the published version of the manuscript.

Funding: This research received no external funding.

Institutional Review Board Statement: Not applicable.

Informed Consent Statement: Not applicable.

Data Availability Statement: The data presented in this study are available on request from the corresponding author.

Conflicts of Interest: The authors declare no conflict of interest.

References

1. Eliot, T.S. The dry salvages. In *Four Quartets*; Reprinted 1971; Houghton Mifflin Harcourt: New York, NY, USA, 1943; pp. 35–48.
2. Chen, J.M.; Rehman, M.U.; Vo, X.V. Clustering commodity markets in space and time: Clarifying returns, volatility, and trading regimes through unsupervised machine learning. *Resour. Policy* **2021**, *73*, 102162. [CrossRef]
3. Chen, J.M.; Rehman, M.U. A pattern new in every moment: The temporal clustering of markets for crude oil, refined fuels, and other commodities. *Energies* **2021**, *14*, 6099. [CrossRef]
4. Aguilera, R.F.; Radetzki, M. The synchronized and exceptional price performance of oil and gold: Explanations and prospects. *Resour. Policy* **2017**, *54*, 81–87. [CrossRef]
5. Reboredo, J.C.; Rivera-Castro, M.A.; Ugolini, A. Downside and upside risk spillovers between exchange rates and stock prices. *J. Bank. Financ.* **2016**, *62*, 76–96. [CrossRef]
6. Rehman, M.U.; Shahzad, S.J.H.; Uddin, G.S.; Hedström, A. Precious metal returns and oil shocks: A time varying connectedness approach. *Resour. Policy* **2018**, *58*, 77–89. [CrossRef]
7. Liu, C.; Naeem, M.A.; Rehman, M.U.; Farid, S.; Shahzad, S.J.H. Oil as hedge, safe-haven, and diversifier for conventional currencies. *Energies* **2020**, *13*, 4354. [CrossRef]
8. Umar, Z.; Jareño, F.; Escribano, A. Oil price shocks and the return and volatility spillover between industrial and precious metals. *Energy Econ.* **2021**, *99*, 105291. [CrossRef]
9. Rubbaniy, G.; Khalid, A.A.; Syriopoulos, K.; Samitas, A. Safe-haven properties of soft commodities during times of COVID-19. *J. Commod. Mark.* **2022**, *27*, 100223. [CrossRef]
10. Talbot, J.M. Tropical commodity chains, forward integration strategies and international inequality: Coffee, cocoa and tea. *Rev. Int. Pol. Econ.* **2002**, *9*, 701–734. [CrossRef]
11. Liu, J.; Han, J. Spectral clustering. In *Data Clustering: Algorithms and Applications*; Aggarwal, C.C., Reddy, C.K., Eds.; Chapman and Hall, CRC Press: Boca Raton, FL, USA, 2014; pp. 177–200. [CrossRef]
12. Liu, R.; Zhang, H. Segmentation of 3D meshes through spectral clustering. In Proceedings of the 12th Pacific Conference on Computer Graphics and Applications, Seoul, Republic of Korea, 6–8 October 2004; pp. 298–305. [CrossRef]
13. Von Luxburg, U. A tutorial on spectral clustering. *Stat. Comput.* **2007**, *17*, 395–416. [CrossRef]
14. Comaniciu, D.; Meer, P. Mean shift: A robust approach toward feature space analysis. *IEEE Trans. Pattern Anal. Mach. Intel.* **2002**, *24*, 603–619. [CrossRef]
15. Frey, B.J.; Dueck, D. Clustering by passing messages between data points. *Science* **2007**, *315*, 972–976. [CrossRef] [PubMed]
16. Bodenhofer, U.; Kothmeier, A.; Hochreiter, S. APCluster: An R package for affinity propagation clustering. *Bioinformatics* **2011**, *27*, 2463–2464. [CrossRef] [PubMed]
17. Shang, F.; Jiao, L.C.; Shi, J.; Wang, F.; Gong, M. Fast affinity propagation clustering: A multilevel approach. *Pattern Recog.* **2012**, *45*, 474–486. [CrossRef]
18. Murtagh, F. A survey of recent advances in hierarchical clustering algorithms. *Comput. J.* **1983**, *26*, 354–359. [CrossRef]
19. Soni, K.G.; Patel, A. Comparative analysis of *k*-means and *k*-medoids algorithm on IRIS data. *Int. J. Comput. Intell. Res.* **2017**, *13*, 899–906. Available online: https://www.ripublication.com/ijcir17/ijcirv13n5_21.pdf (accessed on 28 June 2023).
20. Van der Maaten, L. Accelerating *t*-SNE using tree-based algorithms. *J. Mach. Learn. Res.* **2014**, *15*, 3221–3245. Available online: https://jmlr.org/papers/v15/vandermaaten14a.html (accessed on 28 June 2023).
21. Van der Maaten, L.; Hinton, G. Visualizing high-dimensional data using *t*-SNE. *J. Mach. Learn. Res.* **2008**, *9*, 2579–2605. Available online: https://www.jmlr.org/papers/v9/vandermaaten08a.html (accessed on 28 June 2023).
22. Van der Maaten, L.; Hinton, G. Visualizing non-metric similarities in multiple maps. *Mach. Learn.* **2012**, *87*, 33–55. [CrossRef]
23. Kilian, L. The economic effects of energy price shocks. *J. Econ. Lit.* **2008**, *46*, 871–909. [CrossRef]
24. Federer, J.P. Oil price volatility and the macroeconomy. *J. Macroecon.* **1996**, *18*, 1–26. [CrossRef]
25. Sim, N.; Zhou, H. Oil prices, stock return, and the dependence between their quantiles. *J. Bank. Financ.* **2015**, *55*, 1–8. [CrossRef]
26. Joëts, M.; Mignon, V.; Razafindrabe, T. Does the volatility of commodity prices reflect macroeconomic uncertainty? *Energy Econ.* **2017**, *68*, 313–326. [CrossRef]
27. Hamilton, J.D. What is an oil shock? *J. Econom.* **2003**, *113*, 363–398. [CrossRef]
28. Blair, B.F.; Rezek, J.P. The effects of hurricane Katrina on price pass-through for Gulf Coast gasoline. *Econ. Lett.* **2008**, *98*, 229–234. [CrossRef]
29. Kaiser, M.J.; Yu, Y.; Jablonowski, C.J. Modeling lost production from destroyed platforms in the 2004–2005 Gulf of Mexico hurricane seasons. *Energy* **2009**, *34*, 1156–1171. [CrossRef]

30. Ji, Q.; Guo, J.-F. Oil price volatility and oil-related events: An internet concern study perspective. *Appl. Energy* **2015**, 256–264. [CrossRef]
31. Morgan, C.W.; Rayner, A.J.; Ennew, C.T. Price instability and commodity future markets. *World Dev.* **1994**, *22*, 1729–1736. [CrossRef]
32. Vo, L.H.; Le, T.-H. Eatery, energy environment and economic system, 1970-2017: Understanding volatility spillover patterns a global sample. *Energy Econ.* **2021**, *100*, 105391. [CrossRef]
33. Rafiq, S.; Ruhul Salim, R.; Bloch, H. Impact of crude oil volatility on economic activities: An empirical investigation in the Thai economy. *Resour. Policy* **2009**, *34*, 121–132. [CrossRef]
34. Filis, G.; Degiannakis, S.; Floros, C. Dynamic correlation between stock market and oil prices: The case of oil-importing and oil-exporting countries. *Int. Rev. Financ. Anal.* **2011**, *20*, 152–164. [CrossRef]
35. Guesmi, K.; Fattoum, S. Return and volatility transmission between oil prices and oil-exporting and oil-importing countries. *Econ. Model.* **2014**, *38*, 305–310. [CrossRef]
36. Kumar, S.; Khalfaoui, R.; Tiwari, A.K. Does geopolitical risk improve the directional predictability from oil to stock returns? Evidence from oil-exporting and oil-importing countries. *Resour. Policy* **2021**, *74*, 102253. [CrossRef]
37. Ebrahim, Z.; Inderwildi, O.R.; King, D.A. Macroeconomic impacts of oil price volatility: Mitigation and resilience. *Front. Energy* **2014**, *8*, 9–24. [CrossRef]
38. Guo, H.; Kliesen, K.L. Oil price volatility and U.S. macroeconomic activity. *Fed. Reserve Bank St. Louis Rev.* **2005**, *87*, 669–683. Available online: https://research.stlouisfed.org/publications/review/2005/11/01/oil-price-volatility-and-u-s-macroeconomic-activity (accessed on 28 June 2023). [CrossRef]
39. Mork, K.A. Oil and the macroeconomy when prices go up and down: An extension of Hamilton's results. *J. Pol. Econ.* **1989**, *97*, 740–744. [CrossRef]
40. Borenstein, S.; Cameron, A.C.; Gilbert, R. Do gasoline prices respond asymmetrically to crude oil price changes? *Q. J. Econ.* **1997**, *112*, 305–339. [CrossRef]
41. Douglas, C.; Herrera, A.M. Why are gasoline prices sticky? A test of alternative models of price adjustment. *J. Appl. Econom.* **2010**, *25*, 903–928. [CrossRef]
42. Douglas, C.C.; Herrera, A.M. Dynamic pricing and asymmetries in retail gasoline markets: What can they tell us about price stickiness? *Econ. Lett.* **2014**, *122*, 247–252. [CrossRef]
43. Karrenbrock, J.D. The behavior of retail gasoline prices: Symmetric or not? *Fed. Reserve Bank St. Louis Rev.* **1991**, *73*, 19–29. Available online: https://research.stlouisfed.org/publications/review/1991/07/01/the-behavior-of-retail-gasoline-prices-symmetric-or-not (accessed on 28 June 2023). [CrossRef]
44. Eleftheriou, K.; Nijkamp, P.; Polemis, M.L. Asymmetric price adjustments in US gasoline markets: Impacts of spatial dependence on the "rockets and feathers" hypothesis. *Reg. Stud.* **2019**, *53*, 667–680. [CrossRef]
45. Galeotti, M.; Lanza, A.; Manera, M. Rockets and feathers revisited: An international comparison on European gasoline markets. *Energy Econ.* **2003**, *25*, 175–190. [CrossRef]
46. Tappata, M. Rockets and feathers: Understanding asymmetric pricing. *Rand J. Econ.* **2009**, *40*, 673–687. [CrossRef]
47. Lewis, M.; Noel, M. The speed of gasoline price response in markets with and without Edgeworth cycles. *Rev. Econ. Stat.* **2011**, *93*, 672–682. [CrossRef]
48. Noel, M.D.; Chu, L. Forecasting gasoline prices in the presence of Edgeworth price cycles. *Energy Econ.* **2015**, *51*, 204–214. [CrossRef]
49. Maskin, E.; Tirole, J. A theory of dynamic oligopoly, II: Price competition, kinked demand curves, and Edgeworth cycles. *Econometrica* **1988**, *56*, 571–599. [CrossRef]
50. Fershtman, C.; Fishman, A. Price cycles and booms: Dynamic search equilibrium. *Am. Econ. Rev.* **1992**, *82*, 1221–1233. Available online: https://www.jstor.org/stable/2117475 (accessed on 28 June 2023).
51. Lewis, M.S. Asymmetric price adjustment and consumer search: An examination of the retail gasoline market. *J. Econ. Manag. Strategy* **2011**, *20*, 409–449. [CrossRef]
52. Bachmeier, L.J.; Griffin, J.M. New evidence on asymmetric gasoline price responses. *Rev. Econ. Stat.* **2003**, *85*, 772–776. [CrossRef]
53. Bremmer, D.S.; Kesselring, R.G. The relationship between U.S. retail gasoline and crude oil prices during the Great Recession: "Rockets and feathers" or "balloons and rocks" behavior? *Energy Econ.* **2016**, *55*, 200–210. [CrossRef]
54. Apergis, N.; Vouzavalis, G. Asymmetric pass through of oil prices to gasoline prices: Evidence from a new country sample. *Energy Policy* **2018**, *114*, 519–528. [CrossRef]
55. Koirala, K.H.; Mishra, A.K.; D'Antoni, J.M.; Mehlhorn, J.E. Energy prices and agricultural commodity prices: Testing correlation using copulas method. *Energy* **2015**, *81*, 430–436. [CrossRef]
56. Roman, M.; Górecka, A.; Domagała, J. The linkages between crude oil and food prices. *Energies* **2020**, *13*, 6545. [CrossRef]
57. Serra, T. Volatility spillovers between food and energy markets: A semiparametric approach. *Energy Econ.* **2011**, *33*, 1155–1164. [CrossRef]
58. Kumar, S.; Tiwari, A.K.; Raheem, I.D.; Hille, E. Time-varying dependence structure between oil and agricultural commodity markets: A dependence-switching CoVaR copula approach. *Resour. Policy* **2021**, *72*, 102049. [CrossRef]
59. Lucotte, Y. Co-movements between crude oil and food prices: A post-commodity boom perspective. *Econ. Lett.* **2016**, *147*, 142–147. [CrossRef]

60. Ferrer, R.; Shahzad, S.J.H.; López, R.; Jareño, F. Time and frequency dynamics of connectedness between renewable energy stocks and crude oil prices. *Energy Econ.* **2018**, *76*, 1–20. [CrossRef]
61. Carpio, L.G.T. The effects of oil price volatility on ethanol, gasoline, and sugar price forecasts. *Energy* **2019**, *181*, 1012–1022. [CrossRef]
62. Reboredo, J.C. Do food and oil prices co-move? *Energy Policy* **2012**, *49*, 456–467. [CrossRef]
63. Du, X.; Yu, C.L.; Hayes, D.J. Speculation and volatility spillover in the crude oil and agricultural commodity markets: A Bayesian analysis. *Energy Econ.* **2011**, *33*, 497–503. [CrossRef]
64. Cabrera, B.L.; Schulz, F. Volatility linkages between energy and agricultural commodity prices. *Energy Econ.* **2016**, *54*, 190–203. [CrossRef]
65. Enciso, S.R.A.; Fellmann, T.; Dominguez, I.P.; Santini, F. Abolishing biofuel policies: Possible impacts on agricultural price lev- els, price variability and global food security. *Food Policy* **2016**, *61*, 9–26. [CrossRef]
66. Gardebroek, C.; Hernandez, M.A. Do energy prices stimulate food price volatility? Examining volatility transmission between US oil, ethanol and corn markets. *Energy Econ.* **2013**, *40*, 119–129. [CrossRef]
67. Karyotis, C.; Alijani, S. Soft commodities and the global financial crisis: Implications for the economy, resources and institutions. *Res. Int. Bus. Financ.* **2016**, *37*, 350–359. [CrossRef]
68. McPhail, L.L.; Babcock, B.A. Impact of US biofuel policy on US corn and gasoline price variability. *Energy* **2012**, *37*, 505–513. [CrossRef]
69. Szucs, D.; Ioannidis, J.P.A. When null hypothesis significance testing is unsuitable for research: A reassessment. *Front. Hum. Neurosci.* **2017**, *11*, 390. [CrossRef]
70. Wasserstein, R.L.; Lazar, N.A. The ASA statement on *p*-values: Context, process, and purpose. *Am. Stat.* **2016**, *70*, 129–133. [CrossRef]
71. Cohen, H.W. *P* values: Use and misuse in medical literature. *Am. J. Hypertens.* **2011**, *24*, 18–23. [CrossRef]
72. Quintana, D.S.; Williams, D.R. Bayesian alternatives for common null-hypothesis significance tests in psychiatry: A non-technical guide using JASP. *BMC Psychiatry* **2018**, *18*, 178. [CrossRef]
73. Farasat, A.; Gross, G.; Nagi, R.; Nikolaev, A.G. Social network analysis with data fusion. *IEEE Trans. Comput. Soc. Syst.* **2016**, *3*, 88–99. [CrossRef]
74. Newman, N.C.; Porter, A.L.; Newman, D.; Trumbach, C.C.; Bolan, S.D. Comparing methods to extract technical content for technological intelligence. *J. Eng. Technol. Manag.* **2014**, *32*, 97–109. [CrossRef]
75. Taber, P.; Staes, C.J.; Phengphoo, S.; Rocha, E.; Lam, A.; del Fiol, G.; Maviglia, S.M.; Rocha, R.A. Developing a sampling method and preliminary taxonomy for classifying COVID-19 public health guidance for healthcare organizations and the general public. *J. Biomed. Inform.* **2021**, *120*, 103852. [CrossRef]
76. Salisu, A.; Fasanya, I.O. Modelling oil price volatility with structural breaks. *Energy Policy* **2013**, *52*, 554–562. [CrossRef]
77. Eldredge, N.; Gould, S.J. Punctuated equilibria: An alternative to phyletic gradualism. In *Models in Paleobiology*; Schopf, T.J.M., Ed.; Freeman, Cooper & Co.: San Francisco, CA, USA, 1972; pp. 82–115.
78. Black, F. How to use the holes in Black-Scholes. *J. Appl. Corp. Financ.* **1989**, *1*, 67–73. [CrossRef]
79. Cox, J.C.; Ross, S.A. The valuation of options for alternative stochastic processes. *J. Financ. Econ.* **1976**, *3*, 145–166. [CrossRef]
80. Kou, S.G. A jump-diffusion model for option pricing. *Manag. Sci.* **2002**, *48*, 1086–1101. [CrossRef]
81. Merton, R.C. Option pricing when underlying stock returns are discontinuous. *J. Financ. Econ.* **1976**, *3*, 125–144. [CrossRef]
82. Crosby, J. A multi-factor jump-diffusion model for commodities. *Quant. Financ.* **2008**, *8*, 181–200. [CrossRef]
83. Da Fonseca, J.; Ignatieva, K. Jump activity analysis for affine jump-diffusion models: Evidence from the commodity market. *J. Bank. Financ.* **2019**, *99*, 45–62. [CrossRef]
84. Hilliard, J.E.; Jorge, A.; Reis, J.A. Jump processes in commodity futures prices and options pricing. *Am. J. Agric. Econ.* **1999**, *81*, 273–286. [CrossRef]
85. Schmitz, A.; Wang, Z.; Kimn, J.H. A jump diffusion model for agricultural commodities with Bayesian analysis. *J. Futures Mark.* **2014**, *34*, 235–260. [CrossRef]
86. Grenander, U.; Miller, M.I. Representations of knowledge in complex systems. *J. R. Stat. Soc. B* **1994**, *56*, 549–603. [CrossRef]
87. Li, D.; Liu, J.; Zhao, Y. Forecasting of PM2.5 concentration in Beijing using hybrid deep learning framework based on attention mechanism. *Appl. Sci.* **2022**, *12*, 11155. [CrossRef]
88. Niu, M.; Zhang, Y.; Ren, Z. Deep learning-based PM2.5 long time-series prediction by fusing multisource data—A case study of Beijing. *Atmosphere* **2023**, *14*, 340. [CrossRef]
89. Zhang, S.; Guo, B.; Dong, A.; He, J.; Xu, Z.; Chen, S.X. Cautionary tales on air-quality improvement in Beijing. *Proc. R. Soc. A* **2017**, *473*, 20170457. [CrossRef] [PubMed]
90. Méndez, M.; Merayo, M.G.; Núñez, M. Machine learning algorithms to forecast air quality: A survey. *Artif. Intell. Rev.* **2023**. [CrossRef]
91. Forster, M. Hegel's dialectical method. In *The Cambridge Companion to Hegel*; Beiser, F.C., Ed.; Cambridge University Press: Cambridge, UK, 1993; pp. 130–170.
92. Rosen, M. *Hegel's Dialectic and Its Criticism*; Cambridge University Press: Cambridge, UK, 1982.
93. Chauvet, M. An economic characterization of business cycle dynamics with factor structure and regime switching. *Int. Econ. Rev.* **1998**, *39*, 969–996. [CrossRef]

94. Chauvet, M.; Piger, J.M. Identifying business cycle turning points in real time. *Rev. Fed. Reserve Bank St. Louis* **2003**, *85*, 47–60. [CrossRef]
95. Chauvet, M.; Piger, J. A comparison of the real-time performance of business cycle dating methods. *J. Bus. Econ. Stat.* **2008**, *26*, 42–49. [CrossRef]
96. Costanza, R. Ecological economics: Reintegrating the study of humans and nature. *Ecol. Appl.* **1996**, *6*, 978–990. [CrossRef]
97. Alabert, A.; Berti, A.; Caballero, R.; Ferrante, M. No-free-lunch theorems in the continuum. *Theor. Comput. Sci.* **2015**, *600*, 98–106. [CrossRef]
98. Ho, Y.C.; Pepyne, D.L. Simple explanation of the no-free-lunch theorem and its implications. *J. Optim. Theory Appl.* **2002**, *115*, 549–570. [CrossRef]
99. Wolpert, D.H. The lack of a priori distinctions between learning algorithms. *Neural Comput.* **1996**, *8*, 1341–1390. [CrossRef]
100. Hamming, R.W. The unreasonable effectiveness of mathematics. *Am. Math. Mon.* **1980**, *87*, 81–90. [CrossRef]
101. Wigner, E. The unreasonable effectiveness of mathematics in the natural sciences. *Commun. Pure Appl. Math.* **1960**, *13*, 1–14. [CrossRef]
102. Velupillai, K.V. The unreasonable ineffectiveness of mathematics in economics. *Camb. J. Econ.* **2005**, *29*, 849–872. [CrossRef]
103. Halevy, A.; Norvig, P.; Pereira, F. The unreasonable effectiveness of data. *IEEE Intell. Syst.* **2009**, *24*, 8–12. [CrossRef]

engineering proceedings

MDPI

Proceeding Paper

Probability-Density-Based Energy-Saving Recommendations for Household Refrigerating Appliances †

Francisco Rodríguez-Cuenca [1,*], Eugenio F. Sánchez-Úbeda [1], José Portela [1,2], Antonio Muñoz [1], Víctor Guizien [3], Andrea Veiga Santiago [3] and Alicia Mateo González [3]

1 Instituto de Investigación Tecnológica (IIT), Escuela Técnica Superior de Ingeniería ICAI, Universidad Pontificia Comillas, 28015 Madrid, Spain; eugenio.sanchez@comillas.edu (E.F.S.-Ú.); jportela@comillas.edu (J.P.); amunoz@comillas.edu (A.M.)

2 Facultad de Ciencias Económicas y Empresariales, Universidad Pontificia Comillas, 28015 Madrid, Spain

3 Advanced Analytics Market Iberia, Endesa Energía, 28042 Madrid, Spain ; victor.guizien@enel.com (V.G.); andrea.veiga@enel.com (A.V.S.); alicia.mateo@enel.com (A.M.G.)

* Correspondence: frodriguezc@comillas.edu

† Presented at the 9th International Conference on Time Series and Forecasting, Gran Canaria, Spain, 12–14 July 2023.

Abstract: The power sector is a major contributor to anthropogenic global warming and is responsible for 38% of total energy-related carbon dioxide emissions and 66% of carbon dioxide emission growth in 2018. In OECD member countries, the residential sector consumes a significant amount of electrical energy, with household refrigerating appliances alone accounting for 30–40% of the total consumption. To analyze the energy use of each domestic appliance, researchers have developed Appliance-Level Energy Characterization (ALEC), a set of techniques that provide insights into individual energy consumption patterns. This study proposes a novel methodology that utilizes robust probability density estimation to detect refrigerators with high energy consumption and recommend tailored energy-saving measures. The methodology considers two consumption features: base energy consumption (energy usage without human interaction) and relative energy consumption (energy usage influenced by human interaction). To assess the approach's effectiveness, the methodology was tested on a dataset of 30 different appliances from monitored homes, yielding positive results that support the robustness of the proposed method.

Keywords: household refrigerating appliances; energy-saving recommendations; appliance-level energy characterization

Citation: Rodríguez-Cuenca, F.; Sánchez-Úbeda, E.F.; Portela, J.; Muñoz, A.; Guizien, V.; Santiago, A.V.; González, A.M. Probability-Density-Based Energy-Saving Recommendations for Household Refrigerating Appliances. *Eng. Proc.* **2023**, *39*, 43. https://doi.org/10.3390/engproc2023039043

Academic Editor: Hosin (David) Lee

Published: 4 July 2023

1. Introduction

The power sector is responsible for 38% of total energy-related carbon dioxide emissions and 66% of carbon dioxide emission growth in 2018, making it the main cause of anthropogenic global warming according to [1]. The residential sector is the largest consumer of electricity, accounting for 30–40% of total consumption in OECD member countries [2]. Thus, it is unsurprising that the United Nations' Sustainable Development Goal 12 aims to promote sustainable consumption and production patterns to reduce humanity's impact on the environment [3].

One of the means to pursue that objective is Appliance-Level Energy Characterization (ALEC). ALEC is the technique used for the analysis of consumers' energy use for each domestic appliance [4]. This analysis can be used to provide feedback and energy-saving recommendations to users, which, if provided after energy consumption, can have an impact ranging from 3.8% to 8.4% of energy savings [4].

There are two main ways to provide ALEC: Intrusive Load Monitoring (ILM) and Non-Intrusive Load Monitoring (NILM). Although ILM provides highly accurate results [5], its adoption is limited due to the high cost and difficulty of installing and maintaining individual sensors for each target appliance [6]. To overcome these challenges, NILM uses intelligent techniques to disaggregate data collected from a single sensor, eliminating the need for individual sensors [7].

This paper focuses on household refrigerating appliances, which include refrigerators (fridges), freezers, or a combination of both. These appliances are responsible for an average of 13.4% of the electrical energy consumption of private households in OECD countries, according to [8].

One effective strategy for promoting energy efficiency in household refrigerating appliances is energy labeling. Typically, energy labels use a rating system that ranges from A+++ (very efficient) to G (least efficient). These labels provide essential information, such as the appliance's energy consumption, annual energy cost, and other relevant energy-related data. As a result, consumers can easily compare the energy efficiency of different models of refrigerators and select the most efficient one that meets their needs. By choosing the most efficient model, significant energy savings can be achieved over the lifetime of the appliance. For instance, in the UK, energy labeling has resulted in a 26% reduction in consumption per refrigerator within ten years [9]. However, it is worth considering the age of the product, as research shows that 16-year-old fridges experience, on average, a 27% increase in energy consumption, with the cabinet's thermal resistance being one of the most significant factors contributing to the degradation of efficiency [10,11].

Consumer behavior also plays a significant role in the energy consumption of household refrigerating appliances. For instance, a decrease in temperature by one degree Celsius can cause the energy consumption of a refrigerator to rise by 7.8% [12], and excessive daily openings can further increase it by 10% [13].

Consumption habits of household refrigerating appliances can be classified into two categories: Direct Using Behavior (DUB) and Indirect Using Behavior (IUB). DUB refers to the actions of consumers that directly interact with the appliances, such as opening the door, while IUB pertains to external factors that impact the appliances, such as their installation location [14].

According to a recent study, a commonly used refrigeration appliance's DUB accounts for over 13.5% of its total energy consumption within a year. However, the study found that the IUB had a greater impact on energy consumption. Specifically, the ambient operating temperature in consumer homes was responsible for an average of approximately 18.5% of the appliance's total energy consumption [14].

This study is a crucial component of the RC4ALL project, which aims to provide energy-saving techniques for households by closely monitoring particular appliances in a carefully selected group of homes. The insights and information gleaned from these monitored households will then be utilized to provide personalized recommendations to other households throughout Spain. This study focuses on identifying refrigerators that consume more energy than necessary and can benefit from automated energy-saving recommendations. The study proposes a new method based on robust probability density estimation to select these refrigerators. The selected refrigerators will then receive customized recommendations to reduce their energy usage. The RC4ALL project is a collaboration between Comillas University and Endesa and is being funded by the Ministry of Science and Innovation (MCI) and the State Research Agency (AEI) [15].

The rest of the paper is organized as follows: Section 2 describes the dataset, and Section 3 presents the proposed methodology for the selection of fridges to be notified about with significant recommendations in order to reduce their energy consumption. A case example is presented in Section 4, while conclusions and future work are in Section 5.

2. Fridge Dataset Analysis

We have collected historical energy consumption data from various home appliances at five-minute intervals. We aim to use this information to provide energy-saving recommendations tailored to fridges that need it. Due to clients' installation limitations, most of the fridges' data became fully available in November 2022. For that reason, we only include energy usage recorded between 1 November 2022, and 1 February 2023 in this study. To prepare the dataset, we filtered out those fridges that had less than 60 consumption days in that time period. Consequently, in this section, we analyze the resulting 30 fridges' consumption history. Note that we do not have additional information about the fridges, such as the energy label, their size, or their age.

2.1. Operating Cycle Analysis

The frequency and value of a fridge's operating cycles are among the primary metrics that characterize it [10]. Fridges operate intermittently and are only turned on when necessary, i.e., when the temperature inside rises above a threshold. The first step in analyzing fridges is to count the number of cycles each fridge has per unit of time and compare it to others. To achieve this, we counted the number of cycles for each fridge by tracking the times per day that the energy consumption increased from below average to above average. As depicted in Figure 1, the sampled fridges had an average of approximately 20 cycles per day, with a maximum of fewer than 60 cycles. The box plots also indicate that some fridges, such as Fridges 9 and 10, had a significant variation in cycle counts, while others, such as Fridge 11, remained stable; bear in mind that no data cleansing has been carried out.

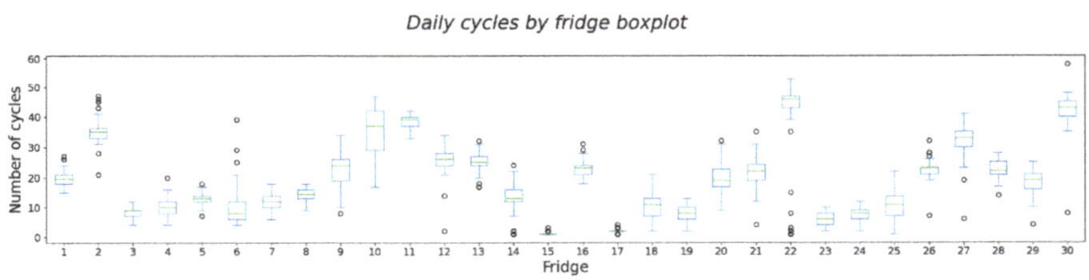

Figure 1. Boxplot of daily cycles per fridge.

2.2. Interval Analysis

To compare fridges' energy consumption, we created consumption profiles for each day by dividing them into 3 h intervals, as can be seen in Figure 2, which would include at least one cycle per interval for all fridges.

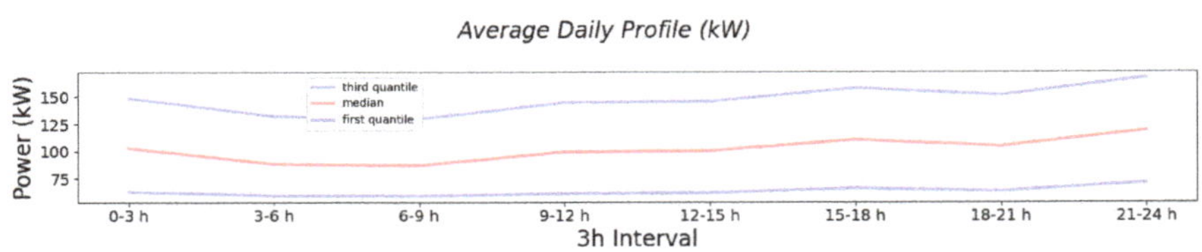

Figure 2. Average daily profile of all fridges.

To estimate the fridge energy consumption without human interaction, we selected the interval with the lowest consumption, which we called the *base consumption*. This approach allows for a fair comparison of energy consumption between different fridges without considering differences in user behavior. We applied a 7-day Centered Moving Average [16] to soften the impact of possible outliers and improve the base consumption estimation. Figure 3 illustrates the benefits of applying the moving average over one of the fridges' energy consumption time series.

To normalize the total energy consumption of each fridge, we used the daily base consumption as a daily reference. This enabled us to calculate the *relative consumption* of each fridge as a percentage of base consumption, allowing us to study energy consumption due to human interaction. Figure 4 illustrates the first percentile, median, and third percentile consumption for all fridges in 3 h intervals in terms of relative consumption. It is evident from the relative energy consumption that, on average, most fridges have a lower energy consumption in the morning between 3 AM and 9 AM, gradually increasing as it approaches the evening. This pattern can be attributed to higher ambient temperatures and more frequent door openings during the daytime hours.

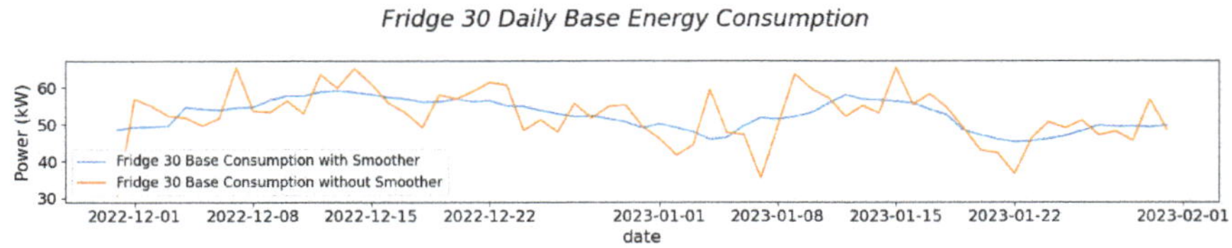

Figure 3. Illustrative example of the daily base consumption estimation for Fridge 30, with and without the centered moving average smoother.

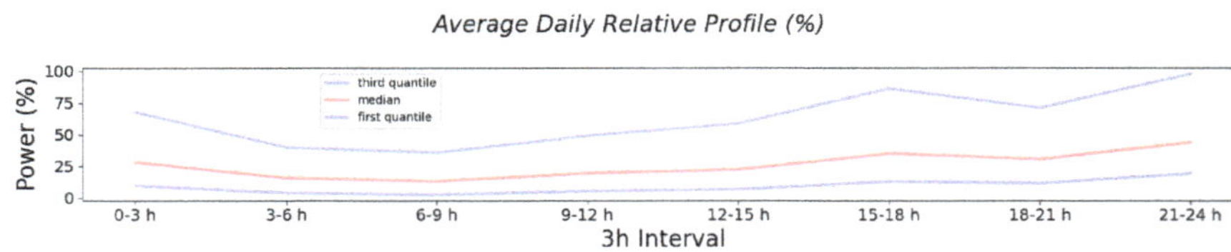

Figure 4. Average daily relative profile of all fridges.

Figure 5 displays the consumption of all 30 fridges in terms of total consumption, base consumption, and relative consumption. There are some crucial observations to note. Firstly, the base consumption for each fridge is relatively stable, with a slightly decreasing trend due to lower temperatures in January than in December. Secondly, relative consumption provides a more realistic comparison between fridges' daily consumption fluctuations than absolute consumption values.

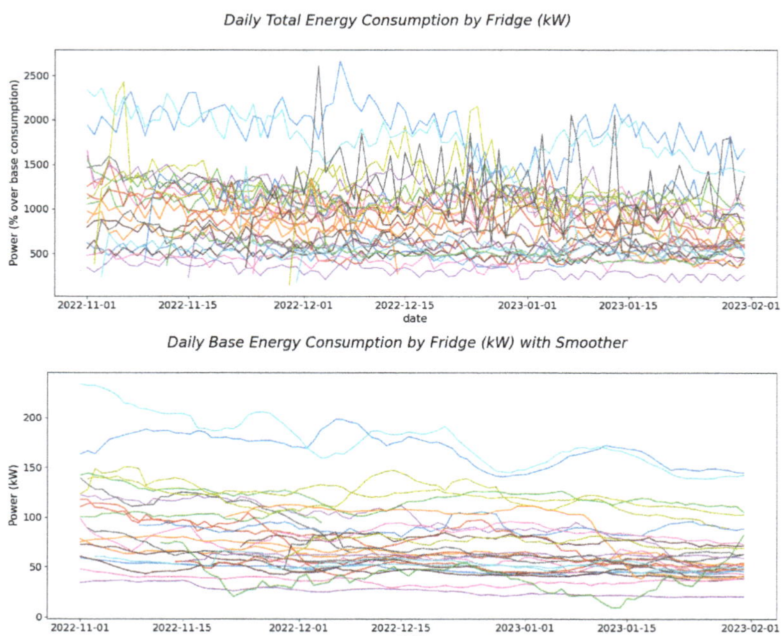

Figure 5. *Cont.*

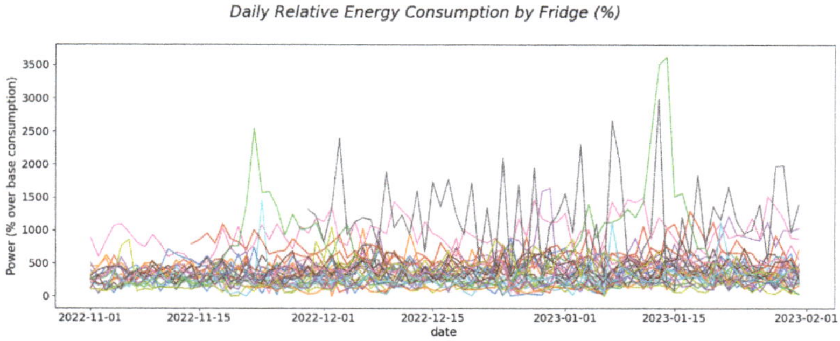

Figure 5. Daily total, base, and relative consumption of each fridge, represented by different colors.

2.3. Variance Analysis

Analysis of Variance (ANOVA) is a widely used statistical method in research [17]. It partitions the observed variance of a variable into components that can be attributed to different groups, providing a measure of confidence in the influence of a particular factor on the resulting variable [18]. In this study, we employed one-way ANOVA with an F-test to assess the degree of influence of the hour interval, day of the week (dow), and month of the year on the resulting relative consumption of each fridge, including their interactions.

As shown in Table 1, the factors of month and interval had, on average, a statistically significant impact on the energy consumption for most fridges (p-value < 0.1), while the day of the week factor and interaction pairs did not. Based on these results, we will primarily use consumption at each three-hour interval to characterize the daily consumption of fridges, as the day of the week is not significant and we only have records for three months out of the year.

Table 1. ANOVA p-value results.

	Month	Dow	Interval	Dow∗Interval	Dow∗Month	Month∗Interval
mean	0.0714	0.4525	0.0287	0.4487	0.6031	0.4908
std	0.1930	0.3398	0.1156	0.3408	0.3611	0.3764
min	0.0000	0.0001	0.0000	0.0000	0.0004	0.0001
max	0.7677	0.9934	0.5846	0.9991	0.9969	0.9844

2.4. Clustering Analysis

To gain more insights into the usage patterns of fridges, we utilized an unsupervised machine learning algorithm. Our approach involved analyzing the daily relative energy consumption profiles of individual fridges using the K-means algorithm, with Dynamic Time Warping as the distance metric [19]. To determine the optimal number of clusters, we used the maximum value of the silhouette as an automated selection approach [20]. The resulting cluster profiles, shown in Figure 6, revealed that while fridges typically exhibit lower energy consumption in the mornings that gradually increases throughout the day, there can be significant variations in usage patterns both within and between fridges.

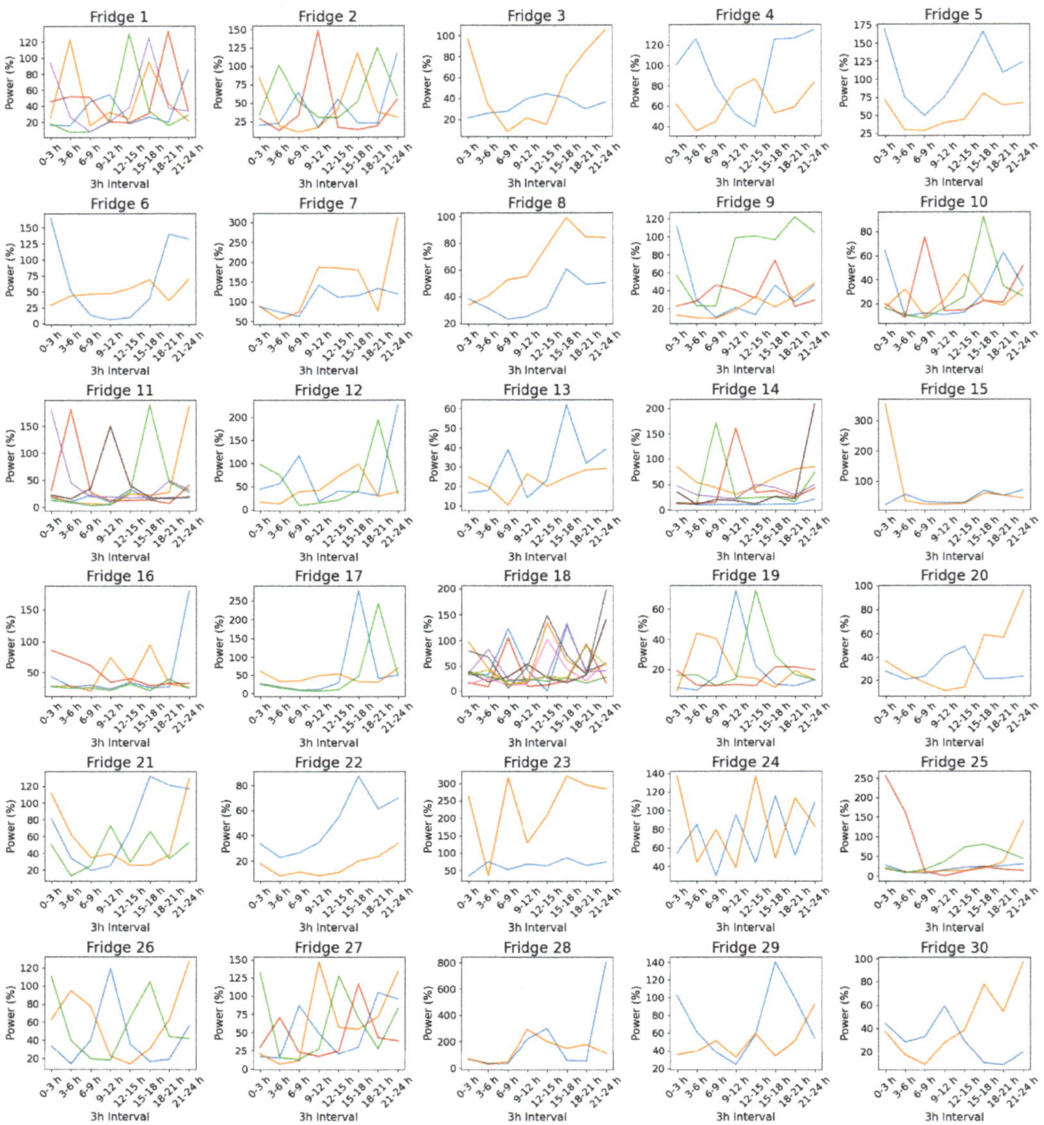

Figure 6. Cluster profiles of relative energy consumption, represented by different colors.

3. Proposed Methodology

This section describes the proposed methodology for selecting refrigerators exhibiting anomalous or exceedingly high consumption records and the assignment of energy-saving recommendations tailored to their needs. As we can see in Figure 7, our method consists of three phases: the extraction of features, the comparison of fridges based on these features, and the elaboration of energy-saving recommendations based on the comparison.

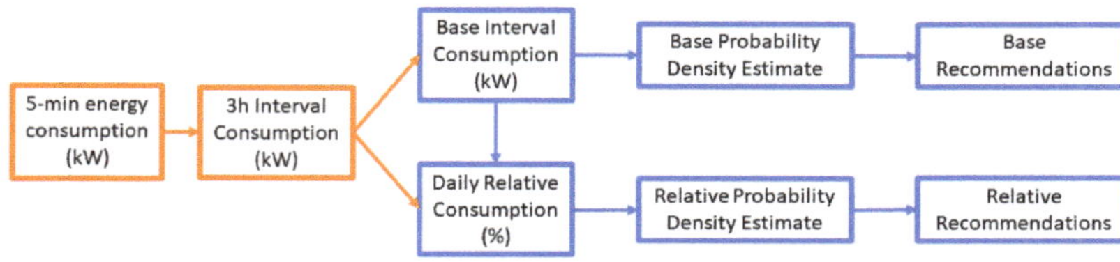

Figure 7. Schema for the proposed methodology.

3.1. Base and Relative Consumption Features

We will use as features the daily base and relative energy consumption for each fridge. The base consumption will be calculated as the three-hour interval of least consumption, as described in the dataset analysis section. The relative consumption for each day is calculated as a proportion of the total daily consumption over the base consumption. Each fridge will then have a daily time series of base and relative consumption, as can be seen in Figure 8. The base energy consumption represents a fridge's energy consumption without users' interaction, while the relative consumption represents the users' impact on energy consumption.

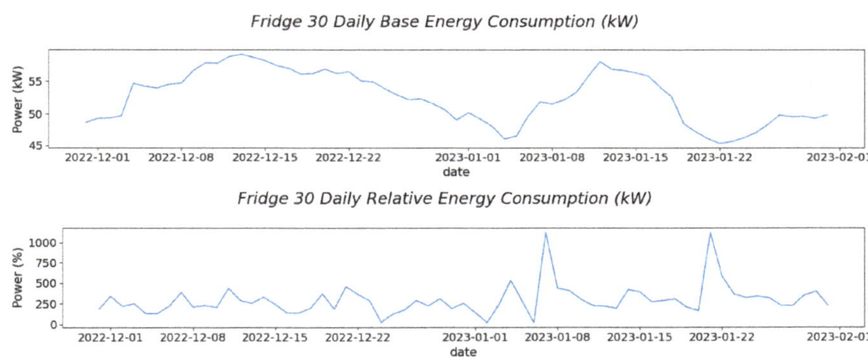

Figure 8. Base and relative consumption history of a sample fridge.

3.2. Comparison of Fridges Based on Features

After characterizing each fridge by its base and relative consumption time series, we aim to compare it against others. To achieve this, we employ a non-parametric approach that estimates the probability density function of daily energy consumption for each fridge and compares it with that of all other fridges [21]. The probability density values for each day will then be aggregated by fridge using the median statistic and will serve as a metric for how probable each fridge's consumption is compared to the rest on its base and relative consumption dimensions. Figure 9 presents an example of this approach, where the probability density estimation of energy consumption of Fridge 30 for each day is represented over the others' distribution and aggregated using the median statistic. The median statistic has been chosen as a measure of central tendency as it is not distorted by outliers or skewed data [22].

3.3. Fridge Recommendations Based on Comparison

Fridges will be shown recommendations according to their probability density scores. These recommendations will depend on if they have a low score on base energy consumption probability density or a low score on relative energy consumption probability density. These recommendations will also be accompanied by their percentile of consumption among other fridges and the expected savings if they follow these guidelines.

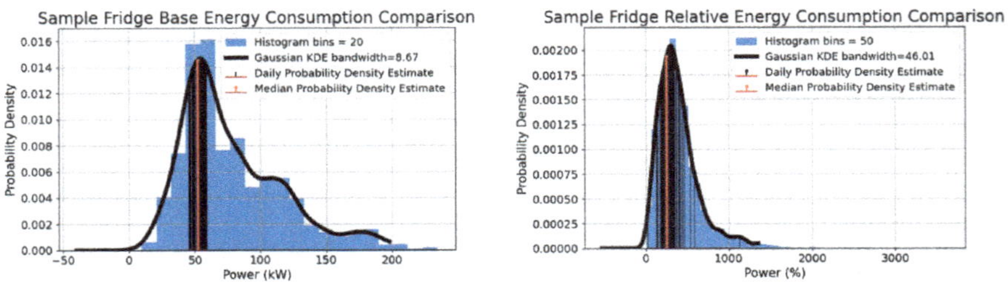

Figure 9. Base and relative consumption for Fridge 30 over the histogram and estimated probability density function of all fridges except itself.

If a fridge's median probability density estimation of base energy consumption is low and falls to the right of the distribution, this indicates that the fridge's energy consumption without human intervention is high compared to the rest.

Alternatively, if a fridge's median probability density estimation of relative energy consumption is low and falls to the right of the distribution, it indicates that the fridge's energy consumption is high due to user behavior.

The methodology involves providing energy-saving recommendations tailored to address whether the high energy consumption is due to high base consumption or high relative consumption. Table 2 shows examples of these recommendations, based on established guidelines provided by the Energy Star agency [23] and research studies cited in [24] and [12].

Table 2. Energy-saving recommendations.

Base Energy Recommendations	Relative Energy Recommendations
Set the appropriate temperature.	Keep the door closed.
Place your fridge in a cool place.	Do not introduce hot food.
Allow air circulation behind the fridge.	Keep the freezer as full as possible.
Check the door seals.	Keep containers closed to lower ambient humidity.

4. Results

In this section, we will apply the proposed methodology to the household refrigerating appliances in our dataset.

First, we calculated the daily base and relative energy consumption for each fridge. The histograms and the probability density functions in Figure 10 show how, once consumption is normalized from total to relative, the distribution of energy consumption is much more transparent and comparable between appliances.

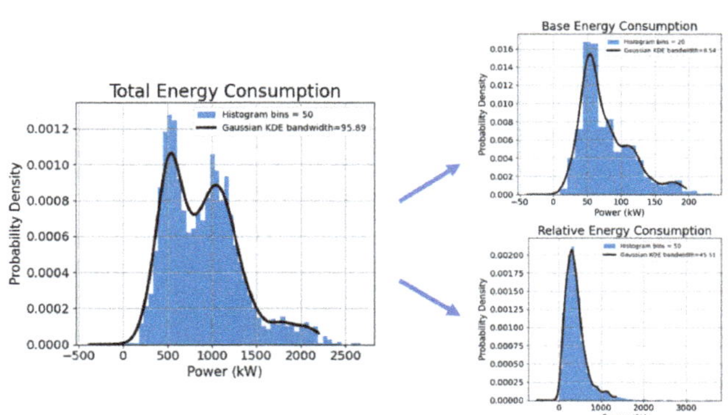

Figure 10. Energy consumption histogram and estimated probability density function for all fridges.

Following the proposed methodology, we then conducted a comparison of the refrigerators. Firstly, we calculated their daily estimated probability density of base and relative energy consumption against the probability density function estimated from all other fridges. The resulting data were then aggregated by fridge using the median statistic. Figure 11 displays the median probability density estimation for each fridge against the probability density function of all fridges for both base and relative energy consumption.

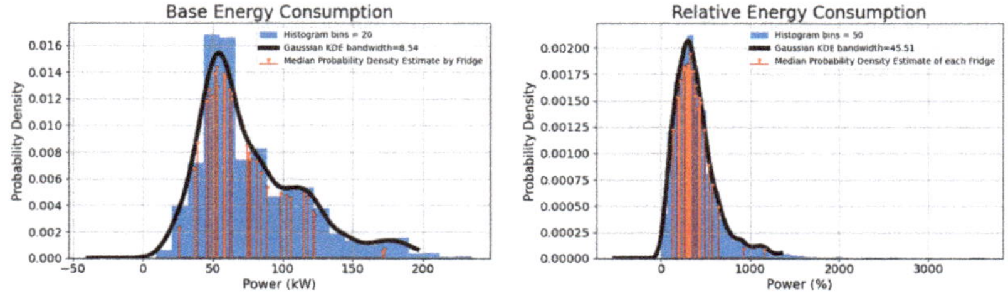

Figure 11. Base and relative consumption histogram and estimated probability density function of all fridges.

After the application of the probability density method to compare the fridges, we identified those in our dataset that exhibited high consumption and low probability density. Figure 12 displays the fridges selected based on their base energy consumption. Notably, Fridges 1 and 10 had a significantly lower probability density and higher consumption compared to the other fridges, making them stand out. We will therefore show the users of these two fridges the recommendations based on the fridge's performance without human interaction.

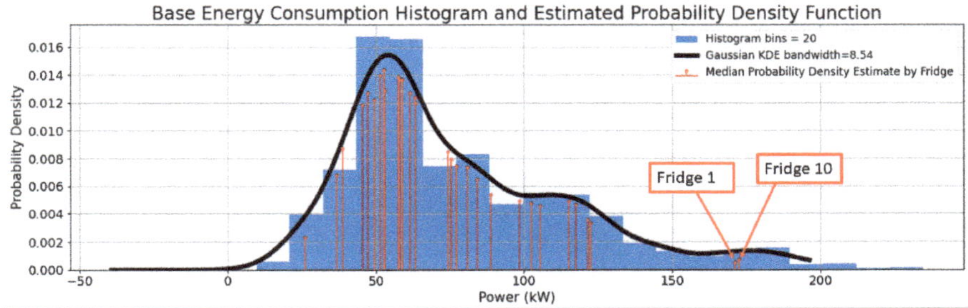

Fridge ID	Median Density of Probability Estimate	Median Base Daily Consumption (kW)
1	0.0005	171.23
10	0.0007	172.48

Figure 12. Fridges selected for base consumption recommendations.

Figure 13 presents the results for relative energy consumption. The probability density estimates revealed that Fridges 7 and 28 consumed energy above the expected levels. We will recommend these fridge users adopt the consumer habits suggested by the experts to reduce their energy consumption.

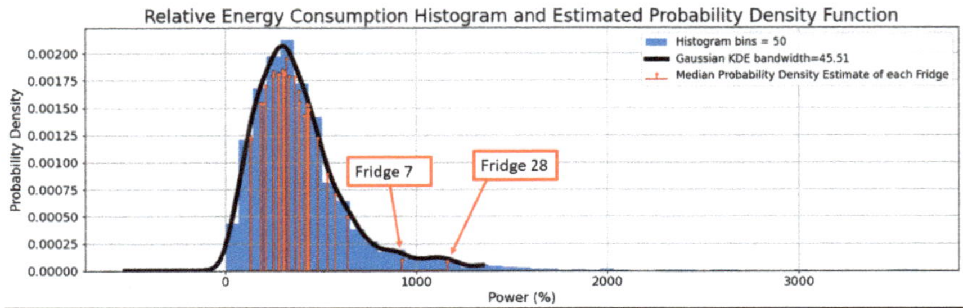

Fridge ID	Median Density of Probability	Median Relative Daily Consumption (% over Base)
28	0.0085	1160.64
7	0.0097	926.02

Figure 13. Fridges selected for relative consumption recommendations.

5. Conclusions and Future Work

We collected energy usage data from 30 household refrigerating appliances between November 2022 and February 2023. Firstly, the data were analyzed to study the frequency of operating cycles. Daily energy profiles were then grouped into 3 h intervals, and the period with the minimum energy consumption was identified as the base consumption. This represents the refrigerator's energy usage in the absence of human interaction. The relative energy consumption was then calculated as a proportion of base consumption to understand the user's impact on energy consumption, which showed that energy consumption was on average lower in the morning and higher in the evening, coinciding with usage hours. We then used ANOVA to analyze the impact of different factors on energy consumption, finding that the month and interval had a significant impact, but not the day of the week. Lastly, a K-Means with DTW clustering analysis was performed to identify different energy consumption patterns of each fridge. It revealed that different days can produce varying energy consumption profiles within and between fridges.

We have proposed a novel methodology to identify high-energy-consuming refrigerators and provide personalized energy-saving suggestions using robust probability density estimation. The process includes three stages: feature extraction, comparison of fridges based on features, and providing tailored recommendations. We use daily base and relative energy consumption as features and compare fridges based on probability density estimates. Recommendations are then provided based on a fridge's position in the distribution of energy consumption and offer guidelines for enhancing energy efficiency.

We tested our methodology on 30 refrigerators. The fridges were compared using their daily energy consumption against the probability density function estimated from all other fridges' daily base and relative energy consumption. The base energy consumption analysis revealed that Fridges 1 and 10 had a high, improbable consumption while the relative energy consumption analysis did the same for Fridges 7 and 28. The proposed methodology proved to be a reliable method to identify fridges with extreme energy behavior.

In future work, we will continue recording consumption to further improve the dataset volume, as it will enable us to study the effect of seasons on fridges' energy consumption. We would also like to track the effectiveness of these recommendations in modifying energy-consuming behavior.

Author Contributions: Conceptualization, F.R.-C. and E.F.S.-Ú.; methodology, F.R.-C. and E.F.S.-Ú.; software, F.R.-C.; validation, F.R.-C., E.F.S.-Ú. and V.G.; formal analysis, F.R.-C. and E.F.S.-Ú.; investigation, F.R.-C.; resources, V.G., A.V.S. and A.M.G.; data curation, F.R.-C. and V.G.; writing—original draft preparation, F.R.-C.; writing—review and editing, F.R.-C., E.F.S.-Ú., J.P., A.M., V.G., A.V.S. and A.M.G.; visualization, F.R.-C.; supervision, E.F.S.-Ú.; project administration, E.F.S.-Ú.; funding acquisition, E.F.S.-Ú., J.P., A.M., A.V.S. and A.M.G. All authors have read and agreed to the published version of the manuscript.

Funding: This work was supported by the RC4ALL project (Agencia Estatal de Investigación, RTC2019-007380-3), which has received funding from the Ministry of Science and Innovation (MCI) of the Government of Spain and the State Research Agency (AEI).

Institutional Review Board Statement: Not applicable.

Informed Consent Statement: Informed consent was obtained from all subjects involved in the study.

Data Availability Statement: Restrictions apply to the availability of these data. The data were obtained from households participating in the RC4ALL project (Agencia Estatal de Investigación, RTC2019-007380-3). The dataset is neither public nor available in the way that it has been used as a source in this article.

Conflicts of Interest: The authors declare no conflict of interest.

References

1. Energy, G. *CO_2 Status Report*; IEA (International Energy Agency): Paris, France, 2019.
2. White, L.V.; Sintov, N.D. Health and financial impacts of demand-side response measures differ across sociodemographic groups. *Nat. Energy* **2020**, *5*, 50–60. [CrossRef]
3. Messerli, P.; Murniningtyas, E.; Eloundou-Enyegue, P.; Foli, E.G.; Furman, E.; Glassman, A.; Hernández Licona, G.; Kim, E.M.; Lutz, W.; Moatti, J.P.; et al. Independent Group of Scientists appointed by the Secretary-General, Global Sustainable Development Report 2019: The Future is Now – Science for Achieving Sustainable Development; United Nations: New York, NY, USA, 2019.
4. Ehrhardt-Martinez, K.; Donnelly, K.A.; Laitner, S. *Advanced Metering Initiatives and Residential Feedback Programs: A Meta-Review for Household Electricity-Saving Opportunities*; American Council for an Energy-Efficient Economy: Washington, DC, USA, 2010.
5. Ridi, A.; Gisler, C.; Hennebert, J. A survey on intrusive load monitoring for appliance recognition. In Proceedings of the 2014 22nd international conference on pattern recognition, Stockholm, Sweden, 24–28 August 2014; pp. 3702–3707.
6. Welikala, S.; Dinesh, C.; Ekanayake, M.P.B.; Godaliyadda, R.I.; Ekanayake, J. Incorporating appliance usage patterns for non-intrusive load monitoring and load forecasting. *IEEE Trans. Smart Grid* **2017**, *10*, 448–461. [CrossRef]
7. Liaqat, R.; Sajjad, I.A.; Waseem, M.; Alhelou, H.H. Appliance level energy characterization of residential electricity demand: Prospects, challenges and recommendations. *IEEE Access* **2021**, *9*, 148676–148697. [CrossRef]
8. *Cool Appliances: Policy Strategies for Energy Efficient Homes*; Organisation for Economic Co-Operation and Development, OECD: Paris, France, 2003.
9. Heap, R. Refrigeration and air conditioning–the response to climate change. *Bull. IIR* **2001**, *5*, 2–16.
10. Paul, A.; Baumhögger, E.; Elsner, A.; Reineke, M.; Hueppe, C.; Stamminger, R.; Hoelscher, H.; Wagner, H.; Gries, U.; Becker, W.; et al. Impact of aging on the energy efficiency of household refrigerating appliances. *Appl. Therm. Eng.* **2022**, *205*, 117992. [CrossRef]
11. Cappelletti, F.; Manes, F.; Rossi, M.; Germani, M. Evaluating the environmental sustainability of durable products through life cycle assessment. The case of domestic refrigerators. *Sustain. Prod. Consum.* **2022**, *34*, 177–189. [CrossRef]
12. Saidur, R.; Masjuki, H.H.; Choudhury, I. Role of ambient temperature, door opening, thermostat setting position and their combined effect on refrigerator-freezer energy consumption. *Energy Convers. Manag.* **2002**, *43*, 845–854. [CrossRef]
13. Liu, D.Y.; Chang, W.R.; Lin, J.Y. Performance comparison with effect of door opening on variable and fixed frequency refrigerators/freezers. *Appl. Therm. Eng.* **2004**, *24*, 2281–2292. [CrossRef]
14. Hueppe, C.; Geppert, J.; Moenninghoff-Juessen, J.; Wolff, L.; Stamminger, R.; Paul, A.; Elsner, A.; Vrabec, J.; Wagner, H.; Hoelscher, H.; et al. Investigating the real life energy consumption of refrigeration appliances in Germany: Are present policies sufficient? *Energy Policy* **2021**, *155*, 112275. [CrossRef]
15. *RC4ALL: Inteligencia Artificial Para un Consumo Responsable*; Endesa: Madrid, Spain, 2020.
16. Kendall, M.; Stuart, A.; Ord, J. *Kendall's Advanced Theory of Statistics, Volume 3: Design and Analysis, and Time Series*; John Wiley & Sons: Hoboken, NJ, USA, 1987.
17. Kim, T.K. Understanding one-way ANOVA using conceptual figures. *Korean J. Anesthesiol.* **2017**, *70*, 22–26. [CrossRef] [PubMed]
18. Fisher, R.A. *Statistical Methods for Research Workers*; Springer: Berlin/Heidelberg, Germany, 1992.
19. Anh, D.T.; Thanh, L.H. An efficient implementation of k-means clustering for time series data with DTW distance. *Int. J. Bus. Intell. Data Min.* **2015**, *10*, 213–232. [CrossRef]
20. Rousseeuw, P.J. Silhouettes: A graphical aid to the interpretation and validation of cluster analysis. *J. Comput. Appl. Math.* **1987**, *20*, 53–65. [CrossRef]
21. Pöthkow, K.; Hege, H.C. Nonparametric models for uncertainty visualization. In *Computer Graphics Forum*; Wiley Online Library: Hoboken, NJ, USA, 2013; Volume 32, pp. 131–140.
22. Wilcox, R.R.; Keselman, H. Modern robust data analysis methods: Measures of central tendency. *Psychol. Methods* **2003**, *8*, 254. [CrossRef] [PubMed]

23. *Refrigerator Best Practices*; Energy Star, U.S. Environment Protection Agency: Washington, DC, USA, 2023.
24. Harrington, L. Prediction of Energy Consumption of Refrigerators During Use. Ph.D. Thesis, Department of Infrastructure Engineering, Melbourne School of Engineering, The University of Melbourne, Melbourne, VIC, Australia, 2018. Available online: https://minerva-access.unimelb.edu.au/handle/11343/213357 (accessed on 18 June 2023).

engineering proceedings

MDPI

Proceeding Paper

Time Series Analysis in Hydrogeological Conceptual Model Upgrading †

Paola Gattinoni

Department of Civil and Environmental Engineering, Politecnico di Milano, 20133 Milan, Italy; paola.gattinoni@polimi.it

† Presented at the 9th International Conference on Time Series and Forecasting, Gran Canaria, Spain, 12–14 July 2023.

Abstract: The modeling of hydrogeological processes often involves a quantitative description of complex systems in which a limited dataset is available, bringing about the formulation of conceptual models able to describe them in a simplified framework. In order to evaluate the reliability of these conceptual models, a statistical description of the elements composing the system can be useful, especially with reference to their mutual interactions. This study shows, through some applicative examples in the hydrogeological field, that the statistical analysis of characterizing the parameters and cause–effect relations arising from time series monitoring data can give useful information about the system dynamic, thus contributing to updating the conceptual model and therefore improving the results of following numerical modeling. Indeed, this dynamic description of the system, with the introduction of the verification and validation processes of the conceptual model, allows the correction of possible errors due to a lack of data or the phenomenon's complexity. This leads to many hydrogeological issues, such as the identification of the most productive aquifer or the one that has the highest vulnerability to pollution, as well as zones of interest in groundwater flow that can trigger slope instability.

Keywords: conceptual model; hydrogeological issues; monitoring data; numerical modeling; time series analysis

1. Introduction

Numerical modeling is often used for describing and forecasting the behavior of complex hydrogeological systems, and has several purposes, from water resources management [1], to pollution analysis and related remediation design [2], as well as in slope stability problems [3].

Nevertheless, in mathematical modeling, not all geological and hydrogeological issues can be fully investigated. In fact, most geological phenomena cannot be described by simple predictable quantities (due to cause–effect relationships) on the basis of boundary conditions or as a function of the values obtained from other variables. On the contrary, many typical hydrogeological parameters can be considered as random variables due to:

- the intrinsic variability of natural processes;
- an inability to understand the physical system in its complexity;
- a lack of sufficient data, both in terms of quality and/or quantity [4].

In the applicative field, the hydrogeological complexity of both the system and the processes, and their limited knowledge, can involve unexpected risk, bringing about the need for a different approach in the system characterization [5].

Consequently, when dealing with hydrogeological risk assessment, a statistics-based conceptual model should be considered in order to take into account the different risk scenarios, and therefore to identify the proper mitigation measures.

Conceptual models are simplified descriptions of complex natural-anthropic systems, involving the identification of geomaterials, the reconstruction of their distribution in space,

Citation: Gattinoni, P. Time Series Analysis in Hydrogeological Conceptual Model Upgrading. *Eng. Proc.* **2023**, *39*, 44. https://doi.org/10.3390/engproc2023039044

Academic Editors: Ignacio Rojas, Hector Pomares, Luis Javier Herrera, Fernando Rojas and Olga Valenzuela

Published: 4 July 2023

as well as their physical and technical characterization; this characterization is aimed at forecasting the system behavior, by taking into account the main processes which rule it. In fact, based on the conceptual model, the numerical or analytical model can be implemented and afterwards validated; the validation process often involves an improvement in the conceptual model, by means of a refinement in available data or an increasing complexity in the processes description. Regardless, the need for such an improvement must be quantitatively assessed, and this assessment can be efficiently based on time series analysis (i.e., by comparing time series of different parameters, or time series of simulated vs. observed parameters).

This paper proposes some typical hydrogeological situations in which a statistical approach to the conceptual model definition was essential for the reliability of the results of subsequent numerical modeling. With this aim, the hydrogeological system is represented through quantitative conceptual models, in which some elements, and their mutual interactions, are analyzed from a statistical point of view, endeavoring to reproduce the system evolution in a dynamic framework and involving back-analysis upgrading.

2. Conceptual Model Update: From Monitoring Data to Error Detection

In order to implement a statistical framework in which the conceptual model can be dynamically improved, it is necessary to quantify the effect of the modeling error on the forecasting and, therefore, proceed with checks and updates of the model itself, based on monitoring data [6]; the goal is to establish if, and to what extent, the model can explain and predict the observed phenomena, and detect any inconsistencies that require model refinement [7]. Indeed, the mismatch in between the time series of observed values compared to modeling results (values obtained by simulating the system behavior based on the defined conceptual model) should be minimized [8], by taking into account the fact that possible errors can arise from:

- modeling errors—because of incorrect hypotheses in the conceptual model, the model fails in reproducing some processes or some cause–effect relationships;
- parametric errors—due to an insufficient knowledge of the parameters describing the system behavior (i.e., connected to measurement errors, system heterogeneity or incorrect working scale).

The simplest method in order to detect conceptual model errors consists of a graphical comparison between the calculated and observed data [9]; for quantifying the capability of the model to fit the observed data, a goodness-of-fit test can be carried out. More specifically, an analysis of the deviations (i.e., the differences between the observed and simulated data) makes it possible to identify errors together with their typology. In fact, conceptual model errors are systematic and their effects on the modeling results depends on the spatial and temporal scale of the simulation. Consequently, a correlation between residuals over time or space means that a trend exists and, therefore, the model is affected by errors [10]. In more detail, if the residual deviation is constant but not null, the parameters have not been properly calibrated whereas, if it shows a trend, the model setting should be improved; only when it is null can the model be considered acceptable.

This approach has been implemented in a dynamic framework for hydrogeological risk assessment, which accounts for a conceptual model to be updated based on the time series analysis of the residuals (Figure 1). A similar approach had already given quite good results in managing the geological risk in civil engineering design [11] and built heritage [12]. In the present study, this framework has been applied in some applicative problems typical of the hydrogeological field, resulting in a significant improvement in risk mitigation and management.

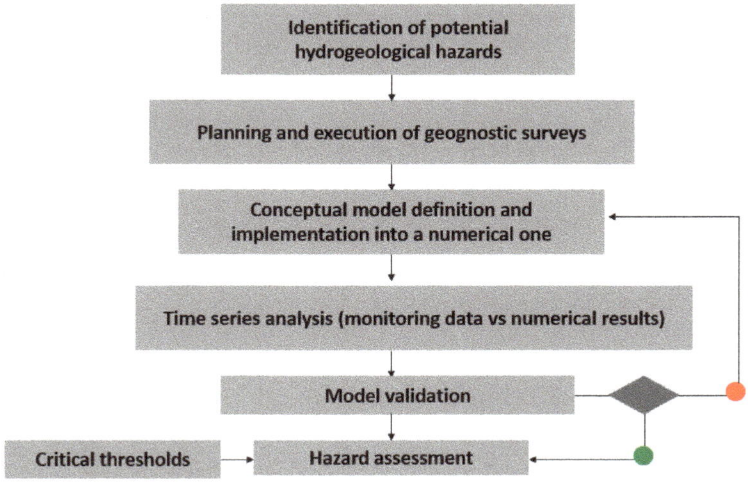

Figure 1. Framework for hydrogeological risk assessment with dynamic conceptual modeling.

3. Examples of Time Series Analysis in Hydrogeological Modeling Improvement

3.1. Hydrogeological Risk in Tunneling

The hydrogeological risk in tunneling [13–15] consists in an inflow risk for the tunnel [16,17], and a water table drawdown with spring depletion for the environment [18–20]; these risks can be analyzed by considering, as a performance function, the water balance over a control volume, and the corresponding elements (i.e., rainfall and permeability) can be statistically described, leading to a stochastic modeling approach. Therefore, modeling results can be compared to monitoring data, typically with references to time series of water table drawdown or spring flow rate; for instance, the differences in between the simulated and observed values of water table (residuals) can be analyzed; if differences are detected, specifically an increasing deviation in residuals (purple line in Figure 2), it means that the conceptual model, based on which the numerical one has been implemented, was too simplified. When dealing with hydrogeological problems, a quite typical error consists in considering the presence of a single aquifer, instead of two different aquifers locally interconnected. In such a case, the performance function describing the conceptual model has to be improved for taking into account the two aquifers and their interaction; nevertheless, some parametrical errors could still be present (blue line in Figure 2 exhibits a constant not zero value), and a further calibration of the modeling parameters is still required.

3.2. Landslide Hydrogeological Risk

The Vajont landslide is a typical example in which several attempts were made to frame and interpret the geological and hydrogeological setting, improving the conceptual model from time to time.

In particular, as far as the hydrogeological conceptual model aimed at describing the "landslide-aquifer-reservoir" is concerned, time series arising from some piezometers of different depths were analyzed and compared to time series of reservoir level and displacements. Based on this comparison, an inconsistency in water table data was detected; in particular, by carrying out an analysis of the linear correlation between the data recorded by piezometers and the reservoir level, one of the piezometers showed data uncorrelated with respect to the others (Figure 3). The first conceptual model, which accounted for a single-layer aquifer, was unable to explain this anomaly; this meant that it contained an error. In fact, it was too simplified for reproducing the real system complexity. Indeed, subsequent studies [21] reconstructed the correct conceptual model, by identifying the existence of a low permeability layer which separates two aquifers: one in the landslide mass and the other one in the underlying bedrock. This updated conceptual model could explain the trend of both shallow and deep piezometers. In fact, the two aquifers had different flow conditions; the shallow one was strongly influenced by the reservoir level (as shown in pz1 and pz3) whereas, in the confined aquifer (pz2), the water table was independent of the reservoir level and strictly correlated to the total rainfall of the previous period [21]. This updated conceptual model allowed the authors to identify a stability threshold for the landslide, as a function of the reservoir level and rainfall in the previous 30 days, related to the measured landslide displacements.

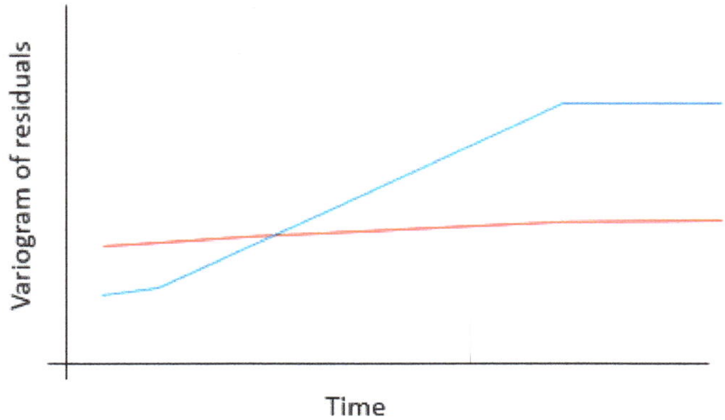

Figure 2. Deviations between modeling results and observed data obtained by a simplified one-aquifer model (in blue) and a more complex multi-aquifer model (in red).

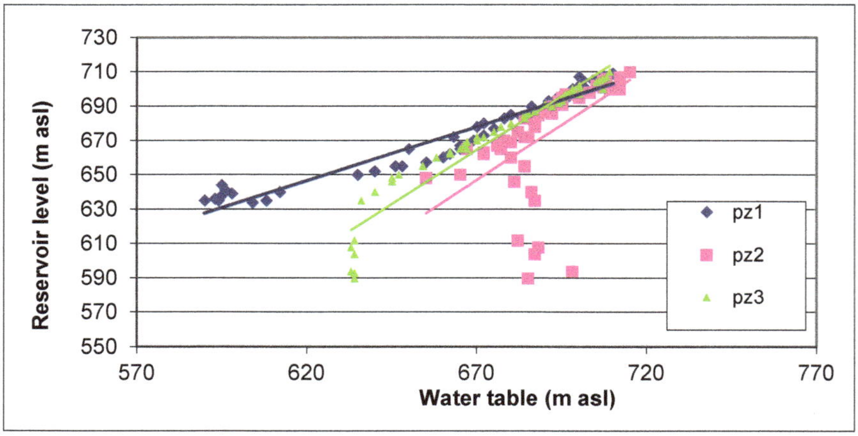

Figure 3. Statistical correlation between water table in piezometers of different depths (pz1 and pz3 are in the shallow aquifer, whereas pz2 is in the confined one) and the reservoir level.

3.3. Groundwater Pollution Risk

The characterization of groundwater pollution involves the statistical analysis of monitoring data both in time and in space (i.e., multivariate analyses, association and correlation indexes on time series, variogram analyses on special distribute data), and it should lead to the identification of the pollution source and processes, as well as to a description of the groundwater quality and vulnerability.

In more detail, the statistical analysis of monitoring data should concern not only the time evolution of pollution, but also the possible relationships between the different contaminants (Figure 4), as well as the relationships between water table and pollution trends. Indeed, an analysis of correlation can be carried out among the time series of all those variables that are representative of cause–effect phenomena; for example, by comparing the time series of water table with the time series of the concentration of some peculiar pollutants, it is possible to verify whether the latter arises from a release of contaminants stored into an aquitard or in non-saturated soils.

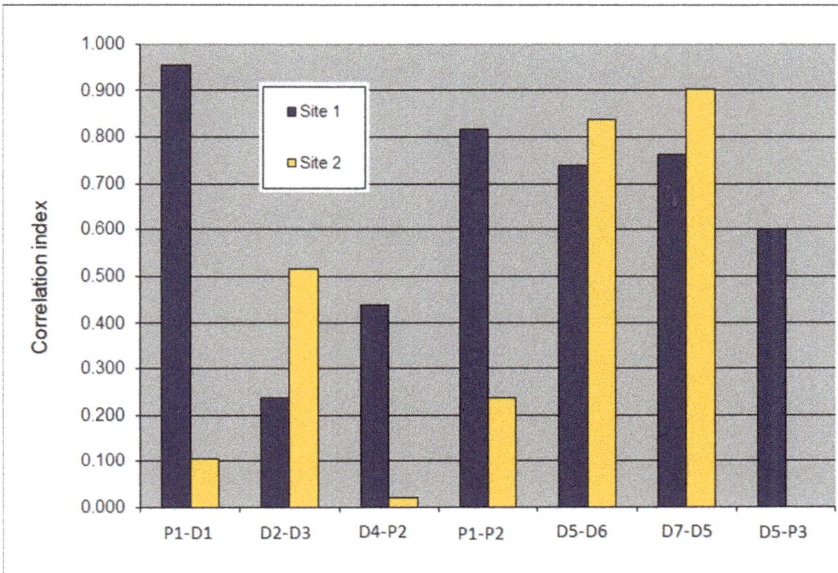

Figure 4. Correlation indexes between the concentrations of some compounds in different polluted sites: in site 1, primary contaminants (Pi) and their degradation products (Di) are predominant and well correlated; in site 2, the correlation between Pi and Di decreases, which involves a different pollution source (afterwards identified with a dump).

Once the processes involved in the pollution event have been identified, a further analysis can be carried out by using variograms in order to identify not only the most polluted areas and their evolution though time, but the type of contamination, too (hot-spot, plume or diffuse; Figure 5).

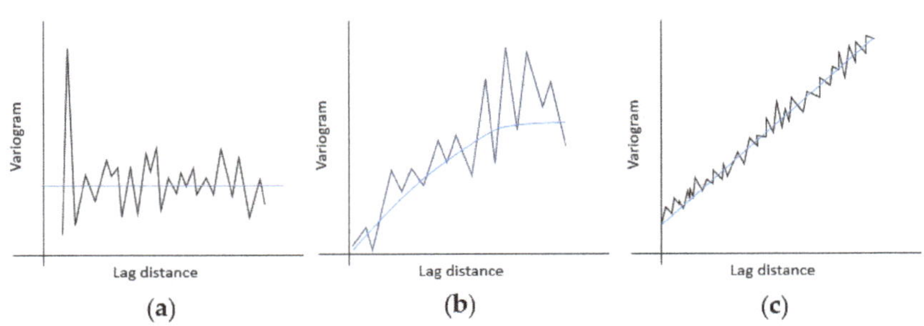

Figure 5. Variograms corresponding to different types of groundwater pollution: (**a**) hot spots bring about a nugget effect, such as for the CVM in an industrial site, (**b**) plume involves a spherical variogram in the groundwater flow direction, such as for chlorides in the same industrial site and (**c**) diffuse contamination usually shows a linear trend, such as for nitrates in a large-scale alluvial aquifer.

4. Conclusions

Groundwater flow and transport are controlled by physical properties that are characterized by a high degree of heterogeneity and by scales of variation that span several orders of magnitude. This complexity can lead to unexpected risk in many applicative fields, from water resources management to civil engineering design and land planning, and therefore requires a strategy for risk mitigation. With this aim, a statistical analysis of the available data and time series can be used in order to detect possible errors in the hydrogeological conceptual model, thus improving its reliability.

The examples proposed in the present paper show the usefulness of this approach in the applicative field of hydrogeology, as it allows, firstly, the upgrading of the conceptual model of the system and then the subsequent numerical modeling results, especially when the hydrogeological setting is very complex. In fact, this approach quantifies the relevance of the different elements involved in the hydrogeological system and the related processes, therefore identifying errors which can be corrected in order to update the conceptual model.

Funding: This research received no external funding.

Institutional Review Board Statement: Not applicable.

Informed Consent Statement: Not applicable.

Data Availability Statement: Data sharing not applicable.

Conflicts of Interest: The authors declare no conflict of interest.

References

1. Boronina, A.; Renard, P.; Balderer, W.; Christodoulides, A. Groundwater resources in the Kouros catchment (Cyprus): Data analysis and numerical modelling. *J. Hydrol.* **2002**, *271*, 130–149. [CrossRef]
2. Francani, V.; Gattinoni, P. Statistical risk analysis of groundwater pollution. In *GeoCongress 2008: Geosustainability and Geohazard Mitigation*; American Society of Civil Engineers: Reston, VA, USA, 2008; Volume 178, pp. 154–161.
3. Gattinoni, P. Parametrical landslide modeling for the hydrogeological susceptibility assessment: From the Crati Valley to the Cavallerizzo landslide (Southern Italy). *Nat. Hazards* **2008**, *50*, 161–178. [CrossRef]
4. Loaiciga, H.A.; Charbeneau, R.J.; Everett, L.G.; Fogg, G.E.; Hobbs, B.F.; Rouhani, S. Review of Ground-Water Quality Monitoring Network Design. *J. Hydraul. Eng.* **1992**, *118*, 11–37. [CrossRef]
5. Makropoulos, C.; Koutsoyiannis, D.; Stanić, M.; Djordjević, S.; Prodanović, D.; Dašić, T.; Prohaska, S.; Maksimović, C.; Wheater, H. A multi-model approach to the simulation of large scale karst flows. *J. Hydrol.* **2008**, *348*, 412–424. [CrossRef]
6. Hassan, A.E. A methodology for validating numerical ground water models. *Groundwater* **2004**, *42*, 347–362. [CrossRef] [PubMed]
7. Straface, S. A procedure of Model Validation with a Bayesian Approach. In *XXX IAHR Congress—Agora Thessaloniki. Calibration and Validation of 3D Numerical Model*; International Association of Hydraulic Engineering and Research: Thessaloniki, Greece, 2003.
8. Gaganis, P.; Smith, L. A Bayesian Approach to the quantification of the effect of model error on the predictions of groundwater models. *Water Resour. Res.* **2001**, *37*, 2309–2322. [CrossRef]
9. Flavelle, P. A quantitative measure of model validation and its potential use for regulatory purposes. *Adv. Water Resour.* **1992**, *15*, 5–13. [CrossRef]
10. Davis, P.A.; Goodrich, M.T. A Proposed Strategy for the Validation of Ground-Water Flow and Solute Transport Models. In *Proceedings from GEOVAL 1990*; Swedish Nuclear Power Inspectorate: Stockholm, Sweden, 1990.
11. Gattinoni, P.; Scesi, L.; Arieni, L.; Canavesi, M.; Zaffaroni, F. A new rating system for hydrogeological risk management along railway infrastructures in Prealpine zone (northern Italy). *Innov. Infrastruct. Solutions* **2021**, *6*, 1–14. [CrossRef]
12. De Finis, E.; Gattinoni, P.; Scesi, L.; Valletta, A. Conceptual and numerical modelling for hydrogeological hazard assessment in the UNESCO site of Castelseprio (northern Italy). In Proceedings of the International Multidisciplinary Scientific GeoConference SGEM 2018, Albena, Bulgaria, 30 June–9 July 2018; Volume 18, pp. 161–176.
13. Gattinoni, P.; Scesi, L. From hydrogeological hazard identification to risk assessment in tunnelling: An example in northern Italy. In Proceedings of the 20th International Multidisciplinary Scientific GeoConference SGEM, Albena, Bulgaria, 16–25 August 2020; pp. 519–530.
14. Loew, S. Groundwater hydraulics and environmental impacts of tunnels in crystalline rocks. In *Meccanica e Ingegneria delle Rocce*; MIR: Torino, Italy, 2002; pp. 201–217.
15. Molinero, J.; Samper, J.; Juanes, R. Numerical modeling of the transient hydrogeological response produced by tunnel construction in fractured bedrocks. *Eng. Geol.* **2002**, *64*, 369–386. [CrossRef]

16. Celestino, T.B.; Gianbastiani, M.; Bortolucci, A.A. Water inflows in tunnels: Back-analysis and role of different lining systems. In *AITES-ITA 2001, Progress in Tunnelling after 2000*; Teuscher, P., Colombo, A., Eds.; Pàtron Editore: Bologna, Italy, 2001; Volume 2, pp. 547–554.
17. Goodman, R.E.; Moye, D.G.; Van Schalkwyk, A.; Javandel, I. Ground water inflow during tunnel driving. *Eng. Geol.* **1965**, *2*, 39–56.
18. Colombo, L.; Gattinoni, P.; Scesi, L. Influence of underground structures and infrastructures on the groundwater level in the urban area of milan, italy. *Int. J. Sustain. Dev. Plan.* **2017**, *12*, 176–184. [CrossRef]
19. Anagnostou, G. The influence of tunnel excavation on the hydraulic head. *Int. J. Numer. Anal. Methods Géoméch.* **1995**, *19*, 725–746. [CrossRef]
20. Dematteis, A.; Kalamaras, G.; Eusebio, A. A systems approach for evaluating springs drawdown due to tunnelling. In *AITES-ITA 2001, Progress in Tunnelling after 2000*; Teuscher, P., Colombo, A., Eds.; Pàtron Editore: Bologna, Italy, 2001; Volume 1, pp. 257–264.
21. Hendron, A.J.; Patton, F.D. *The Vaiont Slide, A Geotechnical Analysis Based on New Geologic Observations of the Failure Surface*; Tech. Rep. GL-85-5; U.S. Army Corps of Engineers: Washington, DC, USA, 1985.

engineering proceedings

MDPI

Proceeding Paper

Nonstationary Frequency Analysis of Extreme Rainfall in the Taihu Lake Basin, China †

Yuting Jin [1], Shuguang Liu [1,2,*], Zhengzheng Zhou [1,*], Qi Zhuang [1] and Guihui Zhong [1]

[1] Department of Hydraulic Engineering, Tongji University, Shanghai 200092, China; 1951555@tongji.edu.cn (Y.J.); 2110026@tongji.edu.cn (Q.Z.); 04098@tongji.edu.cn (G.Z.)
[2] Key Laboratory of Yangtze River Water Environment, Ministry of Education, Tongji University, Shanghai 200092, China
* Correspondence: liusgliu@tongji.edu.cn (S.L.); 19058@tongji.edu.cn (Z.Z.); Tel.: +86-13641853120 (S.L.); +86-13816108758 (Z.Z.)
† Presented at the 9th International Conference on Time Series and Forecasting, Gran Canaria, Spain, 12–14 July 2023.

Abstract: Nonstationary is one of the prominent phenomena in the current hydrological time series due to climate change and urban expansion. In this study, using the long time series rainfall data from rain gauges and satellite rainfall data, the trend and abrupt change of rainfall in the Taihu Lake basin, China, are examined by the Mann–Kendall (MK) test and the Pettitt test, using rain gague data. For seven water conservancy zones in this basin, the intensity–duration–frequency curves (IDFs) are obtained using satellite rainfall and the stochastic storm transposition (SST) method, providing a method for rainfall frequency analysis based on nonstationary assumption. The IDFs results between the conventional frequency analysis method with the stationary assumption and the SST-based method are compared. The results show an overall increasing trend of annual total rainfall in the Taihu Lake Basin, with significant changes at most stations. The SST-based results show a significant difference of IDFs in seven conservancy zones, which are linked to nonstationary changes in rainfall series. Our results provide an important reference for understanding the nonstationary changes and nonstationary frequency analysis of extreme rainfall in the Taihu Lake basin.

Keywords: nonstationary changes; extreme rainfall; stochastic storm transposition; rainfall frequency analysis; Taihu Lake basin

Citation: Jin, Y.; Liu, S.; Zhou, Z.; Zhuang, Q.; Zhong, G. Nonstationary Frequency Analysis of Extreme Rainfall in the Taihu Lake Basin, China. *Eng. Proc.* **2023**, *39*, 45. https://doi.org/10.3390/engproc2023039045

Academic Editors: Ignacio Rojas, Hector Pomares, Luis Javier Herrera, Fernando Rojas and Olga Valenzuela

Published: 4 July 2023

1. Introduction

With global warming and rapid urbanization, extreme hydrometeorological events, such as extreme rainfall/flood occur more frequently, with increasing intensity and severe damage in urban areas [1]. It results in the potential need to update the current design standard for hydraulic infrastructure to protect human life and property [2]. To this end, the nonstationarity-based frequency analysis of hydrological time series is of vital importance. The Taihu Lake basin contains several mega-cities and large and medium-sized cities with highly developed urban economies and highly concentrated populations [3,4]. As the urbanization process in the Taihu Lake basin further accelerates, the scale of cities expands, the area of rivers and lakes decreases dramatically, and the flood storage capacity of the basin is greatly affected. At the same time, flooding caused by extreme precipitation has become one of the key issues affecting the safe and orderly development of the region [5].

Current research on the characteristics of storms in the Taihu Lake basin is mainly based on hydrological data from ground stations [6]. However, due to the fixed location, small number, and uneven distribution of hydrological stations, the actual maximum rainfall is likely to be "missed" and the spatial heterogeneity of rainfall is difficult reflect [7]. Therefore, in the traditional urban design storm, the assumption of a spatially uniform rainfall distribution is often used to simplify the calculation [8]. However, as urbanization has accelerated in recent years, the differences in the spatial and temporal distribution of rainfall have further intensified, and it is necessary to consider the spatial distribution of rainfall to ensure the reliability of the design storm results. In this study, taking the Taihu Lake basin, which has experienced rapid urbanization in recent years, as an example, we examine the nonstationary changes in extreme rainfall time series and then propose a nonstationary-based frequency analysis.

Stochastic storm transposition (SST) is a regional rainfall frequency analysis method based on a combination of regional probabilistic resampling and spatial transformation of storms and is now widely used in various regions of the United States [9–12]. The main principle to delineate a storm-shift area with a target region as the core is to select a series of extreme rainfall events within the area as a "storm catalog", simulate and extend the storm sequence based on a combination of probabilistic resampling and geographic shifts, and then perform frequency analysis. The main advantage of this method is that it avoids the assumption of consistency of time series required by traditional methods and, at the same time, describes the spatial variability of rainfall in addition to its intensity and magnitude, reducing the uncertainty of the design storms.

Several studies have shown that high-resolution precipitation data combined with SST methods can improve the accuracy of calculations in rainfall/flood studies. For example, Zhou [12] used 16a radar data (2000–2015) for an urban area design storm study in the Baltimore metropolitan area, USA; Wright [10] used 10a radar data (2001–2010) for a flood study in the Little Sugar Creek Watershed, USA; and Zhuang [13] used 21a radar data (2000–2020) for an urban area design storm study in Shanghai, China, all of which have been well used. This paper analyzes the spatial and temporal distribution characteristics of each conservancy zone in the Taihu Lake basin based on high-resolution satellite rainfall data and also uses the SST method for storm frequency analysis, the results of which can provide a basis for flood control planning and disaster prevention and mitigation in the basin.

The paper is organized as follows: in Section 2, the scope of the study data and the methodology will be presented. In Section 3, the results and discussions will be introduced. A summary and conclusions are presented in Section 4.

2. Materials and Methods

2.1. Study Area and Data

The Taihu Lake basin is located between 118°55′~122°12′ E and 30°06′~31°53′ N. It has a total area of 36,900 km^2 and is dominated by plains, which account for 4/6 of the total area; water surface accounts for 1/6; and hills account for 1/6. This study is based on the Taihu Lake basin conservancy zone as the study unit.

A total of 9 rain gauges were selected as representative rain gauges for each of the 7 conservancy zones in the Taihu Lake basin, and the 60-year (1961–2020) daily rainfall time series was used for the study. The locations of the rain gauges are shown in Figure 1.

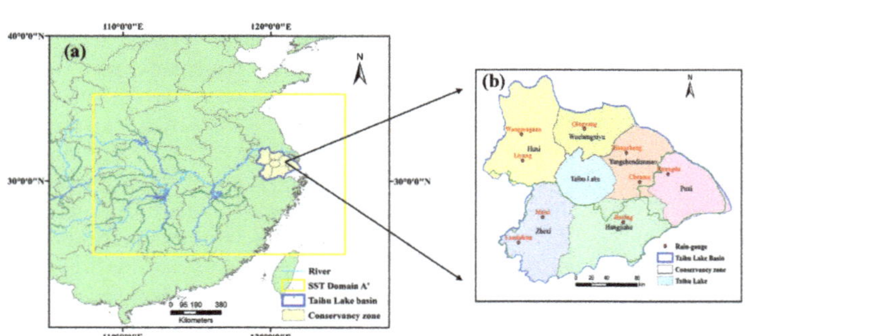

Figure 1. Location of the study region (**a**). The distribution of the seven conservancy zones in Taihu Lake Basin and the location of the 9 representative rain gauges (**b**).

In this paper, the GPM-IMERG gridded satellite rainfall data are provided by NASA were used to study and analyze the data from 1 January 2001 to 31 December 2020, with a temporal resolution of 30 min and a spatial resolution of 0.1° × 0.1°. It has been shown that this dataset has good adaptation in China.

2.2. Research Methods

2.2.1. Nonstationary Analysis

In this paper, the Mann–Kendal test [14,15] and the Pettitt [16] variation point analysis method were used to test the trend and abrupt change points of the rainfall series, respectively, and the spatial heterogeneity and non-coherence analysis of the extreme rainfall indicators were conducted. All the above methods are widely used in the analysis of time series; the specific methods can be seen in references [16–18].

2.2.2. Stochastic Storm Transposition (SST) Method

The main steps in applying the SST method to a regional design storm study are summarized as follows [19]. A more detailed description of the SST method can be found in the literature [20].

(1) A region including the study area is defined as the transposition domain area A′, and it is determined that the transposition of the storm will be carried out within this area. The transposition domain should contain enough storm events, and its shape can be defined as a rectangular area or an arbitrary shape. It is usually determined based on a comprehensive analysis of the regional hydro-meteorological and geographical characteristics. The outline in Figure 2 is the selected transposition domain in this paper.

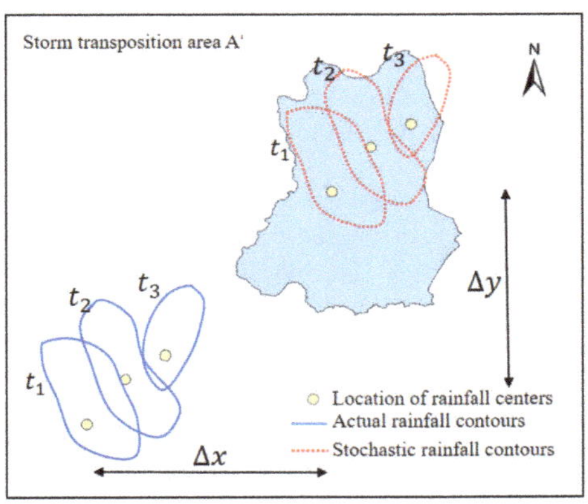

Figure 2. Depiction of stochastic storm transposition procedures for a single storm, taking the Huxi Conservancy zone as an example.

(2) The maximum *m* storms (no overlapping time between rainfall events) in the transposition domain are selected from the n-year satellite rainfall series to form a subset containing spatial and temporal rainfall data as the "Storm catalog". When A′ is a "homogeneous area", which means that the storm has uniform characteristics in all parts of the area, the probability of storm occurrence is the same in all parts of the area. When the displacement area A′ is a "heterogeneity area", it means that there is spatial heterogeneity in the distribution of storms in the area. The probability of storm occurrence in different parts of the transposition domain is different, and the probability magnitude is different. The probability of occurrence of storm events differs from one place to another within the transposition domin, and the probability can be determined by the location (longitude and latitude) of m storms based on the non-parametric estimation method of Gaussian kernel density. In this paper, we calculate the probability of storms shown in the figure based on the premise that the transposition domain is a "heterogeneity area".

(3) Simulation of the "annual maximum storms" sequence. This means that k_i storms are randomly selected from the storm catalog. The random variable k_i is assumed to follow the Poisson distribution. The parameter of the Poisson distribution is $\lambda = m/n$, such as in the storm catalog in the paper; if m = 200 rainfall events and n = 20 years of rainfall data, then $\lambda = 10.0$. The relative movement and evolution of the entire storm field in all periods are not changed; only the spatial location of the storm occurrence is changed (see Figure 2). After shifting, the maximum value of t–h rainfall accumulation in the target watershed A was retained as the "annual maximum storms".

(4) Process (3) is repeated N times to construct the "annual maximum storms" sequence with a duration of t–h for N years. If we assume that the above process is repeated N = 500 times, we can obtain the annual maximum storm sequence of 500a. At this time, the range of the recurrence period is 1~500a. Note that since the shifted storms are randomly selected from the storm catalog, the shifting process is also called "resampling". Therefore, the cumulative rainfall calculated in target area A for each resampling process will not be repeated. After resampling, the "annual maximum storms" sequence is obtained for N years. The annual exceedance probability of each storm i is pi = i/Tmax, and the recurrence period Ti = 1/pi. The calculation results can be plotted as empirical IDF curves or input to hydrological models.

3. Results and Discussion

3.1. Nonstationary Analysis

3.1.1. Annual Rainfall Analysis

From 1960 to 2020, the annual total rainfall of rain gauges in the seven water conservancy zones in the Taihu Lake basin ranged from 521 to 2338 mm, and the average annual rainfall ranged from 1075 to 1580 mm, with a decreasing trend from southwest to northeast. The MK test showed that the annual total rainfall in more than 80% of the water conservancy zones showed non-stationary changes and an overall increasing trend. From the Pettitt test results, the annual total rainfall mainly changed abruptly in 1979 and 2008, among which the Meixi, Jiaxing, Huangdu, and Chenmu rain gauges located in the southeastern part of the Taihu Lake basin all had abrupt changes in 1979, and the Wangmuguan, Liyang, and Laoshikan rain gauges located in the western part of the Taihu Lake basin all had abrupt changes in 2008. The rainfall sequence was tested with a significance level of 0.05, and 37.5% of the rain gauges that passed the significance level test were all urban rain gauges with a high urbanization level (Table 1). The results of the nonstationary analysis are shown in Figure 3.

Table 1. The results of the nonstationary analysis of the annual total rainfall of rain gauges in the seven water conservancy zones in Taihu Lake basin.

Water Conservancy Zone	Rain Gauge	Trend	Change Point	Significance Test
Yangchendianmao	Xiangcheng	increasing	1997	×
	Chenmu	increasing	1979	√
Zhexi	Meixi	increasing	1979	×
	Laoshikan	increasing	2008	×
Wuchengxiyu	Qingyang	increasing	2005	×
Huxi	Wangmuguan	no trend	2008	×
	Liyang	no trend	2008	×
Huangjiahu	Jiaxing	increasing	1979	√
Puxi	Huangdu	increasing	1979	√

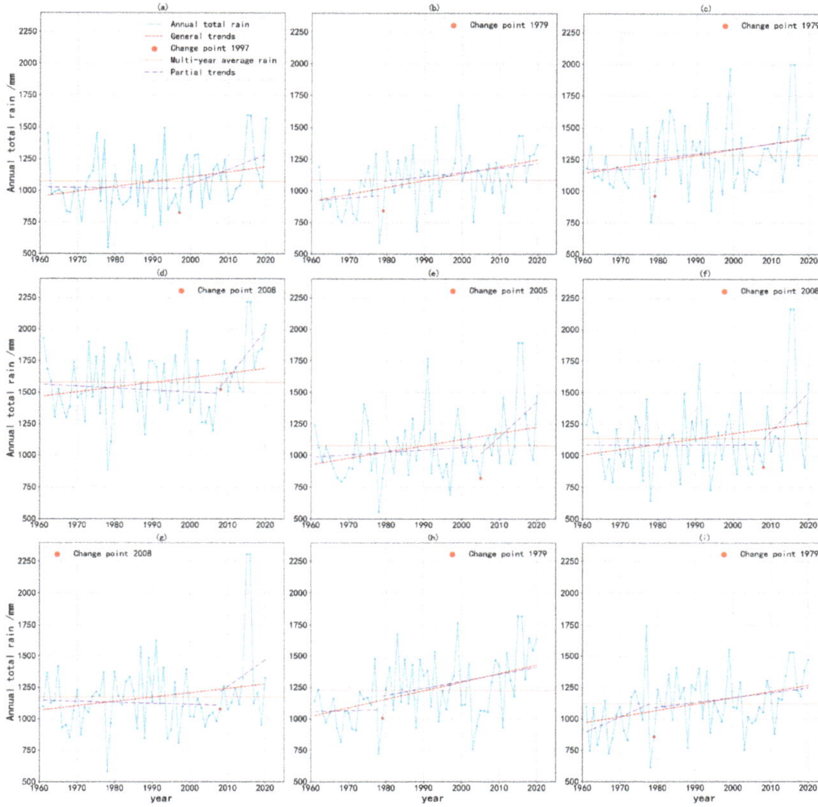

Figure 3. The results of nonstationary analysis of the annual total rainfall of each representative rain gauge in the seven water conservancy zones in Taihu Lake basin. Analysis results for Xiangchen (**a**); Chenmu (**b**) rain-gauges in Yangchendianmao water conservancy zone; Meixi (**c**); Laoshikan (**d**) rain-gauges in Zhexi water conservancy zone; Qingyang (**e**) rain-gaugs in Wuchengxiyu water conservancy zone; Wangmuguan (**f**); Liyang (**g**) rain-gauges in Huxi water conservancy zone; Jiaxing (**h**) rain-gauge in Hangjiahu water conservancy zone; Huangdu (**i**) rain-gauge in Puxi water conservancy zone.

3.1.2. Extreme Rainfall Analysis

Taking the annual maximum 1-day rainfall (Rx1d) as an example, shown in Figure 4, we analyzed the extreme rainfall characteristics of the seven water conservancy zones in the Taihu Lake basin. The Rx1d at each station from 1960 to 2020 ranged from 27 to 176 mm, and comparing the Chenmu, Xiangcheng, and Huangdu stations with the Meixi and Laoshikan stations, we can obtain that the frequency of extreme rainfall is higher in areas with more urbanization. The average frequency and intensity of heavy rainfall and extreme precipitation events in the rest of the water conservancy zones, except Huxi, have shown an overall upward trend over the past 60 years, and a significant upward trend in the past 5 years, with an overall pattern of decreasing extreme rainfall from southwest to northeast, which is consistent with spatial variation in annual total rainfall.

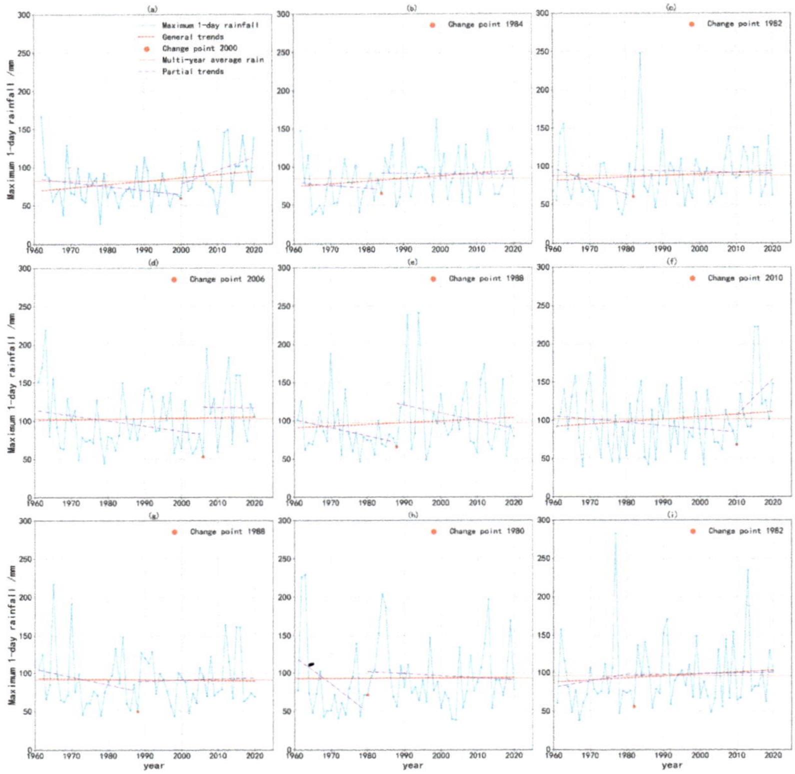

Figure 4. The results of nonstationary analysis of the annual maximum 1-day rainfall of each representative rain gauge in the seven water conservancy zones in Taihu Lake basin, showing average rainfall, change point, the trend of rainfall before and after the change, and the overall trend of rainfall. Analysis results for Xiangchen (**a**); Chenmu (**b**) rain-gauges in Yangchendianmao water conservancy zone; Meixi (**c**); Laoshikan (**d**) rain-gauges in Zhexi water conservancy zone; Qingyang (**e**) rain-gaugs in Wuchengxiyu water conservancy zone; Wangmuguan (**f**); Liyang (**g**) rain-gauges in Huxi water conservancy zone; Jiaxing; (**h**) rain-gauge in Hangjiahu water conservancy zone; Huangdu (**i**) rain-gauge in Puxi water conservancy zone.

3.2. *SST-Based Rainfall Frequency Analysis*

The design storm for various return periods for each conservancy zone was generated based on the storm catalog, with a duration of 3d. The upper and lower curve intervals were plotted using the 95th and 5th percentiles of the 500 estimates generated by the SST (Figure 5). The results were compared with the conventional method of moments. The figure shows that the SST estimates in each water conservancy zone were generally analogous to the moment method but slightly lower than the moment method for large return periods. Similar results have been found in previous studies [12,13,20,21]. A possible reason is that the design storm under the large recurrence period is mainly generated by a few storm events in the storm catalog after transposition; therefore, the estimates may be upper-bounded. Another reason may be due to the quality of the GPM satellite, which had a better ability to capture weak precipitation and less ability to detect heavy precipitation [22]. Therefore, the design results of the SST are reasonable from a comprehensive viewpoint. It is advisable to increase the proportion of extreme storm events in the storm catalog or improve the accuracy of rainfall data to improve the accuracy of the SST for designing storms. In different regions, due to climate variations, each parameter shows different effects on the design results, and further solutions to the problems of SST uncertainty analysis and design storm underestimation are needed for deeper research.

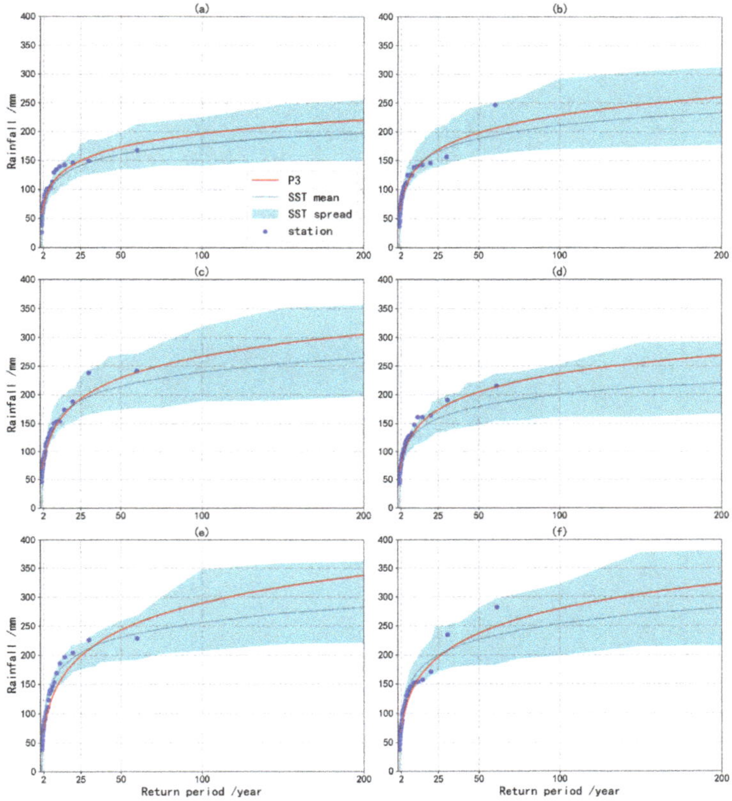

Figure 5. Comparison of IDF estimates between SST and moment method for Yangchendianmao (**a**), Zhexi (**b**), Wuchengxiyu (**c**), Huxi (**d**), Hangjiahu (**e**), and Puxi (**f**) water conservancy zones.

A design storm generated by the conventional method generally only considers the intensity and duration of rainfall, simplifying the spatial distribution of rainfall and the temporal variation, which restricts the accuracy of a design storm/ design flood. The SST method, on the other hand, can estimate the spatial distribution of rainfall (see Figure 6). For design storm calculations in large-scale basins, the heterogeneity of rainfall spatial distribution is significant. Currently, the SST method is used at the watershed level for de-sign storm calculations in major watersheds in the United States; and in China, it is applied mainly at the urban level. Taking 3d rainfall with a return period of 200 years as an example, the design storms are mainly concentrated in urban areas, and the more developed the urbanization is, the more heterogeneous the spatial distribution of the design storms in the water conservancy zones.

The SST design storm calculation for the Taihu Lake basin based on the water conservancy zone unit improves the accuracy of the calculation results as well as the reference value compared with the traditional design storm calculation for the whole basin and has a certain reference value for the design storm calculation for the basin with complex topography, a large area, and uneven economic development status within the basin. It can be seen that there is inhomogeneity in the spatial distribution of design storms, which is more significant under large recurrence periods.

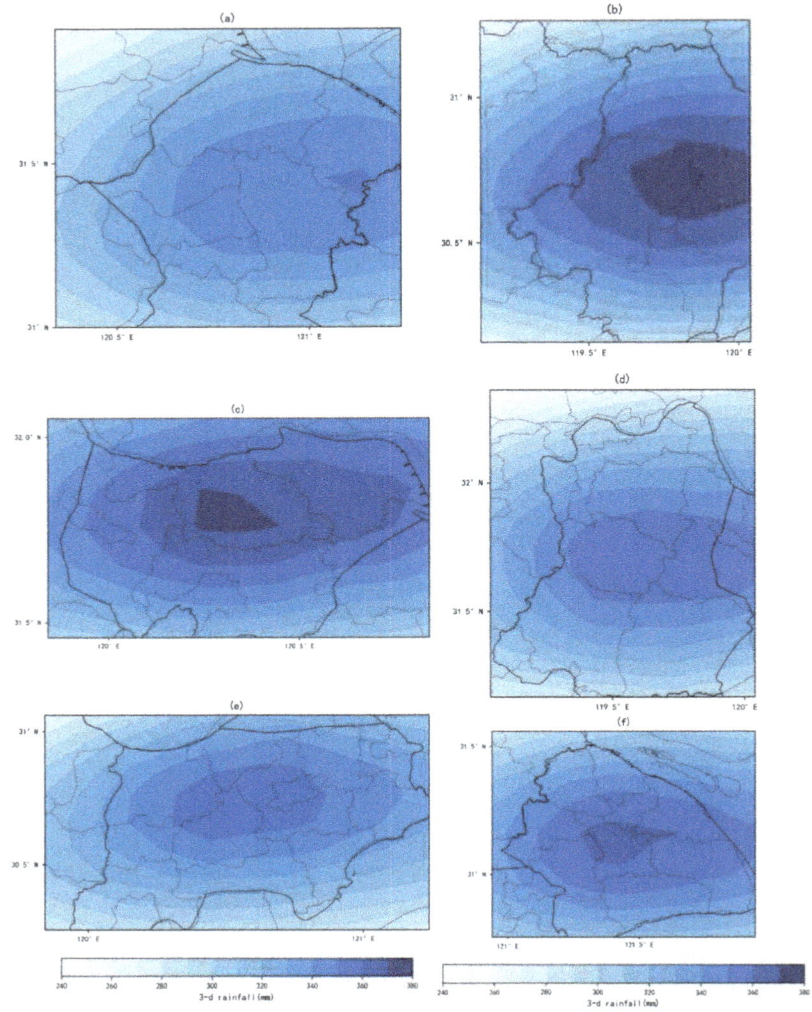

Figure 6. Spatial distribution of 3d rainfall estimation at 200-year return period for Yangchendianmao (**a**), Zhexi (**b**), Wuchengxiyu (**c**), Huxi (**d**), Hangjiahu (**e**), and Puxi (**f**) water conservancy zones.

4. Conclusions

(1) The average annual rainfall in the Taihu Lake basin was approximately 1265 mm for the 1960–2020 period. The rainfall has increased rapidly in the past 10 years. In recent years, the rainfall centers in the Taihu Lake basin have been concentrated in the Puxi and Yangchengdian Mao areas, showing a decreasing trend from Puxi to the surrounding areas.

(2) Rainstorms in the Taihu Lake basin are still mainly localized, concentrated in the more urbanized areas, the southwestern mountainous and southeastern coastal areas, and the northeastern areas, such as Yangchengdianmao. The western part of the lake has relatively few rainstorms. It is recommended that flood control and drainage in the city center be strengthened in the future.

(3) The SST results show that the rainfall in each water conservancy zone is mainly concentrated in the urban area, gradually decreases in the surrounding area, and is influenced by the topography. The spatial distribution variability of rainfall in the long duration

is large, and the heterogeneity in spatial distribution is more significant under the large return periods.

(4) Future work could focus on the error uncertainty analysis and correction method of satellite data in the SST method, further improve the calculation accuracy of the SST method, and combine the SST method with the basin subsurface data for analysis.

Author Contributions: Conceptualization, Y.J. and S.L.; methodology, Z.Z. and Q.Z.; software, Y.J.; validation, Y.J., Z.Z. and Q.Z.; formal analysis, Y.J.; investigation, S.L.; resources, S.L.; data curation, G.Z.; writing—original draft preparation, Y.J.; writing—review and editing, Z.Z.; visualization, Y.J.; supervision, S.L.; project administration, S.L.; funding acquisition, S.L. and G.Z. All authors have read and agreed to the published version of the manuscript.

Funding: This research was funded by the National Natural Science Foundation of China, grant number 42271031.

Institutional Review Board Statement: Not applicable.

Informed Consent Statement: Not applicable.

Data Availability Statement: Data is available at [23].

Conflicts of Interest: The authors declare no conflict of interest. The funders had no role in the design of the study, in the collection, analysis, or interpretation of data, in the writing of the manuscript, or in the decision to publish the results.

References

1. IPCC. *Climate Change 2022: Impacts, Adaptation and Vulnerability*; IPCC: Paris, France, 2022.
2. Zuo, H.; Lu, S.; Hu, Q. Trend analysis of temperature and precipitation in China in the last 50 years. *Plateau Weather* **2004**, *2*, 238–244.
3. Dai, S.; Zhang, Y. Flooding characteristics and mitigation measures in Taihu Lake Basin. In Proceedings of the Taihu Advanced Forum, Shanghai, China, 2 December 2004; pp. 88–93.
4. Taihu Basin Authority of Ministry of Water Resources. *Taihu Lake Basin and Southeast Rivers Water Resources Bulletin 2018*; Taihu Basin Authority of MWR: Shanghai, China, 2019.
5. LI, L.; Hu, Q.; Wang, Y.; Liu, Y.; Zhu, Y.; Wang, L.; Cui, T. Spatial and temporal distribution characteristics of water production changes in the Taihu Lake basin during rapid urbanization. *Lake Sci.* **2019**, *31*, 1289–1298.
6. Zhan, Y. Taihu Lake Basin Wucheng Xiyu District Flood Control Planning. Master's Thesis, Hohai University, Nanjing, China, 2005.
7. Wang, C.; Zhai, P. Preliminary analysis of changes in extreme heavy precipitation events in major cities in China. *Clim. Environ. Stud.* **2009**, *14*, 553–560.
8. Ji, X.; Na, Y.; Xia, C.; Shuai, C.; Yu, B.; Lin, B. Application of the generalized cumulative rainfall percentage method to the study of design storm rain patterns in the Taihu Lake region. *Jiangsu Water Resour.* **2020**, *284*, 29–32. [CrossRef]
9. Smith, J.A.; Baeck, M.L.; Villarini, G.; Welty, C.; Miller, A.J.; Krajewski, W.F. Analyses of a long-term, high-resolution radar rainfall data set for the Baltimore metropolitan region. *Water Resour. Res.* **2012**, 48. [CrossRef]
10. Wright, D.B.; Smith, J.A.; Villarini, G.; Baeck, M.L. Long-term high-resolution radar rainfall fields for urban hydrology. *J. Am. Water Resour. Assoc.* **2014**, *50*, 713–734. [CrossRef]
11. Yang, L.; Tian, F.Q.; Smith, J.A.; Hu, H.P. Urbansignatures in thespatial clusteringof summer heavy rainfall events over the Beijing metropolitan region. *J. Geophys. Res.-Atmos.* **2014**, *119*, 1203–1217. [CrossRef]
12. Zhou, Z.; Smith, J.A.; Wright, D.B.; Baeck, M.L.; Yang, L.; Liu, S. Storm Catalog-Based Analysis of Rainfall Heterogeneity and Frequency in a Complex Terrain. *Water Resour. Res.* **2019**, *55*, 1871–1889. [CrossRef]
13. Zhuang, Q.; Zhou, Z.; Liu, S.; Wright, D.B.; Araruna Junior, J.T.; Makhinov, A.N.; Makhinova, A.F. Bivariate rainfall frequency analysis in an urban Watershed: Combining copula theory with stochastic storm transposition. *J. Hydrol.* **2022**, *615*, 128648. [CrossRef]
14. Kendall, M.G. *Rank Correlation Methods*, 2nd ed.; Springer: Berlin/Heidelberg, Germany, 1955.
15. Liu, X.; Zhou, Y.; Lu, Q.; Zhao, Z.; Wen, W.; Li, C.; Wang, H.; Lu, Y.; Song, W.; Qiao, L. Method for Determining Anti-Floating and Anti-Water level Based on Mann-Kendall Trend Analysis, Involves Defining Mann-Kendall Trend Test Statistic, Calculating Z-Value of Mann-Kendall Trend Analysis Statistical Variable. CN Patent CN111877416-A, 20 November 2020.
16. Pettitt, A.N. A Non-Parametric Approach to the Change-Point Problem. *J. R. Stat. Soc.* **1979**, *28*, 126–135. [CrossRef]
17. Mann, H.B. Nonparametric test against trend. *Econometrica* **1945**, *13*, 245–259. [CrossRef]
18. Mood, A.M. The Distribution Theory of Runs. *Ann. Math. Stat.* **1940**, *11*, 367–392. [CrossRef]
19. Zhou, Z.; Liu, S.; Wright, D.B. Analysis of urban design storm based on stochastic storm transposition. *Adv. Water Sci.* **2020**, *31*, 583–591. [CrossRef]

20. Wright, D.B.; Smith, J.A.; Villarini, G.; Baeck, M.L. Estimating the frequency of extreme rainfall using weather radar and stochastic storm transposition. *J. Hydrol.* **2013**, *488*, 150–165. [CrossRef]
21. Zhuang, Q.; Liu, S.; Zhou, Z. The Change of Spatio-Temporal Precipitation and Rainfall Frequency Analysis in Shanghai. *J. China Hydrol.* **2021**, *41*, 74–80. [CrossRef]
22. Wu, Y.; Zhang, Z.; Jin, Q. Accuracy evaluation of GPM precipitation product in Yangtze River basin. *Yangtze River* **2019**, *50*, 77–85+152. [CrossRef]
23. Huffman, G.J.; Stocker, E.F.; Bolvin, D.T.; Nelkin, E.J.; Tan, J. *GPM IMERG Early Precipitation L3 Half Hourly 0.1 Degree x 0.1 Degree V06*; Goddard Earth Sciences Data and Information Services Center (GES DISC): Greenbelt, MD, USA, 2019.

engineering proceedings

MDPI

Proceeding Paper

Comparison of Privatization in the Republic of Croatia and Selected Former Communist Countries †

Helena Nikolic [1,*] and Jan Horacek [2]

1 Faculty of Economics and Business, University of Zagreb, 10000 Zagreb, Croatia
2 Electrolux d.o.o., 10000 Zagreb, Croatia; janhoracek2@gmail.com
* Correspondence: hnikolic@efzg.hr

† Presented at the 9th International Conference on Time Series and Forecasting, Gran Canaria, Spain, 12–14 July 2023.

Abstract: The paper deals with privatization processes in five selected countries of the communist regime and their comparative analysis. Most countries have historically encountered the need to privatize state-owned enterprises. A closed and inefficient economic system would reach the brink of resilience and change was necessary. Privatization was a conceptual solution. Due to diversified economic systems, internal social and political differences, as well as the complexity of the privatization process itself, the ways in which it has been implemented vary greatly from country to country. However, the aspiration has generally always meant overall economic improvement, and the implementation of rapid, formal, as painless as possible, preferably spontaneous, and transparent privatization. Still, everything took place in several stages and there was a mass, and most often coupon, privatization in one of the phases. It was concluded that each privatization process is specific, but there are still some overlaps. The main distinguishing criteria are related to the approach towards privatization (modular or inflexible) as well as centralization (Croatia, Czechoslovakia, and Poland) and decentralization (Slovenia and Hungary) of the system that implements and controls privatization. In addition, in some countries there has been a lack of public support due to numerous embezzlements, frauds and attempts to exploit positions of power at a given time, while on the other hand orderliness, legitimacy, and innovation have resulted in an in-flow of foreign capital and successful privatization supported by the public.

Keywords: privatization; Republic of Croatia; communist countries

Citation: Nikolic, H.; Horacek, J. Comparison of Privatization in the Republic of Croatia and Selected Former Communist Countries. *Eng. Proc.* **2023**, *39*, 48. https://doi.org/10.3390/engproc2023039048

Academic Editors: Ignacio Rojas, Hector Pomares, Luis Javier Herrera, Fernando Rojas and Olga Valenzuela

Published: 4 July 2023

1. Introduction

Privatization was rarely mentioned in the Eastern Bloc countries until the early 1980s. However, after the fall of the Berlin Wall, it became a major preoccupation of political actors, the economic academic community, and the public. It was necessary to define a strategic plan that would lead to increased competitiveness, a greater degree of internationalization and overall personal and social well-being. At that time, social ownership, as a public good, began to dissolve in most of the former communist countries. Accordingly, it was private ownership that was the conceptual solution since it is the foundation of market business. However, confusing and turbulent struggles for ownership led to a series of problems for companies and employees. One of the main background problems was the desire for manipulation which led to numerous malversations. The complexity of the topic has aroused the interest of many authors to research this issue in the domestic and foreign environment.

The main aim of the paper is to compare the model of privatization in the Republic of Croatia with models of privatization in selected former communist countries (Poland, Hungary, Slovenia and Czechoslovakia). The authors seek to find answers on questions related to the similarities and differences between these five models and to identify and itemize better and worse solutions in the process of resolving the privatization node that has appeared in all European countries whose political systems had been based on communism. In addition, the aim of this paper is to present the privatization processes in Croatia and selected former communist countries.

The article deals with the analysis of secondary data in the domain of privatization and problems related to the privatization processes. A particular focus has been placed on the sources of the most prominent economists dealing with the topic of privatization in selected countries, since privatization issues do not only cover the economic sphere, but the topic of privatization has also been approached with interdisciplinarity. Moreover, it is not only a consequence of economic processes but is a result of a broader legal, political, and economic connotation.

The organization of the paper is as follows. The second chapter deals with the topic of privatization in Croatia. The situation in Yugoslavia in the second half of the 20th century is described and the key moments that brought Croatia closer to privatization and market liberalization in general are discussed. A brief overview of the chronology of Croatian privatization through a comparison of its features is also presented. The third chapter deals with a review of privatization in selected former communist countries. Their historical path to privatization and the characteristics of privatization are described and the effects on citizens and the state are presented. Finally, a comparison of privatization in Croatia with the mentioned countries through certain parameters is made. The conclusion, as the last chapter, summarizes the topics discussed in the previous chapters.

2. Privatization in Croatia

2.1. Political Context and Economic Crisis in Yugoslavia

Yugoslavia was specific in comparison with other socialist countries. It sought to reform the socialist economy through decentralization of leadership and limited introduction of competitive market forces [1]. Self-management was introduced while administrations and work councils ran businesses relatively independently. This enabled a higher sense of belonging of the employees to the companies, and consequently their motivation was at a higher level. The system of social ownership was based on a non-proprietary principle in which the basis of the system was not capital but labor [2].

The problems of the Yugoslav economy began in the early 1980s. Annual inflation had grown steadily, and the government tried to solve the problem by printing money, which led to hyperinflation (1980: 40.7%, 1989: 1252%) [3]. The loans from the Western world (led by the US) which assisted in the development of the Yugoslav economy in order to separate Yugoslav policy from the influence of the USSR (Union Republics of the Soviet Union) became increasingly expensive and increasingly difficult to obtain. Furthermore, the inability to finance production and sustain the economy led to a vicious cycle that bankrupted Yugoslavia in 1982 (though it was never officially published). Yugoslavia's total debt was constantly growing as was its trade deficit. Productivity was low due to a poor organizational structure of self-management and constant neglect of market demands [4]. After the unofficial bankruptcy, Yugoslavia was forced to introduce austerity measures in order to reprogram its debts with the IMF (International Monetary Fund) and obtain new capital. Restrictions on the import of goods were introduced which resulted in a shortage of basic items. Due to the shortage of gasoline, vehicle usage was also restricted (depending on the number of the license plate people were able to drive only on certain days). The economic crisis led to a strong political crisis, strengthening nationalism and resentment of the Member States. Political leaders tried to determine the owners of social property and introduce equality of all forms of ownership. Their main aim was to introduce a capital and labor market instead of the current negotiated preferential business, social planning, credit relations and monopoly. Such a strategic approach resulted in a short-term recovery of the Yugoslav economy. Specifically, inflation was stabilized and living standards and market activity increased [5].

2.2. Specific Features of Croatian Privatization

In the case of privatization, the state took over the entire management mechanism on all essentials related to the company. This form of asset centralization was described by Gregurek (2001) as "original state-level accumulation". The lack of strategic goals, the neglect of the market as a reference in determining the value of the company, and the frequent institutional rules and legal framework changes had affected the uncertainty of potential prospective investors and led to numerous frauds that were not characteristic for Central European countries.

Privatization in Croatia began in 1989 with liberalization as a response to the actual crisis. Transformation of ownership and adoption of the law which regulates the implementation represents the first phase of Croatian privatization (1991–1994). The emphasis was on protecting strategically important state-owned enterprises from privatization. Other companies were subject to sale, and employees were given priority in buying shares on preferential terms. However, the question arose to whom the ownership of individual companies belonged after the collapse of the old system. Good examples are the cases of Hungary and Poland, where the problem of ownership was relatively easy to solve because it was assumed that it was state-owned, while in Yugoslavia it was not known who the real owners of the capital were [1]. The funds had been perceived as social property used by workers. The disadvantage of the first phase was the subjective valorization of the value of enterprises due to the prompt sale and war. The second phase was marked by economic stabilization which affected the competitiveness of the companies. During this period, the companies that avoided the first privatization wave entered the privatization process. The Croatian Privatization Fund and the Pension Fund disposed of their assets [6]. The third wave of privatization was politically inclined. It was characterized by mass coupon privatization and the distribution of shares of questionable quality to the overall public. Specifically, companies facing bankruptcy were for sale. The fourth stage of privatization continued the sale of the bankrupt companies. However, an initiative to revise conversion and privatization emerged on suspicion of numerous frauds and criminal acts [7].

3. Privatization in Selected Former Communist Countries

3.1. Slovenia and Privatization

At the beginning of the transition, the economic situation in Slovenia was significantly different from the situation in other transition countries. The degree of marketability was much higher, and the system of self-government in Yugoslavia and the reforms implemented in the 1980s enabled a high degree of enterprise independence. Central planning was rarely seen, and most business decisions were made independently in enterprises with the limitations of the actual semi-command system. The breakup of Yugoslavia and the war threatened Slovenia's international trade. The reorientation to western markets was not easy, especially because of the high risk of doing business with a country near a war environment. Many companies did not survive the new situation and unemployment increased significantly in the early 1990s. Furthermore, the situation was complicated by the fact that this crisis was taking place in the context of very high inflation. One of the most important instruments in curbing inflation was a restrictive wage policy that would lead to a real decline in purchasing power and living standards of the population [8]. Slovenia was also fighting for international recognition which came rather late. Despite political and economic uncertainty and recession, there were surpluses in macroeconomic balances and a reversal in lowering inflation rates [9].

The period of transition from recession to long-term growth occurred in the mid-1990s, and in those years injections and unemployment began to decline. By 1997, key reforms had been implemented. There was a bank remediation and corporate ownership transformation. The economy functioned as a true market economy and all former socially owned enterprises were transformed into joint-stock companies. The historical context of Slovenian privatization began in the late 1980s [10]. Legislation sought to abolish self-governing and social relations in order to replace them with capitalist property relations.

The Workers' Council had the right to sell the business to private owners, and the gradual transformation of socially owned enterprises into joint ventures was also allowed. However, the government struggled with the implementation of massive, rapid, centralized, and distributive privatization through mutual funds that was intended to break up with any form of political, social and economic heritage from Yugoslavia. More specifically, it was impossible to initially convert all socially owned property into state property and it made no sense to allow politicians to choose new politically committed owners [11]. On the other hand, a decentralized and commercial approach to privatization was proposed in which managers and workers would initiate privatization processes spontaneously and slowly. This would avoid the negative effects of privatization in the transition process, and existing property rights would strengthen managers [12]. The struggle between the two approaches resulted in a crisis of the relatively efficient Slovenian economy. They were losing the market and numerous companies were becoming insolvent and on the verge of bankruptcy, while banks were accumulating a large number of bad loans.

Mencinger (2007) divides enterprises into three groups: (i) large non-profit enterprises most affected by political events and transition, (ii) less-affected enterprises that should be privatized under the provisions of the new ownership-restructuring law, and (iii) utilities that are supposed to remain in state ownership. The Development Fund established corporate governance for the first group of companies with the aim of solving financial problems and creating preconditions for privatization. The ultimate goal of the program was to liquidate or sell the companies over a two-year period. Initially, the program was successful and the results of the first year were impressive, but gradually the program changed, and the Development Fund turned into an institution to provide various forms of non-transparent and often politically motivated state aid to troubled companies.

Finally, at the end of 1992, an alternative solution was found—the Law on Transformation of Social Property was enacted, which had the features of both approaches to privatization—decentralization and gradual privatization and predominantly distributive privatization via vouchers to Slovenian citizens. The proportions by which the shares in the companies were divided were defined: 10% of the social capital belonged to the Pension Fund and the Former Owners' Compensation Fund, and 20% of the capital was transferred to the Development Fund, which sold these shares to newly established private investment funds, and employees in exchange for their certificates of ownership. For the remaining 40% of the capital, four options were available: (i) buy-out provided to employees on preferential terms, (ii) public offer to citizens, (iii) public offer to privatization funds, and (iv) buy-out to a strategic partner. Ownership certificates facilitated the free distribution of shares among employees and citizens.

Since companies were able to choose between different methods of privatization, the most popular was the internal purchase of shares, which included mandatory transfers of 40% of shares to institutional owners (three funds). This was in line with the expectations and legacies of self-management. The choice of this model was enhanced with a 50% discount and the possibility of deferred payments of up to five years. Workers and managers received a majority in 44% of enterprises, but this represented only 22.9% of total capital, while funds accounted for almost 41% in the ownership structure of enterprises after the privatization process [13]. Another popular method was the combination of public auction with internal stock allocation, which was a method preferred by profitable large companies.

In 2005, the entire Slovenian economy was reformed, with effects on the further course of privatization. Specifically, the enterprises which remained the property of the Government after the first phase of mass privatization had to be privatized by scattered commercial privatization. Thus, the period of non-transparent consolidation of ownership at low prices that followed the first stage of privatization was about to end. In the second phase of privatization, the emphasis was on transparency and international actions, as well as on coordinated privatization, from which even the largest state-owned enterprises were not excluded. The patterns by which the state withdrew were strictly defined and its impact on the economy gradually diminished. Sales criteria included maximizing the sales price, working together and constantly evaluating the company's profitability, and finding as many potential customers as possible. Furthermore, given the shallow Slovenian capital market, the goal was to independently increase the value of the portfolio of shareholders in accordance with the rules of the stock market. Finally, there are companies that were not necessarily subject to immediate privatization, as they are large companies that are important to the national economy. Their sale needed to be consistent with the obligations of the funds, the market situation, and the sale of other state-owned enterprises [14].

3.2. Hungary and Privatization

Hungary has been trying to reform the socialist economy since the 1980s through decentralization of leadership and the limited introduction of competitive market forces. The countries of Central and Eastern Europe lost confidence in the Soviet model and began to look for alternative methods of securing conditions for capital accumulation [15]. Hungary sought to secure fresh capital. The strategic approach involved the sale to foreign owners from Western countries. In the late 1980s, a series of laws led to the gradual creation of conditions for the legalization of privatization. The three most important laws that enabled Hungary to begin privatization and transition to a new business system were: (i) the Foreign Investment Act 1988, which created extremely favorable conditions for foreign private companies; (ii) the Corporations Act, which allowed the formation of corporations in the same year; and (iii) the 1989 Conversion Act, which allowed state-owned enterprises and cooperatives to be converted into joint stock companies. Spontaneous privatization arising from these laws is presented as "machinations survival of the elite of the old system" [12].

Privatization in this form was not supported by the public. Therefore, the State Property Agency was established, whose task was to control the privatization process in accordance with the property policy directives. These directives created a legal framework for more transparent privatization in Hungary. The Hungarian model of privatization is different to other countries. A flexible approach has been put in place that implements privatization on a case-by-case basis. In the first two years, this process was created by managers in companies, with virtually no state control. Subsequently, certain laws were enacted, and the State Property Agency was established, but there was never a significant role for the state in centralizing the process due to the Agency's failure to accelerate privatization [1].

Privatization in Hungary can be divided into several categories: the first category is characterized by spontaneous privatization; the second category is related to active privatization where the main objective is to attract investors; the third category is the self-privatization process, which is defined as an independent procedure with the support of an authorized consultant; the fourth category is privatization initiated by investors; and the fifth category is the sale of shares on preferential terms to employees (up to 15% of shares). Regardless of the category, the main objective of the Hungarian privatization was to create conditions for future economic operations to be as efficient as possible and to achieve a privatization price and a satisfactory level of revenue to be used to service the national debt. The privatization process was carried out in accordance with market conditions (the issue of domestic and international demand and purchasing power), and state intervention was only possible in emergency situations. An important element of the strategy was the implementation of various privatization methods that would allow a wide range of investors capable of acquiring companies (discounts and benefits for workers) and the transfer of state ownership without compensation to local authorities, social security, foundations and public welfare associations, but only if this reduced government grants [2].

The first privatization program was launched by the State Property Agency in September 1990 and at that time twenty companies of different sizes and profiles were put up for sale. Although the program initially aroused special interest, mainly among potential Western investors, it eventually proved unsuccessful as most of the enterprises remained state-owned, with most of them under the direct control of the State Property Agency.

The most common form of ownership changes up to that point had been encouraged by managers, and despite the restrictions and legal solutions that the State Property Agency had tried to impose, the maneuvering space of managers, although reduced, still played a large role in the so-called spontaneous privatizations. This situation led to great speculation and management efforts to downplay the value of the company. Thus, the process began of transforming the communist classes to new owners [12].

At the end of June 1993, all state-owned enterprises were required to be converted into business corporations. This decision forced enterprises to undertake rapid and formal privatization. The major problem was the inability to raise funds from local citizens who would invest in privatized companies. Therefore, the government formulated specific schemes to assist the population in obtaining funding. The mechanisms included: renting certain companies, using existing loans, buying property through a term plan, and small investment programs for shares buying. Nevertheless, foreign capital has played an important role in Hungarian privatization in the form of pure investments and through participation in privatization projects [16].

3.3. *Czechoslovakia and Privatization*

For many years, Czechoslovakia was not in favor of reforms towards liberalism since it had a balanced macroeconomy, a few competitive branches, a skilled workforce and negligible indebtedness [17]. Still, privatization was inevitable. The implemented privatization model was one of the most innovative solutions to the problem of mass privatization because it did not represent the classic forms of privatization. Namely, the goal was to carry out mass privatization without prior reconstruction of the company and to transform the country from 100% state ownership into a country with an unusually small, well-defined, state sector and a fast-growing private sector in a short period of time [18]. This was achieved by distributing vouchers to citizens in a fast and transparent manner to prevent a lack of capital among the population. They were allowed, carefree, to buy shares of one of the offered state-owned companies. The purchase could be made independently or through investment privatization funds [19].

An important feature of the Czechoslovak model was the possibility that investors could develop their own variants of company privatization that would compete equally with the management's proposals on the privatization of specific companies. Such a system allowed the competent government institution to select the best of the proposed models for enterprise transformation. Privatization was carried out exclusively by ministries, the government and the national privatization committees. The reason for this was to avoid the uncontrolled growth of managerial power that occurs during spontaneous privatizations. However, there were some exceptions [20,21]. The implementation of this privatization model involved a large number of state-owned enterprises being owned by several thousand small shareholders, domestic citizens. The benefit of such rapid privatization was certainly the awakening of the entrepreneurial spirit in this rigid communist system. However, the shortcomings and risks of this privatization method are reflected in the later functioning of the companies. Due to the huge dispersion of ownership, the management of the company is difficult, and this problem is expressed until the shareholders decide to hand over their shares to a certain investment fund or some other institution. In addition, the downside is the high cost of implementing such privatization, which does not contribute significantly to the state budget in the future. In any case, privatization of this magnitude has not been recorded anywhere else. Analysts estimate that public support was crucial for the success and realization of the proposed conversion concept. In addition, foreign capital, mostly of German origin, also contributed significantly [22].

The first significant problems arose in the post-privatization period. The lack of appropriate regulation contributed to the chaotic reshaping of the portfolio acquired in the initial phase of privatization. Consequently, there was a notable reduction in the concentration of ownership in the first year of the third round of privatization. Moreover, it took more than four years for ownership concentration indices to normalize and to increase in line with initial expectations [13].

3.4. Poland and Privatization

Poland has the longest history of private ownership of all former communist countries when considering "non-state economic activities that have been present since the late 1950s" [19]. The purpose of private companies was to advance the economy through additional capital and new technology. However, legislation on privatization was established subsequently. The 1989 consolidation plan enabled the transformation of enterprises into joint stock companies. A directorial experiment was also conducted. It provided an opportunity for managers of state-owned enterprises to be engaged in economic activities outside the management of their own enterprises [12]. This led to an increase in political and social stability, which is why some of the privatization measures had to be slowed down due to citizens' dissatisfaction with austerity and waiver measures. But the government gained public trust and the Poles were willing to make a temporary "sacrifice" as a pledge for a better future [15].

The formal Privatization Law was established in 1990. It redintegrated a normative framework flexible enough to be able to implement different privatization strategies with an emphasis on centralist regulation and control of privatization. The fundamental goals of privatization included the reduction in the public sector and the inherent burden on the state budget, as well as the generation of revenues. Thus, an efficient corporate governance system was established that visibly improved the performance of the company itself [2]. Although the first exemplary privatizations were officially rated as successful, it soon became apparent that the case-by-case method was too slow and too expensive. After the change of government, the method of urgent mass privatization was supposed to be implemented. The plan was to distribute free vouchers to citizens who would then exchange them for shares in mutual funds. However, due to political instability and opposition interventions, the concept of share division has not been implemented. Privatization continued, on a case-by-case basis, through the capital method and privatization through bankruptcy [19].

According to Polic (2008), two methodological approaches were applied in the Polish privatization model. The first was the privatization of capital (indirect privatization), and the second was the sale of company assets (direct privatization). Leasing dominated as a method of direct privatization. Thus, the ownership was transferred to the employed workers in "buy-out" privatizations. On the other hand, the bankruptcy method was implemented in the case of small and medium enterprises. The company's assets (or part of it) were sold or more often rented to domestic or foreign buyers. The most widespread was the so-called leveraged buy-out, in which investors paid 30% of the appraised value as an investment and repaid the rest during the usual five-year period [12].

A special feature of the Polish privatization process is the sectoral approach. Namely, they are trying to restructure and privatize companies from the same sector because of the similarities that these companies have. Thus, they are able to overcome the concurrent difficulties faced by companies in a particular industry. Privatization is decided by the competent ministry, company management and the workers' council. However, they usually hire foreign consulting firms that prepare feasibility studies for restructuring and privatization. This approach has accelerated the privatization process itself visibly and made it more cost-effective and efficient. Privatization has resulted in the growth of the private sector, the growth of private partnerships and employment in the tertiary and quaternary sectors [22].

4. Comparative Analysis of Privatization Processes in Croatia and Selected Former Communist Countries

Normative goals in all countries regarding privatization were relatively similar. Thus, the political classes strived for economic growth, prosperity of citizens, inclusion in world trade, the desire to become a significant player on the international market, high employment, high standards, etc. The starting positions of the former communist countries and Croatia were different when comparing their openness to the market at a time when they were under communist regimes. Consequently, they all considered privatization as the solution to a number of internal problems, the growth of competitiveness and overall economic progress.

Croatia and Slovenia, which originated from Yugoslavia, were more market-oriented than other socialist countries, especially Slovenia as the most developed Yugoslav republic. Namely, their communism was relatively soft. In that manner, their openness to the market was much higher, especially in the period after the death of Josip Broz Tito [23]. Poland encouraged economic activities and also had a positive experience with private initiatives. Furthermore, Hungary was often considered as the most unfettered country in the Eastern bloc. This was achieved through a series of reforms that gradually moved Hungary away from the Soviet grip and turned it to the West. On the other hand, Czechoslovakia was the most rigorous of all selected countries in terms of economic freedoms. Accordingly, until the beginning of the transition, it showed no affinity for privatization and the international market in general.

Macroeconomic and political goals dominated in the design and selection of privatization models in each country. The unfavorable institutional and legal framework for post-privatization ownership transformation had only constrained effects on the results of post-privatization ownership transformation. The reason why the privatization process had stalled on many occasions was politics and the inability of major political actors to agree on privatization models. The best example of the struggles of different privatization approaches was visible in Slovenia, where the struggle of Mencinger's and Sachs's concept lasted for 16 months, which eventually ended in a compromise and taking certain parts from everyone's program. A similar situation occurred in Poland, where parliamentary parties were reluctant to support the government for the method of urgent mass privatization after it became clear that case-by-case privatization was not fulfilling its task due to slowness. Hungary was characterized by a flexible approach and minimal state interference. State intervention existed exclusively in cases of government rewarding of loyal party members. The awards included key positions in central economic decision-making institutions or on the boards of directors of state-owned enterprises that had undergone a conversion or been "temporarily" placed under state control. The political elites in Croatia and Czechoslovakia had a great influence and it can be said that the process was centralized and in the hands of the state. Czechoslovakia's advantage was that managers did not have much influence or synergy with political leadership. In addition, external investors were given the opportunity to create privatization programs for companies.

Mass privatization is the most attractive phase of privatization for citizens due to the lack of funds for the purchase of shares and ownership of shares in companies. In Croatia, it was most prevalent in the third phase of privatization, where shares were distributed to a wider population despite a portfolio of dubious value. Czechoslovakia had a specific form of mass privatization in which every adult citizen was allowed to participate. On the other hand, in Croatia, vouchers were given only to certain social strata. If we look at the number of coupons for privatized companies, in Czechoslovakia this approach led to the privatization of 1849 companies with private capital of 9.25 billion euros [24], while in Croatia the number of companies was 471, and the amount of privatized capital was 3.6 billion German marks [25]. Furthermore, Slovenia, which only at the end of 1992 passed a law that fully regulated the privatization process, decided to award vouchers and this led to workers and managers having a majority share in 44% of companies.

Citizens' perceptions of privatization processes show differences in the view of privatization, and it is interesting that the most pessimistic about privatization were the citizens of Croatia, 50% of whom considered privatization to be extremely negative. Along with the Croats, there are also the Poles, of whom only 18% believed that privatization is a positive process. On the other hand, the citizens of Czechoslovakia, who are considered the biggest winners of the privatization process in the former communist countries, were the most satisfied with privatization [15].

5. Conclusions

Privatization in selected countries has brought a lot of benefits to consumers, which is reflected in lower prices and better quality of products and services. Internationalization was necessary and there was no alternative. Therefore, it can be concluded that the privatization process has been relatively positive given the historical context in which it took place. An additional aggravating circumstance is inadaptability, as it was very difficult for many to move from one system to another. Most of the "mistakes" made during privatization in all countries can be attributed to beginner and inexperienced omissions due to ignorance of the market mechanism and inadequate adjustment to market competition conditions.

Citizens expected privatization to bring fresh capital, a more efficient management system, and investments to impoverished companies that would contribute to the wider community, but in many cases positions of power were exploited. The Czechoslovak model proved to be a complete success that brought fresh foreign capital with incredible speed and led to the widest privatization, which until the collapse of Czechoslovakia was synonymous with successful privatization.

The fact is that the privatization process in these countries has significantly reduced public debt by increasing budget revenues. In addition, successfully privatized companies have created new jobs through their development and thus contributed to employment growth. In most cases, due to successful privatization, there has been an increase in salaries, and by participating in the purchase of shares, a large number of employees have become owners of privatized companies. This has allowed them to participate in the management of the company as small shareholders, although in certain cases this has proved to be a negative determinant because the dispersion of ownership could lead to the impossibility of running the company efficiently.

Author Contributions: Conceptualization, H.N. and J.H.; methodology, J.H. and H.N.; software, H.N.; validation, H.N. and J.H.; formal analysis, J.H.; investigation, J.H.; resources, J.H.; data curation, J.H.; writing—original draft preparation, J.H.; writing—review and editing, H.N.; visualization, J.H.; supervision, H.N.; project administration, H.N.; funding acquisition, H.N. All authors have read and agreed to the published version of the manuscript.

Funding: This research received no external funding.

Institutional Review Board Statement: Not applicable.

Informed Consent Statement: Not applicable.

Data Availability Statement: Not applicable.

Conflicts of Interest: The authors declare no conflict of interest.

References

1. Njavro, Đ. Privatizacija. *Društvena Istraživanja J. Gen. Soc. Issues* **1993**, *2*, 87–105.
2. Kalogjera, D. Privatizacija u stabilizaciji i razvoju hrvatskog gospodarstva. *Društvena Istraživanja J. Gen. Soc. Issues* **1993**, *2*, 51–86.
3. Allock. *Explaining Yugoslavia*; Hurst& Company: London, UK, 2000.
4. Bennett, C. *Yugoslavia's Bloody Collapse: Causes, Course and Consequences*; NYU Press: New York, NY, USA, 1995.
5. Milanović, T. Kriza 1980-ih u Jugoslaviji i međunarodni pogledi. *Časopis Studenata Povijesti Sveučilišta u Zadru* **2011**, *4*, 85–92.
6. Gregurek, M. Stupanj i učinci privatizacije u Hrvatskoj. *Econ. Rev.* **2001**, *52*, 155–188.
7. Bendeković, J. Privatization in Croatia. *Econ. Rev.* **2000**, *51*, 55–90.
8. Grdešić, M. Tranzicija, sindikati i političke elite u Sloveniji. *Politička Misao* **2006**, *18*, 121–141.
9. Kračun, D. Tranzicija, stabilizacija i ekonomski rast: Iskustvo Slovenije. *Ekon. Pregl.* **2005**, *56*, 145–162.
10. Mencinger, J. Privatization in Slovenia. *Slovenian L. Rev.* **2006**, *3*, 65.
11. Simoneti, M.; Gregoric, A. Menedžersko lastništvo in uspešnost poslovanja privatiziranih podjetij v Sloveniji. *IB Revija* **1992**, *1*, 87–104.
12. Polić, N. Model privatizacije i transformacija vlasništva privatiziranog kapitala. *Ekon. Pregl.* **2008**, *60*, 396–410.

13. Assaf, A.; Cvelbar, K.L. Privatization, market competition, international attractive-ness, management tenure and hotel performance: Evidence from Slovenia. *Int. J. Hosp. Manag.* **2011**, *30*, 391–397. [CrossRef]
14. Dale, G. Tranzicija u srednjoj i istočnoj Europi. *Diskrepancija Studentski Časopis za Društveno-Humanističke Teme* **2011**, *13*, 81–98.
15. Kiss, Y. Paradoksi privatizacije u istočnoj i srednjoj Europi. *Društvena Istraživanja* **1993**, *2*, 3–30.
16. Mejstrik, M. *Privatization in Czechoslovakia*; Privatisation; Routledge: Lodon, UK, 2003; pp. 108–119.
17. Čučković, N. Privatizacija u srednjoj i istočnoj Europi: Program vaučer privatizacije u Češko-slovačkoj. *Društvena Istraživanja J. Gen. Soc. Issues* **1993**, *2*, 139–148.
18. Šokčević, Š. Privatizacija društvenog/ državnog vlasništva—(ne)uspjeli tranzicijski proces? *Diaconvensia Teološki Pril.* **2009**, *15*, 103–154.
19. Carlin, W.; Mayer, C. Restructuring enterprises in eastern Europe. *Econ. Policy* **1992**, *7*, 311–352. [CrossRef]
20. Mládek, J. Initialization of Privatization through Restitution and Small Privatization. In *The Privatization Process in East-Central Europe*; Springer: Boston, MA, USA, 1997; pp. 45–54.
21. Kupka, M. Transformation of ownership in Czechoslovakia. *Sov. Stud.* **1992**, *44*, 25–36. [CrossRef]
22. Aussenegg, W. Privatization versus private sector initial public offerings in Poland. *Multinatl. Financ. J.* **2000**, *4*, 69–99. [CrossRef]
23. Rajšić, P. The Economy of Tito's Yugoslavia: Delaying the Inevitable Collapse. Available online: http://predragrajsic.blogspot.com/2014/03/the-economy-of-titos-yugoslavia.html (accessed on 1 July 2023).
24. Lipton, D.; Sachs, J. Privatization in Eastern Europe: The case of Poland. In *Development Policy*; Palgrave Macmillan: London, UK, 1990; pp. 169–212.
25. Štulhofer, A. *Proces Privatizcije i Hrvatska Javnost 1996–1998*; Privatizacija i javnost; Institut Društvenih Znanosti I. Pilar: Zagreb, Croatia, 1999; pp. 1–31.

engineering proceedings

MDPI

Proceeding Paper

It Can't Get No Worse: Using Twitter Data to Improve GDP Estimates for Developing Countries †

Agustín Indaco

Economics, Carnegie Mellon University in Qatar, Ar-Rayyan P.O. Box 24866, Qatar; aindaco@andrew.cmu.edu

† Presented at the 9th International Conference on Time Series and Forecasting, Gran Canaria, Spain, 12–14 July 2023.

Abstract: This paper shows that we can use social media data to improve the accuracy of GDP estimates at the country level for developing countries. I use all publicly available image tweets from 2012 and 2013 to estimate GDP at the country level for developing countries. First, I find that one can explain 76% of the cross-country variation in GDP with the volume of tweets sent from each country. I then show that the residuals on these Twitter-GDP estimates are significantly larger for countries with allegedly poor data quality. I then use Nigeria as a case study to show that this method delivers much more timely and accurate estimates than those presented by official statistic agencies.

Keywords: national accounts; social media data; nowcasting

Citation: Indaco, A. It Can't Get No Worse: Using Twitter Data to Improve GDP Estimates for Developing Countries. *Eng. Proc.* **2023**, *39*, 49. https://doi.org/10.3390/engproc2023039049

Academic Editors: Ignacio Rojas, Hector Pomares, Luis Javier Herrera, Fernando Rojas and Olga Valenzuela

Published: 4 July 2023

1. Introduction

On 7 April 2014, the Nigerian Bureau of Statistics declared that their 2013 gross domestic product (GDP) estimates were being revised upward from USD 269 billion to USD 510 billion [1]. Overnight, the Nigerian economy had grown by 89 percent and was now the largest economy in Africa, overtaking South Africa in the process.

However, this sudden increase in GDP was not a result of successful economic policies or benevolent external conditions. Instead, it was a product of a national accounting process known as *rebasing*. Until 2014, Nigeria's GDP was constructed by sampling businesses weighed in accordance to the importance each sector had on the Nigerian economy in 1990. Evidently, in the ensuing years, the economy morphed. This made emerging industries vastly underrepresented in the country's GDP estimate, and vice versa. In this sense, the new and updated estimate was expected to more closely reflect the *true* GDP of the Nigerian economy. However, it also meant that up until that point, policymakers, investors, and everyone else making economic decisions based on the *old* GDP estimates were relying on grossly inaccurate economic data.

This example, of which there are several others in the past few years (e.g., Zambia in 2010, Kenya and Tanzania in 2013, and Uganda in 2014 just to name a few examples from other African countries. In each of these cases, the revised figures where 13–28% higher than the previous estimates), sheds light on how complicated it is to put together national statistics and how (oftentimes) inaccurate official GDP estimates are.

The inaccuracy of GDP measurements at the country level tends to be accentuated in developing countries. This is due to several factors. First, statistical offices in developing countries tend to have fewer resources to construct these estimates. Second, given that developing countries tend to have relatively large informal sectors, they are oftentimes included in official estimates [2]. Given that informal companies many times do not keep proper accounting books, this generally complicates matters because informal companies tend to provide inaccurate financial statements. These conditions leave statistical offices in developing countries with a complicated task, to put together a reliable estimate of the size of the countries' economy, which includes a sizeable sector that generally does not count with reliable data and to put together this estimate with limited resources. The inaccuracy of GDP measurements at the country level tends to be accentuated in developing countries. This is due to several factors. First, statistical offices in developing countries tend to have fewer resources to construct these estimates. Second, given that developing countries tend to have relatively large informal sectors, they are oftentimes included in official estimates [2]. Given that informal companies many times do not keep proper accounting books, they tend to provide inaccurate statements of their own finances.

These conditions leave statistical offices in developing countries with a complicated task, to use their limited resources to put together a reliable estimate of the size of the countries' economy from a sample of (mostly) unreliable firm-level data. This leads economists like [2] to conclude that GDP statistics from African countries are "best guesses of aggregate production".

These concerns and limitations have motivated efforts to find proxies that may estimate economic activity. Several authors have proposed using satellite night-light images to estimate GDP at the country level or sub-national level (A thorough overview can be found in [3]). The use of night-lights has motivated economists to look for other proxies to measure economic activity. Ref. [4] estimate the German business cycle at a monthly level by measuring toll activity on important highways by heavy transport vehicles. On their part, Ref. [5] use Google Trends search data to estimate economic activity. Finally, in a paper closely related to this one, Ref. [6] suggests the use of social media data for estimating GDP, both at the national and sub-national level. The paper shows how social media can be used as a supplement to official GDP estimates to improve their accuracy. Given their accuracy and availability, these alternative estimates could also serve as a tool for non-governmental agencies and international organizations to corroborate official GDP estimates.

This paper proposes a way in which statistic agencies and international organization can use social media data to improve the accuracy of their economic measurements. In particular, I use all publicly available tweets with images sent in 2012 and 2013 to estimate GDP at the country level for developing countries. I find that one can accurately estimate GDP at the country level by using the volume of tweets shared from each location. I then gather World Bank data on the quality and fidelity of the official economic data released by each country and find that the residuals of the Twitter-GDP estimates for countries with allegedly poor data quality tend to be larger than for countries that are considered to have more reliable economic data. I finalize by using the aforementioned rebasing case of Nigeria as an example of how social media data offers valuable information in finding the true level of economic activity for a developing country. For 2012 and 2013, I find that the GDP estimate using Twitter data is in fact quite close to the greatly revised estimate.

2. Materials and Methods

The Twitter data for this paper were obtained directly from Twitter. The dataset was awarded via the 2014 Twitter Data Grant submission, which was awarded to the Cultural Analytics Lab directed by Lev Manovich. The dataset contains all Twitter posts containing geo-tagged images between 1 January 2012 and 31 December 2013. As per [7], approximately 20% of tweets are geographically located, while [8] reports that 42% of tweets contain an image. However, the latter analysis was limited to 1 million tweets sent by US West Coast users, which could skew the results. To account for this, the author collected 10,000 tweets randomly in December 2018 using the Twitter API. Among this set, 4.9% of tweets were geo-located, and 22.8% contained images.

The dataset contains 140 million tweets from all over the world, each with a unique Twitter user ID, the latitude and longitude from where the tweet was sent from (with 5 decimal points for a precision of 1.1 m), the tweet's date and time, the image tweeted, and any accompanying text. Figure 1 shows a map indicating the location from where all image tweets were sent from.

Bots that sent over five tweets in a minute were removed to prevent them from biasing the data. However, this did not significantly alter the results presented in Section 3.

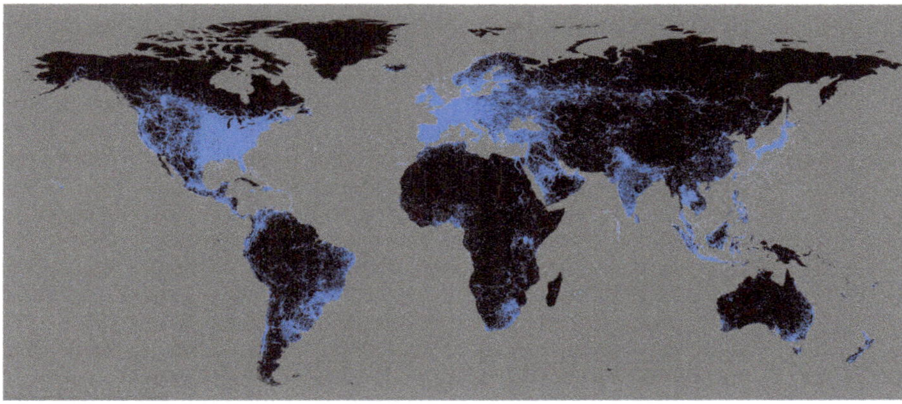

Figure 1. This figure shows the location from where image tweets in our data were sent from. Each blue dot represents an image tweet sent from that location. The map gives an overall indication of the countries and regions with more Twitter activity.

Table 1 summarizes the Twitter data by year and by income group (based on the World Bank's classification) for developing countries. The data show that the average number of tweets per country rose from around 50,000 in 2012 to almost 250,000 in 2013, suggesting an increase in image tweets. Additionally, while countries with higher incomes had more tweets on average, the growth rates of tweet numbers from 2012 to 2013 were higher among lower income countries.

Table 1. Twitter data summary statistics: Mean and SD

	2012	**2013**
Tweets	52,219.1	245,805.3
	(148,147.3)	(782,123.6)
Upper-middle (51)	89,123	406,004.9
	(201,367.9)	(836,887.4)
Lower-middle (43)	37,394.5	209,938.9
	(106,230.9)	(929,686.9)
Low (28)	735.9	3602.5
	(898.9)	(4370.9)

Notes: Top row shows the mean number of tweets per country and standard deviation in brackets for all developing countries in the dataset. The bottom half of the table shows the mean number of tweets and the standard deviation by country in each income group. The number of countries per income group is shown in brackets.

3. Results

3.1. Estimating GDP from Tweets

In order to suggest that social media data can serve as a proxy for estimating economic activity in developing countries, I first show that annual GDP can be accurately estimated using solely the volume of tweets sent from each country.

I use the precise location (latitude–longitude) to geocode the country of origin where each post was sent from, aggregate the volume of tweets by country per year and estimate

$$\ln GDP_{i,t} = \beta_0 + \alpha_t + \beta_1 \ln Population_{i,t} + \beta_2 \ln Tweets_{i,t} + \varepsilon_{i,t}, \quad (1)$$

where the explained variable is the natural log of GDP of country i in year t. The coefficient we are most interested in is β_2, which shows the relevance of the number of image tweets taken from that country in each of those years for estimating GDP. In Equation (1), we control for the population size in each country and include year fixed effects (α_t) to control for any differences in the use of Twitter from one year to the other, as well as changes in global economic conditions.

The corresponding estimates are reported in Table 2. There are 122 developing countries in the dataset with data on GDP, Twitter and population for both years. I also remove countries in which Twitter was banned for a period of time during any of these years, such as China and Iran. In column 1, I regress the natural log of GDP solely on the number of image tweets sent from each country as well as country fixed effects. This is the baseline regression. The coefficient of interest on $ln(Tweets)$ is positive and highly significant and the R^2 is 0.76. Columns 2–4 run the same model separately for countries in each income group. In all cases, the coefficient of interest is positive and statistically significant at the 1% level. The R^2 varies from 0.53 to 0.75 depending on the income group. Column 5 controls for the population estimate for each of these countries in the regression. Column 5 shows this model when we include all developing countries. Relative to column 2, the coefficient on tweets declines but remains positive and statistically significant and the R^2 increases to 0.9. Columns 6–8 run this model separately for countries in each income group. In all cases, the coefficient of interest is positive and statistically significant (either at the 1% level or 5% level for upper-middle income countries) and the R^2 varies between 0.89 and 0.98.

Table 2. Estimating country GDP for developing countries.

	(1)	(2)	(3)	(4)	(5)	(6)	(7)	(8)
Countries:	**All**	**Low**	**Lower-Middle**	**Upper-Middle**	**All**	**Low**	**Lower-Middle**	**Upper-Middle**
ln(Tweets)	0.60 ***	0.42 ***	0.58 ***	0.67 ***	0.13 ***	0.37 ***	0.16 ***	0.05 **
	(0.02)	(0.06)	(0.04)	(0.04)	(0.02)	(0.04)	(0.03)	(0.03)
ln(Population)					0.51 ***	0.82 ***	0.78 ***	0.98 ***
					(0.02)	(0.07)	(0.04)	(0.03)
R^2	0.76	0.53	0.71	0.75	0.90	0.90	0.96	0.98
Adj. R^2	0.76	0.51	0.71	0.74	0.90	0.89	0.95	0.98
Num. obs.	244	56	86	102	244	56	86	102
RMSE	1.01	0.67	1.01	1.29	0.65	0.31	0.40	0.38

*** $p < 0.01$, ** $p < 0.05$. **Notes:** The dependent variable in all columns is the log of GDP. Using the World Bank classification, columns 2–4 and 6–8 estimate the log of GDP for the subset of low, lower-middle, and upper-middle income countries separately. Columns 5–8 control for the log of population in each country. All models include year fixed effects. Standard errors are included in parenthesis.

Table 2 shows that the volume of image tweets sent in a year is a valuable measure for estimating GDP at the country level, being able to explain 76% of the cross-country variation in GDP on its own. Figure 2 plots the residuals of Equation (1) against the fitted values, enabling us to study the distribution of the residuals. The figure indicates that they seem to be randomly distributed around zero.

3.2. Data Quality Issues

Section 1 showed that GDP estimates have been criticized for being inaccurate, particularly in developing countries. If this is the case, it would imply that Equation (1) is fitting the data to the GDP reported by countries, which is not necessarily the *true* and *accurate* GDP of these countries. Hence, it is possible that a portion of the differences between the Twitter-GDP and official GDP estimates arise due to measurement error in official GDP estimates. In this case, data from tweets could be used by statistical agencies as a complementary measure to produce more accurate estimates.

To examine this, I will incorporate a measure of data quality developed by the World Bank. The World Bank's Statistical Capacity Indicator is a composite score assessing the capacity of a country's statistical system. It is based on a diagnostic framework assessing areas including methodology, data sources, and periodicity and timeliness. The overall score is a simple average of all three area scores on a scale of 0–100, where higher values indicate better data quality assessment.

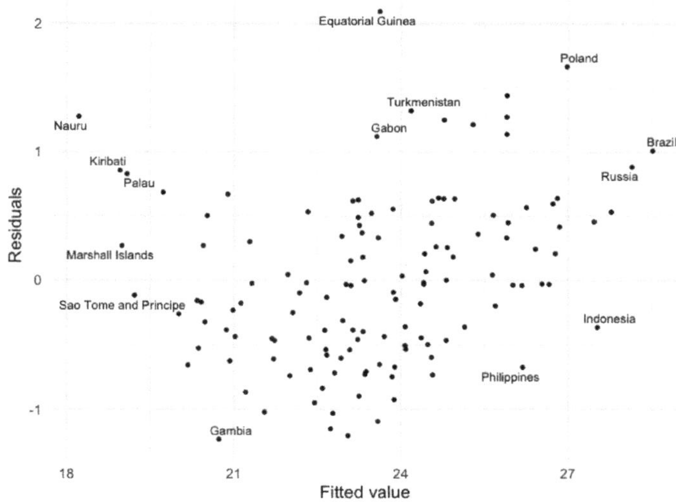

Figure 2. This figure enables us to study the distribution of the residuals from our main specification from Equation (1). The figure plots the residuals of tweets on GDP, against the fitted values of that model. For clarity, only 10% of the observations in the lowest density regions are labeled.

Given that the World Bank works solely with low-income, lower-middle income, and upper-middle income countries, the data available for such measures are restricted to these countries. There are 140 countries for which there is an indicator on the quality of the data, as well as GDP, Twitter, population, and percent of population with access to the Internet.

I will then collect the residuals of Equation (1) using the subset data and run the following regression

$$Residuals_{i,t}^2 = \beta_0 + \beta_1 DataQuality_{i,t} + \varepsilon_{i,t}, \quad (2)$$

where I will regress the squared residuals for country i in year t on the data quality index and GDP. The coefficient of interest is β_1; a negative and statistically significant coefficient would indicate that the residuals in my baseline model in Equation (1) are larger for countries with low data quality, and vice versa. Table 3 shows that the data quality indicator coefficient is in fact negative and statistically significant at the 1% confidence level. This indicates that GDP estimates under the baseline model are more accurate for countries with high-quality data, and vice versa.

3.3. Example: Nigeria's Rebasing

In April 2014, the Nigerian Bureau of Statistics announced that they were changing the base year for their GDP calculations from 1990 to 2010. Authorities explained that the change was made to reflect the new structure of the economy, which currently relied more heavily on the financial and communication sectors, among others [1]. Interestingly though, the revised GDP estimates for 2012 and 2013 (the two most recent years) were larger by roughly 90% each year.

These revisions, which are not infrequent (according to a statement from the Nigeria's Bureau of Statistics, they are considering to rebase their GDP estimates again [9]) enable us to see if GDP estimates could be more accurate if we were to rely on GDP estimates based on volume of tweets.

For this exercise, we gather the coefficients calculated in Equation (1) to estimate Nigeria's GDP for 2012 and 2013 based on the volume of image tweets sent from the country in each of those years. We then compare these estimates to both the original (i.e., the one using the old series) and the revised (i.e., the one using the new series) official GDP estimates for those years.

Table 3. Data quality issues.

Dep. var.:	**Residual^2**
Data Quality	−0.01 ***
	(< 0.01)
R^2	0.04
Adj. R^2	0.03
Num. obs.	244
RMSE	1.30

*** $p < 0.01$. **Notes:** The dependent variable in columns 1–4 is the log of GDP; and in column 5 the residual squared. Using the World Bank classification, columns 1–4 estimate the log of GDP for the subset of low-income, low-middle, and upper-middle income countries combined. In column (5), I will collect the squared residuals of the estimation in column 4 and regress them on the data quality index and the log of tweets.

Table 4 shows these different estimates. We see that the Twitter-GDP estimates are closer to the revised estimates, than the old series estimates are. While the revised estimates are 93 and 90% higher than the original estimates, respectively, they are 45 and 49% higher than the Twitter-GDP estimates, respectively.

Table 4. Data quality issues.

	2012	**2013**
Tweets	64,674	311,557
GDP estimates *(billions of USD)*		
Twitter-GDP	319.9	343.8
Old series	240.3	269.6
New series	453.9	509.9

Notes: First row shows the number of image tweets sent from Nigeria in 2012 and 2013, respectively. The following row shows the estimated GDP using the coefficients calculated in Equation (1) and the number of tweets. The following two rows show the official GDP estimates, using the old series as well as the new revised series. All GDP estimates are in billions of USD.

4. Discussion

The main goal of this paper is to study whether social media data from Twitter could be used as a proxy for estimating GDP for developing countries. In particular, this paper analyzes whether social media data can be used to improve the accuracy of official GDP estimates for low-income countries. First, I find that the volume of image tweets sent from a country, together with the population, can explain roughly 90% of the cross-country variation in GDP for developing countries. This is pretty much in line with what others have found for using social media to estimate economic activity [5,6].

As discussed in Section 1, developing countries tend to have highly inaccurate official GDP estimates. Hence, it is possible that a significant share of the differences between the Twitter-GDP and official GDP estimates arise due to measurement errors in these official GDP estimates. If this is in fact the case, data from tweets could be used by statistical agencies as a complementary measure to produce more accurate estimates. I study this by collecting the residuals in the baseline model and then running a regression on a measure of data quality in each country. The negative coefficient on the data quality index in Equation (2) suggests that there is information to be captured from Twitter data that could help close the gap between estimated GDP and the *true* GDP. Social media data could thus be used as a complement to survey data to increase the accuracy of GDP estimates.

Furthermore, given that the measurement errors stemming from official GDP estimates and Twitter-GDP estimates are not correlated, we can use both measures together to improve their accuracy [10]. This was also one of the motivating factors to push for the use of night-lights to improve GDP measurements in [11].

A word of caution should be expressed before incorporating social media (or related) data to produce official statistics. While these data sources could represent valuable information, it is troublesome to incorporate measures that only represent a proxy. In other words, it is important to understand the underlying mechanism relating economic activity and social media posts. Ref. [6] explores the underlying mechanism between tweets and economic activity and finds evidence to suggest that social media posts are often used by users to showcase consumption of goods and services to their network of followers. Thus, a larger number of posts represents a larger share of consumption, which is a significant part of what drives the economy.

Nonetheless, the frequency with which users post on social media and reasons why they choose to do so can evolve and change quickly. A much deeper understanding of these mechanisms are needed before governments and statistic agencies rely on these measures when putting together official measures. Given that people are generating increasingly large volumes of data on social media applications (and related software), it would be sensible to research these more carefully to see if they can help us obtain more accurate measurements on the state of the economy.

Funding: This research received no external funding.

Institutional Review Board Statement: Not applicable.

Informed Consent Statement: Not applicable.

Data Availability Statement: Restrictions apply to the availability of these data. The data were awarded via the 2014 Twitter Data Grant submission, which was awarded to the Cultural Analytics Lab directed by Lev Manovich. Aggregated data are available from the author upon request.

Acknowledgments: I am grateful to David Jaeger, Francesc Ortega, and Lev Manovich for their useful comments and suggestions.

Conflicts of Interest: The author declares no conflict of interest.

Abbreviations

The following abbreviations are used in this manuscript:

GDP	Gross Domestic Product

References

1. Blas, J.; Wallis, W. Nigeria Almost Doubles GDP in Recalculation. *Financial Times*, 7 April 2014; Volume 7, p. 8.
2. Jerven, M. *Poor Numbers: How We Are Misled by African Development Statistics and What to Do about It*; Cornell University Press: Ithaca, NY, USA, 2013.
3. Donaldson, D.; Storeygard, A. The View from Above: Applications of Satellite Data in Economics. *J. Econ. Perspect.* **2016**, *30*, 171–198. [CrossRef]
4. Askitas, N.; Zimmermann, K.F. Nowcasting Business Cycles Using Toll Data. *J. Forecast.* **2013**, *32*, 299–306. [CrossRef]
5. Woloszko, N. *Tracking Activity in Real Time with Google Trends*; OECD Economics Department Working Papers 1634; OECD Publishing: Paris, France, 2020. [CrossRef]
6. Indaco, A. From twitter to GDP: Estimating economic activity from social media. *Reg. Sci. Urban Econ.* **2020**, *85*, 103591. [CrossRef]
7. Weidemann, C.; Swift, J. Social media location intelligence: The next privacy battle—An ArcGIS add-in and analysis of geospatial data collected from Twitter.com. *Int. J. Geoinform.* **2013**, *9*, 21–27.
8. Lee, K. What Analyzing 1 Million Tweets Taught Us. 2015. Available online: https://thenextweb.com/socialmedia/2015/11/03/what-analyzing-1-million-tweets-taught-us/ (accessed on 2 October 2018).
9. Eboh, C. Nigeria to Rebase GDP to Determine Structure of Economy. Available online: https://www.reuters.com/article/nigeria-economy-rebasing-idAFL8N2K960S (accessed on 2 January 2021)
10. Rao, B.L.S.P. *Identifiability in Stochastic Models: Characterization of Probability Distributions*; Academic Press: Cambridge, MA, USA, 1992.
11. Henderson, J.V.; Storeygard, A.; Weil, D.N. Measuring Economic Growth from Outer Space. *Am. Econ. Rev.* **2012**, *102*, 994–1028. [CrossRef] [PubMed]

engineering proceedings

MDPI

Proceeding Paper

EEG-Based Neural Synchrony Predicts Evaluative Engagement with Music Videos †

Nikki Leeuwis [1,2,*] and Tom van Bommel [2]

1 Tilburg School of Humanities and Digital Sciences, Tilburg University, 5037 AB Tilburg, The Netherlands
2 Unravel Research, 3581 KW Utrecht, The Netherlands
* Correspondence: n.leeuwis@tilburguniversity.edu
† Presented at the 9th International Conference on Time Series and Forecasting, Gran Canaria, Spain, 12–14 July 2023.

Abstract: The use of neuroimaging to predict individual and population-wide behaviors, also known as neuroforecasting, was long applied to estimate movie popularity. Only recently, EEG-based neural synchronization, which is indicative of engagement, was found as a valid predictor of the listening behavior of the population. However, the population's evaluative responses to the songs were not incorporated. To fill this void, this study explored whether neural synchrony can also be related to likes, dislikes and comments for the same songs on YouTube more than two years after their release. In this way, we aimed to separate passive engagement (i.e., listening) from active engagement (evaluating). The results showed that neural synchrony was a significant predictor of the likes and comments on YouTube, even after controlling for explicit liking ratings from the lab study. In contrast, frontal alpha asymmetry did not predict YouTube likes. Thus, engagement as represented by neural synchronization could be a valuable tool for predicting active as well as passive engagement with entertainment products. This underlines the value of neural similarity in predicting the impact of music and videos before their true effect in the crowd can be known.

Keywords: neuroforecasting; neuromarketing; neural synchrony

Citation: Leeuwis, N.; van Bommel, T. EEG-Based Neural Synchrony Predicts Evaluative Engagement with Music Videos. *Eng. Proc.* **2023**, *39*, 50. https://doi.org/10.3390/engproc2023039050

Academic Editors: Ignacio Rojas, Hector Pomares, Luis Javier Herrera, Fernando Rojas and Olga Valenzuela

Published: 4 July 2023

1. Introduction

Traditionally, when investigating human decision-making, an individual's previous choices were the best indicator of future ones. However, more recently, it was supposed that the choices humans make can be linked to changes in brain activity. Due to the continuous advances in brain-imaging design and analysis, the use of brain activity to predict individual and population-wide choices emerged in recent years. This prediction is also called neuroforecasting [1] and can be applied in a variety of domains including consumer behavior, neuromarketing, health-related choices and financial decision-making [2–5]. Researchers tried to pinpoint neural dynamics that predict individual and population-wide behaviors, but with mixed results. In a study by Genevsky et al. [6], the neural activity in the reward center in the brain of only 30 human participants was predictive of the market-level crowdfunding outcomes weeks later, while the behavior of those participants did not provide any predictive value for market-level behavior.

Within this line of research, Berns and Moore [7] examined the brain activity in the reward center of their subjects during listening to music, and they were able to correlate these neural dynamics to the sales data of the music albums. Later studies [8] again found that music popularity could be forecasted with brain activity, but these implemented a metric of brain activity called neural synchrony.

The first notice of neural synchronization in the context of predicting behavior was in 2004, when Hasson et al. [9] identified similar brain responses across the subject pool while they were watching a movie. This similarity was not only found in visual and auditory cortices (where it could be expected due to the audiovisual input of the movie), but frontal and parietal regions also showed synchronized responses. This indicates that brain regions that are associated with narrative and emotional processing were engaged while watching the movie in a similar way across participants. Engagement with the narrative is a common definition of the neural synchronization during natural stimuli [10]. Several studies showed that scrambling a story, movie or music part significantly decreased the similarity in brain patterns between individuals [11,12]. This metric of synchronized brain activity between individuals was shown to be indicative of popularity of movies [13,14], social buzz [15] and recall [13,16]. Other studies, such as Chan et al. [17], showed that this metric is not only related to passive engagement but also evaluative responses.

Next to neural synchrony, another metric related to personal preferences should be considered as well; frontal alpha asymmetry (FAA) is a popular metric in neuromarketing and was linked to personal liking in several studies [18–20]. Moreover, FAA was a significant predictor in the study of Leeuwis et al. [8]; on an individual level, the FAA during a song was predictive of the individual ratings provided afterwards.

Thus, to capture the predictive value of neural measurements of evaluative responses on a population-level, this study presents an accompanying analysis to Leeuwis et al. [8], who showed that the number of plays on Spotify could be predicted by this neural marker. To elaborate the finding that passive engagement in the form of consuming the content could be predicted, we added the predictive value of neural synchrony for active engagement in the form of evaluative responses to the songs; we inferred the number of likes, dislikes and comments on YouTube and investigated whether neural markers of synchronization and frontal alpha asymmetry within a small group of individuals could be related to this evaluative engagement of the crowd as well.

2. Materials and Methods

Data for the experiment were obtained from Leeuwis et al. [8], where 30 participants (23 female, 7 male; Age M $\pm$ SD = 26.87 $\pm$ 10.80) listened to fragments of music tracks on two albums that were just released a few days prior to the experiment. During listening to the 24 s excerpt of each track, brain activity was recorded on nine EEG channels. After listening to a track, participants rated the track on a 1–5 Likert scale before continuing to the next one. The order of the albums was counterbalanced, and the series of tracks were randomized between subjects too.

Passive engagement was quantified by the number of views on YouTube. Four dependent variables (DV) of active engagement on YouTube were evaluated 2 years and 10 months after the release of the albums; likes and comments were directly assessed from YouTube.com. Dislikes were obtained with the Google Chrome extension Return to Dislikes (https://returnyoutubedislike.com/ accessed on 23 February 2023). Additionally, likes of the song YouTube were corrected by the times the song was played on YouTube, such that this dependent variable was reflecting the number of likes per play of the video.

The EEG data were processed by Leeuwis et al. [8]. After pre-processing, power spectral density was calculated in the alpha frequency range (8–12 Hz) and this data in the central electrodes (C3, Cz, and C4) were pairwisely correlated between all participants for each track separately to calculate neural synchrony for each track. Moreover, frontal alpha asymmetry (FAA) was calculated from the F3 and F4 electrodes. More details about the processing can be found in Leeuwis et al. [8].

Outliers were removed when their score on the dependent variable exceeded the boundary of 2.5 SD above or below the mean and by Mahalanobis distance on both the explicit ratings by participants in the lab and neural measurements.

Linear regression models were fitted to assess the predictability of the neural measurements. As noted by Boksem and Smidts [5], when assessing the predictability of neural measures, their value should be above and beyond the traditional methods (e.g., stated preferences) to be relevant. Therefore, for each DV, a baseline model was created as H0 to be compared to models H1 and H2 incorporating brain activity measures:

H0: *DV~Likes Lab + Music Video (y/n)*

H1: *DV~Likes Lab + Music Video (y/n) + Neural Synchrony*

H2: *DV~Likes Lab + Music Video (y/n) + Frontal Alpha Asymmetry*

This meant that the basic (H0) model incorporates the average explicit liking of the respondents in the lab and the fact that a track was accompanied by a music video as a dichotomous factor. It is important to note here that the participants in the lab did not see the music video, they only listened to musical excerpts of the songs while the desktop screen was black. The DV was one of the five YouTube variables mentioned earlier: the likes, dislikes, number of comments, the likes-per-view ratio, or likes-per-dislike ratio.

Consequently, the H0 model was then elaborated by adding neural synchrony or frontal aslpha asymmetry to the model and comparing the explained variance with an ANOVA test between the models. For the DV number of views, only H1 was compared to H0, as the previous research already showed that frontal asymmetry was not predictive of passive engagement in this context [8]. The assumptions underlying linear regression were tested with a Shapiro–Wilk test for normality of residuals and Breusch–Pagan test checking the assumption of constant variance [21]. Linearity assumptions were checked visually. Numerical variables were scaled before inputted to the model.

Investigating four dependent variables, and for each comparing two models to H0, resulted in 8 statistical tests. Moreover, the model for views was compared only on neural synchrony, making the total of 9 tests. Following Bonferroni correction, significance levels for these tests were set at 0.006.

Data acquisition, pre-processing, PSD and FAA calculation was carried out using iMotions (2019). The neural synchrony was calculated using R (R Core Team, 2019), and statistical analysis was also performed in R (R Core Team, 2022).

3. Results

One track exceeded the 2.5 SD boundary on likes and comments and was also identified as an outlier in Mahalanobis distance. This track was removed from analysis, which left 23 tracks in the data for analysis (liking in lab before removal: M = 2.67, SD = 0.28; liking in lab after removal: M = 2.66, SD = 0.28; likes YouTube before removal: M = 64.73, SD = 63.18; likes YouTube after removal: M = 56.58, SD = 50.09). Models without outlier removal did not meet the assumption of normality of errors; thus, all models reported here had the outlier removed.

3.1. YouTube Views

To start with active engagement, we applied the model of Leeuwis et al. [8] on the passive engagement on YouTube. To correct for normality of model errors, the YouTube views were log transformed before they were put into the model. Here, as well, neural synchrony added predictive value to the baseline model ($F(20) = 8.57$, $p = 0.009$). The H0 model predicted 29.05% of the views ($R^2_{adj} = 0.291$, $F(2,20) = 5.50$, $p = 0.012$), while the model incorporating neural measures predicted 48.34% of the views ($R^2_{adj} = 0.483$, $F(2,20) = 7.86$, $p = 0.001$).

3.2. YouTube Likes

The H0 model significantly predicted the number of likes on YouTube ($R^2_{adj} = 0.573$, $F(2,20) = 15.76$, $p < 0.001$), where the fact that the track was released with an accompanying music video was significantly predictive (beta = 1.59, $t(20) = 5.38$, $p < 0.001$) but the average subjective liking of the track by the participants in the lab was not (beta = 0.26, $t(20) = 1.65$, $p = 0.11$), although an increasing line was observed (Figure 1).

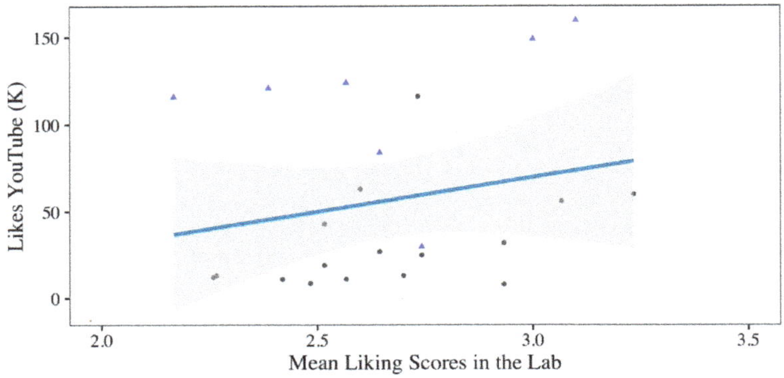

Figure 1. The distribution of liking scores in the lab (1 to 5 Likert scale) and the likes of that track on YouTube (×1000) almost three years later. The blue line indicates the fitted regression line. The gray area indicates the confidence interval. The tracks that are accompanied by a music video on YouTube are indicated by blue triangles (please note that subjects in the lab only listened to an excerpt of the track and did not see the music video).

Adding neural synchrony to this model of YouTube likes increased the predictive value ($F(20) = 9.68$, $p = 0.006$). The predicted variance was improved to 70.21% ($R^2_{adj} = 0.7038$, $F(3,19) = 18.28$, $p < 0.001$). The predictive value of the stated preferences in the lab remained insignificant (beta = 0.13, $t(19) = 1.09$, $p = 0.29$), while the fact that the track was released with an accompanying music video was significantly predictive (beta = 1.46, $t(19) = 5.80$, $p < 0.001$), as well as neural synchrony (beta = 0.38, $t(19) = 3.11$, $p = 0.006$) (Figure 2).

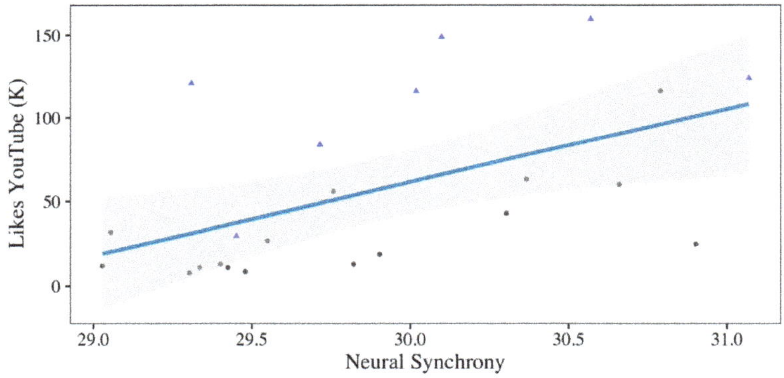

Figure 2. The neural synchrony scores for each track in the lab and the likes (×1000) of that track on YouTube almost three years later. The blue line indicates the fitted regression line. The gray area indicates the confidence interval. The tracks that are accompanied by a music video on YouTube are indicated by blue triangles (please note that subjects in the lab only listened to an excerpt of the track and did not see the music video).

Predicting the evaluative responses on YouTube by the stated preferences in the lab and the accompanying frontal alpha asymmetry measures, provided a significant model ($R^2_{adj} = 0.553$, $F(3,19) = 10.08$, $p < 0.001$), although no improvement in predictive value was achieved beyond the model incorporating only the lab-based liking ($F(20) = 0.120$, $p = 0.73$).

3.3. *YouTube Dislikes*

The same comparisons were also performed for dislikes on YouTube. To meet the assumptions of homogeneity of variance, the dislikes were log transformed. The H0 model significantly predicted the number of dislikes on YouTube ($R^2_{adj} = 0.553$, $F(2,20) = 14.60$, $p < 0.001$), where the fact that the track was released with an accompanying music video was significantly predictive (beta = 1.53, $t(20) = 5.04$, $p < 0.001$) but the average liking of the track by the participants in the lab was not (beta = 0.28, $t(20) = 2.00$, $p = 0.06$). Adding neural synchrony to this model increased the predictive value ($F(20) = 7.39$, $p = 0.014$), although not significantly when considering Bonferroni correction. The predicted variance was improved to 66.11% ($R^2_{adj} = 0.661$, $F(3,19) = 15.30$, $p < 0.001$). The predictive value of the stated preferences in the lab remained insignificant (beta = 0.19, $t(19) = 1.50$, $p = 0.15$), while the fact that the track was released with an accompanying music video was significantly predictive (beta = 1.40, $t(19) = 5.22$, $p < 0.001$), as well as neural synchrony (beta = 0.35, $t(19) = 2.72$, $p = 0.014$). Interestingly, the positive value points out that neural synchrony was positively correlated to likes as well as dislikes. This indicates that it was a predictor of engagement rather than the direction of this engagement. Frontal alpha asymmetry did not add any predictive value to the H0 model ($F(20) = 0.43$, $p = 0.52$; $R^2_{adj} = 0.540$, $F(3,19) = 9.60$, $p < 0.001$).

3.4. YouTube Comments

Similarly, we further examined the predictability of YouTube comments on each video. To meet the assumption of normality of model errors, the comments were log transformed before putting into the model. The H0 model significantly predicted the number of comments on YouTube ($R^2_{adj} = 0.616$, $F(2,20) = 18.7$, $p < 0.001$), where the fact that the track was released with an accompanying music video was significantly predictive (beta = 1.71, $t(20) = 6.09$, $p < 0.001$) but the average liking of the track by the participants in the lab was not (beta = 0.07, $t(20) = 0.50$, $p = 0.62$).

Adding neural synchrony to this model increased the predictive value ($F(20) = 11.23$, $p = 0.003$), where the predicted variance was improved to 74.61% ($R^2_{adj} = 0.746$, $F(3,19) = 22.55$, $p < 0.001$). The predictive value of the stated preferences in the lab remained insignificant (beta = −0.03, $t(19) = -0.28$, $p = 0.78$), while the fact that the track was released with an accompanying music video was significantly predictive (beta = 1.58, $t(19) = 6.79$, $p < 0.001$), as well as neural synchrony (beta = 0.38, $t(19) = 3.35$, $p = 0.003$).

Adding frontal alpha asymmetry to this model did not increase the predictive value ($F(20) = 0.20$, $p = 0.659$), the model only explained 60.0% of the variance ($R^2_{adj} = 0.600$, $F(3,19) = 12.01$, $p < 0.001$).

3.5. Likes Per View Ratio

The H0 model predicting likes-per-view ratio explained only 20.02% of the variance ($R^2_{adj} = 0.202$, $F(2,20) = 3.79$, $p = 0.04$). In this case, adding neural synchrony did not add predictive power ($F(20) = 0.20$, $p = 0.66$), as the model explained 16.88% of the variance ($R^2_{adj} = 0.169$, $F(3,19) = 2.49$, $p = 0.09$). FAA did also not improve the model ($F(20) = 0.36$, $p = 0.56$), as the model explained 17.57% of the variance ($R^2_{adj} = 0.176$, $F(3,19) = 2.56$, $p = 0.09$).

4. Discussion

In investigating the use of neuroimaging techniques to predict population-wide evaluative responses to music on YouTube, we explored the predictive value of neural synchrony and frontal alpha asymmetry. The study presented an analysis of data from a previous experiment by Leeuwis et al. [8], where the neural markers of synchronization and frontal alpha asymmetry in response to music tracks of two new music albums were recorded. The results showed that neural synchrony had predictive value above and beyond stated preferences for evaluative responses by the population. The neural similarity was a positive factor in predicting likes as well as dislikes and comments. This indicates the potential usefulness of neural synchrony not only in predicting passive engagement (consuming the content) but also active engagement (evaluating the content) with music tracks, regardless of the direction of the evaluation. No such results were found for frontal alpha asymmetry.

To further examine the active engagement measurement of likes, we corrected those by the number of times the videos were played and found that the predictive value was not present anymore.

While one could argue that the latter could be expected since the previous analysis by Leeuwis et al. [8] showed that the number of times the song was played was significantly predicted by neural synchrony, we believe this analysis of evaluative responses shows a new aspect of engagement that was not covered in previous explorations of neural synchrony. The previous analyses by Berns and Moore [7] and Leeuwis et al. [8], where the popularity was derived from album sales and online streams, respectively, were relating more to passive engagement. In this dataset, we validated once again that passive engagement was better predicted when neural synchrony was included in the model, although the explained variance was lower than the results observed for Spotify plays [8]. This could have to do with a difference platform engagement or the fact that these data were gathered two years later compared to the Spotify data.

Our study shows that the number of likes was not significantly predicted when corrected for the total number of views, which indicates that the number of views could be the most important contributor to the number of likes on the platform. However, the number of views could not be known beforehand, making it not feasible for popularity prediction. The other factors explored in the data, namely the neural correlates and the fact that the song is released with music video can be established almost immediately after the release of the song, enabling prediction of the evaluative responses straight away.

Previous studies that found that similarity in brain responses was indicative of evaluative responses [17] were more focused on market review panelists. This is different from YouTube video (dis)likes in the sense that for the latter, the likes are provided by individuals that consumed the content on a voluntary basis while the panelists are intentionally exposed to the stimuli. Ideally, in future research, we would correct the likes not for the total number of plays but by the number of individuals that played the video. This may present a more realistic scenario as it may be that a small group of individuals (assuming this will be the likers) is responsible for a great lot of the plays [22].

The fact that frontal alpha asymmetry did not add extra explanatory value could be explained by the fact that this metric is mostly found to be indicative of individual liking [8,18,19]. Moreover, the catchiness of music is not necessarily depending on positive emotions elicited; irritating tunes stick particularly well [23], explaining why approach may not be sufficiently predictive of passive or active engagement with the content. However, some studies also found generalizations out-of-sample to be correlated to FAA measurements; Shestyuk et al. [24] showed that frontal asymmetry in a combination of alpha and beta bands was predictive of TV viewership and social media engagement.

In discussing these measures, it should be taken into account that we discussed the brain activity measures of unimodal stimuli; only sound. For FAA, this was shown to elicit weaker activity patterns than multimodal stimuli that most of the previous studies employed [25]. Neural synchronization may also be diminished due to this unimodal approach, since engagement in general was higher with multimodal design [26].

Another limitation of the study was the fact that only two music albums were incorporated as stimuli, which may not be representative of the wide variety of entertainment media available. Future research could investigate the predictive value of neural metrics in larger and more diverse stimulus sets. A replication with new music could provide a valuable addition to this dataset. Moreover, the mediation variable of music videos, social media campaigns or marketing budget were not accounted for but could have an interesting effect on the predicted values of likes on YouTube.

Future research in this area could focus on tackling these limitations. By creating a full image of all variables impacting entertainment popularity, the undisguised effect of neural measure beyond all other could be explored. Moreover, replication of studies employing neural synchrony is needed to strengthen the validity of this emerging metric. Especially since various calculation methods exist, including different locations, frequency bands, neuroimaging techniques and algorithms in general. Moment-to-moment fluctuations of neural synchrony may also provide an interesting metric for examination of entertainment [11,27]. The field of neuroforecasting may also be expanded by the implementation of more sophisticated prediction algorithms, machine learning or deep learning, which generally improves prediction accuracies [28].

Our results underline that neuroforecasting has an additional value to surveys when determining the success of entertainment types. Applications could include ranking songs on an album to determine the marketing budget to put in each of them. Moreover, the field can be broadened to other branches such as marketing, health and financial decision-making where this metric could be used to indicate the engagement and predict the impact of communication statements.

5. Conclusions

This work provided valuable insights into the potential usefulness of neuroforecasting in predicting human behavior. The research findings suggest that neural synchrony may have predictive value above and beyond stated preferences, indicating that brain activity can be a useful tool in predicting not only population-wide passive engagement behaviors but also active engagements; neural synchrony added value for predicting YouTube likes, dislikes and comments almost three years after the release of the music tracks. Our results showed a distinction between active and passive engagement and the potential of neural synchrony to provide an indication of both types of interaction with the content. Thereby, this work further contributes to the body of literature of neuroforecasting. Further research in this field could have important implications for a variety of domains, including consumer behavior, neuromarketing, health-related choices, and financial decision making.

Author Contributions: Conceptualization, N.L. and T.v.B.; methodology, N.L. and T.v.B.; formal analysis, N.L.; writing—original draft preparation, N.L.; writing—review and editing, T.v.B.; visualization, N.L.; supervision, T.v.B. All authors have read and agreed to the published version of the manuscript.

Funding: This research received no external funding.

Institutional Review Board Statement: Ethical review and approval was not required for the study on human participants in accordance with the local legislation and institutional requirements. Prior to the experiment, participants singed an informed consent in accordance with NMSBA Code of Ethics and EU GDPR guidelines.

Informed Consent Statement: Informed consent was obtained from all participants involved in the study.

Data Availability Statement: The raw data supporting the conclusions of this article will be made available by the authors, without undue reservation.

Conflicts of Interest: Authors were employed by the company Unravel Research. Authors' employment does not depend on the outcomes or publication of this study. The authors declare no financial or otherwise competing conflicts of interest.

References

1. Knutson, B.; Genevsky, A. Neuroforecasting aggregate choice. *Curr. Dir. Psychol. Sci.* **2018**, *27*, 110–115. [CrossRef] [PubMed]
2. Berkman, E.T.; Falk, E.B. Beyond brain mapping: Using neural measures to predict real-world outcomes. *Curr. Dir. Psychol. Sci.* **2013**, *22*, 45–50. [CrossRef]
3. Stallen, M.; Borg, N.; Knutson, B. Brain activity foreshadows stock price dynamics. *J. Neurosci.* **2021**, *41*, 3266–3274. [CrossRef] [PubMed]
4. Falk, E.B.; Berkman, E.T.; Whalen, D.; Lieberman, M.D. Neural activity during health messaging predicts reductions in smoking above and beyond self-report. *Health Psychol.* **2011**, *30*, 177. [CrossRef] [PubMed]
5. Boksem, M.A.S.; Smidts, A. Brain responses to movie trailers predict individual preferences for movies and their population-wide commercial success. *J. Mark. Res.* **2015**, *52*, 482–492. [CrossRef]
6. Genevsky, A.; Yoon, C.; Knutson, B. When brain beats behavior: Neuroforecasting crowdfunding outcomes. *J. Neurosci.* **2017**, *37*, 8625–8634. [CrossRef]
7. Berns, G.S.; Moore, S.E. A neural predictor of cultural popularity. *J. Consum. Psychol.* **2012**, *22*, 154–160. [CrossRef]

8. Leeuwis, N.; Pistone, D.; Flick, N.; van Bommel, T. A Sound Prediction: EEG-Based Neural Synchrony Predicts Online Music Streams. *Front. Psychol.* **2021**, *12*, 672980. [CrossRef]
9. Hasson, U.; Nir, Y.; Levy, I.; Fuhrmann, G.; Malach, R. Intersubject Synchronization of Cortical Activity During Natural Vision. *Science* **2004**, *303*, 1634–1640. [CrossRef]
10. Grall, C.; Tamborini, R.; Weber, R.; Schmälzle, R. Stories collectively engage listeners' brains: Enhanced intersubject correlations during reception of personal narratives. *J. Commun.* **2021**, *71*, 332–355. [CrossRef]
11. Dmochowski, J.P.; Sajda, P.; Dias, J.; Parra, L.C. Correlated components of ongoing EEG point to emotionally laden attention—A possible marker of engagement? *Front. Hum. Neurosci.* **2012**, *6*, 112. [CrossRef] [PubMed]
12. Kaneshiro, B.; Nguyen, D.T.; Norcia, A.M.; Dmochowski, J.P.; Berger, J. Natural music evokes correlated EEG responses reflecting temporal structure and beat. *NeuroImage* **2020**, *214*, 116559. [CrossRef] [PubMed]
13. Barnett, S.B.; Cerf, M. A ticket for your thoughts: Method for predicting content recall and sales using neural similarity of moviegoers. *J. Consum. Res.* **2017**, *44*, 160–181. [CrossRef]
14. Christoforou, C.; Papadopoulos, T.C.; Constantinidou, F.; Theodorou, M. Your brain on the movies: A computational approach for predicting box-office performance from viewer's brain responses to movie trailers. *Front. Neuroinform.* **2017**, *11*, 72. [CrossRef]
15. Dmochowski, J.P.; Bezdek, M.A.; Abelson, B.P.; Johnson, J.S.; Schumacher, E.H.; Parra, L.C. Audience preferences are predicted by temporal reliability of neural processing. *Nat. Commun.* **2014**, *5*, 4567. [CrossRef]
16. Hasson, U.; Furman, O.; Clark, D.; Dudai, Y.; Davachi, L. Enhanced intersubject correlations during movie viewing correlate with successful episodic encoding. *Neuron* **2008**, *57*, 452–462. [CrossRef]
17. Chan, H.-Y.; Smidts, A.; Schoots, V.C.; Dietvorst, R.C.; Boksem, M.A. Neural similarity at temporal lobe and cerebellum predicts out-of-sample preference and recall for video stimuli. *Neuroimage* **2019**, *197*, 391–401. [CrossRef]
18. Davidson, R.J.; Ekman, P.; Saron, C.D.; Senulis, J.A.; Friesen, W.V. Approach-withdrawal and cerebral asymmetry: Emotional expression and brain physiology: I. *J. Personal. Soc. Psychol.* **1990**, *58*, 330. [CrossRef]
19. Vecchiato, G.; Toppi, J.; Astolfi, L.; De Vico Fallani, F.; Cincotti, F.; Mattia, D.; Bez, F.; Babiloni, F.; Spectral, E.E.G. Spectral EEG frontal asymmetries correlate with the experienced pleasantness of TV commercial advertisements. *Med. Biol. Eng. Comput.* **2011**, *49*, 579–583. [CrossRef]
20. Briesemeister, B.B.; Tamm, S.; Heine, A.; Jacobs, A.M. Approach the good, withdraw from the bad—A review on frontal alpha asymmetry measures in applied psychological research. *Psychology* **2013**, *4*, 261. [CrossRef]
21. Zaman, A. Inconsistency of the Breusch-Pagan test. *J. Econ. Soc. Res.* **2000**, *2*, 1–11.
22. Kaimann, D.; Tanneberg, I.; Cox, J. "I will survive": Online streaming and the chart survival of music tracks. *Manag. Decis. Econ.* **2021**, *42*, 3–20. [CrossRef]
23. Beaman, C.P. The literary and recent scientific history of the earworm: A review and theoretical framework. *Audit. Percept. Cogn.* **2018**, *1*, 42–65. [CrossRef]
24. Shestyuk, A.Y.; Kasinathan, K.; Karapoondinott, V.; Knight, R.T.; Gurumoorthy, R. Individual EEG measures of attention, memory, and motivation predict population level TV viewership and Twitter engagement. *PLoS ONE* **2019**, *14*, e0214507. [CrossRef]
25. Sabu, P.; Stuldreher, I.; Kaneko, D.; Brouwer, A.-M. A review on the role of affective stimuli in event-related frontal alpha asymmetry. *Front. Comput. Sci.* **2022**, *4*, 869123. [CrossRef]
26. MitsisMitsis, K.; Zarkogianni, K.; Kalafatis, E.; Dalakleidi, K.; Jaafar, A.; Mourkousis, G.; Nikita, K.S. A multimodal approach for real time recognition of engagement towards adaptive serious games for health. *Sensors* **2022**, *22*, 2472. [CrossRef]
27. Kaneshiro, B.; Nguyen, D.T.; Norcia, A.M.; Dmochowski, J.P.; Berger, J. Inter-subject EEG correlation reflects time-varying engagement with natural music. *bioRxiv* **2021**, 1–7. [CrossRef]
28. Alimardani, M.; Kaba, M. Deep Learning for Neuromarketing; Classification of User Preference using EEG Signals. In Proceedings of the ACM International Conference Proceeding Series, 12th Augmented Human International Conference, Rovaniemi, Finland, 22–24 February 2021; Association for Computing Machinery: New York, NY, USA, 2021; pp. 1–7. [CrossRef]

engineering proceedings

Proceeding Paper

Calculating the Effectiveness of COVID-19 Non-Pharmaceutical Interventions with Interrupted Time Series Analysis via Clustering-Based Counterfactual Country †

Fatemeh Navazi *, Yufei Yuan and Norm Archer

DeGroote School of Business, McMaster University, Hamilton, ON L8S 4M4, Canada
* Correspondence: navazif@mcmaster.ca

† Presented at the 9th International Conference on Time Series and Forecasting, Gran Canaria, Spain, 12–14 July 2023.

Abstract: During the first year of the COVID-19 pandemic, governments only had access to non-pharmaceutical interventions (NPIs) to mitigate the spread of the disease. Various methods have been discussed in the literature for calculating the effectiveness of NPIs. Among these methods, the interrupted time series analysis method is the area of our interest. To study the second wave, we clustered countries based on levels of implemented NPIs, except for the target NPI (X) whose effectiveness wanted to be evaluated. To do so, the COVID-19 Policy Response Tracker data-set gathered by the "Our World in Data" team of Oxford University, and COVID-19 statistical data gathered by the John Hopkins Hospital were used. After clustering, we selected a counterfactual country from the countries that were in the same cluster as the target country, and implemented NPI (X) at its lowest level. Thus, the target country and the counterfactual country were similar in implementation level of other NPIs and only differed in the implementation level of the target NPI (X). Therefore, we can calculate the effectiveness of NPI (X) without being concerned about the impurity of the effectiveness values that might be caused by other NPIs. This allowed us to calculate the effectiveness of NPI (X) using the interrupted time series analysis with the control group. Interrupted time series analysis assesses the effect of different policy-implementation levels by evaluating interruptions caused by policies in trend and level after the policy-implementation date. Before the NPI-implementation date, the implementation levels of NPIs were similar in both selected countries. After this date, the counterfactual country could be treated as a baseline for calculating changes in the trends and levels of COVID-19 cases in the target country. To demonstrate this approach, we used the generalized least square (GLS) method to estimate interrupted time series parameters related to the effectiveness of school closure (the target NPI) in Spain (the target country). The results show that increasing the implementation level of school closure caused a 34% decrease in COVID-19 prevalence in Spain after only 10 days compared to the counterfactual country.

Keywords: interrupted time series analysis; COVID-19; public health; Non-Pharmaceutical Intervention (NPI); counterfactual analysis

Citation: Navazi, F.; Yuan, Y.; Archer, N. Calculating the Effectiveness of COVID-19 Non-Pharmaceutical Interventions with Interrupted Time Series Analysis via Clustering-Based Counterfactual Country. *Eng. Proc.* **2023**, *39*, 51. https://doi.org/10.3390/engproc2023039051

Academic Editors: Ignacio Rojas, Hector Pomares, Luis Javier Herrera, Fernando Rojas and Olga Valenzuela

Published: 5 July 2023

1. Introduction

The highly transmissible COVID-19 virus was announced as a variant of concern (VoC) by the World Health Organization (WHO) on the 11 March 2020 [1]. During the first year of the COVID-19 pandemic, governments used non-pharmaceutical interventions (NPIs) to control the spread of COVID-19, in the absence of medications and vaccines.

Restrictive NPIs can be categorized at the individual level, such as face covering mandates, and the social level, i.e., public transport restrictions, school closure, workplace closure, public event cancelations, restrictions on gatherings, stay-at-home orders, and international travel bans. Researchers have applied various methodologies to assess the effectiveness of these NPIs. Previous research has shown that even after developing and administering the COVID-19 vaccine and before achieving herd immunity, the NPIs remained effective [1].

Navazi et al. [1] analyzed the effects of lockdown on the third wave of COVID-19 in the province of Ontario, Canada, using the interrupted time series analysis (ITSA) method considering vaccination percentage; however, they did not have a counterfactual baseline to calculate lockdown effectiveness and only relied on predicting what would happen in the future, based on pre-intervention trends. Conversely, having a counterfactual country is more realistic. Saki et al. [2] found ITSA to be effective in estimating the effectiveness of social distancing in Iran. However, a limitation of their work was that they assumed the relationship between NPI implementation and COVID-19 confirmed cases to be linear, while it is known that epidemics follow a non-linear trend. This led us to add some non-linear terms to our model.

Auger et al. [3] also used ITSA to estimate the effectiveness of school closure. Although they added some of the other NPIs as covariates to their model, the effectiveness of school closure was not calculated independently and might depend on other NPI implementation levels. Therefore, in this study, we used ITSA with a control group to isolate the effectiveness of the NPI. Thayer et al. [4] used ITSA to investigate the effectiveness of lockdown in India. They considered lockdown implementation levels and estimated the effectiveness of different levels of lockdown. However, their research was limited to a single country, whereas when several countries with similar NPI-implementation levels exist, they can be used for ITSA with a control group. Emeto et al. [5] used ITSA with a matched-control country to study the effectiveness of border closure in Africa. However, the multi-country study by Ballard et al. [6] only considered the region as the ITSA model's input variable, not as a control group. Shah et al. [7] applied ITSA individually to three regions and compared the impact of the first lockdown on non-COVID-19 patient hospital admission in these three regions. In this case, they implemented single-group ITSA three times without an accompanying counterfactual analysis.

There are some challenges in single-group interrupted time series analysis, such as being unable to control for other competing factors [8], poor internal validity [8], or using treatment group pre-intervention trends as counterfactuals [5]. To solve these issues, we will use other ITSA designs, such as using a control group, to obtain valid and causal results.

We developed a clustering-based counterfactual selection integrated with an interrupted time series analysis with a control group to address the research gap. For example, if we consider eight main restrictive NPIs to study, we will cluster countries based on seven NPIs. The remaining NPI (X) is the one whose effectiveness we are interested in studying. The selection mechanism of the control country was based on two criteria: (1) being in the same cluster as the target country and (2) having the greatest difference with the target country in implementing NPI (X) and implementing NPI (X) at a zero or lower level. Having a comparable control group helps to implement a robust approach that considers treatment effects [8]. This way, the change in COVID-19 prevalence resulting from the pattern of a COVID-19 wave, rather than the intervention, is considered, and will not lead to an underestimation or overestimation of the number of COVID-19 cases [8].

Regarding wave selection, if we are interested in studying the effectiveness of both non-pharmaceutical interventions and pharmaceutical interventions such as vaccines, then a wave during 2021 should be selected, whereas if we want to only study the effectiveness of NPIs, we should select the first or second wave of COVID-19 during 2020. As the first COVID-19 wave's statistics have many uncertainties resulting from the shortage of test kits and non-standard COVID-19 diagnosis methods [1], we will focus on a study of the second wave of COVID-19.

2. Materials

This study used time series data from 8 NPIs in the COVID-19 Policy Response Tracker of Oxford University [9]. Effectiveness of the NPI that we are interested in studying is school closure, which means students and teachers are not required to go to school for in-person activities, and educational activities are provided online (except for some lab research) [3]. The 8 NPI implementation levels are as follows:

1. Public event cancelation implemented at 3 levels: no measure (0), recommended cancelation (1), and required cancelation (2).
2. Restrictions on gatherings, which limits the number of people in gatherings to less than 10 (4), 10 to 100 (3), 100 to 1000 (2), limiting only very large gatherings (>1000 people) (1), no limit (0) (five levels).
3. Workplace closure implemented at four levels: no closure (0), recommended closure (1), required closure for some (2), and required closure for all (3).
4. School closure implemented at four levels: no closure (0), recommended closure (1), required closure for some (2), and required closure for all (3).
5. Restriction on internal movement is implemented at three levels: no measure (0), recommend movement restriction (1), and required movement restriction (2).
6. International travel ban implemented at five levels: no measures (0), screening (1), quarantining of arrivals from high-risk regions (2), banning high-risk regions (3), and total border closure (4).
7. Public transport closure implemented at three levels: no measures (0), recommend closing (or significantly reduced volume/route/means of available transport) (1), required closure (or prohibiting most people from using it) (2).
8. Stay-at-home requirements implemented at four levels, including no measure (0), recommending not leaving the house (1); requiring not leaving the house, with exceptions for grocery shopping, essential trips, etc. (2); requiring not leaving the house with minimal exceptions like once every few days, etc. (3).

The COVID-19 statistical data, including COVID-19 prevalence provided by John Hopkins Hospital [10], was also used. Additionally, we calculated a date column showing how many days had passed since the start of the second wave.

3. Methods

Interrupted time series analysis is a statistical methodology used to study intervention effects that can cause both level and trend changes [1]. When only one group is exposed to the policy, researchers use a "pre-post" observational study design. However, if we have two groups and only one is exposed to the policy, we can use a "pre-post with control" observational study design. One group without exposure to the policy works as a control group for the other. This method is also called a difference in differences study. Having a control country helps to capture the change in trend caused by policy instead of the nonlinearity (curvilinear) of the outcome wave. Since adding a control group solidifies the study, we will use this research design.

In order to determine what would have happened if there had been no NPI X (target NPI to study its effectiveness), we needed a counterfactual country that did not implement NPI X but implemented the rest of the NPIs at a similar level with the target country. This method is called counterfactual analysis. Therefore, we developed a hybrid methodology with clustering-based counterfactual selection to find a suitable control group for interrupted time series analysis with control, as hybrid methodologies are well known for covering the shortcomings of a methodology with a second method [11].

First, we clustered all countries with their data based on all studied NPIs except for NPI X. The K-Means clustering algorithm [12] was used for this purpose. This identifies which countries are in the same cluster as the target country. Among these countries, we determined pairs of countries that implemented NPI X at different levels. We then looked at the countries paired with the target country and selected the country which could work as a control group. A control country must have implemented NPI X at zero or lower levels for a period before and after NPI X implementation level changes in the target country. After selecting the control country estimate, we could use interrupted time series with counterfactual analysis for the control group to determine the effectiveness of NPI X. The methodology's steps are illustrated in Figure 1.

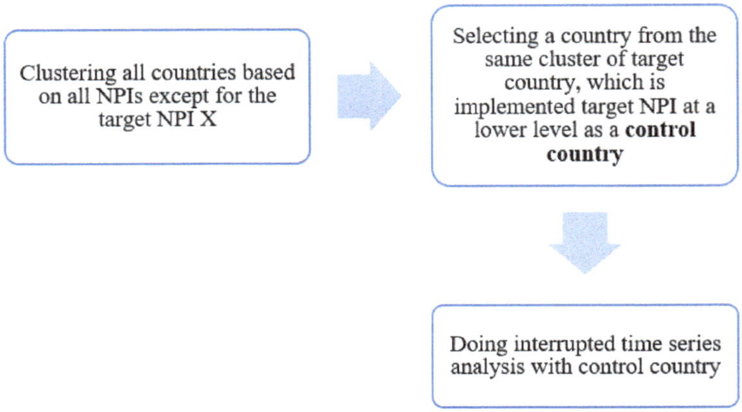

Figure 1. Developed hybrid methodology.

We have data from several populous European countries, including Austria, Belgium, Czechia, Denmark, Finland, France, Germany, Italy, Netherlands, Norway, Poland, Portugal, Romania, Russia, Spain, Switzerland, United Kingdom, available to select the target and control countries.

In this study, the second wave of COVID-19, which occurred at the end of 2020 and the beginning of 2021 in most countries, was studied. This is because COVID-19 vaccines, developed at the beginning of 2021, affected immunization, while we want to study NPIs, not pharmaceutical interventions. Moreover, clustering by considering vaccine administration percentage is difficult given vaccine shortages at the beginning of 2021, countries varied in vaccine administration percentage, and it is hard to cluster countries based on pharmaceutical interventions and NPIs at the same time.

The target country of this study is Spain, and the target NPI whose effectiveness is investigated is school closure. We focused on finding a control country that implemented school closure at a lower level by clustering countries based on other NPIs.

We clustered countries based on NPIs mentioned in the Materials section, except for school closure. Since the data related to each NPI are time series data during the second wave of COVID-19, the problem was clustering time series data. One of the methods for clustering time series data is considering each time point as a data column (feature) for clustering. Figure 2 illustrates how the data frame was transformed for time series clustering purposes. We considered just the first 60 days of the second wave because the performance of K-Means depends on the number of features, and it cannot perform well when the number of features is increased [13]. Moreover, the minimum duration for the second COVID-19 wave was 65 days in Poland.

The time series clustering was coded in Python 3.8, and the K-Means algorithm from Python's Scikit-learn library was used. Since we wanted to cluster time series data, dynamic time wrapping (DTW) distance [13] was used for calculating the distance among points and cluster centroids instead of simple Euclidean distance in the body of the K-Means algorithm. Euclidean distance ignores the time dimension of data and cannot take into account time shifts, whereas dynamic time wrapping can handle these features [13].

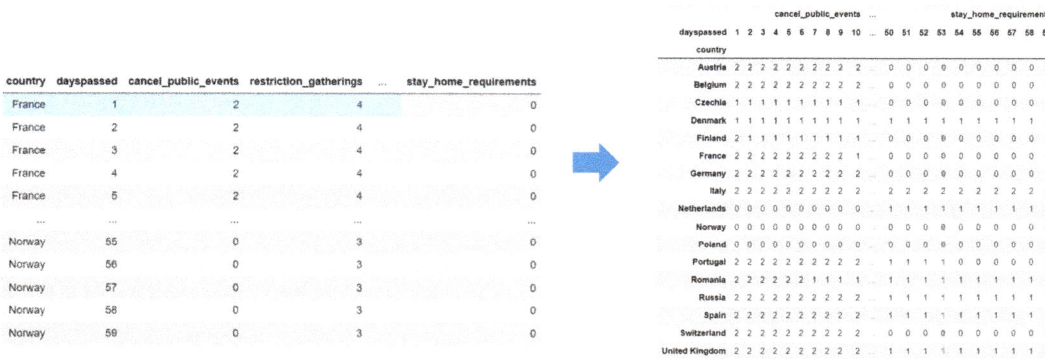

country	dayspassed	cancel_public_events	restriction_gatherings	...	stay_home_requirements
France	1	2	4		0
France	2	2	4		0
France	3	2	4		0
France	4	2	4		0
France	5	2	4		0
...	...	...	...		...
Norway	55	0	3		0
Norway	56	0	3		0
Norway	57	0	3		0
Norway	58	0	3		0
Norway	59	0	3		0

	cancel_public_events										...										stay_home_requirements
dayspassed	1	2	3	4	5	6	7	8	9	10	...	50	51	52	53	54	55	56	57	58	59
country																					
Austria	2	2	2	2	2	2	2	2	2	2	...	0	0	0	0	0	0	0	0	0	0
Belgium	2	2	2	2	2	2	2	2	2	2	...	0	0	0	0	0	0	0	0	0	0
Czechia	1	1	1	1	1	1	1	1	1	1	...	0	0	0	0	0	0	0	0	0	0
Denmark	1	1	1	1	1	1	1	1	1	1	...	1	1	1	1	1	1	1	1	1	1
Finland	2	1	1	1	1	1	1	1	1	1	...	0	0	0	0	0	0	0	0	0	0
France	2	2	2	2	2	2	2	2	2	1	...	0	0	0	0	0	0	0	0	0	0
Germany	2	2	2	2	2	2	2	2	2	2	...	0	0	0	0	0	0	0	0	0	0
Italy	2	2	2	2	2	2	2	2	2	2	...	2	2	2	2	2	2	2	2	2	2
Netherlands	0	0	0	0	0	0	0	0	0	0	...	1	1	1	1	1	1	1	1	1	1
Norway	0	0	0	0	0	0	0	0	0	0	...	0	0	0	0	0	0	0	0	0	0
Poland	0	0	0	0	0	0	0	0	0	0	...	0	0	0	0	0	0	0	0	0	0
Portugal	2	2	2	2	2	2	2	2	2	2	...	1	1	1	1	0	0	0	0	0	0
Romania	2	2	2	2	2	2	1	1	1	1	...	1	1	1	1	1	1	1	1	1	1
Russia	2	2	2	2	2	2	2	2	2	2	...	1	1	1	1	1	1	1	1	1	1
Spain	2	2	2	2	2	2	2	2	2	2	...	1	1	1	1	1	1	1	1	1	1
Switzerland	2	2	2	2	2	2	2	2	2	2	...	0	0	0	0	0	0	0	0	0	0
United Kingdom	2	2	2	2	2	2	2	2	2	2	...	1	1	1	1	1	1	1	1	1	1

Figure 2. Data frame transformation for time series clustering.

One of the limitations of K-Means clustering is that the number of clusters (K) is not determined. Several ways exist to determine the number of clusters, like elbow law or Silhouette score. Here, we used Silhouette score, which measures how well the items (countries) in one cluster are separated from items (countries) in another cluster. The maximum Silhouette score is one, and it assigns higher values to better clustering results [13]. Figure 3 illustrates the Silhouette metrics for different numbers of clusters. Since we had 17 countries to cluster, the maximum number of clusters was set to 9 because we needed at least one pair of countries in each cluster in order to select the control country. Since K = 2 has the highest Silhouette score, we set the number of clusters at 2.

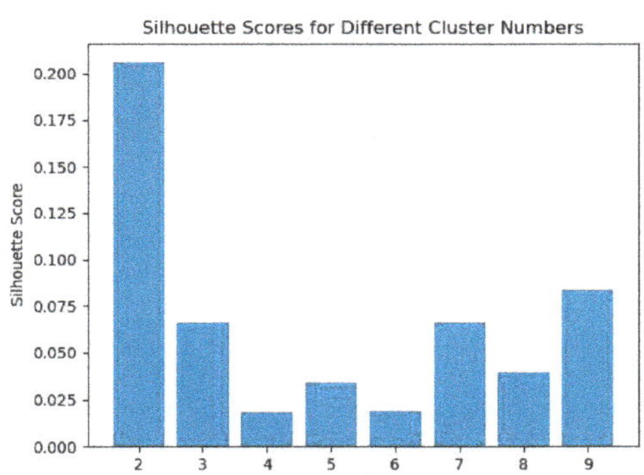

Figure 3. Determining the number of clusters by Silhouette score.

The K-Means algorithm clustered the given countries in 2 clusters, as illustrated in Figure 4.

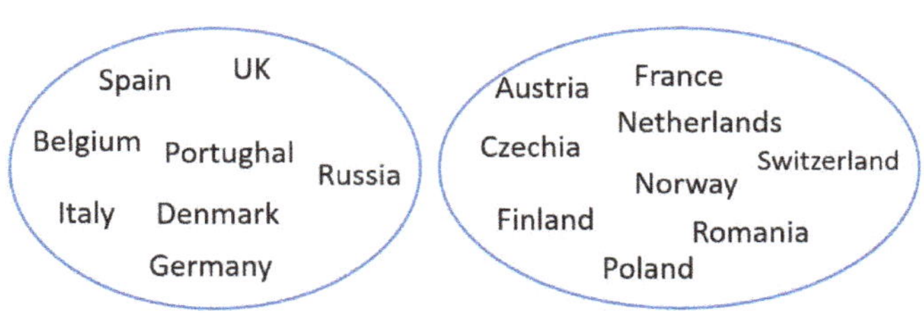

Figure 4. Clustering (K = 2) results for the second wave of COVID-19.

In order to select the counterfactual country from the cluster that Spain is in, a loop was coded in Python that paired all countries in the cluster with Spain. Then, this loop checked for a significant difference between the mean [14] of NPI X implementation levels in each pair. Statistical paired t-test was used to check the mean difference [11]. This helped to select the counterfactual country. Table 1 shows all countries paired with Spain, the mean difference in school closure implementation level, and the *p*-value of the *t*-test.

Table 1. *t*-test results for the mean difference of NPI (X).

Country Pairs	Mean Difference	*p*-Value of *t*-Test
(Spain and Belgium)	1	0
(Spain and Denmark)	1	0
(Spain and Italy)	0.8276	<0.001
(Spain and Germany)	0.6897	<0.001
(Spain and UK)	0.5517	<0.001
(Spain and Portugal)	0	NA
(Spain and Russia)	0	NA

Belgium and Denmark had the highest difference in implementing school closure. Between these two countries, we found Belgium to be the more suitable country as a control group. While Spain changed the school closure implementation level from 2 to 3 on day 101 of the second wave, Belgium changed its level from 1 to 2 on day 109 of the second wave. However, in Denmark, the school closure implementation level changed from 2 to 1 on day 24 of the second wave, again from 1 to 0 on day 105 of the second wave, and from 0 to 2 on day 124 of the second wave. As there was more fluctuation in school closure implementation levels in Denmark, Belgium seemed a better option for counterfactual analysis. Moreover, at least one level difference exists between school closure implementations in Spain and Belgium. This allowed us to investigate the effectiveness of higher-level school closure implementations using interrupted time series analysis with Belgium as a control country.

Interrupted time series analysis was implemented using NLME [15] and CAR [16] libraries of the RStudio software version 1.4.1106. The interrupted time series model that we wanted to analyze is as follows:

$$\begin{aligned} \text{COVID}-19 \quad \text{prevalence} \sim & \text{ time} + \text{TC} + \text{TC} \times \text{time} + \text{level change} \\ & + \text{trend change} + \text{TC} \times \text{level change} + \text{TC} \times \text{trend change} \\ & + (\text{trend change})^2 + \text{TC} \times (\text{trend change})^2 \end{aligned}$$

TC is the Target Country, a binary variable showing whether it is an intervention country (1) or a control country (0). By integrating clustering-based counterfactual analysis with interrupted time series analysis, we could overcome some of the drawbacks of interrupted time series stated in [17], such as the difficulty of isolating one policy's effect. So, it is effective to have a control group.

We found the periods of the autoregressive residual and moving average by doing a preliminary interrupted time series analysis. The ordinary least squares (OLS) method was used for preliminary interrupted time series analysis. R squared of the GLS model was 72%, which meant that 72% of COVID-19 prevalence could be explained by the input variables, including days passed since the beginning of the second COVID-19 wave and the NPI implementation levels. Since OLS assumes the error terms are independent, we found suitable periods for autoregression or moving averages by checking for autocorrelation.

To check for autocorrelation, the auto correlation function (ACF) plot and partial ACF (PACF) plot of the OLS model residuals are illustrated in Figure 5. We considered the maximum 23-day possible lag. Since ACF shows exponential decay, we utilized autoregression (AR). AR means that the error term is related to the error term in previous periods. Since PACF does not show an exponential decay, considering the moving average was not necessary. We examined the PACF plot to determine the AR order and pick the highest violation from the 95% two-way confidence interval (dashed lines). As the PACF for number 7 was larger, AR = 7 was considered for the model.

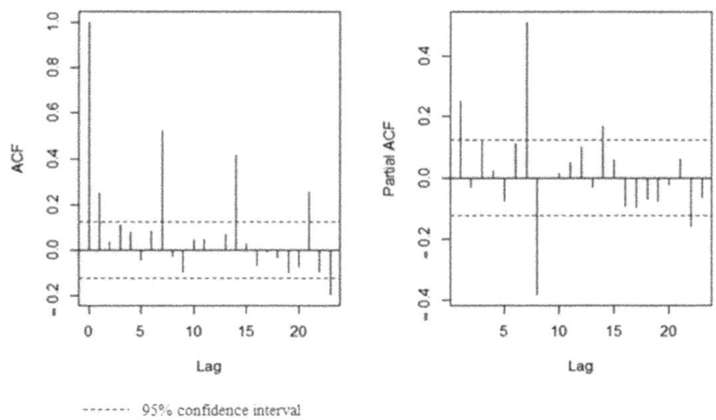

Figure 5. ACF and PACF plot of OLS model residuals.

The second method for checking autocorrelation was the Durbin–Watson test. The Durbin–Watson test which used to check autoregressive residuals confirmed AR (7). The ideal value for the statistical Durbin–Watson test is 2. If the statistics are above 2, it shows a positive correlation; if they are below 2, it shows a negative correlation [16]. Results of the Durbin–Watson test reported in Table 2 confirms 7 days of lag for autoregression with the lowest absolute statistics (0.9090) and zero *p*-value (*p*-value < 0.01).

Table 2. Durbin–Watson test results.

Lag (Days)	Statistics	*p*-Value	Lag (Days)	Statistics	*p*-Value
1	1.4840	**0.000**	8	2.0052	0.806
2	1.9166	0.292	9	2.1426	0.180
3	1.7634	0.024	10	1.8629	0.514
4	1.8283	0.104	11	1.8383	0.394
5	2.0698	0.674	12	1.8948	0.746
6	1.7943	0.082	13	1.7339	0.180
7	0.9090	**0.000**	14	1.0369	**0.000**

4. Results and Discussion

After preliminary analysis, we used the generalized least square (GLS) method, which, is an extension of the OLS model that considers both moving average and autoregressive errors [16]. The GLS model's coefficients and their *p*-values are reported in Table 3. The residual standard error for this model is 217.9.

Table 3. GLS results with AR = 7.

Name of Component	Coefficient Estimates	*p*-Value
Intercept	−130.2954	0.3478
time	8.4291	0 ***
TC	137.2721	0.4841
TC × time	−6.1002	0.0208 *
level change	434.8590	0 ***
trend change	14.9470	0.3280
TC × level change	−511.7353	0 ***
TC × trend change	10.6900	0.6206
$(\text{trend change})^2$	−2.6273	0 ***
TC × $(\text{trend change})^2$	1.9237	0.0132 *

*** at 0.001 level; * at 0.05 level.

In the following, we provide the interpretation of the significant coefficients with a *p*-value less than 0.05. The COVID-19 prevalence in the control group increased (8.4 units per time period). However, in the target group, it was 6.1 units less, which means 2.3 (8.4–6.1) units per time period. In other words, Spain had a −6 cases per million decrease per time period compared to Belgium.

Moreover, the level coefficient shows a significant and positive level change in the control group (434). However, the level change coefficient caused by intervention in the target country was −511. Since it is a differential level change, it means that intervention caused an approximately −77 (−511 + 434) decrease in the level of COVID-19 prevalence in the intervention country.

As COVID-19 prevalence follows an exponential pattern in each wave, the trend^2 term was added to the model, and its coefficients were significant. This means that the COVID-19 prevalence pattern followed an exponential trend. For the control country, this coefficient was −2.6, and for the target country, it was −0.7 (−2.6 + 1.9); although it was less, it was still negative.

Overall, the school closure intervention led to a reduction in the time coefficient, a negative trend^2 coefficient, and a decrease in the level of COVID-19 prevalence; therefore, it was effective. On the other hand, the trend change coefficients in both target and control countries were insignificant.

In Figure 6, the blue line is the Spain data, and the red line is the Belgium data (control group). The dashed blue line shows the counterfactual for Spain based on the control group data (without the time-squared term). The blue line after the vertical black dashed line shows the COVID-19 prevalence after stricter intervention implementation in the target country. As shown in Figure 6, one level stricter school closure decreased the COVID-19 prevalence level and caused a negative trend^2.

In order to approximate the effectiveness of one level increase in school closure implementation, we predicted the COVID-19 prevalence 10 days after the school closer level increase, considering intervention (using the main model) and without considering intervention (using the counterfactual model). Ten days were selected, as it takes at least five days (COVID-19 incubation period [18]) to observe the NPI effect on COVID-19 prevalence. The predicted COVID-19 prevalence was 385.5, while its counterfactual equivalent, considering the trend^2 coefficient, was 587. This means that one level of school closure increase caused an approximately 34% decrease in COVID-19 prevalence after 10 days, which is significant. Therefore, school closure was an effective NPI.

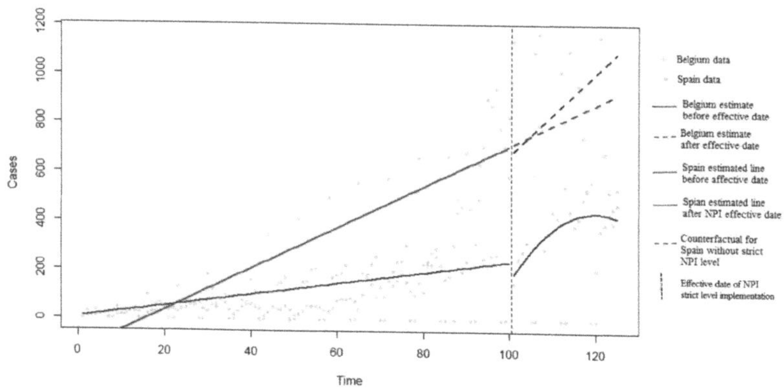

Figure 6. COVID-19 prevalence in target and control countries with and without school closure level increase.

5. Conclusions

Interrupted time series analysis has been used in the COVID-19 literature to investigate the effectiveness of NPIs. However, the single-group interrupted time series analysis has poor internal validity as well as other shortcomings. In this study, a new statistical methodology was developed and shown to be able to handle these shortcomings. By combining a clustering algorithm with conventional single-group interrupted time series analysis, we found an appropriate control country for our target country and changed the research design to ITSA with a control group. This counterfactual country acted as a baseline for the post-intervention period. Here, a K-Means algorithm was used to cluster countries based on all NPIs except the target NPI with two clusters (K = 2), which had the highest Silhouette score. In our case, Belgium was selected as a control group for Spain, to assess the effectiveness of school closure during the second wave of COVID-19, because it was in the same cluster as Spain and had one of the highest differences in implementing school closure compared with Spain. The interrupted time series with control group results showed that increasing one level of the "School closure" NPI effectively reduced the level and time coefficient of COVID-19 prevalence and maintained the negative trend^2 coefficient.

In addition, the ITSA with counterfactual analysis showed that school closure caused a 34% reduction in COVID-19 prevalence 10 days after increasing the level of school closure in Spain. This means that school closure was an effective policy, and adherence to it was important in mitigating the spread of COVID-19. Extending the interrupted time series model to consider adherence to an implemented NPI is an opportunity for future research. Furthermore, improving the performance of the clustering aspect using complex time series clustering methodologies is another future research opportunity. All in all, depending on how many samples are available for clustering, the methodology we developed can also be applied to assessing other public health policies.

Author Contributions: Conceptualization, F.N., Y.Y. and N.A.; methodology, F.N. and N.A.; software, F.N.; validation, F.N.; formal analysis, F.N.; investigation, Y.Y. and N.A.; resources, F.N.; data curation, F.N.; writing—original draft preparation, F.N.; writing—review and editing, N.A. and Y.Y.; visualization, F.N.; supervision, Y.Y. and N.A.; project administration, Y.Y.; funding acquisition, Y.Y. All authors have read and agreed to the published version of the manuscript.

Funding: This research received no external funding.

Institutional Review Board Statement: Not applicable.

Informed Consent Statement: Not applicable.

Data Availability Statement: Data are available at https://ourworldindata.org/policy-responses-COVID.

Conflicts of Interest: The authors declare no conflict of interest.

References

1. Navazi, F.; Yuan, Y.; Archer, N. The effect of the Ontario stay-at-home order on COVID-19 third wave infections including vaccination considerations: An interrupted time series analysis. *PLoS ONE* **2022**, *17*, e0265549. [CrossRef] [PubMed]
2. Saki, M.; Ghanbari, M.K.; Behzadifar, M.; Imani-Nasab, M.H.; Behzadifar, M.; Azari, S.; Bakhtiari, A.; Wu, J.; Bragazzi, N.L. The impact of the social distancing policy on COVID-19 incidence cases and deaths in Iran from february 2020 to january 2021: Insights from an interrupted time series analysis. *Yale J. Biol. Med.* **2021**, *94*, 13–21. [PubMed]
3. Auger, K.A.; Shah, S.S.; Richardson, T.; Hartley, D.; Hall, M.; Warniment, A.; Timmons, K.; Bosse, D.; Ferris, S.A.; Brady, P.W.; et al. Association between Statewide School Closure and COVID-19 Incidence and Mortality in the US. *JAMA-J. Am. Med. Assoc.* **2020**, *324*, 859–870. [CrossRef] [PubMed]
4. Thayer, W.M.; Hasan, M.Z.; Sankhla, P.; Gupta, S. An interrupted time series analysis of the lockdown policies in India: A national-level analysis of COVID-19 incidence. *Health Policy Plan.* **2021**, *36*, 620–629. [CrossRef] [PubMed]
5. Emeto, T.I.; Alele, F.O.; Ilesanmi, O.S. Evaluation of the effect of border closure on COVID-19 incidence rates across nine African countries: An interrupted time series study. *Trans. R. Soc. Trop. Med. Hyg.* **2021**, *115*, 1174–1183. [CrossRef] [PubMed]
6. Ballard, M.; Olsen, H.E.; Millear, A.; Yang, J.; Whidden, C.; Yembrick, A.; Thakura, D.; Nuwasiima, A.; Christiansen, M.; Ressler, D.J.; et al. Continuity of community-based healthcare provision during COVID-19: A multicountry interrupted time series analysis. *BMJ Open* **2022**, *12*, e052407. [CrossRef] [PubMed]
7. Shah, S.A.; Brophy, S.; Kennedy, J.; Fisher, L.; Walker, A.; Mackenna, B.; Curtis, H.; Inglesby, P.; Davy, S.; Bacon, S.; et al. Articles Impact of fi rst UK COVID-19 lockdown on hospital admissions: Interrupted time series study of 32 million people. *eClinicalMedicine* **2022**, *49*, 101462. [CrossRef] [PubMed]
8. Linden, A. Challenges to validity in single-group interrupted time series analysis. *J. Eval. Clin. Pract.* **2017**, *23*, 413–418. [CrossRef] [PubMed]
9. Hale, T.; Angrist, N.; Goldszmidt, R.; Kira, B.; Petherick, A.; Phillips, T.; Webster, S.; Cameron-Blake, E.; Hallas, L.; Majumdar, S.; et al. A global panel database of pandemic policies (Oxford COVID-19 Government Response Tracker). *Nat. Hum. Behav.* **2021**, *5*, 529–538. [CrossRef] [PubMed]
10. Johns Hopkins University. Johns Hopkins Coronavirus Resource Center. 2021. Available online: https://coronavirus.jhu.edu/ (accessed on 26 October 2021).
11. Navazi, F.; Sazvar, Z.; Tavakkoli-Moghaddam, R. A sustainable closed-loop location-routing-inventory problem for perishable products. *Sci. Iran.* **2023**, *30*, 757–783. [CrossRef]
12. Hartigan, J.A.; Wong, M.A. Algorithm AS 136: A K-Means Clustering Algorithm. *J. R. Stat. Soc. Ser. C Appl. Stat.* **1979**, *28*, 100–108. [CrossRef]
13. Verleysen, M.; François, D. The curse of dimensionality in data mining and time series prediction. In *Computational Intelligence and Bioinspired Systems, Proceedings of the 8th International Work-Conference on Artificial Neural Networks, IWANN 2005, Vilanova i la Geltrú, Spain, 8–10 June 2005*; Springer: Berlin/Heidelberg, Germany, 2005; pp. 758–770.
14. Memari, P.; Navazi, F.; Jolai, F. Hybrid wind-municipal solid waste biomass power plant location selection considering waste collection problem: A case study. *Energy Sources Part B Econ. Plan. Policy* **2021**, *16*, 719–739. [CrossRef]
15. Pinheiro, J.; Bates, D.; DebRoy, S.; Sarkar, D.; R Core Team. *nlme: Linear and Nonlinear Mixed Effects Models, R Package version 3*; R Core Team: Vienna, Austria, 2021; pp. 1–153. Available online: https://cran.r-project.org/package=nlme (accessed on 30 May 2023).
16. Fox, J.; Weisberg, S. Time-series regression and generalized least squares: An appendix. In *An R Companion to Applied Regression*; SAGE: Thousand Oaks, CA, USA, 2018; pp. 1–10. Available online: http://tinyurl.com/carbook (accessed on 28 June 2021).
17. Navazi, F.; Yuan, Y.; Archer, N. A review of big data analytics models for assessing nonpharmaceutical interventions for COVID-19 pandemic management. in press.
18. Li, Q.; Guan, X.; Wu, P.; Wang, X.; Zhou, L.; Tong, Y.; Ren, R.; Leung, K.S.M.; Lau, E.H.Y.; Wong, J.Y.; et al. Early Transmission Dynamics in Wuhan, China, of Novel Coronavirus–Infected Pneumonia. *N. Engl. J. Med.* **2020**, *382*, 1199–1207. [CrossRef] [PubMed]

engineering proceedings

Proceeding Paper

A Novel Unconstrained Geometric BINAR(1) Model †

Sunecher Yuvraj [1,*] and Mamode Khan Naushad [2]

1 Department of Accounting, Finance and Economics, Faculty of Business Management and Finance, University of Technology Mauritius, Port-Louis 11108, Mauritius
2 Department of Economics and Statistics, University of Mauritius, Reduit 80834, Mauritius; n.mamodekhan@uom.ac.mu
* Correspondence: ysunecher@utm.ac.mu

† Presented at the 9th International Conference on Time Series and Forecasting, Gran Canaria, Spain, 12–14 July 2023.

Abstract: Modelling the non-stationary unconstrained bivariate integer-valued autoregressive of order 1 (NSUBINAR(1)) model is challenging due to the complex cross-correlation relationship between the counting series. Hence, this paper introduces a novel non-stationary unconstrained BINAR(1) with geometric marginals (NSUBINAR(1)GEOM) based on the assumption that the counting series are both influenced by the same time-dependent explanatory variables. The generalized quasi-likelihood (GQL) estimation method is used to estimate the regression and dependence parameters. Monte Carlo simulations and an application to a real-life accident series data are presented.

Keywords: non-stationary; unconstrained; BINAR(1); GQL; geometric

1. Introduction

In the literature, several researchers have developed first-order bivariate integer-valued autoregressive (BINAR(1)) models to analyse bivariate time series of counts. Originally, Pedeli and Karlis [1,2] developed two constrained BINAR(1) models with Poisson (CBINAR(1)P) and negative binomial (NB) (CBINAR(1)NB) innovations by extending the classical INAR(1) model of McKenzie [3] based on the binomial thinning mechanism [4]. These two models were developed under stationary moment assumptions only and the cross-correlation between the bivariate series was induced by the correlated Poisson and NB innovations, hence implying a constrained relationship. By the same token, Pedeli and Karlis [5] extended the CBINAR(1)P model to an unconstrained BINAR(1) model with Poisson innovations (UBINAR(1)P) under the same condition of stationarity. In this latter model, the cross-correlation between the series was induced by the correlated Poisson innovation terms and the relationship between the observations of each counting series with previous-lagged observations of the other series.

Likewise, Ristic, Nastic, Jayakumar and Bakouch [6] and Nastic, Ristic and Popovic [7] developed a stationary UBINAR(1) model with geometric marginals (UBINAR(1)GEOM) with independent mixed geometric innovations. Hence, the cross-correlation relationship was induced only by the relation of the current observations with previous-lagged observations of the other series via the negative binomial (NB) thinning operator. Interestingly, Nastic et al. [7] showed in their paper that the UBINAR(1)GEOM yields better AICs than the above over-dispersed BINAR(1) models. However, it is worth mentioning that the UBINAR(1)GEOM model was developed only for stationary time series and hence, cannot be used to analyse non-stationary real-life over-dispersed series.

As for non-stationary time series of counts, few BINAR(1) models have been developed. Mamodekhan, Sunecher and Jowaheer [8] developed a CBINAR(1) model with Poisson innovations under non-stationarity assumption (NSCBINAR(1)P) induced by time-dependent explanatory variables. In a similar context, Sunecher, Mamodekhan and

Citation: Yuvraj, S.; Naushad, M.K. A Novel Unconstrained Geometric BINAR(1) Model. *Eng. Proc.* **2023**, *39*, 52. https://doi.org/10.3390/engproc2023039052

Academic Editors: Ignacio Rojas, Hector Pomares, Luis Javier Herrera, Fernando Rojas and Olga Valenzuela

Published: 5 July 2023

Jowaheer [9] developed a non-stationary CBINAR(1) model with NB innovations for over-dispersed time series of counts (NSCBINAR(1)NB). However, an unconstrained BINAR(1) model under the non-stationarity context that can model over-dispersed counting series has not yet been developed in the literature. Hence, in this paper we propose to develop a non-stationary unconstrained BINAR(1) model with geometric innovations (NSUBINAR(1)GEOM) similar to Pedeli and Karlis [5]. However, such model development poses some computational challenges in estimating the unknown model parameters as it is rather difficult to specify the joint generating function [10].

In the paper by Pedeli and Karlis [2,5], the conditional maximum likelihood (CML) approach was compared with the method of moments (MoM), where the authors concluded that CML yields far better estimates than MoM, but at huge computational costs. A similar conclusion was drawn by Nastic et al. [7] who used the least-square (LS) technique, as an alternative to CML, which omits the likelihood function. In the same way, due to the computational challenges of the CML, Mamodekhan et al. [8] and Sunecher et al. [9] developed the generalized quasi-likelihood (GQL) approach in the non-stationary bivariate context. Mamodekhan et al. [8] compared the GQL with CML, where it was shown that GQL yields asymptotically equally efficient estimates as CML. Hence, based on the above findings, the unknown parameters of the NSUBINAR(1)GEOM will be estimated using the GQL.

The organization of the paper is as follows: In the next section, the NSUBINAR(1) model is developed. In Section 3, the GQL approach is developed under the non-stationarity bivariate context. Section 4 focuses on the simulation part where BINAR(1) data with geometric marginals are generated and the GQL approach is used to estimate the model parameters. In Section 5, the model is applied to the accident data in Mauritius. The conclusion is provided in the final section.

2. The Non-Stationary Unconstrained BINAR(1) with Geometric Marginals (NSUBINAR(1)GEOM)

The UBINAR(1) model is specified as:

$$Y_t^{[1]} = \rho_{11} * Y_{t-1}^{[1]} + \rho_{12} * Y_{t-1}^{[2]} + R_t^{[1]} \tag{1}$$

$$Y_t^{[2]} = \rho_{21} * Y_{t-1}^{[1]} + \rho_{22} * Y_{t-1}^{[2]} + R_t^{[2]} \tag{2}$$

based on the following assumptions:

(a) $Y_t^{[k]}$ is geometric such that $Y_t^{[k]} \sim \text{Geom}(\frac{\mu_t^{[k]}}{1+\mu_t^{[k]}})$. Hence, $\text{E}(Y_t^{[k]}) = \mu_t^{[k]}$ and $\text{Var}(Y_t^{[k]}) = \mu_t^{[k]}(1+\mu_t^{[k]})$, where $\mu_t^{[k]} = \exp(\boldsymbol{x}_t'\boldsymbol{\beta}^{[k]})$ with $\boldsymbol{x}_t = [x_{t1}, x_{t2}, \ldots, x_{tj}, \ldots, x_{tp}]'$ is a $(p \times 1)$ vector of covariates influencing both $Y_t^{[1]}$ and $Y_t^{[2]}$, with corresponding regression coefficients $\boldsymbol{\beta}^{[k]} = [\beta_1^{[k]}, \beta_2^{[k]}, \ldots, \beta_j^{[k]}, \ldots, \beta_p^{[k]}]'$ for $t = 1, 2, \ldots, T$ and $k \in \{1, 2\}$.

(b) $*$ is the binomial thinning operator [4] such that $\rho_{ij} * Y_{t-1}^{[j]} = \sum_{m=1}^{Y_{t-1}^{[j]}} Z_m$ with $Z_m \sim \text{Geom}(\frac{\rho_{ij}}{1+\rho_{ij}})$. Hence, $E(\rho_{ij} * Y_{t-1}^{[j]}) = \rho_{ij} E(Y_{t-1}^{[j]})$ and $\text{Var}(\rho_{ij} * Y_{t-1}^{[j]}) = \rho_{ij}(1+\rho_{ij})E(Y_{t-1}^{[j]}) + \rho_{ij}^2 \text{Var}(Y_{t-1}^{[j]})$.

(c)

$$\text{Corr}(R_t^{[1]}, R_{t'}^{[2]}) = \begin{cases} \kappa_{12,t} & t = t', \\ 0 & t \neq t'. \end{cases}$$

$$\begin{aligned} E(Y_t^{[1]}) &= E(\rho_{11} * Y_{t-1}^{[1]}) + E(\rho_{12} * Y_{t-1}^{[2]}) + E(R_t^{[1]}) \\ \mu_t^{[1]} &= \rho_{11}\mu_{t-1}^{[1]} + \rho_{12}\mu_{t-1}^{[2]} + E(R_t^{[1]}). \end{aligned} \tag{3}$$

Re-arranging Equation (3), we have

$$E(R_t^{[1]}) = \lambda_t^{[1]} = \mu_t^{[1]} - \rho_{11}\mu_{t-1}^{[1]} - \rho_{12}\mu_{t-1}^{[2]}. \tag{4}$$

Similarly,

$$E(R_t^{[2]}) = \lambda_t^{[2]} = \mu_t^{[2]} - \rho_{21}\mu_{t-1}^{[1]} - \rho_{22}\mu_{t-1}^{[2]}. \tag{5}$$

$$\begin{aligned}
\mathrm{Var}(Y_t^{[1]}) &= \mathrm{Var}(\rho_{11} * Y_{t-1}^{[1]} + \rho_{12} * Y_{t-1}^{[2]} + R_t^{[1]}) \\
\mathrm{Var}(Y_t^{[1]}) &= \rho_{11}(1+\rho_{11})E(Y_{t-1}^{[1]}) + \rho_{11}^2\mathrm{Var}(Y_{t-1}^{[1]}) + \rho_{12}(1+\rho_{12})E(Y_{t-1}^{[2]}) + \rho_{12}^2\mathrm{Var}(Y_{t-1}^{[2]}) \\
&\quad + 2\rho_{11}\rho_{12}\mathrm{Cov}(Y_{t-1}^{[1]}, Y_{t-1}^{[2]}) + \mathrm{Var}(R_t^{[1]}) \\
\mu_t^{[1]} + \mu_t^{[1]2} &= \rho_{11}(1+\rho_{11})\mu_{t-1}^{[1]} + \rho_{11}^2(\mu_{t-1}^{[1]} + {\mu_{t-1}^{[1]}}^2) + \rho_{12}(1+\rho_{12})\mu_{t-1}^{[2]} + \rho_{12}^2(\mu_{t-1}^{[2]} + {\mu_{t-1}^{[2]}}^2) \\
&\quad + 2\rho_{11}\rho_{12}\mathrm{Cov}(Y_{t-1}^{[1]}, Y_{t-1}^{[2]}) + \mathrm{Var}(R_t^{[1]})
\end{aligned} \tag{6}$$

Re-arranging Equation (6), we have

$$\begin{aligned}
\mathrm{Var}(R_t^{[1]}) &= \mu_t^{[1]}(1+\mu_t^{[1]}) - \rho_{11}(1+\rho_{11})\mu_{t-1}^{[1]} - \rho_{11}^2\mu_{t-1}^{[1]}(1+\mu_{t-1}^{[1]}) - \rho_{12}(1+\rho_{12})\mu_{t-1}^{[2]} \\
&\quad - \rho_{12}^2\mu_{t-1}^{[2]}(1+\mu_{t-1}^{[2]}) - 2\rho_{11}\rho_{12}\mathrm{Cov}(Y_{t-1}^{[1]}, Y_{t-1}^{[2]})
\end{aligned} \tag{7}$$

and similarly,

$$\begin{aligned}
\mathrm{Var}(R_t^{[2]}) &= \mu_t^{[2]}(1+\mu_t^{[2]}) - \rho_{21}(1+\rho_{21})\mu_{t-1}^{[1]} - \rho_{21}^2\mu_{t-1}^{[1]}(1+\mu_{t-1}^{[1]}) - \rho_{22}(1+\rho_{22})\mu_{t-1}^{[2]} \\
&\quad - \rho_{22}^2\mu_{t-1}^{[2]}(1+\mu_{t-1}^{[2]}) - 2\rho_{21}\rho_{22}\mathrm{Cov}(Y_{t-1}^{[1]}, Y_{t-1}^{[2]}),
\end{aligned} \tag{8}$$

The above moments clearly indicate that the marginal distribution of $R_t^{[k]}$ is rather complex to derive. To facilitate the derivation of the cross-covariances, we write Equations (1) and (2) in vector form as follows:

$$\boldsymbol{Y}_t = \boldsymbol{A} * \boldsymbol{Y}_{t-1} + \boldsymbol{R}_t \tag{9}$$

with $\boldsymbol{Y_t} = [Y_t^{[1]}, Y_t^{[2]}]'$, $\boldsymbol{R_t} = [R_t^{[1]}, R_t^{[2]}]'$, $\boldsymbol{A} = \begin{pmatrix} \rho_{11} & \rho_{12} \\ \rho_{21} & \rho_{22} \end{pmatrix}$.

Assuming $\boldsymbol{\Sigma}_{h,t} = \begin{bmatrix} \mathrm{Cov}(Y_t^{[1]}, Y_{t+h}^{[1]}) & \mathrm{Cov}(Y_{t+h}^{[1]}, Y_t^{[2]}) \\ \mathrm{Cov}(Y_t^{[1]}, Y_{t+h}^{[2]}) & \mathrm{Cov}(Y_t^{[2]}, Y_{t+h}^{[2]}) \end{bmatrix}$, from Pedeli and Karlis [5] and Ristic et al. [6], it was shown that $\boldsymbol{\Sigma}_{h,t} = \boldsymbol{A}^h\boldsymbol{\Sigma}_{0,t}$ and hence

$$\begin{aligned}
\mathrm{Cov}(Y_t^{[1]}, Y_t^{[2]}) &= \mathrm{Cov}(\rho_{11} * Y_{t-1}^{[1]} + \rho_{12} * Y_{t-1}^{[2]} + R_t^{[1]}, \rho_{21} * Y_{t-1}^{[1]} + \rho_{22} * Y_{t-1}^{[2]} + R_t^{[2]}) \\
&= (\rho_{11}\rho_{22} + \rho_{12}\rho_{21})\mathrm{Cov}(Y_{t-1}^{[1]}, Y_{t-1}^{[2]}) + \rho_{11}\rho_{21}(\mu_{t-1}^{[1]} + {\mu_{t-1}^{[1]}}^2) \\
&\quad + \rho_{22}\rho_{12}(\mu_{t-1}^{[2]} + {\mu_{t-1}^{[2]}}^2) + [\kappa_{12,t}\sqrt{\mathrm{Var}(R_t^{[1]})}\sqrt{\mathrm{Var}(R_t^{[2]})}].
\end{aligned} \tag{10}$$

Note that if $\alpha = \rho_{11} = \rho_{12}$, $\gamma = \rho_{21} = \rho_{22}$, $\mu = \mu_{t-1}^{[1]} = \mu_{t-1}^{[2]}$ and $\kappa_{12,t} = 0$, Equation (10) simply reduces to $\mathrm{Cov}(Y_t^{[1]}, Y_t^{[2]}) = \frac{2\alpha\gamma}{1-2\alpha\gamma}\mu(1+\mu)$, which is the same as in Ristic et al. [6].

3. Estimation Method

The GQL equation to estimate the regression parameters is specified as:

$$\boldsymbol{D_\beta}'\boldsymbol{\Sigma_\beta}^{-1}(\boldsymbol{f} - \boldsymbol{\mu}) = 0 \tag{11}$$

with score vector $\boldsymbol{f} = [\boldsymbol{f}_1, \boldsymbol{f}_2, \ldots, \boldsymbol{f}_t, \ldots, \boldsymbol{f}_{t+h}, \ldots, \boldsymbol{f}_T]$ with $\boldsymbol{f}_t = [Y_t^{[1]}, Y_t^{[2]}]'$ and $\boldsymbol{\mu} = [\boldsymbol{\mu}_1, \boldsymbol{\mu}_2, \ldots, \boldsymbol{\mu}_t, \ldots, \boldsymbol{\mu}_T]$ with corresponding mean $\boldsymbol{\mu}_t = [\mu_t^{[1]}, \mu_t^{[2]}]'$ for $t = 1, 2, \ldots, T$.

The covariance matrix $\boldsymbol{\Sigma}_\beta$ is a $(2T \times 2T)$. The derivative matrix $\boldsymbol{D}_\beta$ is denoted by $\boldsymbol{D}_\beta = [\boldsymbol{D}_1, \boldsymbol{D}_2, \ldots, \boldsymbol{D}_t, \ldots, \boldsymbol{D}_T]'$ with

$$\boldsymbol{D}_t = \begin{pmatrix} \frac{\partial \mu_t^{[1]}}{\partial \beta^{[1]}} & 0 \\ 0 & \frac{\partial \mu_t^{[2]}}{\partial \beta^{[2]}} \end{pmatrix}_{2p \times 2}$$

where $\frac{\partial \mu_t^{[k]}}{\partial \beta_j^{[k]}} = \mu_t^{[k]} x'_{tj}$.

The Newton–Raphson iterative technique is used to estimate the regression parameters as follows:

$$\begin{pmatrix} \hat{\boldsymbol{\beta}}_{r+1}^{[1]} \\ \hat{\boldsymbol{\beta}}_{r+1}^{[2]} \end{pmatrix} = \begin{pmatrix} \hat{\boldsymbol{\beta}}_{r}^{[1]} \\ \hat{\boldsymbol{\beta}}_{r}^{[2]} \end{pmatrix} + [\boldsymbol{D}_\beta{}' \boldsymbol{\Sigma}_\beta{}^{-1} \boldsymbol{D}_\beta]_r^{-1} [\boldsymbol{D}_\beta{}' \boldsymbol{\Sigma}_\beta{}^{-1} (\boldsymbol{f} - \boldsymbol{\mu})]_r \tag{12}$$

where $\hat{\boldsymbol{\beta}}_r^{[k]}$ are the estimates at the rth iteration and $[.]_r$ are the values of the expression at the rth iteration.

For an initial value of $[\hat{\rho}_{11}, \hat{\rho}_{12}, \hat{\rho}_{21}, \hat{\rho}_{22}, \hat{\kappa}_{12,t}, \boldsymbol{\beta}]$, we solve the iterative Equation (12) until convergence. These estimates are consistent and under mild regulatory conditions, $(\hat{\boldsymbol{\beta}} - \boldsymbol{\beta})'$ is asymptotically normal with a mean of 0 and a covariance matrix of $[\boldsymbol{D}_\beta{}' \boldsymbol{\Sigma}_\beta{}^{-1} \boldsymbol{D}_\beta]^{-1}$ as shown in [8,9,11].

A second GQL is specified to estimate the dependence parameter $\boldsymbol{\psi} = [\rho_{11}, \rho_{12}, \rho_{21}, \rho_{22}]$ as follows:

$$\boldsymbol{D}_\psi{}' \boldsymbol{\Sigma}_\psi{}^{-1} (\boldsymbol{Y}_\psi - \boldsymbol{\mu}_\psi) = 0, \tag{13}$$

with $\boldsymbol{Y}_\psi = [Y_1^{[1]} Y_1^{[2]} | Y_0^{[1]}, Y_0^{[2]}, Y_2^{[1]} Y_2^{[2]} | Y_1^{[1]}, Y_1^{[2]}, \ldots, Y_t^{[1]} Y_t^{[2]} | Y_{t-1}^{[1]}, Y_{t-1}^{[2]}, \ldots,$ $Y_T^{[1]} Y_T^{[2]} | Y_{T-1}^{[1]}, Y_{T-1}^{[2]}]'_{T \times 1}$ and $\boldsymbol{\mu}_\psi = E(\boldsymbol{Y}_\psi)$.

The $(T \times T)$ covariance matrix $\boldsymbol{\Sigma}_\psi$ comprises of $\mathrm{Var}(Y_t^{[1]} Y_t^{[2]} \mid Y_{t-1}^{[1]}, Y_{t-1}^{[2]})$ along the diagonal and $\mathrm{Cov}(Y_t^{[1]} Y_t^{[2]} Y_{t+h}^{[1]} Y_{t+h}^{[2]} \mid Y_{t-1}^{[1]}, Y_{t-1}^{[2]}, Y_{t+h-1}^{[1]}, Y_{t+h-1}^{[2]})$ in the off-diagonal entries. All the entries are of higher-order moments and hence, the 'working' multivariate normality assumption structure is used to compute these entries as in [9,12].

As for the $(T \times 4)$ derivative matrix $\boldsymbol{D}_\psi$,

$$\begin{aligned} &E(Y_t^{[1]} Y_t^{[2]} \mid Y_{t-1}^{[1]}, Y_{t-1}^{[2]}) \\ &= [\kappa_{12,t} [\mu_t^{[1]} (1 + \mu_t^{[1]}) - \rho_{11} (1 + \rho_{11}) \mu_{t-1}^{[1]} - \rho_{11}^2 \mu_{t-1}^{[1]} (1 + \mu_{t-1}^{[1]}) - \rho_{12} (1 + \rho_{12}) \mu_{t-1}^{[2]} \\ &- \rho_{12}^2 \mu_{t-1}^{[2]} (1 + \mu_{t-1}^{[2]})]^{\frac{1}{2}} [\mu_t^{[2]} (1 + \mu_t^{[2]}) - \rho_{21} (1 + \rho_{21}) \mu_{t-1}^{[1]} - \rho_{21}^2 \mu_{t-1}^{[1]} (1 + \mu_{t-1}^{[1]}) \\ &- \rho_{22} (1 + \rho_{22}) \mu_{t-1}^{[2]} - \rho_{22}^2 \mu_{t-1}^{[2]} (1 + \mu_{t-1}^{[2]})]^{\frac{1}{2}} + (\rho_{11} Y_{t-1}^{[1]} + \rho_{12} Y_{t-1}^{[2]} + \mu_t^{[1]} - \rho_{11} \mu_{t-1}^{[1]} \\ &- \rho_{12} \mu_{t-1}^{[2]}) \times (\rho_{21} Y_{t-1}^{[1]} + \rho_{22} Y_{t-1}^{[2]} + \mu_t^{[2]} - \rho_{21} \mu_{t-1}^{[1]} - \rho_{22} \mu_{t-1}^{[2]})] \end{aligned} \tag{14}$$

As for the estimates of $\kappa_{12,t}$, they are estimated using the method of moments as follows:

$$\kappa_{12,t} = \frac{\tilde{\mathrm{Cov}}(Y_t^{[1]}, Y_t^{[2]}) - (\hat{\rho}_{11} \hat{\rho}_{22} + \hat{\rho}_{12} \hat{\rho}_{21}) \tilde{\mathrm{Cov}}(Y_{t-1}^{[1]}, Y_{t-1}^{[2]}) - \hat{\rho}_{11} \hat{\rho}_{21} (\hat{\mu}_{t-1}^{[1]} + \hat{\mu}_{t-1}^{[1]}{}^2) - \hat{\rho}_{22} \hat{\rho}_{12} (\hat{\mu}_{t-1}^{[2]} + \hat{\mu}_{t-1}^{[2]}{}^2)}{\sqrt{\mathrm{Var}(R_t^{[1]})} \sqrt{\mathrm{Var}(R_t^{[2]})}} \tag{15}$$

where $\hat{\mu}_0^{[k]} = \hat{\mu}_1^{[k]}$, $\tilde{\mathrm{Cov}}(Y_t^{[1]}, Y_t^{[2]}) = \frac{1}{T} \sum_{t=1}^{T} (y_t^{[1]} - \hat{\mu}_t^{[1]})(y_t^{[2]} - \hat{\mu}_t^{[2]})$ and $\tilde{\mathrm{Cov}}(Y_{t-1}^{[1]}, Y_{t-1}^{[2]}) = \frac{1}{T-1} \sum_{t=2}^{T} (y_{t-1}^{[1]} - \hat{\mu}_{t-1}^{[1]})(y_{t-1}^{[2]} - \hat{\mu}_{t-1}^{[2]})$.

The Newton–Raphson iteration for the second GQL yields

$$\left(\hat{\boldsymbol{\psi}}_{r+1} \right) = \left(\hat{\boldsymbol{\psi}}_{r} \right) + [\boldsymbol{D}_{\psi}{}'\hat{\boldsymbol{\Sigma}}_{\psi}{}^{-1}\boldsymbol{D}_{\psi}]_r^{-1}[\boldsymbol{D}_{\psi}{}'\hat{\boldsymbol{\Sigma}}_{\psi}{}^{-1}(\boldsymbol{Y}_{\psi} - \boldsymbol{\mu}_{\psi})]_r \tag{16}$$

where $\hat{\boldsymbol{\psi}}_r$ are the estimates at the rth iteration and $[.]_r$ are the values of the expression at the rth iteration.

The regression estimates $\hat{\boldsymbol{\beta}}$ obtained from Equation (12) are used to solve Equation (16) until convergence. These updated values of $\hat{\boldsymbol{\psi}}$ are in turn replaced in Equation (12) to obtain a new set of regression parameters, which are again used to obtain a new set of dependence parameters and this cycle continues until convergence of the two sets of parameters. These estimates are consistent and under mild regulatory conditions, $(\hat{\boldsymbol{\psi}} - \boldsymbol{\psi})'$ is asymptotically normal with a mean of 0 and a covariance matrix $[\boldsymbol{D}_{\psi}{}'\boldsymbol{\Sigma}_{\psi}{}^{-1}\boldsymbol{D}_{\psi}]^{-1}$.

Forecasting Equations

The forecasting equations are derived as follows:

Given $Y_t^{[k]}$, the forecasting function is expressed as

$$E(Y_{t+1}^{[1]}|Y_t^{[1]}, Y_t^{[2]}) = \hat{\mu}_{t+1}^{[1]} + \hat{\rho}_{11}(Y_t^{[1]} - \hat{\mu}_t^{[1]}) + \hat{\rho}_{12}(Y_t^{[2]} - \hat{\mu}_t^{[2]}), \tag{17}$$

$$E(Y_{t+1}^{[2]}|Y_t^{[1]}, Y_t^{[2]}) = \hat{\mu}_{t+1}^{[2]} + \hat{\rho}_{21}(Y_t^{[1]} - \hat{\mu}_t^{[1]}) + \hat{\rho}_{22}(Y_t^{[2]} - \hat{\mu}_t^{[2]}) \tag{18}$$

4. Simulation Study

In this section we generate BINAR(1) time series data with geometric marginals under the following time-varying covariate design:

$$x_{t1} = \begin{cases} -\cos(2\pi t) + 0.01 & (t = 1, \ldots, T/4) \\ \sin(2\pi t) + 0.05 & (t = (T/4) + 1, \ldots, 3T/4) \\ \cos(2\pi t) + 0.10 & (t = (3T/4) + 1, \ldots, T) \end{cases}$$

$$x_{t2} = \begin{cases} (1/t) & (t = 1, \ldots, T/4) \\ (-1/t) & (t = (T/4) + 1, \ldots, 3T/4) \\ t & (t = (3T/4) + 1, \ldots, T) \end{cases}$$

where $\mu_t^{[k]} = \exp(x_{t1}\beta_1^{[k]} + x_{t2}\beta_2^{[k]})$. Assuming $[\rho_{12}, \rho_{21}] = [0.5, 0.5]$, $[\rho_{11}, \rho_{22}] = [0.9, 0.9], [0.3, 0.9], [0.3, 0.3]$, $\beta^{[1]} = 0.5$ and $\beta^{[2]} = 0.9$ for $t = 1, 2, \ldots, T = 100, 500, 1000$, we generate $R_t^{[k]}$ using the inverse transformation method as in [13]. A total of 5000 Monte Carlo replications are made under the above combinations and the simulated mean estimates are shown below:

From Table 1, we observe that the GQL estimates are consistent and that the cross-correlation parameter $\kappa_{12,1}$ is close to unity. In addition, as the time points increase, we notice a decrease in the standard errors, with GQL yielding low standard errors as also demonstrated in [14]. Some details on the number of non-convergent simulations include: for $\rho_{11} = \rho_{22} = 0.9$ under GQL, around 360 simulations failed for $T = 100$, 300 for $T = 500$ and 220 for $T = 1000$. For $\rho_{11} = 0.3$ and $\rho_{22} = 0.9$, around 340 GQL simulations failed for $T = 100$, 275 for $T = 500$ and 190 for $T = 1000$. However, when $\rho_{11} = \rho_{22} = 0.3$, the GQL algorithms failed in 315 simulations for $T = 100$, 215 for $T = 500$ and 170 for $T = 1000$. The failures were mainly due to either an ill-conditioned covariance matrix or the Hessian structure in Equation (12). Hence, it is concluded from this section that GQL yields far superior estimates than GLS and GMM, and constitutes of a slightly better non-convergent computational problem.

Table 1. GQL estimates of the parameters and standard errors under the non-stationary geometric BINAR(1) model.

ρ_{11}	ρ_{22}	T	Methods	$\hat{\beta}_1^{[1]}$	$\hat{\beta}_2^{[1]}$	$\hat{\beta}_1^{[2]}$	$\hat{\beta}_2^{[2]}$	$\hat{\rho}_{11}$	$\hat{\rho}_{22}$	$\hat{\rho}_{12}$	$\hat{\rho}_{21}$	$\hat{\kappa}_{12,1}$
0.9	0.9	100	GQL	0.4823 (0.0910)	0.4870 (0.0931)	0.8876 (0.0977)	0.8840 (0.0946)	0.8819 (0.1163)	0.8847 (0.1115)	0.4859 (0.1125)	0.4847 (0.1153)	0.9815
		500	GQL	0.4923 (0.0517)	0.4914 (0.0512)	0.8950 (0.0547)	0.8917 (0.0585)	0.8942 (0.0649)	0.8920 (0.0625)	0.4945 (0.0623)	0.4940 (0.0630)	0.9909
		1000	GQL	0.4991 (0.0122)	0.4988 (0.0171)	0.8960 (0.0194)	0.8981 (0.0146)	0.8984 (0.0271)	0.8994 (0.0224)	0.5004 (0.0209)	0.5006 (0.0237)	0.9979
0.3	0.9	100	GQL	0.4894 (0.0969)	0.4891 (0.0977)	0.8870 (0.0915)	0.8840 (0.0959)	0.2856 (0.1133)	0.8868 (0.1172)	0.4899 (0.1128)	0.4888 (0.1119)	0.9826
		500	GQL	0.4926 (0.0526)	0.4927 (0.0556)	0.8944 (0.0507)	0.8959 (0.0512)	0.2915 (0.0681)	0.8928 (0.0694)	0.4965 (0.0671)	0.4940 (0.0631)	0.9918
		1000	GQL	0.4995 (0.0137)	0.4975 (0.0175)	0.8969 (0.0111)	0.8994 (0.0141)	0.2988 (0.0213)	0.8963 (0.0293)	0.5008 (0.0251)	0.5011 (0.0221)	0.9995
0.3	0.3	100	GQL	0.4823 (0.0981)	0.4870 (0.0911)	0.8804 (0.0931)	0.8896 (0.0928)	0.2812 (0.1169)	0.2834 (0.1160)	0.4854 (0.1135)	0.4835 (0.1197)	0.9819
		500	GQL	0.4929 (0.0594)	0.4942 (0.0589)	0.8910 (0.0562)	0.8935 (0.0580)	0.2964 (0.0614)	0.2931 (0.0621)	0.4920 (0.0677)	0.4915 (0.0681)	0.9913
		1000	GQL	0.4956 (0.0152)	0.4992 (0.0125)	0.8987 (0.0135)	0.8990 (0.0133)	0.2988 (0.0241)	0.2980 (0.0208)	0.5004 (0.0219)	0.5001 (0.0211)	0.9966

5. Analysing the Time Series of Day and Night Road Accidents in Mauritius

In this section we analyse the monthly day and night accident series data in Mauritius collected from January 2011 to January 2020 that connects the capital city of Port-Louis and the tourist zone Grand-Bay, Mauritius, totalling 109 bivariate time series data. With a sample cross-correlation of 0.3267, it is rationale to believe that there exists a cross-correlation between the two series since both sets of data were collected on the same route. The summary statistics illustrate that day and night accidents have means (variance) of 8.6422 (27.5652) and 4.2110 (13.1310), respectively. Given the significant over-dispersion, the NSUBINAR(1)GEOM in Section 2 is applied to analyse the time series. The covariates we consider are: number of speed cameras (SC) in this area, the number of police officers deployed on street patrol in the different police stations in this area (PO), the number of times the streets in the area have been re-maintained during the years (NS) and number of roundabouts (RA) from Port-Louis to Grand-Bay.

Tables 2 and 3 show the regression estimates and the serial and dependence estimates of the in-sample accident data from January 2011 to August 2019, totalling 104 paired observations, while the out-sample data from September 2019 to January 2020 were used to validate the model. Note, the in-sample data were also analysed using the NSCBINAR(1)NB from [9] and the estimates were compared with the NSCBINAR(1)GEOM and NSUBINAR(1)GEOM.

From Table 2, using the NSUBINAR(1)GEOM, we can note an expected decrease in the number of day and night accidents by 9 and 5%, respectively, if there is an installation of an additional speed camera along the motorway M2. As more police patrols are re-enforced in the area, the number of accidents is expected to decrease by 9% during the day and 7% during the night. Similarly, better road maintenance contributes to a decrease in day accidents by 7% and 6% in night accidents. Roundabout construction must be carefully monitored as this factor leads to an expected increase in the number of day accidents by 6% and 8% during the night. In comparison with the NSCBINAR(1)NB, the signs estimated effects are the same as in NSUBINAR(1)GEOM with little fluctuation in the estimates and their corresponding standard errors, but far better than NSCBINAR(1)GEOM. Using the corresponding forecasting in Equations (17) and (18) and in [9], and based on the out-sample observations from September 2019 to January 2020, Table 4 displays the RMSEs and mean absolute deviation (MAD) under the NSCBINAR(1)NB, NSCBINAR(1)GEOM and NSUBINAR(1)GEOM.

Table 2. Monthly day and night accidents: GQL estimates of the regression parameters.

Model	Time Series	Intercept	NS	SC	PO	RA	$\hat{c}$
NSCBINAR(1)NB	Day Accidents	2.5353	−0.0815	−0.0942	−0.0934	0.0760	1.8475
	s.e	(0.0633)	(0.0396)	(0.0456)	(0.0397)	(0.0289)	(0.0948)
	Night Accidents	0.9272	−0.0790	−0.0671	−0.0824	0.0943	0.8965
	s.e	(0.0742)	(0.0245)	(0.0213)	(0.0329)	(0.0330)	(0.0980)
NSUBINAR(1)GEOM	Day Accidents	2.4445	−0.0824	−0.0952	−0.0955	0.0668	
	s.e	(0.0560)	(0.0273)	(0.0314)	(0.0215)	(0.0164)	
	Night Accidents	0.9106	−0.0714	−0.0572	−0.0747	0.0817	
	s.e	(0.0710)	(0.0180)	(0.0157)	(0.0260)	(0.0205)	
NSCBINAR(1)GEOM	Day Accidents	2.413	−0.0929	−0.0866	−0.0961	0.0852	
	s.e	(0.0958)	(0.0353)	(0.0385)	(0.0419)	(0.0342)	
	Night Accidents	0.9623	−0.0876	−0.0894	−0.0785	0.0899	
	s.e	(0.0952)	(0.0367)	(0.0387)	(0.0375)	(0.0345)	

Table 3. Monthly day and night accidents: GQL estimates of the dependence parameters.

Model	Time Series	$\hat{\rho}$ − Serial	$\hat{\rho}$ − Cross	$\hat{\kappa}_{12,1}$
NSCBINAR(1)NB	Day Accidents	0.2620		0.0065
	s.e			
	Night Accidents	0.2941		
	s.e			
NSUBINAR(1)GEOM	Day Accidents	0.2748	0.0728	0.0025
	s.e	(0.0346)	(0.0234)	
	Night Accidents	0.2438	0.0563	
	s.e	(0.0388)	(0.0191)	
NSCBINAR(1)GEOM	Day Accidents	0.2145		0.0034
	s.e			
	Night Accidents	0.2442		
	s.e			

Table 4. RMSE and MAD for the one step-ahead forecast for the number of monthly day and night accidents.

Model	RMSE $Y_t^{[1]}$	RMSE $Y_t^{[2]}$	MAD $Y_t^{[1]}$	MAD $Y_t^{[2]}$
NSCBINAR(1)NB	0.132	0.141	0.109	0.120
NSUBINAR(1)GEOM	0.120	0.129	0.098	0.104
NSCBINAR(1)GEOM	0.196	0.189	0.155	0.144

The standard errors, RMSE and MAD, illustrate that NSUBINAR(1)GEOM yields almost the same measures as NSCBINAR(1)NB, but better estimates than the NSCBINAR(1)GEOM.

6. Conclusions

In this paper, the unconstrained non-stationary BINAR(1) model with geometric marginals was modelled. However, during the development of this model, it is observed that the joint probability function of the innovation series is rather complex to derive and this limits the construction of a conditional likelihood function to estimate the unknown model parameters. Hence, this paper proposes an alternative GQL estimation approach that only requires the correct specification of the score and moment vectors. As for the derivation of the higher-order entries of the auto-covariance matrix, the multivariate normality assumption was used. With regard to the simulation study, it is shown that the GQL approach provides consistent parameter estimates and statistically more efficient estimates. In the analysis of the accident data, reliable estimates of the different covariates were obtained and comparable to NSCBINAR(1)NB and NSCBINAR(1)GEOM. The RMSE and MAD show that NSUBINAR(1)GEOM yields better forecasts than the other two competing models, a similar conclusion illustrated in [7]. This model is commendable to analyse over-dispersed series under non-stationary setups characterized by time-dependent effects.

Author Contributions: Conceptualization, S.Y. and M.K.N.; methodology, S.Y.; software, M.K.N.; validation, S.Y. and M.K.N.; formal analysis, S.Y.; investigation, S.Y.; resources, M.K.N.; data curation, M.K.N.; writing—original draft preparation, S.Y.; writing—review and editing, M.K.N.; visualization, S.Y. and M.K.N.; supervision, M.K.N.; project administration, S.Y. All authors have read and agreed to the published version of the manuscript.

Funding: This research received no external funding.

Institutional Review Board Statement: Not applicable.

Informed Consent Statement: Not applicable.

Data Availability Statement: https://cso.gov.tt/subject/traffic-statistics (accessed on 27 June 2023).

Conflicts of Interest: The authors declare no conflict of interest.

References

1. Pedeli, X.; Karlis, D. *Bivariate INAR(1) Models*; Technical Report; Athens University of Economics: Athens, Greece, 2009.
2. Pedeli, X.; Karlis, D. A bivariate INAR(1) process with application. *Stat. Model. Int. J.* **2011**, *11*, 325–349. [CrossRef]
3. McKenzie, E. Autoregressive moving-average processes with Negative Binomial and geometric marginal distributions. *Adv. Appl. Probab.* **1986**, *18*, 679–705. [CrossRef]
4. Steutel, F.; Van Harn, K. Discrete analogues of self-decomposability and stability. *Ann. Probab.* **1979**, *7*, 3893–3899. [CrossRef]
5. Pedeli, X.; Karlis, D. Some properties of multivariate INAR(1) processes. *Comput. Stat. Data Anal.* **2013**, *67*, 213–225. [CrossRef]
6. Ristic, M.; Nastic, A.; Jayakumar, K.; Bakouch, H. A bivariate INAR(1) time series model with geometric marginals. *Appl. Math. Lett.* **2012**, *25*, 481–485. [CrossRef]
7. Nastic, A.; Ristic, M.; Popovic, P. Estimation in a bivariate integer-valued autoregressive process. *Commun. -Stat.-Theory Methods* **2016**, *45*, 5660–5678. [CrossRef]
8. Mamode Khan, N.; Sunecher, Y.; Jowaheer, V. Modelling a Non-Stationary BINAR(1) Poisson Process. *J. Stat. Comput. Simul.* **2016**, *86*, 3106–3126. [CrossRef]
9. Sunecher, Y.; Mamodekhan, N.; Jowaheer, V. A GQL estimation approach for analysing non-stationary over-dispersed BINAR(1) time series. *J. Stat. Comput. Simul.* **2017**, *87*, 1911–1924. [CrossRef]
10. Nastic, A.; Laketa, P.; Ristic, M. Random environment integer-valued autoregressive process. *J. Time Ser. Anal.* **2016**, *37*, 267–287. [CrossRef]
11. Sutradhar, B.; Jowaheer, V.; Rao, P. Remarks on asymptotic efficient estimation for regression effects in stationary and non-stationary models for panel count data. *Braz. J. Probab. Stat.* **2014**, *28*, 241–254. [CrossRef]
12. Jowaheer, V.; Mamode Khan, N.; Sunecher, Y. A non-stationary BINAR(1) process with negative binomial innovations for modeling the number of goals in the first and second half: the case study of Arsenal Football Club. *Commun. -Stat.-Case Stud. Data Anal. Appl.* **2017**, *2*, 21–33. [CrossRef]
13. Ristic, M.; Bakouch, H.; Nastic, A. A new geometric first-order integer-valued autoregressive (NGINAR(1)) process. *J. Stat. Plan. Inference* **2009**, *136*, 2218–2226. [CrossRef]
14. Sunecher, Y.; Mamodekhan, N.; Jowaheer, V. Estimating the parameters of a BINMA Poisson model for a non-stationary bivariate time series. *Commun. Stat. Simul. Comput.* **2016**, *46*, 6803–6827. [CrossRef]

MDPI
St. Alban-Anlage 66
4052 Basel
Switzerland
www.mdpi.com

Engineering Proceedings Editorial Office
E-mail: engproc@mdpi.com
www.mdpi.com/journal/engproc

www.ingramcontent.com/pod-product-compliance
Lightning Source LLC
Chambersburg PA
CBHW061500220926
51850CB00004B/137
* 9 7 8 3 0 3 6 5 9 7 2 8 7 *